Lecture Notes in Computer Science 13805

More information about this series at https://link.springer.com/bookseries/558

Leonid Karlinsky · Tomer Michaeli ·
Ko Nishino (Eds.)

Computer Vision – ECCV 2022 Workshops

Tel Aviv, Israel, October 23–27, 2022
Proceedings, Part V

 Springer

Editors
Leonid Karlinsky
IBM Research - MIT-IBM Watson AI Lab
Massachusetts, USA

Tomer Michaeli 🄳
Technion – Israel Institute of Technology
Haifa, Israel

Ko Nishino 🄳
Kyoto University
Kyoto, Japan

ISSN 0302-9743   ISSN 1611-3349 (electronic)
Lecture Notes in Computer Science
ISBN 978-3-031-25071-2   ISBN 978-3-031-25072-9 (eBook)
https://doi.org/10.1007/978-3-031-25072-9

This Springer imprint is published by the registered company Springer Nature Switzerland AG
The registered company address is: Gewerbestrasse 11, 6330 Cham, Switzerland

Foreword

Organizing the European Conference on Computer Vision (ECCV 2022) in Tel-Aviv during a global pandemic was no easy feat. The uncertainty level was extremely high, and decisions had to be postponed to the last minute. Still, we managed to plan things just in time for ECCV 2022 to be held in person. Participation in physical events is crucial to stimulating collaborations and nurturing the culture of the Computer Vision community.

There were many people who worked hard to ensure attendees enjoyed the best science at the 17th edition of ECCV. We are grateful to the Program Chairs Gabriel Brostow and Tal Hassner, who went above and beyond to ensure the ECCV reviewing process ran smoothly. The scientific program included dozens of workshops and tutorials in addition to the main conference and we would like to thank Leonid Karlinsky and Tomer Michaeli for their hard work. Finally, special thanks to the web chairs Lorenzo Baraldi and Kosta Derpanis, who put in extra hours to transfer information fast and efficiently to the ECCV community.

We would like to express gratitude to our generous sponsors and the Industry Chairs Dimosthenis Karatzas and Chen Sagiv, who oversaw industry relations and proposed new ways for academia-industry collaboration and technology transfer. It's great to see so much industrial interest in what we're doing!

Authors' draft versions of the papers appeared online with open access on both the Computer Vision Foundation (CVF) and the European Computer Vision Association (ECVA) websites as with previous ECCVs. Springer, the publisher of the proceedings, has arranged for archival publication. The final version of the papers is hosted by SpringerLink, with active references and supplementary materials. It benefits all potential readers that we offer both a free and citeable version for all researchers, as well as an authoritative, citeable version for SpringerLink readers. Our thanks go to Ronan Nugent from Springer, who helped us negotiate this agreement. Last but not least, we wish to thank Eric Mortensen, our publication chair, whose expertise made the process smooth.

October 2022

Rita Cucchiara
Jiří Matas
Amnon Shashua
Lihi Zelnik-Manor

Preface

Welcome to the workshop proceedings of the 17th European Conference on Computer Vision (ECCV 2022). This year, the main ECCV event was accompanied by 60 workshops, scheduled between October 23–24, 2022. We received 103 workshop proposals on diverse computer vision topics and unfortunately had to decline many valuable proposals because of space limitations. We strove to achieve a balance between topics, as well as between established and new series. Due to the uncertainty associated with the COVID-19 pandemic around the proposal submission deadline, we allowed two workshop formats: hybrid and purely online. Some proposers switched their preferred format as we drew near the conference dates. The final program included 30 hybrid workshops and 30 purely online workshops. Not all workshops published their papers in the ECCV workshop proceedings, or had papers at all. These volumes collect the edited papers from 38 out of the 60 workshops. We sincerely thank the ECCV general chairs for trusting us with the responsibility for the workshops, the workshop organizers for their hard work in putting together exciting programs, and the workshop presenters and authors for contributing to ECCV.

October 2022

Tomer Michaeli
Leonid Karlinsky
Ko Nishino

Organization

General Chairs

Rita Cucchiara   University of Modena and Reggio Emilia, Italy
Jiří Matas   Czech Technical University in Prague,
    Czech Republic
Amnon Shashua   Hebrew University of Jerusalem, Israel
Lihi Zelnik-Manor   Technion – Israel Institute of Technology, Israel

Program Chairs

Shai Avidan   Tel-Aviv University, Israel
Gabriel Brostow   University College London, UK
Giovanni Maria Farinella   University of Catania, Italy
Tal Hassner   Facebook AI, USA

Program Technical Chair

Pavel Lifshits   Technion – Israel Institute of Technology, Israel

Workshops Chairs

Leonid Karlinsky   IBM Research - MIT-IBM Watson AI Lab, USA
Tomer Michaeli   Technion – Israel Institute of Technology, Israel
Ko Nishino   Kyoto University, Japan

Tutorial Chairs

Thomas Pock   Graz University of Technology, Austria
Natalia Neverova   Facebook AI Research, UK

Demo Chair

Bohyung Han   Seoul National University, South Korea

Social and Student Activities Chairs

Tatiana Tommasi   Italian Institute of Technology, Italy
Sagie Benaim   University of Copenhagen, Denmark

Diversity and Inclusion Chairs

Xi Yin   Facebook AI Research, USA
Bryan Russell   Adobe, USA

Communications Chairs

Lorenzo Baraldi   University of Modena and Reggio Emilia, Italy
Kosta Derpanis   York University and Samsung AI Centre Toronto,
                                Canada

Industrial Liaison Chairs

Dimosthenis Karatzas   Universitat Autònoma de Barcelona, Spain
Chen Sagiv   SagivTech, Israel

Finance Chair

Gerard Medioni   University of Southern California and Amazon,
                                USA

Publication Chair

Eric Mortensen   MiCROTEC, USA

Workshops Organizers

W01 - AI for Space

Tat-Jun Chin   The University of Adelaide, Australia
Luca Carlone   Massachusetts Institute of Technology, USA
Djamila Aouada   University of Luxembourg, Luxembourg
Binfeng Pan   Northwestern Polytechnical University, China
Viorela Ila   The University of Sydney, Australia
Benjamin Morrell   NASA Jet Propulsion Lab, USA
Grzegorz Kakareko   Spire Global, USA

W02 - Vision for Art

Alessio Del Bue   Istituto Italiano di Tecnologia, Italy
Peter Bell   Philipps-Universität Marburg, Germany
Leonardo L. Impett   École Polytechnique Fédérale de Lausanne
                                (EPFL), Switzerland
Noa Garcia   Osaka University, Japan
Stuart James   Istituto Italiano di Tecnologia, Italy

W03 - Adversarial Robustness in the Real World

Angtian Wang	Johns Hopkins University, USA
Yutong Bai	Johns Hopkins University, USA
Adam Kortylewski	Max Planck Institute for Informatics, Germany
Cihang Xie	University of California, Santa Cruz, USA
Alan Yuille	Johns Hopkins University, USA
Xinyun Chen	University of California, Berkeley, USA
Judy Hoffman	Georgia Institute of Technology, USA
Wieland Brendel	University of Tübingen, Germany
Matthias Hein	University of Tübingen, Germany
Hang Su	Tsinghua University, China
Dawn Song	University of California, Berkeley, USA
Jun Zhu	Tsinghua University, China
Philippe Burlina	Johns Hopkins University, USA
Rama Chellappa	Johns Hopkins University, USA
Yinpeng Dong	Tsinghua University, China
Yingwei Li	Johns Hopkins University, USA
Ju He	Johns Hopkins University, USA
Alexander Robey	University of Pennsylvania, USA

W04 - Autonomous Vehicle Vision

Rui Fan	Tongji University, China
Nemanja Djuric	Aurora Innovation, USA
Wenshuo Wang	McGill University, Canada
Peter Ondruska	Toyota Woven Planet, UK
Jie Li	Toyota Research Institute, USA

W05 - Learning With Limited and Imperfect Data

Noel C. F. Codella	Microsoft, USA
Zsolt Kira	Georgia Institute of Technology, USA
Shuai Zheng	Cruise LLC, USA
Judy Hoffman	Georgia Institute of Technology, USA
Tatiana Tommasi	Politecnico di Torino, Italy
Xiaojuan Qi	The University of Hong Kong, China
Sadeep Jayasumana	University of Oxford, UK
Viraj Prabhu	Georgia Institute of Technology, USA
Yunhui Guo	University of Texas at Dallas, USA
Ming-Ming Cheng	Nankai University, China

W06 - Advances in Image Manipulation

Radu Timofte	University of Würzburg, Germany, and ETH Zurich, Switzerland
Andrey Ignatov	AI Benchmark and ETH Zurich, Switzerland
Ren Yang	ETH Zurich, Switzerland
Marcos V. Conde	University of Würzburg, Germany
Furkan Kınlı	Özyeğin University, Turkey

W07 - Medical Computer Vision

Tal Arbel	McGill University, Canada
Ayelet Akselrod-Ballin	Reichman University, Israel
Vasileios Belagiannis	Otto von Guericke University, Germany
Qi Dou	The Chinese University of Hong Kong, China
Moti Freiman	Technion, Israel
Nicolas Padoy	University of Strasbourg, France
Tammy Riklin Raviv	Ben Gurion University, Israel
Mathias Unberath	Johns Hopkins University, USA
Yuyin Zhou	University of California, Santa Cruz, USA

W08 - Computer Vision for Metaverse

Bichen Wu	Meta Reality Labs, USA
Peizhao Zhang	Facebook, USA
Xiaoliang Dai	Facebook, USA
Tao Xu	Facebook, USA
Hang Zhang	Meta, USA
Péter Vajda	Facebook, USA
Fernando de la Torre	Carnegie Mellon University, USA
Angela Dai	Technical University of Munich, Germany
Bryan Catanzaro	NVIDIA, USA

W09 - Self-Supervised Learning: What Is Next?

Yuki M. Asano	University of Amsterdam, The Netherlands
Christian Rupprecht	University of Oxford, UK
Diane Larlus	Naver Labs Europe, France
Andrew Zisserman	University of Oxford, UK

W10 - Self-Supervised Learning for Next-Generation Industry-Level Autonomous Driving

Xiaodan Liang	Sun Yat-sen University, China
Hang Xu	Huawei Noah's Ark Lab, China

Fisher Yu   ETH Zürich, Switzerland
Wei Zhang   Huawei Noah's Ark Lab, China
Michael C. Kampffmeyer   UiT The Arctic University of Norway, Norway
Ping Luo   The University of Hong Kong, China

W11 - ISIC Skin Image Analysis

M. Emre Celebi   University of Central Arkansas, USA
Catarina Barata   Instituto Superior Técnico, Portugal
Allan Halpern   Memorial Sloan Kettering Cancer Center, USA
Philipp Tschandl   Medical University of Vienna, Austria
Marc Combalia   Hospital Clínic of Barcelona, Spain
Yuan Liu   Google Health, USA

W12 - Cross-Modal Human-Robot Interaction

Fengda Zhu   Monash University, Australia
Yi Zhu   Huawei Noah's Ark Lab, China
Xiaodan Liang   Sun Yat-sen University, China
Liwei Wang   The Chinese University of Hong Kong, China
Xiaojun Chang   University of Technology Sydney, Australia
Nicu Sebe   University of Trento, Italy

W13 - Text in Everything

Ron Litman   Amazon AI Labs, Israel
Aviad Aberdam   Amazon AI Labs, Israel
Shai Mazor   Amazon AI Labs, Israel
Hadar Averbuch-Elor   Cornell University, USA
Dimosthenis Karatzas   Universitat Autònoma de Barcelona, Spain
R. Manmatha   Amazon AI Labs, USA

W14 - BioImage Computing

Jan Funke   HHMI Janelia Research Campus, USA
Alexander Krull   University of Birmingham, UK
Dagmar Kainmueller   Max Delbrück Center, Germany
Florian Jug   Human Technopole, Italy
Anna Kreshuk   EMBL-European Bioinformatics Institute,
                                   Germany
Martin Weigert   École Polytechnique Fédérale de Lausanne
                                   (EPFL), Switzerland
Virginie Uhlmann   EMBL-European Bioinformatics Institute, UK

Peter Bajcsy   National Institute of Standards and Technology,
                           USA
Erik Meijering   University of New South Wales, Australia

W15 - Visual Object-Oriented Learning Meets Interaction: Discovery, Representations, and Applications

Kaichun Mo   Stanford University, USA
Yanchao Yang   Stanford University, USA
Jiayuan Gu   University of California, San Diego, USA
Shubham Tulsiani   Carnegie Mellon University, USA
Hongjing Lu   University of California, Los Angeles, USA
Leonidas Guibas   Stanford University, USA

W16 - AI for Creative Video Editing and Understanding

Fabian Caba   Adobe Research, USA
Anyi Rao   The Chinese University of Hong Kong, China
Alejandro Pardo   King Abdullah University of Science and
                           Technology, Saudi Arabia
Linning Xu   The Chinese University of Hong Kong, China
Yu Xiong   The Chinese University of Hong Kong, China
Victor A. Escorcia   Samsung AI Center, UK
Ali Thabet   Reality Labs at Meta, USA
Dong Liu   Netflix Research, USA
Dahua Lin   The Chinese University of Hong Kong, China
Bernard Ghanem   King Abdullah University of Science and
                           Technology, Saudi Arabia

W17 - Visual Inductive Priors for Data-Efficient Deep Learning

Jan C. van Gemert   Delft University of Technology, The Netherlands
Nergis Tömen   Delft University of Technology, The Netherlands
Ekin Dogus Cubuk   Google Brain, USA
Robert-Jan Bruintjes   Delft University of Technology, The Netherlands
Attila Lengyel   Delft University of Technology, The Netherlands
Osman Semih Kayhan   Bosch Security Systems, The Netherlands
Marcos Baptista Ríos   Alice Biometrics, Spain
Lorenzo Brigato   Sapienza University of Rome, Italy

W18 - Mobile Intelligent Photography and Imaging

Chongyi Li   Nanyang Technological University, Singapore
Shangchen Zhou   Nanyang Technological University, Singapore

Ruicheng Feng	Nanyang Technological University, Singapore
Jun Jiang	SenseBrain Research, USA
Wenxiu Sun	SenseTime Group Limited, China
Chen Change Loy	Nanyang Technological University, Singapore
Jinwei Gu	SenseBrain Research, USA

W19 - People Analysis: From Face, Body and Fashion to 3D Virtual Avatars

Alberto Del Bimbo	University of Florence, Italy
Mohamed Daoudi	IMT Nord Europe, France
Roberto Vezzani	University of Modena and Reggio Emilia, Italy
Xavier Alameda-Pineda	Inria Grenoble, France
Marcella Cornia	University of Modena and Reggio Emilia, Italy
Guido Borghi	University of Bologna, Italy
Claudio Ferrari	University of Parma, Italy
Federico Becattini	University of Florence, Italy
Andrea Pilzer	NVIDIA AI Technology Center, Italy
Zhiwen Chen	Alibaba Group, China
Xiangyu Zhu	Chinese Academy of Sciences, China
Ye Pan	Shanghai Jiao Tong University, China
Xiaoming Liu	Michigan State University, USA

W20 - Safe Artificial Intelligence for Automated Driving

Timo Saemann	Valeo, Germany
Oliver Wasenmüller	Hochschule Mannheim, Germany
Markus Enzweiler	Esslingen University of Applied Sciences, Germany
Peter Schlicht	CARIAD, Germany
Joachim Sicking	Fraunhofer IAIS, Germany
Stefan Milz	Spleenlab.ai and Technische Universität Ilmenau, Germany
Fabian Hüger	Volkswagen Group Research, Germany
Seyed Ghobadi	University of Applied Sciences Mittelhessen, Germany
Ruby Moritz	Volkswagen Group Research, Germany
Oliver Grau	Intel Labs, Germany
Frédérik Blank	Bosch, Germany
Thomas Stauner	BMW Group, Germany

W21 - Real-World Surveillance: Applications and Challenges

| Kamal Nasrollahi | Aalborg University, Denmark |
| Sergio Escalera | Universitat Autònoma de Barcelona, Spain |

Radu Tudor Ionescu	University of Bucharest, Romania
Fahad Shahbaz Khan	Mohamed bin Zayed University of Artificial Intelligence, United Arab Emirates
Thomas B. Moeslund	Aalborg University, Denmark
Anthony Hoogs	Kitware, USA
Shmuel Peleg	The Hebrew University, Israel
Mubarak Shah	University of Central Florida, USA

W22 - Affective Behavior Analysis In-the-Wild

Dimitrios Kollias	Queen Mary University of London, UK
Stefanos Zafeiriou	Imperial College London, UK
Elnar Hajiyev	Realeyes, UK
Viktoriia Sharmanska	University of Sussex, UK

W23 - Visual Perception for Navigation in Human Environments: The JackRabbot Human Body Pose Dataset and Benchmark

Hamid Rezatofighi	Monash University, Australia
Edward Vendrow	Stanford University, USA
Ian Reid	University of Adelaide, Australia
Silvio Savarese	Stanford University, USA

W24 - Distributed Smart Cameras

Niki Martinel	University of Udine, Italy
Ehsan Adeli	Stanford University, USA
Rita Pucci	University of Udine, Italy
Animashree Anandkumar	Caltech and NVIDIA, USA
Caifeng Shan	Shandong University of Science and Technology, China
Yue Gao	Tsinghua University, China
Christian Micheloni	University of Udine, Italy
Hamid Aghajan	Ghent University, Belgium
Li Fei-Fei	Stanford University, USA

W25 - Causality in Vision

Yulei Niu	Columbia University, USA
Hanwang Zhang	Nanyang Technological University, Singapore
Peng Cui	Tsinghua University, China
Song-Chun Zhu	University of California, Los Angeles, USA
Qianru Sun	Singapore Management University, Singapore
Mike Zheng Shou	National University of Singapore, Singapore
Kaihua Tang	Nanyang Technological University, Singapore

W26 - In-Vehicle Sensing and Monitorization

Jaime S. Cardoso	INESC TEC and Universidade do Porto, Portugal
Pedro M. Carvalho	INESC TEC and Polytechnic of Porto, Portugal
João Ribeiro Pinto	Bosch Car Multimedia and Universidade do Porto, Portugal
Paula Viana	INESC TEC and Polytechnic of Porto, Portugal
Christer Ahlström	Swedish National Road and Transport Research Institute, Sweden
Carolina Pinto	Bosch Car Multimedia, Portugal

W27 - Assistive Computer Vision and Robotics

Marco Leo	National Research Council of Italy, Italy
Giovanni Maria Farinella	University of Catania, Italy
Antonino Furnari	University of Catania, Italy
Mohan Trivedi	University of California, San Diego, USA
Gérard Medioni	Amazon, USA

W28 - Computational Aspects of Deep Learning

Iuri Frosio	NVIDIA, Italy
Sophia Shao	University of California, Berkeley, USA
Lorenzo Baraldi	University of Modena and Reggio Emilia, Italy
Claudio Baecchi	University of Florence, Italy
Frederic Pariente	NVIDIA, France
Giuseppe Fiameni	NVIDIA, Italy

W29 - Computer Vision for Civil and Infrastructure Engineering

Joakim Bruslund Haurum	Aalborg University, Denmark
Mingzhu Wang	Loughborough University, UK
Ajmal Mian	University of Western Australia, Australia
Thomas B. Moeslund	Aalborg University, Denmark

W30 - AI-Enabled Medical Image Analysis: Digital Pathology and Radiology/COVID-19

Jaime S. Cardoso	INESC TEC and Universidade do Porto, Portugal
Stefanos Kollias	National Technical University of Athens, Greece
Sara P. Oliveira	INESC TEC, Portugal
Mattias Rantalainen	Karolinska Institutet, Sweden
Jeroen van der Laak	Radboud University Medical Center, The Netherlands
Cameron Po-Hsuan Chen	Google Health, USA

Diana Felizardo   IMP Diagnostics, Portugal
Ana Monteiro   IMP Diagnostics, Portugal
Isabel M. Pinto   IMP Diagnostics, Portugal
Pedro C. Neto   INESC TEC, Portugal
Xujiong Ye   University of Lincoln, UK
Luc Bidaut   University of Lincoln, UK
Francesco Rundo   STMicroelectronics, Italy
Dimitrios Kollias   Queen Mary University of London, UK
Giuseppe Banna   Portsmouth Hospitals University, UK

W31 - Compositional and Multimodal Perception

Kazuki Kozuka   Panasonic Corporation, Japan
Zelun Luo   Stanford University, USA
Ehsan Adeli   Stanford University, USA
Ranjay Krishna   University of Washington, USA
Juan Carlos Niebles   Salesforce and Stanford University, USA
Li Fei-Fei   Stanford University, USA

W32 - Uncertainty Quantification for Computer Vision

Andrea Pilzer   NVIDIA, Italy
Martin Trapp   Aalto University, Finland
Arno Solin   Aalto University, Finland
Yingzhen Li   Imperial College London, UK
Neill D. F. Campbell   University of Bath, UK

W33 - Recovering 6D Object Pose

Martin Sundermeyer   DLR German Aerospace Center, Germany
Tomáš Hodaň   Reality Labs at Meta, USA
Yann Labbé   Inria Paris, France
Gu Wang   Tsinghua University, China
Lingni Ma   Reality Labs at Meta, USA
Eric Brachmann   Niantic, Germany
Bertram Drost   MVTec, Germany
Sindi Shkodrani   Reality Labs at Meta, USA
Rigas Kouskouridas   Scape Technologies, UK
Ales Leonardis   University of Birmingham, UK
Carsten Steger   Technical University of Munich and MVTec,
                                 Germany
Vincent Lepetit   École des Ponts ParisTech, France, and TU Graz,
                                 Austria
Jiří Matas   Czech Technical University in Prague,
                                 Czech Republic

W34 - Drawings and Abstract Imagery: Representation and Analysis

Diane Oyen	Los Alamos National Laboratory, USA
Kushal Kafle	Adobe Research, USA
Michal Kucer	Los Alamos National Laboratory, USA
Pradyumna Reddy	University College London, UK
Cory Scott	University of California, Irvine, USA

W35 - Sign Language Understanding

Liliane Momeni	University of Oxford, UK
Gül Varol	École des Ponts ParisTech, France
Hannah Bull	University of Paris-Saclay, France
Prajwal K. R.	University of Oxford, UK
Neil Fox	University College London, UK
Ben Saunders	University of Surrey, UK
Necati Cihan Camgöz	Meta Reality Labs, Switzerland
Richard Bowden	University of Surrey, UK
Andrew Zisserman	University of Oxford, UK
Bencie Woll	University College London, UK
Sergio Escalera	Universitat Autònoma de Barcelona, Spain
Jose L. Alba-Castro	Universidade de Vigo, Spain
Thomas B. Moeslund	Aalborg University, Denmark
Julio C. S. Jacques Junior	Universitat Autònoma de Barcelona, Spain
Manuel Vázquez Enríquez	Universidade de Vigo, Spain

W36 - A Challenge for Out-of-Distribution Generalization in Computer Vision

Adam Kortylewski	Max Planck Institute for Informatics, Germany
Bingchen Zhao	University of Edinburgh, UK
Jiahao Wang	Max Planck Institute for Informatics, Germany
Shaozuo Yu	The Chinese University of Hong Kong, China
Siwei Yang	Hong Kong University of Science and Technology, China
Dan Hendrycks	University of California, Berkeley, USA
Oliver Zendel	Austrian Institute of Technology, Austria
Dawn Song	University of California, Berkeley, USA
Alan Yuille	Johns Hopkins University, USA

W37 - Vision With Biased or Scarce Data

Kuan-Chuan Peng	Mitsubishi Electric Research Labs, USA
Ziyan Wu	United Imaging Intelligence, USA

W38 - Visual Object Tracking Challenge

Matej Kristan	University of Ljubljana, Slovenia
Aleš Leonardis	University of Birmingham, UK
Jiří Matas	Czech Technical University in Prague, Czech Republic
Hyung Jin Chang	University of Birmingham, UK
Joni-Kristian Kämäräinen	Tampere University, Finland
Roman Pflugfelder	Technical University of Munich, Germany, Technion, Israel, and Austrian Institute of Technology, Austria
Luka Čehovin Zajc	University of Ljubljana, Slovenia
Alan Lukežič	University of Ljubljana, Slovenia
Gustavo Fernández	Austrian Institute of Technology, Austria
Michael Felsberg	Linköping University, Sweden
Martin Danelljan	ETH Zurich, Switzerland

Contents – Part V

W20 - Safe Artificial Intelligence for Automated Driving

W21 - Real-World Surveillance: Applications and Challenges

W18 - Challenge on Mobile Intelligent Photography and Imaging

W18 - Challenge on Mobile Intelligent Photography and Imaging

This was the first workshop on Mobile Intelligent Photography and Imaging (MIPI). The workshop aimed to emphasize the integration of novel image sensors and imaging algorithms. Together with the workshop, we organized five exciting challenge tracks, including RGB+ToF Depth Completion, Quad-Bayer Re-mosaic, RGBW Sensor Re-mosaic, RGBW Sensor Fusion, and Under-display Camera Image Restoration. The challenges attracted hundreds of participations. The workshop also received high-quality papers and the winnering challenge teams and the authors of the best workshop papers were invited to present their work. We also invited renowned keynote speakers from both industry and academia to share their insights and recent work. The workshop provided a fertile ground for researchers, scientists, and engineers from around the world to disseminate their research outcomes and push forward the frontiers of knowledge within novel image sensors and imaging systems-related areas.

October 2022

Chongyi Li
Shangchen Zhou
Ruicheng Feng
Jun Jiang
Wenxiu Sun
Chen Change Loy
Jinwei Gu

MIPI 2022 Challenge on RGB+ToF Depth Completion: Dataset and Report

Wenxiu Sun[1,3], Qingpeng Zhu[1], Chongyi Li[4(✉)], Ruicheng Feng[4],
Shangchen Zhou[4], Jun Jiang[4], Qingyu Yang[2], Chen Change Loy[4],
Jinwei Gu[2,3], Dewang Hou[5], Kai Zhao[6], Liying Lu[7], Yu Li[8], Huaijia Lin[7],
Ruizheng Wu[9], Jiangbo Lu[9], Jiaya Jia[7], Qiang Liu[10], Haosong Yue[10],
Danyang Cao[10], Lehang Yu[10], Jiaxuan Quan[10], Jixiang Liang[10], Yufei Wang[11],
Yuchao Dai[11], Peng Yang[11], Hu Yan[12], Houbiao Liu[12], Siyuan Su[12],
Xuanhe Li[12], Rui Ren[13], Yunlong Liu[13], Yufan Zhu[13], Dong Lao[14],
Alex Wong[14,15], and Katie Chang[15]

[1] SenseTime Research and Tetras.AI, Beijing, China
{sunwx,zhuqingpeng}@tetras.ai
[2] SenseBrain, San Jose, USA
[3] Shanghai AI Laboratory, Shanghai, China
[4] Nanyang Technological University, Singapore, Singapore
chongyi.li@ntu.edu.sg
[5] Peking University, Beijing, China
dewh@pku.edu.cn
[6] Tsinghua University, Beijing, China
[7] The Chinese University of Hong Kong, Shatin, Hong Kong
lylu@cse.cuhk.edu.hk
[8] International Digital Economy Academy (IDEA), Shenzhen, China
[9] SmartMore, Shatin, Hong Kong
[10] BeiHang University, Beijing, China
[11] School of Electronics and Information, Northwestern Polytechnical University,
Xi'an, China
[12] Amlogic, Shanghai, China
hu.yan@amlogic.com
[13] Xidian University, Xi'an, China
[14] UCLA, Los Angeles, USA
lao@cs.ucla.edu
[15] Yale University, New Haven, USA

Abstract. Developing and integrating advanced image sensors with novel algorithms in camera systems is prevalent with the increasing demand for computational photography and imaging on mobile platforms. However, the lack of high-quality data for research and the rare opportunity for in-depth exchange of views from industry and academia constrain the development of mobile intelligent photography and imaging (MIPI). To bridge the gap, we introduce the first MIPI challenge including five tracks focusing on novel image sensors and imaging algorithms. In this paper, RGB+ToF Depth Completion, one of the five tracks, working on the fusion of RGB sensor and ToF sensor (with spot

---

MIPI 2022 challenge website: http://mipi-challenge.org/MIPI2022/.

L. Karlinsky et al. (Eds.): ECCV 2022 Workshops, LNCS 13805, pp. 3–20, 2023.
https://doi.org/10.1007/978-3-031-25072-9_1

illumination) is introduced. The participants were provided with a new dataset called TetrasRGBD, which contains 18k pairs of high-quality synthetic RGB+Depth training data and 2.3k pairs of testing data from mixed sources. All the data are collected in an indoor scenario. We require that the running time of all methods should be real-time on desktop GPUs. The final results are evaluated using objective metrics and Mean Opinion Score (MOS) subjectively. A detailed description of all models developed in this challenge is provided in this paper. More details of this challenge and the link to the dataset can be found at https://github.com/mipi-challenge/MIPI2022

Keywords: RGB+ToF · Depth completion · MIPI challenge

1   Introduction

RGB+ToF Depth Completion uses sparse ToF depth measurements and a pre-aligned RGB image to obtain a complete depth map. There are a few advantages of using sparse ToF depth measurements. First, the hardware power consumption of sparse ToF (with spot illumination) is low compared to full-field ToF (with flood illumination) depth measurements, which is very important for mobile applications to prevent overheating and fast battery drain. Second, the sparse depth has higher precision and long-range depth measurements due to the focused energy in each laser bin. Third, the Multi-path Interference (MPI) problem that usually bothers full-field iToF measurement is greatly diminished in the sparse ToF depth measurement. However, one obvious disadvantage of sparse ToF depth measurement is the depth density. Take iPhone 12 Pro [21] for example, which is equipped with an advanced mobile Lidar (dToF), the maximum raw depth density is only 1.2%. If directly use as is, the depth would be too sparse to be applied to typical applications like image enhancement, 3D reconstruction, and AR/VR applications.

In this challenge, we intend to fuse the pre-aligned RGB and sparse ToF depth measurement to obtain a complete depth map. Since the power consumption and processing time are still important factors to be balanced, in this challenge, the proposed algorithm is required to process the RGBD data and predict the depth in real-time, i.e. reaches speeds of 30 frames per second, on a reference platform with a GeForce RTX 2080 Ti GPU. The solution is not necessarily deep learning solution, however, to facilitate the deep learning training, we provide a high-quality synthetic depth training dataset containing 20,000 pairs of RGB and ground truth depth images of 7 indoor scenes. We provide a data loader to read these files and a function to simulate sparse depth maps that are close to the real sensor measurements. The participants are also allowed to use other public-domain dataset, for example, NYU-Depth v2 [28], KITTI Depth Completion dataset [32], Scenenet [23], Arkitscenes [1], Waymo open dataset [30], etc. A baseline code is available as well to understand the whole pipeline and to wrap up quickly. The testing data comes from mixed sources, including synthetic data, spot-iToF, and samples that are manually subsampled from iPhone 12 Pro processed depth which uses dToF. The algorithm performance will be ranked by objective metrics: relative mean absolute error (RMAE), edge-weighted mean absolute error (EWMAE), relative depth shift (RDS), and relative temporal standard deviation (RTSD). Details of the metrics are described in the Evaluation section. For the final evaluation, we will also evaluate subjectively in metrics that could not be measured effectively using objective metrics, for example, XY-resolution, edge sharpness, smoothness on flat surfaces, etc.

This challenge is a part of the Mobile Intelligent Photography and Imaging (MIPI) 2022 workshop and challenges which emphasize the integration of novel image sensors and imaging algorithms, which is held in conjunction with ECCV 2022. It consists of five competition tracks:

1. RGB+ToF Depth Completion uses sparse, noisy ToF depth measurements with RGB images to obtain a complete depth map.
2. Quad-Bayer Re-mosaic converts Quad-Bayer RAW data into Bayer format so that it can be processed with standard ISPs.
3. RGBW Sensor Re-mosaic converts RGBW RAW data into Bayer format so that it can be processed with standard ISPs.
4. RGBW Sensor Fusion fuses Bayer data and a monochrome channel data into Bayer format to increase SNR and spatial resolution.
5. Under-display Camera Image Restoration improves the visual quality of image captured by a new imaging system equipped with under-display camera.

2 Challenge

To develop an efficient and high-performance RGB+ToF Depth Completion solution to be used for mobile applications, we provide the following resources for participants:

- A high-quality and large-scale dataset that can be used to train and test the solution;
- The data processing code with data loader that can help participants to save time to accommodate the provided dataset to the depth completion task;
- A set of evaluation metrics that can measure the performance of a developed solution;
- A suggested requirement of running time on multiple platforms that is necessary for real-time processing.

2.1 Problem Definition

Depth completion [3, 6, 8, 9, 11–13, 15, 16, 20, 22, 24, 25] aims to recover dense depth from sparse depth measurements. Earlier methods concentrate on retrieving dense depth maps only from the sparse ones. However, these approaches are limited and not able to recover depth details and semantic information without the availability of multi-modal data. In this challenge, we focus on the RGB+ToF sensor fusion, where a pre-aligned RGB image is also available as guidance for depth completion. In our evaluation, the depth resolution and RGB resolution are fixed at 256 × 192, and the input depth map sparsity ranges from 1.0% to 1.8%. As a reference, the KITTI depth completion dataset [32] has sparsity around 4%–5%. The target of this challenge is to predict a dense depth map given the sparsity depth map and a pre-aligned RGB image at the allowed running time constraint (please refer to Sect. 2.5 for details).

2.2   Dataset: TetrasRGBD

The training data contains 7 image sequences of aligned RGB and ground-truth dense depth from 7 indoor scenes (20,000 pairs of RGB and depth in total). For each scene, the RGB and the ground-truth depth are rendered along a smooth trajectory in our created 3D virtual environment. RGB and dense depth images in the training set have a resolution of 640 × 480 pixels. We also provide a function to simulate the sparse depth maps that are close to the real sensor measurements[1]. A visualization of an example frame of RGB, ground-truth depth, and simulated sparse depth is shown in Fig. 1.

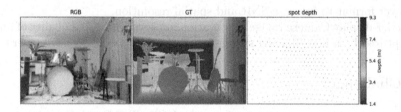

Fig. 1. Visualization of an example training data. Left, middle, and right images correspond to RGB, ground-truth depth, and simulated spot depth data, respectively.

The testing data contains, a) *Synthetic*: a synthetic image sequence (500 pairs of RGB and depth in total) rendered from an indoor virtual environment that differs from the training data; b) *iPhone dynamic*: 24 image sequences of dynamic scenes collected from an iPhone 12Pro (600 pairs of RGB and depth in total); c) *iPhone static*: 24 image sequences of static scenes collected from an iPhone 12Pro (600 pairs of RGB and depth in total); d) *Modified phone static*: 24 image sequences of static scenes (600 pairs of RGB and depth in total) collected from a modified phone. Please note that depth noises, missing depth values in low reflectance regions, and mismatch of field of views between RGB and ToF cameras could be observed from this real data. RGB and dense depth images in the entire testing set have the resolution of 256 × 192 pixels. RGB and spot depth data from the testing set are provided and the GT depth are not available to participants. The depth data in both training and testing sets are in meters.

---

[1] https://github.com/zhuqingpeng/MIPI2022-RGB-ToF-depth-completion.

2.3   Challenge Phases

The challenge consisted of the following phases:

1. Development: The registered participants get access to the data and baseline code, and are able to train the models and evaluate their running time locally.
2. Validation: The participants can upload their models to the remote server to check the fidelity scores on the validation dataset, and to compare their results on the validation leaderboard.
3. Testing: The participants submit their final results, code, models, and fact-sheets.

2.4   Scoring System

Objective Evaluation. We define the following metrics to evaluate the performance of depth completion algorithms.

– Relative Mean Absolute Error (RMAE), which measures the relative depth error between the completed depth and the ground truth, i.e.

$$\text{RMAE} = \frac{1}{M \cdot N} \sum_{m=1}^{M} \sum_{n=1}^{N} \left| \frac{\hat{D}(m,n) - D(m,n)}{D(m,n)} \right|, \tag{1}$$

where M and N denote the height and width of depth, respectively. D and $\hat{D}$ represent the ground-truth depth and the predicted depth, respectively.
– Edge Weighted Mean Absolute Error (EWMAE), which is a weighted average of absolute error. Regions with larger depth discontinuity are assigned higher weights. Similar to the idea of Gradient Conduction Mean Square Error (GCMSE) [19], EWMAE applies a weighting coefficient $G(m,n)$ to the absolute error between pixel (m,n) in ground-truth depth D and predicted depth $\hat{D}$, i.e.

$$\text{EWMAE} = \frac{1}{M \cdot N} \left| \frac{\sum_{m=1}^{M} \sum_{n=1}^{N} G(m,n) \cdot [\hat{D}(m,n) - D(m,n)]}{\sum_{m=1}^{M} \sum_{n=1}^{N} G(m,n)} \right|, \tag{2}$$

where the weight coefficient G is computed in the same way as in [19].
– Relative Depth Shift (RDS), which measures the relative depth error between the completed depth and the input sparse depth on the set of pixels where there are valid depth values in the input sparse depth, i.e.

$$\text{RDS} = \frac{1}{n(S)} \sum_{s \in S} \left| \frac{\hat{D}(s) - X(s)}{X(s)} \right|, \tag{3}$$

where X represents the input sparse depth. S denotes the set of the coordinates of all spot pixels, i.e. pixels with valid depth value. $n(S)$ denotes the cardinality of S.
– Relative Temporal Standard Deviation (RTSD), which measures the temporal standard deviation normalized by depth values for static scenes, i.e.

$$\text{RTSD} = \frac{1}{M \cdot N} \left(\sqrt{\frac{\sum_{f=1}^{F} \left(\hat{D}(m,n) - \sum_{f=1}^{F} \hat{D}(m,n) \right)}{F}} \middle/ \sum_{f=1}^{F} \hat{D}(m,n) \right), \tag{4}$$

8 W. Sun et al.

where F denotes the number of frames.

RMAE, EWMAE, and RDS will be measured on the testing data with GT depth. RTSD will be measured on the testing data collected from static scenes. We will rank the proposed algorithms according to the score calculated by the following formula, where the coefficients are designed to balance the values of different metrics,

$$\text{Objective score} = 1 - 1.8 \times \text{RMAE} - 0.6 \times \text{EWMAE} - 3 \times \text{RDS} - 4.6 \times \text{RTSD}. \quad (5)$$

For each dataset, we report the average results over all the processed images belonging to it.

Subjective Evaluation. For subjective evaluation, we adapt the commonly used Mean Opinion Score (MOS) with blind evaluation. The score is on a scale of 1 (bad) to 5 (excellent). We invited 16 expert observers to watch videos and give their subjective score independently. The scores of all subjects are averaged as the final MOS.

2.5 Running Time Evaluation

The proposed algorithms are required to be able to process the RGB and sparse depth sequence in real-time. Participants are required to include the average run time of one pair of RGB and depth data using their algorithms and the information in the device in the submitted readme file. Due to the difference of devices for evaluation, we set different requirements of running time for different types of devices according to the AI benchmark data from the website[2]. Although the running time is still far from real-time and low power consumption on edge computing devices, we believe this could set up a good starting point for researchers to further push the limit in both academia and industry.

3 Challenge Results

From 119 registered participants, 18 teams submitted their results in the validation phase, 9 teams entered the final phase and submitted the valid results, code, executables, and factsheets. Table 1 summarizes the final challenge results. Team 5 (ZoomNeXt) shows the best overall performance, followed by Team 1 (GAMEON) and Team 4 (Singer). The proposed methods are described in Sect. 4 and the team members and affiliations are listed in Appendix A.

---

[2] https://ai-benchmark.com/ranking_deeplearning_detailed.html.

Table 1. MIPI 2022 RGB+ToF Depth Completion challenge results and final rankings. Team ZoomNeXt is the challenge winner.

Team No.	Team name/User name	RMAE	EWMAE	RDS	RTSD	Objective	Subjective	Final
5	ZoomNeXt/Devonn, k-zha14	0.02935	0.13928	0.00004	0.00997	0.81763	3.53125	**0.76194**
1	GAMEON/hail_hydra	**0.02183**	**0.13256**	0.00736	0.01127	0.80723	**3.55469**	0.75908
4	Singer/Yaxiong_Liu	0.02651	0.13843	0.00270	0.01012	0.81457	3.32812	0.74010
0	NPU-CVR/Arya22	0.02497	0.13278	0.00011	0.00747	**0.84071**	3.14844	0.73520
2	JingAM/JingAM	0.03767	0.13826	0.00459	0.00000	0.83545	2.96094	0.71382
6	NPU-CVR/jokerWRN	0.02547	0.13418	0.00101	0.00725	0.83729	2.75781	0.69443
8	Anonymous/anonymous	0.03015	0.13627	0.00002	0.01716	0.78497	2.76562	0.66905
3	MainHouse113/renruixdu	0.03167	0.14771	0.01103	0.01162	0.76781	2.71875	0.65578
7	UCLA Vision Lab/laid	0.03890	0.14731	0.00014	0.00028	0.83990	2.09375	0.62933

To analyze the performance on different testing dataset, we also summarized the objective score (RMAE) and the subjective score (MOS) per dataset in Table 2, namely *Synthetic*, *iPhone dynamic*, *iPhone static*, and *Modified phone static*. Note that due to there is no RTSD score for dynamic datasets and the RDS in the submitted results is usually very small if participants used hard depth replacement, we only present the RMAE score in this table as an objective indicator. Team ZoomNeXt performs the best in the *Modified phone static* subset, and moderate in other subsets. Team GAMEON performs the best in the subset of *Synthetic*, *iPhone dynamic*, and *iPhone static*, however, obvious artifacts could be observed in the *Modified phone static* subset.

Figure 2 shows a single frame visualization of all the submitted results in the test dataset. Top-left pair is from *Synthetic* subset, top-right pair is from *iPhone dynamic* subset, bottom-left pair is from *iPhone static* subset, and bottom-right pair is from *Modified phone static* subset. It can be observed that all the models could reproduce a semantically meaningful dense depth map. Team 1 (GAMEON) shows the best XY-resolution, where the tiny structures (chair legs, fingers, etc.) could be correctly estimated. Team 5 (ZoomNeXt) and Team 4 (Singer) show the most stable cross-dataset performance, especially in the modified phone dataset.

Table 2. Evalutions of RMAE and MOS in the testing dataset.

Team No.	Team Name/User Name	Synthetic		iPhone dynamic		iPhone static		Mod. phone static	
		RMAE	MOS	RMAE	MOS	RMAE	MOS	RMAE	MOS
5	ZoomNeXt/Devonn, k-zha14	0.06478	3.37500	0.01462	3.53125	0.01455	3.40625	/	**3.8125**
1	GAMEON/hail_hydra	**0.05222**	4.18750	**0.00919**	3.96875	**0.00915**	3.84375	/	2.21875
4	Singer/Yaxiong_Liu	0.06112	3.34375	0.01264	3.68750	0.01154	2.71875	/	3.56250
0	NPU-CVR/Arya22	0.05940	3.50000	0.01099	3.09375	0.01026	3.25000	/	2.75000
2	JingAM/JingAM	0.0854	2.50000	0.01685	3.50000	0.01872	3.28125	/	2.56250
6	NPU-CVR/jokerWRN	0.06061	3.15625	0.01116	2.62500	0.01050	2.87500	/	2.37500
8	Anonymous/anonymous	0.06458	3.87500	0.01589	2.75000	0.01571	2.53125	/	1.90625
3	MainHouse113/renruixdu	0.06928	2.90625	0.01718	3.06250	0.01482	2.03125	/	2.87500
7	UCLA Vision Lab/laid	0.08677	1.71875	0.01997	2.18750	0.01793	2.12500	/	2.34375

Fig. 2. Visualization of results. The set of images shown in the top left is from the Synthetic subset including RGB image, sparse depth, ground-truth depth (if available), and results of 9 teams. Similarly, the set of images shown in the top right is from iPhone dynamic subset, the set of images shown in the bottom left is from the iPhone static subset, and the set of images shown in the bottom right is from the Modified phone subset.

The running time of submitted methods on their individual platforms is summarized in Table 3. All methods satisfied the real-time requirements when converted to a reference platform with a GeForce RTX 2080 Ti GPU.

Table 3. Running time of submitted methods. All of the methods satisfied the real-time inference requirement.

Team No.	Team name/User name	Inference Time	Testing Platform	Upper Limit
5	ZoomNeXt/Devonn, k-zha14	18 ms	Tesla V100 GPU	34 ms
1	GAMEON/hail_hydra	33 ms	GeForce RTX 2080 Ti	33 ms
4	Singer/Yaxiong_Liu	33 ms	GeForce RTX 2060 SUPER	50 ms
0	NPU-CVR/Arya22	23 ms	GeForce RTX 2080 Ti	33 ms
2	JingAM/JingAM	10 ms	GeForce RTX 3090	/
6	NPU-CVR/jokerWRN	24 ms	GeForce RTX 2080 Ti	33 ms
8	Anonymous/anonymous	7 ms	GeForce RTX 2080 Ti	33 ms
3	MainHouse113/renruixdu	28 ms	GeForce 1080 Ti	48 ms
7	UCLA Vision Lab/laid	81 ms	GeForce 1080 Max-Q	106 ms

4   Challenge Methods

In this section, we describe the solutions submitted by all teams participating in the final stage of MIPI 2022 RGB+ToF Depth Completion Challenge. A brief taxonomy of all the methods is in Table 4.

Table 4. A taxonomy of the all the methods in the final stage.

Team name	Fusion	Multi-scale	Refinement	Inspired from	Ensemble	Additional data
ZoomNeXt	Late	No	SPN series	FusionNet [33]	No	No
GAMEON	Early	Yes	SPN series	NLSPN [24]	No	ARKitScenes [1]
Singer	Late	Yes	SPN series	Sehlnet [18]	No	No
NPU-CVR	Mid	No	Deformable Conv	GuideNet [31]	No	No
JingAM	Late	No	No	FusionNet [33]	No	No
Anonymous	Early	No	No	FCN	Yes (Self)	No
MainHouse113	Early	Yes	No	MobileNet [10]	No	No
UCLA Vision Lab	Late	No	Bilateral Filter	ScaffNet [34], FusionNet [33]	Yes (Network)	SceneNet [23]

4.1   ZoomNeXt

Team ZoomNeXt proposes a lightweight and efficient multimodal depth completion (EMDC) model, shown in Fig. 3, which carries their solution to better address the following three key problems.

1. How to fuse multi-modality data more effectively? For this problem, the team adopted a Global and Local Depth Prediction (GLDP) framework [11,33]. For the confidence maps used by the fusion module, they adjusted the pathways of the features to have relative certainty of global and local depth predictions. Furthermore, the losses calculated on the global and local predictions are adaptively weighted to avoid model mismatch.

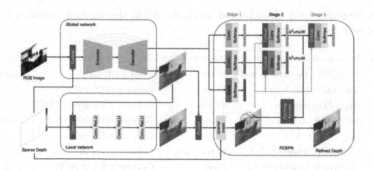

Fig. 3. Model architecture of ZoomNeXt team.

2. How to reduce the negative effects of missing values regions in sparse modality? They first replaced the traditional upsampling in the U-Net in global network with pixel-shuffle [27], and also removed the batch normalization in local network, as they are fragile to features with anisotropic distribution and degrade the results.
3. How to better recover scene structures for both objective metrics and subjective quality? They proposed key innovations in terms of SPN [4,5,12,17,24] structure and loss function design. First, they proposed the funnel convolutional spatial propagation network (FCSPN) for depth refinement. FCSPN can fuse the point-wise results from large to small dilated convolutions in each stage, and the maximum dilation at each stage is designed to be gradually smaller, thus forming a funnel-like structure stage by stage. Second, they also proposed a corrected gradient loss to handle the extreme depth (0 or inf) in the ground-truth.

4.2 GAMEON

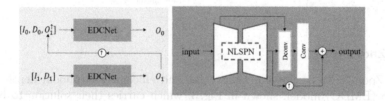

Fig. 4. Model architecture of GAMEON team.

GAMEON team proposed a multi-scale architecture to complete the sparse depth map with high performance and fast speed. Knowledge distillation method is used to distill information from large models. Besides, they proposed a stability constraint to increase the model stability and robustness.

They adopt NLSPN [24] as their baseline, and improve it in the following several aspects. As shown in the left figure of Fig. 4, they adopted a multi-scale scheme to produce the final result. At each scale, the proposed EDCNet is used to generate the dense depth map at the current resolution. The detailed architecture of EDCNet is shown in the right figure of Fig. 4, where NLSPN is adopted as the base network to produce the dense depth map at 1/2 resolution of the input, then two convolution layers are used to refine and generate the full resolution result. To enhance the performance, they further adopt knowledge distillation to distill information from Midas (DPT-Large) [26]. They distill information from the penultimate layer of Midas to the penultimate layer of the depth branch of NLSPN. A convolution layer of kernel size 1 is used to solve the channel dimension gap. Furthermore, they also propose a stability constraint to increase the model stability and robustness. In particular, given a training sample, they adopt the thin plate splines (TPS) [7] transformation (very small perturbations) on the input and ground-truth to generate the transformed sample. After getting the outputs for these two samples, they apply the same transformation to the output of the original sample, which is constrained to be the same as the output of the transformed sample.

4.3   Singer

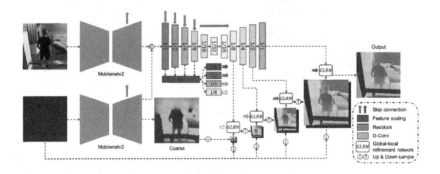

Fig. 5. Model architecture of Singer team.

Team Singer proposed a depth completion approach that separately estimates high- and low-frequency components to address the problem of how to sufficiently utilize of multimodal data. Based on their previous work [18], they proposed a novel Laplacian pyramid-based [2,14,29] depth completion network, which estimates low-frequency components from sparse depth maps by downsampling and contains a Laplacian pyramid decoder that estimates multi-scale residuals to reconstruct complex details of the scene. The overall architecture of the network is shown in Fig. 5.

They use two independent mobilenetv2 to extract features from RGB and Sparse ToF depth. The features from two decoders are fused in a resnet-based encoder. To recover high-frequency scene structures while saving computational cost, the proposed Laplacian pyramid representation progressively adds information on different frequency bands so that the scene structure at different scales can be hierarchically recovered during reconstruction. Instead of the simple upsampling and summation of the two parts in two frequency bands, they proposed a global-local refinement network (GLRN) to fully use different levels of features from the encoder to estimate residuals at various scales and refine them. To save the computational cost of using spatial propagation networks, they introduced a dynamic mask for the fixed kernel of CSPN termed as Affinity Decay Spatial Propagation Network (AD-SPN), which is used to refine the estimated depth maps at various scales through spatial propagation.

4.4 NPU-CVR

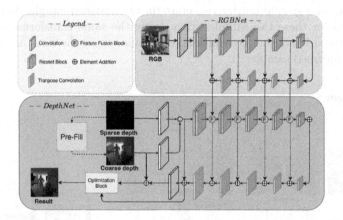

Fig. 6. Model architecture of NPU-CVR team.

Team NPU-CVR submitted two results with a difference in network structural parameter setting. To efficiently fill the sparse and noisy depth maps captured by the ToF camera into dense depth maps, they propose an RGB-guided depth completion method. The overall framework of their method shown in Fig. 6 is based on residual learning. Given a sparse depth map, they first fill it into a coarse dense depth map by pre-processing. Then, they obtain the residual by the network based on the proposed guided feature fusion block, and the residual is added to the coarse depth map to obtain the fine depth map. In addition, they also propose a depth map optimization block that enables deformable convolution to further improve the performance of the method.

4.5 JingAM

Team JingAM improves on the FusionNet proposed in the paper [33]. This paper takes RGB map and sparse depth map as input. The network structure is divided into global network and local network. The global information is obtained through the global network. In addition, the confidence map is used to combine the two inputs according to the uncertainty in the later fusion method. On this basis, they added the skip-level structure and more bottlenecks, and canceled the guidance map and replaced it with the global depth prediction. The loss function weights of the prediction map, global map, and local map are 1, 0.2, and 0.1 respectively. In terms of data enhancement, the brightness, saturation, and contrast were randomly adjusted from 0.1 to 0.8. The sampling step size of the depth map is randomly adjusted between 5 and 12, and change the input size from fixed size to random size. The series of strategies they adopted significantly improved RMSE and MAE compared to the paper [33].

4.6   Anonymous

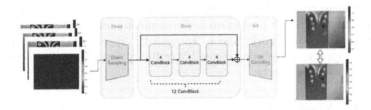

Fig. 7. Model architecture of Anonymous team.

Team Anonymous proposed their method based on a fully convolutional neural network (FCN) which consists of 2 downsampling convolutional layers, 12 residual blocks, and 2 upsampling convolutional layers to produce the final result as shown in Fig. 7. During the training, 25 sequential images and their sparse depth maps from one scene of the provided training dataset will be selected as inputs, then the FCN will predict 25 dense depth maps. The predicted dense depth maps and their ground truth will be used to calculate element-wise training loss consisting of L1 loss, L2 loss, and RMAE loss. Furthermore, RTSD will be calculated with the predicted dense depth maps and regarded as one item of the training loss. During the evaluation, the input RGB image and the sparse depth map will also be concatenated along the channel dimension and sent to the FCN to predict the final dense depth map.

4.7   MainHouse113

As shown in Fig. 8, Team Mainhouse113 uses a multi-scale joint prediction network (MSPNet) that inputs RGB image and sparse depth image, and simultaneously predicts the dense depth image and uncertainty map. They introduce the uncertainty-driven loss to guide network training and leverage a course-to-fine strategy to train the model. Second, they use an uncertainty attention residual

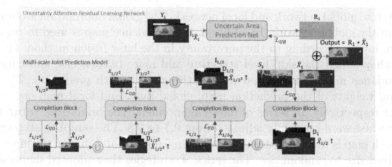

Fig. 8. Model architecture of Mainhouse113 team.

learning network (UARNet), which inputs RGB image, sparse depth image and dense depth image from MSPNet, and outputs residual dense depth image. The uncertainty map from MSPNet serves as an attention map when training the UARNet. The final depth prediction is the element-sum of the two dense depth images from the two networks above. To meet the speed requirement, they use depthwise separable convolution instead of ordinary convolution in the first step, which may lose a few effects.

4.8   UCLA Vision Lab

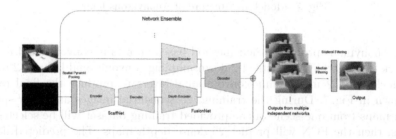

Fig. 9. Model architecture of UCLA team.

Team UCLA Vision Lab proposed a method that aims to learn a prior on the shapes populating the scene from only sparse points. This is realized as a light-weight encoder-decoder network (ScaffNet) [34] and because it is only conditioned on the sparse points, it generalizes across domains. A second network, also an encoder-decoder network (FusionNet) [31], takes the putative depth map and the RGB colored image as input using separate encoder branches and outputs the residual to refine the prior. Because the two networks are extremely light-weight (13ms per forward pass), they are able to instantiate an ensemble of them to yield robust predictions. Finally, they perform an image-guided bilateral filtering step on the output depth map to smooth out any spurious predictions. The overall model architecture during inference time is shown in Fig. 9.

5   Conclusions

In this paper, we summarized the RGB+ToF Depth Completion challenge in the first Mobile Intelligent Photography and Imaging workshop (MIPI 2022) held in conjunction with ECCV 2022. The participants were provided with a high-quality training/testing dataset, which is now available for researchers to download for future research. We are excited to see the new progress contributed by the submitted solutions in such a short time, which are all described in this paper. The challenge results are reported and analyzed. For future works, there is still plenty room for improvements including dealing with depth outliers/noises, precise depth boundaries, high depth resolution, dark scenes, as well as low latency and low power consumption, etc.

Acknowledgements. We thank Shanghai Artificial Intelligence Laboratory, Sony, and Nanyang Technological University to sponsor this MIPI 2022 challenge. We thank all the organizers and all the participants for their great work.

A   Teams and Affiliations

ZoomNeXt Team
Title: Learning An Efficient Multimodal Depth Completion Model
Members: [1]Dewang Hou (dewh@pku.edu.cn), [2]Kai Zhao
Affiliations: [1]Peking University, [2]Tsinghua University

GAMEON Team
Title: A multi-scale depth completion network with high stability
Members: [1]Liying Lu (lylu@cse.cuhk.edu.hk), [2]Yu Li, [1]Huaijia Lin, [3]Ruizheng Wu, [3]Jiangbo Lu, [1]Jiaya Jia
Affiliations: [1]The Chinese University of Hong Kong, [2]International Digital Economy Academy (IDEA), [3]SmartMore

Singer Team
Title: Depth Completion Using Laplacian Pyramid-Based Depth Residuals
Members: Qiang Liu (476582539@qq.com), Haosong Yue, Danyang Cao, Lehang Yu, Jiaxuan Quan, Jixiang Liang
Affiliations: BeiHang University

NPU-CVR Team
Title: An efficient residual network for depth completion of sparse Time-of-Flight depth maps
Members: Yufei Wang (wangyufei1951@gmail.com), Yuchao Dai, Peng Yang
Affiliations: School of Electronics and Informa-

tion, Northwestern Polytechnical University

JingAM Team
Title: Depth completion with RGB guidance and confidence
Members: Hu Yan (hu.yan@amlogic.com), Houbiao Liu, Siyuan Su, Xuanhe Li
Affiliations: Amlogic, Shanghai, China

Anonymous Team
Title: Prediction consistency is learning from yourself

MainHouse113 Team
Title: Uncertainty-based deep learning framework with depthwise separable convolution for depth completion
Members: Rui Ren (1019479834@qq.com), Yunlong Liu, Yufan Zhu
Affiliations: Xidian University

UCLA Vision Lab Team
Title: Learning shape priors from synthetic data for depth completion
Members: [1]Dong Lao (lao@cs.ucla.edu), Alex Wong, [1]Katie Chang
Affiliations: [1]UCLA, [2]Yale University

References

1. Baruch, G., et al.: Arkitscenes-a diverse real-world dataset for 3D indoor scene understanding using mobile RGB-D data. arXiv preprint arXiv:2111.08897 (2021)
2. Chen, X., Chen, X., Zhang, Y., Fu, X., Zha, Z.J.: Laplacian pyramid neural network for dense continuous-value regression for complex scenes. IEEE Trans. Neural Netw. Learn. Syst. **32**(11), 5034–5046 (2020)
3. Chen, Z., Badrinarayanan, V., Drozdov, G., Rabinovich, A.: Estimating depth from RGB and sparse sensing. In: Proceedings of the European Conference on Computer Vision (ECCV), pp. 167–182 (2018)
4. Cheng, X., Wang, P., Guan, C., Yang, R.: CSPN++: learning context and resource aware convolutional spatial propagation networks for depth completion. In: Proceedings of the AAAI Conference on Artificial Intelligence, vol. 34, pp. 10615–10622 (2020)
5. Cheng, X., Wang, P., Yang, R.: Depth estimation via affinity learned with convolutional spatial propagation network. In: Proceedings of the European Conference on Computer Vision (ECCV), pp. 103–119 (2018)
6. Cheng, X., Wang, P., Yang, R.: Learning depth with convolutional spatial propagation network. IEEE Trans. Pattern Anal. Mach. Intell. **42**(10), 2361–2379 (2019)
7. Duchon, J.: Splines minimizing rotation-invariant semi-norms in sobolev spaces. In: Schempp, W., Zeller, K. (eds.) Constructive Theory of Functions of Several Variables, pp. 85–100. Springer, Heidelberg (1977). https://doi.org/10.1007/BFb0086566
8. Eldesokey, A., Felsberg, M., Holmquist, K., Persson, M.: Uncertainty-aware CNNs for depth completion: uncertainty from beginning to end. In: Proceedings of the IEEE/CVF Conference on Computer Vision and Pattern Recognition, pp. 12014–12023 (2020)
9. Eldesokey, A., Felsberg, M., Khan, F.S.: Confidence propagation through CNNs for guided sparse depth regression. IEEE Trans. Pattern Anal. Mach. Intell. **42**(10), 2423–2436 (2019)
10. Howard, A.G., et al.: MobileNets: efficient convolutional neural networks for mobile vision applications. arXiv preprint arXiv:1704.04861 (2017)
11. Hu, J., et al.: Deep depth completion: a survey. arXiv preprint arXiv:2205.05335 (2022)
12. Hu, M., Wang, S., Li, B., Ning, S., Fan, L., Gong, X.: PENet: towards precise and efficient image guided depth completion. In: 2021 IEEE International Conference on Robotics and Automation (ICRA), pp. 13656–13662. IEEE (2021)
13. Imran, S., Long, Y., Liu, X., Morris, D.: Depth coefficients for depth completion. In: Proceedings of the IEEE/CVF Conference on Computer Vision and Pattern Recognition, pp. 12438–12447. IEEE (2019)
14. Jeon, J., Lee, S.: Reconstruction-based pairwise depth dataset for depth image enhancement using CNN. In: Proceedings of the European Conference on Computer Vision (ECCV), pp. 422–438 (2018)
15. Lee, B.U., Lee, K., Kweon, I.S.: Depth completion using plane-residual representation. In: Proceedings of the IEEE/CVF Conference on Computer Vision and Pattern Recognition, pp. 13916–13925 (2021)
16. Li, A., Yuan, Z., Ling, Y., Chi, W., Zhang, C., et al.: A multi-scale guided cascade hourglass network for depth completion. In: Proceedings of the IEEE/CVF Winter Conference on Applications of Computer Vision, pp. 32–40 (2020)

17. Lin, Y., Cheng, T., Zhong, Q., Zhou, W., Yang, H.: Dynamic spatial propagation network for depth completion. arXiv preprint arXiv:2202.09769 (2022)

18. Liu, Q., Yue, H., Lyu, Z., Wang, W., Liu, Z., Chen, W.: SEHLNet: separate estimation of high-and low-frequency components for depth completion. In: 2022 International Conference on Robotics and Automation (ICRA), pp. 668–674. IEEE (2022)

19. López-Randulfe, J., Veiga, C., Rodríguez-Andina, J.J., Farina, J.: A quantitative method for selecting denoising filters, based on a new edge-sensitive metric. In: 2017 IEEE International Conference on Industrial Technology (ICIT), pp. 974–979. IEEE (2017)

20. Lopez-Rodriguez, A., Busam, B., Mikolajczyk, K.: Project to adapt: domain adaptation for depth completion from noisy and sparse sensor data. In: Proceedings of the Asian Conference on Computer Vision (2020)

21. Luetzenburg, G., Kroon, A., Bjørk, A.A.: Evaluation of the apple iPhone 12 pro lidar for an application in geosciences. Sci. Rep. 11(1), 1–9 (2021)

22. Ma, F., Cavalheiro, G.V., Karaman, S.: Self-supervised sparse-to-dense: self-supervised depth completion from lidar and monocular camera. In: 2019 International Conference on Robotics and Automation (ICRA), pp. 3288–3295. IEEE (2019)

23. McCormac, J., Handa, A., Leutenegger, S., Davison, A.J.: Scenenet RGB-D: can 5M synthetic images beat generic imagenet pre-training on indoor segmentation? In: Proceedings of the IEEE International Conference on Computer Vision, pp. 2678–2687 (2017)

24. Park, J., Joo, K., Hu, Z., Liu, C.-K., So Kweon, I.: Non-local spatial propagation network for depth completion. In: Vedaldi, A., Bischof, H., Brox, T., Frahm, J.-M. (eds.) ECCV 2020. LNCS, vol. 12358, pp. 120–136. Springer, Cham (2020). https://doi.org/10.1007/978-3-030-58601-0_8

25. Qu, C., Nguyen, T., Taylor, C.: Depth completion via deep basis fitting. In: Proceedings of the IEEE/CVF Winter Conference on Applications of Computer Vision, pp. 71–80 (2020)

26. Ranftl, R., Lasinger, K., Hafner, D., Schindler, K., Koltun, V.: Towards robust monocular depth estimation: mixing datasets for zero-shot cross-dataset transfer. IEEE Trans. Pattern Anal. Mach. Intell. 44(3), 1623–1637 (2020)

27. Shi, W., et al.: Real-time single image and video super-resolution using an efficient sub-pixel convolutional neural network. In: Proceedings of the IEEE Conference on Computer Vision and Pattern Recognition, pp. 1874–1883 (2016)

28. Silberman, N., Hoiem, D., Kohli, P., Fergus, R.: Indoor segmentation and support inference from RGBD images. In: Fitzgibbon, A., Lazebnik, S., Perona, P., Sato, Y., Schmid, C. (eds.) ECCV 2012. LNCS, vol. 7576, pp. 746–760. Springer, Heidelberg (2012). https://doi.org/10.1007/978-3-642-33715-4_54

29. Song, M., Lim, S., Kim, W.: Monocular depth estimation using laplacian pyramid-based depth residuals. IEEE Trans. Circuits Syst. Video Technol. 31(11), 4381–4393 (2021)

30. Sun, P., et al.: Scalability in perception for autonomous driving: Waymo open dataset. In: Proceedings of the IEEE/CVF Conference on Computer Vision and Pattern Recognition, pp. 2446–2454 (2020)

31. Tang, J., Tian, F.P., Feng, W., Li, J., Tan, P.: Learning guided convolutional network for depth completion. IEEE Trans. Image Process. 30, 1116–1129 (2020)

32. Uhrig, J., Schneider, N., Schneider, L., Franke, U., Brox, T., Geiger, A.: Sparsity invariant CNNs. In: International Conference on 3D Vision (3DV) (2017)
33. Van Gansbeke, W., Neven, D., De Brabandere, B., Van Gool, L.: Sparse and noisy lidar completion with RGB guidance and uncertainty. In: 2019 16th International Conference on Machine Vision Applications (MVA), pp. 1–6. IEEE (2019)
34. Wong, A., Cicek, S., Soatto, S.: Learning topology from synthetic data for unsupervised depth completion. IEEE Robot. Autom. Lett. **6**(2), 1495–1502 (2021)

MIPI 2022 Challenge on Quad-Bayer Re-mosaic: Dataset and Report

Qingyu Yang[1(✉)], Guang Yang[1], Jun Jiang[1], Chongyi Li[4], Ruicheng Feng[4],
Shangchen Zhou[4], Wenxiu Sun[2,3], Qingpeng Zhu[2], Chen Change Loy[4],
Jinwei Gu[1,3], Zhen Wang[5], Daoyu Li[5], Yuzhe Zhang[5], Lintao Peng[5],
Xuyang Chang[5], Yinuo Zhang[5], Yaqi Wu[6], Xun Wu[7], Zhihao Fan[8],
Chengjie Xia[9], Feng Zhang[6,7,8,9], Haijin Zeng[10], Kai Feng[11], Yongqiang Zhao[11],
Hiep Quang Luong[10], Jan Aelterman[10], Anh Minh Truong[10],
Wilfried Philips[10], Xiaohong Liu[12], Jun Jia[12], Hanchi Sun[12], Guangtao Zhai[12],
Longan Xiao[13], Qihang Xu[13], Ting Jiang[14], Qi Wu[14], Chengzhi Jiang[14],
Mingyan Han[14], Xinpeng Li[14], Wenjie Lin[14], Youwei Li[14], Haoqiang Fan[14],
Shuaicheng Liu[14], Rongyuan Wu[15], Lingchen Sun[15], and Qiaosi Yi[15,16]

[1] SenseBrain, San Jose, USA
{yangqingyu,jiangjun}@sensebrain.site
[2] SenseTime Research and Tetras.AI, Beijing, China
[3] Shanghai AI Laboratory, Shanghai, China
[4] Nanyang Technological University, Singapore, Singapore
[5] Beijing Institute of Technology, Beijing, China
wzhstruggle@bit.edu.cn
[6] Harbin Institute of Technology, Harbin 150001, China
[7] Tsinghua University, Beijing 100084, China
[8] University of Shanghai for Science and Technology, Shanghai 200093, China
[9] Zhejiang University, Hangzhou 310027, China
[10] IMEC & Ghent University, Ghent, Belgium
Haijin.Zeng@imec.be
[11] Northwestern Polytechnical University, Xi'an, China
[12] Shanghai Jiao Tong University, Shanghai, China
[13] Transsion, Shenzhen, China
[14] Megvii Technology, Beijing, China
jiangting@megvii.com
[15] OPPO Research Institute, Shenzhen, China
[16] East China Normal University, Shanghai, China

Abstract. Developing and integrating advanced image sensors with novel algorithms in camera systems is prevalent with the increasing demand for computational photography and imaging on mobile platforms. However, the lack of high-quality data for research and the rare opportunity for in-depth exchange of views from industry and academia constrain the development of mobile intelligent photography and imaging

---

Q. Yang, J. Jiang, C. Li, S. Zhou, R. Feng, W. Sun, Q. Zhu, C. C. Loz, J. Gu are the MIPI 2022 challenge organizers.
The other authors participated in the challenge. Please refer to Appendix A for details.
MIPI 2022 challenge website: http://mipi-challenge.org/.

(MIPI). To bridge the gap, we introduce the first MIPI challenge including five tracks focusing on novel image sensors and imaging algorithms. In this paper, Quad Joint Remosaic and Denoise, one of the five tracks, working on the interpolation of Quad CFA to Bayer at full-resolution is introduced. The participants were provided with a new dataset including 70 (training) and 15 (validation) scenes of high-quality Quad and Bayer pair. In addition, for each scene, Quad of different noise level were provided at 0 dB, 24 dB and 42 dB. All the data were captured using a Quad sensor in both outdoor and indoor conditions. The final results are evaluated using objective metrics including PSNR, SSIM [6], LPIPS [10] and KLD. A detailed description of all models developed in this challenge is provided in this paper. More details of this challenge and the link to the dataset can be found in https://github.com/mipi-challenge/MIPI2022.

Keywords: Quad · Remosaic · Bayer · Denoise · MIPI challenge

1   Introduction

Quad is a popular CFA pattern (Fig. 1) used widely for smartphone cameras. Its binning mode is used under low light to achieve enhanced image quality by averaging four pixels within a 2×2 neighborhood. While SNR is improved in the binning mode, the spatial resolution is reduced as a tradeoff. To allow the output Bayer to have the same spatial resolution as the input Quad under normal lighting conditions, we need an interpolation procedure to convert Quad to a Bayer pattern. The interpolation process is usually referred to as remosaic. A good remosaic algorithm should be able to get the Bayer output from Quad with least artifacts, such as moire pattern, false color, and so forth.

The remosaic problem becomes more challenging when the input Quad becomes noisy. A joint remosaic and denoise task is thus in demand for real world applications.

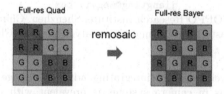

Fig. 1. The Quad remosaic task.

In this challenge, we intend to remosaic the Quad input to obtain a Bayer at the same spatial resolution. The solution is not necessarily deep-learning. However, to facilitate the deep learning training, we provide a dataset of high-quality Quad and Bayer pair (70 scenes for training, 15 for validation, and 15 for test). We provide a data-loader to read these files and a simple ISP in Fig. 2

to visualize the RGB output from the Bayer and to calculate loss functions. The participants are also allowed to use other public-domain dataset. The algorithm performance is evaluated and ranked using objective metrics: Peak Signal-to-Noise Ratio (PSNR), Structural Similarity Index (SSIM) [6], Learned Perceptual Image Patch Similarity (LPIPS) [10] and KL-divergence (KLD). The objective metrics of a baseline method is available as well to provide a benchmark.

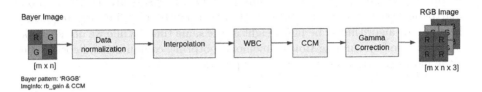

Fig. 2. An ISP to visualize the output Bayer and to calculate the loss function.

This challenge is a part of the Mobile Intelligent Photography and Imaging (MIPI) 2022 workshop and challenges which emphasizing the integration of novel image sensors and imaging algorithms, which is held in conjunction with ECCV 2022. It consists of five competition tracks:

1. RGB+ToF Depth Completion uses sparse, noisy ToF depth measurements with RGB images to obtain a complete depth map.
2. Quad-Bayer Re-mosaic converts Quad-Bayer RAW data into Bayer format so that it can be processed with standard ISPs.
3. RGBW Sensor Re-mosaic converts RGBW RAW data into Bayer format so that it can be processed with standard ISPs.
4. RGBW Sensor Fusion fuses Bayer data and a monochrome channel data into Bayer format to increase SNR and spatial resolution.
5. Under-display Camera Image Restoration improves the visual quality of image captured by a new imaging system equipped under-display camera.

2   Challenge

To develop high quality Quad Remosaic solution, we provide the following resources for participants:

– A high-quality Quad and Bayer dataset; As far as we know, this is the first and only dataset consisting of aligned Quad and Bayer pair, relieving the pain of data collection to develop learning-based remosaic algorithms;
– A data processing code with dataloader to help participants get familiar with the provided dataset;
– A simple ISP including basic ISP blocks to visualize the algorithm output and to calculate the loss function on RGB results;
– A set of objective image quality metrics to measure the performance of a developed solution;

2.1 Problem Definition

Quad remosaic aims to interpolate the input Quad CFA pattern to obtain a Bayer of the same resolution. The remosaic task is needed mainly because current camera ISPs usually cannot process CFAs other than the Bayer pattern. In addition, the remosaic task becomes more challenging when the noise level gets higher, thus requiring more advanced algorithms to avoid image quality artifacts. In addition to the image quality requirement, Quad sensors are widely used on smartphones with limited computational budget and battery life, thus requiring the remosaic algorithm to be lightweight at the same time. While we do not rank solutions based on running time or memory footprint, computational cost is one of the most important criteria in real applications.

2.2 Dataset: Tetras-Quad

The training data contains 70 scenes of aligned Quad (input) and Bayer (ground-truth) pair. For each scene, noise is sythesized on the 0 dB Quad input to provide the noisy Quad input at 24 dB and 42 dB respectively. The synthesized noise consists of read noise and shot noise, and the noise model is measured on a Quad sensor. The data generation steps are shown in Fig. 3. The testing data contains Quad input of 15 scenes at 0 dB, 24 dB and 42 dB, but the GT Bayer results are not available to participants.

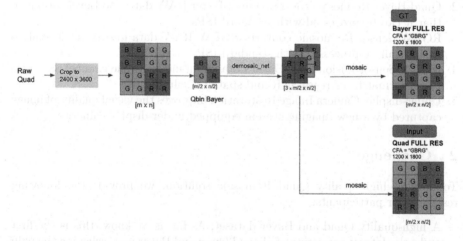

Fig. 3. Data generation of the Quad remosaic task. The Quad raw data is captured using a Quad sensor and cropped to be 2400 × 3600. A Qbin Bayer is obtained by averaging a 2 × 2 block of the same color, and we demosaic the Bayer to get a RGB using the demosaic-net. The RGB image is in turn mosaiced to get the input Quad and aligned ground truth Bayer.

2.3   Challenge Phases

The challenge consisted of the following phases:

1. Development: The registered participants get access to the data and baseline code, and are able to train the models and evaluate their run-time locally.
2. Validation: The participants can upload their models to the remote server to check the fidelity scores on the validation dataset, and to compare their results on the validation leaderboard.
3. Testing: The participants submit their final results, codes, models, and factsheets.

2.4   Scoring System

Objective Evaluation. The evaluation consists of (1) the comparison of the remosaic output (Bayer) with the reference ground truth Bayer, and (2) the comparison of RGB from the predicted and ground truth Bayer using a simple ISP (the code of the simple ISP is provided). We use

1. Peak Signal To Noise Ratio (PSNR)
2. Structural Similarity Index Measure (SSIM) [6]
3. KL divergence (KLD)
4. Learned perceptual image patch similarity (LPIPS) [10]

to evaluate the remosaic performance. The PNSR, SSIM and LPIPS will be applied to the RGB from the Bayer using the provided simple ISP code, while KL divergence is evaluated on the predicted Bayer directly.

A metric weighting PSNR, SSIM, KL divergence, and LPIPS is used to give the final ranking of each method, and we will report each metric separately as well. The code to calculate the metrics is provided. The weighted metric is shown below. The M4 score is between 0 and 100, and the higher the score, the better the overall image quality.

$$M4 = PSNR \cdot \text{SSIM} \cdot 2^{1-\text{LPIPS}-\text{KLD}}, \tag{1}$$

For each dataset we report the average results over all the processed images belonging to it.

3   Challenge Results

Six teams submitted their results in final phase, and their results have been verified using their submitted code as well. Table 1 summarizes the final challenge results. **op-summer-po**, **JHC-SJTU** and **IMEC-IPI & NPU-MPI** are the top three teams ranked by M4 by Eq. 1, and **op-summer-po** shows the best overall performance. The proposed methods are described in Sect. 4 and the team members and affiliations are listed in Appendix A.

Table 1. MIPI 2022 Joint Quad Remosaic and Denoise challenge results and final rankings. PSNR, SSIM, LPIPS and KLD are calculated between the submitted results from each team and the ground truth data. A weighted metric, M4, by Eq. 1 is used to rank the algorithm performance, and the team with the highest M4 is the winner. The M4 of the top 3 teams are highlighted.

Team name	PSNR	SSIM	LPIPS	KLD	M4
BITSpectral	37.2	0.96	0.11	0.03	66
HITZST01	37.2	0.96	0.11	0.06	64.82
IMEC-IPI & NPU-MPI	37.76	0.96	0.1	0.014	**67.95**
JHC-SJTU	37.64	0.96	0.1	0.007	**67.99**
MegNR	36.08	0.95	0.095	0.023	64.1
op-summer-po	37.93	0.965	0.104	0.019	**68.03**

To learn more about the algorithm performance, we evaluated the qualitative image quality in addition to the objective IQ metrics in Fig. 4 and 5 respectively. While all teams in Table 1 have achieved high PSNR and SSIM, the detail and texture loss can be found on the Teddy bear in Fig. 4 and trees in Fig. 5 when the input has a high amount of noise. Oversmoothing tends to yield higher PSNR at the cost of detail loss perceptually.

Fig. 4. Qualitative image quality (IQ) comparison. The results of one of the test scenes (42 dB) are shown. While the top three remosaic methods achieve high objective IQ metrics in Table 1, details and texture loss are noticeable on the Teddy bear.

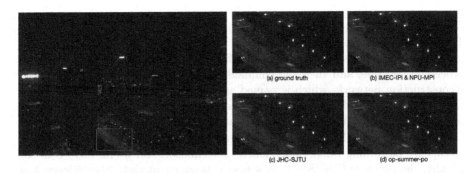

Fig. 5. Qualitative image quality (IQ) comparison. The results of one of the test scenes (42 dB) are shown. Oversmoothing in the top three methods in Table 1 can be found when compared with the ground truth. The trees in the image can barely be identified in (b),(c) and (d).

In addition to benchmarking the image quality of remosaic algorithms, computational efficiency is evaluated because of wide adoptions of Quad sensors on smartphones. We measured the running time of the remosaic solutions of the top three teams (based on M4 by Eq. 1) in Table 2. While running time is not employed in the challenge to rank remosaic algorithms, the computational cost is of critical importance when developing algorithms for smartphones. JHC-SJTU achieved the shortest running time among the top three solutions on a workstation GPU (NVIDIA Tesla V100-SXM2-32GB). With sensor resolution of mainstream smartphones reaching 64M or even higher, power-efficient remosaic algorithms are highly desirable.

Table 2. Running time of the top three solutions ranked by Eq. 1 in the 2022 Joint Quad Remosaic and Denoise challenge. The running time of input of 1200 × 1800 was measured, while the running time of a 64M input Quad was based on estimation. The measurement was taken on a NVIDIA Tesla V100-SXM2-32GB GPU.

Team name	1200 × 1800 (measured)	64M (estimated)
IMEC-IPI & NPU-MPI	6.1 s	180 s
JHC-SJTU	**1 s**	**29.6 s**
op-summer-po	4.4 s	130 s

4   Challenge Methods

In this section, we describe the solutions submitted by all teams participating in the final stage of MIPI 2022 Joint Quad Remosaic and Denoise Challenge.

4.1 MegNR

The overall pipeline is shown in Fig. 6. The Quad input is firstly split into independent channels by the pixel-unshuffle(PU) [4], which is served as a pre-processing module, and then the channels are fed into the Quad remosaic and reconstruction network, which is named as HAUformer, and it is similar to the Uformer [5]. The difference between them lies in the fact that we construct two modules Hybrid Attention Local-Enhanced Block(HALEB) and Overlapping Cross-Attention Block(OCAB) to build group to replace the LeWin Blocks [5] to capture more long-range dependencies information and useful local context. Finally, the pixel-shuffle(PS) [3] as a post-processing module, the independent channels output by the network are restored to the standard bayer output.

The main contribution can be summarized as follow:

- MegNR introduces channel attention and overlapping cross-attention module to Transformer to better aggregate input and cross-window information. This simple design significantly improves the remosaic and restoration quality.
- A new loss function (Eq. 2) is proposed to constrain the network in multiple dimensions and significantly improve the network effect. **M** denotes our loss function, $\mathbf{O_R}$ is the RGB output and **GT** is the ground truth. **PSNR**, **SSIM** and **LPIPS** denote the visual perception metric.

$$\mathbf{M} = -\mathbf{PSNR}(\mathbf{O_R}, \mathbf{GT}) \cdot \mathbf{SSIM}(\mathbf{O_R}, \mathbf{GT}) \cdot 2^{1-\mathbf{LPIPS}(\mathbf{O_R}, \mathbf{GT})} \tag{2}$$

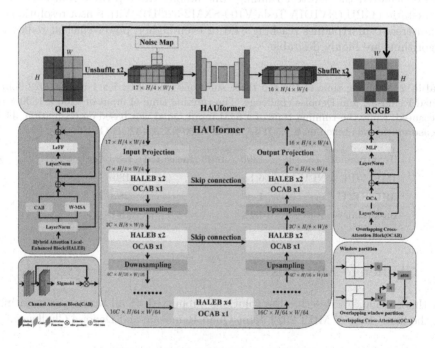

Fig. 6. Model architecture of MegNR.

4.2   IMEC-IPI & NPU-MPI

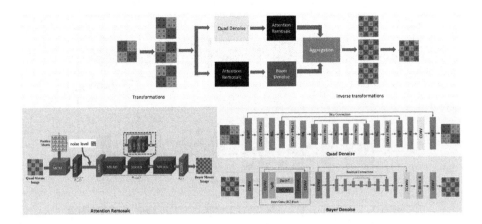

Fig. 7. Model architecture of IMEC-IPI & NPU-MPI.

IMEC-IPI & NPU-MPI proposed a joint denoise and remosaic model consisting of two transformation invariance guided parallel pipelines with independent Swin-Conv-UNet, wavelet based denoising module and mosaic attention based remosaicking module. The overall architecture is in Fig. 7.

Firstly, to avoid aliasing and artifacts in remosaicing, we utilize joint spatial-channel correlation in a raw Quad Bayer image by using a mosaic convolution module (MCM) and a mosaic attention module (MAM).

Secondly, for the hard cases that the proposed naive mode trained by using existing training datasets does not give satisfactory results, we proposed to identify difficult patches and create a better re-training set, leading to enhanced performance on the hard cases.

Thirdly, for the denoising module, we employed the SCUNet [8] and DWT [1] to address the realistic noise in this task. The SCUNet is used to denoise the Quad Bayer image, and we do not use the pre- trained mode, and only train it on the provided images. The DWT is used to denoise the remosaiced Bayer image.

Finally, to solve the question of how to combine denoising and remosaicking, i.e., denoising first or remosaicking first, we proposed a new parallel strategy and found the best way to reconstruct Bayer images from a noisy Quad Bayer mosaic, as shown in Fig. 7. In addition, to further improve the model robustness and generality, we consider transformation invariance: a rigid transformation of the input should lead to an output with the same transformation. Here, we enforce this invariance through the network design instead of increasing the training dataset by direct rotating the training images.

4.3   HITZST01

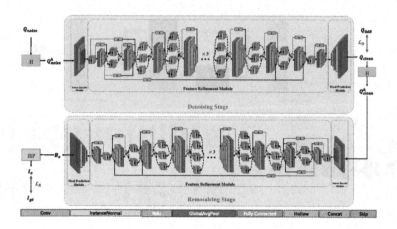

Fig. 8. Model architecture of HITZST.

HITZST proposed a two-stage paradigm that reformulate this task into denoise first and then remosaic. By introducing two supervisory signals: clean QBC Q_{clean} for denoise stage and RGB images I_{gt} generated from clean Bayer B_{gt} for remosaic stage, our full framework is trained in a two-stage supervised manner in Fig. 8.

In general, our proposed network consists of two stages: denoise stage and remosaic stage. For each stage, we use the same backbone (JQNet). Specifically, our network framework, as shown consists of three sub-modules: the source encoder module, the feature refinement module and the final prediction module.

Source Encoder Module consists of two convolution layers. Its role is to reconstruct image texture information from mosaic image as the input (Iraw) of feature refinement module. **Feature Refinement Module** takes Iraw as input to fully select, refine and enhance useful information in mosaic domain. It is formed by sequential connection of MCCA [2] [11] blocks of different scale. **And Final Prediction Module** after getting a great representation Irefine of the mosaic image from the feature refinement network, final prediction module takes the Irefine as input and perform demosaicking to reconstruct a full-channel hyperspectral image.

4.4 BITSpectral

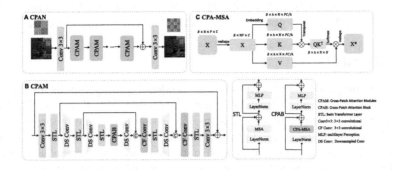

Fig. 9. Model architecture of BITSpectral.

BITSpectral proposed a transformer- based network called CPAN in Fig. 9. The proposed Cross-Patch Attention Network (CPAN) follows the residual learning structure. The input convolutional layer extracts the embeddings of input images. The embeddings are next delivered into the Cross-Patch Attention Modules (CPAM). The reported network includes 5 CPAMs in total. Different from conventional transformer modules, CPAM directly exploits the global attention between feature patches. CPAM is a U-shape sub-network which can reduce the computational complexity. We utilize the Swin Transformer Layer (STL) to extract the attention within feature patches in each stage, and Cross-Patch Attention Block (CPAB) to directly obtain the global attention between patches for the innermost stage.

CPAB differs from STL in that it enhances the global field of perception by cross-patch attention. Current transformers in vision, such as Swin Transformer and ViT, perform self-attention within patches. They does not sufficiently consider the connection between patches, and thus the global feature extraction capability is limited. CPAB consists of a Cross-Patch Attention Multi head Self-attention(CPA-MSA) and a MLP layer. CPA-MSA is an improvement on traditional MSA, focusing on the relationship between different patches. The input of CPA-MSA is X ($X \in \mathbb{R}^{B \times N \times P \times C}$), where B denotes the batch size of training, N is the number of patches cut per feature map, P is the number of pixels per patch, and C indicates the number of channels. Firstly, X is reshaped to $B \times NP \times C$. Then X is linearly transformed to obtain Q, K, V, whose dimensions are $\mathbb{R}^{B \times h \times N \times \frac{PC}{h}}$. We can get $QK^T \in \mathbb{R}^{B \times h \times N \times N}$ and compute $Attention(Q, K, V) \in \mathbb{R}^{B \times h \times N \times \frac{PC}{h}}$. Finally, they are reshaped to be the output $X_{out} \in \mathbb{R}^{B \times N \times P \times C}$.

In the implementation, we divide the dataset into the training set (65 images) and validation set (5 images). The images in training set are cropped into patches of 128×128 pixels. To improve inference efficiency, The images in the testing set and validation set are not cropped. We utilized the Adam optimizer with $\beta_1 = 0.9, \beta_2 = 0.999, e = 10^{-6}$ for training. In order to accelerate the convergence rate, we applied the cosine annealing learning rate strategy, whose starting learning rate is set at 4×10^{-4}. On a single 3090 GPU, the training procedure took around 20 h to complete 100 epochs. The CPAM-Net needs 0.08 s to reconstruct an image of 1200×1800 pixels in both validation and testing datasets.

4.5   JHC-SJTU

JHC-SJTU proposed a joint remosaicing and denoising pipeline for raw data in quad-Bayer pattern. The pipeline consists of two stages as shown in Fig. 10. In the first stage, a multi-resolution feature extraction and fusion network is used to jointly remosaic the quad Bayer to standard Bayer (RGGB) and denoise it. The ground truth raw data in RGGB pattern is used for supervision. In the second stage, the simple ISP provided by the organizers is concatenated to the output of the network. We finetune the network in the first stage using the RGB

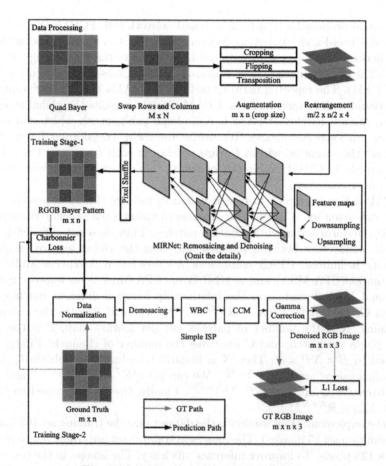

Fig. 10. Model architecture of JHC-SJTU.

images generated by the ISP without updating its parameters, i.e. the demosaicing network. The jointly remosaicing and denosing network is MIRNet [7]. The main contributions of the proposed solution include:

– We propose a preprocessing method: for quad-Bayer raw data, we first swap the second column and the third column in each 4×4 quad-Bayer unit, and then swap the second row and the third row of each unit. After that, the quad-Bayer raw data is converted to a RGGB-alike Bayer. Then, we decompose the RGGB map into four channel maps that are the R channel, G1 channel, G2 channel, and B channel, respectively. With these preprocessing, we can extract the channel information collected from different sensors.

– After the training in the first stage, we concatenate the trained ISP to the jointly remosaicing and denoising network. The output of our network and the ground truth RGGB raw data are processed to generate the RGB images. Then, the generated ground-truth images are used as a color-supervision to finetune our network. In this stage, we freeze the parameters of ISP. Compared to raw data, RGB images contain more information, which can further improve the quality of the generated raw data. Our experiments demonstrate the improvement in PSNR is about 0.2 dB.

– We also use data augmentation to expand the training samples. In training, cropping, flipping, and transposition are randomly applied to the raw data. In addition, we design a cropping-based method to unify the Bayer pattern after these augmentations. The experimental results show these augmentation methods can increase the PSNR value by about 0.4 dB.

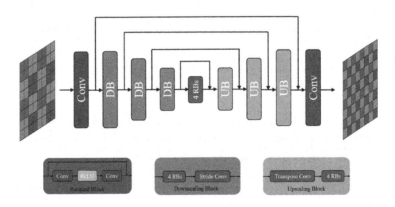

Fig. 11. Model architecture of the op-summer-po team.

4.6 Op-summer-po

Op-summer-po proposed to do the Quad remosaic and denoise using DRUNet [9]. The overall model architecture is shown in Fig. 11. The DRUNet has four scales with 64, 128, 256, and 512 channels respectively. Each scale has an identity skip connection between the stride convolution (stride = 2) and the transpose convolution. This connection concatenates encoder and decoder features. Each encoder or decoder includes four residual blocks.

To avoid detail loss during denoising, LPIPS [10] is used both on the raw and RGB in the training. No pre-trained models or external datasets were used for the task. During training, the resolution of HR patches is 128 × 128. We implemented the experiment on 1 * V100 with PyTorch and the batch size is 16 per GPU. The optimizer is Adam. The learning rate is initially set as 0.0001, and gets 0.5 decays every 100 epochs.

5 Conclusions

In this paper, we summarized the Joint Quad Remosaic and Denoise challenge in the first Mobile Intelligent Photography and Imaging workshop (MIPI 2022) held in conjunction with ECCV 2022. The participants were provided with a high-quality training/testing dataset, which is now available for researchers to download for future researches. We are excited to see the so many submissions within such a short period, and we look forward for more researches in this area.

Acknowledgements. We thank Shanghai Artificial Intelligence Laboratory, Sony, and Nanyang Technological University to sponsor this MIPI 2022 challenge. We thank all the organizers and participants for their great work.

A Teams and Affiliations

BITSpectral
Title: Cross-Patch Attention Network for Quad Joint Remosaic and Denoise
Members: Zhen Wang, (wzhstruggle@bit.edu.cn), Daoyu Li, Yuzhe Zhang, Lintao Peng, Xuyang Chang, Yinuo Zhang
Affiliations: Beijing Institute of Technology

HITZST01
Title: Multi-scale convolution network (JDRMCANet) with joint denoise and remosaic.
Members: [1]Yaqi Wu (titimasta@163.com), [2]Xun Wu, [3]Zhihao Fan, [4]Chengjie Xia, Feng Zhang,
Affiliations: [1]Harbin Institute of Technology, Harbin, 150001, China, [2]Tsinghua University, Beijing, 100084, China, [3]University of Shanghai for Science and Technology, Shanghai, 200093, China, [4]Zhejiang University, Hangzhou, 310027, China

IMEC-IPI & NPU-MPI
Title: Hard Datasets and Invariance guided Parallel Swin-Conv-Attention Network for Quad Joint Remosaic and Denoise.
Members: [1]Haijin Zeng (Haijin.Zeng@imec.be), [2]Kai Feng, [2]Yongqiang Zhao, [1]Hiep Quang Luong, [1]Jan Aeltferman, [1]Anh Minh Truong, and [1]Wilfried Philips.
Affiliations: [1]IMEC & Ghent University [2]Northwestern Polytechnical University

JHC-SJTU
Title: Jointly Remosaicing and Denoising for Quad Bayer with Multi-Resolution Feature Extraction and Fusion.
Members: [1]Xiaohong Liu (xiaohongliu@sjtu.edu.cn), [1]Jun Jia, [1]Hanchi Sun, [1]Guangtao Zhai, [2]Anlong Xiao, [2]Qihang Xu
Affiliations: [1]Shanghai Jiao Tong University, [2]Transsion

MegNR
Title: HAUformer: Hybrid Attention-guided U-shaped Transformer for Quad Remosaic Image Restoration.
Members: Ting Jiang (jiangting@megvii.com), Qi Wu, Chengzhi Jiang, Mingyan Han, Xinpeng Li, Wenjie Lin, Youwei Li, Haoqiang Fan and Shuaicheng Liu
Affiliations: Megvii Technology

op-summer-po
Title: Two LPIPS Functions in Raw and RGB domains for Quad-Bayer Joint Remosaic and Denoise.
Members: [1]Rongyuan Wu (1104138645@qq.com), [1]Lingchen Sun, [1,2]Qiaosi Yi
Affiliations: [1]OPPO Research Institute, [2]East China Normal University

References

1. Abdelhamed, A., Afifi, M., Timofte, R., Brown, M.S.: NTIRE 2020 challenge on real image denoising: dataset, methods and results. In: Proceedings of the IEEE/CVF Conference on Computer Vision and Pattern Recognition (CVPR) Workshops (2020)
2. Kim, B.-H., Song, J., Ye, J.C., Baek, J.H.: PyNET-CA: enhanced PyNET with channel attention for end-to-end mobile image signal processing. In: Bartoli, A., Fusiello, A. (eds.) ECCV 2020. LNCS, vol. 12537, pp. 202–212. Springer, Cham (2020). https://doi.org/10.1007/978-3-030-67070-2_12
3. Shi, W., et al.: Real-time single image and video super-resolution using an efficient sub-pixel convolutional neural network. In: Proceedings of the IEEE Conference on Computer Vision and Pattern Recognition, pp. 1874–1883 (2016)
4. Sun, B., Zhang, Y., Jiang, S., Fu, Y.: Hybrid pixel-unshuffled network for lightweight image super-resolution. arXiv preprint arXiv:2203.08921 (2022)
5. Wang, Z., Cun, X., Bao, J., Zhou, W., Liu, J., Li, H.: Uformer: a general U-shaped transformer for image restoration. In: Proceedings of the IEEE/CVF Conference on Computer Vision and Pattern Recognition, pp. 17683–17693 (2022)
6. Wang, Z., Bovik, A.C., Sheikh, H.R., Simoncelli, E.P.: Image quality assessment: from error visibility to structural similarity. IEEE Trans. Image Process. **13**(4), 600–612 (2004)
7. Zamir, S.W., et al.: Learning enriched features for real image restoration and enhancement. In: Vedaldi, A., Bischof, H., Brox, T., Frahm, J.-M. (eds.) ECCV 2020. LNCS, vol. 12370, pp. 492–511. Springer, Cham (2020). https://doi.org/10.1007/978-3-030-58595-2_30
8. Zhang, K., et al.: Practical blind denoising via swin-conv-unet and data synthesis. arXiv preprint (2022)
9. Zhang, K., Li, Y., Zuo, W., Zhang, L., Van Gool, L., Timofte, R.: Plug-and-play image restoration with deep denoiser prior. IEEE Trans. Pattern Anal. Mach. Intell. (2021)
10. Zhang, R., Isola, P., Efros, A.A., Shechtman, E., Wang, O.: The unreasonable effectiveness of deep features as a perceptual metric. In: Proceedings of the IEEE Conference on Computer Vision and Pattern Recognition (CVPR) (2018)
11. Zhang, Y., Li, K., Li, K., Wang, L., Zhong, B., Fu, Y.: Image super-resolution using very deep residual channel attention networks. In: Ferrari, V., Hebert, M., Sminchisescu, C., Weiss, Y. (eds.) ECCV 2018. LNCS, vol. 11211, pp. 294–310. Springer, Cham (2018). https://doi.org/10.1007/978-3-030-01234-2_18

MIPI 2022 Challenge on RGBW Sensor Re-mosaic: Dataset and Report

Qingyu Yang[1(✉)], Guang Yang[1,2,3,4,5,6,7], Jun Jiang[1], Chongyi Li[4],
Ruicheng Feng[4], Shangchen Zhou[4], Wenxiu Sun[2,3], Qingpeng Zhu[2],
Chen Change Loy[4], Jinwei Gu[1,3], Lingchen Sun[5], Rongyuan Wu[5], Qiaosi Yi[5,6],
Rongjian Xu[7], Xiaohui Liu[7], Zhilu Zhang[7], Xiaohe Wu[7], Ruohao Wang[7],
Junyi Li[7], Wangmeng Zuo[7], and Faming Fang[6]

[1] SenseBrain, San Jose, USA
{yangqingyu,jiangjun}@sensebrain.site
[2] SenseTime Research and Tetras.AI, Beijing, China
[3] Shanghai AI Laboratory, Shanghai, China
[4] Nanyang Technological University, Singapore, Singapore
[5] OPPO Research Institute, Shenzhen, China
slcbbd111@sina.com
[6] East China Normal University, Shanghai, China
[7] Harbin Institute of Technology, Harbin, China

Abstract. Developing and integrating advanced image sensors with novel algorithms in camera systems is prevalent with the increasing demand for computational photography and imaging on mobile platforms. However, the lack of high-quality data for research and the rare opportunity for in-depth exchange of views from industry and academia constrain the development of mobile intelligent photography and imaging (MIPI). To bridge the gap, we introduce the first MIPI challenge including five tracks focusing on novel image sensors and imaging algorithms. In this paper, RGBW Joint Remosaic and Denoise, one of the five tracks, working on the interpolation of RGBW CFA to Bayer at full-resolution is introduced. The participants were provided with a new dataset including 70 (training) and 15 (validation) scenes of high-quality RGBW and Bayer pair. In addition, for each scene, RGBW of different noise level were provided at 0 dB, 24 dB and 42 dB. All the data were captured using a RGBW sensor in both outdoor and indoor conditions. The final results are evaluated using objective metrics including PSNR, SSIM [5], LPIPS [7] and KLD. A detailed description of all models developed in this challenge is provided in this paper. More details of this challenge and the link to the dataset can be found in https://github.com/mipi-challenge/MIPI2022.

Keywords: RGBW · Remosaic · Bayer · Denoise · MIPI challenge

Q. Yang, J. Jiang, C. Li, S. Zhou, R. Feng, W. Sun, Q. Zhu, C. C. Loy, J. Gu are the MIPI 2022 challenge organizers
The other authors participated in the challenge. Please refer to Appendix A for details.
MIPI 2022 challenge website: http://mipi-challenge.org/.

L. Karlinsky et al. (Eds.): ECCV 2022 Workshops, LNCS 13805, pp. 36–45, 2023.
https://doi.org/10.1007/978-3-031-25072-9_3

1 Introduction

RGBW is a new type of CFA pattern (Fig. 1) designed for image quality enhancement under low light conditions. Thanks to the higher optical transmittance of white pixels over conventional red, green and blue pixels, the signal-to-noise (SNR) ratio of the sensor output becomes significantly improved, thus boosting the image quality especially under low light conditions. Recently several phone OEMs, including Transsion, Vivo and Oppo have adopted RGBW sensors in their flagship smartphones to improve the camera image quality [1–3].

On the other hand, conventional camera ISPs can only work with Bayer patterns, thereby requiring an interpolation procedure to convert RGBW to a Bayer pattern. The interpolation process is usually referred to as remosaic, and a good remosaic algorithm should be able (1) to get a Bayer output from RGBW with least artifacts, and (2) to fully take advantage of the SNR and resolution benefit of white pixels.

The remosaic problem becomes more challenging when the input RGBW becomes noisy, especially under low light conditions. A joint remosaic and denoise task is thus in demand for real world applications.

Fig. 1. The RGBW remosaic task.

In this challenge, we intend to remosaic the RGBW input to obtain a Bayer at the same spatial resolution. The solution is not necessarily deep-learning. However, to facilitate the deep learning training, we provide a dataset of high-quality RGBW and Bayer pair (70 scenes for training, 15 for validation, and 15 for test). We provide a data-loader to read these files and a simple ISP in Fig. 2 to visualize the RGB output from the Bayer and to calculate loss functions. The participants are also allowed to use other public-domain dataset. The algorithm performance is evaluated and ranked using objective metrics: Peak Signal-to-Noise Ratio (PSNR), Structural Similarity Index (SSIM) [5], Learned Perceptual Image Patch Similarity (LPIPS) [7] and KL-divergence (KLD). The objective metrics of a baseline method is available as well to provide a benchmark.

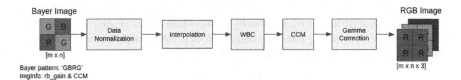

Fig. 2. An ISP to visualize the output Bayer and to calculate the loss function.

This challenge is a part of the Mobile Intelligent Photography and Imaging (MIPI) 2022 workshop and challenges which emphasizing the integration of novel image sensors and imaging algorithms, which is held in conjunction with ECCV 2022. It consists of five competition tracks:

1. RGB+ToF Depth Completion uses sparse, noisy ToF depth measurements with RGB images to obtain a complete depth map.
2. Quad-Bayer Re-mosaic converts Quad-Bayer RAW data into Bayer format so that it can be processed with standard ISPs.
3. RGBW Sensor Re-mosaic converts RGBW RAW data into Bayer format so that it can be processed with standard ISPs.
4. RGBW Sensor Fusion fuses Bayer data and a monochrome channel data into Bayer format to increase SNR and spatial resolution.
5. Under-display Camera Image Restoration improves the visual quality of image captured by a new imaging system equipped under-display camera.

2   Challenge

To develop high quality RGBW Remosaic solution, we provide the following resources for participants:

- A high-quality RGBW and Bayer dataset; As far as we know, this is the first and only dataset consisting of aligned RGBW and Bayer pair, relieving the pain of data collection to develop learning-based remosaic algorithms;
- A data processing code with dataloader to help participants get familiar with the provided dataset;
- A simple ISP including basic ISP blocks to visualize the algorithm output and to calculate the loss function on RGB results;
- A set of objective image quality metrics to measure the performance of a developed solution;

2.1   Problem Definition

RGBW remosaic aims to interpolate the input RGBW CFA pattern to obtain a Bayer of the same resolution. The remosaic task is needed mainly because current camera ISPs usually cannot process CFAs other than the Bayer pattern. In addition, the remosaic task becomes more challenging when the noise level gets higher, thus requiring more advanced algorithms to avoid image quality artifacts. In addition to the image quality requirement, RGBW sensors are widely used on smartphones with limited computational budget and battery life, thus requiring the remosaic algorithm to be lightweight at the same time. While we do not rank solutions based on running time or memory footprint, computational cost is one of the most important criteria in real applications.

2.2 Dataset: Tetras-RGBW-RMSC

The training data contains 70 scenes of aligned RGBW (input) and Bayer (ground-truth) pair. For each scene, noise is sythesized on the 0 dB RGBW input to provide the noisy RGBW input at 24 dB and 42 dB respectively. The synthesized noise consists of read noise and shot noise, and the noise model is measured on a RGBW sensor. The data generation steps are shown in Fig. 3. The testing data contains RGBW input of 15 scenes at 0 dB, 24 dB and 42 dB, but the GT Bayer results are not available to participants.

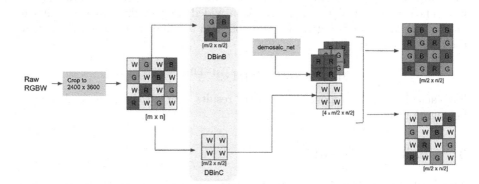

Fig. 3. Data generation of the RGBW remosaic task. The RGBW raw data is captured using a RGBW sensor and cropped to be 2400 × 3600. A Bayer (DbinB) and white (DbinC) image are obtained by averaging the same color in the diagonal direction within a 2 × 2 block. We demosaic the Bayer (DbinB) to get a RGB using the demosaic-net. The white (DbinC) is concatenated to the RGB image to have RGBW for each pixel, which is in turn mosaiced to get the input RGBW and aligned ground truth Bayer.

2.3 Challenge Phases

The challenge consisted of the following phases:

1. Development: The registered participants get access to the data and baseline code, and are able to train the models and evaluate their run-time locally.
2. Validation: The participants can upload their models to the remote server to check the fidelity scores on the validation dataset, and to compare their results on the validation leaderboard.
3. Testing: The participants submit their final results, codes, models, and fact-sheets.

2.4 Scoring System

Objective Evaluation. The evaluation consists of (1) the comparison of the remosaic output (Bayer) with the reference ground truth Bayer, and (2) the comparison of RGB from the predicted and ground truth Bayer using a simple ISP (the code of the simple ISP is provided). We use

1. Peak Signal To Noise Ratio (PSNR)
2. Structural Similarity Index Measure (SSIM) [5]
3. KL divergence (KLD)
4. Learned perceptual image patch similarity (LPIPS) [7]

to evaluate the remosaic performance. The PNSR, SSIM and LPIPS will be applied to the RGB from the Bayer using the provided simple ISP code, while KL divergence is evaluated on the predicted Bayer directly.

A metric weighting PSNR, SSIM, KL divergence, and LPIPS is used to give the final ranking of each method, and we will report each metric separately as well. The code to calculate the metrics is provided. The weighted metric is shown below. The M4 score is between 0 and 100, and the higher the score, the better the overall image quality.

$$M4 = PSNR \cdot \text{SSIM} \cdot 2^{1-\text{LPIPS}-\text{KLD}}, \tag{1}$$

For each dataset we report the average results over all the processed images belonging to it.

3   Challenge Results

The results of the top three teams are shown in Table 1 in the final test phase after we verified their submission using their code. **op-summer-po**, **HIT-IIL** and **Eating, Drinking, and Playing** are the top three teams ranked by M4 by Eq. 1, and **op-summer-po** shows the best overall performance. The proposed methods are described in Sect. 4 and the team members and affiliations are listed in Appendix A.

Table 1. MIPI 2022 Joint RGBW Remosaic and Denoise challenge results and final rankings. PSNR, SSIM, LPIPS and KLD are calculated between the submitted results from each team and the ground truth data. A weighted metric, M4, by Eq. 1 is used to rank the algorithm performance, and the top three teams with the highest M4 are included in the table.

Team name	PSNR	SSIM	LPIPS	KLD	M4
op-summer-po	36.83	0.957	0.115	0.018	64.89
HIT-IIL	36.34	0.95	0.129	0.02	63.12
Eating, Drinking, and Playing	36.77	0.957	0.132	0.019	63.98

To learn more about the algorithm performance, we evaluated the qualitative image quality in addition to the objective IQ metrics in Fig. 4 and 5 respectively. While all teams in Table 1 have achieved high PSNR and SSIM, the detail and texture loss can be found on the book cover in Fig. 4 and on the mesh in Fig. 5 when the input has a large amount of noise. Oversmoothing tends to yield higher PSNR at the cost of detail loss perceptually.

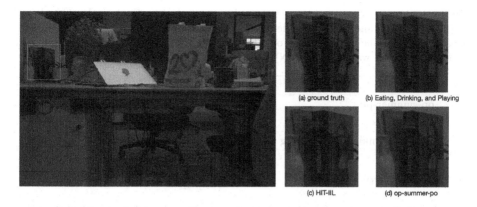

Fig. 4. Qualitative image quality (IQ) comparison. The results of one of the test scenes (42 dB) are shown. While the top three remosaic methods achieve high objective IQ metrics in Table 1, details and texture loss are noticeable on the book cover. The texts on the book are barely interpretable in (b), (c) and (d).

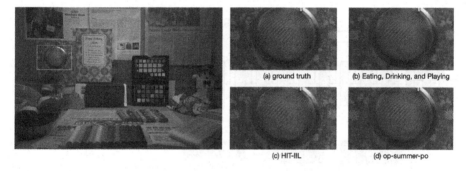

Fig. 5. Qualitative image quality (IQ) comparison. The results of one of the test scenes (42 dB) are shown. Oversmoothing in the top three methods in Table 1 can be found when compared with the ground truth. The text under the mesh can barely be recognized, and the mesh texture become distorted in (b),(c) and (d).

In addition to benchmarking the image quality of remosaic algorithms, computational efficiency is evaluated because of wide adoptions of RGBW sensors on smartphones. We measured the running time of the remosaic solutions of the top three teams (based on M4 by Eq. 1) in Table 2. While running time is not employed in the challenge to rank remosaic algorithms, the computational cost is of critical importance when developing algorithms for smartphones. HIT-IIL achieved the shortest running time among the top three solutions on a workstation GPU (NVIDIA Tesla V100-SXM2-32GB). With sensor resolution of mainstream smartphones reaching 64M or even higher, power-efficient remosaic algorithms are highly desirable.

Table 2. Running time of the top three solutions ranked by Eq. 1 in the 2022 Joint RGBW Remosaic and Denoise challenge. The running time of input of 1200×1800 was measured, while the running time of a 64M input RGBW was based on estimation. The measurement was taken on a NVIDIA Tesla V100-SXM2-32GB GPU.

Team name	1200×1800 (measured)	64M (estimated)
op-summer-po	6.2 s	184 s
HIT-IIL	**4.1 s**	**121.5 s**
Eating, Drinking and Playing	10.4 s	308 s

4 Challenge Methods

In this section, we describe the solutions submitted by all teams participating in the final stage of MIPI 2022 Joint RGBW Remosaic and Denoise Challenge.

4.1 Op-summer-po

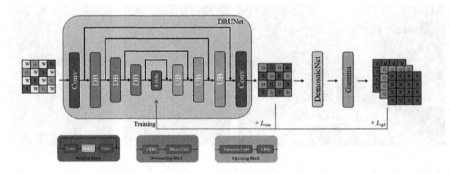

Fig. 6. Model architecture of op-summer-po.

op-summer-po proposed a framework based on DRUNet [6] as shown in Fig. 6, which has four scales with 64, 128, 256, and 512 channels respectively. Each scale has an identity skip connection between the stride convolution (stride = 2) and the transpose convolution. This connection concatenates encoder and decoder features. Each encoder or decoder includes four residual blocks.

The input of the framework is the raw RGBW image, and the output is the estimated raw Bayer image. Then the estimated raw Bayer image is sent to DemosaicNet and Gamma Transform to get the full-resolution RGB image. Moreover, to get a better perception quality, two LPIPS [7] functions in both raw and rgb domains are used.

4.2 HIT-IIL

HIT-IIL proposed an end-to-end method to learn RGBW re-mosaicing and denoising jointly. We split RGBW images into 4-channel while maintaining image size, as shown in Fig. 7. Specifically, we first repeat the RGBW image as 4 channels, then make each channel represent a color type (i.e., one of white, green, blue and red). When the pixel color type is different from the color represented by the channel, its value is set to 0. Such input mode provides the positional information of the color for the network and significantly improves performance in experiments.

As for the end-to-end network, We adopt NAFNet [4] to re-mosaic and denoise RGBW images. NAFNet contains the 4-level encoder-decoder and bottleneck. For the encoder part, the numbers of NAFNet's blocks for each level are 2, 4, 8, and 24. For the decoder part, the numbers of NAFNet's blocks for 4 levels are all 2. The number of NAFNet's blocks for the bottleneck is 12.

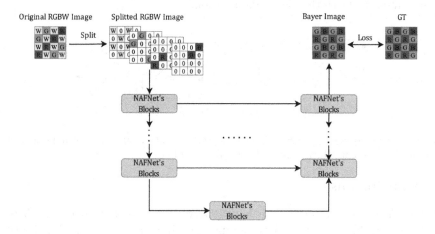

Fig. 7. Model architecture of HIT-IIL.

4.3 Eating, Drinking and Playing

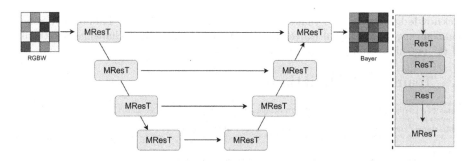

Fig. 8. Model architecture of Eating, Drinking and Playing.

Eating, Drinking and Playing proposed a UNet-Transformer for RGBW Joint Remosaic and Denoise, and its overall architecture is in Fig. 8. To be more specific, we aim to estimate a raw Bayer image O from the corresponding raw

RGBW image I. To achieve this, we decrease the distance between the output O and the ground truth raw Bayer image GT by minimizing L1 loss.

Similar to DRUNet [6], the UNet-Transformer adopts an encode-decode structure and has four levels with 64, 128, 256, and 512 channels respectively. The main difference between UNet-Transformer and DRUNet is that UNet-Transformer adopts Multi-ResTransformer (MResT) blocks rather than residual convolution blocks in each level of the encode and decode. As the main component of UNet-Transformer, the MResT cascades multiple Res-Transformer (ResT) blocks, which can study local and global information simultaneously. The structure of Rest is illustrated in Fig. 9. In the ResT, the first two convolution layer is adopted to study the local information and then we employ a self-attention to learn the global information.

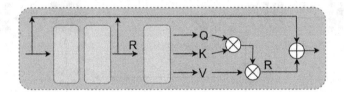

Fig. 9. The structure of the Res-Transformer(ResT) block. R is the reshape and the blue block is the convolution layer. (Color figure online)

5   Conclusions

In this paper, we summarized the Joint RGBW Remosaic and Denoise challenge in the first Mobile Intelligent Photography and Imaging workshop (MIPI 2022) held in conjunction with ECCV 2022. The participants were provided with a high-quality training/testing dataset, which is now available for researchers to download for future researches. We are excited to see so many submissions within such a short period, and we look forward for more researches in this area.

Acknowledgements. We thank Shanghai Artificial Intelligence Laboratory, Sony, and Nanyang Technological University to sponsor this MIPI 2022 challenge. We thank all the organizers and participants for their great work.

A   Teams and Affiliations

op-summer-po
Title: Two LPIPS Functions in Raw and RGB domains for RGBW Joint Remosaic and Denoise
Members: [1]Lingchen Sun, (slcbbd111@sina.com), [1]Rongyuan Wu, [1,2]Qiaosi Yi
Affiliations: [1]OPPO Research Institute, [2]East China Normal University

HIT-IIL
Title: Data Input and Augmentation Strategies for RGBW Image Re-Mosaicing
Members: Rongjian Xu (ronjon.xu@gmail.com), Xiaohui Liu, Zhilu Zhang, Xiaohe Wu, Ruohao Wang, Junyi Li, Wangmeng Zuo
Affiliations: Harbin Institute of Technology

Eating, Drinking, and Playing
Title: UNet-Transformer for RGBW Joint Remosaic and Denoise
Members: [1,2]Qiaosi Yi (51205901027@stu.ecnu.edu.cn), [1]Rongyuan Wu, [1]Lingchen Sun, [2]Faming Fang
Affiliations: [1]OPPO Research Institute, [2]East China Normal University

References

1. Camon 19 Pro. https://www.tecno-mobile.com/phones/product-detail/product/camon-19-pro-5g
2. OPPO unveils multiple innovative imaging technologies. https://www.oppo.com/en/newsroom/press/oppo-future-imaging-technology-launch/
3. Vivo x80 is the only vivo smartphone with a Sony imx866 sensor: The world's first RGBW bottom sensors. https://www.vivoglobal.ph/vivo-X80-is-the-only-vivo-smartphone-with-a-Sony-IMX866-Sensor-The-Worlds-First-RGBW-Bottom-Sensors
4. Chen, L., Chu, X., Zhang, X., Sun, J.: Simple baselines for image restoration. arXiv preprint arXiv:2204.04676 (2022)
5. Wang, Z., Bovik, A.C., Sheikh, H.R., Simoncelli, E.P.: Image quality assessment: from error visibility to structural similarity. IEEE Trans. Image Process. **13**(4), 600–612 (2004)
6. Zhang, K., Li, Y., Zuo, W., Zhang, L., Van Gool, L., Timofte, R.: Plug-and-play image restoration with deep denoiser prior. IEEE Trans. Pattern Anal. Mach. Intell. (2021)
7. Zhang, R., Isola, P., Efros, A.A., Shechtman, E., Wang, O.: The unreasonable effectiveness of deep features as a perceptual metric. In: Proceedings of the IEEE Conference on Computer Vision and Pattern Recognition (CVPR) (2018)

MIPI 2022 Challenge on RGBW Sensor Fusion: Dataset and Report

Qingyu Yang[1]([✉]), Guang Yang[1,2,3,4], Jun Jiang[1], Chongyi Li[4],
Ruicheng Feng[4], Shangchen Zhou[4], Wenxiu Sun[2,3], Qingpeng Zhu[2],
Chen Change Loy[4], Jinwei Gu[1,3], Zhen Wang[5], Daoyu Li[5], Yuzhe Zhang[5],
Lintao Peng[5], Xuyang Chang[5], Yinuo Zhang[5], Liheng Bian[5], Bing Li[6],
Jie Huang[6], Mingde Yao[6], Ruikang Xu[6], Feng Zhao[6], Xiaohui Liu[7],
Rongjian Xu[7], Zhilu Zhang[7], Xiaohe Wu[7], Ruohao Wang[7], Junyi Li[7],
Wangmeng Zuo[7], Zhuang Jia[8], DongJae Lee[9], Ting Jiang[10], Qi Wu[10],
Chengzhi Jiang[10], Mingyan Han[10], Xinpeng Li[10], Wenjie Lin[10], Youwei Li[10],
Haoqiang Fan[10], and Shuaicheng Liu[10]

[1] SenseBrain, San Jose, USA
{yangqingyu,jiangjun}@sensebrain.site
[2] SenseTime Research and Tetras.AI, Beijing, China
[3] Shanghai AI Laboratory, Shanghai, China
[4] Nanyang Technological University, Singapore, Singapore
[5] Beijing Institute of Technology, Beijing, China
[6] University of Science and Technology of China, Hefei, China
frigid@mail.ustc.edu.cn
[7] Harbin Institute of Technology, Harbin, China
[8] Xiaomi, Beijing, China
jiazhuang@xiaomi.com
[9] KAIST, Daejeon, South Korea
jhtwosun@kaist.ac.kr
[10] Megvii Technology, Beijing, China
jiangting@megvii.com

Abstract. Developing and integrating advanced image sensors with novel algorithms in camera systems is prevalent with the increasing demand for computational photography and imaging on mobile platforms. However, the lack of high-quality data for research and the rare opportunity for in-depth exchange of views from industry and academia constrain the development of mobile intelligent photography and imaging (MIPI). To bridge the gap, we introduce the first MIPI challenge including five tracks focusing on novel image sensors and imaging algorithms. In this paper, RGBW Joint Fusion and Denoise, one of the five tracks, working on the fusion of binning-mode RGBW to Bayer at half resolution is introduced. The participants were provided with a new dataset including 70 (training) and 15 (validation) scenes of high-quality RGBW

Q. Yang, J. Jiang, C. Li, S. Zhou, R. Feng, W. Sun, Q. Zhu, C. C. Loy, J. Gu, are the MIPI 2022 challenge organizers. The other authors participated in the challenge. Please refer to Appendix A for details.
MIPI 2022 challenge website: http://mipi-challenge.org/.

and Bayer pair. In addition, for each scene, RGBW of 24 dB and 42 dB are provided. All the data were captured using a RGBW sensor in both outdoor and indoor conditions. The final results are evaluated using objective metrics including PSNR, SSIM [11], LPIPS [15] and KLD. A detailed description of all models developed in this challenge is provided in this paper. More details of this challenge and the link to the dataset can be found in https://github.com/mipi-challenge/MIPI2022

Keywords: RGBW · Fusion · Bayer · Denoise · MIPI challenge

1   Introduction

RGBW is a new type of CFA pattern (Fig. 1(a)) designed for image quality enhancement under low light conditions. Thanks to the higher optical transmittance of white pixels over conventional red, green and blue pixels, the signal-to-noise (SNR) ratio of the sensor output becomes significantly improved, thus boosting the image quality especially under low light conditions. Recently several phone OEMs, including Transsion, Vivo and Oppo have adopted RGBW sensors in their flagship smartphones to improve the camera image quality [1–3].

The binning mode of RGBW is mainly used in the camera preview mode and video mode, in which pixels of the same color are averaged in the diagonal direction within a 2×2 window in RGBW to further improve the image quality and to reduce the noise. A fusion algorithm is thereby needed to take the input of a diagonal-binning-bayer (DBinB) and a diagonal-binning-white (DBinC) to obtain a Bayer of better signal-to-noise (SNR) ratio in Fig. 1(b). A good fusion algorithm should be able (1) to get a Bayer output from RGBW with least artifacts, and (2) to fully take advantage of the SNR and resolution benefit of white pixels.

The RGBW fusion problem becomes more challenging when the input DbinB and DbinC become noisy especially under low light conditions. A joint fusion and denoise task is thus in demand for real world applications.

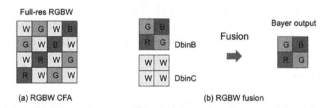

Fig. 1. The RGBW Fusion task: (a) the RGBW CFA. (b) In the binning mode, DbinB and DbinC are obtained by diagonal averaging of pixels of the same color within a 2×2 window. The joint fusion and denoise algorithm takes DbinB and DbinC as input to get a high-quality Bayer.

In this challenge, we intend to fuse the RGBW inputs (DBinB and DBinC in Fig. 1(b)) to denoise and improve the Bayer at half resolution. The solution is not necessarily deep-learning. However, to facilitate the deep learning training, we provide a dataset of high-quality binning-mode RGBW (DBinB and DBinC) and the output Bayer pair (70 scenes for training, 15 for validation, and 15 for test). We provide a data-loader to read these files and a simple ISP in Fig. 2 to visualize the RGB output from the Bayer and to calculate loss functions. The participants are also allowed to use other public-domain dataset. The algorithm performance is evaluated and ranked using objective metrics: Peak Signal-to-Noise Ratio (PSNR), Structural Similarity Index (SSIM) [11], Learned Perceptual Image Patch Similarity (LPIPS) [15] and KL-divergence (KLD). The objective metrics of a baseline method is available as well to provide a benchmark.

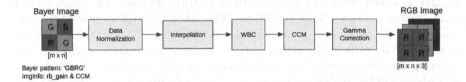

Fig. 2. An ISP to visuailize the output Bayer and to calculate the loss function.

This challenge is a part of the Mobile Intelligent Photography and Imaging (MIPI) 2022 workshop and challenges which emphasizing the integration of novel image sensors and imaging algorithms, which is held in conjunction with ECCV 2022. It consists of five competition tracks:

1. RGB+ToF Depth Completion uses sparse, noisy ToF depth measurements with RGB images to obtain a complete depth map.
2. Quad-Bayer Re-mosaic converts Quad-Bayer RAW data into Bayer format so that it can be processed with standard ISPs.
3. RGBW Sensor Re-mosaic converts RGBW RAW data into Bayer format so that it can be processed with standard ISPs.
4. RGBW Sensor Fusion fuses Bayer data and a monochrome channel data into Bayer format to increase SNR and spatial resolution.
5. Under-display Camera Image Restoration improves the visual quality of image captured by a new imaging system equipped under-display camera.

2 Challenge

To develop high quality RGBW fusion solution, we provide the following resources for participants:

- A high-quality RGBW (DbinB and DbinC) and Bayer dataset; As far as we know, this is the first and only dataset consisting of aligned RGBW and Bayer pair, relieving the pain of data collection to develop learning-based fusion algorithms;

- A data processing code with dataloader to help participants get familiar with the provided dataset;
- A simple ISP including basic ISP blocks to visualize the algorithm output and to calculate the loss function on RGB results;
- A set of objective image quality metrics to measure the performance of a developed solution;

2.1 Problem Definition

The RGBW fusion task aims to fuse the DBinB and DBinC of RGBW (Fig. ?? (b)) to improve the image quality of the Bayer output at half resolution. By incorporating the white pixels (DBinC) of higher spatial resolution and higher SNR, the output Bayer potentially would have better image quality. In addition, the RGBW binning mode is mainly used for the preview and video modes on smartphones, thus requiring the fusion algorithm to be lightweight and power-efficient. While we do not rank solutions based on the running time or memory footprint, the computational cost is one of the most important criteria in real applications.

2.2 Dataset: Tetras-RGBW-Fusion

The training data contains 70 scenes of aligned RGBW (DBinB and DBinC input) and Bayer (ground-truth) pair. For each scene, DBinB at 0 dB is used as the ground truth. Noise is sythesized on the 0 dB half-resolution DbinB and DbinC data to provide the noisy input at 24 dB and 42 dB respectively. The synthesized noise consists of read noise and shot noise, and the noise model is measured on a RGBW sensor. The data generation steps are shown in Fig. 3. The testing data contains DbinB and DbinC inputs of 15 scenes at 24 dB and 42 dB, but the GT Bayer results are not available to participants.

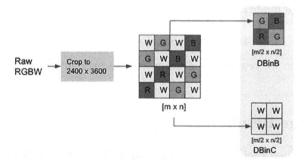

Fig. 3. Data generation of the RGBW fusion task. The RGBW raw data is captured using a RGBW sensor and cropped to be 2400 × 3600. A Bayer (DbinB) and white (DbinC) image are obtained by averaging the same color in the diagonal direction within a 2 × 2 block.

2.3   Challenge Phases

The challenge consisted of the following phases:

1. Development: The registered participants get access to the data and baseline code, and are able to train the models and evaluate their run-time locally.
2. Validation: The participants can upload their models to the remote server to check the fidelity scores on the validation dataset, and to compare their results on the validation leaderboard.
3. Testing: The participants submit their final results, codes, models, and fact-sheets.

2.4   Scoring System

Objective Evaluation. The evaluation consists of (1) the comparison of the fused output (Bayer) with the reference ground truth Bayer, and (2) the comparison of RGB from the predicted and ground truth Bayer using a simple ISP (the code of the simple ISP is provided). We use

1. Peak Signal To Noise Ratio (PSNR)
2. Structural Similarity Index Measure (SSIM) [11]
3. KL divergence (KLD)
4. Learned perceptual image patch similarity (LPIPS) [15]

to evaluate the fusion performance. The PNSR, SSIM and LPIPS will be applied to the RGB from the Bayer using the provided simple ISP code, while KL divergence is evaluated on the predicted Bayer directly.

A metric weighting PSNR, SSIM, KL divergence, and LPIPS is used to give the final ranking of each method, and we will report each metric separately as well. The code to calculate the metrics is provided. The weighted metric is shown below. The M4 score is between 0 and 100, and the higher the score, the better the overall image quality.

$$M4 = PSNR \cdot \text{SSIM} \cdot 2^{1-\text{LPIPS}-\text{KLD}}, \tag{1}$$

For each dataset we report the average results over all the processed images belonging to it.

3   Challenge Results

Six teams submited their results in final phase. Table 1 summarizes the results in final test phase after we verified their submission using their code. **LLCKP**, **MegNR** and **jzsherlock** are the top three teams ranked by M4 by Eq. 1, and **LLCKP** shows the best overall performance. The proposed methods are described in Sect. 4 and the team members and affiliations are listed in Appendix A.

To learn more about the algorithm performance, we evaluated the qualitative image quality in addition to the objective IQ metrics in Fig. 4 and 5 respectively. While all teams in Table 1 have achieved high PSNR and SSIM, the detail and texture loss can be found on the yellow box in Fig. 4 and on the test chart in Fig. 5 when the input has a large amount of noise and the scene is under low light condition. Oversmoothing tends to yield higher PSNR at the cost of detail loss perceptually.

In addition to benchmarking the image quality of fusion algorithms, computational efficiency is evaluated because of wide adoptions of RGBW sensors on smartphones. We measured the running time of the RGBW fusion solutions of the top three teams in Table 2. While running time is not employed in the challenge to rank fusion algorithms, the computational cost is of critical importance when developing algorithms for smartphones. jzsherlock achieved the shortest running

Table 1. MIPI 2022 Joint RGBW Fusion and Denoise challenge results and final rankings. PSNR, SSIM, LPIPS and KLD are calculated between the submitted results from each team and the ground truth data. A weighted metric, M4, by Eq. 1 is used to rank the algorithm performance, and the top three teams with the highest M4 are highlighted.

Team name	PSNR	SSIM	LPIPS	KLD	M4
BITSpectral	36.53	0.958	0.126	0.027	63.27
BIVLab	35.09	0.94	0.174	0.0255	57.98
HIT-IIL	36.66	0.958	0.128	0.02196	63.62
jzsherlock	37.05	0.958	0.132	0.29	**63.84**
LLCKP	36.89	0.952	0.054	0.017	**67.07**
MegNR	36.98	0.96	0.098	0.0156	**65.55**

(a) ground truth   (b) jzsherlock

(c) LLCKP   (d) MegNR

Fig. 4. Qualitative image quality (IQ) comparison. The results of one of the test scenes (42 dB) are shown. While the top three fusion methods achieve high objective IQ metrics in Table 1, details and texture loss are noticeable on the yellow box. The texts on the box are barely interpretable in (b), (c) and (d) (Color figure online).

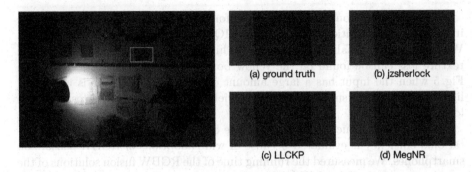

(a) ground truth   (b) jzsherlock

(c) LLCKP   (d) MegNR

Fig. 5. Qualitative image quality (IQ) comparison. The results of one of the test scenes (42 dB) are shown. Oversmoothing in the top three methods in Table 1 can be found when compared with the ground truth. The test chart becomes distorted in (b), (c), and (d).

time among the top three solutions on a workstation GPU (NVIDIA Tesla V100-SXM2-32 GB). With sensor resolution of mainstream smartphones reaching 64M or even higher, power-efficient fusion algorithms are highly desirable.

Table 2. Running time of the top three solutions ranked by Eq. 1 in the 2022 Joint RGBW Fusion and Denoise challenge. The running time of input of 1200×1800 was measured, while the running time of a 64M RGBW sensor was based on estimation (the binning-mode resolution of a 64M RGBW sensor is 16M). The measurement was taken on a NVIDIA Tesla V100-SXM2-32 GB GPU.

Team name	1200×1800 (measured)	16M (estimated)
jzsherlock	**3.7 s**	**27.4 s**
LLCKP	7.1 s	52.6 s
MegNR	12.4 s	91.9 s

4   Challenge Methods

In this section, we describe the solutions submitted by all teams participating in the final stage of MIPI 2022 RGBW Joint Fusion and Denoise Challenge.

4.1 BITSpectral

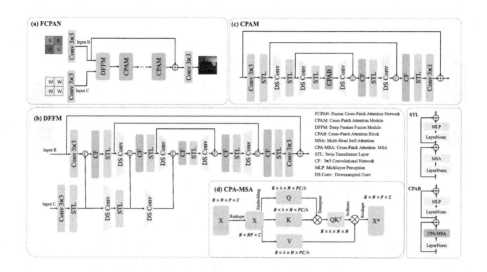

Fig. 6. Model architecture of BITSpectral.

BITSpectral developed a transformer-based network, Fusion Cross-Patch Attention Network (FCPAN), for this joint fusing and denoising task. The FCPAN is presented in Fig. 6(a) that consists of a Deep Feature Fusion Module (DFFM) and several Cross-Patch Attention Modules (CPAM). The input of DFFM contains a RGGB bayer pattern and a W channel. The output of DFFM is the fused features of RGBW, which is fed to CPAM for depth feature extraction. CPAM is a U-shape network with spatial downsampling to reduce the computational complexity. We propose to use 4 CPAMs in our network.

Figure 6 also includes details of Swin Transformer Layer [6] (STL), the Cross-Patch Attention Block (CPAB), and Cross-Patch Attention Multi-Head Self-Attention (CPA-MSA). We used STL to extract the attention within feature patches in each stage, and CPAB to obtain the global attention among patches for the innermost stage directly. Compared with STL, CPAB has extended range of perception due to the cross-patch attention.

4.2   BIVLab

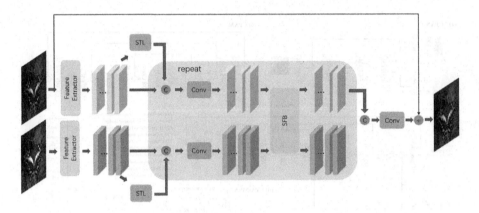

Fig. 7. Model architecture of BIVLab.

BIVLab proposed a Self-Guided Spatial-Frequency Complement Network(SG-SFCN) for RGBW joint fusion and denoise task. As shown in Fig. 7, the swin transformer layer (STL) [6] is adopted to extract rich features from DbinB and DbinC bayer separately. SpaFre blocks (SFB) [12] is then fusing the DbinB and DbinC bayer in complementary spatial and frequency domains. In order to handle the different noise levels, the features extracted by the STL, which contain the noise-level information, are applied to each SFB as a guidance. Finally, the denoised bayer is obtained by adding the predicted bayer residual to the original DbinB bayer. During the training, all the images are cropped to patches with size 720 × 720 in order to guarantee essential global information.

4.3   HIT-IIL

HIT-IIL proposed a NAFNet [4] based model for RGBW Joint Fusion and Denoise task. As shown in Fig. 8, the entire framework consists of 4-level encoder-decoder and bottleneck module. For the encoder part, the numbers of NAFNet's

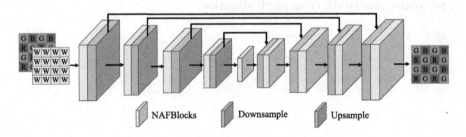

Fig. 8. Model architecture of HIT-IIL.

blocks for each level are 2, 2, 4, and 8. For the decoder part, the numbers of NAFNet's blocks are set to 2 for all 4 levels. In addition, the bottleneck module contains 24 NAFNet's blocks. Unlike the original NAFNet design, the skip connection between the input and the output is removed in our method.

During the training, we also use two data augmentation strategies. The first one is mixup, which generate synthesised images as:

$$\hat{\mathbf{x}} = \mathbf{a} * \mathbf{x}_{24} + (\mathbf{1} - \mathbf{a}) * \mathbf{x}_{42}, \tag{2}$$

In which $\mathbf{x}_{24}$ and $\mathbf{x}_{42}$ denote images of the same scene with noise levels of 24 dB and 42 dB. A random variable $\mathbf{a}$ is selected between 0 and 1 to generate the synthesised image $\hat{\mathbf{x}}$. Our second augmentation strategy is the image flip proposed by [5].

4.4 Jzsherlock

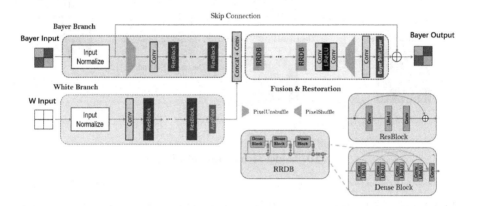

Fig. 9. Model architecture of jzsherlock.

Jzsherlock proposed a dual-branch network for the RGBW joint fusion and denoise task. The entire architecture, consisting of bayer branch and white branch, is shown in Fig. 9. The bayer branch's input is a normalized noisy bayer image and output the denoised result. After pixel unshuffle operation with scale = 2, the bayer image is converted to GBRG channels. We use stacked Res-Blocks without BatchNorm (BN) layers to extract the feature maps of noisy bayer image. On the other hand, the white branch extracts the features from corresponding white image using stacked ResBlocks as well. An average pooling layer rescale the white image features to the same size as bayer branch for feature fusion. Several Residual-in-Residual Dense Blocks (RRDB) [9] are applied on fused feature maps for the restoration. After the RRDB blocks, a Conv+LeakyReLU+Conv structure is applied to enlarge the feature map channels by scale of 4. Then pixel shuffle with scale = 2 is applied to upscale the feature maps to the input size. A Conv layer is used to convert the output to the GBRG 4 channels. Finally, a skip connection is applied to add the input bayer to form the final denoised result.

The network is trained by L1 loss in the normalized domain. The final selected input normalize method is min-max normalize with min = 64 and max = 1023, with values out of the range clipped.

4.5 LLCKP

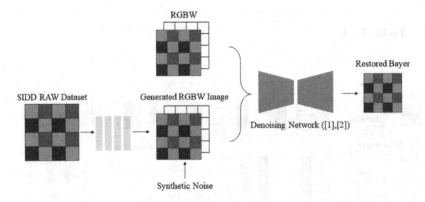

Fig. 10. Model architecture of LLCKP.

LLCKP proposed a denoising method that based on existing image restoration model [4,14]. As shown in Fig. 10, we synthesize RGBW images from GT GBRG images and with additional synthetic noise and real-noise pair (noisy images provided by challenge). We also used 20,000 pairs of RAW Image from SIDD with normal exposure and synthesize RGBW image as extra data. During the training, the Restormer model's [14] weights are pre-trained on SIDD RGB images. Data augmentation [5] and cutmix [13] are applied during the training phase.

4.6 MegNR

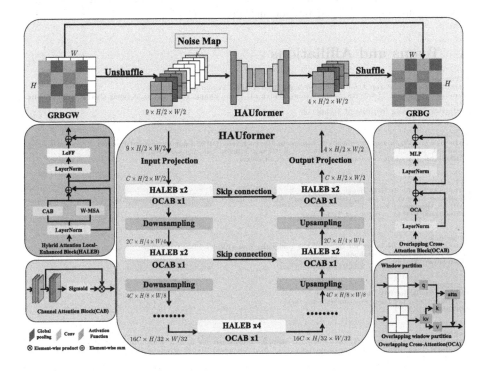

Fig. 11. Model architecture of MegNR.

MegNR proposed a pipeline for the RGBW Joint Fusion and Denoise task. The overall diagram is shown in Fig. 11. The pixel-unshuffle(PU) [8] is firstly applied to RGBW images to split it into independent channels. Inspired by Uformer [10], we develop our RGBW fusion and reconstruction network, HAUformer. We replace the LeWin Blocks [10] in Uformer's original design and includes two modules Hybrid Attention Local-Enhanced Block(HALEB) and Overlapping Cross-Attention Block(OCAB) to capture more long-range dependencies information and useful local context. Finally, the pixel-shuffle(PS) [7] module restores output to the standard bayer format.

5 Conclusions

In this paper, we summarized the Joint RGBW Fusion and Denoise challenge in the first Mobile Intelligent Photography and Imaging workshop (MIPI 2022) held in conjunction with ECCV 2022. The participants were provided with a high-quality training/testing dataset, which is now available for researchers to download for future researches. We are excited to see so many submissions within such a short period, and we look forward for more researches in this area.

Acknowledgements. We thank Shanghai Artificial Intelligence Laboratory, Sony, and Nanyang Technological University to sponsor this MIPI 2022 challenge. We thank all the organizers and participants for their great work.

A   Teams and Affiliations

BITSpectral
Title: Fusion Cross-Patch Attention Network for RGBW Joint Fusion and Denoise
Members: Zhen Wang (wzhstruggle@163.com), Daoyu Li, Yuzhe Zhang, Lintao Peng, Xuyang Chang, Yinuo Zhang, Liheng Bian
Affiliations: Beijing Institute of Technology

BIVLab
Title: Self-Guided Spatial-Frequency Complement Network for RGBW Joint Fusion and Denoise
Members: Bing Li (frigid@mail.ustc.edu.cn), Jie Huang, Mingde Yao, Ruikang Xu, Feng Zhao
Affiliations: University of Science and Technology of China

HIT-IIL
Title: NAFNet for RGBW Image Fusion
Members: Xiaohui Liu (xh720199@gmail.com), Xiaohui Liu, Rongjian Xu, Zhilu Zhang, Xiaohe Wu, Ruohao Wang, Junyi Li, Wangmeng Zuo
Affiliations: Harbin Institute of Technology

jzsherlock
Title: Dual Branch Network for Bayer Image Denoising Using White Pixel Guidance
Members: Zhuang Jia (jiazhuang@xiaomi.com)
Affiliations: Xiaomi

LLCKP
Title: Synthetic RGBW image and noise
Members: DongJae Lee (jhtwosun@kaist.ac.kr)
Affiliations: KAIST

MegNR
Title: HAUformer: Hybrid Attention-guided U-shaped Transformer for RGBW Fusion Image Restoration
Members: Ting Jiang (jiangting@megvii.com), Qi Wu, Chengzhi Jiang, Mingyan Han, Xinpeng Li, Wenjie Lin, Youwei Li, Haoqiang Fan, Shuaicheng Liu
Affiliations: Megvii Technology

References

1. Camon 19 pro. https://www.tecno-mobile.com/phones/product-detail/product/camon-19-pro-5g
2. Oppo unveils multiple innovative imaging technologies. https://www.oppo.com/en/newsroom/press/oppo-future-imaging-technology-launch/
3. vivo x80 is the only vivo smartphone with a sony imx866 sensor: the world's first RGBW bottom sensors. https://www.vivoglobal.ph/vivo-X80-is-the-only-vivo-smartphone-with-a-Sony-IMX866-Sensor-The-Worlds-First-RGBW-Bottom-Sensors/
4. Chen, L., Chu, X., Zhang, X., Sun, J.: Simple baselines for image restoration. arXiv preprint arXiv:2204.04676 (2022)
5. Liu, J., et al.: Learning raw image denoising with bayer pattern unification and bayer preserving augmentation. In: Proceedings of the IEEE/CVF Conference on Computer Vision and Pattern Recognition Workshops (2019)
6. Liu, Z., et al.: Swin transformer: hierarchical vision transformer using shifted windows. In: Proceedings of the IEEE/CVF International Conference on Computer Vision, pp. 10012–10022 (2021)
7. Shi, W., et al.: Real-time single image and video super-resolution using an efficient sub-pixel convolutional neural network. In: Proceedings of the IEEE Conference on Computer Vision and Pattern Recognition, pp. 1874–1883 (2016)

8. Sun, B., Zhang, Y., Jiang, S., Fu, Y.: Hybrid pixel-unshuffled network for lightweight image super-resolution. arXiv preprint arXiv:2203.08921 (2022)

9. Wang, X., et al.: Esrgan: enhanced super-resolution generative adversarial networks. In: Proceedings of the European Conference on Computer Vision (ECCV) Workshops (2018)

10. Wang, Z., Cun, X., Bao, J., Zhou, W., Liu, J., Li, H.: Uformer: a general u-shaped transformer for image restoration. In: Proceedings of the IEEE/CVF Conference on Computer Vision and Pattern Recognition, pp. 17683–17693 (2022)

11. Wang, Z., Bovik, A.C., Sheikh, H.R., Simoncelli, E.P.: Image quality assessment: from error visibility to structural similarity. IEEE Trans. Image Process. **13**(4), 600–612 (2004)

12. Xu, S., Zhang, J., Zhao, Z., Sun, K., Liu, J., Zhang, C.: Deep gradient projection networks for pan-sharpening. In: Proceedings of the IEEE/CVF Conference on Computer Vision and Pattern Recognition, pp. 1366–1375 (2021)

13. Yun, S., Han, D., Oh, S.J., Chun, S., Choe, J., Yoo, Y.: Cutmix: Regularization strategy to train strong classifiers with localizable features. In: Proceedings of the IEEE/CVF International Conference on Computer Vision, pp. 6023–6032 (2019)

14. Zamir, S.W., Arora, A., Khan, S., Hayat, M., Khan, F.S., Yang, M.H.: Restormer: efficient transformer for high-resolution image restoration. In: Proceedings of the IEEE/CVF Conference on Computer Vision and Pattern Recognition, pp. 5728–5739 (2022)

15. Zhang, R., Isola, P., Efros, A.A., Shechtman, E., Wang, O.: The unreasonable effectiveness of deep features as a perceptual metric. In: Proceedings of the IEEE Conference on Computer Vision and Pattern Recognition (CVPR) (2018)

MIPI 2022 Challenge on Under-Display Camera Image Restoration: Methods and Results

Ruicheng Feng[1]([⊠]), Chongyi Li[1], Shangchen Zhou[1], Wenxiu Sun[3,4],
Qingpeng Zhu[4], Jun Jiang[2], Qingyu Yang[2], Chen Change Loy[1], Jinwei Gu[2,3],
Yurui Zhu[5], Xi Wang[5], Xueyang Fu[5], Xiaowei Hu[3], Jinfan Hu[6], Xina Liu[6],
Xiangyu Chen[3,6,14], Chao Dong[3,6], Dafeng Zhang[7], Feiyu Huang[7],
Shizhuo Liu[7], Xiaobing Wang[7], Zhezhu Jin[7], Xuhao Jiang[8], Guangqi Shao[8],
Xiaotao Wang[8], Lei Lei[8], Zhao Zhang[9], Suiyi Zhao[9], Huan Zheng[9],
Yangcheng Gao[9], Yanyan Wei[9], Jiahuan Ren[9], Tao Huang[10],
Zhenxuan Fang[10], Mengluan Huang[10], Junwei Xu[10], Yong Zhang[11],
Yuechi Yang[11], Qidi Shu[12], Zhiwen Yang[11], Shaocong Li[11], Mingde Yao[5],
Ruikang Xu[5], Yuanshen Guan[5], Jie Huang[5], Zhiwei Xiong[5], Hangyan Zhu[13],
Ming Liu[13], Shaohui Liu[13], Wangmeng Zuo[13], Zhuang Jia[8], Binbin Song[14],
Ziqi Song[15], Guiting Mao[15], Ben Hou[15], Zhimou Liu[15], Yi Ke[15],
Dengpei Ouyang[15], Dekui Han[15], Jinghao Zhang[5], Qi Zhu[5], Naishan Zheng[5],
Feng Zhao[5], Wu Jin[16], Marcos Conde[17], Sabari Nathan[18], Radu Timofte[17],
Tianyi Xu[19], Jun Xu[19], P. S. Hrishikesh[20], Densen Puthussery[20], C. V. Jiji[21],
Biao Jiang[22], Yuhan Ding[22], WanZhang Li[22], Xiaoyue Feng[22], Sijing Chen[22],
Tianheng Zhong[22], Jiyang Lu[23], Hongming Chen[23], Zhentao Fan[23],
and Xiang Chen[24]

[1] Nanyang Technological University, Singapore, Singapore
{ruicheng002,chongyi.li}@ntu.edu.sg
[2] SenseBrain, San Jose, USA
[3] Shanghai AI Laboratory, Shanghai, China
[4] SenseTime Research and Tetras.AI, Beijing, China
[5] University of Science and Technology of China, Hefei, China
[6] Shenzhen Institutes of Advanced Technology, Chinese Academy of Sciences,
Shenzhen, China
[7] Samsung Research China, Beijing, China
[8] Xiaomi, Beijing, China
[9] Hefei University of Technology, Hefei, China
[10] School of Artificial Intelligence, Xidian University, Xi'an, China
[11] School of Remote Sensing and Information Engineering, Wuhan University,
Wuhan, China

R. Feng, C. Li, S. Zhou, W. Sun, Q. Zhu, J. Jiang, Q. Yang, C. C. Loy and J. Gu—MIPI 2022 challenge organizers. The other authors participated in the challenge. Please refer to Appendix for details. MIPI 2022 challenge website: http://mipi-challenge.org.

Supplementary Information The online version contains supplementary material available at https://doi.org/10.1007/978-3-031-25072-9_5

[12] State Key Laboratory of Information Engineering in Surveying, Mapping and Remote Sensing, Wuhan University, Wuhan, China
[13] Harbin Institute of Technology, Harbin, China
[14] University of Macau, Macau, China
[15] Changsha Research Institute of Mining and Metallurgy, Changsha, China
[16] Tianjin University, Tianjin, China
[17] Computer Vision Lab, University of Wurzburg, Würzburg, Germany
[18] Couger Inc., Tokyo, Japan
[19] School of Statistics and Data Science, Nankai University, Tianjin, China
[20] Founding Minds Software, Thiruvananthapuram, India
[21] Department of Electronics and Communication, SRM University AP, Amaravati, India
[22] Fudan University, Shanghai, China
[23] Shenyang Aerospace University, Shenyang, China
[24] Nanjing University of Science and Technology, Nanjing, China

Abstract. Developing and integrating advanced image sensors with novel algorithms in camera systems is prevalent with the increasing demand for computational photography and imaging on mobile platforms. However, the lack of high-quality data for research and the rare opportunity for in-depth exchange of views from industry and academia constrain the development of mobile intelligent photography and imaging (MIPI). To bridge the gap, we introduce the first MIPI challenge including five tracks focusing on novel image sensors and imaging algorithms. In this paper, we summarize and review the Under-Display Camera (UDC) Image Restoration track on MIPI 2022. In total, 167 participants were successfully registered, and 19 teams submitted results in the final testing phase. The developed solutions in this challenge achieved state-of-the-art performance on Under-Display Camera Image Restoration. A detailed description of all models developed in this challenge is provided in this paper. More details of this challenge and the link to the dataset can be found at https://github.com/mipi-challenge/MIPI2022.

Keywords: Under-Display Camera · Image restoration · MIPI challenge

1   Introduction

The demand for smartphones with full-screen displays has drawn interest from manufacturers in a newly-defined imaging system, Under-Display Camera (UDC). In addition, it also demonstrates practical applicability in other scenarios, *e.g.*, for videoconferencing with more natural gaze focus as cameras are placed at the center of the displays.

Under-Display Camera (UDC) is an imaging system whose camera are placed underneath a display. However, a widespread commercial productions of UDC are prevented by poor imaging quality caused by diffraction artifacts. Such artifacts are unique to UDC, caused by the gaps between display pixels that act as an aperture and induces diffraction artifacts in the captured image. Typical diffraction artifacts include flare, saturated blobs, blur, haze, and noise. Therefore, while bringing better user experience, UDC may sacrifice image quality, and affect other downstream vision tasks. The complex and diverse distortions make the reconstruction problem extremely challenging. Zhou *et al.* [41,42] pioneer the attempt of the UDC image restoration and propose a Monitor Camera Imaging System (MCIS) to capture paired data. However, their work only simulates incomplete degradation. To alleviate this problem, Feng *et al.* [9] reformulate the image formation model and synthesis the UDC image by considering the diffraction flare of the saturated region in the high-dynamic-range (HDR) images. This challenge are based on the dataset proposed in [9], and aims to restore UDC images with complicated degradations. More details will be discussed in the following sections.

We hold this image restoration challenge in conjunction with MIPI Challenge which will be held on ECCV 2022. We are seeking an efficient and high-performance image restoration algorithm to be used for recovering under-display camera images. MIPI 2022 consists of five competition tracks:

- RGB+ToF Depth Completion uses sparse, noisy ToF depth measurements with RGB images to obtain a complete depth map.
- Quad-Bayer Re-mosaic converts Quad-Bayer RAW data into Bayer format so that it can be processed with standard ISPs.
- RGBW Sensor Re-mosaic converts RGBW RAW data into Bayer format so that it can be processed with standard ISPs.
- RGBW Sensor Fusion fuses Bayer data and a monochrome channel data into Bayer format to increase SNR and spatial resolution.
- Under-Display Camera Image Restoration improves the visual quality of image captured by a new imaging system equipped under-display camera.

2 MIPI 2022 Under-Display Camera Image Restoration

To facilitate the development of efficient and high-performance UDC image restoration solution, we provide a high-quality dataset to be used to training and testing and a set of evaluation metrics that can measure the performance of developed solutions. This challenge aims to advance research on UDC image restoration.

2.1 Datasets

The dataset is collected and synthesized using a model-based simulation pipeline as introduced in [9]. The training split contains 2016 pairs of $800 \times 800 \times 3$ images. Image values are ranging from [0, 500] and constructed in '.npy' form. The validation set is a sub-set of the testing set in [9], and contains 40 pairs of images. The testing set consists of another 40 pairs of images. The input images from validation and testing set are provided and the GT are not available to participants. Both input and Ground Truth data are high dynamic range. For evaluation, all measurements are computed in tonemapped images (Modified Reinhard). The tone mapping operation can be expressed as $f(x) = \frac{x}{x+0.25}$.

2.2   Evaluation

The evaluation measures the objective fidelity and the perceptual quality of the UDC images with reference ground truth images. We use the standard Peak Signal To Noise Ratio (PSNR) and the Structural Similarity (SSIM) index as often employed in the literature. In addition, Learned Perceptual Image Patch Similarity (LPIPS) [37] will be used as a complement. All measurements are computed in tone-mapped images (Modified Reinhard [23]). For the final ranking, we choose PSNR as the main measure, however the top ranked solutions are expected to also achieve an above average performance on SSIM and LPIPS. For the dataset we report the average results over all the processed images.

2.3   Challenge Phase

The challenge consisted of the following phases:

1. Development: The registered participants get access to the data and baseline code, and are able to train the models and evaluate their run-time locally.
2. Validation: The participants can upload their models to the remote server to check the fidelity scores on the validation dataset, and to compare their results on the validation leaderboard.
3. Testing: The participants submit their final results, codes, models, and factsheets.

3   Challenge Results

Among 167 registered participants, 19 teams successfully submitted their results, codes, factsheets in the final test phase. Table 1 reports the final test results and rankings of the teams. The methods evaluated in Table 1 are briefly described in Sect. 4 and the team members are listed in Appendix. We have the following observations. First, the USTC_WXYZ team is the first place winner of this challenge, while XPixel Group and SRC-B team win the second place and overall third place, respectively. Second, most methods achieve high PSNR performance (over 40 dB). This indicates most degradations, e.g., glare and haze, are easy to restore. Only three teams train their models with extra data, and several top-ranked teams apply ensemble strategies (self-ensemble [26], model ensemble, or both).

Table 1. Results of MIPI 2022 challenge on UDC image restoration. 'Runtime' for per image is tested and averaged across the validation datasets, and the image size is 800×800. 'Params' denotes the total number of learnable parameters.

Team name	User name	Metric			Params (M)	Runtime (s)	Platform	Extra data	Ensemble
		PSNR	SSIM	LPIPS					
USTC_WXYZ	YuruiZhu	$48.48_{(1)}$	$0.9934_{(1)}$	$0.0093_{(1)}$	16.85	0.27	Nvidia A100	Yes	model
XPixel Group	JFHu	$47.78_{(2)}$	$0.9913_{(5)}$	$0.0122_{(6)}$	14.06	0.16	Nvidia A6000	–	–
SRC-B	xiaozhazha	$46.92_{(3)}$	$0.9929_{(2)}$	$0.0098_{(2)}$	23.56	0.13	RTX 3090	–	self-ensemble + model
MIALGO	Xhjiang	$46.12_{(4)}$	$0.9892_{(10)}$	$0.0159_{(13)}$	7.93	9	Tesla V100	–	–
LVGroup_HFUT	HuanZheng	$45.87_{(5)}$	$0.9920_{(3)}$	$0.0109_{(4)}$	6.47	0.92	GTX 2080Ti	–	–
GSM	Zhenxuan_Fang	$45.82_{(6)}$	$0.9917_{(4)}$	$0.0106_{(3)}$	11.85	1.88	RTX 3090	–	–
Y2C	y2c	$45.56_{(7)}$	$0.9912_{(6)}$	$0.0129_{(8)}$	/	/	GTX 3090	–	self-ensemble
VIDAR	Null	$44.05_{(8)}$	$0.9908_{(7)}$	$0.0120_{(5)}$	33.8	0.47	RTX 3090Ti	–	model ensemble
IILLab	zhuhy	$43.45_{(9)}$	$0.9897_{(9)}$	$0.0125_{(7)}$	82.2	2.07	RTX A6000	–	self-ensemble
jzsherlock	jzsherlock	$43.44_{(10)}$	$0.9899_{(8)}$	$0.0133_{(9)}$	14.82	1.03	Tesla A100	–	self-ensemble
Namecantbenull	Namecantbenull	$43.13_{(11)}$	$0.9872_{(13)}$	$0.0144_{(10)}$	/	/	Tesla A100	/	/
MeVision	Ziqi_Song	$42.95_{(12)}$	$0.9892_{(11)}$	$0.0150_{(11)}$	/	0.08	NVIDIA 3080ti	Yes	–
BIVLab	Hao-mk	$42.04_{(13)}$	$0.9873_{(12)}$	$0.0155_{(12)}$	2.98	2.52	Tesla V100	–	self-ensemble
RushRushRush	stillwaters	$39.52_{(14)}$	$0.9820_{(14)}$	$0.0216_{(14)}$	5.9	0.63	RTX 2080Ti	/	–
JMU-CVLab	nanashi	$37.46_{(15)}$	$0.9773_{(16)}$	$0.0370_{(16)}$	2	0.48	Tesla P100	–	–
eye3	SummerinSummer	$36.70_{(16)}$	$0.9783_{(15)}$	$0.0326_{(15)}$	26.13	1.07	Tesla V100	–	–
FMS Lab	hrishikeshps94	$35.77_{(17)}$	$0.9719_{(17)}$	$0.0458_{(18)}$	4.40	0.04	Tesla V100	–	–
EDLC2004	jiangbiao	$35.50_{(18)}$	$0.9616_{(18)}$	$0.0453_{(17)}$	31	0.99	RTX 2080Ti	Yes	–
SAU_LCFC	chm	$32.75_{(19)}$	$0.9591_{(19)}$	$0.0566_{(19)}$	0.98	0.92	Tesla V100	–	–

4   Challenge Methods and Teams

USTC_WXYZ Team. Inspired by [6,10,34], this team design an enhanced multi-inputs multi-outputs network, which mainly consists of dense residual blocks and the cross-scale gating fusion modules. The overall architecture of the network is depicted in Fig. 1. The training phase could be divided into three stages: i) Adopt the Adam optimizer with a batch size of 3 and the patch size of 256×256. The initial learning rate is 2×10^{-4} and is adjusted with the Cosine Annealing scheme, including 1000 epochs in total. ii) Adopt the Adam optimizer with a batch size of 1 and the patch size of 512×512. The initial learning rate is 2×10^{-5} and is adjusted with the Cosine Annealing scheme, including 300 epochs in total. iii) Adopt the Adam optimizer with a batch size of 2 and the patch size of 800×800. The initial learning rate is 8×10^{-6} and is adjusted with the Cosine Annealing scheme, including 150 epochs in total. During inference, the team adopt model ensemble strategy averaging the parameters of multiple models trained with different hyperparameters, which brings around 0.09 dB increase on PSNR.

XPixel Group. This team designs a UNet-like structure (see Fig. 2), making full use of the hierarchical multi-scale information from low-level features to high-level features. To ease the training procedure and facilitate the information flow, several residual blocks are utilized in the base network, and the dynamic kernels in the skip connection bring better flexibility. Since UDC images have different holistic brightness and contrast information, the team applies the condition network with spatial feature transform (SFT) to deal with input images with location-specific and image-specific operations (mentioned in HDRUNet

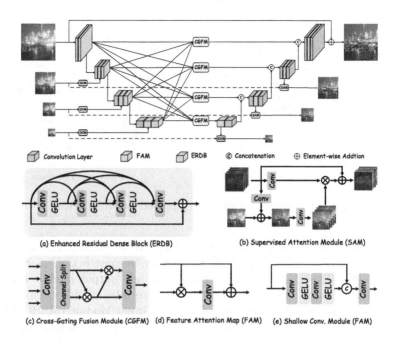

Fig. 1. The network architecture proposed by USTC_WXYZ Team.

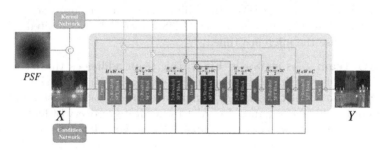

Fig. 2. The network architecture proposed by XPixel Group.

[4]), which could provide spatially variant manipulations. In addition, they incorporate the information of point spread function (PSF) provided in DISCNet [9], which has been demonstrated its effectiveness through extensive experiments on both synthetic and real UDC data. For training, the authors randomly cropped 256×256 patches from the training images as inputs. The mini-batch size is set to 32 and the whole network is trained for 6×10^5 iterations. The learning rate is initialized as 2×10^{-4}, decayed with a CosineAnnealing schedule, and restarted at $[50K, 150K, 300K, 450K]$ iterations.

SRC-B Team. This team proposes Multi-Refinement Network (MRNet) for Image Restoration on Under-Display Camera, as shown in Fig. 3. In this

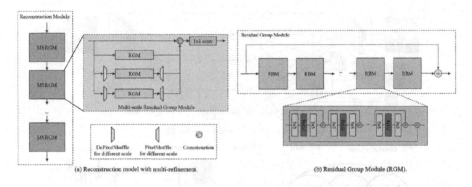

(a) Reconstruction model with multi-refinement.   (b) Residual Group Module (RGM).

Fig. 3. The network architecture proposed by SRC-B Team.

challenge, the authors modify and improve the MRNet [1], which mainly composed of 3 modules: shallow Feature Extraction, reconstruction module and output module. The shallow Feature Extraction and output module only use one convolution layer. Multi-refinement is the main idea of the reconstruction module, which includes N Multi-scale Residual Group Module (MSRGM) and progressively refine the features. MSRGM is also used to fuse information from three scales to improve the representation ability and robustness of model, where each scale is composed of several residual group module (RGM). RGM contains G residual block module (RBM). In addition, the authors propose to remove Channel Attention (CA) module in RBM [38], since it brings limited improvement but increase the inference time.

MIALGO Team. As shown in Fig. 4, this team addresses the UDC image restoration problem with a Residual Dense Network (RDN) [39] as the backbone. The authors also adopt the strategy of multi-resolution feature fusion in HRNet [25] to improve the performance. In order to facilitate recovery of high dynamic range images, the authors generate 6 non-HDR images with different exposures range from the given HDR images, where the value ranges of each split are: [0, 0.5], [0, 2], [0, 8], [0, 32], [0, 128], [0, 500], and normalize them and then stack into 18 channels of data for training. According to the dynamic range of the data, 2000 images in the training datasets (the remaining 16 are used as validation sets) are divided into two datasets: i) Easy sample dataset: 1318 images whose intensities lie in [0, 32]. ii) Hard sample dataset: easy samples plus doubled remaining hard samples. They use two datasets to train two models, each of which is first trained with 500k iterations. Then they fine-tune the easy sample model and the hard sample model by 200K iterations. In the validation/test phase, the 18 channel data is merged into 3 channel HDR images and fed to the network.

LVGroup_HFUT Team. Considering the specificity of the UDC image restoration task (mobile deployment), this team design a lightweight and real-time model for UDC image restoration using a simple UNet architecture with magic modifications as shown in Fig. 5. Specifically, they combine the full-resolution network FRC-Net [40] and the classical UNet [24] to construct the

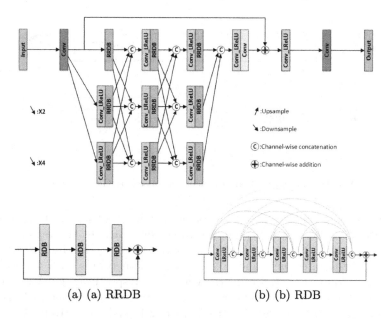

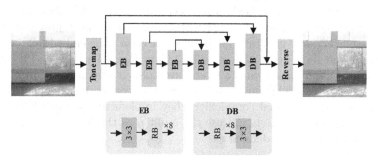

Fig. 4. The network architecture proposed by MIALGO Team.

Fig. 5. The network architecture proposed by LVGroup_HFUT Team.

model, which presents two following advantages: 1) directly stacking the residual blocks at the original resolution to ensure that the model learns sufficient spatial structure information, and 2) stacking the residual blocks at the downsampled resolution to ensure that the model learns sufficient semantic-level information.

During training, they first perform tone mapping, and then perform a series of data augmentation sequentially, including: 1) random crop to 384×384; 2) vertical flip with probability 0.5; 3): horizontal flip with probability 0.5. They train the model for 3000 epochs on provided training dataset with initial learning rate 1e−4 and batch size 4.

GSM Team. This team reformulates UDC image restoration as a Maximum a Posteriori estimation problem with the learned Gaussian Scale Mixture (GSM) models. Specifically, the x can be solved by

$$x^{(t+1)} = x^{(t)} - 2\delta\{\mathbf{A}^T(\mathbf{A}x^{(t)} - y) + w^{(t)}(x^{(t)} - u^{(t)})\}, \qquad (1)$$

68   R. Feng et al.

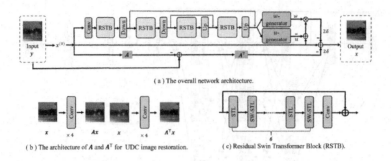

(a) The overall network architecture.

(b) The architecture of A and A^T for UDC image restoration.   (c) Residual Swin Transformer Block (RSTB).

Fig. 6. The network architecture proposed by GSM Team.

where $\mathbf{A}^T$ denotes the transposed version of $\mathbf{A}$, δ denotes the step size, t denotes the tth iteration for iteratively optimizing $\boldsymbol{x}$, $\boldsymbol{w}^{(t)}$ denotes the regularization parameters, and $\boldsymbol{u}^{(t)}$ denotes the mean of the GSM model. In [11], a UNet was used to estimated $\boldsymbol{w}^{(t)}$ and $\boldsymbol{u}^{(t)}$ and two sub-networks with 4 Resblocks and 2 Conv layers were used to learn $\mathbf{A}$ and $\mathbf{A}^T$, respectively. For UDC image restoration, they develop a network based on Swin Transformer to learn the GSM prior (*i.e.*, $\boldsymbol{w}^{(t)}$ and $\boldsymbol{u}^{(t)}$) and use 4 Conv layers, respectively. As shown in Fig. 6, the team constructs an end-to-end network for UDC image restoration. The transformer-based GSM prior network contains an embedding layer, 4 Residual Swin Transformer Blocks (RSTB) [16], two downsampling layers, two upsampling layers, a w-generator, and an u-generator. The embedding layer, the w-generator and the u-generator are a 3×3 Conv layer. The RSTB [16] contains 6 Swin Transformer Layers and a Conv layer plus a skip connection. The features of the first two RSTBs are reused by two skip connections, respectively. For reducing the computational complexity, they implement the united framework Eq. 1 with only one iteration and use the input $\boldsymbol{y}$ as the initial value $\boldsymbol{x}^{(0)}$.

The loss function L is defined as

$$
\begin{aligned}
L_1 &= \sqrt{\|\boldsymbol{x} - \hat{\boldsymbol{x}}\|^2 + \epsilon}, \\
L_2 &= \sqrt{\|\Delta(\boldsymbol{x}) - \Delta(\hat{\boldsymbol{x}})\|^2 + \epsilon}, \\
L_3 &= \|\mathcal{FFT}(\boldsymbol{x}) - \mathcal{FFT}(\hat{\boldsymbol{x}})\|_1, \\
L &= L_1 + 0.05 \times L_2 + 0.01 \times L_3,
\end{aligned}
\tag{2}
$$

where $\boldsymbol{x}$ and $\hat{\boldsymbol{x}}$ denote the label and the output of the network, Δ denotes the Laplacian operator, and $\mathcal{FFT}$ is the FFT operation.

Y2C Team. This team takes model [6,35] as backbone network and introduces the multi-scale design. Besides, they reconstruct each color channel by a complete branch instead of processing them together. The overall architecture of the Multi-Scale and Separable Channels reconstruction Network (MSSCN) for image restoration of under-display camera is shown in Fig. 7. The network takes the multi-scale degraded images (1x, 0.5x, 0.25x) as inputs, and the initial feature maps of each color channel in each scale are extracted respectively. Before extracting feature maps, they use the discrete wavelet transform (DWT) [17] to

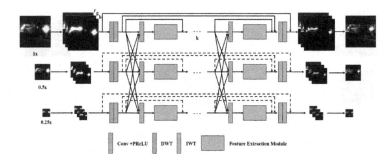

Fig. 7. The network architecture proposed by Y2C Team.

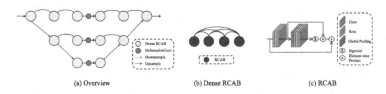

Fig. 8. The network architecture proposed by VIDAR Team.

reduce the resolution in each scale and improve the efficiency of reconstruction. In the fusion module, this team fuses the feature maps in different scales in the same spatial resolution, concatenates the feature maps from each color channel and each scale, and processes them using a depth-wise convolution. In the feature extraction module, any efficient feature extraction block can be used to extract feature maps. Residual group (RG) with multiple residual channel attention blocks (RCAB) [38] is used as the feature extraction module. To improve the learning capacity of recovering images, they repeat the feature fusion module and feature extraction module k times. Then the reconstructed images are generated by the inverse discrete wavelet transform (IWT).

During training phase, random rotation and flip are used for data augmentation. All models are trained with multi-scale L1 loss and frequency reconstruction loss. During testing phase, they adopt self-ensemble strategy, bringing around 0.9 dB increase on PSNR.

VIDAR Team. As shown in Fig. 8, this team designs a dense U-Net for under-display camera image restoration. The architecture aggregates and fuses feature information with Dense RCAB [38]. During training, batch size is set to 4 with the patch of 384×384 and the optimizer is ADAM [14] by setting $\beta_1 = 0.9$, $\beta_2 = 0.999$. The initial learning rate is 1×10^{-4}. They use Charbonnier loss to train models and stop training when no notable decay of training loss is observed.

IILLab Team. This team presents a Wiener Guided Coupled Dynamic Filter Network, which takes advantage of both DWDN [8] and BPN [31]. As shown in Fig. 9, on the U-net based backbone, they first used feature level wiener for global restoration with the estimated degradation kernel, then used the coupled dynamic filter network for local restoration. The U-net based backbone obtained

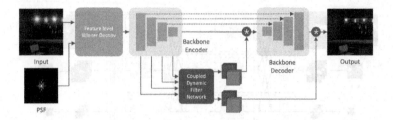

Fig. 9. The network architecture proposed by IILLab Team.

the global and local information from the feature level wiener deconvolution and the coupled dynamic filter network, the UDC degradation was modeled with higher accuracy, thus obtained a better restoration result. Wiener Guided Coupled Dynamic Filter Network was jointly trained under the supervision of L1 loss and (VGG-based) perceptual loss [13].

jzsherlock Team. This team proposes the U-RRDBNet for the task of UDC image restoration. Since UDC images are suffering from the strong glare caused by optical structure of OLED display, which usually occurs when light source (*e.g.*street lamp, sun) in the view and degrade a large area around the light source, larger receptive field is required to deal with this task. The proposed architecture and processing pipeline is illustrated in Fig. 10. Though the original U-net and its variants have achieved promising results on segmentation tasks, the capability of representation learning of the U-net is still limited for the dense prediction in low-level tasks. So the authors apply the Residual-in-Residual Dense Block (RRDB) [28] to substitute simple Conv layers in U-net after each down/up-sampling operation to increase the network capability. They first perform the modified Reinhard tone-mapping, and then feed the tone-mapped image into the end-to-end U-RRDBNet for restoration, and produces the image in tonemapped domain with artifacts removed. The output of network is then transformed back to HDR domain using the inverse of Reinhard tonemapping. The model is first trained using L_1 loss and perceptual loss [13] in tonemapped domain, and then finetuned with MSE loss for higher PSNR performance.

Namecantbenull Team. This team designs a deep learning model for the UDC image restoration based on the U-shaped network. In this network, they incorporate the convolution operation and attention mechanism into the transformer block architecture, which has been proved effective in image restoration tasks in [2]. Since there is no need to calculate the self-attention coefficient matrices, the memory cost and computation complexity can be reduced significantly. The overall network structure is shown in Fig. 11. Specifically, they substitute the simplified channel attention module with the spatial and channel attention module [3] to additionally consider the correlation of the spatial pixels. They train the model with L1 and perceptual loss on the tone-mapped domain.

MeVision Team. This team follow [15] and presents a two-branch network to restore UDC images. As shown in Fig. 12, the input images are processed and

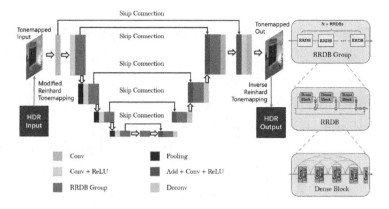

Fig. 10. The network architecture proposed by jzsherlock Team.

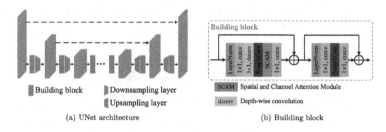

(a) UNet architecture   (b) Building block

Fig. 11. The network architecture proposed by Namecantbenull Team.

fed to the restoration network. The original resolution image is processed in one branch while the blurred image with noise is processed in another one. Finally, the two branches are connected in series by the affine transformation connection. Then, they follow [20] and modify the high-frequency reconstruction branch with encoder-decoder structure to reduce parameters. And They converted the smoothed diluted residual block into IMD block [12] so that the signals can propagate directly from skip connections to bottom layers.

BIVLab Team. This team develops a Self-Guided Coarse-to-Fine Network (SG-CFN) to progressively reconstruct the degraded under-display camera images. As shown in Fig. 13, the proposed SG-CFN consists of two branches, restoration branch and condition branch. The restoration branch is constructed by Feature Extraction Module (FEM) based on improved RSTBs [16], which incorporates the paralleled central difference convolution (CDC) with swin transfomer layer (STL) to extract rich features at multiple scales. The condition branch is constructed by fast Fourier convolution (FFC) blocks [5], that endowed with global receptive field to model the distribution of the large-scale point-spread function (PSF), which is indeed the culprit of the degradation. Furthermore, the multi-scale representations extracted from restoration branch are processed via the Degradation Correction Module (DCM) to restore clean features, guided by the corresponding condition information. To fully exploit the restored multi-scale clean features,

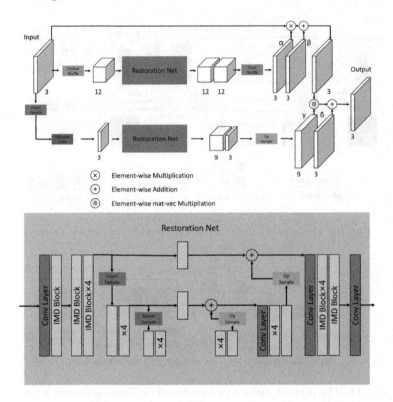

Fig. 12. The network architecture proposed by MeVision Team.

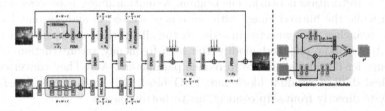

Fig. 13. The network architecture proposed by BIVlab Team.

they enable the asymmetric feature fusion inspired by [6] to facilitate the flexible information flow. Images captured through under-display camera typically suffer from diffraction degradation caused by large-scale PSF, resulting in blurred and detail attenuated images. Thereby, (1) incorporating CDC into restoration branch can great help to avoid over-smooth feature extraction and enrich the representation capability, (2) constructing the condition branch with FFC blocks endow the global receptive field to capture the degradation information, (3) correcting the degraded features from both local adaptation and global modulation makes the process of restoration more effective. During testing phase, they adopt self-ensemble strategy and it brings 0.63 dB performance gain on PSNR.

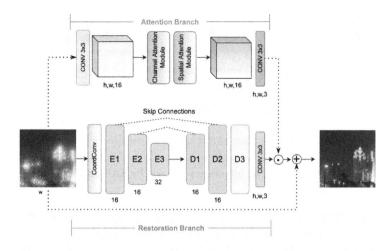

Fig. 14. The network architecture proposed by JMU-CVLab Team.

RushRushRush Team. This team reproduced the MIRNetV2 [36] on the UDC dataset. The network utilize both spatially-precise high-resolution representations and contextual information from the low-resolution representations. The multi-scale residual block contains: (i) parallel multi-resolution convolution streams to extract multi-scale features, (ii) information exchange across streams, (iii) non-local attention mechanism to capture contextual information, and (iv) attention based multi-scale feature aggregation. The approach learns an enriched set of features that combines contextual information from multiple scales, while simultaneously preserving the high-resolution spatial details.

JMU-CVLab Team. This team proposes a dual-branch lightweight neural network. Inspired by recent camera modelling approaches [7], they design a dual-branch model, depending on the tradeoff between resources-performance. As shown in Fig. 14, the method combines ideas from deblurring [21] and HDR [19] networks, and attention methods [18,29]. This does not rely on metadata or extra information as the PSF. The main image restoration branch with a Dense Residual UNet architecture [24,39]. They use an initial CoordConv [18] layer to encode positional information. Three Encoder blocks, each formed by 2 dense residual layers (DRL) [39] followed by the corresponding downsampling pooling. The decoder blocks D_1 and D_2 consists on a bilinar upsampling layer and 2 DRL The decoder block D_3 does not have an upsampling layer, consists on 2 DRL and a series of convolutions with 7,5, and 1 kernel size, the final convolution (cyan color) produces the residual using a tanh activation. Similar to [19], the additional attention branch aims to generate an attention map (per channel) to control the hallucination on overexposed areas. The attention map is generated after applying a CBAM block [29] on the features of the original image, and the final convolution is activated using a sigmoid function.

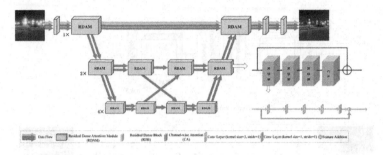

Fig. 15. The network architecture proposed by SAU-LCFC Team.

eye3 Team. This team implements the network architecture by Restormer [33]. Instead of exploring new architecture for under-display camera restoration, they attempt to mine the potential of the existed method to advance this field. Following the scheme of [33], they train an Restormer in a progressive fashion with L_1 loss for $300k$ iterations, with $92k$, $64k$, $48k$, $36k$, $36k$ and $24k$ iterations for patch size 128, 160, 192, 256, 320, 384, respectively. Then they fix the patch size to 384×384 and use a mask loss strategy [30] to train the model for another $36k$ iterations, where saturated pixels are masked out. After that, inspired by self-training [32], they add Gaussian noise to the well trained model and repeat the training process.

FMS Lab Team. Dual Branch Wavelet Net (DBWN) [17] that is proposed to restore the images degraded in an under-display camera imaging system. The overall network structure is similar to [22], where the network branches restore the high and low spatial frequency components of the image separately.

EDLC2004 Team. This team adopted TransWeather [27] model to deal with image restoration problem for UDC images. Specifically, they combine different loss functions including L_1 loss, $L2$ loss. They train the model from scratch and it took approximately 17 h with two 2080Ti GPUs.

SAU_LCFC Team. The team proposes a Hourglass-Structured Fusion Network (HSF-Net) as shown in Fig. 15. They start from a coarse-scale stream, and then gradually repeat top-down and bottom-up processing to form more coarse-to-fine scale streams one by one. In each stream, the residual dense attention modules (RDAMs) are introduced as the basic component to deeply learn the features of under-display camera images. Each RDAM contains 3 dense residual blocks (RDBs) in series with channel-wise attention (CA). In each RDB, the first 4 convolutional layers are adopted to elevate the number of feature maps, while the last convolutional layer is employed to aggregate feature maps. The growth rate of RDB is set to 16.

5   Conclusions

In this report, we review and summarize the methods and results of MIPI 2022 challenge on Under-Display Camera Image Restoration. All the proposed methods are based on deep CNNs and most of them share similar U-shape backbone to boost performance.

Acknowledgements. We thank Shanghai Artificial Intelligence Laboratory, Sony, and Nanyang Technological University to sponse this MIPI 2022 challenge. We thank all the organizers and all the participants for their great work.

References

1. Abuolaim, A., Timofte, R., Brown, M.S.: NTIRE 2021 challenge for defocus deblurring using dual-pixel images: methods and results. In: Proceedings of the IEEE/CVF Conference on Computer Vision and Pattern Recognition, pp. 578–587 (2021)
2. Chen, L., Chu, X., Zhang, X., Sun, J.: Simple baselines for image restoration. arXiv preprint arXiv:2204.04676 (2022)
3. Chen, L., et al.: SCA-CNN: spatial and channel-wise attention in convolutional networks for image captioning. In: Proceedings of the IEEE Conference on Computer Vision and Pattern Recognition, pp. 5659–5667 (2017)
4. Chen, X., Liu, Y., Zhang, Z., Qiao, Y., Dong, C.: HDRUNet: single image HDR reconstruction with denoising and dequantization. In: Proceedings of the IEEE/CVF Conference on Computer Vision and Pattern Recognition, pp. 354–363 (2021)
5. Chi, L., Jiang, B., Mu, Y.: Fast Fourier convolution. Adv. Neural. Inf. Process. Syst. **33**, 4479–4488 (2020)
6. Cho, S.J., Ji, S.W., Hong, J.P., Jung, S.W., Ko, S.J.: Rethinking coarse-to-fine approach in single image deblurring. In: Proceedings of the IEEE/CVF International Conference on Computer Vision, pp. 4641–4650 (2021)
7. Conde, M.V., McDonagh, S., Maggioni, M., Leonardis, A., Pérez-Pellitero, E.: Model-based image signal processors via learnable dictionaries. In: Proceedings of the AAAI Conference on Artificial Intelligence, vol. 36, pp. 481–489 (2022)
8. Dong, J., Roth, S., Schiele, B.: Deep wiener deconvolution: Wiener meets deep learning for image deblurring. Adv. Neural. Inf. Process. Syst. **33**, 1048–1059 (2020)
9. Feng, R., Li, C., Chen, H., Li, S., Loy, C.C., Gu, J.: Removing diffraction image artifacts in under-display camera via dynamic skip connection networks. In: IEEE Conference on Computer Vision and Pattern Recognition (CVPR) (2021)
10. Huang, G., Liu, Z., Van Der Maaten, L., Weinberger, K.Q.: Densely connected convolutional networks. In: CVPR, vol. 1, p. 3 (2017)
11. Huang, T., Dong, W., Yuan, X., Wu, J., Shi, G.: Deep gaussian scale mixture prior for spectral compressive imaging. In: Proceedings of the IEEE/CVF Conference on Computer Vision and Pattern Recognition, pp. 16216–16225 (2021)
12. Hui, Z., Gao, X., Yang, Y., Wang, X.: Lightweight image super-resolution with information multi-distillation network. In: Proceedings of the 27th ACM International Conference on Multimedia, pp. 2024–2032 (2019)

13. Johnson, J., Alahi, A., Fei-Fei, L.: Perceptual losses for real-time style transfer and super-resolution. In: Leibe, B., Matas, J., Sebe, N., Welling, M. (eds.) ECCV 2016. LNCS, vol. 9906, pp. 694–711. Springer, Cham (2016). https://doi.org/10.1007/978-3-319-46475-6_43

14. Kingma, D.P., Ba, J.: Adam: a method for stochastic optimization. CoRR abs/1412.6980 (2015)

15. Koh, J., Lee, J., Yoon, S.: BNUDC: a two-branched deep neural network for restoring images from under-display cameras. In: Proceedings of the IEEE/CVF Conference on Computer Vision and Pattern Recognition, pp. 1950–1959 (2022)

16. Liang, J., Cao, J., Sun, G., Zhang, K., Van Gool, L., Timofte, R.: SwinIR: image restoration using Swin transformer. In: Proceedings of the IEEE/CVF International Conference on Computer Vision, pp. 1833–1844 (2021)

17. Liu, P., Zhang, H., Zhang, K., Lin, L., Zuo, W.: Multi-level wavelet-CNN for image restoration. In: Proceedings of the IEEE Conference on Computer Vision and Pattern Recognition Workshops, pp. 773–782 (2018)

18. Liu, R., et al.: An intriguing failing of convolutional neural networks and the CoordConv solution. In: Advances in Neural Information Processing Systems, vol. 31 (2018)

19. Liu, Y.L., et al.: Single-image HDR reconstruction by learning to reverse the camera pipeline. In: Proceedings of the IEEE/CVF Conference on Computer Vision and Pattern Recognition, pp. 1651–1660 (2020)

20. Mao, X., Shen, C., Yang, Y.B.: Image restoration using very deep convolutional encoder-decoder networks with symmetric skip connections. In: Advances in Neural Information Processing Systems, vol. 29 (2016)

21. Nah, S., Son, S., Lee, S., Timofte, R., Lee, K.M.: NTIRE 2021 challenge on image deblurring. In: Proceedings of the IEEE/CVF Conference on Computer Vision and Pattern Recognition, pp. 149–165 (2021)

22. Panikkasseril Sethumadhavan, H., Puthussery, D., Kuriakose, M., Charangatt Victor, J.: Transform domain pyramidal dilated convolution networks for restoration of under display camera images. In: Bartoli, A., Fusiello, A. (eds.) ECCV 2020. LNCS, vol. 12539, pp. 364–378. Springer, Cham (2020). https://doi.org/10.1007/978-3-030-68238-5_28

23. Reinhard, E., Stark, M., Shirley, P., Ferwerda, J.: Photographic tone reproduction for digital images. In: Proceedings of the 29th Annual Conference on Computer Graphics and Interactive Techniques, pp. 267–276 (2002)

24. Ronneberger, O., Fischer, P., Brox, T.: U-Net: convolutional networks for biomedical image segmentation. In: Navab, N., Hornegger, J., Wells, W.M., Frangi, A.F. (eds.) MICCAI 2015. LNCS, vol. 9351, pp. 234–241. Springer, Cham (2015). https://doi.org/10.1007/978-3-319-24574-4_28

25. Sun, K., Xiao, B., Liu, D., Wang, J.: Deep high-resolution representation learning for human pose estimation. In: Proceedings of the IEEE/CVF Conference on Computer Vision and Pattern Recognition, pp. 5693–5703 (2019)

26. Timofte, R., Rothe, R., Van Gool, L.: Seven ways to improve example-based single image super resolution. In: Proceedings of the IEEE Conference on Computer Vision and Pattern Recognition, pp. 1865–1873 (2016)

27. Valanarasu, J.M.J., Yasarla, R., Patel, V.M.: TransWeather: transformer-based restoration of images degraded by adverse weather conditions. In: Proceedings of the IEEE/CVF Conference on Computer Vision and Pattern Recognition, pp. 2353–2363 (2022)

28. Wang, X., et al.: ESRGAN: enhanced super-resolution generative adversarial networks. In: Leal-Taixé, L., Roth, S. (eds.) ECCV 2018. LNCS, vol. 11133, pp. 63–79. Springer, Cham (2019). https://doi.org/10.1007/978-3-030-11021-5_5

29. Woo, S., Park, J., Lee, J.-Y., Kweon, I.S.: CBAM: convolutional block attention module. In: Ferrari, V., Hebert, M., Sminchisescu, C., Weiss, Y. (eds.) ECCV 2018. LNCS, vol. 11211, pp. 3–19. Springer, Cham (2018). https://doi.org/10.1007/978-3-030-01234-2_1

30. Wu, Y., et al.: How to train neural networks for flare removal. In: Proceedings of the IEEE/CVF International Conference on Computer Vision, pp. 2239–2247 (2021)

31. Xia, Z., Perazzi, F., Gharbi, M., Sunkavalli, K., Chakrabarti, A.: Basis prediction networks for effective burst denoising with large kernels. In: Proceedings of the IEEE/CVF Conference on Computer Vision and Pattern Recognition, pp. 11844–11853 (2020)

32. Xie, Q., Luong, M.T., Hovy, E., Le, Q.V.: Self-training with noisy student improves ImageNet classification. In: Proceedings of the IEEE/CVF Conference on Computer Vision and Pattern Recognition, pp. 10687–10698 (2020)

33. Zamir, S.W., Arora, A., Khan, S., Hayat, M., Khan, F.S., Yang, M.H.: Restormer: efficient transformer for high-resolution image restoration. In: Proceedings of the IEEE/CVF Conference on Computer Vision and Pattern Recognition, pp. 5728–5739 (2022)

34. Zamir, S.W., et al.: Multi-stage progressive image restoration. In: Proceedings of the IEEE/CVF Conference on Computer Vision and Pattern Recognition, pp. 14821–14831 (2021)

35. Zamir, S.W., et al.: Learning enriched features for fast image restoration and enhancement. arXiv preprint arXiv:2205.01649 (2022)

36. Zamir, S.W., et al.: Learning enriched features for fast image restoration and enhancement. IEEE Trans. Pattern Anal. Mach. Intell. (TPAMI) (2022)

37. Zhang, R., Isola, P., Efros, A.A., Shechtman, E., Wang, O.: The unreasonable effectiveness of deep features as a perceptual metric. In: Proceedings of the IEEE Conference on Computer Vision and Pattern Recognition, pp. 586–595 (2018)

38. Zhang, Y., Li, K., Li, K., Wang, L., Zhong, B., Fu, Y.: Image super-resolution using very deep residual channel attention networks. In: Ferrari, V., Hebert, M., Sminchisescu, C., Weiss, Y. (eds.) ECCV 2018. LNCS, vol. 11211, pp. 294–310. Springer, Cham (2018). https://doi.org/10.1007/978-3-030-01234-2_18

39. Zhang, Y., Tian, Y., Kong, Y., Zhong, B., Fu, Y.: Residual dense network for image super-resolution. In: The IEEE Conference on Computer Vision and Pattern Recognition (CVPR) (2018)

40. Zhang, Z., Zheng, H., Hong, R., Fan, J., Yang, Y., Yan, S.: FRC-Net: a simple yet effective architecture for low-light image enhancement (2022)

41. Zhou, Y., et al.: UDC 2020 challenge on image restoration of under-display camera: methods and results. arXiv preprint arXiv:2008.07742 (2020)

42. Zhou, Y., Ren, D., Emerton, N., Lim, S., Large, T.: Image restoration for under-display camera. arXiv preprint arXiv:2003.04857 (2020)

Continuous Spectral Reconstruction from RGB Images via Implicit Neural Representation

Ruikang Xu[1], Mingde Yao[1], Chang Chen[2], Lizhi Wang[3], and Zhiwei Xiong[1]([✉])

[1] University of Science and Technology of China, Hefei, China
{xurk,mdyao}@mail.ustc.edu.cn, zwxiong@ustc.edu.cn
[2] Huawei Noah's Ark Lab, Beijing, China
chenchang25@huawei.com
[3] Beijing Institute of Technology, Beijing, China
wanglizhi@bit.edu.cn

Abstract. Existing spectral reconstruction methods learn discrete mappings from spectrally downsampled measurements (e.g., RGB images) to a specific number of spectral bands. However, they generally neglect the continuous nature of the spectral signature and only reconstruct specific spectral bands due to the intrinsic limitation of discrete mappings. In this paper, we propose a novel continuous spectral reconstruction network with implicit neural representation, which enables spectral reconstruction of arbitrary band numbers for the first time. Specifically, our method takes an RGB image and a set of wavelengths as inputs to reconstruct the spectral image with arbitrary bands, where the RGB image provides the context of the scene and the wavelengths provide the target spectral coordinates. To exploit the spectral-spatial correlation in implicit neural representation, we devise a spectral profile interpolation module and a neural attention mapping module, which exploit and aggregate the spatial-spectral correlation of the spectral image in multiple dimensions. Extensive experiments demonstrate that our method not only outperforms existing discrete spectral reconstruction methods but also enables spectral reconstruction of arbitrary and even extreme band numbers beyond the training samples.

Keywords: Computational photography · Hyperspectral image reconstruction · Implicit neural representation

1 Introduction

Spectral images record richer spectrum information than traditional RGB images, which have been proven useful in various vision-based applications, such as anomaly detection [21], object tracking [50], and segmentation [16]. To acquire spectral images, conventional spectral imaging technology relies on either spatial or spectral scanning for capturing the spectral signature to a number of bands,

L. Karlinsky et al. (Eds.): ECCV 2022 Workshops, LNCS 13805, pp. 78–94, 2023.
https://doi.org/10.1007/978-3-031-25072-9_6

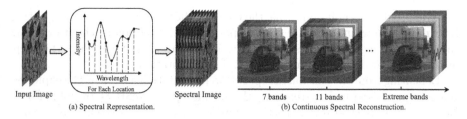

Input Image   Spectral Image   7 bands   11 bands   Extreme bands
                (a) Spectral Representation.   (b) Continuous Spectral Reconstruction.

Fig. 1. (a) The main difference of spectral reconstruction between previous methods and NeSR. Previous methods learn a discrete mapping to reconstruct the spectral image as the "blue dashes". We learn a continuous spectral representation and sample it to the target number of spectral bands as the "red curve". (b) Continuoue spectral reconstruction. (Color figure online)

which is of high complexity and time-consuming [10,17,43,45,53]. As an alternative, spectral reconstruction from RGB images is regarded as an attractive solution owning to the easy acquisition of RGB images [3,30,38,60].

Existing methods [23,31,38,56,58] for spectral reconstruction aim to learn a discrete mapping from an RGB image to a spectral image, which directly generates a specific number of spectral bands from three bands, as illustrated by the "blue dashes" in Fig. 1(a). However, this kind of representation ignores the continuous nature of the spectral signature. In the physical world, the spectral signature is in a continuous form where the high-dimension correlation is naturally hidden in the continuous representation [9,22]. To approximate the natural representation of the spectral signature, we reformulate the spectral reconstruction process, where a number of spectral bands are resampled from a continuous spectral curve, as illustrated by the "red curve" in Fig. 1(a).

Recent works adopt the concept of implicit neural representation for super-resolution [14] and 3D reconstruction [26,35] tasks, which obtain the continuous representation of signals with high fidelity. Inspired by this line of works, we aim to learn the continuous representation of the spectral signature for spectral reconstruction by leveraging a continuous and differentiable function [24,27]. However, it is non-trivial to exploit existing continuous representation methods for spectral images directly, since they only focus on the spatial correlation of RGB images [14,34]. The high-dimension spatial-spectral correlation, which plays a vital role for high-fidelity spectral reconstruction, remains unexplored [42,46].

To fill this gap, we propose Neural Spectral Reconstruction (NeSR) to continuously represent the spectral signature by exploiting the spatial-spectral correlation. Specifically, we first adopt a feature encoder to extract deep features from the input RGB image for representing the context information of the scene. We then take the target wavelengths as the coordinate information to learn the projection from the deep features to the spectral intensities of the corresponding wavelengths. To exploit the spatial-spectral correlation for continuous spectral reconstruction, we devise a Spectral Profile Interpolation (SPI) module and a Neural Attention Mapping (NAM) module. The former encodes the spatial-spectral correlation to the deep features leveraging the vertical and horizontal spectral profile interpolation, while the latter further enriches the spatial-spectral information using an elaborate spatial-spectral-wise attention mechanism. With the above two modules, the spatial-spectral correlation is exploited in the deep features for learning a continuous spectral representation.

Benefiting from the continuous spectral representation as well as the exploitation of spatial-spectral correlation, the advantage of NeSR is twofold. First, NeSR enables spectral reconstruction of arbitrary and even extreme band numbers beyond the training samples for the first time, as illustrated in Fig. 1(b). Second, NeSR notably improves the performance over state-of-the-art methods for spectral image reconstruction, bringing 12% accuracy improvement with only 2% parameter increase than the strongest baseline on the NTIRE2020 challenge dataset.

The contributions of this paper are summarized as follows:

(1) For the first time, we propose NeSR to reconstruct spectral images with an arbitrary number of spectral bands while keeping high accuracy via implicit neural representation.
(2) We propose the SPI and NAM modules to exploit the spatial-spectral correlation of depth features extracted from input RGB images for continuous spectral representation.
(3) Extensive experiments demonstrate that NeSR outperforms state-of-the-art methods in reconstructing spectral images with a specific number of bands.

2   Related Work

2.1   Implicit Neural Representation

Implicit neural representation aims to model an object as a continuous and differentiable function that maps coordinates and deep features to the corresponding signal, which is parameterized by a deep neural network. Recent works demonstrate its potential for modeling surfaces [11,20,35], shapes [6] [25], and the appearance of 3D objects [28,29]. Mildenhall *et al.* first introduce the implicit neural representation for synthesizing novel views of complex scenes using a sparse set of input views, named Neural Radiance Filed (NeRF) [26]. Compared with explicit 3D representations, implicit neural representation can capture subtle details. With the great success of NeRF [26], a large number of works extend implicit neural representation to other applications, such as 3D-aware generalization [12,39], pose estimation [37,55], relighting [7,36].

To get a more general representation, recent works estimate latent codes and share an implicit space for different objects or scenes [14,15,34,52], instead of learning an independent implicit neural representation for each object. The implicit space can be generated with an auto-encoder architecture. For example, Sitzmann *et al.* [33] propose a meta-learning-based method for sharing the implicit space. Mescheder *et al.* [24] propose to generate a global latent space of given images as input and use an occupancy function conditioning to perform the 3D reconstruction. Sitzmann *et al.* [34] replace ReLU with periodic activation functions and demonstrate that it can model the signals in higher quality. Chen *et al.* [14] utilize local implicit image function for representing natural and complex images. Yang *et al.* [52] propose a Transformer-based implicit neural representation for screen image super-resolution. Different from these works, we propose NeSR to address the continuous spectral reconstruction from RGB images by fully exploring the spatial-spectral correlation.

2.2 Spectral Reconstruction from RGB Images

Most existing spectral imaging systems rely on either spatial or spectral scanning [17,43,45]. They need to capture the spectral information of a single point or a single band separately, and then scan the whole scene to get a full spectral image. Due to the difficulty of measuring information from scenes with moving content, they are unsuitable for real-time operation. Moreover, these spectral imaging systems remain prohibitively expensive for consumer-grade usage. As an alternative, reconstructing spectral images from RGB images is a relatively low-cost and convenient approach to acquire spectral images. However, it is a severely ill-posed problem, since much information is lost after integrating the spectral radiance into RGB values.

Many methods have been proposed for spectral reconstruction from RGB images [32,44]. Early works leverage sparse coding to recover the lost spectral information from RGB images. Arad *et al.* [3] first exploit the spectral prior from a vast amount of data to create a sparse dictionary, which facilitates the spectral reconstruction. Later, Aeschbacher *et al.* [1] further improve the performance of Arad's method leveraging the A+ framework [40] from super-resolution. Alternatively, Akhtar *et al.* [2] utilize Gaussian processes and clustering instead of dictionary learning. With the great success of convolution neural networks in low-level computer vision, this task has received increasing attention from the deep learning direction [38,56,60]. Xiong *et al.* [51] propose a unified deep learning framework HSCNN for spectral reconstruction from both RGB and compressive measurements. Shi *et al.* [31] propose two improvements to boost the performance of HSCNN. Li *et al.* [23] propose an adaptive weighted attention network (AWAN) to explore the camera spectral sensitivity prior for improving the reconstruction accuracy. Zhao *et al.* [58] propose a hierarchical regression network (HRNet) for reconstructing spectral images from RGB images. Cai *et al.* [8] propose a Transformer-based model for spectral reconstruction, which utilize the mask-guided attention mechanism to capture the global correlation for reconstruction. However, these methods reconstruct the spectral images by the discrete mapping, which fixes the number of output spectral bands and ignores the continuous nature of the spectral signature. In this work, we propose NeSR to learn the continuous representation for spectral reconstruction from RGB images, which improves accuracy for reconstructing spectral images with a specific number of bands and enables spectral reconstruction with an arbitrary and extreme number of bands.

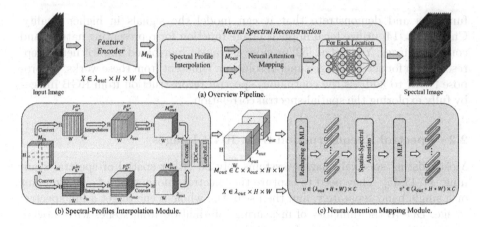

Fig. 2. The proposed Neural Spectral Reconstruction.

3   Neural Spectral Reconstruction

3.1   Overview

The overview of NeSR is shown in Fig. 2(a). Given an input RGB image, we cascade a feature encoder to extract the deep feature $M_{in} \in \mathbb{R}^{\lambda_{in} \times H \times W}$ as one input. For learning the continuous spectral representation, we take the 3D coordinate $X \in \mathbb{R}^{\lambda_{out} \times H \times W}$ as the other input. Specifically, for the spectral dimension, we use the normalized spectral wavelengths to reconstruct the spectral intensities of the corresponding wavelengths. Then, NeSR takes the deep feature M_{in} and the coordinate X to reconstruct the spectral image with the target of bands as the output, denoted as

$$Y = \mathcal{F}(M_{in}, X), \qquad (1)$$

where $\mathcal{F}$ stands for the overall processing of NeSR.

Specifically, given the deep feature M_{in} and the coordinate X, NeSR upsamples the deep feature to the target number of spectral bands for subsequent coordinate information fusion at first. To this end, the SPI module is designed to upsample the deep feature and encode the spatial-spectral correlation leveraging the vertical and horizontal spectral profile interpolation. Then, the NAM module is designed to further exploit the spatial-spectral correlation of the fused feature by global attention in spatial and spectral dimensions, which computes the similarity of different channels according to the spatial-spectral-wise attention mechanism.

After exploiting the spatial-spectral correlation and fusing the coordinate information, a new deep feature v^* is generated. Finally, each latent code of this feature is fed to a multi-layer perception (MLP) to predict the corresponding spectral signature. Iterating over all locations of the target spectral image, NeSR can reconstruct the spectral image with the desired number of bands while keeping high reconstruction fidelity.

3.2   Spectral Profile Interpolation

The SPI module upsamples the deep feature to the target number of spectral bands and encodes the spatial-spectral correlation of the spectral image. The spatial-spectral correlation is an essential characteristic of spectral representation [42,46]. To upsample the feature while exploiting this correlation, we propose the concept of Spectral Profile (SP) inspired by other high-dimension image reconstruction tasks [48,49,61].

We first give the definition of SP as follows. Consider a spectral image as a 3D volume $I(h, w, \lambda)$, where h and w stands for the spatial dimensions and λ stands for the spectral dimension, the vertical SP $P_{w^*}(h, \lambda)$ and the horizontal SP $P_{h^*}(w, \lambda)$ are the slices generated when $w = w^*$ and $h = h^*$, respectively. Since much information is lost after integrating the spectral radiance into RGB values, we exploit the spatial-spectral correlation of the spectral image in the feature domain. The flow diagram of the SPI module is shown in Fig. 2(b).

In the SPI module, given the deep feature $M_{in} \in \mathbb{R}^{\lambda_{in} \times H \times W}$, we first convert it into the vertical and horizontal SPs to exploit the correlations in the spatial and spectral dimensions, denoted as $P_{w^*}^{in}(h, \lambda) \in \mathbb{R}^{H \times \lambda_{in}}$ and $P_{h^*}^{in}(w, \lambda) \in \mathbb{R}^{W \times \lambda_{in}}$. Then, the upsampled vertical SP $P_{w^*}^{sr}(h, \lambda) \in \mathbb{R}^{H \times \lambda_{out}}$ and the upsampled horizontal SP $P_{h^*}^{sr}(w, \lambda) \in \mathbb{R}^{W \times \lambda_{out}}$ can be generated as

$$
\begin{aligned}
P_{w^*}^{sr}(h, \lambda) &= Up\left(P_{w^*}^{in}(h, \lambda)\right), \\
P_{h^*}^{sr}(w, \lambda) &= Up\left(P_{h^*}^{in}(w, \lambda)\right),
\end{aligned}
\tag{2}
$$

where $Up(\cdot)$ stands for the upsampling operation. After converting the above upsampled SPs back to the feature maps, we obtain the intermediate results $M_{out}^h, M_{out}^w \in \mathbb{R}^{\lambda_{out} \times H \times W}$. To generate the upsampled deep feature for the coordinate information fusion, we fuse the intermediate results with the concatenation operation and utilize the 3D convolution layers with LeakyReLU, which can be denoted as

$$
M_{out} = Conv_{3D}\left(\left[M_{out}^h, M_{out}^w\right]\right),
\tag{3}
$$

where $M_{out} \in \mathbb{R}^{C \times \lambda_{out} \times H \times W}$ stands for the output feature of the SPI module, which can be viewed as $\lambda_{out} \times H \times W$ latent codes corresponding to the coordinate of the spectral image. $Conv_{3D}(\cdot)$ and $[\cdot, \cdot]$ denote the 3D convolution layers with LeakyReLU and the concatenation operation, respectively.

3.3   Neural Attention Mapping

The NAM module further exploits the spatial-spectral correlation for continuous spectral representation by a new attention mechanism. Inspired by the success of self-attention-based architectures, such as Transformer, in natural language processing [41] and computer vision [13,47,59], we propose the spatial-spectral-wise attention mechanism, which captures the interactions of different dimensions to exploit the spectral-spatial correlation. The flow diagram of the NAM module is shown in Fig. 2(c).

Different from the standard Transformer block that receives a 1D token embedding [18] as input, the NAM module takes a 3D tensor as input. To handle the 3D tensor, we first reshape the 3D feature M_{out} and the 3D coordinate X to the 1D token embedding, and then map the embedding with an MLP, denoted as

$$v = MLP\left(Reshape([M_{out}, X])\right),\qquad(4)$$

where $MLP(\cdot)$ and $Reshape(\cdot)$ stand for the MLP and the reshape operation, respectively. $v \in \mathbb{R}^{(\lambda_{out} \cdot H \cdot W) \times C}$ denotes the fused token embedding, which fuses the information of the coordinate and the deep feature. Then, v is fed to the spatial-spectral attention block to exploit the spatial-spectral correlation of the fused token embedding.

In computing attention, we first map v into key and memory embeddings for modeling the deep correspondences. Unlike the self-attention mechanism that only focuses on a single dimension, we compute the attention maps of different dimensions to capture the interactions among dimensions for exploring the correlation of the fused token embedding. Benefiting from that v compresses the spectral-spatial information to one dimension, the attention maps fully exploit the spectral-spectral correlation to compute the similarity among channels, denoted as

$$\begin{aligned} Q, K, V &= vW_q, vW_k, vW_v, \\ v^* &= MLP(V \otimes \text{SoftMax}(Q^{\mathrm{T}} \otimes K)), \end{aligned}\qquad(5)$$

where $Q, K, V \in \mathbb{R}^{(\lambda_{out} \cdot H \cdot W) \times C}$ stand for *query*, *key* and *value* to generate the attention, and W_q, W_k, W_v stand for the corresponding weights, respectively. $\otimes$ and T denote the batch-wise matrix multiplication and the batch-wise transposition, respectively. Finally, we generate the output token embedding of the NAM module $v^* \in \mathbb{R}^{(\lambda_{out} \cdot H \cdot W) \times C}$ by an MLP.

3.4   Loss Function

We adopt the Mean Relative Absolute Error (MRAE) as the loss function, which is defined as

$$\mathcal{L} = \frac{1}{N} \sum_{i=1}^{N} (Y^{(i)} - I_{GT}^{(i)} / (I_{GT}^{(i)} + \varepsilon)),\qquad(6)$$

where $Y^{(i)}$ and $I_{GT}^{(i)}$ stand for the ith ($i = 1, ..., N$) pixel of the reconstructed and ground truth spectral images, respectively. We set $\varepsilon = 1 \times 10^{-3}$ due to zero points in the ground truth spectral image.

4   Experiments

4.1   Comparison to State-of-the-Art Methods

To quantitatively evaluate the effectiveness of the proposed method, we first compare NeSR with state-of-the-art methods in reconstructing spectral images with a specific number of spectral bands.

Table 1. Quantitative comparison on the NTIRE2020, CAVE and NTIRE2018 datasets for 31-band spectral reconstruction from RGB images. Red and blue indicate the best and the second best performance, respectively.

Methods	NTIRE2020 clean		NTIRE2020 real		CAVE		NTIRE2018 clean		NTIRE2018 real		Param (M)
	MRAE	RMSE	MRAE	RMSE	MRAE	RMSE	MRAE	RMSE	MRAE	RMSE	
BI	0.16566	0.04551	0.17451	0.04307	5.74425	0.16886	0.12524	0.01941	0.15862	0.02375	-
HSCNN-R	0.04128	0.01515	0.07278	0.01892	0.19613	0.03535	0.01943	0.00389	0.03437	0.00621	1.20
HSCNN-R + NeSR	0.03832	0.01323	0.06978	0.01814	0.15353	0.03304	0.01611	0.00383	0.03171	0.00584	1.34
HRNet	0.04007	0.01401	0.06756	0.01821	0.17219	0.02987	0.01521	0.00368	0.02985	0.00571	31.70
HRNet + NeSR	0.03701	0.01284	0.06691	0.01797	0.16431	0.02977	0.01466	0.00351	0.02895	0.00557	31.88
AWAN	0.03441	0.01215	0.06883	0.01711	0.19156	0.03752	0.01226	0.00255	0.03121	0.00577	28.59
AWAN + NeSR	0.02996	0.00989	0.06661	0.01632	0.13226	0.02789	0.01159	0.00229	0.03003	0.00552	29.29

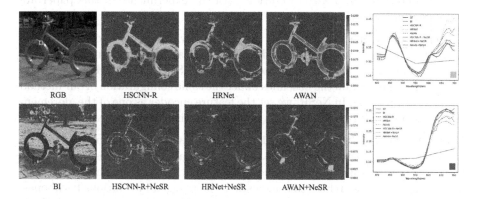

Fig. 3. Visualization of the error maps of different methods of spectral reconstruction from RGB images on "ARAD_HS_0465" of the NTIRE2020 "Clean" dataset. We show the spectral signatures of selected pixels and mark them by red and yellow rectangles in the RGB image. Please zoom in to see the difference of spectral intensity.

Datasets. In this work, we select three datasets as the benchmark for training and evaluation, including NTIRE2020 [5], CAVE [54] and NTIRE2018 [4]. NTIRE2020 and NTIRE2018 are the benchmarks for the hyperspectral reconstruction challenges in NTIRE2020 and NTIRE2018, respectively. Both datasets consist of two tracks: "Clean" and "Real World", in which each spectral image consists of 31 successive spectral bands ranging from 400 nm to 700 nm with a 10 nm increment. CAVE is a 16-bit spectral image dataset containing 32 scenes with 31 successive spectral bands ranging from 400 nm to 700 nm with a 10 nm increment. For training, the image pairs of the three datasets are randomly cropped to 64 × 64 and normalized to range [0, 1]. The training sets of the NTIRE2020 and NTIRE2018 datasets and the randomly picked 28 scenes from the CAVE dataset are used for training. We use the validation sets of the NTIRE2020 and NTIRE2018 datasets and the remaining 4 scenes from the CAVE dataset as the test datasets. We utilize the MRAE and Root Mean Square Error (RMSE) as metrics to quantitatively evaluate the reconstructed spectral images.

Implementation Details. Our proposed method is trained on spectral-RGB images on the three datasets separately. We take $\lambda_{min} = 400$ nm and $\lambda_{max} = 700$ nm as the minimum and maximum spectral coordinates since the input RGB images are synthesized in such a visible light range. We then normalize the target spectral coordinate λ_i to $[-1, 1]$ as $2 \times (\lambda_i - \lambda_{min})/(\lambda_{max} - \lambda_{min}) - 1$. We employ a 4-layer MLP with ReLU activation to reconstruct spectral images at the end of NeSR, and the hidden dimensions are (128, 128, 256, and 256). We take the bilinear interpolation as the upsampling operation in the SPI module. Adam optimizer is utilized with parameters $\beta_1 = 0.9$ and $\beta_2 = 0.999$. The learning rate is initially set to 1×10^{-4} and is later downscaled by a factor of 0.5 after every 2×10^4 iterations till 3×10^5 iterations. All the experiments in this paper are conducted in PyTorch 1.6.

Comparison Methods. To verify the superiority of our method, we compare NeSR with different baselines, including one classical method (bilinear interpolation, BI) and three deep-learning-based methods (HSCNN-R [31], HRNet [58] and AWAN [23]). We retrain the baselines under the same configuration of our method for a fair comparison. In NeSR, we set the feature encoder as one of the aforementioned deep-learning-based methods by removing the output layer, respectively.

Quantitative Evaluation. Table 1 shows the results of different methods on all benchmarks. It can be seen that, on the one hand, NeSR boosts the performance of the corresponding baseline with a slight parameter increase. On the other hand, NeSR is architecture-agnostic to be plugged into various backbones, which is effective for both simple (e.g., HSCNN-R) and complex (e.g., AWAN) architectures. Specifically, HRNet+NeSR achieves 6.9% decrease in MRAE than HRNet with only 0.5% parameters increase on the validation set of NTIRE2020 "Clean", and it shows the best performance on the validation set of NTIRE2018 "Real World". AWAN+NeSR achieves the best performance on the validation set of two tracks of NTIRE2020, which brings 12.9% decrease in MRAE than AWAN with 2.4% parameters increase.

Qualitative Evaluation. To evaluate the perceptual quality of spectral reconstruction, we show the error maps of different methods on the NTIRE2020 "Clean" validation set. As visualized in Fig. 3, we can observe a significant error decrease brought by NeSR. Besides, the spectral signatures reconstructed by NeSR for selected pixels are closer to the ground truth than the corresponding baselines (the yellow and red boxes). Therefore, NeSR can boost all the corresponding baselines in terms of spatial and spectral reconstruction accuracy.

4.2 Continuous Spectral Reconstruction

Previous methods of spectral reconstruction fix the number of output bands. Different from existing methods, by leveraging the continuous spectral representation, NeSR can reconstruct the spectral images with an arbitrary number of spectral bands and maintain high fidelity using a single model.

Dataset and Implementation. We build a spectral dataset based on ICVL dataset [3] for *continuous spectral reconstruction*. We remove duplicate scenes

Table 2. Quantitative comparison on the ICVL dataset for spectral reconstruction from RGB images with arbitrary bands. Red and <u>blue</u> indicate the best and the second best performance, respectively.

Methods	31 bands		16 bands		11 bands		7 bands	
	MRAE	RMSE	MRAE	RMSE	MRAE	RMSE	MRAE	RMSE
BI	0.15573	0.02941	0.16611	0.03052	0.19012	0.03293	0.20214	0.03276
Sparse coding	0.04843	0.01116	0.05722	0.01287	0.07681	0.01451	0.09841	0.02079
AWAN (-D)	0.02298	0.00534	0.02313	0.00544	0.02412	0.00565	0.02535	0.00578
AWAN (-S)	<u>0.02212</u>	<u>0.00530</u>	<u>0.02241</u>	<u>0.00543</u>	<u>0.02347</u>	<u>0.00567</u>	<u>0.02484</u>	<u>0.00571</u>
AWAN + NeSR	0.02001	0.00498	0.02024	0.00505	0.02129	0.00522	0.02119	0.00541

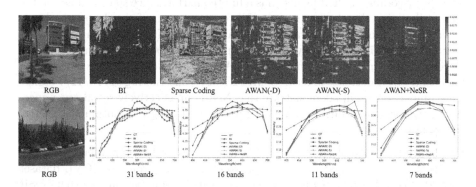

Fig. 4. Top: Visualization of the error maps of different methods of spectral reconstruction from RGB images on "BGU_HS_00044" of the ICVL dataset with 31 bands. Bottom: Spectral curves of reconstructed images of spectral reconstruction from RGB images with different band numbers on "BGU_HS_00045" of ICVL.

from the ICVL dataset and randomly choose 80 scenes for training and 10 scenes for testing. We process raw data of the ICVL dataset to generate the spectral images from 400 nm to 700 nm with different band numbers. Specifically, 31/16/11/7 spectral bands are used to validate the effectiveness of NeSR, which can be replaced by other band numbers. The implementation details are the same as in Sect. 4.1.

Comparison Methods. Since there are few available methods that could reconstruct spectral images with different bands using a single model, we set up three methods as baselines in this section (See Table 2). BI: We interpolate RGB images following the prior work [56,60], which operates the interpolation function along the spectral dimension, to obtain spectral images with the desired band numbers. Sparse coding: The sparse coding method [3] reconstructs spectral images leveraging a sparse dictionary of spectral signatures and their corresponding RGB projections, which can reconstruct spectral images with desired output band numbers. Deep-learning-based: We select AWAN [23] as a representative deep-learning-based baseline for comprehensive comparisons. Because the current deep models cannot reconstruct spectral images with an arbitrary number of bands, we separately train different models for different band numbers (7/11/16/31 bands), denoted as AWAN(-S). We also design a two-step strategy, denoted as AWAN(-D), by first reconstructing spectral images with a large number of spectral bands (e.g., 61 bands) and then downsampling them to the target number of spectral bands.

Quantitative Evaluation. We compare the quantitative results between NeSR and the baselines as mentioned above. As shown in Table 2, reconstructing the spectral image with an arbitrary number of bands by BI shows poor performance because the RGB image is integrated on the spectral dimension, and direct interpolation between RGB channels cannot correspond to any spectral wavelength. The sparse coding method [3] also has a limited performance since it relies on the sparse dictionary but lacks the powerful representation of the deep neural networks. Our continuous spectral representation (AWAN+NeSR) can reconstruct spectral images into arbitrary bands with a single model, which surpasses the aforementioned methods. Compared with AWAN(-S), NeSR achieves 9.5%, 9.6%, 9.2%, and 14.5% decrease in terms of MRAE for reconstructing 31/16/11/7 bands. The comparison results demonstrate that NeSR can effectively reconstruct spectral images with an arbitrary number of bands leveraging the continuous spectral representation.

Qualitative Evaluation. To give a visual comparison of different methods, we display the error maps of one representative scene from the ICVL dataset with 31 spectral bands in Fig. 4. From the visualization, we can see that NeSR provides higher reconstruction fidelity than other methods. We also give the spectral curves of different methods for reconstructing different spectral band numbers, which shows that our method has a lower spectral error than other methods. The qualitative comparison demonstrates that NeSR can reconstruct spectral images with an arbitrary number of spectral bands while keeping high fidelity.

4.3 Extreme Spectral Reconstruction

Our NeSR can also generate spectral images with an extreme band number beyond the training samples. We conduct experiments on the ICVL dataset to validate this point. During the training stage, NeSR takes an RGB image as the input, and the target spectral image is randomly sampled in the range of 7 to 31 bands. In the inference stage, NeSR is able to reconstruct spectral images with an extreme band number (e.g., 61 bands), which have larger spectral resolution and are not contained in the training samples. We show the quantitative and qualitative comparison results in Table 3 and Fig. 5.

It can be seen that NeSR has superior performance on spectral image reconstruction with extreme band numbers and significantly outperforms the baselines. It is worth noting that the reconstructed spectral images (41 bands to 61 bands) are out of the distribution of training samples, and NeSR has never seen them in the training stage, which indicates that NeSR empowers the practical continuous representation for spectral reconstruction.

Table 3. Left: Quantitative results of reconstructing spectral images from RGB images on the extreme spectral reconstruction. Right: Quantitative results of reconstructing high spectral-resolution images from low spectral-resolution images.

Methods	61 bands		51 bands		41 bands	
	MRAE	RMSE	MRAE	RMSE	MRAE	RMSE
BI	0.14844	0.02868	0.14962	0.02978	0.14914	0.03169
AWAN + BI	0.03593	0.00848	0.03375	0.00791	0.03332	0.00771
AWAN + NeSR	0.02689	0.00728	0.02655	0.00722	0.02677	0.00704

Methods	16-31 bands		16-61 bands	
	MRAE	RMSE	MRAE	RMSE
BI	0.01521	0.00578	0.01837	0.00576
AWAN	0.01285	0.00271	0.01346	0.00296
AWAN + NeSR	0.01232	0.00245	0.01312	0.00278

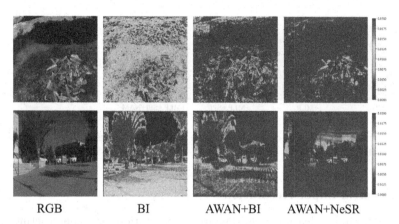

RGB   BI   AWAN+BI   AWAN+NeSR

Fig. 5. Error maps of reconstructed spectral images from RGB images on extreme spectral reconstruction. Top: Reconstructed results with 61 bands on "BGU_HS_00177". Bottom: Reconstructed results with 51 bands on "BGU_HS_00062".

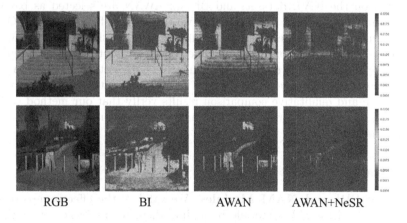

RGB   BI   AWAN   AWAN+NeSR

Fig. 6. Error maps of reconstructed spectral images on ICVL dataset. Top: Reconstructed results from 16 bands to 31 bands on "BGU_HS_00061". Bottom: Reconstructed results from 16 bands to 61 bands on "BGU_HS_00075".

Table 4. Left: Ablation on SPI and NAM modules. Right: Ablation on the attention mechanism. Both experiments are performed on the NTIRE2020 dataset.

Methods	MRAE	RMSE
AWAN	0.03441	0.01215
AWAN+MLP	0.03387	0.01211
AWAN+MLP+SPI	0.03139	0.01102
AWAN+MLP+SPI+NAM	0.02996	0.00989

Methods	MRAE	RMSE
Spectral-Wise	0.03214	0.01001
Spatial-Wise	0.03222	0.01032
Spatial-Spectral-Wise	0.02996	0.00989

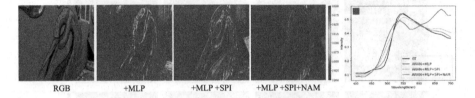

RGB   +MLP   +MLP +SPI   +MLP +SPI+NAM

Fig. 7. Visualization results of ablation on the proposed modules on "ARAD_HS_0464" of the NTIRE2020 dataset.

4.4 Spectral Super Resolution

Since NeSR aims to learn the continuous representation for reconstructing spectral images, it can be also applied to reconstruct high spectral-resolution images from low spectral-resolution images. Specifically, we reconstruct the spectral image with 16 bands to 31 and 61 bands by a single model. Experiments are conducted on the ICVL dataset, and BI and AWAN are selected as baseline methods, in which AWAN needs to train two times for different settings. Other implementation details are the same as in Sect. 4.1. As shown in Table 3, our method (+NeSR) outperforms the baseline methods in both settings. To give a visual comparison, we also show the error maps in both settings in Fig. 6. As can be seen, our method recovers the spectral information with a lower error. The quantitative and qualitative comparison results verify that our method can be utilized for spectral super-resolution leveraging the continuous representation of spectral images.

4.5 Ablation Studies

Impact of SPI and NAM Modules. We validate the effectiveness of the SPI module and the NAM module by adding them to the basic network step by step. We use AWAN cascaded with an MLP as the basic network to ensure that it has a similar parameter amount to our method, and conduct an experiment on the NTIRE2020 "Clean" dataset. The results of the ablation experiment are shown in Table 4. Since the deep feature lacks the spatial-spectral correlation, the baseline shows a limited reconstruction fidelity without the SPI and NAM modules. When inserting the SPI module into the baseline, the MRAE decreases from 0.03387 to 0.03139, and the NAM module also contributes to

0.00123 MRAE decrease. Moreover, from the error maps and the spectral curves in Fig. 7, we can see that the reconstruction error decreases when inserting the SPI and NAM modules into the basic network. The quantitative and qualitative comparison results demonstrate that our proposed modules are effective in enriching the spectral information and exploring the spatial-spectral correlation for learning the continuous spectral representation.

Impact of the Attention Mechanism. To validate the effectiveness of our spectral-spatial-wise attention, we compare it with spectral-wise and spatial-wise attention [19,57]. We use AWAN as the feature encoder and maintain the SPI module, and conduct an experiment on the NTIRE2020 "Clean" dataset. For each model, we only replace the attention mechanism of the NAM module. The results of the ablation experiment are shown in Table 4. Since the spatial-wise and spectral-wise attention cannot fully exploit the spectral-spatial correlation, they show a limited reconstruction accuracy. Compared with the spatial-wise/spectral-wise attention, our spectral-spatial-wise attention decreases the MRAE from 0.03222/0.03214 to 0.02996. The experiment results demonstrate that our nwe attention mechanism is effective in capturing the interactions of different channels for boosting performance.

5 Conclusion

In this paper, we propose NeSR to enable spectral reconstruction of arbitrary band numbers for the first time by learning the continuous spectral representation. NeSR inputs an RGB image and a set of wavelengths, as the context of the scene and the target spectral coordinates, to control the output band numbers of the spectral image. For high-fidelity reconstruction, we devise the SPI and NAM modules to exploit the spectral-spatial correlation in implicit neural representation. Extensive experiments demonstrate that NeSR significantly improves the reconstruction accuracy over the corresponding baselines with little parameter increase. Moreover, NeSR can effectively reconstruct spectral images with an arbitrary and even extreme number of bands leveraging the continuous spectral representation.

Acknowledgments. This work was supported in part by the National Natural Science Foundation of China under Grants 62131003 and 62021001.

References

1. Aeschbacher, J., Wu, J., Timofte, R.: In defense of shallow learned spectral reconstruction from RGB images. In: ICCVW (2017)
2. Akhtar, N., Mian, A.: Hyperspectral recovery from RGB images using gaussian processes. IEEE Trans. Pattern Anal. Mach. Intell. **42**(1), 100–113 (2018)
3. Arad, B., Ben-Shahar, O.: Sparse recovery of hyperspectral signal from natural RGB images. In: Leibe, B., Matas, J., Sebe, N., Welling, M. (eds.) ECCV 2016. LNCS, vol. 9911, pp. 19–34. Springer, Cham (2016). https://doi.org/10.1007/978-3-319-46478-7_2

4. Arad, B., Ben-Shahar, O., Timofte, R.: NTIRE 2018 challenge on spectral reconstruction from RGB images. In: CVPRW (2018)
5. Arad, B., Timofte, R., Ben-Shahar, O., Lin, Y.T., Finlayson, G.D.: NTIRE 2020 challenge on spectral reconstruction from an RGB image. In: CVPRW (2020)
6. Atzmon, M., Lipman, Y.: SAL: sign agnostic learning of shapes from raw data. In: CVPR (2020)
7. Boss, M., Braun, R., Jampani, V., Barron, J.T., Liu, C., Lensch, H.P.: NeRD: neural reflectance decomposition from image collections. In: ICCV (2021)
8. Cai, Y., et al.: Mask-guided spectral-wise transformer for efficient hyperspectral image reconstruction. In: CVPR (2022)
9. Cao, G., Bachega, L.R., Bouman, C.A.: The sparse matrix transform for covariance estimation and analysis of high dimensional signals. IEEE Trans. Image Process. **20**(3), 625–640 (2010)
10. Cao, X., Du, H., Tong, X., Dai, Q., Lin, S.: A prism-mask system for multispectral video acquisition. IEEE Trans. Pattern Anal. Mach. Intell. **33**(12), 2423–2435 (2011)
11. Chabra, R., et al.: Deep local shapes: learning local SDF priors for detailed 3D reconstruction. In: Vedaldi, A., Bischof, H., Brox, T., Frahm, J.-M. (eds.) ECCV 2020. LNCS, vol. 12374, pp. 608–625. Springer, Cham (2020). https://doi.org/10.1007/978-3-030-58526-6_36
12. Chan, E.R., Monteiro, M., Kellnhofer, P., Wu, J., Wetzstein, G.: Pi-GAN: periodic implicit generative adversarial networks for 3D-aware image synthesis. In: CVPR (2021)
13. Chen, H., et al.: Pre-trained image processing transformer. In: CVPR (2021)
14. Chen, Y., Liu, S., Wang, X.: Learning continuous image representation with local implicit image function. In: CVPR (2021)
15. Chen, Z., Zhang, H.: Learning implicit fields for generative shape modeling. In: CVPR (2019)
16. Dao, P.D., Mantripragada, K., He, Y., Qureshi, F.Z.: Improving hyperspectral image segmentation by applying inverse noise weighting and outlier removal for optimal scale selection. ISPRS J. Photogramm. Remote. Sens. **171**, 348–366 (2021)
17. Descour, M., Dereniak, E.: Computed-tomography imaging spectrometer: experimental calibration and reconstruction results. Appl. Opt. **34**(22), 4817–4826 (1995)
18. Dosovitskiy, A., et al.: An image is worth 16x16 words: transformers for image recognition at scale. In: ICLR (2020)
19. Fu, J., et al.: Dual attention network for scene segmentation. In: CVPR (2019)
20. Jiang, C., Sud, A., Makadia, A., Huang, J., Nießner, M., Funkhouser, T., et al.: Local implicit grid representations for 3D scenes. In: CVPR (2020)
21. Jiang, K., Xie, W., Lei, J., Jiang, T., Li, Y.: LREN: low-rank embedded network for sample-free hyperspectral anomaly detection. In: AAAI (2021)
22. Kuybeda, O., Malah, D., Barzohar, M.: Rank estimation and redundancy reduction of high-dimensional noisy signals with preservation of rare vectors. IEEE Trans. Signal Process. **55**(12), 5579–5592 (2007)
23. Li, J., Wu, C., Song, R., Li, Y., Liu, F.: Adaptive weighted attention network with camera spectral sensitivity prior for spectral reconstruction from RGB images. In: CVPRW (2020)
24. Mescheder, L., Oechsle, M., Niemeyer, M., Nowozin, S., Geiger, A.: Occupancy networks: learning 3D reconstruction in function space. In: CVPR (2019)
25. Michalkiewicz, M., Pontes, J.K., Jack, D., Baktashmotlagh, M., Eriksson, A.: Implicit surface representations as layers in neural networks. In: CVPR (2019)

26. Mildenhall, B., Srinivasan, P.P., Tancik, M., Barron, J.T., Ramamoorthi, R., Ng, R.: NeRF: representing scenes as neural radiance fields for view synthesis. In: ECCV (2020)
27. Niemeyer, M., Geiger, A.: Giraffe: representing scenes as compositional generative neural feature fields. In: CVPR (2021)
28. Niemeyer, M., Mescheder, L., Oechsle, M., Geiger, A.: Differentiable volumetric rendering: learning implicit 3D representations without 3D supervision. In: CVPR (2020)
29. Oechsle, M., Mescheder, L., Niemeyer, M., Strauss, T., Geiger, A.: Texture fields: learning texture representations in function space. In: CVPR (2019)
30. Robles-Kelly, A.: Single image spectral reconstruction for multimedia applications. In: ACM MM (2015)
31. Shi, Z., Chen, C., Xiong, Z., Liu, D., Wu, F.: HSCNN+: advanced CNN-based hyperspectral recovery from RGB images. In: CVPRW (2018)
32. Shi, Z., Chen, C., Xiong, Z., Liu, D., Zha, Z.J., Wu, F.: Deep residual attention network for spectral image super-resolution. In: ECCVW (2018)
33. Sitzmann, V., Chan, E.R., Tucker, R., Snavely, N., Wetzstein, G.: MetaSDF: meta-learning signed distance functions. In: NIPS (2020)
34. Sitzmann, V., Martel, J., Bergman, A., Lindell, D., Wetzstein, G.: Implicit neural representations with periodic activation functions. In: NIPS (2020)
35. Sitzmann, V., Zollhöfer, M., Wetzstein, G.: Scene representation networks: continuous 3D-structure-aware neural scene representations. In: NIPS (2019)
36. Srinivasan, P.P., Deng, B., Zhang, X., Tancik, M., Mildenhall, B., Barron, J.T.: NeRV: neural reflectance and visibility fields for relighting and view synthesis. In: CVPR (2021)
37. Su, S.Y., Yu, F., Zollhoefer, M., Rhodin, H.: A-NeRF: surface-free human 3D pose refinement via neural rendering. arXiv:2102.06199 (2021)
38. Sun, B., Yan, J., Zhou, X., Zheng, Y.: Tuning IR-cut filter for illumination-aware spectral reconstruction from RGB. In: CVPR (2021)
39. Tancik, M., et al.: Learned initializations for optimizing coordinate-based neural representations. In: CVPR (2021)
40. Timofte, R., De Smet, V., Van Gool, L.: A+: adjusted anchored neighborhood regression for fast super-resolution. In: ACCV (2014)
41. Vaswani, A., et al.: Attention is all you need. In: NIPS (2017)
42. Wang, L., Sun, C., Fu, Y., Kim, M.H., Huang, H.: Hyperspectral image reconstruction using a deep spatial-spectral prior. In: CVPR (2019)
43. Wang, L., Xiong, Z., Gao, D., Shi, G., Zeng, W., Wu, F.: High-speed hyperspectral video acquisition with a dual-camera architecture. In: CVPR (2015)
44. Wang, L., Xiong, Z., Huang, H., Shi, G., Wu, F., Zeng, W.: High-speed hyperspectral video acquisition by combining nyquist and compressive sampling. IEEE Trans. Pattern Anal. Mach. Intell. $41(4)$, 857–870 (2019)
45. Wang, L., Xiong, Z., Shi, G., Wu, F., Zeng, W.: Adaptive nonlocal sparse representation for dual-camera compressive hyperspectral imaging. IEEE Trans. Pattern Anal. Mach. Intell. $39(10)$, 2104–2111 (2017)
46. Wang, P., Wang, L., Leung, H., Zhang, G.: Super-resolution mapping based on spatial-spectral correlation for spectral imagery. IEEE Trans. Geosci. Remote Sens. $59(3)$, 2256–2268 (2020)
47. Weng, W., Zhang, Y., Xiong, Z.: Event-based video reconstruction using transformer. In: ICCV (2021)
48. Wu, G., et al.: Light field image processing: an overview. IEEE J. Sel. Top. Signal Process. $11(7)$, 926–954 (2017)

49. Xiao, Z., Xiong, Z., Fu, X., Liu, D., Zha, Z.J.: Space-time video super-resolution using temporal profiles. In: ACM MM (2020)
50. Xiong, F., Zhou, J., Qian, Y.: Material based object tracking in hyperspectral videos. IEEE Trans. Image Process. **29**, 3719–3733 (2020)
51. Xiong, Z., Shi, Z., Li, H., Wang, L., Liu, D., Wu, F.: HSCNN: CNN-based hyperspectral image recovery from spectrally undersampled projections. In: ICCVW (2017)
52. Yang, J., Shen, S., Yue, H., Li, K.: Implicit transformer network for screen content image continuous super-resolution. In: NIPS (2021)
53. Yao, M., Xiong, Z., Wang, L., Liu, D., Chen, X.: Spectral-depth imaging with deep learning based reconstruction. Opt. Express **27**(26), 38312–38325 (2019)
54. Yasuma, F., Mitsunaga, T., Iso, D., Nayar, S.K.: Generalized assorted pixel camera: postcapture control of resolution, dynamic range, and spectrum. IEEE Trans. Image Process. **19**(9), 2241–2253 (2010)
55. Yen-Chen, L., Florence, P., Barron, J.T., Rodriguez, A., Isola, P., Lin, T.Y.: iNeRF: inverting neural radiance fields for pose estimation. arXiv:2012.05877 (2020)
56. Zhang, L., et al.: Pixel-aware deep function-mixture network for spectral super-resolution. In: AAAI (2020)
57. Zhang, Y., Li, K., Li, K., Wang, L., Zhong, B., Fu, Y.: Image super-resolution using very deep residual channel attention networks. In: ECCV (2018)
58. Zhao, Y., Po, L.M., Yan, Q., Liu, W., Lin, T.: Hierarchical regression network for spectral reconstruction from RGB images. In: CVPRW (2020)
59. Zhu, X., Su, W., Lu, L., Li, B., Wang, X., Dai, J.: Deformable DETR: deformable transformers for end-to-end object detection. In: ICLR (2020)
60. Zhu, Z., Liu, H., Hou, J., Zeng, H., Zhang, Q.: Semantic-embedded unsupervised spectral reconstruction from single RGB images in the wild. In: ICCV (2021)
61. Zuckerman, L.P., Naor, E., Pisha, G., Bagon, S., Irani, M.: Across scales and across dimensions: temporal super-resolution using deep internal learning. In: Vedaldi, A., Bischof, H., Brox, T., Frahm, J.-M. (eds.) ECCV 2020. LNCS, vol. 12352, pp. 52–68. Springer, Cham (2020). https://doi.org/10.1007/978-3-030-58571-6_4

Event-Based Image Deblurring
with Dynamic Motion Awareness

Patricia Vitoria(ORCID), Stamatios Georgoulis, Stepan Tulyakov[✉],
Alfredo Bochicchio, Julius Erbach, and Yuanyou Li

Huawei Technologies, Zurich Research Center, Zürich, Switzerland
{patricia.vitoria.carrera,stamatios.georgoulis,stepan.tulyakov,
alfredo.bochicchio,julius.erbach,liyuanyou1}@huawei.com

Abstract. Non-uniform image deblurring is a challenging task due to
the lack of temporal and textural information in the blurry image itself.
Complementary information from auxiliary sensors such event sensors
are being explored to address these limitations. The latter can record
changes in a logarithmic intensity asynchronously, called events, with
high temporal resolution and high dynamic range. Current event-based
deblurring methods combine the blurry image with events to jointly estimate per-pixel motion and the deblur operator. In this paper, we argue
that a divide-and-conquer approach is more suitable for this task. To this
end, we propose to use modulated deformable convolutions, whose kernel
offsets and modulation masks are dynamically estimated from events to
encode the motion in the scene, while the deblur operator is learned from
the combination of blurry image and corresponding events. Furthermore,
we employ a coarse-to-fine multi-scale reconstruction approach to cope
with the inherent sparsity of events in low contrast regions. Importantly,
we introduce the first dataset containing pairs of real RGB blur images
and related events during the exposure time. Our results show better
overall robustness when using events, with improvements in PSNR by
up to 1.57 dB on synthetic data and 1.08 dB on real event data.

Keywords: Image deblurring · Dataset · Event-based vision ·
Deformable convolution

1 Introduction

Conventional cameras operate by accumulating light over the exposure time and
integrating this information to produce an image. During this time, camera or
scene motion may lead to a mix of information across a neighborhood of pixels,
resulting in blurry images. To account for this effect, some cameras incorporate
an optical image stabilizer (OIS), which tries to actively counteract the camera
shake. However, OIS can deal just with small motions and it still fails in the case
of scene motion. To reduce the impact of motion blur, one can alternatively adapt

---

Supplementary Information The online version contains supplementary material
available at https://doi.org/10.1007/978-3-031-25072-9_7.

L. Karlinsky et al. (Eds.): ECCV 2022 Workshops, LNCS 13805, pp. 95–112, 2023.
https://doi.org/10.1007/978-3-031-25072-9_7

the exposure time according to the motion observed in the scene [8], albeit at the potential loss of structural information. In the inevitable case where an image is already blurred, image deblurring methods try to recover the sharp image behind the motion-blurred scene. However, restoring a photorealistic sharp image from a single blurry image is an ill-posed problem due to the loss of temporal and textural information.

In the last years, image-based deblurring algorithms have shown impressive advances [3,4,11,14,17,20,21,25,33,38,42,51,52,54,55], but they still struggle while reconstructing high-frequency details. Recently, together with the intensity camera, a new breed of bio-inspired sensors containing high-frequency details, called event cameras, have been employed in image deblurring. Event cameras can asynchronously detect local log intensity changes, called *events*, at each pixel independently during the exposure time of a conventional camera with very low latency (1 μs) and high dynamic range (≈ 140 dB). Understandably, due to their capacity to capture high temporal resolution and fine-grained motion information, event cameras pose the deblurring problem in a more tractable way. Despite their great potential, current event-based image deblurring methods come with their own limitations. *(1)* They are restricted to low-resolution Active Pixel Sensor (APS) intensity frames provided by the event camera [13,16,31,32, 40] or *(2)* work just with grayscale images or using color events by treating each channel separately, failing to exploit the intra-correlations between image channels [13,16,31,32,40]. *(3)* They are either based on physical models [16,31, 32] that are sensitive to the lack of events in low-contrast regions as well as their noisy nature, or they use the event data just as an additional input to the image without explicitly exploiting the information provided by the events [13,56]. *(4)* They do not generalize well to real data [13,56].

To overcome the above-mentioned issues, in this work, we propose a divide-and-conquer deblurring approach that decouples the estimation of the per-pixel motion from the deblur operator. Due to their inherent temporal information, we use the input events to dynamically estimate per-pixel trajectories in terms of kernel offsets (and modulation masks) that correspond to the per-pixel motion. On the other hand, the deblur operator is learned from a set of blurry images and corresponding event features and stays fixed after training. This decoupling of tasks is achieved via the use of modulated deformable convolutions [58]; see Fig. 2 for an illustration. Moreover, we frame our approach as a multi-scale, coarse-to-fine learning task to deal with the inherent sparsity of events. In particular, coarser scales can take broader contextual information into account, while finer scales are encouraged to preserve the high-frequency details of the image, thereby alleviating the noisy and sparse nature of events. Notably, our algorithm works directly on high-resolution (HR) RGB images using only a monochrome event camera, thus exploiting the intra-correlation between channels and avoiding the intrinsic limitations of using color event cameras. An overview of our approach can be seen in Fig. 1.

Due to the lack of real-world datasets containing HR RGB images with their corresponding events during the exposure time, and to demonstrate the ability

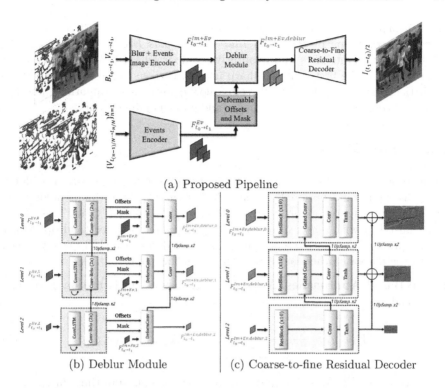

(a) Proposed Pipeline

(b) Deblur Module | (c) Coarse-to-fine Residual Decoder

Fig. 1. Proposed Architecture. (a) Two encoders (blue and red) are used to compute blurry image and events features $\{\mathbf{F}_{t_0 \to t_1}^{Im+Ev}\}_{l=0}^{L}$ and events features $\{\mathbf{F}_{t_0 \to t_1}^{Ev}\}_{l=0}^{L}$ that are used to compute the deblur operator and motion-related kernel offsets together with modulated masks, respectively. Then the (b) deblur module utilizes a set of modulated deformable convolutions to estimate the deblur $\{\mathbf{F}_{t_0 \to t_1}^{Im+Ev,deblur}\}_{l=0}^{L}$ from the features $\{\mathbf{F}_{t_0 \to t_1}^{Im+Ev}\}_{l=0}^{L}$ at different scales by using offsets and mask computed from $\{\mathbf{F}_{t_0 \to t_1}^{Ev}\}_{l=0}^{L}$. The sharp image is decoded from the deblurred features by progressively reconstructing the residual image from coarser-to-finer scales. (Color figure online)

of our method to generalize to real data, we collected a new dataset, called RGBlur+E, composed of real HR blurry images and events.

Contributions of this work are as follows

1. We propose the first divide-and-conquer deblurring approach that decouples the estimation of per-pixel motion from the deblur operator. To explicitly exploit the dense motion information provided by the events, we employ modulated deformable convolutions. The latter allows us to estimate the integration locations (i.e. kernel offsets and modulation masks), which are solely related to motion, directly from events.
2. To deal with the inherent sparsity of events as well as their lack in low contrast regions, we propose to employ a multi-scale, coarse-to-fine approach, that will exploit high frequency local details of events and broad motion information.

3. To evaluate and facilitate future research, we collect the first real-world dataset consisting of *HR RGB Blur Images plus Events* (RGBlur+E). To allow both quantitative and qualitative comparisons, RGBlur+E consists of two subsets: the first contains sharp images, corresponding events, and synthetically generated blurry images to allow quantitative evaluations, and the second consists of real blurry images and corresponding events.

2   Related Work

In this section, we discuss prior work on *image deblurring*, i.e. given a blurry image as input (and optionally events during the exposure time), predict the sharp latent image at a certain timestamp of the exposure time. This is not to be mistaken with *video deblurring*, i.e. from a blurry video (and optionally events) produce the sharp latent video, which is a different task (typically solved in conjunction with other tasks) not studied in this paper.

Image-Based Approaches. A generic image-based blur formation model can be defined as

$$\mathbf{B}(x,y) = \frac{1}{t_1 - t_0} \sum_{t=t_0}^{t_1} \mathbf{I}_t(x,y) = \langle \mathbf{I}_{nn}(x,y), \mathbf{k}(x,y) \rangle \tag{1}$$

where $\mathbf{B} \in \mathbb{R}^{H \times W \times C}$ is the blurry image over the exposure time $[t_0, t_1]$, $\mathbf{I}_t \in \mathbb{R}^{H \times W \times C}$ is the sharp image at time $t \in [t_0, t_1]$, $\mathbf{I}_{nn}(x,y)$ is a window of size $K \times K$ around pixel (x,y) and $\mathbf{k}$ is a per-pixel blur kernel. Blind deblurring formulates the problem as one of blind deconvolution where the goal is to recover the sharp image $\mathbf{I}_t$ without knowledge of the blur kernel $\mathbf{k}$ [19]. Most traditional methods apply a two-step approach, namely blur kernel estimation followed by non-blind deconvolution [5,10,23,29]. Additionally, assumptions about the scene prior are made to solve the problem. Examples of such priors are the Total Variation (TV) [2], sparse image priors [23,49], color channel statistics [30,50], patch recurrence [24], and outlier image signals [9]. Nonetheless, most of those methods are limited to uniform blur. To handle spatially-varying blur, more flexible blur models have been proposed based on a projective motion path model [41,45,53], or a segmentation of the image into areas with different types of blur [7,15,22,28]. However, those methods are computationally heavy (e.g. computation time of 20 mins [39] or an hour [15] per image) and successful results will depend on the accuracy of the blur models.

Image deblurring has benefited from the advances of deep convolutional neural networks (CNNs) by directly estimating the maping from blurry images to sharp. Several network designs have been proposed that increase the receptive field [55], use multi-scale strategies [3,4,11,25,27,38,42,54], learn a latent presentation [26,43], use deformable convolutions [14,51], progressively refine the result [33,46,52], or use perceptual metrics [20,21]. Also, some works combined model-based and learning-based strategies, by first estimating the motion kernel field using a CNN, and then applying a non-blind deconvolution [1,12,39].

However, despite advances in the field, image-based approaches struggle to reconstruct structure or perform accurate deblurring in the presence of large motions, also in real case scenarios and in the case of model-based approaches, they are bound by the employed model of the blur kernel.

Event-Based Approaches. An event camera outputs a sequence of *events*, denoted by (x, y, t, σ), which record log intensity changes. (x, y) are image coordinates, t is the time the event occurred, and $\sigma = \pm 1$, called the polarity, denotes the sign of the intensity change. Due to their high temporal resolution, and hence reliably encoded motion information, event sensors have been used for image deblurring. Several event-based methods perform image deblurring on grayscale images, or alternatively on color images by processing color events channel-wise[1]. In [32], the authors model the relationship between the blurry image, events, and latent frames through an Event-based Double Integral (EDI) model defined as,

$$\mathbf{B} = \frac{1}{t_1 - t_0} \sum_{t=t_0}^{t_1} \mathbf{I}_t = \frac{\mathbf{I}_{t_0}}{t_1 - t_0} \odot \sum_{t=t_0}^{t_1} \exp\left(c\mathbf{E}_{t_0 \rightarrow t}\right) \qquad (2)$$

where $\mathbf{E}_{t_0 \rightarrow t} \in \mathbb{R}^{H \times W}$ is the sum of events between t_0 and t, and c is a constant. $\mathbf{E}_{t_0 \rightarrow t}$ can be computed as the integral of events between t_0 and t. However, the noisy and lossy nature of event data often results in strongly accumulated noise, and loss of details and contrast in the scene. In [31], the authors extend the EDI model to the multiple EDI (mEDI) model to get smoother results based on multiple images and their events. [16] combines the optimization approach of the EDI model with an end-to-end learning strategy. In particular, results obtained by [32] are refined by exploiting computed optical flow information and boundary guidance maps. Similarly, [44] proposes to incorporate an iterative optimization method into a deep neural network that performs denoising, deblurring, and super-resolution simultaneously.

Direct one-to-one mapping from a blurry image plus events has been performed by [13,40,48,56]. [13] utilize two consecutive U-Nets to perform image deblurring followed by high frame rate video generation. In [40], two encoders-based Unet are used where image and events features are combined using cross-modal attention and the skip connections across stages are guided by an events mask. Additionally, they propose an alternative representation to voxel grid that has into account the image deblurring formation process. A drawback of [13,40,48] methods is that either the method just work on grayscale images or color events cameras are needed. In [56], the authors use the information of a monochrome event camera to compute an event representation through E2VID [35]. Then, the event representation is used as an additional input to an RGB image-based network [54]. However, direct one-to-one mapping approaches often fail to generalize to real data. To overcome this issue, [48] use a self-supervised approach based on optical flow estimation where real-world events are exploited to alleviate the performance degradation caused by data inconsistency.

---

[1] Note that, the use of color events comes with its drawbacks. Small effective resolution and reduction of light per pixel, the need for demosaicking at the event level, and the lack of good commercial color event cameras, to mention a few.

In this work, instead of learning all the deblurring process at once, we propose to decouple the learning of per-pixel motion from the deblur operators, and retain only the essential information for each sub-task. One of the benefits of the proposed method is that while the implicit deblur operator is learnt and fixed, the per-pixel motion is dynamically learned for each single example adapting the sampling of the kernel in the deblur module to the motion of each specific blur image. Moreover, to solve common artifacts of event-based methods, we use a multi-scale coarse-to-fine approach that handles the lack of events in low contrast regions and their noisy nature. Additionally, the proposed approach works directly with a combination of HR RGB images and monochrome event cameras and we propose the first dataset that combines HR RGB images with real events.

Deformable Convolutions. A major limitation of current (image- or event-based) deblurring methods is that by using standard convolutions, they end up applying the same deblur operators, learned during training, with a fixed receptive field across the entire image, regardless of the motion pattern that caused the blur. In contrast, deformable convolutions [6] learn kernel offsets per pixel, and thus can effectively change the sampling locations of the deblur operators. This allows for more 'motion-aware' deblur operators with the ability to adapt their receptive field according to the motion pattern. In [58], the authors propose to also learn a modulation mask in conjunction with kernel offsets, i.e. modulated deformable convolutions, to attenuate the contribution of each sampling location. See Fig. 2 for an illustration of the sampling process in a modulated deformable convolution. In image deblurring, modulated deformable convolutions have been used to adapt the receptive field to the blur present in the scene [14,51]. In this paper, we propose to use it differently. Since events contain fine-grained motion information per pixel, we can estimate 'motion-aware' deblur operators by conditioning the learning of kernel offsets and modulation masks solely to encoded event features. This allows the receptive field of deblur operators to dynamically adapt according to the underlying motion pattern.

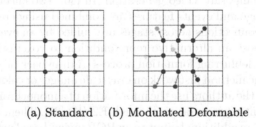

(a) Standard   (b) Modulated Deformable

Fig. 2. Standard vs Deformable Convolution. Illustration of kernel sampling locations in a 3 × 3 standard and modulated deformable convolution [58]. (a) Regular sampling in a standard convolution. (b) In a modulated deformable convolution, the sampling deforms in the direction of the learnable offsets (red lines) and is weighted by the learnable modulation mask. (Color figure online)

3 Method

Problem formulation. Given a blurry RGB HR image $\mathbf{B}_{t_0 \to t_1}$ and the corresponding event stream $\mathbf{E}_{t_0 \to t_1}$ containing all asynchronous events triggered during exposure time $[t_0, t_1]$, the proposed method reconstructs a sharp latent image $\mathbf{I}_t$ corresponding to the mid exposure timestamp $t = (t_1 + t_0)/2$.

3.1 System Overview

Figure 1 illustrates the proposed event-based image deblurring architecture, which consists of: an RGB blurry image and events encoder, a separate recurrent events encoder, a *deblur* module, and a multi-scale residual decoder.

The first encoder computes a multi-scale feature representation, $\{\mathbf{F}^{Im+Ev}_{t_0 \to t_1}\}^L_{l=0}$, from the blurry image $\mathbf{B}_{t_0 \to t_1}$ and the stream of events $\mathbf{E}_{t_0 \to t_1}$ during the exposure time represented as a *voxel grid* with five equally-sized temporal bins.

Similarly, the second encoder computes a multi-scale feature representation $\{\mathbf{F}^{Ev}_{t_0 \to t_1}\}^L_{l=0}$, but solely from the events. Differently, the stream of events $\mathbf{E}_{t_0 \to t_1}$ is split into fixed-duration chunks $\{\mathbf{V}_{t_{(n-1)/N} \to t_{n/N}}\}^N_{n=1}$, with each chunk containing all events within a time window $1/N$ represented as a voxel grid with five equally-sized temporal bins. The event chunks are integrated over time using a ConvLSTM module.

The deblur module exploits the features from the second encoder $\{\mathbf{F}^{Ev}_{t_0 \to t_1}\}^L_{l=0}$ to dynamically estimate kernel offsets and modulation masks per pixel that adapt the sampling locations of the modulated deformable convolutions according to the per-pixel motion pattern of the input image. Sequentially, the modulated deformable convolutions are applied to the first encoder features $\{\mathbf{F}^{Im+Ev}_{t_0 \to t_1}\}^L_{l=0}$, at each scale, essentially performing the deblur operators at each feature level.

Finally, the deblurred features $\{\mathbf{F}^{Im+Ev,deblur}_{t_0 \to t_1}\}^L_{l=0}$ are passed through a multi-scale decoder that progressively reconstructs the residual image in a coarse-to-fine manner. By doing so, our method exploits both image- and event-based information. Coarser scales will encode broad contextual information that helps deblurring if events are missing, while original resolution scales will estimate finer details by exploiting image and event structural information.

Section 3.2 and 3.3 detail the deblur module and coarse-to-fine reconstruction respectively.

3.2 Deblur Module

A standard 2D convolution consists of two steps: 1) sampling on a regular grid $\mathcal{R}$ over the input feature map $\mathbf{x}$; and 2) summation of sampled values weighted by w. The grid $\mathcal{R}$ defines the receptive field, and dilation. As illustrated in Fig. 2(a), for deformable convolutions the regular grid $\mathcal{R}$ is augmented with offsets $\{\Delta x_k | n = 1, ..., N\}$, and in the case of modulated deformable convolutions additionally modulated scalars Δm_k weight each position.

Blurry images by itself do not contain explicit temporal information or motion cues during the image acquisition, thus, being bad candidates as a clue to estimate kernel offsets and modulation masks. In contrast, due to their high temporal resolution, events contain all fine-grained motion cues during the exposure time that led to the blurry image, allowing for one-to-one mapping solutions. Motivated by this observation, in this work, we exploit the inherent motion information in the event chunks to learn the offsets Δx_k and modulated scalars Δm_k that 1) adapt the sampling grid of the modulated deformable convolution to the per-pixel motion of the scene, and 2) weight every sampling position according to its actual contribution. The estimated offsets and masks are directly applied to $\mathbf{F}^{Im+Ev}(x + x_k + \Delta x_k)$ to perform deblurring at the feature level. In particular, the modulated deformable convolutions can be expressed as:

$$\mathbf{F}^{Im+Ev,deblur}(x) = \sum_{k=1}^{K} w_k \cdot \mathbf{F}^{Im+Ev}(x + x_k + \Delta x_k) \cdot \Delta m_k. \qquad (3)$$

with w_k being the kernel convolutional weights. To address inherent limitations of event data, i.e. noisy events or lack of events in low-contrast regions, we perform the feature deblurring in a multi-scale fashion (see Sect. 3.3. Specifically, as shown in Fig. 1(b), to generate deblur features at the l-th scale we re-use the features computed at the $(l - 1)$-th scale. Note that, in order to compute the offsets and mask we first integrate the chunk of events using a ConvLSTM module [47] followed by two Conv+ReLU blocks that return the offsets and mask at each scale. The ConvLSTM layer will encode the overall temporal information and enable the network to learn the integration of events, as done analytically in the EDI formulation [32] or recently is done similarly by the SCER preprocessing module in [40].

3.3   Multi-scale Coarse-to-Fine Approach

By nature, event information is inherently sparse, noisy and they lack in low contrast regions. To address this problem, we propose to, first, perform feature deblurring in a multi-scale fashion and, second, progressively estimate the sharp image by computing residuals at different scales from multi-scale deblur features $\{\mathbf{F}^{Im+Ev,deblur}_{t_0 \to t_1}\}^L_{l=0}$.

Coarse-to-fine reconstruction has been widely used in image-based restoration methods [33,46,52]. While the motivation of those methods is to recover the clear image progressively by diving into the problem in small sub-tasks, in this work, the goal of this approach is to deal with the sparsity and lack of events in some regions. Coarser scales will learn broader contextual information where motion can be estimated even in locations where there are no events and being then more robust to noisy or missing events in low-contrast areas. On the other hand, scales closer to the original resolution will exploit the fine details contained in the event data translating to details preservation in the final output image.

Specifically, as shown in Fig. 1(b–c), deblurred features are estimated at different scales $\{\mathbf{F}^{Im+Ev,deblur}_{t_0 \to t_1}\}^L_{l=0}$ as explained in Sect. 3.2 and used to perform a coarse-to-fine residual reconstruction using an individual decoder block at each scale. In particular, we employ a decoder at each scale composed of ten residual blocks, each block consisting of Conv+ReLU+Conv, an *attention mechanism* (except in the last scale) that combines features from the current scale l and the previous scale $l - 1$, and a convolutional layer. The attention mechanism uses gating to select the most informative features from each scale. Note that, the residual estimation at scale l will be added to the estimation at the previous scale $l - 1$.

3.4 Loss Function

To train the network, a sum over all the scales l of a combination of an $L1$ image reconstruction loss L_{rec}^l, and a perceptual loss [57] L_{percep}^l is used.

Reconstruction Loss. The $L1$ reconstruction loss at a level l is defined as

$$L_{rec}^l = ||\mathbf{I} \downarrow_{2l} - \tilde{\mathbf{I}}_l||_1 \tag{4}$$

where $\mathbf{I} \downarrow_{2l}$ denotes the sharp image downsampled by a factor of N, and $\tilde{\mathbf{I}}_l$ the estimated sharp image equals to the sum of the blur image $\mathbf{B}_{t_0 \to t_1} \downarrow_{2l}$ and the learned residual $\mathbf{R}_l$.

Perceptual Loss. The perceptual loss computes a feature representation of the reconstructed image and the target image by using a VGG network [37] that was trained on ImageNet [36], and averages the distances between VGG features across multiple layers as follows

$$L_{percep}^l = \sum_n \frac{1}{H_n W_n} \sum_{h,w} ||\omega_n \odot (\Phi_n(\mathbf{I} \downarrow_{2l})_{h,w} - \Phi_n(\tilde{\mathbf{I}}_l)_{h,w})||_2^2, \tag{5}$$

where H_l and W_l are the height and width of feature map Φ_n at layer n and ω_n are weights for each features. Note that, features are unit-normalized in the channel dimension. By using the perceptual loss (see Eq. 5), our network effectively learns to endow the reconstructed images with natural statistics (i.e. with features close to those of natural images).

4   RGBlur+E Dataset

Due to a lack of datasets for evaluating event-based deblurring algorithms on RGB HR images, we capture the first dataset containing real RGB blur images together with the corresponding events during exposure time, called *RGBlur+E*.

As shown in Table 3, we built a hybrid setup that uses a high-speed FLIR BlackFly S RGB GS camera and a Prophesee Gen4 monochrome event camera mounted on a rigid case with a 50/50 one-way mirror to share incoming light. The final resolution of both image and event data is 970×625. Both have been hardware synchronized and built into a beamsplitter setup. We have recorded a wide range of scenarios (cars, text, crossroads, people, close objects, buildings) and covered different motion patterns (static, pan, rotation, walk, hand-held camera, 3D motion). More details are provided in Table 3.

The dataset is composed by two subsets: *RGBlur+E-HS* and *RGBlur+E-LS*. In RGBlur+E-HS, we use high-speed video sequences at 180 fps to generate synthetic blurry images by averaging a fixed number of frames with ground truth sharp images and events during the exposure time. We collect a total of 32 video sequences. We split the dataset into 18 sequences for training and 14 for testing. After synthesizing the blur images, the dataset contains 7600 pairs or frames for training and 3586 for testing. In RGBlur+E-LS, we recorded 29 sequences at 28 fps with a total of 6590 frames with a longer exposure time containing a combination of blurry images and events during the exposure time. However, there is no ground truth data available.

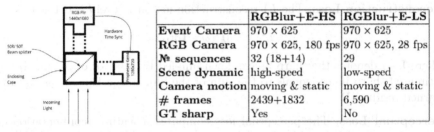

	RGBlur+E-HS	RGBlur+E-LS
Event Camera	970 × 625	970 × 625
RGB Camera	970 × 625, 180 fps	970 × 625, 28 fps
№ sequences	32 (18+14)	29
Scene dynamic	high-speed	low-speed
Camera motion	moving & static	moving & static
# frames	2439+1832	6,590
GT sharp	Yes	No

(a) Beamsplitter setup   (b) Dataset Specifications

Fig. 3. RGBlur+E setup and specifications. Image acquisition diagram and specifications summary of the proposed RGBlur+E dataset.

5   Experiments

5.1   Experimental Settings

Data Preparation

GoPro. The GoPro dataset is widely adopted for image- and event-based deblurring. To synthesize blurry images we average between 7 and 13 frames for training. For testing, we fix the number of frames to 11. Events are synthesized using the ESIM simulator [34].

RGBlur+E Dataset. To close the gap between simulated and real events, we finetune the pretrained model on the GoPro on the training set of the training subset of the proposed RGBlur+E-HS.

Training Setting. All experiments in this work are done using the PyTorch framework. For training, we use the Adam optimizer [18] with standard settings, batches of size 6 and learning rate 10^{-4} decreased by a factor 2 every 10 epoch. We train the full pipeline for 50 epochs. The total number of parameters to optimize is 5.7M with an inference time of 1.8 s. Notice the reduced number of parameters compared to other state of the art methods.

5.2   Ablation of the Effectiveness of the Components

Table 1 shows an ablation study that analyzes the contribution of each component on the final performance of the algorithm. We see, (1) that event data eases the problem of image deblurring by boosting the performance by +3.18 dB, (2) that adapting the receptive field of the convolutions by using a second encoder to compute offsets and masks in the deblur module helps in the performance of the algorithm (+1.18 dB) (3) that pre-processing the event by using a ConvLSTM is more informative for image deblurring than just using the voxel grid representation (+1.69 dB) (4) the benefits coarse-to-fine approach rather than computing the residual image directly at the finer scale (+0.29 dB).

Table 1. Components analysis on the GoPro dataset [25]. Im. stands for image input, Events for their use as an additional input, DM for Deblur Module computed from events and C2F for coarse-to-fine reconstruction.

Im	C2F	Events	DM	LSTM	PSNR ↑	SSIM ↑	LPIPS ↓
✓					28.28	0.86	0.19
✓	✓				28.57	0.87	0.18
✓	✓	✓			31.46	0.92	0.15
✓	✓	✓	✓		32.64	0.93	0.13
✓	✓	✓	✓	✓	**34.33**	**0.94**	**0.11**

5.3   Comparison with State-of-the-Art Models

We perform an extensive comparison with state-of-the-art *image* deblurring methods. We re-ran experiments if code is available for the best-performing methods. Our comparison includes model-based [12,15,45,49], learning-based [3,4,11,14,17,20,21,25,33,38,42,51,52,54,55] and event-based methods [13,16, 31,32,56]. The comparison is twofold: first, as widely adopted in the literature, we compare quantitatively and qualitatively with the GoPro synthetic dataset. Second, we compare quantitatively with the first subset of the RGBlur+E dataset and qualitatively with the second subset.

Comparison on the GoPro Dataset. Table 2 reports the results on the synthetic GoPro dataset. Gray values are directly extracted from papers. Our approach achieves the best performance in terms of both PSNR and SSIM improving the PSNR by +5, 94, +1, 57, and +2, 08 dB compared with model-, learning-, and event-based methods, respectively. We show in Fig. 4 a visual comparison between our method and several state-of-the-art methods. The proposed method recovers the sharpest details in various challenging situations, *e.g.* details on the

Table 2. Quantitative comparisons on the synthetic GoPro dataset [25]. Chen et al. ([1]) computes the values by averaging 7 frames instead of 11. Gray values are extracted directly from the paper.

Method	Events	PSNR ↑	SSIM ↑
Model-based			
Xu et al. [49]	✗	21.00	0.741
Hyun et al. [15]	✗	23.64	0.824
Whyte et al. [45]	✗	24.60	0.846
Gong et al. [12]	✗	26.40	0.863
Carbajal et al. [1]	✗	28.39	0.82
Learning-based			
Jin et al. [17]	✗	26.98	0.892
Kupyn et al. [20]	✗	28.70	0.858
Nah et al. [25]	✗	29.08	0.913
Zhang et al. [55]	✗	29.18	0.931
Kupyn et al. v2 [21]	✗	29.55	0.934
Yuan et al. [51]	✗	29.81	0.937
Zhang et al. [54]	✗	30.23	0.900
Tao et al. [42]	✗	30.27	0.902
Gao et al. [11]	✗	31.17	0.916
Park et al. [33]	✗	31.15	0.916
Suin et al. [38]	✗	31.85	0.948
Huo et al. [14]	✗	32.09	0.931
Cho et al. [4]	✗	32.45	0.934
Zamir et al. [52]	✗	32.65	0.937
Chen et al. [3]	✗	32.76	0.937
Event-based			
Pan et al. [32]	Color	29.06	0.943
Pan et al. [31]	Color	30.29	0.919
Jiang et al. [16]	Color	31.79	0.949
Chen et al. [13][1]	Color	32.99[1]	0.935[1]
Zhang et al. [56]	Gray	32.25	0.929
Ours	Gray	**34.33**	**0.944**

Table 3. RGBlur+E-HS performance.

Method	Zamir et al. [52]	Chen et al. [3]	Pan et al. [32]	Ours	Ours finetuned
PSNR ↑	33.16	32.67	24.20	31.95	**34.24**
SSIM ↑	0.86	0.86	0.76	0.85	**0.87**

Fig. 4. Deblurred examples on the synthetic GoPro dataset [25]. First row: each blur image example. Second to fifth row: results obtained by [3,4,14,52] image-based methods where [14] uses deformable convolutions and [3,4,52] use a multi-scale approach, [32] event-based method and our method.

faces, like eyes, nose, and eyebrows or the car plate numbers. Also, our method can handle low-contrast situations where no events are present, *e.g.* the fingers in the moving hand in the first example, or severe blur situations.

Comparison on the Proposed RGBlur+E Dataset. One of the main drawbacks of state-of-the-art methods is the lack of generalization to real blur and events. In this section, we evaluate the performance in real case scenarios. We use RGBlur+E-HS to quantitatively evaluate the performance of event-based methods to real event data. Note that, this set contains real events but synthetically generated blurry images by averaging 11 frames. To fairly compare image-based method where the entire input is all synthetic, we finetune the our model with the training subset of RGBlur+E-HS. We report quantitative results in Table 3. Our model outperform all other methods while encountering real events. Examples of this set can be found in the last row in Fig. 5 and in the Supp. Mat.

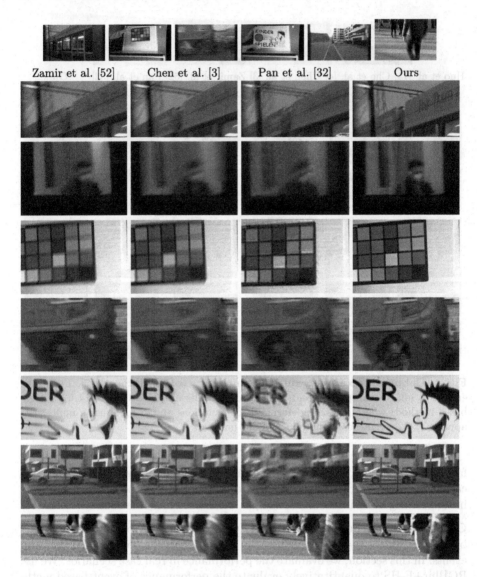

Fig. 5. Representative examples of state-of-the-art methods on the proposed *RGBlur+E-LS* (six first examples) and the *RGBlur+E-HS* dataset (last example). First row: each blur image example. Second to eighth, from left to right: results of Zamir et al. [52], Chen et al. [3], Pan et al. [32] and our result.

6   Conclusions

In this work, we propose a novel deep learning architecture that decouples the estimation of the per-pixel motion from the deblur operator. First, we dynamically learn offsets and masks containing motion information from events. Those are used by modulated deformable convolutions that will deblur image and event features at different scales. Moreover, we reconstruct the sharp image in a coarse-to-fine multi-scale approach to overcome the inherent sparsity and noisy nature of events. Additionally, we introduce the *RGBlur+E* dataset, the first dataset containing real HR RGB images plus events. Extensive evaluations show that the proposed approach achieves superior performance compared to state-of-the-art image and event-based methods on both synthetic and real-world datasets.

References

1. Carbajal, G., Vitoria, P., Delbracio, M., Musé, P., Lezama, J.: Non-uniform blur kernel estimation via adaptive basis decomposition. arXiv preprint arXiv:2102.01026 (2021)
2. Chan, T.F., Wong, C.K.: Total variation blind deconvolution. IEEE Trans. Image Process. **7**(3), 370–375 (1998)
3. Chen, L., Lu, X., Zhang, J., Chu, X., Chen, C.: HINet: half instance normalization network for image restoration. In: Proceedings of the IEEE/CVF Conference on Computer Vision and Pattern Recognition, pp. 182–192 (2021)
4. Cho, S.J., Ji, S.W., Hong, J.P., Jung, S.W., Ko, S.J.: Rethinking coarse-to-fine approach in single image deblurring. In: Proceedings of the IEEE/CVF International Conference on Computer Vision, pp. 4641–4650 (2021)
5. Cho, S., Lee, S.: Fast motion deblurring. In: ACM SIGGRAPH Asia 2009 papers, pp. 1–8 (2009)
6. Dai, J., et al.: Deformable convolutional networks. In: Proceedings of the IEEE International Conference on Computer Vision, pp. 764–773 (2017)
7. Dai, S., Wu, Y.: Removing partial blur in a single image. In: 2009 IEEE Conference on Computer Vision and Pattern Recognition, pp. 2544–2551. IEEE (2009)
8. Delbracio, M., Kelly, D., Brown, M.S., Milanfar, P.: Mobile computational photography: a tour. Annu. Rev. Vis. Sci. **7**, 571–604 (2021). Annual Reviews
9. Dong, J., Pan, J., Su, Z., Yang, M.H.: Blind image deblurring with outlier handling. In: Proceedings of the IEEE International Conference on Computer Vision, pp. 2478–2486 (2017)
10. Fergus, R., Singh, B., Hertzmann, A., Roweis, S.T., Freeman, W.T.: Removing camera shake from a single photograph. In: ACM SIGGRAPH 2006 Papers, pp. 787–794 (2006)
11. Gao, H., Tao, X., Shen, X., Jia, J.: Dynamic scene deblurring with parameter selective sharing and nested skip connections. In: Proceedings of the IEEE/CVF Conference on Computer Vision and Pattern Recognition, pp. 3848–3856 (2019)
12. Gong, D., et al.: From motion blur to motion flow: a deep learning solution for removing heterogeneous motion blur. In: Proceedings of the IEEE Conference on Computer Vision and Pattern Recognition, pp. 2319–2328 (2017)
13. Haoyu, C., Minggui, T., Boxin, S., YIzhou, W., Tiejun, H.: Learning to deblur and generate high frame rate video with an event camera. arXiv preprint arXiv:2003.00847 (2020)
14. Huo, D., Masoumzadeh, A., Yang, Y.H.: Blind non-uniform motion deblurring using atrous spatial pyramid deformable convolution and deblurring-reblurring consistency. arXiv preprint arXiv:2106.14336 (2021)
15. Hyun Kim, T., Ahn, B., Mu Lee, K.: Dynamic scene deblurring. In: Proceedings of the IEEE International Conference on Computer Vision, pp. 3160–3167 (2013)

16. Jiang, Z., Zhang, Y., Zou, D., Ren, J., Lv, J., Liu, Y.: Learning event-based motion deblurring. In: Proceedings of the IEEE/CVF Conference on Computer Vision and Pattern Recognition, pp. 3320–3329 (2020)
17. Jin, M., Meishvili, G., Favaro, P.: Learning to extract a video sequence from a single motion-blurred image. In: Proceedings of the IEEE Conference on Computer Vision and Pattern Recognition, pp. 6334–6342 (2018)
18. Kingma, D.P., Ba, J.: Adam: a method for stochastic optimization. arXiv preprint arXiv:1412.6980 (2014)
19. Kundur, D., Hatzinakos, D.: Blind image deconvolution. IEEE Signal Process. Mag. 13(3), 43–64 (1996)
20. Kupyn, O., Budzan, V., Mykhailych, M., Mishkin, D., Matas, J.: Deblurgan: blind motion deblurring using conditional adversarial networks. In: Proceedings of the IEEE Conference on Computer Vision and Pattern Recognition, pp. 8183–8192 (2018)
21. Kupyn, O., Martyniuk, T., Wu, J., Wang, Z.: Deblurgan-v2: deblurring (orders-of-magnitude) faster and better. In: Proceedings of the IEEE/CVF International Conference on Computer Vision, pp. 8878–8887 (2019)
22. Levin, A.: Blind motion deblurring using image statistics. Adv. Neural. Inf. Process. Syst. 19, 841–848 (2006)
23. Levin, A., Weiss, Y., Durand, F., Freeman, W.T.: Understanding and evaluating blind deconvolution algorithms. In: 2009 IEEE Conference on Computer Vision and Pattern Recognition, pp. 1964–1971. IEEE (2009)
24. Michaeli, T., Irani, M.: Blind deblurring using internal patch recurrence. In: Fleet, D., Pajdla, T., Schiele, B., Tuytelaars, T. (eds.) ECCV 2014. LNCS, vol. 8691, pp. 783–798. Springer, Cham (2014). https://doi.org/10.1007/978-3-319-10578-9_51
25. Nah, S., Hyun Kim, T., Mu Lee, K.: Deep multi-scale convolutional neural network for dynamic scene deblurring. In: Proceedings of the IEEE Conference on Computer Vision and Pattern Recognition, pp. 3883–3891 (2017)
26. Nimisha, T.M., Kumar Singh, A., Rajagopalan, A.N.: Blur-invariant deep learning for blind-deblurring. In: Proceedings of the IEEE International Conference on Computer Vision, pp. 4752–4760 (2017)
27. Noroozi, M., Chandramouli, P., Favaro, P.: Motion deblurring in the wild. In: Roth, V., Vetter, T. (eds.) GCPR 2017. LNCS, vol. 10496, pp. 65–77. Springer, Cham (2017). https://doi.org/10.1007/978-3-319-66709-6_6
28. Pan, J., Hu, Z., Su, Z., Lee, H.Y., Yang, M.H.: Soft-segmentation guided object motion deblurring. In: Proceedings of the IEEE Conference on Computer Vision and Pattern Recognition, pp. 459–468 (2016)
29. Pan, J., Hu, Z., Su, Z., Yang, M.H.: Deblurring text images via l0-regularized intensity and gradient prior. In: Proceedings of the IEEE Conference on Computer Vision and Pattern Recognition, pp. 2901–2908 (2014)
30. Pan, J., Sun, D., Pfister, H., Yang, M.H.: Blind image deblurring using dark channel prior. In: Proceedings of the IEEE Conference on Computer Vision and Pattern Recognition, pp. 1628–1636 (2016)
31. Pan, L., Hartley, R., Scheerlinck, C., Liu, M., Yu, X., Dai, Y.: High frame rate video reconstruction based on an event camera. IEEE Trans. Pattern Anal. Mach. Intell. (2020)
32. Pan, L., Scheerlinck, C., Yu, X., Hartley, R., Liu, M., Dai, Y.: Bringing a blurry frame alive at high frame-rate with an event camera. In: Proceedings of the IEEE/CVF Conference on Computer Vision and Pattern Recognition, pp. 6820–6829 (2019)

33. Park, D., Kang, D.U., Kim, J., Chun, S.Y.: Multi-temporal recurrent neural networks for progressive non-uniform single image deblurring with incremental temporal training. In: Vedaldi, A., Bischof, H., Brox, T., Frahm, J.-M. (eds.) ECCV 2020. LNCS, vol. 12351, pp. 327–343. Springer, Cham (2020). https://doi.org/10.1007/978-3-030-58539-6_20
34. Rebecq, H., Gehrig, D., Scaramuzza, D.: ESIM: an open event camera simulator. In: Conference on Robot Learning, pp. 969–982. PMLR (2018)
35. Rebecq, H., Ranftl, R., Koltun, V., Scaramuzza, D.: Events-to-video: bringing modern computer vision to event cameras. In: Proceedings of the IEEE/CVF Conference on Computer Vision and Pattern Recognition, pp. 3857–3866 (2019)
36. Russakovsky, O., et al.: Imagenet large scale visual recognition challenge. Int. J. Comput. Vision **115**(3), 211–252 (2015)
37. Simonyan, K., Zisserman, A.: Very deep convolutional networks for large-scale image recognition. arXiv preprint arXiv:1409.1556 (2014)
38. Suin, M., Purohit, K., Rajagopalan, A.: Spatially-attentive patch-hierarchical network for adaptive motion deblurring. In: Proceedings of the IEEE/CVF Conference on Computer Vision and Pattern Recognition, pp. 3606–3615 (2020)
39. Sun, J., Cao, W., Xu, Z., Ponce, J.: Learning a convolutional neural network for non-uniform motion blur removal. In: Proceedings of the IEEE Conference on Computer Vision and Pattern Recognition, pp. 769–777 (2015)
40. Sun, L., et al.: Mefnet: multi-scale event fusion network for motion deblurring. arXiv preprint arXiv:2112.00167 (2021)
41. Tai, Y.W., Tan, P., Brown, M.S.: Richardson-lucy deblurring for scenes under a projective motion path. IEEE Trans. Pattern Anal. Mach. Intell. **33**(8), 1603–1618 (2010)
42. Tao, X., Gao, H., Shen, X., Wang, J., Jia, J.: Scale-recurrent network for deep image deblurring. In: Proceedings of the IEEE Conference on Computer Vision and Pattern Recognition, pp. 8174–8182 (2018)
43. Tran, P., Tran, A.T., Phung, Q., Hoai, M.: Explore image deblurring via encoded blur kernel space. In: Proceedings of the IEEE/CVF Conference on Computer Vision and Pattern Recognition, pp. 11956–11965 (2021)
44. Wang, B., He, J., Yu, L., Xia, G.-S., Yang, W.: Event enhanced high-quality image recovery. In: Vedaldi, A., Bischof, H., Brox, T., Frahm, J.-M. (eds.) ECCV 2020. LNCS, vol. 12358, pp. 155–171. Springer, Cham (2020). https://doi.org/10.1007/978-3-030-58601-0_10
45. Whyte, O., Sivic, J., Zisserman, A., Ponce, J.: Non-uniform deblurring for shaken images. Int. J. Comput. Vision **98**(2), 168–186 (2012)
46. Wieschollek, P., Hirsch, M., Scholkopf, B., Lensch, H.: Learning blind motion deblurring. In: Proceedings of the IEEE International Conference on Computer Vision, pp. 231–240 (2017)
47. Xingjian, S., Chen, Z., Wang, H., Yeung, D.Y., Wong, W.K., Woo, W.C.: Convolutional LSTM network: a machine learning approach for precipitation nowcasting. In: Advances in Neural Information Processing Systems, pp. 802–810 (2015)
48. Xu, F., et al.: Motion deblurring with real events. In: Proceedings of the IEEE/CVF International Conference on Computer Vision, pp. 2583–2592 (2021)
49. Xu, L., Zheng, S., Jia, J.: Unnatural l0 sparse representation for natural image deblurring. In: Proceedings of the IEEE Conference on Computer Vision and Pattern Recognition, pp. 1107–1114 (2013)
50. Yan, Y., Ren, W., Guo, Y., Wang, R., Cao, X.: Image deblurring via extreme channels prior. In: Proceedings of the IEEE Conference on Computer Vision and Pattern Recognition, pp. 4003–4011 (2017)

51. Yuan, Y., Su, W., Ma, D.: Efficient dynamic scene deblurring using spatially variant deconvolution network with optical flow guided training. In: Proceedings of the IEEE/CVF Conference on Computer Vision and Pattern Recognition, pp. 3555–3564 (2020)
52. Zamir, S.W., et al.: Multi-stage progressive image restoration. In: Proceedings of the IEEE/CVF Conference on Computer Vision and Pattern Recognition, pp. 14821–14831 (2021)
53. Zhang, H., Wipf, D.: Non-uniform camera shake removal using a spatially-adaptive sparse penalty. In: Advances in Neural Information Processing Systems, pp. 1556–1564. Citeseer (2013)
54. Zhang, H., Dai, Y., Li, H., Koniusz, P.: Deep stacked hierarchical multi-patch network for image deblurring. In: Proceedings of the IEEE/CVF Conference on Computer Vision and Pattern Recognition, pp. 5978–5986 (2019)
55. Zhang, J., et al.: Dynamic scene deblurring using spatially variant recurrent neural networks. In: Proceedings of the IEEE Conference on Computer Vision and Pattern Recognition, pp. 2521–2529 (2018)
56. Zhang, L., Zhang, H., Chen, J., Wang, L.: Hybrid deblur net: deep non-uniform deblurring with event camera. IEEE Access 8, 148075–148083 (2020)
57. Zhang, R., Isola, P., Efros, A.A., Shechtman, E., Wang, O.: The unreasonable effectiveness of deep features as a perceptual metric. In: CVPR (2018)
58. Zhu, X., Hu, H., Lin, S., Dai, J.: Deformable convnets V2: more deformable, better results. In: Proceedings of the IEEE/CVF Conference on Computer Vision and Pattern Recognition, pp. 9308–9316 (2019)

UDC-UNet: Under-Display Camera Image Restoration via U-shape Dynamic Network

Xina Liu[1,2], Jinfan Hu[1,2], Xiangyu Chen[1,3,4], and Chao Dong[1,4(✉)]

[1] Shenzhen Key Lab of Computer Vision and Pattern Recognition, SIAT-SenseTime Joint Lab, Shenzhen Institutes of Advanced Technology, Chinese Academy of Sciences, Beijing, China
{xn.liu,jf.hu1,chao.dong}@siat.ac.cn, chxy95@gmail.com
[2] University of Chinese Academy of Sciences, Beijing, China
[3] University of Macau, Zhuhai, China
[4] Shanghai AI Laboratory, Shanghai, China

Abstract. Under-Display Camera (UDC) has been widely exploited to help smartphones realize full-screen displays. However, as the screen could inevitably affect the light propagation process, the images captured by the UDC system usually contain flare, haze, blur, and noise. Particularly, flare and blur in UDC images could severely deteriorate the user experience in high dynamic range (HDR) scenes. In this paper, we propose a new deep model, namely UDC-UNet, to address the UDC image restoration problem with an estimated PSF in HDR scenes. Our network consists of three parts, including a U-shape base network to utilize multi-scale information, a condition branch to perform spatially variant modulation, and a kernel branch to leverage the prior knowledge of the PSF. According to the characteristics of HDR data, we additionally design a tone mapping loss to stabilize network optimization and achieve better visual quality. Experimental results show that the proposed UDC-UNet outperforms the state-of-the-art methods in quantitative and qualitative comparisons. Our approach won second place in the UDC image restoration track of the MIPI challenge. Codes and models are available at https://github.com/J-FHu/UDCUNet.

Keywords: Under display camera · Image restoration

1 Introduction

As a new design for smartphones, Under-Display Camera (UDC) systems can provide a bezel-less and notch-free viewing experience, and attracts much attention from the industry. However, it is difficult to preserve the full functionality of the imaging sensor under a display. As the propagation of light in the imaging process is inevitably affected, various forms of optical diffraction and interference would be generated. The image captured by UDC often suffers from noise, flare,

---

X. Liu and J. Hu—Equal contributions,

© The Author(s), under exclusive license to Springer Nature Switzerland AG 2023
L. Karlinsky et al. (Eds.): ECCV 2022 Workshops, LNCS 13805, pp. 113–129, 2023.
https://doi.org/10.1007/978-3-031-25072-9_8

Fig. 1. Visual results of under-display camera image restoration. Affected by the UDC imaging mechanism, artifacts including flare, blur, and haze will appear in UDC images. Our method generates visually pleasing results by alleviating the artifacts.

haze, and blurry artifacts. Besides, UDC images are often captured under a high dynamic range in real scenes. Therefore, severe over-saturation could occur in highlight regions, as shown in Fig. 1.

Image restoration task aim to restore the clean image from its degraded version, such as denoising [16,51,52], deraining [29,30,56], deblurring [23,27], super-resolution [4,7,54], and HDR reconstruction [3,10]. Similar to these tasks, UDC image restoration aims at reconstructing the degraded image generated by the UDC system. To model the complicated degradation process of the UDC imaging system, existing works [11,55] propose to utilize a particular diffraction blur kernel, i.e. Point Spread Function (PSF). Then UDC image restoration can be regarded as an inversion problem for a measured PSF. Though existing methods have made significant progress in this task, their performance is still limited.

In this work, we propose UDC-UNet to restore images captured by UDC in HDR scenes. The framework consists of a base network, a condition branch, and a kernel branch. First, we exploit the commonly used U-shape structure to build the base network for utilizing the multi-scale information hierarchically. Then, to achieve the spatially variant modulation for different regions with different exposures, we adopt several spatial feature transform (SFT) layers [46] to build the condition branch as the design in [3]. Additionally, the prior knowledge of an accurately measure PSF has been proven to be useful for the restoration process [11]. Hence, we also add a kernel branch that uses an estimated PSF to refine the intermediate features. Besides, considering the data characteristic of HDR images, we design a new tone mapping loss to normalize image values into [0,1]. It can not only balance the effects of pixels with different intensities but also stabilize the training process. Experimental results show that our method surpasses the state-of-the-art methods in both quantitative performance and visual quality. Our approach can generate visually pleasing results without artifacts even in the over-saturated regions, as shown in Fig. 1. The proposed UDC-UNet won second place in the MIPI UDC image restoration challenge.

In summary, our contributions are three-fold:

1) We propose a new deep network, UDC-UNet, to address the UDC image restoration problem. 2) We design a tone mapping loss to address the UDC image restoration problem in HDR scenes. 3) Experiments demonstrate the superiority of our method in quantitative performance and visual quality.

2   Related Work

2.1   UDC Restoration

Under-display camera is a new imaging system that mounts a display screen on top of a traditional digital camera lens. While providing a better user experience of viewing and interaction, the UDC system may sacrifice the quality of photography. Due to low light transmission rate and diffraction effects in the UDC system, significant quality degradations may appear in UDC images, including flare, haze, blur, and noise. Some works [39,45] have analyzed the imaging principle of the UDC system and attempted to restore UDC images with deconvolutional-based methods like Wiener Filter [14]. Qin et al. [38] demonstrated the issue of see-through image blurring of transparent organic light-emitting diodes display. Kwon et al. [24] paid attention to the analysis of transparent displays. Zhou et al. [55] recovered the original signal from the point-spread-function-convoluted image. Feng et al. [11] treated the UDC image restoration as a non-blind inverse problem for an accurately measured PSF.

Until now, there is only few works to directly solve the problem of UDC image restoration. MCIS [55] modeled the image formation pipeline of UDC and proposed to address the image restoration of UDC using a Deconvolution-based Pipeline (DeP) and data-driven learning-based methods. Although this variant of UNet solved many problems of UDC restoration, it lacks consideration of HDR in data generation and PSF measurement. A newly proposed structure named DISCNet [11] considered high dynamic range (HDR) and measured the real-world PSF of the UDC system. Besides, DISCNet regarded PSF as useful domain knowledge and added a separate branch to fuse its information. Experiments have shown the effectiveness of doing so in removing diffraction image artifacts.

However, in HDR scenes, existing methods still could not perform well in both blurring of unsaturated regions and flare in over-saturated areas. UDC images in this scenario have different exposures and brightness. Previous networks apply the same filter weights across all regions, whose capability are no longer powerful enough in this task. Spatial feature transform (SFT) layer is a excellent design provided in SFTGAN [46] that can achieve affine transformation to the intermediate features and perform a spatially variant modulation. A lot of works [15,25,42] have proven the effectiveness of the SFT layer, and especially, HDRUNet [3] demonstrated the effectiveness of the module for HDR reconstruction. Thus, we adopt exploit SFT in our network to process the various patterns across the image. Additionally, since Feng et al. [11] presented that the dynamic convolutions generated by the measured PSF could also benefit the UDC image restoration task in HDR scenes, we utilize this mechanism in our design to further refine the intermediate features.

2.2   Image Restoration

Image restoration aims to reconstruct a high-quality image from its degraded version. Prior to the deep-learning era, many traditional image processing methods [1,17,19,36,43,44] have been proposed for image restoration. Since the first deep learning method was successfully applied in low-level vision tasks [8], a flurry of approaches have been developed for various image restoration problems, including super-resolution [4,9,22,54], denoising [16,51,52], deraining [29,30,56], deblurring [23,27], and HDR reconstruction [3,10].

As one of the most important task in the field of image restoration, image super-resolution attracts the attention of numerous researchers, and thus a series of methods [4,9,31,32,47,54] have been proposed and made great progress. Denoising is a fundamental task for image restoration, and some good methods have been designed to deal with this problem [16,26,33,51,52]. Deblurring, deraining and HDR reconstruction are also practical problems in the image restoration field. Deblurring methods aim to address the image blur caused by camera shake, object motion, and out-of-focus [6,13,23,27,37,53]. Deraining aims at removing the rain streaks from the degraded images, which is also a widely studied problem in image restoration [12,29,30,40,56]. Reconstructing LDR images to their HDR version can greatly improve the visual quality of images and viewing experience for users. There are also many works focusing on this task [3,5,10,20,34].

Since UNet [41] first proposed the U-shape network for biomedical image segmentation, this design is widely used in image restoration task. For example, Cho et al. [6] introduced a U-shape network to deal with image deblurring. Chen et al. [3,10] also proposed UNet-style networks for single image HDR reconstruction. As U-shape network can better utilize multi-scale information and save computations by deploying multiple down-sampling and up-sampling operations, we also exploit this structure for UDC image restoration task. With the development of large models and the pre-training technique, many works propose to address image restoration problems using Transformer [32,48,49] and try to deal with multiple image restoration task simultaneously [2,28]. However, there are still few methods for UDC image restoration.

3   Methodology

3.1   Problem Formulation

As mentioned in [11], a real-world UDC image formation model with multiple types of degradations can be formulated as

$$\hat{y} = \phi[C(x * k + n)], \tag{1}$$

where x represents the real HDR scene irradiance. k is the known convolution kernel, commonly referred to as the PSF for this task. "$*$" denotes the convolution operation and n represents the camera noise. $C(\cdot)$ denotes the clipping operation for the digital sensor r with limited dynamic range and $\phi(\cdot)$ represents the non-linear tone-mapping function.

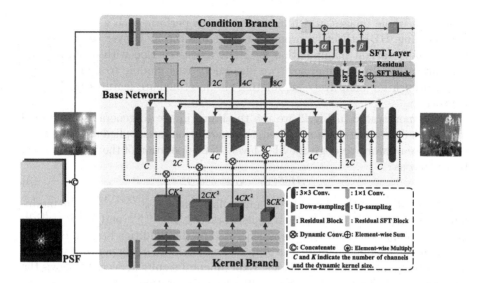

Fig. 2. The overall architecture of UDC-UNet. It consists of a U-shape base network, a condition branch, and a kernel branch.

Following [11], we treat this task as a non-blind image restoration problem, where the degraded image $\hat{y}$ and the estimated PSF k are given to restore the clean image. Hence, our proposed network f can be defined as

$$x' = f(\hat{y}, k; \theta), \tag{2}$$

where x' denotes the predicted image and θ represents the network parameters.

3.2 Network Structure

The overall architecture of the proposed network consists of three components – a base network, a condition branch, and a kernel branch as shown in Fig. 2,

Base Network. A U-shape structure is adopted to build the base network. Its effectiveness has been proven in many image restoration methods [3,10,48,50]. This kind of structure helps the network make full use of the hierarchical multiscale information from low-level features to high-level features. The shallow layers in the network can gradually extract the features with a growing receptive field and map the input image to high-dimensional representation. For the deep layers, features can be reconstructed step by step from the encoded representation. Benefiting from the skip connections, shallow features, and deep features can be combined organically. For the basic blocks in the base network, we simply exploit the common residual block with two 3×3 convolutions.

Condition Branch. The key to UDC image restoration in HDR scenes is to deal with the blurring of unsaturated regions and address the flare in oversaturated areas. Traditional convolution filters apply the same filter weights

118   X. Liu et al.

across the whole images, which are inefficient to perform such a spatially variant mapping. HDR-UNet [3] proposes to leverage the SFT layer [46] to achieve
the spatially variant manipulation for HDR reconstruction with denoising and
dequantization. In our approach, we also adopt this module to build the condition branch for adaptively modulating the intermediate features in the base
network.

As demonstrated in Fig. 2, we use the input image to generate conditional
maps with different resolutions. Then, these conditions are applied to the features
in the base network through the SFT layers. The operation of the SFT layer can
be written as

$$SFT(x) = \alpha \odot x + \beta, \tag{3}$$

where "$\odot$" denotes the element-wise multiplication. $x \in \mathbb{R}^{C \times H \times W}$ is the modu-
lated intermediate features in the base network. $\alpha \in \mathbb{R}^{C \times H \times W}$ and $\beta \in \mathbb{R}^{C \times H \times W}$
are two modulation coefficient maps generated by the condition branch. By using
such a modulation mechanism, the proposed network can easily perform spatially
variant mapping for different regions. Experimental results in Sect. 4.2 shows the
effectiveness of this branch for the UDC image restoration task.

Kernel Branch. Feng et al. [11] have demonstrated that the PSF can provide
useful prior knowledge for UDC image restoration. Thus, we exploit a similar
approach to leverage the PSF to further refine the intermediate features in our
base network. Following [11,15], we first project the PSF onto a b-dimensional
vector as the kernel code. Then, we stretch the kernel code into $H \times W \times b$
features and concatenate it with the degraded image of size $H \times W \times C$ to
obtain the conditional input of size $H \times W \times (b+C)$. Through the kernel branch,
the conditional inputs are mapped into scale-specific kernel features. For the
modulated features of $h \times w \times c$, the generated kernel feature is of size $h \times w \times ck^2$.
Then, the dynamic convolutions are performed on each pixel as

$$F_{i,j,c} = K_{i,j,c} \cdot M_{i,j,c}, \tag{4}$$

where $K_{i,j,c}$ represents the $k \times k$ filter reshaped from the kernel feature at position
(i, j, c) with the size of $1 \times 1 \times k^2$. $M_{i,j,c}$ and $F_{i,j,c}$ denote the $k \times k$ patch centered
at (i, j, c) of the modulated features and the element at (i, j, c) of the output
feature. " $\cdot$ " means the inner product operation. Note that we directly use the
measured PSF provided by [11] in this paper.

3.3   Loss Function

Traditional loss functions for image restoration task, such as ℓ_1 and ℓ_2 loss, assign
the same weight to all pixels regardless of intensity. But for images captured in
HDR scenes, such loss function will make the network focus on those pixels with
large intensity, resulting in an unstable training process and poor performance.
A common practice is to design a specific loss function according to the tone
mapping function [3,35]. Inspired by these works, we propose a $Mapping\_\ell_1$
loss to optimize the network, which is formulated as

$$Mapping\_\ell_1(Y, X) = |Mapping(Y) - Mapping(X)|, \qquad (5)$$

where Y represents the predicted result and X means the corresponding ground truth. For the tone mapping function, $Mapping(I) = I/(I + 0.25)$.

4 Experiments

4.1 Experimental Setup

Dataset. UDC images and their corresponding ground-truth images in real scenes are currently difficult to obtain. Therefore, Feng *et al.* in [11] propose to use the PSF to simulate the degraded UDC images from the clean images. In the MIPI UDC image restoration challenge[1], 132 HDR panorama images with the spatial size of 8196 × 4096 from the HDRI Haven dataset[2] are exploited to generate the synthetic data. This dataset contains images captured in various indoor and outdoor, day and night, nature and urban scenes. The images are first projected to the regular perspective and then cropped into 800 × 800 image patches. Each of these patches is degraded by using Eq. (1) to obtain the simulated UDC images. As the result, 2016 image pairs are available for training. There are also 40 image pairs left as validation set and testing set.

Implementation Details. During the training phase, 256 × 256 patches randomly cropped from the original images are fed into the network. The mini-batch size is set to 32 and the whole network is trained for 6×10^5 iterations. Both the number of residual blocks in the condition branch and kernel branch are set to 2, while the number of residual SFT blocks in the base network is set to [2,8], respectively. We use the exponential moving average strategy during the training. The number of channels C is set to 32 in our UDC-UNet. We also provide a small version of our method, UDC-UNet$_S$, by reducing the channel number C to 20. The initialization strategy [18] and Adam optimizer [21] are adapted for training. The learning rate is initialized as 2×10^{-4} and decayed with a cosine annealing schedule, where $\eta_{min} = 1 \times 10^{-7}$ and $\eta_{max} = 2 \times 10^{-4}$. Furthermore, the learning rate is restarted at $[5 \times 10^4, 1.5 \times 10^5, 3 \times 10^5, 4.5 \times 10^5]$ iterations. For the evaluation metrics, PSNR, SSIM, and LPIPS are calculated in the tone-mapped images through the tone mapping function ($Mapping(I) = I/(I+0.25)$) provided by the challenge.

4.2 Ablation Study

In this part, we first conduct experiments to explore the effectiveness of the key modules. Then we investigate the influence of different conditional inputs on the kernel branch. Finally, we study the effects of different loss functions. Note that for fast validation, we use the UDC-UNet and set the batch size as 6 for all experiments in this section.

---

[1] https://codalab.lisn.upsaclay.fr/competitions/4874#participate.

[2] https://hdrihaven.com/hdris/.

Table 1. Ablation study of the key modules in UDC-UNet.

Models	(a)	(b)	(c)	(d)	(e) (Ours)
Skip connections	✗	✔	✔	✔	✔
Condition branch	✗	✗	✗	✔	✔
Kernel branch	✗	✗	✔	✗	✔
PSNR (dB)	42.19	44.50	44.58	45.23	**45.37**
SSIM	0.9884	0.9897	0.9893	0.9897	**0.9898**
LPIPS	0.0164	**0.0155**	0.0157	0.0166	0.0162

Effectiveness of Key Modules. As our approach consists of three components – the base network, the condition branch, and the kernel branch, we conduct ablation experiments to demonstrate the effectiveness of these modules. In addition, we emphasize that the skip connections (all the dotted lines in the base network shown in Fig. 2) in the base network have significant positive effects on performance. Our base network without any skip connections and extra branches can obtain the quantitative performance of more than 42 dB as shown in Table 1, demonstrating the effectiveness of the U-shape structure for this task. By comparing models (a) and (b), we can observe that the skip connections bring more than 2 dB performance gain. It presents the significance of the skip connection in the base network. The models with the condition branch (d,e) outperform the models without this branch (b,c) by more than 0.7 dB, illustrating the necessity of the condition branch. The kernel branch also brings a considerable performance gain by comparing the models (c,e) and models (b,d). Combining these key modules, our method achieves the highest PSNR and SSIM.

Table 2. Quantitative results of different conditional inputs for the kernel branch.

Conditional Inputs	PSNR(dB)	SSIM	LPIPS
(a) None	45.23	0.9897	0.0166
(b) Image	45.17	0.9896	**0.0162**
(c) PSF	45.26	0.9895	0.0166
(d) Image+PSF	**45.37**	**0.9898**	**0.0162**

Different Conditional Inputs of the Kernel Branch. An intuitive method for introducing the PSF prior is to directly use the estimated PSF to generate the dynamic filters in the kernel branch. Nonetheless, experiments in [11] show that utilizing the combination of the PSF and the degraded image to generate filters can bring more performance improvement. Thus, we also conduct experiments to investigate the different conditional inputs of the kernel branch. As shown in Table 2, using only the degraded image or the PSF does not bring obvious performance gain and even causes a little performance drop. However, when using the image and the PSF simultaneously as the conditional input, the model obtains a nonnegligible performance improvement.

Table 3. Quantitative results of our method using different loss functions.

Losses	PSNR (dB)	SSIM	LPIPS
ℓ_1	41.30	0.9812	0.0301
$Mapping\_\ell_2$	40.19	0.9838	0.0238
$Mapping\_\ell_1$	**45.37**	**0.9898**	**0.0162**

Effects of Different Loss Functions. We also conduct experiments to show the effects of different loss functions, including the traditional ℓ_1 loss, the proposed $Mapping\_\ell_1$ loss, and ℓ_2 loss calculated after the same tone mapping function, denoting as the $Mapping\_\ell_2$ loss. As depicted in Table 3, we can find that our $Mapping\_\ell_1$ brings a large quantitative performance improvement compared to the traditional ℓ_1 loss and also outperforms the $Mapping\_\ell_2$ loss by a large margin. The results show that the loss function significantly affects the optimization of the network for this task.

We also provide the visual comparisons of models using different loss functions, as shown in Fig. 3. We can observe that ℓ_1 loss and $Mapping\_\ell_2$ loss make the network focus more on the over-saturated regions, resulting in severe artifacts at these areas. We visualize the tone mapping function in Fig. 4 to further illustrate the effect of our proposed loss. Since the pixel intensities of the used HDR data are in the range [0,500], the large values could significantly affect network optimization and easily make the network focus on the over-saturated regions. After the tone mapping function, the pixel values can be normalized to the range of [0,1). An intuitive manifestation is that values greater than 1 will be compressed to (0.8,1] through the tone mapping function. Hence, the proposed $Mapping\_\ell_1$ loss can balance the influence of different regions with different exposures for network optimization.

4.3 Comparison with State-of-the-Art Methods

To evaluate the performance of our UDC-UNet, we conduct experiments on the simulated data and make comparisons with several state-of-the-art methods. Since the task of UDC image restoration is a newly-proposed task, there are not many related works available to compare. Therefore, we select three representative methods for comparison, including DISCNet [11] for UDC image restoration, Uformer [48] for general image restoration tasks, and HDRUNet [3] for HDR image reconstruction. Specifically, we retrain the compared methods using the official codes on the same data with the same $Mapping\_\ell_1$ loss. Note

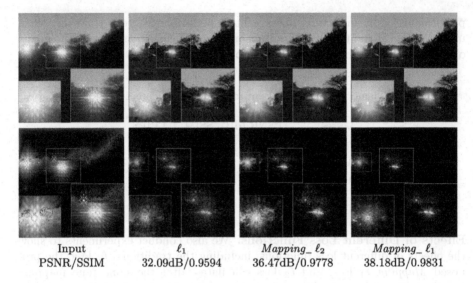

| Input
PSNR/SSIM | ℓ_1
32.09dB/0.9594 | $Mapping\_\ell_2$
36.47dB/0.9778 | $Mapping\_\ell_1$
38.18dB/0.9831 |

Fig. 3. The visual comparisons of models using different loss functions. The bottom row presents the residual maps, $E = Mapping(|Y - X|)$, where Y means the generated output and X indicates the ground truth.

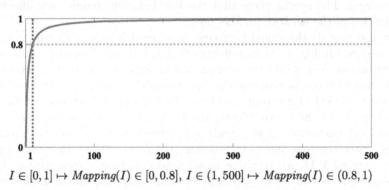

$$I \in [0, 1] \mapsto Mapping(I) \in [0, 0.8], \ I \in (1, 500] \mapsto Mapping(I) \in (0.8, 1)$$

Fig. 4. Visualization of the tone mapping function, *i.e.*, $Mapping(I) = I/(I + 0.25)$.

that we use the small version of Uformer with channel number 32 to save computations. To further demonstrate the effectiveness of our method, we also provide the results of the small version of our method, UDC-UNet$_S$, by reducing the channel number C to 20.

Quantitative Results. The quantitative results are provided in Table 4. We can observe that UDC-UNet achieves the best performance and significantly outperforms the compared state-of-the-art methods by more than 7 dB. Considering the computations, we also provide the results of a smaller variant UDC-UNet$_S$. We can observe that UDC-Net$_S$ also obtains considerable performance

Table 4. Quantitative comparisons of our method with state-of-the-art approaches. The best values are in **bold**, and the second best values are underlined.

Methods	PSNR (dB)	SSIM	LPIPS	Params	GMACs
Uformer [48]	37.97	0.9784	0.0285	20.0M	490
HDRUNet [3]	40.23	0.9832	0.0240	**1.7M**	229
DISCNet [11]	39.89	0.9864	0.0152	3.8M	367
UDC-UNet$_S$ (ours)	45.98	0.9913	0.0128	5.7M	**169**
UDC-UNet (ours)	**47.18**	**0.9927**	**0.0100**	14.0M	402

with relatively fewer parameters and the least computation. The GMACs of all the methods are calculated on the input size of 800 × 800. Note that due to more downsampling operations, our methods have much fewer GMACs than the compared approaches.

Visual Comparison. We provide the visual results of our method and state-of-the-art approaches in Fig. 5. Compared to other methods, our UDC-UNet obtains visually pleasing results with clean textures and little artifacts, especially on the over-saturated regions. Besides, the results generated by UDC-UNet$_S$ also have relatively fewer artifacts compared to other methods. To further demonstrate the superiority of UDC-UNet, we also provide visual comparisons with current models on the testing data. It can be seen that our method obtains better visual quality than the several compared approaches.

5 Results of the MIPI UDC Image Restoration Challenge

We participate in the mobile intelligent photography and imaging (MIPI) challenge[3] and won second place in the UDC image restoration track. The top 5 results are depicted in Table 5.

Table 5. Top 5 results of the MIPI UDC image restoration challenge.

Teams	PSNR(dB)	SSIM	LPIPS
USTC_WXYZ	48.4776	0.9934	0.0093
XPixel Group (ours)	47.7821	0.9913	0.0122
SRC-B	46.9227	0.9929	0.0098
MIALGO	46.1177	0.9892	0.0159
LVGroup_HFUT	45.8722	0.9920	0.0109

---

[3] http://mipi-challenge.org/.

124   X. Liu et al.

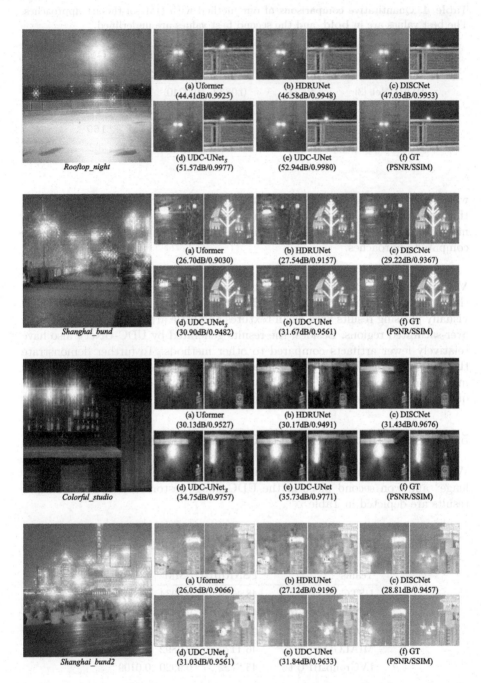

Fig. 5. Visual comparisons of our method with state-of-the-art approaches on the validation set.

Fig. 6. Visual comparisons of our method with state-of-the-art approaches on the testing set.

6   Conclusions

In this paper, we propose UDC-UNet to address UDC image restoration task in HDR scenes. The designed network consists of a U-shape base network to utilize multi-scale information, a condition branch to perform spatially variant modulation, and a kernel branch to leverage the prior knowledge of the measured PSF. In addition, we design a tone mapping loss to balance the effects of pixels with different intensities. Experiments show that our method obtains the best quantitative performance and visual quality compared to state-of-the-art approaches. We participated in the MIPI challenge and won second place in the UDC image restoration track.

Acknowledgements. This work is partially supported by National Natural Science Foundation of China (61906184, U1913210), and the Shanghai Committee of Science and Technology, China (Grant No. 21DZ1100100).

References

1. Calvetti, D., Reichel, L., Zhang, Q.: Iterative solution methods for large linear discrete ill-posed problems. In: Datta, B.N. (eds.) Applied and Computational Control, Signals, and Circuits, pp. 313–367. Springer, Boston (1999). https://doi.org/10.1007/978-1-4612-0571-5_7

2. Chen, H., et al.: Pre-trained image processing transformer. In: Proceedings of the IEEE/CVF Conference on Computer Vision and Pattern Recognition, pp. 12299–12310 (2021)

3. Chen, X., Liu, Y., Zhang, Z., Qiao, Y., Dong, C.: HdruNet: single image HDR reconstruction with denoising and dequantization. In: Proceedings of the IEEE/CVF Conference on Computer Vision and Pattern Recognition, pp. 354–363 (2021)

4. Chen, X., Wang, X., Zhou, J., Dong, C.: Activating more pixels in image super-resolution transformer. arXiv preprint arXiv:2205.04437 (2022)

5. Chen, X., Zhang, Z., Ren, J.S., Tian, L., Qiao, Y., Dong, C.: A new journey from SDRTV to HDRTV. In: Proceedings of the IEEE/CVF International Conference on Computer Vision., pp. 4500–4509 (2021)

6. Cho, S.J., Ji, S.W., Hong, J.P., Jung, S.W., Ko, S.J.: Rethinking coarse-to-fine approach in single image deblurring. In: Proceedings of the IEEE/CVF International Conference on Computer Vision, pp. 4641–4650 (2021)

7. Dong, C., Loy, C.C., He, K., Tang, X.: Learning a deep convolutional network for image super-resolution. In: Fleet, D., Pajdla, T., Schiele, B., Tuytelaars, T. (eds.) ECCV 2014. LNCS, vol. 8692, pp. 184–199. Springer, Cham (2014). https://doi.org/10.1007/978-3-319-10593-2_13

8. Dong, C., Loy, C.C., He, K., Tang, X.: Learning a deep convolutional network for image super-resolution. In: Fleet, D., Pajdla, T., Schiele, B., Tuytelaars, T. (eds.) ECCV 2014. LNCS, vol. 8692, pp. 184–199. Springer, Cham (2014). https://doi.org/10.1007/978-3-319-10593-2_13

9. Dong, C., Loy, C.C., Tang, X.: Accelerating the super-resolution convolutional neural network. In: Leibe, B., Matas, J., Sebe, N., Welling, M. (eds.) ECCV 2016. LNCS, vol. 9906, pp. 391–407. Springer, Cham (2016). https://doi.org/10.1007/978-3-319-46475-6_25

10. Eilertsen, G., Kronander, J., Denes, G., Mantiuk, R.K., Unger, J.: HDR image reconstruction from a single exposure using deep CNNs. ACM Trans. Graphics **36**(6), 1–15 (2017)

11. Feng, R., Li, C., Chen, H., Li, S., Loy, C.C., Gu, J.: Removing diffraction image artifacts in under-display camera via dynamic skip connection network. In: Proceedings of the IEEE/CVF Conference on Computer Vision and Pattern Recognition, pp. 662–671 (2021)

12. Fu, X., Huang, J., Zeng, D., Huang, Y., Ding, X., Paisley, J.: Removing rain from single images via a deep detail network. In: Proceedings of the IEEE Conference on Computer Vision and Pattern Recognition, pp. 3855–3863 (2017)

13. Gao, H., Tao, X., Shen, X., Jia, J.: Dynamic scene deblurring with parameter selective sharing and nested skip connections. In: Proceedings of the IEEE/CVF Conference on Computer Vision and Pattern Recognition, pp. 3848–3856 (2019)

14. Goldstein, J.S., Reed, I.S., Scharf, L.L.: A multistage representation of the wiener filter based on orthogonal projections. IEEE Trans. Inf. Theory **44**(7), 2943–2959 (1998)

15. Gu, J., Lu, H., Zuo, W., Dong, C.: Blind super-resolution with iterative kernel correction. In: Proceedings of the IEEE/CVF Conference on Computer Vision and Pattern Recognition, pp. 1604–1613 (2019)

16. Guo, S., Yan, Z., Zhang, K., Zuo, W., Zhang, L.: Toward convolutional blind denoising of real photographs. In: Proceedings of the IEEE/CVF Conference on Computer Vision and Pattern Recognition, pp. 1712–1722 (2019)

17. Hansen, P.C.: Regularization tools: a Matlab package for analysis and solution of discrete ill-posed problems. Num. Algorithms **6**(1), 1–35 (1994)

18. He, K., Zhang, X., Ren, S., Sun, J.: Delving deep into rectifiers: surpassing human-level performance on imagenet classification. In: Proceedings of the IEEE International Conference on Computer Vision, pp. 1026–1034 (2015)

19. Jauch, W.: The maximum-entropy method in charge-density studies. II. general aspects of reliability. Acta Crystallographica Sect. A Found. Crystallogr. **50**(5), 650–652 (1994)

20. Kim, S.Y., Oh, J., Kim, M.: Deep SR-ITM: joint learning of super-resolution and inverse tone-mapping for 4k UHD HDR applications. In: Proceedings of the IEEE/CVF International Conference on Computer Vision, pp. 3116–3125 (2019)

21. Kingma, D.P., Ba, J.: Adam: a method for stochastic optimization. arXiv preprint arXiv:1412.6980 (2014)

22. Kong, X., Liu, X., Gu, J., Qiao, Y., Dong, C.: Reflash dropout in image super-resolution. In: Proceedings of the IEEE/CVF Conference on Computer Vision and Pattern Recognition, pp. 6002–6012 (2022)

23. Kupyn, O., Martyniuk, T., Wu, J., Wang, Z.: Deblurgan-v2: Deblurring (orders-of-magnitude) faster and better. In: Proceedings of the IEEE/CVF International Conference on Computer Vision, pp. 8878–8887 (2019)

24. Kwon, H.J., Yang, C.M., Kim, M.C., Kim, C.W., Ahn, J.Y., Kim, P.R.: Modeling of luminance transition curve of transparent plastics on transparent OLED displays. Electr. Imaging **2016**(20), 1–4 (2016)

25. Lee, C.H., Liu, Z., Wu, L., Luo, P.: Maskgan: towards diverse and interactive facial image manipulation. In: Proceedings of the IEEE/CVF Conference on Computer Vision and Pattern Recognition, pp. 5549–5558 (2020)

26. Lehtinen, J., Munkberg, J., Hasselgren, J., Laine, S., Karras, T., Aittala, M., Aila, T.: Noise2noise: learning image restoration without clean data. arXiv preprint arXiv:1803.04189 (2018)

27. Li, L., Pan, J., Lai, W.S., Gao, C., Sang, N., Yang, M.H.: Blind image deblurring via deep discriminative priors. Int. J. Comput. Vision **127**(8), 1025–1043 (2019)
28. Li, W., Lu, X., Lu, J., Zhang, X., Jia, J.: On efficient transformer and image pre-training for low-level vision. arXiv preprint arXiv:2112.10175 (2021)
29. Li, X., Wu, J., Lin, Z., Liu, H., Zha, H.: Recurrent squeeze-and-excitation context aggregation net for single image deraining. In: Ferrari, V., Hebert, M., Sminchisescu, C., Weiss, Y. (eds.) ECCV 2018. LNCS, vol. 11211, pp. 262–277. Springer, Cham (2018). https://doi.org/10.1007/978-3-030-01234-2_16
30. Li, Y., Tan, R.T., Guo, X., Lu, J., Brown, M.S.: Rain streak removal using layer priors. In: Proceedings of the IEEE Conference on Computer Vision and Pattern Recognition, pp. 2736–2744 (2016)
31. Li, Z., et al.: Blueprint separable residual network for efficient image super-resolution. In: Proceedings of the IEEE/CVF Conference on Computer Vision and Pattern Recognition, pp. 833–843 (2022)
32. Liang, J., Cao, J., Sun, G., Zhang, K., Van Gool, L., Timofte, R.: Swinir: image restoration using swin transformer. In: Proceedings of the IEEE/CVF International Conference on Computer Vision, pp. 1833–1844 (2021)
33. Liu, P., Zhang, H., Zhang, K., Lin, L., Zuo, W.: Multi-level wavelet-CNN for image restoration. In: Proceedings of the IEEE Conference on Computer Vision and Pattern Recognition Workshops, pp. 773–782 (2018)
34. Liu, Y.L., et al.: Single-image HDR reconstruction by learning to reverse the camera pipeline. In: Proceedings of the IEEE/CVF Conference on Computer Vision and Pattern Recognition, pp. 1651–1660 (2020)
35. Liu, Z., et al.: AdNet: attention-guided deformable convolutional network for high dynamic range imaging. In: Proceedings of the IEEE/CVF Conference on Computer Vision and Pattern Recognition, pp. 463–470 (2021)
36. O'Sullivan, J.A., Blahut, R.E., Snyder, D.L.: Information-theoretic image formation. IEEE Trans. Inf. Theory **44**(6), 2094–2123 (1998)
37. Park, D., Kang, D.U., Kim, J., Chun, S.Y.: Multi-temporal recurrent neural networks for progressive non-uniform single image deblurring with incremental temporal training. In: Vedaldi, A., Bischof, H., Brox, T., Frahm, J.-M. (eds.) ECCV 2020. LNCS, vol. 12351, pp. 327–343. Springer, Cham (2020). https://doi.org/10.1007/978-3-030-58539-6_20
38. Qin, Z., Tsai, Y.H., Yeh, Y.W., Huang, Y.P., Shieh, H.P.D.: See-through image blurring of transparent organic light-emitting diodes display: calculation method based on diffraction and analysis of pixel structures. J. Display Technol. **12**(11), 1242–1249 (2016)
39. Qin, Z., Xie, J., Lin, F.C., Huang, Y.P., Shieh, H.P.D.: Evaluation of a transparent display's pixel structure regarding subjective quality of diffracted see-through images. IEEE Photon. J. **9**(4), 1–14 (2017)
40. Ren, D., Zuo, W., Hu, Q., Zhu, P., Meng, D.: Progressive image deraining networks: a better and simpler baseline. In: Proceedings of the IEEE/CVF Conference on Computer Vision and Pattern Recognition, pp. 3937–3946 (2019)
41. Ronneberger, O., Fischer, P., Brox, T.: U-Net: convolutional networks for biomedical image segmentation. In: Navab, N., Hornegger, J., Wells, W.M., Frangi, A.F. (eds.) MICCAI 2015. LNCS, vol. 9351, pp. 234–241. Springer, Cham (2015). https://doi.org/10.1007/978-3-319-24574-4_28
42. Shao, Y., Li, L., Ren, W., Gao, C., Sang, N.: Domain adaptation for image dehazing. In: Proceedings of the IEEE/CVF Conference on Computer Vision and Pattern Recognition, pp. 2808–2817 (2020)

43. Starck, J.L., Murtagh, F.: Astronomical Image and Data Analysis. Springer Berlin (2007). https://doi.org/10.1007/978-3-662-04906-8
44. Starck, J.L., Pantin, E., Murtagh, F.: Deconvolution in astronomy: a review. Publ. Astron. Soc. Pac. **114**(800), 1051 (2002)
45. Tang, Q., Jiang, H., Mei, X., Hou, S., Liu, G., Li, Z.: 28-2: study of the image blur through ffs LCD panel caused by diffraction for camera under panel. In: SID Symposium Digest of Technical Papers, vol. 51, pp. 406–409. Wiley Online Library (2020)
46. Wang, X., Yu, K., Dong, C., Loy, C.C.: Recovering realistic texture in image super-resolution by deep spatial feature transform. In: Proceedings of the IEEE Conference on Computer Vision and Pattern Recognition, pp. 606–615 (2018)
47. Wang, X., et al.: ESRGAN: enhanced super-resolution generative adversarial networks. In: Leal-Taixé, L., Roth, S. (eds.) ECCV 2018. LNCS, vol. 11133, pp. 63–79. Springer, Cham (2019). https://doi.org/10.1007/978-3-030-11021-5_5
48. Wang, Z., Cun, X., Bao, J., Zhou, W., Liu, J., Li, H.: UforMer: a general u-shaped transformer for image restoration. In: Proceedings of the IEEE/CVF Conference on Computer Vision and Pattern Recognition, pp. 17683–17693 (2022)
49. Zamir, S.W., Arora, A., Khan, S., Hayat, M., Khan, F.S., Yang, M.H.: Restormer: efficient transformer for high-resolution image restoration. In: Proceedings of the IEEE/CVF Conference on Computer Vision and Pattern Recognition, pp. 5728–5739 (2022)
50. Zamir, S.W., Arora, A., Khan, S., Hayat, M., Khan, F.S., Yang, M.H.: RestorMer: efficient transformer for high-resolution image restoration. In: Proceedings of the IEEE/CVF Conference on Computer Vision and Pattern Recognition, pp. 5728–5739 (2022)
51. Zhang, K., Zuo, W., Chen, Y., Meng, D., Zhang, L.: Beyond a gaussian denoiser: Residual learning of deep CNN for image denoising. IEEE Trans. Image Process. **26**(7), 3142–3155 (2017)
52. Zhang, K., Zuo, W., Zhang, L.: FFDNet: toward a fast and flexible solution for CNN-based image denoising. IEEE Trans. Image Process. **27**(9), 4608–4622 (2018)
53. Zhang, K., et al.: Deblurring by realistic blurring. In: Proceedings of the IEEE/CVF Conference on Computer Vision and Pattern Recognition, pp. 2737–2746 (2020)
54. Zhang, Y., Li, K., Li, K., Wang, L., Zhong, B., Fu, Y.: Image super-resolution using very deep residual channel attention networks. In: Ferrari, V., Hebert, M., Sminchisescu, C., Weiss, Y. (eds.) ECCV 2018. LNCS, vol. 11211, pp. 294–310. Springer, Cham (2018). https://doi.org/10.1007/978-3-030-01234-2_18
55. Zhou, Y., Ren, D., Emerton, N., Lim, S., Large, T.: Image restoration for under-display camera. In: Proceedings of the IEEE/CVF Conference on Computer Vision and Pattern Recognition, pp. 9179–9188 (2021)
56. Zhu, L., Fu, C.W., Lischinski, D., Heng, P.A.: Joint bi-layer optimization for single-image rain streak removal. In: Proceedings of the IEEE International Conference on Computer Vision, pp. 2526–2534 (2017)

Enhanced Coarse-to-Fine Network
for Image Restoration
from Under-Display Cameras

Yurui Zhu[1,2], Xi Wang[1,2], Xueyang Fu[1,2(✉)], and Xiaowei Hu[1,2(✉)]

¹ Universityf China, Hefei, China
{zyr,wangxxi}@mail.ustc.edu.cn,
xyfu@ustc.edu.cn, huxiaowei@pjlab.org.cn
² Shanghai AI Laboratory, Shanghai, China

Abstract. New sensors and imaging systems are the indispensable foundation for mobile intelligent photography and imaging. Nowadays, under-display cameras (UDCs) demonstrate practical applicability in smartphones, laptops, tablets, and other scenarios. However, the images captured by UDCs suffer from complex image degradation issues, such as flare, haze, blur, and noise. To solve the above issues, we present an Enhanced Coarse-to-Fine Network (ECFNet) to effectively restore the UDC images, which takes the multi-scale images as the input and gradually generates multi-scale results from coarse to fine. We design two enhanced core components, *i.e.*, Enhanced Residual Dense Block (ERDB) and multi-scale Cross-Gating Fusion Module (CGFM), in the ECFNet, and we further introduce progressive training and model ensemble strategies to enhance the results. Experimental results show superior performance against the existing state-of-the-art methods both qualitatively and visually, and our ECFNet achieves the best performance in terms of all the evaluation metrics in the MIPI-challenge 2022 Under-display Camera Image Restoration track. Our source codes are available at this repository.

Keywords: Under-display cameras · Deep neural networks · And image restoration

1 Introduction

Under-Display Cameras (UDC) is an emerging imaging system technology, which has been applied to various scenarios, such as mobile phones, laptops, and video-conferencing with UDC TV. This novel imaging system can achieve the demands of full-screen devices with larger screen-to-body ratios, which is favored by many consumers and manufacturers. However, the display panel in front of the camera inevitably brings incoming light loss and diffraction, which in turn causes unwanted noise, blurring, fogging, and vertigo effects. The UDCs degradation process in the high dynamic scenes [7] is formulated as

$$\hat{y} = \varphi\left[C(x * k + n)\right] , \tag{1}$$

Y. Zhu—This work was done during his internship at Shanghai AI Laboratory.

where x, $\hat{y}$ are the original images and the UDC degradation images; $\varphi(\cdot)$ represents the non-linear tone mapping operation to reproduce the appearance of high dynamic range images on a display with more constricted dynamic range; $C(\cdot)$ is the truncation function, which limits the range of values for imaging; k refers to the Point Spread Function (PSF), which mainly brings diffraction artifacts; $*$ indicates the 2D convolution operation; n denotes the noise information from the camera device.

To remove undesired image artifacts, lots of works based on deep neural networks (DNNs) have been proposed for UDC image restoration. For example, MSUNET [45] attempts to analyze the UDC degradation processes based on two common displays, *e.g.*, the Transparent OLED (T-OLED) and the phone Pentile OLED (P-OLED). However, they just employ a U-net-based structure to explore the mapping relationship between the UDC distorted images and clean images. Feng *et al.* [7] integrate PSF prior and degradation knowledge to construct the dynamic network [14] to recover the desired high-quality images. Based on the optical properties of OLEDs, BNUDC [17] develops a two-branch structure network to achieve independent high- and low-frequency component reconstruction. Although the existing methods show remarkable performance on the restoration results in terms of multiple evaluation metrics, the local high-frequency artifacts (*e.g.*, flare caused by strong light) are still a challenging task, which largely degrades the visual results.

As mentioned before, images captured by UDC usually introduce many degradation types (*e.g.*, blurry and diffraction). There are already a large number of techniques for image restoration tasks. For example, attention and adaptive gating mechanisms [2,3,6,8,19,23,27,35,40,43] have been widely used in low-level vision tasks, which helps the restoration network to pay extra attention to local regions. NAFNet [4,20] explores the applications and variants of nonlinear activation functions in their networks. Although these methods are not specifically designed for UDC restoration, they also provide potential inspiration for designing more effective UDC restoration frameworks.

In this paper, inspired by [6], we design an effective Enhanced Coarse-to-Fine Network (ECFNet) to restore images from UDCs. To be specific, we maintain the multi-input encoder and multi-output decoder UNet-based framework, which gradually restores latent clean results in different scales in a coarse-to-fine manner. Moreover, we further explore several enhanced strategies in our ECFNet to better restore the images captured by UDC. Firstly, considering the larger resolution size of UDC images, we adopt the images with multiple resolutions as the inputs of the network and output the restored image with the different resolutions. Secondly, we present an enhanced Residual Dense Block (ERDB) as the basic block of our backbone and explore the effect of different non-linear activation functions. Moreover, the multi-scale Cross-Gating Fusion Module (CGFM) is devised to merge cross-scale features to transfer the information from the multi-input encoder into the multi-output decoder with the cross-gating mechanism. Additionally, we equip with progressive training and model ensemble strategies to further improve the restoration performance. Finally, we conduct various experiments to show the effectiveness of our method.

The main contributions of our paper are summarized as follows:

- We present an effective Enhanced Coarse-to-Fine Network (ECFNet) to restore images captured by Under-Display Cameras (UDCs) with various types of degradation.
- We devise two core enhanced modules, *i.e.*, Enhanced Residual Dense Block (ERDB) and multi-scale Cross-Gating Fusion Module (CGFM), to enhance the network by fully exploring the effects of the non-linear activation functions and the feature interaction strategies among different scales.
- Experimental results demonstrate that our method significantly outperforms other UDC image restoration solutions. In the MIPI-challenge 2022 Under-display Camera Image Restoration track, our ECFNet achieves first place in terms of all the evaluation scores ($PSNR$, $SSIM$, and $LPIPS$) and outperforms the others by a large margin.

2   Related Work

UDCs Restoration

For the development of full-screen smartphones, under-display cameras (UDCs) are a crucial technology. Due to the light loss and diffraction introduced by the front display panel, the image captured by the UDC usually suffers some degree of degradation. The UDC system has been analyzed in several previous works. An Edge Diffusion Function model for transparent OLEDs is developed by Kwon *et al.* [18]. Based on diffraction theory, Qin *et al.* [25] propose an accurate method for calculating the perspective image and demonstrate that image blur can be suppressed by modifying the pixel structure of transparent OLED displays. Qin *et al.* [26] and Tang *et al.* [29] describe and analyze the diffraction effects of the UDC system. However, these methods do not recover the already existing blurred UDC images. [44,45] are the first work to perform UDC image restoration on publicly released datasets, by modeling the degraded images captured by the UDC, Zhou *et al.* [45] design an MCI to capture paired images, and propose a data synthesis pipeline for generating UDC images to be generated from natural images. Then based on these synthetic data, they use a variation of Unet to achieve UDC image restoration. However, they only solve the single degradation of images and are difficult to apply to UDC images with real degradation. DAGF [28] proposes a Deep Atrous Guided Filter for image restoration in UDC systems. Panikkasseril *et al.* [24] propose two different networks to recover blurred images that are captured using two types of UDC techniques. Yang *et al.* [33] use residual dense networks to achieve UDC image restoration. Kwon *et al.* [18] propose a novel controllable image restoration framework for UDC. Feng *et al.* [7] achieve restoration of multiple PSFs by adjusting the PSF of the UDC image. BNUDC [17] proposes a two-branch DNN architecture for recovering the high-frequency and low-frequency components of an image, respectively.

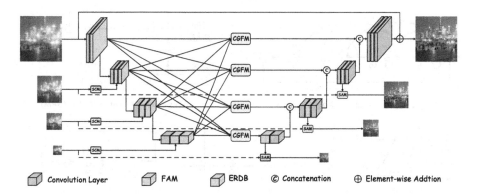

Fig. 1. The architecture of our proposed Enhanced Coarse-to-Fine Network (ECFNet) to restore the images captured by under-display cameras.

Coarse-to-Fine Networks

In many image restoration tasks [6, 9, 21, 30, 38], the coarse-to-fine network design has shown to be effective. For example, MIMOUet [6] rethinks the coarse-to-fine strategy and presents a UNet-like structure network to progressively recover clean images at different scales. And our network is built based on MIMOUet. However, we have made further adjustments from the basic network module, hierarchical feature interaction manner, and network levels to accommodate to the UDCs restoration task.

3   Method

Our proposed Enhanced Coarse-to-Fine Network (ECFNet) is built with multi-input encoder and multi-output decoder UNet-based framework, which gradually generates desired high-quality results of multi-scale sizes in a coarse to fine manner. The specific architecture of our ECFNet is present in Fig. 1, which is based on the MIMO-UNet [6]. Different from MIMO-UNet, we expand the scales of the network and devise the enhanced basic block to improve the network capability. Furthermore, we build the Cross-Gating Fusion Module (CGFM) to transfer the information flow among the encoders and decoders of cross scales. The specific components of our ECFNet will be thoroughly described in the following subsections.

3.1   Enhanced Encoder

The enhanced Encoder consists of four sub-networks, which take the different images with different resolutions as input. Different from MIMO-UNet [6], we further enhance the basic blocks to enhance the capacity of our model. Figure 2 depicts our proposed Enhanced Residual Dense Block, which comprises three convolution layers with kernel size of 3 × 3. Additionally, there are several skip

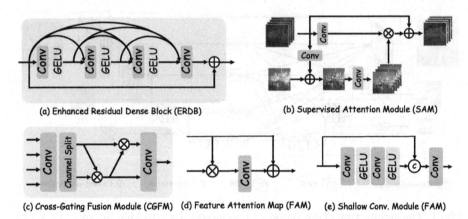

(a) Enhanced Residual Dense Block (ERDB)   (b) Supervised Attention Module (SAM)

(c) Cross-Gating Fusion Module (CGFM)   (d) Feature Attention Map (FAM)   (e) Shallow Conv. Module (FAM)

Fig. 2. The structures of sub-modules in the main architecture.

connections across the various levels to construct a dense network structure. Such residual dense connection manner has been proven to be effective in many computer vision tasks [13,41,42], which could extract richer contextual information and transfer the information flow from shallow to deep layers.

To build more effective residual dense block, we further exploit the effect non-linear activation functions. Inspired by [4], it seems that GELU [12] has more effective than other activation functions, $e.g.$, ReLU [1], PReLU [10]. Refer to [12], GELU could be approximately written as

$$GELU(x) \approx 0.5x \left(1 + tanh \left[\sqrt{2/\pi} \left(x + 0.044715x^3 \right) \right] \right) \qquad (2)$$

Specifically, we replace the plain ReLU with GELU in our Enhanced Residual Dense block (ERDB), which brings obvious performance gain. Hence, we utilize the ERDB as basic block of our enhanced encoder. At each scale of the enhanced encoder, there are six ERDB in total.

3.2 Enhanced Decoder

Similar to the enhanced encoder, Enhanced Decoder (ED) also has sub-networks and takes ERDB as the basic block. Moreover, ED generate four high-quality results with different sizes and we also impose the corresponding intermediate constraints.

Besides, we further introduce the Supervised attention module (SAM) [36] between each two scales of ED, enabling for a considerable performance improvement. The specific structure of SAM is shown in Fig. 2(b).

3.3 Cross-Gating Fusion Module

For many UNet-based image restoration methods [33,34], they usually propagate the feature information flow of the corresponding size in the encoder to the

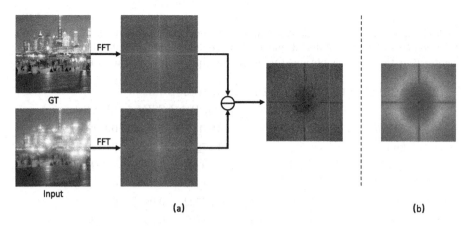

Fig. 3. Analysis of frequency domain. (a) Amplitude difference acquisition pipeline of one pair image. (b) Average amplitude difference map on the evaluation dataset.

decoder of the corresponding size to increase the diversity of features to further improve the performance of the model. Furthermore, some methods [6,15,37] cascade the different scale features to allow the information flow between the encoder and the decoder in a top-to-bottom and bottom-to-top manner.

In this paper, not only we adopt the dense concatenation cross features with different scales, but also we introduce the gating mechanisms to further enhance the capacity of our model. As shown in Fig. 2, we propose multi-scale Cross-Gating Fusion Module (CGFM) enhance the information flow between the encoder and the decoder. CGFM first collects the features at four scales through the concatenation operation, then adjusts the number of feature channels through 1×1 convolutional layer. Then the features are evenly divided into two parts and performs gating operation [34] on the two parts sequentially. Finally, enhanced feature through the concatenation operation could be obtained. Therefore, CGFM at the smallest scale could be expressed as

$$
\begin{aligned}
F &= Conv_{1\times1}(Concat([(EEB_1^{out})^{\downarrow},,(EEB_2^{out})^{\downarrow},(EEB_3^{out})^{\downarrow},EEB_4^{out}])), \\
F_1, F_2 &= Split(F), \\
F_1 &= F_2 \otimes F_1, \\
F_2 &= F_1 \otimes F_2, \\
F_e &= Concat([F_1, F_2]),
\end{aligned}
$$

$$(3)$$

where $EEB_i^{out}, i = 1,2,3,4$ denotes the output of the nth scale enhanced encoder blocks; $Concat(\dots)$ denotes the concatenation operation along the channel dimension; $Conv_{1\times1}(\dots)$ denotes the convolution operation with kernel of 1×1; $\downarrow$ denotes the down-sampling operation; $Split(\dots)$ denotes the split operation along the channel dimension; $\otimes$ denotes the element-wise multiplication.

Table 1. Quantitative comparisons of methods on the official testing datasets of the MIPI-challenge 2022 Under-display Camera Image Restoration track. The best and the second results are boldfaced and underlined, respectively. Note that ECFNet-s3 denotes that the model is obtained with the multiple training strategies.

MIPI 2022 UDC Challenge Track	Metrics		
	PSNR ↑	SSIM ↑	LPIPS ↓
ECFNet-s3 (Ours)	**48.4776**	**0.9934**	**0.0093**
2nd	<u>47.7821</u>	0.9913	0.0122
3rd	46.9227	<u>0.9929</u>	<u>0.0098</u>
4th	46.1177	0.9892	0.0159
5th	45.8722	0.9920	0.0109
6th	45.8227	0.9917	0.0106
7th	45.5613	0.9912	0.0129
8th	44.0504	0.9908	0.0120
9th	43.4514	0.9897	0.0125
10th	43.4360	0.9899	0.0133
11th	43.1300	0.9872	0.0144
12th	42.9475	0.9892	0.0150
13th	42.0392	0.9873	0.0155
14th	39.5205	0.9820	0.0216
15th	37.4569	0.9973	0.0370
16th	36.6961	0.9783	0.0326
17th	35.7726	0.9719	0.0458
18th	35.5031	0.9616	0.0453
19th	32.7496	0.9591	0.0566

3.4   Loss Functions

Because ECFNet has multi-scale outputs, we naturally combine the Charbonnier loss [31] and the multi-scale content losses [22] to optimize our network, which is written as follows:

$$\mathcal{L}_{content} = \sum_{k=1}^{K} \sqrt{\left\| I_{pre}^k - I_{gt}^k \right\|_2 + \epsilon^2},\qquad(4)$$

where K is the number of scales, which equals to 4 as default; I_{pre}^k is the predicted result of scale k; I_{gt}^k is the ground truth of scale k; ϵ is set to 0.0001 as default.

Besides the content loss, we also exploit the frequency domain information to provide auxiliary loss for our network. As shown in Fig. 6, we first provide the visualization of amplitude difference map in frequency domain, which indicates the various degradations brought by UDCs mainly affect the information in the

Table 2. Quantitative comparisons of methods on the official evaluation and testing datasets of TOLED [5]. Note that we reduce the size of the original model similar to BNUDC [17] for fair comparisons. The average inference time (IT) is obtained with TOLED dataset on the NVIDIA 1080Ti GPU device.

TOLED	PARAM(M: 10^6)	IT(seconds)	TEST SET			VAL. SET		
			PSNR↑	SSIM↑	LPIPS↓	PSNR↑	SSIM↑	LPIPS↓
MSUNET [45]	8.9	0.33	37.40	0.976	0.109	38.25	0.977	0.117
IPIUer [44]	24.7	0.43	38.18	0.979	0.113	39.03	0.981	0.121
BAIDU [44]	20.0	0.18	38.23	0.980	0.102	39.06	0.981	0.111
DAGF [28]	1.1	1.30	37.27	0.973	0.141	37.46	0.943	0.090
BNUDC [17]	4.6	0.36	38.22	0.979	0.099	39.09	0.981	0.107
ECFNet-small (ours)	4.1	0.04	37.79	0.953	0.061	38.49	0.955	0.060

Table 3. Ablation study of basic blocks and modules.

Models	Experiment Setting			#Param (M: 10^6)	Metrics	
					PSNR ↑	SSIM ↑
Model-1	Basic Block	Resdiual Block	w LeakeyReLU	16.82 M	34.92	0.976
Model-2			w GELU	16.82 M	35.09	0.981
Model-3		Dense Residual Block	w ReLU	16.85 M	43.33	0.992
Model-4			w LeakeyReLU	16.85 M	44.33	0.993
Model-5			w PReLU	16.85 M	42.98	0.992
Default			w GELU	16.85 M	45.08	0.994
Model-6	Modules	w./o SAM		16.65 M	44.83	0.993
Default		w. SAM		16.85 M	45.08	0.994

mid- and high-frequency regions. Such information loss patterns are not only obvious on a single pair of images, but also on the entire evaluation dataset. Hence, we further apply multi-scale frequency loss to optimize our network, which is defined as follows:

$$\mathcal{L}_{frequency} = \sum_{k=1}^{K} \left\| \mathcal{F}(I_{pre}^k) - \mathcal{F}(I_{gt}^k) \right\|_1, \tag{5}$$

where $\mathcal{F}(\cdot)$ indicates the Fast Fourier Transform (FFT). Finally, the total loss could be defined as

$$\mathcal{L}_{total} = \mathcal{L}_{content} + \lambda \mathcal{L}_{frequency}, \tag{6}$$

where λ denotes the balanced weight and we empirically set λ to 0.5 as default.

4 Experiments

4.1 Implementation Details

We implement our proposed UDC image restoration network via the PyTorch 1.8 platform. Adam optimizer [16] with parameters $\beta_1 = 0.9$ and $\beta_2 = 0.99$ is

Table 4. Ablation study of the losses.

Losses			Metrics	
L1 Loss	Charbonnier L1 Loss	Frequency Loss	PSNR ↑	SSIM ↑
✓			42.26	0.989
	✓		42.50	0.990
✓		✓	44.83	0.993
	✓	✓	45.08	0.994

Table 5. Ablation study of the connection manner between Enhanced Encoder (EE) and Enhanced Decoder (ED).

Connection Manner Between EE and ED	Metrics	
	PSNR ↑	SSIM ↑
Simple Concatentation	44.00	0.991
Single Gating Fusion	44.87	0.993
Cross-Gating Fusion Module (CGFM) (default)	45.08	0.994

adopted to optimize our network. Additionally, motivated by [34], we introduce the progressive training strategy and the specific training phase of our network could be divided into three stages:

(1) We adopt the Adam optimizer with a batch size of three and the patch size of 256×256. The initial learning rate is 2×10^{-4} and is adjusted with the Cosine Annealing scheme, including 1000 epochs in total. The first stage performs on the NVIDIA 1080Ti GPU. We obtain the best model at this stage as the initialization of the second stage.
(2) We adopt the Adam optimizer with a batch size of one and the patch size of 512 × 512. The initial learning rate is 1×10^{-5} and is adjusted with the Cosine Annealing scheme, including 300 epochs in total. The second stage performs on the NVIDIA 1080Ti GPU. We obtain the best model at this stage as the initialization of the next stage.
(3) We adopt the Adam optimizer with a batch size of two and the patch size of 800 × 800. The initial learning rate is 8×10^{-6} and is adjusted with the Cosine Annealing scheme, including 150 epochs in total. The third stage performs on the NVIDIA A100 GPU. In this stage, we additionally expand the training dataset with the testing set of SYNTH Dataset [7]. Finally, We adopt the model ensemble strategy, which averages the parameters of multiple models trained with different training iterations.

Therefore, in order to better distinguish the results of the models, we record the results of the three stages as ECFNet-s1, ECFNet-s2 and ECFNet-s3.

In the testing phase, we adopt the model after fine-tuning to achieve the best performance. Moreover, we utilize a model-ensemble strategy to obtain better

Table 6. Ablation study of the numbers of network inputs and outputs.

Numbers of Network Inputs and Outputs	Metrics	
	PSNR ↑	SSIM ↑
Single Input, Single Output	44.49	0.992
Three Inputs, Three Outputs	44.68	0.993
Four Inputs, Four Outputs (default)	45.08	0.994

Table 7. Ablation study of training strategies.

Results with different strategy	Metrics	
	PSNR ↑	SSIM ↑
ECFNet-s1	45.08	0.994
ECFNet-s2 + w./o External Data	46.23	0.995
ECFNet-s2 + w. External Data	47.80	0.996
ECFNet-s3 + w. External Data	49.04	0.995
ECFNet-s3 + w. External Data + Model Ensemble (default)	49.13	0.996

results. 11G GPU memory is enough to infer our model, and we use one NVIDIA 1080Ti GPU with 11G memory for testing.

4.2 Dataset

We conduct the experiments strictly following the setting of the MIPI-challenge 2022 Under-display Camera Image Restoration track. There are 2016 pairs of $800 \times 800 \times 3$ images in the training split. Image values are produced in ".npy" format and range from $[0, 500]$. There are 40 image pairings in the validation set and 40 image pairs in the testing dataset, respectively. Note that the corresponding ground truths of testing dataset are not publicly available. Besides the above dataset at the training phase, we also exploit an additional 320 pairs of test datasets in dataset SYNTH [7] to augment the training dataset. However, our network never requires the Point Spread Function (PSF) as the external prior to predict the final results, which is more flexible than DISCNet [7]. Only for the challenge dataset, we adopt the additional dataset to improve the restoration performance of our model.

Moreover, in order to validate the effectiveness of our method, we also employ the TOLED dataset [5] to training our network. TOLED dataset totally consists of 300 image pairs, which has been divided into 240 image pairs for training, 30 image pairs for evaluation, and 30 image pairs for testing. Note that the models trained on these two different datasets are different, and then applied to the corresponding test datasets for testing.

4.3   Evaluation Metrics

Similar to previous methods [7,17], we employ three reference-based metrics to verify the effectiveness of our method: Peak Signal-to-Noise Ratio (PSNR), the structural similarity (SSIM) [32], and Learned Perceptual Image Patch Similarity (LPIPS) [39]. For the PSNR and SSIM metrics, higher is better. For the LPIPS metric, lower means better.

Input   GT   Residual of Input   Residual of BNUDC   Residual of Ours

Fig. 4. Visual comparison results of UDC Image Restoration on the evaluation dataset of MIPI-challenge 2022 track. Note that brighter means bigger error.

4.4   Comparations

In Table 1, we report the comparison results among different solutions on the challenge track. Obviously, our method performs best performance in terms of all the evaluation metrics, even 0.7 dB higher than the second-place method. Besides, in Table 2 we achieve the comparable performance on the TOLED evaluation and testing dataset. For fair comparison, note that we reduce the size of the original model similar to BNUDC [17]. And our method is significantly seventeen times faster than BNUDC on the TOLED dataset. In addition, in

Fig. 5. Visual comparison results of UDC Image Restoration on the testing dataset of MIPI-challenge 2022 track.

Fig. 4, 5, 6, to demonstrate that the images generated by our network have better visual quality, we show the residual map of other methods and our on the UDC dataset to increase the visual differentiation. Note that the residual map is the difference between the estimated results and the ground truth. It is clear that our method achieves better visual results, with better recovery for edge flare and less difference between the generated images and the ground truth.

4.5 Ablation Study

We conduct extensive experiments to verify the effect of each components of our method, *e.g.*, modules, strategies and losses. Note that in the ablation study we only use the first stage training manner (refer to ECFNet-s1) for convenience.

Effects of Basic Blocks and Modules. AS reported in Table 3, we compare the performance of models with different non-linear activation functions, including ReLU, LeakeyReLU, PReLU, and GELU. Obviously, using GELU performs better than other activation functions both embedding in the standard residual block [11] and standard dense residual block [13]. For example, compared to using LeakeyReLU, using GELU brings 0.75db performance gains.

In Table 3, we further conduct experiments to verify the effectiveness of the SAM. Using SAM [36] aims at bridging the relation between the different scales, which facilitates achieving 0.25 dB performance gain.

Effects of the Coarse-to-Fine Strategy. Our restoration framework is built on the coarse-to-fine strategy to gradually recover different scales latent clean results. We further report the related results of the model with different numbers of inputs and outputs in Table 6, which could validate the effectiveness of the coarse-to-fine strategy. Obviously, using four inputs and four outputs achieve the best results. Compared with baseline, using multiple outputs to optimize our model brings about 0.59 dB performance improvement.

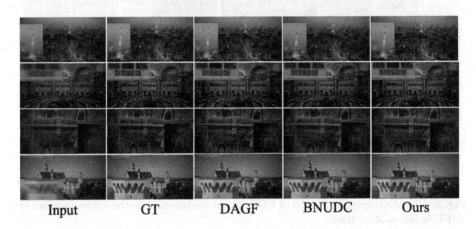

Fig. 6. Visual comparison results of UDC Image Restoration on the TOLED dataset.

Effects of the Loss Functions. Except for the content losses, we additionally introduce the $\mathcal{L}_{frequency}$ to optimize the network from frequency domain. The network performance with different losses is reported in Table 4. Furthermore, we provide the visualization comparisons when models are optimized by different loss combinations in Fig. 7. Combining Charbonnier L1 Loss and Frequency Loss brings the best model performance.

Effects of the Training Strategies. Following Restormer [34], we additionally adopt the progressive training strategy to enhance the model performance. As shown in Table 7, the models trained at different stages are marked as ECFNet-s1, ECFNet-s2, and ECFNet-s3. Experiments show that training with progressively larger patches often leads to better generalization performance gains

Model ensemble strategy, whose results are obtained by linearly combine several model parameters with different training iterations. This brings around a 0.09dB+ (PSNR) increase on the evaluation dataset with help of the model ensemble strategy.

Inference Time. The average inference time per image of our standard ECFNet is 0.2665s, which is obtained with the image resolution of 800 × 800 on the NVIDIA 1080Ti GPU device.

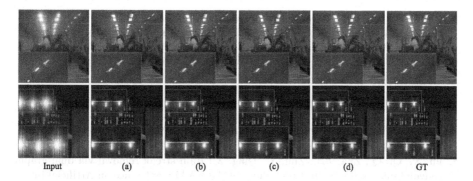

Fig. 7. Model results with different loss combinations. (a) w. L1 Loss; (b) w. Charbonnier L1 Loss; (c) w. L1 Loss and FFT loss; (d) w. Charbonnier L1 Loss and FFT loss (default).

5 Conclusions

In this paper, we design an effective network for restoring images from under-display cameras, named Enhanced Coarse-to-Fine Network (ECFNet). It inherits the previous classic coarse-to-fine network framework and further formulates two enhanced core modules, *i.e.*,, Enhanced Residual Dense Block (ERDB) and multi-scale Cross-Gating Fusion Module (CGFM). Besides, we introduce the progressive training and model ensemble strategy to further improve our model performance. Finally, ECFNet achieves the best performance in terms of all the evaluation metrics in the MIPI-challenge 2022 UDC Image Restoration track.

Acknowledgement. This work was supported by the National Natural Science Foundation of China (NSFC) under Grant 61901433 and in part by the USTC Research Funds of the Double First-Class Initiative under Grant YD2100002003. This work is partially supported by the Shanghai Committee of Science and Technology (Grant No. 21DZ1100100).

References

1. Agarap, A.F.: Deep learning using rectified linear units (ReLU). arXiv preprint arXiv:1803.08375 (2018)
2. Anwar, S., Barnes, N.: Real image denoising with feature attention. In: Proceedings of the IEEE/CVF International Conference on Computer Vision, pp. 3155–3164 (2019)
3. Chen, D., et al.: Gated context aggregation network for image dehazing and deraining. In: 2019 IEEE Winter Conference on Applications of Computer Vision (WACV), pp. 1375–1383. IEEE (2019)
4. Chen, L., Chu, X., Zhang, X., Sun, J.: Simple baselines for image restoration. In: Avidan, S., Brostow, G., Cisse, M., Farinella, G.M., Hassner, T. (eds.) Computer Vision – ECCV 2022. LNCS, vol. 13667. Springer, Cham. https://doi.org/10.1007/978-3-031-20071-7_2

5. Cheng, C.J., et al.: P-79: evaluation of diffraction induced background image quality degradation through transparent OLED display. In: SID Symposium Digest of Technical Papers, vol. 50, pp. 1533–1536. Wiley Online Library (2019)

6. Cho, S.J., Ji, S.W., Hong, J.P., Jung, S.W., Ko, S.J.: Rethinking coarse-to-fine approach in single image deblurring. In: Proceedings of the IEEE/CVF International Conference on Computer Vision, pp. 4641–4650 (2021)

7. Feng, R., Li, C., Chen, H., Li, S., Loy, C.C., Gu, J.: Removing diffraction image artifacts in under-display camera via dynamic skip connection network. In: Proceedings of the IEEE/CVF Conference on Computer Vision and Pattern Recognition, pp. 662–671 (2021)

8. Fu, X., Qi, Q., Zha, Z.J., Zhu, Y., Ding, X.: Rain streak removal via dual graph convolutional network. In: Proceedings of the AAAI Conference on Artificial Intelligence, vol. 35, pp. 1352–1360 (2021)

9. Gao, H., Tao, X., Shen, X., Jia, J.: Dynamic scene deblurring with parameter selective sharing and nested skip connections. In: Proceedings of the IEEE/CVF Conference on Computer Vision and Pattern Recognition, pp. 3848–3856 (2019)

10. He, K., Zhang, X., Ren, S., Sun, J.: Delving deep into rectifiers: surpassing human-level performance on ImageNet classification. In: Proceedings of the IEEE International Conference on Computer Vision, pp. 1026–1034 (2015)

11. He, K., Zhang, X., Ren, S., Sun, J.: Deep residual learning for image recognition. In: Proceedings of the IEEE Conference on Computer Vision and Pattern Recognition, pp. 770–778 (2016)

12. Hendrycks, D., Gimpel, K.: Gaussian error linear units (GELUs). arXiv preprint arXiv:1606.08415 (2016)

13. Huang, G., Liu, Z., Van Der Maaten, L., Weinberger, K.Q.: Densely connected convolutional networks. In: Proceedings of the IEEE Conference on Computer Vision and Pattern Recognition, pp. 4700–4708 (2017)

14. Jia, X., De Brabandere, B., Tuytelaars, T., Gool, L.V.: Dynamic filter networks. In: 29th Proceedings of Conference on Advances in Neural Information Processing System (2016)

15. Kim, S.-W., Kook, H.-K., Sun, J.-Y., Kang, M.-C., Ko, S.-J.: Parallel feature pyramid network for object detection. In: Ferrari, V., Hebert, M., Sminchisescu, C., Weiss, Y. (eds.) ECCV 2018. LNCS, vol. 11209, pp. 239–256. Springer, Cham (2018). https://doi.org/10.1007/978-3-030-01228-1_15

16. Kingma, D.P., Ba, J.: Adam: a method for stochastic optimization. arXiv preprint arXiv:1412.6980 (2014)

17. Koh, J., Lee, J., Yoon, S.: BNUDC: a two-branched deep neural network for restoring images from under-display cameras. In: Proceedings of the IEEE/CVF Conference on Computer Vision and Pattern Recognition, pp. 1950–1959 (2022)

18. Kwon, H.J., Yang, C.M., Kim, M.C., Kim, C.W., Ahn, J.Y., Kim, P.R.: Modeling of luminance transition curve of transparent plastics on transparent OLED displays. Electr. Imaging **2016**(20), 1–4 (2016)

19. Liu, D., Wen, B., Fan, Y., Loy, C.C., Huang, T.S.: Non-local recurrent network for image restoration. In: 31st Proceedings of Conference on Advances in Neural Information Processing Systems (2018)

20. Liu, Z., Mao, H., Wu, C.Y., Feichtenhofer, C., Darrell, T., Xie, S.: A convnet for the 2020s. In: Proceedings of the IEEE/CVF Conference on Computer Vision and Pattern Recognition, pp. 11976–11986 (2022)

21. Ma, Y., Liu, X., Bai, S., Wang, L., He, D., Liu, A.: Coarse-to-fine image inpainting via region-wise convolutions and non-local correlation. In: IJCAI, pp. 3123–3129 (2019)

22. Nah, S., Hyun Kim, T., Mu Lee, K.: Deep multi-scale convolutional neural network for dynamic scene deblurring. In: Proceedings of the IEEE Conference on Computer Vision and Pattern Recognition, pp. 3883–3891 (2017)
23. Pan, X., Zhan, X., Dai, B., Lin, D., Loy, C.C., Luo, P.: Exploiting deep generative prior for versatile image restoration and manipulation. IEEE Trans. Pattern Anal. Mach. Intell. **44**, 7474–7489 (2021)
24. Panikkasseril Sethumadhavan, H., Puthussery, D., Kuriakose, M., Charangatt Victor, J.: Transform domain pyramidal dilated convolution networks for restoration of under display camera images. In: Bartoli, A., Fusiello, A. (eds.) ECCV 2020. LNCS, vol. 12539, pp. 364–378. Springer, Cham (2020). https://doi.org/10.1007/978-3-030-68238-5_28
25. Qin, Z., Tsai, Y.H., Yeh, Y.W., Huang, Y.P., Shieh, H.P.D.: See-through image blurring of transparent organic light-emitting diodes display: calculation method based on diffraction and analysis of pixel structures. J. Display Technol. **12**(11), 1242–1249 (2016)
26. Qin, Z., Xie, J., Lin, F.C., Huang, Y.P., Shieh, H.P.D.: Evaluation of a transparent display's pixel structure regarding subjective quality of diffracted see-through images. IEEE Photonics J. **9**(4), 1–14 (2017)
27. Ren, W., et al.: Gated fusion network for single image dehazing. In: Proceedings of the IEEE Conference on Computer Vision and Pattern Recognition, pp. 3253–3261 (2018)
28. Sundar, V., Hegde, S., Kothandaraman, D., Mitra, K.: Deep Atrous guided filter for image restoration in under display cameras. In: Bartoli, A., Fusiello, A. (eds.) ECCV 2020. LNCS, vol. 12539, pp. 379–397. Springer, Cham (2020). https://doi.org/10.1007/978-3-030-68238-5_29
29. Tang, Q., Jiang, H., Mei, X., Hou, S., Liu, G., Li, Z.: 28-2: study of the image blur through FFS LCD panel caused by diffraction for camera under panel. In: SID Symposium Digest of Technical Papers, vol. 51, pp. 406–409. Wiley Online Library (2020)
30. Wang, L., Li, Y., Wang, S.: Deepdeblur: fast one-step blurry face images restoration. arXiv preprint arXiv:1711.09515 (2017)
31. Wang, X., Chan, K.C., Yu, K., Dong, C., Change Loy, C.: EDVR: video restoration with enhanced deformable convolutional networks. In: Proceedings of the IEEE/CVF Conference on Computer Vision and Pattern Recognition Workshops (2019)
32. Wang, Z., Bovik, A.C., Sheikh, H.R., Simoncelli, E.P.: Image quality assessment: from error visibility to structural similarity. IEEE Trans. Image Process. **13**(4), 600–612 (2004)
33. Yang, Q., Liu, Y., Tang, J., Ku, T.: Residual and dense UNet for under-display camera restoration. In: Bartoli, A., Fusiello, A. (eds.) ECCV 2020. LNCS, vol. 12539, pp. 398–408. Springer, Cham (2020). https://doi.org/10.1007/978-3-030-68238-5_30
34. Zamir, S.W., Arora, A., Khan, S., Hayat, M., Khan, F.S., Yang, M.H.: Restormer: efficient transformer for high-resolution image restoration. In: Proceedings of the IEEE/CVF Conference on Computer Vision and Pattern Recognition, pp. 5728–5739 (2022)
35. Zamir, S.W., et al.: Multi-stage progressive image restoration. In: Proceedings of the IEEE/CVF Conference on Computer Vision and Pattern Rrecognition, pp. 14821–14831 (2021)
36. Zamir, S.W., et al.: Multi-stage progressive image restoration. In: CVPR (2021)

37. Zhang, H., Dai, Y., Li, H., Koniusz, P.: Deep stacked hierarchical multi-patch network for image deblurring. In: Proceedings of the IEEE/CVF Conference on Computer Vision and Pattern Recognition, pp. 5978–5986 (2019)
38. Zhang, K., Tao, D., Gao, X., Li, X., Li, J.: Coarse-to-fine learning for single-image super-resolution. IEEE Trans. Neural Netw. Learn. Syst. **28**(5), 1109–1122 (2016)
39. Zhang, R., Isola, P., Efros, A.A., Shechtman, E., Wang, O.: The unreasonable effectiveness of deep features as a perceptual metric. In: Proceedings of the IEEE Conference on Computer Vision and Pattern Recognition, pp. 586–595 (2018)
40. Zhang, Y., Li, K., Li, K., Wang, L., Zhong, B., Fu, Y.: Image super-resolution using very deep residual channel attention networks. In: Ferrari, V., Hebert, M., Sminchisescu, C., Weiss, Y. (eds.) ECCV 2018. LNCS, vol. 11211, pp. 294–310. Springer, Cham (2018). https://doi.org/10.1007/978-3-030-01234-2_18
41. Zhang, Y., Tian, Y., Kong, Y., Zhong, B., Fu, Y.: Residual dense network for image super-resolution. In: Proceedings of the IEEE Conference on Computer Vision and Pattern Recognition, pp. 2472–2481 (2018)
42. Zhang, Y., Tian, Y., Kong, Y., Zhong, B., Fu, Y.: Residual dense network for image restoration. IEEE Trans. Pattern Anal. Mach. Intell. **43**(7), 2480–2495 (2020)
43. Zhao, H., Kong, X., He, J., Qiao, Yu., Dong, C.: Efficient image super-resolution using pixel attention. In: Bartoli, A., Fusiello, A. (eds.) ECCV 2020. LNCS, vol. 12537, pp. 56–72. Springer, Cham (2020). https://doi.org/10.1007/978-3-030-67070-2_3
44. Zhou, Y., et al.: UDC 2020 challenge on image restoration of under-display camera: methods and results. In: Bartoli, A., Fusiello, A. (eds.) ECCV 2020. LNCS, vol. 12539, pp. 337–351. Springer, Cham (2020). https://doi.org/10.1007/978-3-030-68238-5_26
45. Zhou, Y., Ren, D., Emerton, N., Lim, S., Large, T.: Image restoration for under-display camera. In: Proceedings of the IEEE/CVF Conference on Computer Vision and Pattern Recognition, pp. 9179–9188 (2021)

Learning to Joint Remosaic and Denoise in Quad Bayer CFA via Universal Multi-scale Channel Attention Network

Xun Wu[1], Zhihao Fan[2], Jiesi Zheng[3], Yaqi Wu[4(✉)], and Feng Zhang[5]

[1] Tsinghua University, Beijing 100084, China
wuxun21@mails.tsinghua.edu.cn
[2] University of Shanghai for Science and Technology, Shanghai 200093, China
203590822@st.usst.edu.cn
[3] Zhejiang University, Hangzhou 310027, China
Jaszheng@zju.edu.cn
[4] Harbin Institute of Technology, Harbin 150001, China
wuyaqi930@foxmail.com
[5] Shanghai AI Laboratory, Shanghai, China
zhangfeng@pjlab.org.cn

Abstract. The color filter array widely used in smart phones is mainly Quad Bayer and Bayer. Quad Bayer color filter array (QBC) is a filter shared by four pixels, which can improve the image quality by averaging four pixels in the 2×2 neighborhood under low light conditions. From low-resolution Bayer to full-resolution Bayer has become a very challenging research, especially in the presence of noise. Considering denoise and remosaic, we propose a general two-stage framework JRD-QBC (Joint Remosaic and Denoise in Quad Bayer CFA), including denoise and remosaic. To begin with, for the denoise phase, in order to ensure the difference of each color channel recovery, we convert the input to hollow QBC, and then enter our backbone network, including source encoder module, feature refinement module and final prediction module. After that, get a clean QBC and then use the same network structure to remosaic to generate Bayer. Extensive experiments demonstrate the proposed two-stage method has a good effect in quantitative indicators and subjective vision.

Keywords: Quad Bayer CFA · Remosaic · Denoise · Raw · Channel attention

1 Introduction

As higher resolution and better nighttime imaging capability of smart camera is growing desired over the past few years, many smartphones like Galaxy S20 FE, Xiaomi Redmi Note8 Pro et al. [1] start to use Quad Bayer CFA (QBC)

X. Wu and Z. Fan—Both authors contributed equally to this research.

© The Author(s), under exclusive license to Springer Nature Switzerland AG 2023
L. Karlinsky et al. (Eds.): ECCV 2022 Workshops, LNCS 13805, pp. 147–160, 2023.
https://doi.org/10.1007/978-3-031-25072-9_10

recently. Different from Standard Bayer CFA, four pixels of the same color are grouped in each 2 × 2 Quad Bayer sensor cells. Sensors equipped with QBC use pixel binning under low light to achieve enhanced image quality by averaging four pixels within a 2 × 2 neighborhood. While Signal Noise Ratio (SNR) is improved in the binning mode, the spatial resolution is reduced unfortunately as a tradeoff. When it comes to full image resolution imaging, the performance degrades due to the non-uniformly type of sampling. To allow the output Bayer to have the same spatial resolution as the input QBC under normal lighting conditions while maintaining the nighttime imaging capability, the interpolation process (usually referred to as *Remosaic*, as shown in Fig. 1) is used to convert QBC to a Bayer pattern.

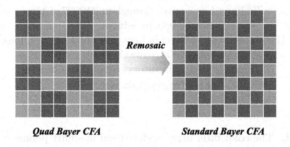

Quad Bayer CFA   Standard Bayer CFA

Fig. 1. Structures of Quad Bayer CFA and Standard Bayer CFA and Remosaic process illustration.

The remosaic problem becomes more challenging when the input QBC becomes noisy. A joint remosaic and denoise task is thus in demand for real world applications. Although QBC has been put to good use, related exploration is very rare, thus designing a method to solve Quad Joint Remosaic and Denoise problem is essential.

In this work, we aim at breaking the predicament of lacking related research and improving the model performance for joint remosaic and denoise task. By learning from Joint Demosaic and Denoise methods on Standard Bayer CFA, we propose a general two-stage framework solution for Quad Joint Remosaic and Denoise task named JQNet, which yielded remarkable result in Quad joint remosaic and denoise challenge held on ECCV'22. We especially design a strong backbone deep-learning networks for Quad Joint Remosaic and Denoise, which universally acts as a basic module in both remosaic and denoise stage. With channel attention block which has shown its efficiency in image enhancement tasks [25], we replace commonly used residual modules with MultiConv Channel Attention (MCCA) block [15]. Our architecture performs well in joint remosaic and denoise task by applying multi-scale feature extraction and selection on noisy QBC to produce the final clean Quad pattern image, and then generates the Bayer pattern through the remosaic module. Our backbone network makes the process of denoise and remosaic transparent and explicable. Our JQNet achieves great performance that makes less false color, noise and better color boundary. In summary, our main contribution are listed below.

- We firstly propose a basic universal module which can be easily appended to other Quad Joint Remosaic and Denoise method as a plug-in backbone to handle both remosaic and denoise task.
- We design a two stage deep learning solution JQNet targeted for joint remosaic and denoise task in Quad Bayer CFA.
- Remarkable results on a comprehensive set of quantitative and qualitative experiments verified the effectiveness of our proposed network structure with parameters reduce sharply.

2   Related Work

2.1   Color Filter Arrays

The design and performance characteristics of CFA mainly depend on the type of color system and the arrangement of color filters in CFA [18,23]. These two basic CFA characteristics specify the requirements for constructing a demosaic solution, which affects its efficiency and cost. The more commonly used CFA in recent years are Bayer and QBC.

Kodak [17] has developed the Bayer pattern which is a technique whereby instead of requiring each pixel to have its own individual three color planes (requiring 24 bits in total), an array of color filters is placed over the camera sensor and this limits the requirement for data bits per pixel to a single 8-bit byte. The Bayer pattern is repeated on the image in a fixed configuration to standardize this process. For Bayer pattern CFA arrays, because human vision is more sensitive than green, green channels are sampled more densely than red and blue channels. A common method is to reconstruct green channels first, and then use the gradient of green channels when rebuilding red and blue channels.

Quad Bayer color filter array (QBC) is four pixels sharing a filter, with an ability to treat a group of four pixels sharing a color filter block as a sensor or a separate sensor. If one of the four pixels on the Quad Bayer sensor captures the noise, only 25% of the information is lost, and the noise does not reduce the sharpness of the image.

2.2   Denoise Raw Images

Image denoising is a basic task in image processing and computer vision. Classical methods usually rely on prior information of images, such as non-local mean [5], sparse coding [3,10,19], BM3D [9], etc. With the development of convolutional neural networks, training end-to-end denoising networks has received considerable attention. Some recent advanced network architectures have produced a large number of cnn-based denoising methods that apply in RGB domain [7,14,16]. Since the shape and distribution of noise are quite different between RAW domain and RGB domain, performance degrades when directly apply these methods to RAW images.

Among the recently proposed datasets for image denoising in the public raw domain [2,4,6], there are also some convolutional neural networks [6,13,14] that have achieved promising results in these datasets. However, it is a tedious work to obtain the noise image and the real image pair. Therefore, synthesizing more realistic raw domain noise data has become a hot topic, including gauss-poisson distribution noise [11,22], gaussian mixture model [26], in-camera process simulation and so on. Foi et al. [22] proposed a method to construct Gauss-Poisson noise model and a network to achieve denoising.

3   Proposed Method

3.1   Problem Formulation

Quad joint Remosaic and Denoise problem is performed on single QBC Q_{noise} in different noisy level to recover corresponding noise-free Bayer B. Given a noise QBC Q_{noise}, the denoise framework can recover a clean QBC Q_{clean}. After that, a remosaic framework is applied to generate the final Bayer output B. The problem can be formulated in Eq. 1.

$$B = \Omega(f(Q_{noise}|\theta_f)|\theta_\Omega) \tag{1}$$

where $f(\cdot|\cdot)$ denotes a denoise framework and $\Omega(\cdot|\cdot)$ is remosaic framework. θ_f and θ_Ω denotes the learnable parameters in $f(\cdot|\cdot)$ and $\Omega(\cdot|\cdot)$, respectively.

3.2   Network Structure

In this section, we introduce the details of our proposed **JQNet** (**J**oint Remosaic and Denoise in **Q**uad Bayer CFA **Net**work).

The full backbone network is composed of three sub-modules: source encoder module, feature refinement module and final prediction module.

Source Encoder Module consists of two convolution layers with a relu in the middle, aims to reconstruct rough texture information I_{rough} from input $I_{QBC} \in \mathbb{R}^{H \times W \times C}$ as the input of feature refinement module:

$$I_{rough} = Conv(Relu(Conv(I_{QBC}))), \tag{2}$$

where $Conv$ denotes convolution and $Relu$ indicates relu activation function.

Feature Refinement Module takes I_{rough} as input to fully select, refine and enhance useful information I_{refine}. It is formed by sequential connection of MCCA blocks of different scales [15], which consist by multi-DoubleConv and channel attention operation(as shown in Fig. 2). MCCA can extract feature at different scales and then use channel attention to perform feature selection for the next stage. In details, we specify four types of MCCA (MCCA3, MCCA5, MCCA7, MCCA9) [15] with the reduction ratio set to 1, 2, 3, 4 respectively and apply numbers of MCCA modules with residual-connect technique to make up the feature refinement network in a particular sequence. The sequence of the operation is clearly marked in Fig. 2:

$$I_{refine} = F_{refine}([I_{rough}]), \qquad (3)$$

where F_{refine} indicates feature refinement module.

Final Prediction Module after getting a great representation I_{refine} of the raw image from the feature refinement network, final prediction module $F_{prediction}$ takes the I_{refine} as input through three convolutional layers and nested two relu to reconstruct the output I_o:

$$I_o = Conv_3(Relu(Conv_2(Relu(Conv_1(I_{refine})))), \qquad (4)$$

where the number of convolutional layers in $Conv_1$, $Conv_2$ and $Conv_3$ is progressively decreasing to the channel of the final output I_o.

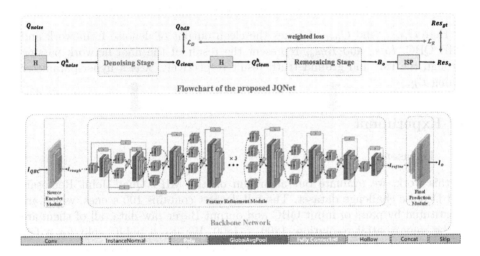

Fig. 2. The architecture of proposed JQNet for Joint Remosaic and Denoise in QBC problem. Our full algorithm is composed of two stage: Denoise stage and Remosaic stage. In the two-phase process, the same backbone network is used. ×3 denotes repeating the MCCA9 unit 3 times. Red arrows ↑ indicates calculated MAE loss, the final loss expression is: $\mathcal{L} = \mathcal{L}_D + \lambda \mathcal{L}_R$, more details are shown in 5.

3.3   Overall Framework

As discussed in Sect. 3.1, we propose a two-stage paradigm that reformulate this task into denoise first and then remosaic. By introducing two supervisory signals: clean QBC Q_{clean} for denoise stage and RGB images I_{gt} generated from clean Bayer B_{gt} for remosaic stage, our full framework is trained in a two-stage supervised manner.

In general, our proposed network consists of two stages: denoise stage and remosaic stage. For each stage, we use the same backbone (JQNet) mentioned above. To be specific, for denoise stage, hollow operation (short for "H") denotes separating channels (more details are discussed in Sect. 4.3). For the first stage, we apply hollow operation to noise QBC Q_{noise} to get Q_{noise}^h as the input of denoise stage, aiming at getting clean QBC Q_{clean}. For the second stage, by applying hollow operation to Q_{clean} as input, we get the recovered Bayer B_o from remosaic framework.

3.4 Loss Function

In many image reconstruction and enhancement tasks, the mean absolute error (MAE) loss is widely used. In this work, our final loss function can be defined by combine the MAE loss in denoise stage and remosaic stage:

$$\mathcal{L} = \mathcal{L}_D + \lambda \mathcal{L}_R = \|Q_{clean} - Q_{0dB}\|_1 + \lambda \|Res_o - Res_{gt}\|_1 \tag{5}$$

Here Q_{clean} and Q_{0dB} denotes the clean output of denoise framework and 0 dB QBC. Res_o and Res_{gt} represent the result of the final network passing through the ISP and ground truth result respectively. λ is a hyper-parameter tuning $\mathcal{L}_R$.

4   Experiment

4.1   Datasets

In this work, we evaluate our algorithm on the MIPI Quad Joint Remosaic and Denoise challenge dataset. The full dataset contains 100 scenes, which are constituted by pairs of input QBC and output Bayer raw data, all of them are of the same spatial resolution: 1200×1800. We use a public split for a fair comparison of 70, 15, and 15 scenes pairs for training, validation, and testing, respectively. For each scene in the training, validation and testing split, there are three noise levels: 0 dB, 24 dB and 42 dB. Some examples in this datasets are shown in Fig. 3.

4.2   Evaluation Metrics

The evaluation consists of two parts: the comparison of the recovered bayer with the ground truth bayer and comparison of the RGB results generated from bayer results mentioned above by a simple ISP. As for the former, we use KL divergence (KLD) as the evaluation metric, and for the latter, we adopt PSNR, SSIM and Learned perceptual image patch similarity (LIPIS) [24] for evaluation. In order to measure the effectiveness of the algorithm and the overall image quality intuitively, we following the setting in MIPI challenge by using $M4$ score as the overall evaluation metric:

$$M4 = PSNR \times SSIM \times 2^{1-LPIPS-KLD} \qquad (6)$$

The M4 score is between 0 and 100, and the higher the score, the better the overall image quality.

Fig. 3. Visualization of MIPI 2022 Challenge on Quad Joint Remosaic and Denoise dataset. This dataset contains multiple scenes (i.e., natural scene, dark scene).

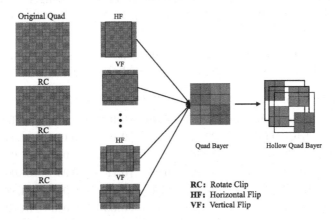

Fig. 4. Details of data augmentation and preprocessing. At first flipping and rotation is apply to the input original QBC, then clipping is adopted to keep the augmented data in the same spatial form (RGGB pattern). At last, hollow operation is applied to split different color channels.

4.3   Implementation Details

Data Augmentation because the one-to-one correspondence between the spatial positions of QBC and Bayer, some data augmentation methods such as flipping and rotation can not be directly performed. However, as shown in Eq. 5, our monitor information for optimizing two stage models are computed in the domain of QBC-to-QBC and QBC-to-RGB, respectively. So these data augmentation methods can be introduced into our data processing pipeline. As shown in Fig. 4, we adopt flipping and rotation directly in QBC-to-RGB domain.

Data Preprocessing at first, to provide pixel location information to enhance the generalization. Considering hollow operation can make network learning more targeted for color channel recovery, so we sample the input QBC into hollow QBC by processing the pixels in the same channel into one channel. After that, original QBC $Q \in \mathbb{R}^{H \times W}$ is processed into hollow QBC $Q_h \in \mathbb{R}^{H \times W \times 3}$ (shown in Fig. 4). Then we sample the full spatial hollow QBC Q_h (1200 × 1800) into some 64 × 64 patches as final input.

Training details for all experiments, our model is implemented in pytorch and runs on 4 Nvidia Titan A100 graphical processing units (GPU). The model is optimized in an Adam optimizer as $\beta_1 = 0.9$, $\beta_2 = 0.99$, learning rate $= 1e - 4$ with a batch size of 56. We set the number of channels in MCCA blocks as 64. To keep two sub-losses in Eq. 5 balance, we set $\lambda = 1.0$.

4.4   Testing Results of MIPI 2022 Challenge on Quad Joint Remosaic and Denoise

The proposed algorithm ranked 5th in *MIPI 2022 Challenge on Quad Joint Remosaic and Denoise*. The final comparison results on testing set are summarized in Table 1. We can observe that our method achieves great performance.

Table 1. The final testing results of MIPI 2022 Challenge on Quad Joint Remosaic and Denoise. The best value is in bold and our results are highlighted in gray.

Rank	Team	PSNR↑	SSIM↑	LPIPS↓	KLD↓	M4↑
1	**op-summer-po**	**37.93**	**0.965**	0.104	0.019	**68.03**
2	JHC-SJTU	37.64	0.96	0.1	**0.0068**	67.99
3	IMEC-IPI	37.76	0.96	0.1	0.014	67.95
4	BITSpectral	37.20	0.96	0.11	0.03	66.00
5	HITZST01	37.20	0.96	0.11	0.06	64.82
6	MegNR	36.08	0.95	**0.095**	0.023	64.10

4.5 Ablation Study

In this section, we perform ablation studies to verify the effectiveness of our proposed two-stage solution, and then, we perform ablation studies by replacing our designed JQNet with other backbones to verify the generality and effectiveness of our proposed JQNet on these two tasks: denoise task and remosaic task. Because of lacking the ground truth of test set, all results reported in this section are computed in valid set.

Effectiveness of Proposed Two-Stage Solution

In this paper, we introduce a two-stage solution for Quad Joint Remosaic and Denoise problem, which consists of denoising stage and remosaic stage. We empirically validate the effectiveness of two-stage solution by adopting one-stage solution for this problem. That is, taking noise QBC Q_{noise}^h as input and directly output clean Bayer B_o. The expression of loss for one-stage solution is:

$$\mathcal{L} = \mathcal{L}_R = \|I_o - I_{gt}\|_1 \tag{7}$$

Table 2. Ablation study of proposed two-stage solution. σ denotes the noise level. The best value is in bold.

Method	σ	PSNR↑	SSIM↑	LPIPS↓	KLD↓	M4↑
Residual[1] [12]	24	34.65	0.9033	0.22044	0.00049	53.72
KPN[1] [20]		34.92	0.9132	0.21034	0.00051	55.11
Unet[1] [21]		34.93	0.9125	0.21213	0.00053	55.02
JQNet[1]		35.06	0.9153	0.21258	0.00043	55.38
JQNet[2]		**35.71**	**0.9174**	**0.20772**	0.00047	**56.71**
Residual[1] [12]	42	29.29	0.8021	0.38641	0.00175	35.90
KPN[1] [20]		30.03	0.8460	0.33287	0.00164	40.29
Unet[1] [21]		30.23	0.8450	0.33605	0.00160	40.43
JQNet[1]		30.07	**0.8624**	0.33598	0.00147	41.05
JQNet[2]		**31.10**	0.8597	**0.32312**	**0.00144**	**42.70**
Residual[1] [12]	Average	31.97	0.8528	0.30343	0.00112	44.15
KPN[1] [20]		32.48	0.8796	0.27161	0.00108	47.29
Unet[1] [21]		32.58	0.8788	0.27409	0.00107	47.32
JQNet[1]		32.57	**0.8889**	0.27430	**0.00095**	47.85
JQNet[2]		**33.41**	0.8885	**0.26542**	0.00096	**49.35**

To simplify the presentation, we use Residual[1], KPN[1], Unet[1], JQNet[1] represent one-stage solution mentioned above with different backbone (i.e., ResNet [12], KPN [20], Unet [21]), and JQNet[2] denotes our proposed two-stage solution. We only report results of 24dB and 42dB because there is no noise in 0dB. As shown in Table 2, our two-stage solution brings a large improvement to the one-stage solution. The improvement in all indicators validate that our two-stage solution can break the gap between different domains by introducing more monitoring information.

Effectiveness and Generalization of proposed JQNet

1) *Denoise.* we empirically validate the effectiveness for denoise task by replacing JQNet in denoise stage with traditional denoise algorithm (i.e., BM3D [8]) and some existing strong vision backbones (i.e., ResNet [12], Unet [21], KPN [20]). BM3D-JQNet, ResNet-JQNet, Unet-JQNet and KPN-JQNet are short for denoising backbone replacement modes of mentioned above, respectively. JQNet^2 denotes our proposed framework in this paper, which consist of two adjacent JQNet. The results shown in Table 3 verifies the effectiveness of our JQNet on denoise task.

Table 3. Ablation study to validate the effectiveness of JQNet for denoise task. σ denotes the noise level. The best value is in bold.

Method	σ	PSNR↑	SSIM↑	LPIPS↓	KLD↓	M4↑
BM3D-JQNet	24	32.08	0.8703	0.28730	0.00109	45.73
ResNet-JQNet		35.38	0.9131	0.21059	0.00048	55.82
Unet-JQNet		34.78	0.9075	0.20985	0.00051	54.56
KPN-JQNet		35.23	0.9032	0.21007	0.00053	55.00
JQNet^2		**35.71**	**0.9174**	**0.20772**	**0.00047**	**56.71**
BM3D-JQNet	42	28.45	0.7955	0.42848	0.00198	33.59
ResNet-JQNet		31.02	0.8575	**0.32213**	**0.00142**	42.51
Unet-JQNet		30.28	0.8490	0.32792	0.00152	40.91
KPN-JQNet		30.97	0.8473	0.32305	0.00156	41.90
JQNet^2		**31.10**	**0.8597**	0.32312	0.00144	**42.70**
BM3D-JQNet	Average	30.27	0.8329	0.35790	0.00154	39.30
ResNet-JQNet		33.20	0.8853	0.26636	**0.00095**	48.84
Unet-JQNet		32.53	0.8782	0.26889	0.00102	47.39
KPN-JQNet		33.10	0.8753	0.26656	0.00105	48.13
JQNet^2		**33.41**	**0.8885**	**0.26542**	0.00096	**49.35**

2) **Remosaic.** To validate the effectiveness for remosaic task, we replace JQNet in remosaic stage with some strong vision backbones (i.e., ResNet [12], Unet [21], KPN [20]). JQNet-ResNet, JQNet-Unet and JQNet-KPN are short for remosaic backbone replacement modes of mentioned above. As shown in Table 4, JQNet can outperform most of the comparisons.

Extensive ablation experimental results shown above validate the proposed JQNet can be applied to these two tasks without any modification, and greatly improve the performance in contrast to baseline backbones.

4.6   Qualitative Evaluation

In this section, we qualitatively illustrate the effectiveness of our proposed two-stage solution and specific designed JQNet. To simplify the comparison, we use the best model besides our $JQNet^2$ mentioned in Sect. 4.5, that is, JQNet-ResNet, ResNet-JQNet and $JQNet^1$. As shown in Fig. 5, our $JQNet^2$ produces cleaner results and avoids lost of color boundary artifacts compared to these baseline methods.

Table 4. Ablation study to validate the effectiveness of JQNet for remosaic task. σ denotes the noise level. The best value is in bold.

Method	σ	PSNR↑	SSIM↑	LPIPS↓	KLD↓	M4↑
ResNet	0	40.07	0.9652	0.05094	**0.00013**	74.67
Unet		40.11	0.9640	0.05913	0.00014	74.19
KPN		40.10	0.9641	0.05122	**0.00013**	74.62
JQNet		**40.13**	**0.9682**	**0.05011**	0.00014	**75.06**
JQNet-ResNet	24	35.57	0.9155	0.21004	**0.00044**	56.28
JQNet-Unet		35.07	0.9127	0.21065	0.00049	55.30
JQNet-KPN		34.84	0.9105	0.20993	0.00052	54.83
$JQNet^2$		**35.71**	**0.9174**	**0.20772**	0.00047	**56.71**
JQNet-ResNet	42	30.88	0.8558	0.32226	0.00143	42.23
JQNet-Unet		29.62	0.8422	0.33741	0.00177	39.44
JQNet-KPN		31.04	0.8588	**0.31934**	**0.00139**	42.68
$JQNet^2$		**31.10**	**0.8597**	0.32312	0.00144	**42.70**
JQNet-ResNet	Average	35.51	0.9122	**0.19441**	**0.00067**	56.68
JQNet-Unet		34.93	0.9063	0.20240	0.00080	54.99
JQNet-KPN		35.32	0.9111	0.19350	0.00068	56.26
$JQNet^2$		**35.65**	**0.9151**	0.19366	0.00068	**57.02**

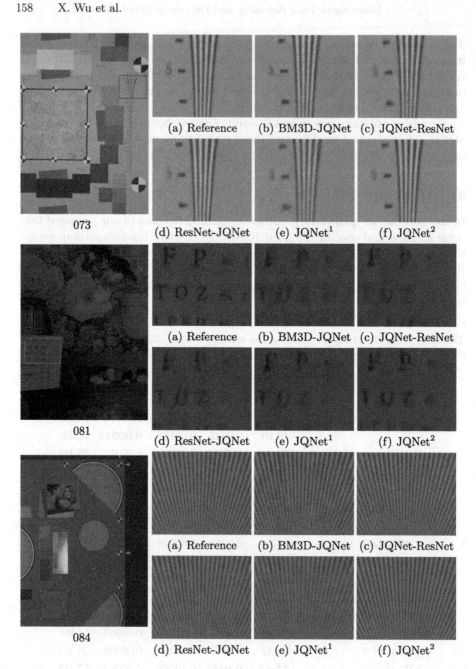

Fig. 5. Qualitative comparison of JDR-QBC results for noise level $\sigma = 42$, our result has better perceptual results with less noise and false color.

5   Conclusion

In this paper, we propose a JRD-QBC network architecture for remosaic and denoise based on Quad Bayer color filter array. We split the original problem into two subtasks, remosaic and denoise. In both subtasks, we use the same network backbone and experiment by using different network backbone structures, the results of which surface the versatility and effectiveness of the network backbone we use.

References

1. A Sharif, S., Naqvi, R.A., Biswas, M.: Beyond joint demosaicking and denoising: an image processing pipeline for a pixel-bin image sensor. In: Proceedings of the IEEE/CVF Conference on Computer Vision and Pattern Recognition, pp. 233–242 (2021)
2. Abdelhamed, A., Lin, S., Brown, M.S.: A high-quality denoising dataset for smartphone cameras. In: Proceedings of the IEEE Conference on Computer Vision and Pattern Recognition, pp. 1692–1700 (2018)
3. Aharon, M., Elad, M., Bruckstein, A.: K-SVD: an algorithm for designing overcomplete dictionaries for sparse representation. IEEE Trans. Signal Process. **54**(11), 4311–4322 (2006)
4. Anaya, J., Barbu, A.: Renoir-a dataset for real low-light image noise reduction. J. Vis. Commun. Image Represent. **51**, 144–154 (2018)
5. Buades, A., Coll, B., Morel, J.M.: A non-local algorithm for image denoising. In: 2005 IEEE Computer Society Conference on Computer Vision and Pattern Recognition (CVPR 2005), vol. 2, pp. 60–65. IEEE (2005)
6. Chen, C., Chen, Q., Xu, J., Koltun, V.: Learning to see in the dark. In: Proceedings of the IEEE Conference on Computer Vision and Pattern Recognition, pp. 3291–3300 (2018)
7. Chen, Y., Pock, T.: Trainable nonlinear reaction diffusion: a flexible framework for fast and effective image restoration. IEEE Trans. Pattern Anal. Mach. Intell. **39**(6), 1256–1272 (2016)
8. Dabov, K., Foi, A., Katkovnik, V., Egiazarian, K.: Image denoising by sparse 3-D transform-domain collaborative filtering. IEEE Trans. Image Process. **16**(8), 2080–2095 (2007)
9. Dabov, K., Foi, A., Katkovnik, V., Egiazarian, K.: Image restoration by sparse 3D transform-domain collaborative filtering. In: Image Processing: Algorithms and Systems VI, vol. 6812, pp. 62–73. SPIE (2008)
10. Elad, M., Aharon, M.: Image denoising via sparse and redundant representations over learned dictionaries. IEEE Trans. Image Process. **15**(12), 3736–3745 (2006)
11. Foi, A., Trimeche, M., Katkovnik, V., Egiazarian, K.: Practical poissonian-gaussian noise modeling and fitting for single-image raw-data. IEEE Trans. Image Process. **17**(10), 1737–1754 (2008)
12. He, K., Zhang, X., Ren, S., Sun, J.: Deep residual learning for image recognition. In: Proceedings of the IEEE Conference on Computer Vision and Pattern Recognition, pp. 770–778 (2016)
13. Hirakawa, K., Parks, T.W.: Joint demosaicing and denoising. IEEE Trans. Image Process. **15**(8), 2146–2157 (2006)

14. Jain, V., Seung, S.: Natural image denoising with convolutional networks. In: Advances in Neural Information Processing Systems, vol. 21 (2008)
15. Kim, B.-H., Song, J., Ye, J.C., Baek, J.H.: PyNET-CA: enhanced PyNET with channel attention for end-to-end mobile image signal processing. In: Bartoli, A., Fusiello, A. (eds.) ECCV 2020. LNCS, vol. 12537, pp. 202–212. Springer, Cham (2020). https://doi.org/10.1007/978-3-030-67070-2_12
16. Lehtinen, J., et al.: Noise2noise: learning image restoration without clean data. arXiv preprint arXiv:1803.04189 (2018)
17. Lukac, R., Plataniotis, K.N.: Color filter arrays: design and performance analysis. IEEE Trans. Consum. Electron. **51**(4), 1260–1267 (2005)
18. Lukac, R., Plataniotis, K.N.: Universal demosaicking for imaging pipelines with an RGB color filter array. Pattern Recogn. **38**(11), 2208–2212 (2005)
19. Mairal, J., Bach, F., Ponce, J., Sapiro, G., Zisserman, A.: Non-local sparse models for image restoration. In: 2009 IEEE 12th International Conference on Computer Vision, pp. 2272–2279. IEEE (2009)
20. Mildenhall, B., Barron, J.T., Chen, J., Sharlet, D., Ng, R., Carroll, R.: Burst denoising with kernel prediction networks. In: Proceedings of the IEEE Conference on Computer Vision and Pattern Recognition, pp. 2502–2510 (2018)
21. Ronneberger, O., Fischer, P., Brox, T.: U-Net: convolutional networks for biomedical image segmentation. In: Navab, N., Hornegger, J., Wells, W.M., Frangi, A.F. (eds.) MICCAI 2015. LNCS, vol. 9351, pp. 234–241. Springer, Cham (2015). https://doi.org/10.1007/978-3-319-24574-4_28
22. Wang, Y., Huang, H., Xu, Q., Liu, J., Liu, Y., Wang, J.: Practical deep raw image denoising on mobile devices. In: Vedaldi, A., Bischof, H., Brox, T., Frahm, J.-M. (eds.) ECCV 2020. LNCS, vol. 12351, pp. 1–16. Springer, Cham (2020). https://doi.org/10.1007/978-3-030-58539-6_1
23. Wilson, P.: Bayer pattern. https://www.sciencedirect.com/topics/engineering/bayer-pattern
24. Zhang, R., Isola, P., Efros, A.A., Shechtman, E., Wang, O.: The unreasonable effectiveness of deep features as a perceptual metric. In: Proceedings of the IEEE Conference on Computer Vision and Pattern Recognition, pp. 586–595 (2018)
25. Zhang, Y., Li, K., Li, K., Wang, L., Zhong, B., Fu, Y.: Image super-resolution using very deep residual channel attention networks. In: Proceedings of the European Conference on Computer Vision (ECCV), pp. 286–301 (2018)
26. Zhu, F., Chen, G., Heng, P.A.: From noise modeling to blind image denoising. In: Proceedings of the IEEE Conference on Computer Vision and Pattern Recognition, pp. 420–429 (2016)

Learning an Efficient Multimodal Depth Completion Model

Dewang Hou[1], Yuanyuan Du[2], Kai Zhao[3], and Yang Zhao[4(✉)]

[1] Peking University, Beijing, China
dewh@pku.edu.cn
[2] Chongqing University, Chongqing, China
[3] Tsinghua University, Beijing, China
zhaok18@tsinghua.org.cn
[4] Hefei University of Technology, Hefei, China
yzhao@hfut.edu.cn

Abstract. With the wide application of sparse ToF sensors in mobile devices, RGB image-guided sparse depth completion has attracted extensive attention recently, but still faces some problems. First, the fusion of multimodal information requires more network modules to process different modalities. But the application scenarios of sparse ToF measurements usually demand lightweight structure and low computational cost. Second, fusing sparse and noisy depth data with dense pixel-wise RGB data may introduce artifacts. In this paper, a light but efficient depth completion network is proposed, which consists of a two-branch global and local depth prediction module and a funnel convolutional spatial propagation network. The two-branch structure extracts and fuses cross-modal features with lightweight backbones. The improved spatial propagation module can refine the completed depth map gradually. Furthermore, corrected gradient loss is presented for the depth completion problem. Experimental results demonstrate the proposed method can outperform some state-of-the-art methods with a lightweight architecture. The proposed method also wins the championship in the MIPI2022 RGB+TOF depth completion challenge.

Keywords: Depth completion · Sparse ToF · RGB guidance

1 Introduction

Depth information plays an important role in various vision applications, such as autonomous driving, robotics, augmented reality, and 3D mapping. In past decades, many depth sensors have been developed and applied to obtain depth information, such as time-of-flight (ToF) and light detection and ranging (LiDAR) sensors. With the rapid development of smart mobile devices, sparse ToF depth measurements have attracted extensive attention recently due to their unique advantages. For example, sparse ToF sensors can effectively avoid multipath interference problem that usually appears in full-field ToF depth measurements. In addition, the power consumption of sparse ToF depth measurement is

L. Karlinsky et al. (Eds.): ECCV 2022 Workshops, LNCS 13805, pp. 161–174, 2023.
https://doi.org/10.1007/978-3-031-25072-9_11

much lower than that of full-field ToF sensors, which is very suitable for mobile devices and edge devices to reduce battery consumption and heating. However, there is a proverb that says you can't have your cake and eat it too. The sparse ToF depth measurement also has significant disadvantages. First, sparse ToF cannot provide a dense pixel-wise depth map due to hardware limitations such as the small amount of laser pulse. Second, sparse ToF depth data easily suffers from noise. To overcome these limitations, depth completion (DC), which converts a sparse depth measurement to a dense pixel-wise depth map, becomes very useful due to industrial demands.

Single modal sparse depth completion is a quite difficult and ill-posed problem because dense pixel-wise depth values are severely missing. Fortunately, aligned RGB images usually can be obtained in these practical scenarios. Therefore, recent depth completion methods often utilize RGB guidance to boost depth completion and show significantly better performance than the unguided way. The straightforward reason is that natural images provide plentiful semantic cues and pixel-wise details, which are critical for filling missing values in sparse depth modality.

For traditional full-field ToF depth maps, many RGB image guided depth super-resolution or enhancement methods have been proposed, e.g., traditional learning-based joint enhancement models [17,40], and recent deep neural network (DNN)-based joint depth image restoration methods [4,25,27,33,41]. For the sparse ToF depth measurements, Ma et al. [25] proposed a single depth regression network to estimate dense scene depth by means of RGB guidance. Tang et al. [33] presented a learnable guided convolution module and applied it to fuse cross-modality information between sparse depth and corresponding image contents.

These RGB-guided depth completion methods can produce better depth maps by comparing them with single modal depth completion. However, current depth completion methods with RGB guidance still face the following two challenges. First, multi-modality usually leads to more computational cost, because different network modules are required for these modalities with different distributions. Note that sparse depth sensors are usually applied in mobile devices or other edge devices with finite computing power and real-time requirement. As a result, the depth completion models not only need to extract, map and fuse multimodal information effectively but also need to meet lightweight architecture and real-time demand. Second, different to full-field depth maps with pixel-wise details, the sparse depth data may introduce extra artifacts during the fusion process. Moreover, the noise contained in sparse depth data may further intensify the negative effects.

In this paper, we proposed a light but efficient multimodal depth completion network based on the following three aspects: fusing multi-modality data more effectively; reducing the negative effects of missing values regions in sparse modality; recovering finer geometric structures for both objective metrics and subjective quality. Overall, the contributions are summarized as follows: **1)** We present a lightweight baseline, which can achieve state-of-the-art results while being computationally efficient. More specifically, the two-branch framework is first used to estimate global and local predictions. Features are exchanged between these two branches to fully make use of cross-modal information. A fusion module with relative confidence maps is presented after two branches, which can further optimize the fused results and accelerate convergence. **2)** we propose a funnel convolutional spatial propagation network, which can gradually improve the depth features with reweighting strategy. **3)** In addition, corrected gradient loss is designed for depth completion problems correspondingly. Experimental results show that the proposed method can achieve superior performance, but also run very fast in the inference stage.

The remainder of this paper is organized as follows. Related works are briefly reviewed in Sect. 2. Section 3 introduces the proposed method in detail. Experimental results are analyzed in Sect. 4, and Sect. 5 concludes the paper.

2 Related Work

2.1 Unguided Depth Completion

Unguided DC methods tend to estimate dense depth map from a sparse depth map directly. Uhrig et al. [34] first applied a sparsity invariant convolutional neural network (CNN) for DC task. Thereafter, many DC networks have been proposed by using the strong learning capability of CNNs [7,8]. Moreover, some unguided DC methods [23,24] use RGB images for auxiliary training, but still perform unguided DC in the inference stage. Unguided methods don't rely on corresponding RGB images, but their performance is naturally worse than RGB-guided DC methods [13]. Because RGB guidance can provide more semantic information and pixel-wise local details. This paper focuses on sparser ToF depth data in real-world applications, and thus mainly discusses the related RGB-guided DC methods.

2.2 RGB-Guided Depth Completion

Currently, different types of guided DC methods have been proposed to employ corresponding RGB images to improve completed depth maps. These RGB-guided DC methods can be roughly divided into four categories, i.e., early fusion model, late fusion model, explicit 3D representation model, and spatial propagation model.

Early fusion methods [6,12,15,25,31,37] directly concatenate depth map and RGB image or their shallow features before feeding them into the main body of networks. These methods usually adopt encoder-decoder structure [25,31] or two-stage coarse to refinement prediction [6,12]. Note that the coarse-to-refinement structure is also frequently used in subsequent residual depth map prediction models [11,19]. Compared to early fusion, late fusion model uses two sub-network to extract features from RGB image and input depth data, and then fuses cross-modal information in the immediate or late stage. Since two modalities are used, this type of method usually adopts dual-branch architecture, such as dual-encoder with one decoder [9,16,43], dual-encoder-decoder [30,33, 39], and global and local depth prediction [18,35]. In addition, there are some 3D representation models learn 3D representations to guide depth completion, such as 3D-aware convolution [2,42] and surface normal representation method [28].

Spatial propagation network (SPN) [21] is also a typical model widely used in DC tasks by treating DC as a depth regression problem. Cheng et al. [5] first applied SPN to the DC task by proposing a convolutional SPN (CSPN). Then, they presented CSPN++ [3] by means of context-aware and resource-aware CSPN. To take advantage of both SPN and late fusion models, Hu et al. [14] further proposed a precise and efficient DC network (PENet). Subsequently, deformable SPN (DSPN) [38] and attention-based dynamic SPN (DySPN) [20] were also applied for RGB-guided DC problem, and achieved impressive performance on KITTI depth completion benchmark [34].

3   Methods

In this paper, we thoroughly study the key components of depth completion methods to derive an Efficient Multimodal Depth Completion network (EMDC), shown in Fig. 1, which tends to solve the above mentioned problems. In the following, we introduce each module of EMDC in detail.

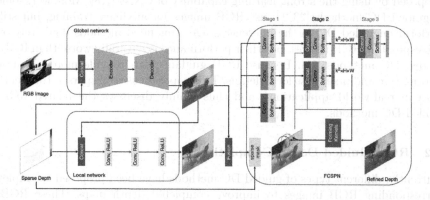

Fig. 1. Network architecture of the proposed EMDC. We introduce a pipeline that consists of two-branch global and local network, fusion module and FCSPN. As an illustrative example, we show the second stage in which FCSPN generates two convolution kernels.

3.1   Global and Local Depth Prediction with Fusion

Motivated by [13,18,35], a global and local depth prediction (GLDP) framework is adopted to fuse the information from two modalities efficiently. As illustrated in Fig. 1, different architectures are designed for the global and local depth estimation sub-networks, respectively. For the global branch, which needs to estimate depth data with the help of RGB image, a typical U-Net structure with large respective fields is used. In addition, pretrained MobileNetV2 [29] is adopted as the encoder in the global branch to obtain strong learning capability with lightweight structures. For the local branch that processes simple depth prediction and homogeneous sparse depth, we merely design a lightweight CNN module to reproduce depth map. Finally, the output depth maps of these two branches are fused via a relative fusion module. For the confidence maps used by the fusion module, we think they should reflect the relative certainty of global and local depth predictions, the so-called relative confidence maps. As shown in Fig. 2, we adjust the pathways of the features to involve this idea directly. Furthermore, we improve the relative fusion module by zero-initializing the relative confidence maps of local depth prediction with the zero-initialization method [1], which can achieve faster convergence and better test performance in our experiments. Related experimental results can be found from the ablation results in Table 2.

To reduce the negative effects introduced by the distribution characteristics of sparse modality, we use pixel-shuffle (PS) [32] to replace the traditional upsampling layer in the global branch. Compared to normal upsampling, PS operation doesn't magnify the non-ideal missing information. In addition, the batch normalization layers in the local branch are all removed, as they are fragile to features with anisotropic spatial distribution and slightly degrade the results.

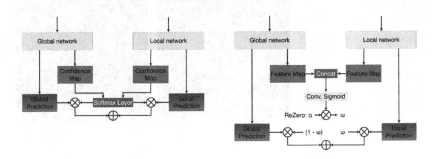

Fig. 2. Left: fusion module proposed in [35], Right: our proposed fusion module.

3.2 Funnel Convolutional Spatial Propagation Network

The most important target of the DC task is to recover better scene structures for both objective metrics and subjective quality. Therefore, to alleviate the problem of recovering scene structures, we propose a Funnel Convolutional Spatial Propagation Network (FCSPN) for depth refinement. As shown in Fig. 1, the FCSPN inherits the basic structure of the spatial propagation network (SPN) [3,4,14,20, 27], which is a popular refinement technique in the DC field. FCSPN can fuse the point-wise results from large to small dilated convolutions in each stage. The maximum dilation of the convolution at each stage is reduced gradually, thus forming a funnel-like structure stage by stage. Compared with CSPN++ [3], the FCSPN generates a new set of kernels, termed reweighting, at each stage to increase the representation capability of SPNs. The FCSPN embraces dynamic filters with kernel reweighting which has a more substantial adaptive capability than adjusting them via attention mechanism [20]. Moreover, the kernel reweighting mechanism merely relies on the kernels in the previous stage and the current depth map in the refinement process, thus reducing the computation complexity considerably. Experimental results show that FCSPN not only has superior performance, but also runs fast in inference, as shown in Table 3.

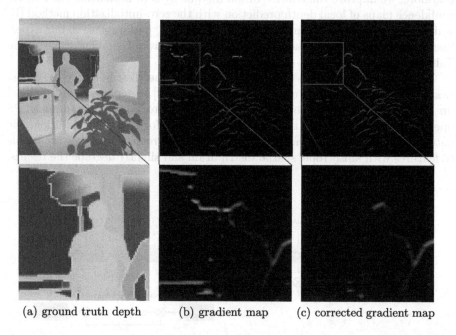

(a) ground truth depth   (b) gradient map   (c) corrected gradient map

Fig. 3. Gradient maps obtained by filtering a depth map. And for simplicity, we only show normalized y-gradient from Sobel-Feldman operator in the vertical direction. And (b) is a naive gradient map and (c) is produced by our method. (Color figure online)

3.3 Loss Functions

Sophisticated depth completion algorithms often focused on minimizing pixel-wise errors which typically leads to over-smoothed results [36]. Degradation of structural information would impede their practical applications, such as safety issues in autonomous driving. To handle this problem, we proposed a corrected gradient loss for the depth completion task. Gradient loss [26] was originally proposed to preserve the structure of natural images. Intuitively, applying it for depth map regression problems may also help to refine the blurry predictions obtained from MAE or MSE losses. However, it is difficult to obtain a complete precisely annotated depth dataset. One of the problems brought about by this is that the gradients, such as those extracted with the Sobel filtering, between depth points that are unreachable by depth measurement and other depth points are inaccurate. We elaborate on this phenomenon in Fig. 3. The dark red regions in the distance in Fig. 3 are actually unreachable by depth measurement. It can be clearly seen that the "annotated" depth of these regions is all set to zero, so the gradient between other areas and them is also unreliable, and even reversed. To handle this problem, we propose the corrected gradient loss. We formulate it as follows:

$$\mathcal{L}_{cgdl}(\hat{Y}, Y) = \left\| F(\hat{Y}) - F(Y) \right\|_p * E(sgn(Y \in \nu)), \tag{1}$$

where $F(\cdot)$ denotes a specific gradient operator or set of gradient operators. ν denotes the valid value range for depth measurement. $sgn(\cdot)$ denotes signum function that extracts the sign of its inputs. $E(\cdot)$ denotes the erosion operation, which is a morphology-based image processing method. $E(sgn(Y \in \nu)$ can also be understood as playing the role of online data cleaning. The loss $\mathcal{L}_{cgdl}$ can be easily introduced to the existing DC models and the final loss function in our work is:

$$\mathcal{L}(\hat{Y}, Y) \quad = \mathcal{L}_{l1}(\hat{Y}, Y) + \lambda_1 * \mathcal{L}_{l1}(\hat{Y}_{global}, Y) \tag{2}$$

$$+ \lambda_2 * \mathcal{L}_{l1}(\hat{Y}_{local}, Y) + \lambda_3 * \mathcal{L}_{cgdl}(\hat{Y}, Y), \tag{3}$$

where

$$\lambda_1 = \mathcal{L}_{l1}(\hat{Y}, Y) / \mathcal{L}_{l1}(\hat{Y}_{global}, Y), \tag{4}$$

$$\lambda_2 = \mathcal{L}_{l1}(\hat{Y}, Y) / \mathcal{L}_{l1}(\hat{Y}_{local}, Y), \tag{5}$$

$$\lambda_3 = 0.7 * \mathcal{L}_{l1}(\hat{Y}, Y) / \mathcal{L}_{cgdl}(\hat{Y}, Y) \tag{6}$$

We also utilize an adaptive loss weight adjustment strategy to search for optimum fusion in the training process. Its implementation can be seen from Eqs. 4 and Eqs. 5. The losses calculated on the results of the two-branch network are strictly constructed to be the same value. Because these losses should be optimized synchronously rather than alternately to match the fusion strategy in the GLDP framework and avoid mode collapse caused by model mismatch [10].

4   Experiments

4.1   Datasets and Metrics

Before Mobile Intelligent Photography and Imaging (MIPI) challenge, there lacks a high-quality sparse depth completion dataset, which hindered related research for both industry and academia. To this end, the RGBD Challenge of MIPI provided a high-quality synthetic DC dataset (MIPI dataset), which contains 20,000 pairs of pre-aligned RGB and ground truth depth images of 7 indoor scenes. For each scene, the RGB and the ground-truth depth are rendered along a smooth trajectory in a created 3D virtual environment.

For testing, the testing data in MIPI dataset comes from mixed sources, including synthetic data, spot-iToF, and some samples subsampled from dToF in iPhone. These mixed testing data can verify the robustness and generalization ability of different DC methods in real-world scenarios.

Four metrics are employed to evaluate the performance of depth completion algorithms, i.e., Relative Mean Absolute Error (RMAE), Edge Weighted Mean Absolute Error (EWMAE), Relative Depth Shift (RDS), and Relative Temporal Standard Deviation (RTSD). Details of these metrics can be found in the technical reports of the MIPI RGBD challenge. The final score combines these four metrics as follows:

$$score = 1 - 1.8 \times RMAE - 0.6 \times EWMAE - 3 \times RDS - 4.6 \times RTSD \quad (7)$$

4.2 Implementational Details

In training stage, we use the AdamW optimizer [22] with $\beta_1 = 0.9$, $\beta_2 = 0.999$. We set the initial learning rate as 0.001, and use a learning scheduler of cosine annealing with a warm start. After a linear increase in the first 10 epochs, the learning rate gradually decreases along the cosine annealing curve after each mini-batch within one period. The proposed EMDC is trained with a mini-batch size of 10 in a total of 150 epochs. The training samples are further augmented via random flip and color jitter. In addition, we use the exponential moving average (EMA) technique to obtain a more stable model. Our algorithm is implemented in PyTorch. A full training of our network takes about 12 h on an NVIDIA Tesla A6000 GPU. The source codes are available along with the pretrained models to facilitate reproduction.[1]

4.3 Experimental Results

Table 1 lists the experimental results of different DC methods, i.e., CSPN [5], PENet [14], FusionNet [35], and the proposed EMDC. From Table 1, we can find that the proposed EMDC can achieve better performance than efficient CSPN, PENet and FusionNet. In addition, Table 1 also lists the performance of SOTA methods of other top teams. Owing to the well-designed architectures, the proposed EMDC can produce better results on EWMAE, RTSD and overall scores.

Figure 4 and Fig. 5 illustrate some subjective results of different DC methods. Overall, the proposed EMDC can reproduce shaper edges, smoother flat surfaces, and clearer details. As shown in Fig. 4, the EMDC can reproduce a more complete flat surface, but other methods cannot well handle different colors on the same surface. By comparing the tiny lines in Fig. 5, the proposed method can reproduce better details than other methods.

---

[1] https://github.com/dwHou/EMDC-PyTorch.

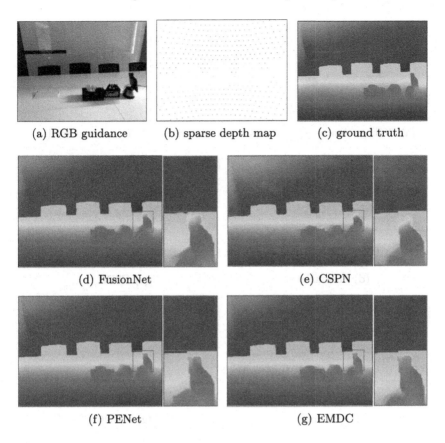

(a) RGB guidance   (b) sparse depth map   (c) ground truth

(d) FusionNet   (e) CSPN

(f) PENet   (g) EMDC

Fig. 4. Depth completion results of different methods.

Table 1. Objective results of different DC methods on MIPI dataset. Red/Blue text: best/second-best.

Method	Overall score↑	RMAE↓	EWMAE↓	RDS↓	RTSD↓
FusionNet [35]	0.795	0.019	0.094	0.009	0.019
CSPN [5]	0.811	0.015	0.090	0.007	0.019
PENet [14]	0.840	0.014	0.087	0.003	0.016
2rd in MIPI	0.846	0.009	0.085	0.002	0.017
3rd in MIPI	0.842	0.012	0.087	0.000	0.018
EMDC (ours)	0.855	0.012	0.084	0.002	0.015

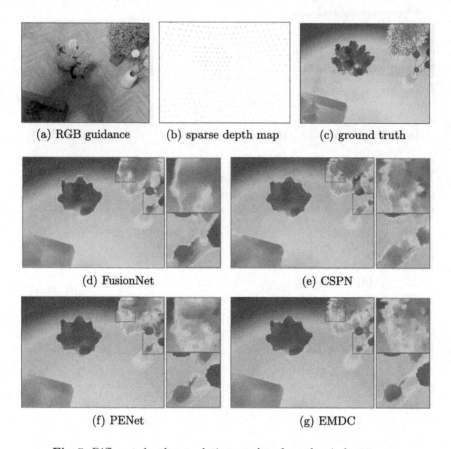

(a) RGB guidance   (b) sparse depth map   (c) ground truth

(d) FusionNet   (e) CSPN

(f) PENet   (g) EMDC

Fig. 5. Different depth completion results of another indoor scene.

Ablation Studies. For ablation testing, we gradually change various settings of the proposed EMDC. All ablation tests are conducted under the same experiment setup, and results are reported in Table 2. Ablation results in Table 2 demonstrate the effectiveness of these modules or settings. The relative certainty fusion module and corrected gradient loss can significantly improve the performance. Using Pixel-Shuffle layer, removing BN layers, and applying re-zero initialization are also helpful to sparse DC tasks. In addition, FCSPN with 9 stages outperforms that with 6 stages, but more iterations also lead to more time-cost.

Inference Time. As sparse ToF sensors are mainly applied in power-limited mobile devices nowadays, the power consumption and processing time are important factors in DC tasks. Hence, we compared the running time of CSPN and FCSPN on the same NVIDIA Tesla V100 GPU. In Table 3, EMDC (FCSPN$_{s=6}$) consists of 6 stages with a total of 15 iterations, and EMDC (FCSPN$_{s=9}$) consists

Table 2. Ablation study on the proposed methods. The results are evaluated by the scoring method mentioned in Sect. 4.3.

Method	(a)	(b)	(c)	(d)	(e)	(f)	(g)	(h)
FCSPN$_{s=6}$		✓						
FCSPN$_{s=9}$			✓	✓	✓	✓	✓	✓
Pixel-Shuffle $_{global\ network}$				✓	✓	✓	✓	✓
Remove BN $_{local\ network}$					✓	✓	✓	✓
Relative Certainty						✓	✓	✓
ReZero Init							✓	✓
Corrected GDL	✓	✓	✓	✓	✓	✓		✓
MIPI overall score ↑	0.798	0.819	0.826	0.834	0.838	0.849	0.847	0.855

Table 3. Runtime of CSPN and the proposed EMDC on 256×192 resolution.

Method	SPN part	Iterations	Propagation time
UNet+CSPN [4]	CSPN	21	25.3 ms
EMDC	FCSPN$_{s=6}$	15	15.7 ms
EMDC	FCSPN$_{s=9}$	21	18.2 ms

of 9 stages with 21 iterations. The improved FCSPN structure can run faster than efficient CSPN. Note that, by comparing the overall scores, although the number of propagation of EMDC (FCSPN$_{s=6}$) is less than CSPN, it can perform better than CSPN on the MIPI Dataset.

5 Conclusions

This paper proposed a lightweight and efficient multimodal depth completion (EMDC) model for sparse depth completion tasks with RGB image guidance. The EMDC consists of a two-branch global and local depth prediction (GLDP) module and a funnel convolutional spatial propagation network (FCSPN). The global branch and local branch in GLDP extract depth features from pixel-wise RGB image and sparse depth data according to the characteristics of different modalities. An improved fusion module with relative confidence maps is then applied to fuse multimodal information. The FCSPN module can refine both the objective and subjective quality of depth maps gradually. A corrected gradient loss is also presented for the depth completion problem. Experimental results show the effectiveness of the proposed method. In addition, the proposed method wins the first place in the RGB+ToF Depth Completion track in Mobile Intelligent Photography and Imaging (MIPI) challenge.

References

1. Bachlechner, T., Majumder, B.P., Mao, H., Cottrell, G., McAuley, J.: Rezero is all you need: fast convergence at large depth. In: Uncertainty in Artificial Intelligence, pp. 1352–1361. PMLR (2021)
2. Chen, Y., Yang, B., Liang, M., Urtasun, R.: Learning joint 2D–3D representations for depth completion. In: Proceedings of the IEEE/CVF International Conference on Computer Vision (ICCV), pp. 10023–10032 (2019)
3. Cheng, X., Wang, P., Guan, C., Yang, R.: CSPN++: learning context and resource aware convolutional spatial propagation networks for depth completion. In: Proceedings of the AAAI Conference on Artificial Intelligence (AAAI), vol. 34, pp. 10615–10622 (2020)
4. Cheng, X., Wang, P., Yang, R.: Depth estimation via affinity learned with convolutional spatial propagation network. In: Ferrari, V., Hebert, M., Sminchisescu, C., Weiss, Y. (eds.) ECCV 2018. LNCS, vol. 11220, pp. 108–125. Springer, Cham (2018). https://doi.org/10.1007/978-3-030-01270-0_7
5. Cheng, X., Wang, P., Yang, R.: Learning depth with convolutional spatial propagation network. IEEE Trans. Pattern Anal. Mach. Intell. **42**(10), 2361–2379 (2019)
6. Dimitrievski, M., Veelaert, P., Philips, W.: Learning morphological operators for depth completion. In: Blanc-Talon, J., Helbert, D., Philips, W., Popescu, D., Scheunders, P. (eds.) ACIVS 2018. LNCS, vol. 11182, pp. 450–461. Springer, Cham (2018). https://doi.org/10.1007/978-3-030-01449-0_38
7. Eldesokey, A., Felsberg, M., Holmquist, K., Persson, M.: Uncertainty-aware CNNs for depth completion: uncertainty from beginning to end. In: Proceedings of the IEEE/CVF Conference on Computer Vision and Pattern Recognition (CVPR), pp. 12014–12023 (2020)
8. Eldesokey, A., Felsberg, M., Khan, F.S.: Propagating confidences through CNNs for sparse data regression. arXiv preprint arXiv:1805.11913 (2018)
9. Fu, C., Dong, C., Mertz, C., Dolan, J.M.: Depth completion via inductive fusion of planar LiDAR and monocular camera. In: IEEE/RSJ International Conference on Intelligent Robots and Systems (IROS), pp. 10843–10848 (2020)
10. Goodfellow, I., et al.: Generative adversarial nets. In: Proceedings on the International Conference on Neural Information Processing Systems (NIPS), vol. 27 (2014)
11. Gu, J., Xiang, Z., Ye, Y., Wang, L.: DenseLiDAR: a real-time pseudo dense depth guided depth completion network. IEEE Robot. Autom. Lett. **6**(2), 1808–1815 (2021)
12. Hambarde, P., Murala, S.: S2DNet: depth estimation from single image and sparse samples. IEEE Trans. Comput. Imaging **6**, 806–817 (2020)
13. Hu, J., et al.: Deep depth completion: a survey. arXiv preprint arXiv:2205.05335 (2022)
14. Hu, M., Wang, S., Li, B., Ning, S., Fan, L., Gong, X.: PENet: towards precise and efficient image guided depth completion. In: IEEE International Conference on Robotics and Automation (ICRA), pp. 13656–13662 (2021)
15. Imran, S., Long, Y., Liu, X., Morris, D.: Depth coefficients for depth completion. In: Proceedings of the IEEE/CVF Conference on Computer Vision and Pattern Recognition (CVPR), pp. 12438–12447 (2019)
16. Jaritz, M., De Charette, R., Wirbel, E., Perrotton, X., Nashashibi, F.: Sparse and dense data with CNNs: depth completion and semantic segmentation. In: IEEE International Conference on 3D Vision (3DV), pp. 52–60 (2018)

17. Kwon, H., Tai, Y.W., Lin, S.: Data-driven depth map refinement via multi-scale sparse representation. In: Proceedings of the IEEE/CVF Conference on Computer Vision and Pattern Recognition (CVPR), pp. 159–167 (2015)

18. Lee, S., Lee, J., Kim, D., Kim, J.: Deep architecture with cross guidance between single image and sparse LiDAR data for depth completion. IEEE Access **8**, 79801–79810 (2020)

19. Liao, Y., Huang, L., Wang, Y., Kodagoda, S., Yu, Y., Liu, Y.: Parse geometry from a line: monocular depth estimation with partial laser observation. In: IEEE International Conference on Robotics and Automation (ICRA), pp. 5059–5066 (2017)

20. Lin, Y., Cheng, T., Zhong, Q., Zhou, W., Yang, H.: Dynamic spatial propagation network for depth completion. arXiv preprint arXiv:2202.09769 (2022)

21. Liu, S., De Mello, S., Gu, J., Zhong, G., Yang, M.H., Kautz, J.: Learning affinity via spatial propagation networks. In: Proceedings on the International Conference on Neural Information Processing Systems (NIPS), vol. 30 (2017)

22. Loshchilov, I., Hutter, F.: Decoupled weight decay regularization. arXiv preprint arXiv:1711.05101 (2017)

23. Lu, K., Barnes, N., Anwar, S., Zheng, L.: From depth what can you see? Depth completion via auxiliary image reconstruction. In: Proceedings of the IEEE/CVF Conference on Computer Vision and Pattern Recognition (CVPR), pp. 11306–11315 (2020)

24. Lu, K., Barnes, N., Anwar, S., Zheng, L.: Depth completion auto-encoder. In: IEEE/CVF Winter Conference on Applications of Computer Vision Workshops (WACVW), pp. 63–73 (2022)

25. Ma, F., Karaman, S.: Sparse-to-Dense: depth prediction from sparse depth samples and a single image. In: IEEE International Conference on Robotics and Automation (ICRA), pp. 4796–4803 (2018)

26. Mathieu, M., Couprie, C., LeCun, Y.: Deep multi-scale video prediction beyond mean square error. arXiv preprint arXiv:1511.05440 (2015)

27. Park, J., Joo, K., Hu, Z., Liu, C.-K., So Kweon, I.: Non-local spatial propagation network for depth completion. In: Vedaldi, A., Bischof, H., Brox, T., Frahm, J.-M. (eds.) ECCV 2020. LNCS, vol. 12358, pp. 120–136. Springer, Cham (2020). https://doi.org/10.1007/978-3-030-58601-0_8

28. Qiu, J., et al.: DeepLiDAR: deep surface normal guided depth prediction for outdoor scene from sparse LiDAR data and single color image. In: Proceedings of the IEEE/CVF Conference on Computer Vision and Pattern Recognition (CVPR), pp. 3313–3322 (2019)

29. Sandler, M., Howard, A., Zhu, M., Zhmoginov, A., Chen, L.C.: MobileNetV2: inverted residuals and linear bottlenecks. In: Proceedings of the IEEE/CVF Conference on Computer Vision and Pattern Recognition (CVPR), pp. 4510–4520 (2018)

30. Schuster, R., Wasenmuller, O., Unger, C., Stricker, D.: SSGP: sparse spatial guided propagation for robust and generic interpolation. In: Proceedings of the IEEE/CVF Winter Conference on Applications of Computer Vision (WACV), pp. 197–206 (2021)

31. Senushkin, D., Romanov, M., Belikov, I., Patakin, N., Konushin, A.: Decoder modulation for indoor depth completion. In: IEEE/RSJ International Conference on Intelligent Robots and Systems (IROS), pp. 2181–2188 (2021)

32. Shi, W., et al.: Real-time single image and video super-resolution using an efficient sub-pixel convolutional neural network. In: Proceedings of the IEEE/CVF Conference on Computer Vision and Pattern Recognition (CVPR), pp. 1874–1883 (2016)

33. Tang, J., Tian, F.P., Feng, W., Li, J., Tan, P.: Learning guided convolutional network for depth completion. IEEE Trans. Image Process. **30**, 1116–1129 (2020)
34. Uhrig, J., Schneider, N., Schneider, L., Franke, U., Brox, T., Geiger, A.: Sparsity invariant CNNs. In: IEEE International Conference on 3D Vision (3DV), pp. 11–20 (2017)
35. Van Gansbeke, W., Neven, D., De Brabandere, B., Van Gool, L.: Sparse and noisy LiDAR completion with RGB guidance and uncertainty. In: IEEE International Conference on Machine Vision Applications (MVA), pp. 1–6 (2019)
36. Wang, Z., Bovik, A.C.: Mean squared error: love it or leave it? A new look at signal fidelity measures. IEEE Sig. Process. Mag. **26**(1), 98–117 (2009)
37. Xu, Y., Zhu, X., Shi, J., Zhang, G., Bao, H., Li, H.: Depth completion from sparse LiDAR data with depth-normal constraints. In: Proceedings of the IEEE/CVF International Conference on Computer Vision (ICCV), pp. 2811–2820 (2019)
38. Xu, Z., Yin, H., Yao, J.: Deformable spatial propagation networks for depth completion. In: IEEE International Conference on Image Processing (ICIP), pp. 913–917 (2020)
39. Yan, Z., et al.: RigNet: repetitive image guided network for depth completion. arXiv preprint arXiv:2107.13802 (2021)
40. Yang, J., Ye, X., Li, K., Hou, C., Wang, Y.: Color-guided depth recovery from RGB-D data using an adaptive autoregressive model. IEEE Trans. Image Process. **23**(8), 3443–3458 (2014)
41. Yang, Y., Wong, A., Soatto, S.: Dense depth posterior (DDP) from single image and sparse range. In: Proceedings of the IEEE/CVF Conference on Computer Vision and Pattern Recognition (CVPR), pp. 3353–3362 (2019)
42. Zhao, S., Gong, M., Fu, H., Tao, D.: Adaptive context-aware multi-modal network for depth completion. IEEE Trans. Image Process. **30**, 5264–5276 (2021)
43. Zhong, Y., Wu, C.Y., You, S., Neumann, U.: Deep RGB-D canonical correlation analysis for sparse depth completion. In: Proceedings on the International Conference on Neural Information Processing Systems (NIPS), vol. 32 (2019)

Learning Rich Information for Quad Bayer Remosaicing and Denoising

Jun Jia[1], Hanchi Sun[1], Xiaohong Liu[2(✉)], Longan Xiao[3], Qihang Xu[3], and Guangtao Zhai[1(✉)]

[1] Institute of Image Communication and Network Engineering, Shanghai Jiao Tong University, Shanghai, China
{jiajun0302,shc15522,zhaiguangtao}@sjtu.edu.cn
[2] John Hopcroft Center for Computer Science, Shanghai Jiao Tong University, Shanghai, China
xiaohongliu@sjtu.edu.cn
[3] Shanghai Transsion Information Technology, Shanghai, China
{longan.xiao1,qihang.xu}@transsion.com

Abstract. In this paper, we propose a DNNs-based solution to jointly remosaic and denoise the camera raw data in Quad Bayer pattern. The traditional remosaic problem can be viewed as an interpolation process that converts the Quad Bayer pattern to a normal CFA pattern, such as the RGGB one. However, this process becomes more challenging when the input Quad Bayer data is noisy. In addition, the limited amount of data available for this task is not sufficient to train neural networks. To address these issues, we view the remosaic problem as a bayer reconstruction problem and use an image restoration model to remove noises while remosaicing the Quad Bayer data implicitly. To make full use of the color information, we propose a two-stage training strategy. The first stage uses the ground-truth RGGB Bayer map to supervise the reconstruction process, and the second stage leverages the provided Image Signal Processor (ISP) to generate the RGB images from our reconstructed bayers. With the use of color information in the second stage, the quality of reconstructed bayers is further improved. Moreover, we propose a data pre-processing method including data augmentation and bayer rearrangement. The experimental results show it can significantly benefit the network training. Our solution achieves the best KLD score with one order of magnitude lead, and overall ranks the second in Quad Joint Remosaic and Denoise @ MIPI-challenge.

Keywords: Quad Bayer · Remosaicing · Denoising · Data augmentation

1 Introduction

In recent years, the increasing demand for the smartphone camera performance has accelerated the high imaging quality of image sensors for smartphones. One

L. Karlinsky et al. (Eds.): ECCV 2022 Workshops, LNCS 13805, pp. 175–191, 2023.
https://doi.org/10.1007/978-3-031-25072-9_12

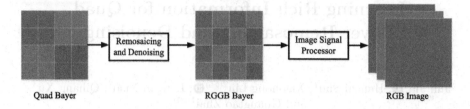

Fig. 1. The overall pipeline of this paper. The input of our model is a Quad Bayer map and the output is a RGGB Bayer map.

of the trends is the multi-pixel, which improves the image resolution by reducing the size of each pixel and arranging more pixels. However, there is a trade-off relationship between pixel miniaturization and decrease in sensitivity. For example, when capturing photos in low-illuminance environments, the weak sensitivities of sensors may decrease the imaging quality. Under this background, a new Color Filter Array (CFA) pattern called Quad Bayer is invented to achieve a good trade-off between the pixel size and the sensitivities of sensors. The Quad Bayer can minimize the decrease in the sensitivities of sensors even if the pixel size is small, which can improve the imaging quality under low light.

Compared to the normal CFA patterns, such as the RGGB Bayer pattern, the four adjacent pixels of a Quad Bayer are clustered with the same color filters. As shown in Fig. 1, the Quad Bayer data has three kinds of color filters and the pixels within a 2 × 2 neighborhood have the same color filters. The Quad Bayer has two modes for low and normal light. When capturing photos under low light, the binning mode enhances the sensitivities of sensors by averaging the four pixels within a 2 × 2 neighborhood, which can improve the imaging quality. As a tradeoff, the spatial resolution is halved. When capturing photos under normal light, the output bayer is supposed to have the same spatial resolution as the input Quad Bayer data. Thus, the original Quad Bayer data needs to be converted to a normal CFA pattern and then fed to the Image Signal Processor (ISP). This converting is an interpolation process called remosaic. The traditional remosaic algorithms are implemented based on hardware. For example, Sony Semiconductor Solutions Corporation (SSS) handles remosaic by installing an array conversion circuit on the image sensor chip[1]. Compared to hardware-based algorithms, software-based remosaic algorithms can be more flexibly applied to different devices. A good remosaic algorithm should be able to get the normal bayer output from the Quad Bayer data with least artifacts, such as moire pattern, false color, and so forth.

However, there are two challenges when designing a remosaic algorithm. The first challenge is that the remosaic problem is difficult when the input Quad Bayer data becomes noisy. Thus, the solution of jointly remosaicing and denoising is in demand for real-world applications. However, denoising and remosaicing are two separate tasks, which makes it difficult to combine them into one algorithm. To address this challenge, we view this process as a reconstruction problem from the

---

[1] https://www.sony-semicon.com/en/technology/mobile/quad-bayer-coding.html.

noisy Quad Bayer map to the clean RGGB Bayer map. Inspired by the success of deep neural networks (DNNs) in image reconstruction tasks, we propose to use a DNNs-based model to remove noises while implicitly rearranging the Quad Bayer map. We present the overall pipeline in real applications in Fig. 1. In addition, we propose a two-stage training strategy to make full use of the color information in both the bayer domain and the RGB domain. As shown in Fig. 2, the first stage uses the ground-truth RGGB Bayer maps to supervise the training of the reconstruction process. After that, the second stage applies the ISP provided by organizers to generate RGB images from the reconstructed bayers. This fine-tuning stage can further improve the quality of our reconstruction.

To train a robust DNNs-based model, we need sufficient training data. However, there is no public dataset that currently contains the paired noisy Quad Bayer data and clean RGGB Bayer data. Thus, the limited amount of data available for this task is the second challenge. Although the organizers of Quad Joint Remosaic and Denoise @MIPI-challenge provide a training set that includes the 210 paired noisy Quad Bayer data and clean RGGB Bayer data, they are not sufficient for training. To address this challenge, we propose a data pre-processing method that employs data augmentation and bayer rearrangement to expand and unify the training samples. The experimental results show this pre-processing can significantly benefit the network training.

After developing and validating our solution, we submit the trained model to Quad Joint Remosaic and Denoise @MIPI-challenge. Our solution is ranked second in the final test phase and achieves the best KLD score. To summarize, our contributions include:

- We propose a DNNs-based model with a two-stage training strategy to jointly remosaic and denoise for Quad Bayer data. By leveraging the two-stage training strategy, the model can make full use of the color information in both the bayer domain and the RGB domain, and the reconstruction quality can be further improved.
- We propose a data pre-processing method including data augmentation and bayer rearrangement. The experimental results show this pre-processing can significantly benefit the network training.
- We submit our solution to Quad Joint Remosaic and Denoise @MIPI-challenge. Our solution is ranked the second in the final test phase and achieves the best KLD score. Codes: https://github.com/jj199603/MIPI2022-QuadBayer

2   Related Works

2.1   Denoising

Image denoising has been greatly concerned in the past few decades. Currently, the image denoising can be classified as two categories, *i.e.,* traditional methods and deep learning based methods.

In traditional image denoising methods, image analysis and processing are usually based on transcendental images. Common methods are 3D transform-domain filtering (BM3D) [10], non-local means (NLM) [4], sparse coding [2], etc. The non-local similarity approach [4] used a non-local algorithm with shared similarity patterns, and the same strategy is applied to [15]. The application of image denoising was implemented using weighted nuclear norm minimization in [18], statistical properties of images were used to remove noise in [43], and scale mixtures of Gaussians were used in the wavelet domain for denoising in [41]. Dictionary learning methods [13] relied on sparse learning [33] from images to obtain a complete dictionary. Traditional denoising methods have certain denoising effectiveness based on reasonable use of image information but the traditional image denoising method is limited because it cannot be extended to all real scenes.

With the development and application of data-driven deep learning, image denoising is given a new processing method. A growing number of researchers are designing novel network architectures based on CNN and transformers, improving the accuracy and versatility of image denoising. [5] first introduced the multi-layer perceptron (MLP) in image denoising and achieves the comparable performance to BM3D. To mimic the real images, many synthetic noise methods are proposed, such as Poissonian-Gaussian noise model [16], Gaussian mixture model [54], camera process simulation [46], and genetic algorithm [8]. Since then, several large physical noise datasets have been generated, such as DND [40] and SIDD [1].

In addition, the real image denoising method is also considered. Researchers first tried the methods previously applied to synthetic noisy datasets on real datasets with model adaptation and tuning [53]. Among them, AINDNet [24] adopted the transfer learning from comprehensive denoising to real denoising, and achieved satisfactory results.

The VDN [48] network architecture based on U-Net [42] was proposed, which assumed that the noise follows an inverse gamma distribution. However, the distribution of noise in the real world is often more complex, so this hypothesis did not apply to many application scenarios. Subsequently, DANet [50] abandoned the hypothesis of noise distribution and used the GAN framework to train the model. Two parallel branches were also used in the structure: one for noise removal and the other for noise generation. A potential limitation was that the training of the GAN-based model was unstable and thus took longer to converge [3]. DANet also used the U-Net architecture in the parallel branch.

Zhang *et al.* [53] recently proposed FFDNet, a denoising network using supervised learning, which connected noise levels as a mapping of noisy images and demonstrated the spatially invariant denoising of real noise with oversmoothed details. MIRNet [51] proposed a general network architecture for image enhancement, such as denoising and super-resolution, with many new building blocks that extracted, exchanged, and exploited multi-scale feature information. InvDN [31] transformed noisy inputs into low-resolution clean images and implicit representations containing noise. To remove noise and restore a clear image,

InvDN replaced noise implicit representation with another implicit representation extracted from previous distribution during noise reduction.

2.2 ISP and Demosaicing

A typical camera ISP assembly line uses a large number of image processing blocks to reconstruct sRGB images from raw sensor data. In another study [27], a two-stage depth network was proposed to replace the camera ISP. The entire ISP of the Huawei P20 smartphone was proposed to be replaced by a combination of deep mode and extensive global functional operations [21]. In [44], the authors proposed a CNN model to suppress image noise and exposure correction to the images captured by smartphone cameras.

Image demosaicing is considered as a low-level ISP task aimed at recreating the CFA pattern of RGB images. However, in practical applications, the image sensors can be affected by noises, which can also lead to the corruption of the final image reconstruction results during demosaicing [29]. In recent work, therefore, the focus has been on the need for a combination of demagnetization and denoising, rather than traditional sequential operations.

In the last four decades, signal processing methods have been widely used for the demosaicing problem [37,47] or resort to the frequency information to improve zipper effect [11,34]. Early methods used frequency method to design the aliasing-free filters [12]. In order to improve near-edge performance, a median chromatic aberration filter [20] was performed and a gradient based approach [39] was widely used. While many methods used chromatic aberrations, Monno recommends using color residuals, starting with a bilinear interpolation of the G channel and then improving the red and blue residuals [38].

However, traditional image processing method can not produce good image quality, easy to produce visual artifacts. More and more demosaicing methods exploit the technology of machine learning; see Energy-based Minimization in [25] or Heide's Complete ISP Modeling [19].

In the last five years, deep learning has become more and more important in low-level visual tasks [23,30,45] than human cognitive abilities. The work related to image demosaicing is summarized below.

Garbi *et al.* presented Bayer with the first end-to-end solution that combined noise reduction and demosaicing [17]. After receiving Bayer images, they extracted four RGGB channels and linked them to a estimated noise channel. The five channels were used as the low-resolution inputs for CNN. They then used a simple network structure similar to VDSR with stacked convolution and global residual paths. Before the final convolution, they also connected the original spliced Bayer plane to the feature mapping of the previous sample. Their main contribution is a data-driven approach to demosaicing the data and publishing a new training dataset by hard patch mining using HDR-VDP2 and moiré detection metric to detect artifact-prone patches [35].

Liu *et al.* [29] proposed an approach based on deep learning, supplemented by density maps and green channel guidance. In [26], the majority-minimization method were merged into a residual denoising network. A deep network was trained using thousands of images to achieve state-of-the-art results [17]. In addition to these supervised learning methods, [14] attempted to address JDD through the unsupervised learning of large numbers of images.

The planar codec structure with symmetric skip connections (RED-Net) was proposed by Mao *et al.* in [36]. RED-Net used skip connections to connect encoder and counter encoder components, but their network is simple and not multi-resolution. They tried different depths of the network, including deeper ones.

Long *et al.* [32] proposed an image segmentation for a full convolutional networks, and the improved version had multi-scale features that captured the background of U-Net [42] at different resolutions. It is demonstrated that the U-Net architecture performed better than the DnCNN [22] network.

3   Proposed Method

In this Section, we first describe the details of the Quad Bayer pre-processing method including data augmentation and bayer rearrangement in Sect. 3.1. Then, the architecture of the model is presented in Sect. 3.2. Finally, we describe the proposed effective two-stage training strategy in Sect. 3.3.

3.1   Quad Bayer Pre-processing

Bayer Rearrangement. As shown in Fig. 2, a raw image in the Quad Bayer pattern consists of multiple 4×4 Quad Bayer units, and each Quad Bayer unit consists of 4 red units, 8 green units, and 4 blue units. Inspired by the success of the raw image processing methods [7,28], a Quad Bayer map is supposed to be decomposed into four channels: R channel, G1 channel, G2 channel, and B channel. For the convenience of channel decomposition, we first swap the second column and the third column in each 4×4 Quad Bayer unit and then swap the second row and the third row of each unit. After that, the Quad Bayer map is converted to a RGGB-alike bayer map and we decompose the RGGB-alike map into four channel maps that are the R channel, the G_1 channel, the G_2 channel, and the B channel, respectively. This process is presented in the first row of Fig. 2.

The above swapping process only includes simple spatial swapping of Quad Bayer units. Thus, the converted RGGB-alike maps still contain noises. Since the provided official ISP does not support a raw image in the Quad Bayer pattern as the input, we also apply this swapping process to the original Quad Bayer data containing noises for visualization in the remaining sections of this paper.

Data Augmentation. We use data augmentation to expand the training samples. The data augmentation processing is applied to the RGGB-alike map after the spatial swapping other than the original Quad Bayer map. In training, the horizontal flip, the vertical flip, and the transposition (permutation of rows and columns) are randomly applied to the RGGB-alike map. To maintain the RGGB

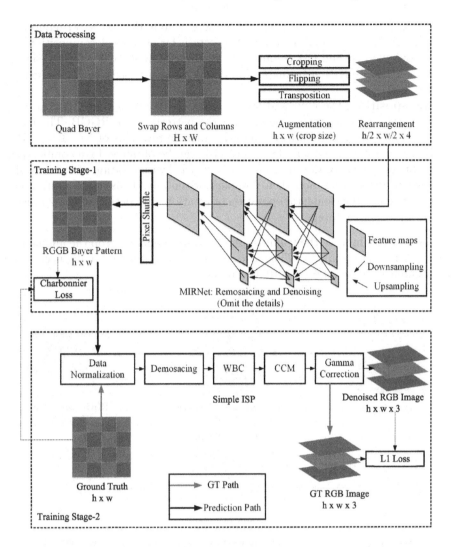

Fig. 2. The solution pipeline of this paper. In the data pre-processing stage, we use data augmentation to expand training samples and convert a Quad Bayer map to four channel RGGB maps. In the first stage, the ground-truth RGGB Bayer maps are used for supervision. In the second stage, the ground-truth RGB image generated by the ISP system is used for supervision. (Color figure online)

pattern after applying augmentations, a cropping-based post-processing method inspired by [28] is used to unify the CFA pattern after these augmentations.

- **vertical flip:** after vertically flipping, a RGGB-alike map is converted to a GBRG-alike map. If we directly decompose the GBRG-alike map into four channels, the channel order will be changed. Thus, we remove the first and last rows of the flipped GBRG-alike map to convert it to a new RGGB-alike map. The details of this process are presented in Fig. 3(a).

182   J. Jia et al.

- **horizontal flip:** after horizontally flipping, a RGGB-alike map is converted to a GRBG-alike map. To unify the pattern for channel decomposition, we remove the first and last columns of the flipped GRBG-alike map to convert it to a new RGGB-alike map. The details of this process are presented in Fig. 3(b).
- **transposition:** the rows and the columns of the original RGGB-alike map are permuted after transposition, but the pattern is still RGGB. The traditional data augmentation methods also include rotations of 90°, 180°, 270°. However, rotations of 90°, 180°, and 270° can be obtained by a combination of flip and transposition. For instance, the clockwise rotation of 90° is equivalent to transposing and then flipping horizontally. Thus, transposition can be viewed as the elemental operation of rotation. The details of transposition are shown in Fig. 3(c).

In addition, to improve the training efficiency, we randomly crop the augmented RGGB-alike map to h × w. The starting coordinates of the cropping need to be even numbers to maintain the RGGB pattern.

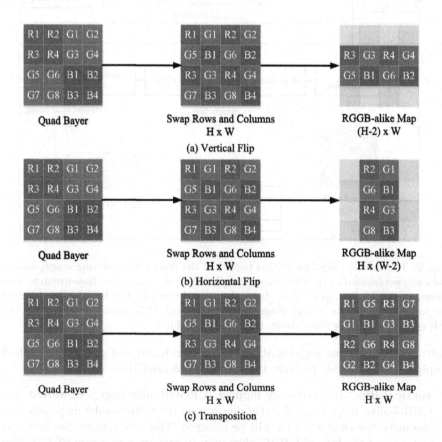

Fig. 3. The details of the data augmentations used in this paper.

3.2   Network for Jointly Remosaicing and Denoising

In this paper, we view the remosaicing problem as a reconstruction process from a Quad Bayer map to a RGGB Bayer map. Inspired by the success of image restoration methods based on deep neural networks [9,49,51,52], we use a DNNs-based neural network to remove noises while remosaicing the Quad pattern to the RGGB pattern implicitly.

We select MIRNet [51] to remove noises and reconstruct RGGB Bayer maps. MIRNet [51] consists of multi-scale feature extraction paths. Feature fusion is carried out among feature maps of different scales to fully learn the multi-scale features of the image. Compared to the models based on auto-encoder [9], MIR-Net contains a high-resolution feature path where the feature maps are not down-sampled, thus more detailed information is preserved. Through the feature fusion among the path of different scales, the high-level semantic features extracted with small resolutions and the low-level features extracted with large resolutions are fully fused and learned. MIRNet exploits multiple attention mechanisms to fuse the multi-scale features, such as spatial attention and channel attention. We present the overall architecture of this model in Fig. 2, omitting the details such as the attention modules.

The input of MIRNet is a $\frac{h}{2} \times \frac{w}{2} \times 4$ map generated from data augmentation and bayer rearrangement. The output of MIRNet is also a $\frac{h}{2} \times \frac{w}{2} \times 4$ map. We apply PixelShuffle operation to the output to generate a h × w × 1 RGGB map which represents the reconstructed RGGB Bayer data.

3.3   Two-Stage Training Strategy

We propose an effective two-stage training strategy to improve the reconstruction quality. As shown in Fig. 2, in the first stage, the model described in Sect. 3.2 is used to jointly remosaic the Quad Bayer data to the RGGB Bayer data and denoise it. The ground-truth RGGB Bayer map is used for supervision.

After the training in the first stage, we concatenate the ISP provided by the challenge organizers to the jointly remosaicing and denoising network. The provided ISP applies normalization, demosaicing, white balance correction, color correction, and gamma correction to a raw image. The output of our network and the ground-truth RGGB Bayer map are processed to generate the corresponding RGB images. Then, the generated ground-truth RGB image is used as a color supervision to finetune our network. In this stage, we freeze the parameters of ISP. Compared to the bayer data, the RGB image contains more color information, which can further improve the quality of the reconstructed raw data.

Loss Functions. In the first training stage, we use *Charbonnier* loss [6] to optimize the parameters of our model, which is defined as:

$$L_{Charbon} = \sqrt{||R_{rec} - R_{gt}||^2 + \epsilon^2} \tag{1}$$

where R_{rec} represents the reconstructed raw data in the RGGB Bayer pattern, R_{gt} represents the ground-truth RGGB Bayer map, and ϵ is hyper-parameter which is set to 10^{-3}. In the second training stage, we compute the L_1 loss between the RGB images generated from the reconstructed raw data and the ground-truth raw data.

4   Experimental Results

4.1   Dataset

The dataset used in experiments is provided by the organizers of the MIPI2022 challenge. The dataset in the development phase is divided into a training set and a validation set. Both the training set and the validation set include samples of three noise levels: 0 dB, 24 dB, and 42 dB. The training set of each noise level includes 70 Quad Bayer files and the corresponding ground-truth files in the RGGB pattern. The validation set of each noise level only includes 15 Quad Bayer files without the corresponding ground-truth files. Thus, the training set includes 270 Quad Bayer samples and the validation set includes 45 Quad Bayer samples. The resolutions of these samples are 1200×1800. The RGB thumbnails of these 85 raw images are also provided for the convenience of visualization.

In experiments, we use all the 270 training samples to train our model and evaluate our model on the 45 validation samples. Because the distribution of the noise model is unknown, no additional datasets are used for training to prevent overfitting.

4.2   Implementation Details

We trained three models for three noise levels (0 dB, 24 dB, and 42 dB), respectively. The training sample is the cropped raw data with a size of 256×256. The hyper-parameter ϵ of Charbonnier loss is set to 10^{-3}. We use Adam to optimize the network. The initial learning rate is 2×10^{-4}, decaying by $1/10$ every 235,200 iterations. The batch size of training is 2 and the average training time of one model is about 36 h in one NVIDIA-2080 Ti. When validating and testing, original Quad Bayer data without any cropping and the output of the model is the corresponding RGGB Bayer map that has the same resolution as the input.

4.3   Evaluation Metrics

The official metrics used by Quad Joint Remosaic and Denoise @MIPI-challenge are Peak Signal To Noise Ratio (PSNR), Structural Similarity Index Measure (SSIM), Learned perceptual image patch similarity (LIPIS), and KL divergence (KLD). The final results and rankings are evaluated by the M_4 score:

$$M_4 = PSNR \times SSIM \times 2^{1-LPIPS-KLD} \qquad (2)$$

In the development and validation phase, we evaluate our model on the validation set. Since the ground-truth RGGB Bayer maps of the validation set are not provided, we can only compute the accurate M_4 scores through the official website. However, the submission number is limited, we cannot compute the accurate M_4 for all experimental results. For the convenience of development and validation, we use PSNR to evaluate some of the experiments.

4.4   Ablation Study

In this section, we analyze the effects of the data augmentation methods and the proposed two-stage training strategy.

Ablation Study of Data Augmentations. To analyze the importances of the data augmentation methods, we first train three models without data augmentations for three noise levels and then train three models with the data augmentations described in Sect. 3.1. During training, we randomly select one augmentation from vertical flips, horizontal flips, and transpositions with a 25% probability, and do not apply any augmentations with a 25% probability. The ablation results on validation set are presented in Table 1. Table 1 shows that using data augmentations can improve the scores of PSNR, SSIM, and LPIPS.

Table 1. The ablation results of data augmentations

Model	PSNR	SSIM	LPIPS	KLD	M4
Without aug	36.24	0.954	0.127	0.00496	64.18
With aug	**36.65**	**0.956**	**0.121**	**0.00628**	**65.15**

Ablation Study of Two-Stage Training Strategy. To analyze the importance of the data augmentation methods, we first train three models for the three noise levels without the second fine-tuning stage. When the PSNR values converge on the validation set, we fine tune these three models on base of the first stage. We use PSNR to evaluate the improvement of the second stage. For 0 dB, the PSNR value increases from 40.91 to 41.17. For 24 dB, the PSNR value increases from 36.22 to 36.43. For 42 dB, the PSNR value increases from 32.27 to 32.36. These results show that fine-tuning the model with RGB images can further improve the visual quality of the reconstructed images. Compared to the raw data, RGB images contain richer color information since the channel number is three times that of raw images.

4.5   Model Complexity and Runtime

The total number of trainable parameters is 31,788,571. When validating and testing, the resolution of the input is 1200×1800 and the average test time in one NVIDIA-2080 Ti is about 0.7 s. Although we use a single model for jointly remosaicing and denoising, our model is not light-weighted, but can be further optimized to adapt real-time application.

186   J. Jia et al.

Table 2. The results and rankings of Quad Joint Remosaic and Denoise @MIPI-challenge

Rank	Team	PSNR	SSIM	LPIPS	KLD	M4
1	op-summer-po	**37.93**	**0.965**	0.104	0.019	**68.03**
2	**JHC-SJTU (ours)**	37.64	0.96	0.1	**0.0068**	67.99
3	IMEC-IPI & NPU	37.76	0.96	0.1	0.014	67.95
4	BITSpectral	37.2	0.96	0.11	0.03	66
5	HITZST01	37.2	0.96	0.11	0.06	64.82
6	MegNR	36.08	0.95	**0.095**	0.023	64.1

Ground Truth   Input   Result   Ground Truth   Input   Result

Fig. 4. The qualitative results of normal brightness images on Quad Joint Remosaic and Denoise @MIPI-challenge validation set.

Ground Truth   Input   Result   Ground Truth   Input   Result

Fig. 5. The qualitative results of low brightness images on Quad Joint Remosaic and Denoise @MIPI-challenge validation set.

4.6 Challenge Submission

After validating the proposed solution, we submit the trained model in Quad Joint Remosaic and Denoise @MIPI-challenge. The dataset in the final test phase includes 15 Quad Bayer files and the resolution of each test sample is 1200×1800. The ranking of our solution is the second as shown in Table 2. Table 2 shows that

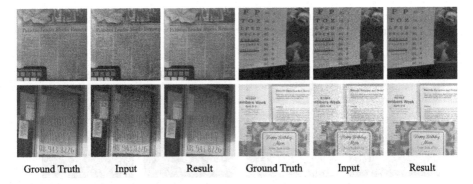

Fig. 6. The qualitative results of text region reconstruction on Quad Joint Remosaic and Denoise @MIPI-challenge validation set.

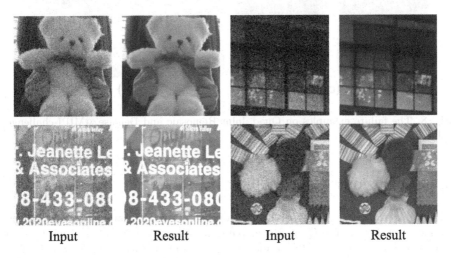

Fig. 7. The qualitative results of text region reconstruction on Quad Joint Remosaic and Denoise @MIPI-challenge test set.

our solution achieves the best KLD score. We present some qualitative results in Fig. 4, Fig. 5, Fig. 6, and Fig. 7. Figure 4 shows the representative reconstruction results of normal brightness images on Quad Joint Remosaic and Denoise @MIPI-challenge validation set, Fig. 5 shows the representative results of low brightness images, Fig. 6 shows the representative results of text regions, and Fig. 7 shows the results on the test set. The noise level of these presented examples is 42 dB.

4.7 Limitations

In experiments, we find that our model has limitations in two scenes. The first scene is the low brightness image as shown in Fig. 8(a)–(c). The second scene is the texture region containing the high noise level as shown in Fig. 8(d), which suffers color errors in the reconstruction result.

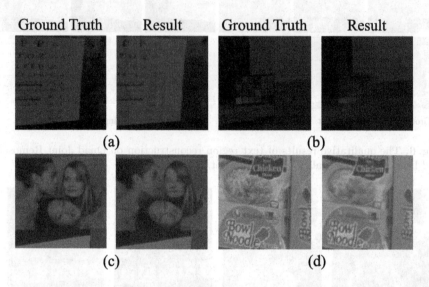

Fig. 8. The failure examples on Quad Joint Remosaic and Denoise @MIPI-challenge validation set.

5   Conclusions

The paper proposes a novel solution of jointly remosaicing and denoising for the camera raw data in the Quad Bayer pattern. We use a DNNs-based multi-scale model to remove noises while remosaicing the Quad Bayer map to the RGGB Bayer map. An effective data pre-processing method is proposed to augment and rearrange the original Quad Bayer data. To make full use of the color information, we propose a two-stage training strategy to fine-tune the model with the corresponding RGB images. The experimental results show that the data pre-processing method and the two-stage training strategy can significantly benefit the network training. We submit our solution to Quad Joint Remosaic and Denoise @MIPI-challenge, and achieve the second rank and the best KLD scores on the final test set. Our solution still has three limitations at this stage: (1) the reconstruction quality of low brightness images needs to be improved, (2) the reconstruction quality is not good enough for high-level noises, and (3) our model has a slightly large number of parameters, which needs to be further optimized for real-time applications. We will solve these problems in the future.

Acknowledgements. This work was supported by the National Key R&D Program of China 2021YFE0206700, NSFC 61831015.

References

1. Abdelhamed, A., Lin, S., Brown, M.S.: A high-quality denoising dataset for smart-phone cameras. In: IEEE/CVF Conference on Computer Vision & Pattern Recognition (2018)
2. Aharon, M., Elad, M., Bruckstein, A.: K-SVD: an algorithm for designing over-complete dictionaries for sparse representation. IEEE Trans. Sig. Process. **54**(11), 4311–4322 (2006). https://doi.org/10.1109/TSP.2006.881199
3. Arjovsky, M., Chintala, S., Bottou, L.: Wasserstein generative adversarial networks. In: International Conference on Machine Learning (2017)
4. Buades, A., Coll, B., Morel, J.M.: A non-local algorithm for image denoising. In: 2005 IEEE Computer Society Conference on Computer Vision and Pattern Recognition, CVPR 2005 (2005)
5. Burger, H.C., Schuler, C.J., Harmeling, S.: Image denoising: can plain neural networks compete with BM3D? In: 2012 IEEE Conference on Computer Vision and Pattern Recognition (CVPR) (2012)
6. Charbonnier, P., Blanc-Feraud, L., Aubert, G., Barlaud, M.: Two deterministic half-quadratic regularization algorithms for computed imaging. In: Proceedings of 1st International Conference on Image Processing, vol. 2, pp. 168–172 (1994). https://doi.org/10.1109/ICIP.1994.413553
7. Chen, C., Chen, Q., Xu, J., Koltun, V.: Learning to see in the dark. In: Proceedings of the IEEE Conference on Computer Vision and Pattern Recognition, pp. 3291–3300 (2018)
8. Chen, J., Chen, J., Chao, H., Ming, Y.: Image blind denoising with generative adversarial network based noise modeling. In: 2018 IEEE/CVF Conference on Computer Vision and Pattern Recognition (CVPR) (2018)
9. Cheng, S., Wang, Y., Huang, H., Liu, D., Fan, H., Liu, S.: NBNet: noise basis learning for image denoising with subspace projection. In: Proceedings of the IEEE/CVF Conference on Computer Vision and Pattern Recognition, pp. 4896–4906 (2021)
10. Dabov, K., Foi, A., Katkovnik, V., Egiazarian, K.: Image denoising by sparse 3-D transform-domain collaborative filtering. IEEE Trans. Image Process. **16**(8), 2080–2095 (2007)
11. Dai, L., Liu, X., Li, C., Chen, J.: AWNet: attentive wavelet network for image ISP. In: Bartoli, A., Fusiello, A. (eds.) ECCV 2020. LNCS, vol. 12537, pp. 185–201. Springer, Cham (2020). https://doi.org/10.1007/978-3-030-67070-2_11
12. Alleysson, D., Süsstrunk, S., Hérault, J.: Linear demosaicing inspired by the human visual system. IEEE Trans. Image Process. **14**(4), 439–449 (2005)
13. Dong, W., Xin, L., Lei, Z., Shi, G.: Sparsity-based image denoising via dictionary learning and structural clustering. In: 2011 IEEE Conference on Computer Vision and Pattern Recognition (CVPR) (2011)
14. Ehret, T., Davy, A., Arias, P., Facciolo, G.: Joint demosaicking and denoising by fine-tuning of bursts of raw images. In: 2019 IEEE/CVF International Conference on Computer Vision (ICCV) (2019)
15. Foi, A., Katkovnik, V., Egiazarian, K.: Pointwise shape-adaptive DCT for high-quality denoising and deblocking of grayscale and color images. IEEE Trans. Image Process. **16**, 1395–1411 (2007)
16. Foi, A., Trimeche, M., Katkovnik, V., Egiazarian, K.: Practical Poissonian-Gaussian noise modeling and fitting for single-image raw-data. IEEE Trans. Image Process. **17**(10), 1737–1754 (2008)

17. Gharbi, M., Chaurasia, G., Paris, S., Durand, F.: Deep joint demosaicking and denoising. ACM Trans. Graph. **35**(6), 191 (2016)
18. Gu, S., Lei, Z., Zuo, W., Feng, X.: Weighted nuclear norm minimization with application to image denoising. In: 2014 IEEE Conference on Computer Vision and Pattern Recognition (CVPR) (2014)
19. Heide, F., et al.: FlexISP: a flexible camera image processing framework. In: International Conference on Computer Graphics and Interactive Techniques (2014)
20. Hirakawa, K., Parks, T.W.: Adaptive homogeneity-directed demosaicing algorithm. IEEE Trans. Image Process. **14**(3), 360 (2005)
21. Ignatov, A., Gool, L.V., Timofte, R.: Replacing mobile camera ISP with a single deep learning model. In: Computer Vision and Pattern Recognition (2020)
22. Kai, Z., Zuo, W., Chen, Y., Meng, D., Lei, Z.: Beyond a Gaussian denoiser: residual learning of deep CNN for image denoising. IEEE Trans. Image Process. **26**(7), 3142–3155 (2016)
23. Kim, I., Song, S., Chang, S., Lim, S., Guo, K.: Deep image demosaicing for sub-micron image sensors. J. Imaging Sci. Technol. **63**(6), 060410-1–060410-12 (2019)
24. Kim, Y., Soh, J.W., Gu, Y.P., Cho, N.I.: Transfer learning from synthetic to real-noise denoising with adaptive instance normalization. In: 2020 IEEE/CVF Conference on Computer Vision and Pattern Recognition (CVPR) (2020)
25. Klatzer, T., Hammernik, K., Knobelreiter, P., Pock, T.: Learning joint demosaicing and denoising based on sequential energy minimization. In: 2016 IEEE International Conference on Computational Photography (ICCP) (2016)
26. Kokkinos, F., Lefkimmiatis, S.: Deep image demosaicking using a cascade of convolutional residual denoising networks. In: Ferrari, V., Hebert, M., Sminchisescu, C., Weiss, Y. (eds.) Computer Vision – ECCV 2018. LNCS, vol. 11218, pp. 317–333. Springer, Cham (2018). https://doi.org/10.1007/978-3-030-01264-9_19
27. Liang, Z., Cai, J., Cao, Z., Zhang, L.: CameraNet: a two-stage framework for effective camera ISP learning. IEEE Trans. Image Process. **30**, 2248–2262 (2019)
28. Liu, J., et al.: Learning raw image denoising with bayer pattern unification and bayer preserving augmentation. In: Proceedings of the IEEE/CVF Conference on Computer Vision and Pattern Recognition Workshops (2019)
29. Liu, L., Jia, X., Liu, J., Tian, Q.: Joint demosaicing and denoising with self guidance. In: 2020 IEEE/CVF Conference on Computer Vision and Pattern Recognition (CVPR) (2020)
30. Liu, X., Shi, K., Wang, Z., Chen, J.: Exploit camera raw data for video super-resolution via hidden Markov model inference. IEEE Trans. Image Process. **30**, 2127–2140 (2021)
31. Liu, Y., et al.: Invertible denoising network: a light solution for real noise removal (2021)
32. Long, J., Shelhamer, E., Darrell, T.: Fully convolutional networks for semantic segmentation. In: Proceedings of the IEEE conference on computer vision and pattern recognition, pp. 3431–3440 (2015)
33. Mairal, J., Bach, F., Ponce, J., Sapiro, G., Zisserman, A.: Non-local sparse models for image restoration. In: 2009 IEEE 12th International Conference on Computer Vision (ICCV) (2010)
34. Malvar, H.S., He, L.W., Cutler, R.: High-quality linear interpolation for demosaicing of bayer-patterned color images. In: 2004 IEEE International Conference on Acoustics, Speech, and Signal Processing (2004)
35. Mantiuk, R., Kim, K.J., Rempel, A.G., Heidrich, W.: HDR-VDP-2: a calibrated visual metric for visibility and quality predictions in all luminance conditions. ACM Trans. Graph. **30**(4), 1–14 (2011)

36. Mao, X.J., Shen, C., Yang, Y.B.: Image restoration using very deep convolutional encoder-decoder networks with symmetric skip connections (2016)
37. Menon, D., Calvagno, G.: Color image demosaicking. Sig. Process. Image Commun. **26**(8–9), 518–533 (2011)
38. Monno, Y., Kiku, D., Tanaka, M., Okutomi, M.: Adaptive residual interpolation for color image demosaicking. In: IEEE International Conference on Image Processing (2015)
39. Pekkucuksen, I., Altunbasak, Y.: Gradient based threshold free color filter array interpolation. In: 2010 17th IEEE International Conference on Image Processing (ICIP) (2010)
40. Plotz, T., Roth, S.: Benchmarking denoising algorithms with real photographs, pp. 2750–2759 (2017)
41. Portilla, J., Strela, V., Wainwright, M.J., Simoncelli, E.P.: Image denoising using scale mixtures of Gaussians in the wavelet domain. IEEE Trans. Image Process. **12**(11), 1338–1351 (2003)
42. Ronneberger, O., Fischer, P., Brox, T.: U-Net: convolutional networks for biomedical image segmentation. In: Navab, N., Hornegger, J., Wells, W.M., Frangi, A.F. (eds.) MICCAI 2015. LNCS, vol. 9351, pp. 234–241. Springer, Cham (2015). https://doi.org/10.1007/978-3-319-24574-4_28
43. Rudin, L.I., Osher, S., Fatemi, E.: Nonlinear total variation based noise removal algorithms. Physica D **60**(1–4), 259–268 (1992)
44. Schwartz, E., Giryes, R., Bronstein, A.M.: DeepISP: towards learning an end-to-end image processing pipeline. IEEE Trans. Image Process. **28**(2), 912–923 (2018)
45. Sharif, S., Naqvi, R.A., Biswas, M.: Beyond joint demosaicking and denoising: an image processing pipeline for a pixel-bin image sensor (2021)
46. Shi, G., Yan, Z., Kai, Z., Zuo, W., Lei, Z.: Toward convolutional blind denoising of real photographs (2018)
47. Xin, L., Gunturk, B., Lei, Z.: Image demosaicing: a systematic survey. In: Proceedings of SPIE - The International Society for Optical Engineering, vol. 6822 (2008)
48. Yue, Z., Yong, H., Zhao, Q., Zhang, L., Meng, D.: Variational denoising network: toward blind noise modeling and removal (2019)
49. Yue, Z., Yong, H., Zhao, Q., Meng, D., Zhang, L.: Variational denoising network: toward blind noise modeling and removal. In: Advances in Neural Information Processing Systems, vol. 32 (2019)
50. Yue, Z., Zhao, Q., Zhang, L., Meng, D.: Dual adversarial network: toward real-world noise removal and noise generation. In: Vedaldi, A., Bischof, H., Brox, T., Frahm, J.-M. (eds.) ECCV 2020. LNCS, vol. 12355, pp. 41–58. Springer, Cham (2020). https://doi.org/10.1007/978-3-030-58607-2_3
51. Zamir, S.W., et al.: Learning enriched features for real image restoration and enhancement. In: Vedaldi, A., Bischof, H., Brox, T., Frahm, J.-M. (eds.) ECCV 2020. LNCS, vol. 12370, pp. 492–511. Springer, Cham (2020). https://doi.org/10.1007/978-3-030-58595-2_30
52. Zamir, S.W., et al.: Learning enriched features for fast image restoration and enhancement (2022)
53. Zhang, K., Zuo, W., Zhang, L.: FFDNet: toward a fast and flexible solution for CNN based image denoising. IEEE Trans. Image Process. **27**(9), 4608–4622 (2017)
54. Zhu, F., Chen, G., Heng, P.A.: From noise modeling to blind image denoising. In: 2016 IEEE Conference on Computer Vision and Pattern Recognition (CVPR) (2016)

Depth Completion Using Laplacian Pyramid-Based Depth Residuals

Haosong Yue, Qiang Liu$^{(\boxtimes)}$, Zhong Liu, Jing Zhang, and Xingming Wu

School of Automation Science and Electrical Engineering, Beihang University,
Beijing 100191, China
4765825391q@gmail.com

Abstract. In this paper, we propose a robust and efficient depth completion network based on residuals. Unlike previous methods that directly predict a depth residual, we reconstruct high-frequency information in complex scenes by exploiting the efficiency of the Laplacian pyramid in representing multi-scale content. Specifically, the framework can be divided into two stages: sparse-to-coarse and coarse-to-fine. In the sparse-to-coarse stage, we only recover depth from the sparse depth map without using any additional color image, and downsample the result to filter out unreliable high-frequency information from the sparse depth measurement. In the coarse-to-fine stage, we use features extracted from both data modalities to model high-frequency components as a series of multi-scale depth residuals via a Laplacian pyramid. Considering the wide distribution of high-frequency information in the frequency domain, we propose a Global-Local Refinement Network (GLRN) to estimate depth residuals separately at each scale. Furthermore, to compensate for the structural information lost by coarse depth map downsampling and further optimize the results with the color image, we propose a novel and efficient Affinity decay spatial propagation network (AD-SPN), which is used to refine the depth estimation results at each scale. Extensive experiments on indoor and outdoor datasets demonstrate that our approach achieves state-of-the-art performance.

Keywords: Depth completion · Laplacian pyramid · Spatial propagation network

1 Introduction

With the rapid development of computer vision, depth perception has become a fundamental problem. It plays an essential role in the research and application of robotics [15], autonomous driving [14], and virtual reality [16]. Many sensors are used to obtain reliable depth information of the scene, such as RGB cameras, LiDAR, or ToF cameras. However, none of them are perfect. Color images obtained from RGB cameras are inexpensive and informative, but their depth values are unobserved. Depth sensors can produce accurate depth measurements but are too sparse and costly. To overcome these challenges, there

L. Karlinsky et al. (Eds.): ECCV 2022 Workshops, LNCS 13805, pp. 192–207, 2023.
https://doi.org/10.1007/978-3-031-25072-9_13

have been extensive works on depth completion, a technique for obtaining dense depth measurements at a low cost, aiming at inferring dense depth maps from sparse depth maps and corresponding high-resolution color images.

Depth completion is a regression task that uses regularization to determine a rule to extrapolate from the observable depth to nearby unobserved depths. The development of neural networks provides a practical tool for data-driven regularization. The early depth completion methods [17,18] mainly take sparse depth maps as input without additional data. However, the performance of these single-model approaches is unsatisfactory due to aliasing in the frequency domain caused by sparse sampling and no other information as a reference. Soon, multi-modal methods [19,20], commonly guided by color images, attract more attention and achieve more excellent results than single-model. Recent image-guided approaches are mainly based on deep convolutional neural networks. The network structure of depth completion has developed from single-modal single-model to multi-modal multi-model. In general, depth completion can be divided into two major strategies: one is ensemble, and the other is refinement. For example, FusionNet [26] proposes a dual-branch network to estimate local and global results separately. DeepLidar [21] introduces the surface normal input and additionally builds a normal path that focuses on the surface normal information. PENet [2] proposes a depth branch and takes the depth map estimated by the color branch as its input. SemAttNet [29] introduces semantic information and builds an additional semantic branch. These methods achieve accurate depth completion results by constructing different branches that enable the model to focus on various scene components and synthesize the advantages of these branches. Different from ensemble methods, CSPN [9,10], NLSPN [12], and CSPN++ [11], use affinity matrices to refine coarse depth completion results in a spatially propagated manner. All of them achieve outstanding performance. However, the predicted dense depth naturally prefers sparse input with blurry details. Under-utilized color information leads to unreliable details and sub-optimal solutions for depth completion.

To address these problems, SEHLNet [1] has proposed a depth completion method that separately estimates high and low-frequency components. However, it is too complex and has low real-time performance, which limits its application. In this paper, we propose a novel Laplacian pyramid-based depth completion network, which estimates low-frequency components from sparse depth maps by downsampling and contains a Laplacian pyramid decoder that estimates multi-scale residuals to reconstruct complex details of the scene. Since the high-frequency information in the scene is widely distributed in the frequency domain, and this distribution has both global generality and local specificity, we propose a global-local residual estimation strategy to estimate the residuals at each scale. Furthermore, the down-sampling operation obtains low-frequency components at a small cost but loses some important geometric information contained in the sparse depth map. So we also propose a novel Affinity decay spatial propagation network, which is used to refine the estimated depth maps at various scales through spatial propagation. AD-SPN not only compensates for the loss of information caused by downsampling but also further optimizes the results by utilizing the detailed information from color images.

The main contributions of our approach can be summarized as follows:

1. We propose adopting the Laplacian pyramid to resolve the depth completion problem. The proposed method successfully restores local details and global structure by recovering depth residuals in different levels of the Laplacian pyramid and summing up those predicted results progressively.
2. In view of the wide distribution of high-frequency residual information in the frequency domain, we propose a global-local refinement network to estimate the global residual and local residual separately at each scale, which is more efficient than directly regressing the residual through the neural network. Furthermore, we propose a novel affinity decay spatial propagation network for refining the estimated depth at each scale.
3. We demonstrate various experimental results on benchmark datasets constructed under complicated indoor and outdoor environments, and show the efficiency and robustness of the proposed method compared to state-of-the-art methods.

2   Related Work

Image-Guided Depth Completion. Early depth completion methods [17,18] only take sparse depth maps as input and do not require additional data. This single-modal approach performs poorly because the depth sensor loses a lot of scene information during sparse sampling. Image-guided depth completion uses additional information during depth completion, such as RGB images. It has attracted much attention and outperforms previous methods because the extra information can compensate for the loss of sparse sampling and contains a lot of surface, edge, or semantic information.

We roughly divide image-guided depth completion methods into two strategies. One of the strategies is ensemble, which builds different multi-branch models through various methods, including changing branch structures [26], input modalities [21,29], multi-scale feature extraction and fusion mechanisms [22,24,28], result ensemble strategies [2], etc. Different branches focus on different depth components and finally fuse their results through adaptive fusion, showing better performance. These methods have achieved good results, but the multi-branch structure often leads to complex models and large memory consumption. Therefore, how to construct suitable different branches and integrate their results remains an open question.

Another strategy is refinement, which optimizes coarse depth estimates. The most commonly used methods are Residual Networks and Spatial Propagation Networks. For example, FCFR-NET [23] designs an energy-based fusion operation to fuse high-frequency information of different modal information and constructs a residual network to learn the residuals of coarse depth maps directly. Although this method is simple and efficient, due to the wide distribution of high-frequency information in the frequency domain, it is difficult for a single residual network

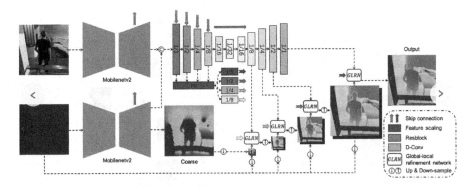

Fig. 1. Network architecture. Our network consists of two mobilenetv2 and a feature fusion encoder, a Laplacian pyramid decoder. The results from the sparse depth network are downsampled as the input to the Laplacian pyramid, and the corresponding residuals are regressed at each scale to refine the depth map. At the same time, the feature maps from the feature fusion encoder are scaled to the same scale, and they can guide the Global-local refinement network to estimate global and local residuals.

(often approaching a band-pass filter) to learn all the high-frequency information. Spatial Propagation Network differs from the methods mentioned above. It uses a deep CNN to learn an affinity matrix that represents the spatial variation of depth values and iteratively refines the coarse depth map. SPN builds a row/column linear propagation model that can be easily applied to many advanced vision tasks. CSPN [9] further improves the linear propagation model and adopts recursive convolution operation to improve the efficiency and accuracy of the model. CSPN++ [11] improves its accuracy and efficiency by learning adaptive convolution kernel size and the number of iterations of propagation. NLSPN [12] effectively excludes irrelevant neighbors in spatial propagation and focuses on the confidence of different regions. DSPN [13,25] uses deformable convolutions to build adaptive receptive fields. In fact, simple spatial propagation models have high efficiency but low accuracy, and complex spatial propagation models have high accuracy but require more resources for training and inference.

Deep Neural Networks in Pyramid Framework. The Laplace pyramid was first proposed by Burt et al. [7]. The purpose of the Laplacian pyramid is to decompose the source image into different spatial frequency bands. The fusion process is performed for each spatial frequency layer separately to extract different features and details of different frequency bands of the decomposition layer, and use different fusion operators to highlight the characteristics and details of specific frequency bands. It is widely used in numerous computer vision tasks, including texture synthesis, image compression, noise removal, and image blending. Lai et al. [8] proposed a deep Laplacian pyramid network for fast and accurate image super-resolution. For depth estimation, [3] designs a Laplacian pyramid neural network, which includes a Laplacian pyramid decoder for signal reconstruction and an adaptive dense feature fusion module for

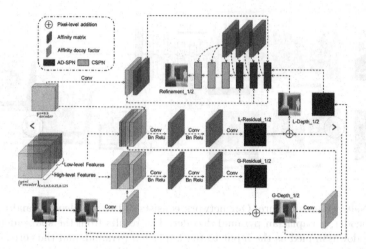

Fig. 2. Global-local refinement network. The feature maps from the feature fusion encoder are divided into high-level and low-level features according to the original scale. High-level features are obtained by upsampling lower-resolution feature maps, which contain more global structural information, and are first used to regress the global residuals. Then, the local residuals are regressed with low-level features and features from the decoder. The final depth is refined through the spatial propagation network, the first three layers are AD-SPN, and the last three are CSPN.

fusing features from input images. [4] incorporates the Laplacian pyramid into the decoder architecture to reconstruct the final depth map from coarse to fine.

This paper introduces the Laplacian pyramid into the depth completion task. We can think of Laplacian pyramids as a series of bandpass filters that are ideal for accurately estimating depth boundaries and global layout.

3   Our Approach

In this section, we introduce the proposed Laplacian pyramid-based deep completion network. We will first briefly describe the network structure of the model. Then we introduce the proposed global-local refinement network, which improves the residual regression capability at each pyramid level. Finally, we introduce the proposed affinity decay spatial propagation network. Overall, our model is a fusion of residual and refinement networks. The Laplacian Pyramid extends the regression range of the residual network for high-frequency information, and the spatial propagation network introduces a new information aggregation method that the standard CNN convolution does not have.

3.1   Network Atructure

The overall architecture of the network is shown in Fig. 1. Due to the pyramid structure of the encoder, the abstraction level of the feature map is negatively related to its resolution. We do not directly perform early or late fusion of features on the two modal data but use two independent mobilenetv2 to extract features from them, respectively. Finally, the features from two decoders are fused in a resnet-based encoder.

Since understanding the rich texture, brightness, and semantic information of color images is a complex matter, estimating coarse depth maps directly from sparse depth maps and color images will generate a lot of noise. We tend only to estimate the coarse depth from the features of the sparse depth map and filter out some unreliable details by downsampling. In this way, the network only needs to fit the depth distribution of the sparse depth map, which significantly reduces the requirements to the network, and a lightweight model can meet it.

For high-frequency information, although the previous method [1] based on spatially variant convolution has achieved remarkable performance on average pixel metrics, it is computationally expensive, and the scene structure of the target image cannot be well recovered. The reason is that it uses complex network architecture to directly regress the residuals while ignoring that high-frequency information is widely distributed in the frequency domain and is challenging to map through one single network. In comparison, the Laplacian pyramid representation progressively adds information of different frequency bands so that the scene structure at different scales can be hierarchically recovered during reconstruction. This enables the predicted final residuals to have average pixel-wise accuracy and preserve scene structures at various scales. Specifically, we first obtain a coarse low-frequency depth map $D_{coarse}^{s=0.125}$ by down-sampling the output of the sparse depth decoder to a 1/8 scale. Then, we progressively estimate the different scale residuals R^s, where $s = 1, 0.5, 0.25, 0.125$ represents the reduction factor relative to the original scale, and add them to the upsampled map $up(D^s)$ from the lower resolution. In each layer of the pyramid, the signal D^s could be constructed by $R^s + up(D^{s/2})$ and refined by a spatial propagation network. Finally, the output of the last layer of the pyramid is the result of depth completion.

3.2 Global-Local Refinement Network

Instead of the simple upsampling and summation of the two parts in two frequency bands, we propose a global-local refinement network to fully use different levels of features from the encoder to estimate residuals at various scales and refine them.

GLRN is shown in Fig. 2. In the Laplacian pyramid at a specific scale s, we scale feature maps with different scales extracted in the encoder to the same scale s. The feature maps with the original scale less than s are regarded as high-level feature maps, which can provide the global structure information of the scene, and the feature maps with the original scale greater than or equal to s and the feature maps from the decoder are regarded as low-level feature maps, which can provide local details information. We first estimate the global residual R_g^s and then predict the local residual R_l^s on this basis. Both the residuals are obtained through three-layer convolution. The final depth is refined through the spatial

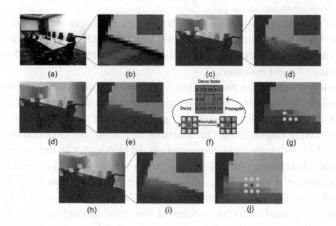

Fig. 3. Compare the performance of CSPN and AD-SPN. (c), (d), (h) represent the depth map before refinement, the depth map refined by AD-SPN, and the depth map refined by CSPN. (f) is the process of affinity decay, the affinity with smaller decay factor tends to zero after three iterations, such as (g). (j) is the fixed propagation neighborhoods in CSPN.

propagation network, the first three layers are our proposed AD-SPN which will be introduced later, and the last three are CSPN.

3.3 Affinity Decay Spatial Propagation Network

Currently, spatial propagation networks (SPNs) [10] are the most popular affinity-based methods for depth refinement. Convolutional spatial propagation network (CSPN) [9] is simple and effective. However, the propagation of each pixel occurs in a fixed receptive field. This may not be optimal for refinement since different pixel needs different local context. CSPN++ [11] further improves its effectiveness and efficiency by learning adaptive convolutional kernel sizes and the number of iterations for the propagation, but its real-time performance is poor. Deformable spatial propagation network (DSPN) [13] uses deformable convolution to replace traditional convolution to achieve a variable receptive field, and the effect is outstanding. However, the learning process of deformable convolution is complicated.

In this work, we need to refine the depth map at each layer of the Laplacian pyramid. So a simple, fast, and efficient spatial propagation network becomes the primary goal to guarantee real-time performance. None of the above methods can meet these requirements. Therefore, we propose a novel Affinity decay spatial propagation network, as shown in Fig. 3. Assuming $d^0 \in R^{m \times n}$ is the initial depth map and $d^N \in R^{m \times n}$ is the refined depth map after N iterations. We can write the spatial propagation as a process of diffusion evolution by reshape $d^t \in R^{m \times n}$ as a column-first one-dimensional vector $H^t \in R^{mn}$, where $t \in 0, 1, 2...N$.

$$H^{t+1} = \begin{bmatrix} 1-\lambda_{0,0} & \kappa_{0,0}(1,0) & \dots & \kappa_{0,0}(m,n) \\ \kappa_{1,0}(0,0) & 1-\lambda_{1,0} & \dots & \kappa_{1,0}(m,n) \\ \vdots & \vdots & \ddots & \vdots \\ \kappa_{m,n}(0,0) & \kappa_{m,n}(1,0) & \dots & 1-\lambda_{m,n} \end{bmatrix} H^t = GH^t \quad (1)$$

where $G \in R^{mn \times mn}$ is a transformation matrix, $\kappa_{i,j}(m,n)$ describes the affinity weight of pixel (i,j) with its neighbors (a,b) and $\lambda_{i,j} = \sum_{a \neq i, b \neq j} \kappa_{i,j}(a,b)$ means that the sum of each row of G is 1. The diffusion process expressed with a partial differential equation (PDE) can be derived as follows,

$$\begin{aligned} H^{t+1} = GH^t &= (I - D + A)H^t \\ H^{t+1} - H^t = GH^t &= -(D - A)H^t \\ \partial H^{t+1} &= -LH^t \end{aligned} \quad (2)$$

where L is the Laplacian matrix, D is the diagonal matrix containing all the $\lambda_{i,j}$, and A is the affinity matrix containing all the $\kappa_{i,j}(a,b)$.

Considering the computing power of the hardware, A is usually a sparse matrix, which can be written as $A = BW$, where W is the weight of affinity matrix and B is a binary matrix whose non-zero elements are distributed according to the existence of affinity. For CSPN, the fixed propagation neighborhood results in B being a fixed matrix with $m \times n \times k \times k$ non-zero elements. The learning of deformable convolution kernels by DSPN is essentially to learn a binary matrix B with a dynamic distribution of non-zero elements. In most cases, we want the spatial propagation to occur in local regions to refine tiny geometries, so inferring B from the global is unnecessary. We introduce a dynamic mask M for the fixed matrix B of CSPN, A = MBW, so that the fixed convolution kernel has spatial heterogeneity. In the implementation, we first estimated an affinity decay matrix $A^d(0 \sim 1)$ with the same shape as the affinity matrix and then multiplied the affinity matrix with A^d and normalized it at each iteration. After several iterations, the affinity with a relatively small factor in A^d tends to 0, while the other affinities are retained, and the result is close to applying the dynamic mask M. Finally, we found that too many decay iterations will cause the model to degenerate, all the affinities except the affinity with the largest factor in A^d are decayed, so we only use A^d in the first three layers of refinement.

3.4 The Training Loss

We train our network with ℓ_1 and ℓ_2 loss in each scale as follows:

$$L^s(D) = \sum_{\rho=1,2} \left(\frac{1}{n} \sum_{p \in Q} |D_{gt}^p - D^p|^\rho \right), \quad (3)$$

where $s = 1, 0.5, 0.25, 0.125$, D and D_{gt} mean the predicted depth map and ground truth at scale s, Q is a set of points which depth value is not zero in

D_{gt}, and n denotes the number of points in Q. The whole loss function can be formulated as:

$$Loss = \lambda_1 L^{0.125}(D) + \lambda_2 L^{0.25}(D) + \lambda_3 L^{0.5}(D) + \lambda_4 L^{1.0}(D), \qquad (4)$$

where λ_1, λ_2, λ_3 and λ_4 are hyper-parameters.

4   Experiments

In this section, we first explain the implementation details and the datasets used in our experiments. Following this, we show qualitative and quantitative comparisons of the proposed method with existing algorithms on the outdoor KITTI dataset. Finally, we perform ablation experiments on the indoor ToF synthetic dataset provided by the MIPI Challenge to verify the effectiveness of each component of the proposed algorithm.

4.1   Implementation Details

Our method was implemented using the PyTorch [27] framework and trained with 4 TITAN RTX GPUs. We use the ADAM optimizer with $\beta_1 = 0.9$, $\beta_2 = 0.99$, and the weight decay is 10^{-6}. We adopt a hierarchical training strategy to train the network sequentially from low resolution to high resolution. We first train the depth estimation results at scale 0.125, where $\lambda_1 = 1$, $\lambda_2, \lambda_3, \lambda_4 = 0$. After the model converges, we start training at scale 0.25, where $\lambda_1, \lambda_2 = 1$, $\lambda_3, \lambda_4 = 0$. And so on until the end of all scale training. Every scale is trained for 50 epochs, and the learning rate starts from 0.001 at each stage, delay by 0.5 every 10 epochs. A total of 200 epochs were trained. The batch size for the indoor ToF synthetic dataset is set to 20. For the KITTI dataset, the batch size is 8. Besides, data augmentation techniques are used, including horizontal random flip and color jitter.

4.2   Datasets

KITTI Dataset is a large outdoor autonomous driving data set, which also is the main benchmark for depth completion. It consists of over 85k color images with corresponding sparse depth maps and semi-dense ground truth maps for training, 6k for validation, and 1k for testing. We use officially selected 1k validation images to evaluate and crop all images to 256×1216 during training.

ToF Synthetic dataset is officially provided by MIPI Challenge. The data contains seven image sequences of aligned RGB and ground-truth dense depth from 7 indoor scenes. For each scene, the RGB and the ground-truth depth are rendered along a smooth trajectory in the created 3D virtual environment. RGB and dense depth images in the dataset have a resolution of 640×480 pixels. During the training process, we use the officially provided sampling method to generate a pseudo-TOF sparse depth map, and conduct experiments according to the officially divided training set and validation set.

4.3   Metrics

For the KITTI DC Dataset, we adopt error metrics same as the dataset benchmark, including RMSE, mean absolute error (MAE), inverse RMSE (RMSE), and inverse MAE (iMAGE). For the ToF Synthetic dataset, mean absolute relative error (REL), and percentage of pixels satisfying δ_τ are additionally selected as the evaluation metrics. All the metrics for evaluation are shown as follows:

RMSE (mm): $\sqrt{\frac{1}{v}\sum_x (\hat{h}_x - h_x)^2}$

MAE (mm): $\frac{1}{v}\sum_x \left| \hat{h}_x - h_x \right|$

iRMSE (1/km): $\sqrt{\frac{1}{v}\sum_x (\frac{1}{\hat{h}_x} - \frac{1}{h_x})^2}$

iMAE (1/km): $\frac{1}{v}\sum_x \left| \frac{1}{\hat{h}_x} - \frac{1}{h_x} \right|$

REL: $\frac{1}{v}\sum_x \left| \frac{\hat{h}_x - h_x}{h_x} \right|$

δ_τ: $max(\frac{\hat{h}_x}{h_x} - \frac{h_x}{\hat{h}_x}) < \tau, \tau \in \left\{ 1.25, 1.25^2, 1.25^3 \right\}$

Table 1. Quantitative comparison with state of the art methods on KITTI Depth Completion testing set.

Method	RMSE mm	MAE mm	iRMSE 1/km	iMAE 1/km
Fusion	772.87	215.02	2.19	0.93
DSPN	766.74	220.36	2.47	1.03
DeepLiDAR	758.38	226.50	2.56	1.15
FuseNet	752.88	221.19	2.34	1.14
ACMNet	744.91	206.09	2.08	0.90
CSPN++	743.69	209.28	2.07	0.90
NLSPN	741.68	199.59	1.99	0.84
GuideNet	736.24	218.83	2.25	0.99
FCFRNet	735.81	217.15	2.20	0.98
PENet	730.08	210.55	2.17	0.94
SEHLNet	714.71	207.51	2.07	0.91
RigNet	712.66	203.25	2.08	0.90
SemAttNet	709.41	205.49	2.03	0.90
DySPN	709.12	192.71	1.88	0.82
LpNet (ours)	710.74	212.51	2.18	0.96

4.4   Evaluation on KITTI Dataset

Table 1 shows the quantitative performance of our proposed method, and compares it with other state-of-the-art methods on the KITTI depth completion benchmark. The results show that our method outperforms most of the previous methods in metrics. Note that though we train our network from scratch

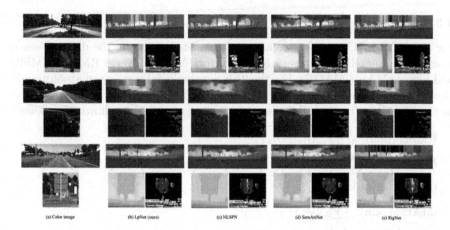

Fig. 4. Qualitative comparisons with state-of-the-art methods on KITTI depth completion test set. Left to right: (a) color image, results of (b) LpNet (ours), (c) NLSPN, (d) SemAttNet, (e) RigNer, respectively. We show dense depth maps and error maps from KITTI leaderboard for better comparison. (Color figure online)

without any additional data, it is still better than those employing additional data. Such as DeepLidar, synthetics surface normal data to train the surface normal prediction network. FusionNet, which uses a pre-trained network trained on CityScapes. SemAttNet uses additional semantic information from the KITTI dataset.

Figure 4 shows a qualitative comparison with several SOTA works. We can find that the depth maps obtained by our proposed method achieve better performance on object boundaries and detail regions. This is because the Laplacian pyramid fits the detailed information in a wider frequency domain by learning multi-scale residuals, and the spatial propagation network further refines the results through spatial changes, which is the ability that general standard convolution does not have. Our proposed AD-SPN is simple and efficient compared to other spatial propagation networks.

4.5 Ablation Study on ToF Synthetic Dataset

We conduct ablation experiments on the ToF synthetic dataset to verify the effectiveness of various parts in our proposed network, including the Laplacian pyramid, the global-local residual estimation strategy, and the affinity decay spatial propagation network. Furthermore, we compare the difference between the staged and direct training strategies, and the results show that our proposed training strategy is reliable.

Effectiveness of the Laplacian Pyramid. We first demonstrate the effectiveness of the proposed Laplacian Pyramid on the depth completion task by

Table 2. Ablation experiments on ToF Synthetic dataset

Method	RMSE	MAE	iRMSE	iMAE	REL	$\delta_{1.25}$	$\delta_{1.25^2}$	$\delta_{1.25^3}$
Baseline	0.5087	0.2073	0.1780	0.0375	0.0820	0.9045	0.9582	0.9800
Residual Net	0.4678	0.1857	0.1550	0.0340	0.0808	0.9161	0.9646	0.9832
Laplace Net	0.4238	0.1612	0.1517	0.0291	0.0732	0.9339	0.9730	0.9879
AD-SPN Net	0.4005	0.1425	0.1726	0.0254	0.0643	0.9378	0.9756	0.9887
CSPN Net	0.4196	0.1578	0.1471	0.0284	0.0684	0.9304	0.9735	0.9882
GLR Net	0.3780	0.1291	0.1410	0.0229	0.0573	0.9452	0.9782	0.9899

comparing the performance in three cases, i.e., (1) Baseline: we use only one mobilenetv2 to build a depth completion network, where the sparse depth map is concatenated with the color image and fed directly into it. (2) Residual net: We first extract the features of the sparse depth map and the color image through two mobilenetv2 respectively, then use the features from the sparse depth map to estimate the coarse depth map, and use the features from the two modalities to estimate the residual. The final result is the sum of the rough depth map and the residual. The network structure is similar to Fig. 1 but without the Laplacian pyramid. (3) Laplace net: It introduces the Laplace pyramid into the residual network. At each layer, we first estimate the residual with the feature map from the decoder, and then sum the residual with the upsampled depth map from the previous layer. As shown in Table 2, Residual net outperforms Baseline, which demonstrates the effectiveness of residual learning. Separately estimating the coarse depth map and its residuals is more efficient than directly estimating depth. This is equivalent to separately estimating the low-frequency and high-frequency components of the depth map in the frequency domain. Laplace net outperforms Baseline and Residual net, demonstrating the Laplace Pyramid's effectiveness. Compared with the single residual network in the Residual net, the pyramid builds a multi-scale residual model, which improves the ability of the network to fit high-frequency components.

Effectiveness of Affinity Decay Spatial Propagation Networks. The purpose of building an affinity decaying spatial propagation network is to improve the performance of the simplest spatial propagation model at a minimal cost. Compared with other complex spatial propagation networks, AD-SPN dramatically enhances the performance of CSPN by adding only one affinity decay matrix. As shown in Table 2, AD-SPN net and CSPN net refine the output of each layer of the Laplacian pyramid with AD-SPN and CSPN, and AD-SPN outperforms CSPN net.

Table 3. Comparison of different training strategies. L1 means that we only supervise the final output depth map of the Laplacian pyramid. L2 indicates that we supervise the output of all pyramid layers but only train 50 epochs. L3 is our proposed staged training strategy, training 200 epochs in 4 stages.

Method	RMSE	MAE	iRMSE	iMAE	REL	$\delta_{1.25}$	$\delta_{1.25^2}$	$\delta_{1.25^3}$
Residual Net	0.4678	0.1857	0.1550	0.0340	0.0808	0.9161	0.9646	0.9832
Laplace Net+L1	0.4654	0.1816	0.1555	0.0331	0.0801	0.9172	0.9651	0.9835
Laplace Net+L2	0.4512	0.1775	0.1544	0.0326	0.0792	0.9210	0.9664	0.9840
Laplace Net+L3	0.4238	0.1612	0.1517	0.0291	0.0732	0.9339	0.9730	0.9879

Color   Baseline   Residual Net   Laplace Net   AD-SPN Net   GLR Net   Ground truth

Fig. 5. Qualitative comparison of ablation experiments.

Effectiveness of Global Local Residual Estimation Strategies. As shown in Table 2, we finally demonstrate the effectiveness of the global-local residual estimation strategy. GLR net is our final model, which adds a global-local residual estimation network to the AD-SPN net and performs best. Instead of directly estimating the residuals at each level of the pyramid with features at the corresponding scale from the feature decoder, GLR net comprehensively references features from the encoder at all scales. We estimate global residuals from high-level low-resolution features and local residuals from low-level high-resolution features. This further improves the model's ability to estimate high-frequency components at each scale. The results of the qualitative comparison are shown in Fig. 5. It can be seen that the results of GLR net have achieved obvious advantages in terms of the overall structure and geometric details, which is in line with the quantitative experimental results. AD-SPN net can significantly improve the model's ability to estimate the scene's geometric contours, demonstrating the effectiveness of our proposed affinity decay. Since AD-SPN refines the output of the Laplacian pyramid, its accuracy is affected by residuals. GLR net can regress high-frequency information over a wider spectrum by separating global and local residuals, so the results are better. To give a more intuitive understanding of our method, we present the outputs of each stage in depth completion in Fig. 6. We did not estimate the global-local residuals on the 1/8 scale because we believe that low frequencies dominate the residuals at this scale, and it is not meaningful to estimate the global and local residuals separately.

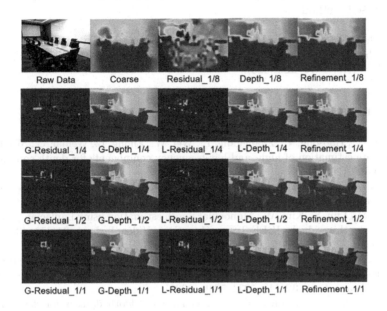

Fig. 6. Outputs of each stage in our model.

Effectiveness of Staged Training Strategies. Finally, we verify the effectiveness of the proposed staged training strategy. As shown in Table 3, directly supervising the final results of the network or the results of each scale output does not significantly improve the performance of the model. We speculate that this is because the convergence rates of residual networks at different pyramid scales are quite different. Direct training will cause the Laplacian pyramid to estimate the residuals when the previous scale does not converge, eventually resulting in a significant degree of overlap between the residuals at different scales, failing to exploit the potential of the Laplacian pyramid fully.

5   Conclusion

We have proposed a robust and efficient residual-based deep completion network. The proposed method reconstructs high-frequency information in complex scenes by exploiting the efficiency of the Laplacian pyramid to represent multi-scale content. Considering the wide distribution of high-frequency information in the frequency domain, we propose a global local refinement network to estimate depth residuals at each scale separately. Furthermore, to compensate for the structural information lost by coarse depth map downsampling and further optimize the results with color image information, we propose a novel and efficient affinity decaying spatial propagation network to refine the depth estimation results at each scale. Our experimental results demonstrated the superiority of the proposed method.

Acknowledgments. This research was funded by National Natural Science Foundation of China (No. 61603020) and the Fundamental Research Funds for the Central Universities (No. YWF-22-L-923).

References

1. Qiang, L., Haosong, Y., Zhanggang, L., Wei, W., Zhong, L., Weihai, C.: SEHLNet: separate estimation of high- and low-frequency components for depth completion. In: ICRA (2022)
2. Hu, M., Wang, S., Li, B., Ning, S., Fan, L., Gong, X.: PENet: towards precise and efficient image guided depth completion. In: ICRA (2021)
3. Song, M., Lim, S., Kim, W.: Monocular depth estimation using Laplacian pyramid-based depth residuals. IEEE Trans. Circ. Syst. Video Technol. **31**(11), 4381–4393 (2021)
4. Chen, X., Chen, X., Zhang, Y., Fu, X., Zha, Z.J.: Laplacian pyramid neural network for dense continuous-value regression for complex scenes. IEEE Trans. Neural Netw. Learn. Syst. **32**(11), 5034–5046 (2021)
5. Jeon, J., Lee, S.: Reconstruction-based pairwise depth dataset for depth image enhancement using CNN. In: Ferrari, V., Hebert, M., Sminchisescu, C., Weiss, Y. (eds.) ECCV 2018. LNCS, vol. 11220, pp. 438–454. Springer, Cham (2018). https://doi.org/10.1007/978-3-030-01270-0_26
6. Li, D., et al.: Involution: inverting the inherence of convolution for visual recognition (2021)
7. Burt, P.J., Adelson, E.H.: The Laplacian pyramid as a compact image code. In: Readings in Computer Vision, vol. 31, no. 4, pp. 671–679 (1987)
8. Lai, W.S., Huang, J.B., Ahuja, N., Yang, M.H.: Fast and accurate image super-resolution with deep Laplacian pyramid networks. IEEE Trans. Pattern Anal. Mach. Intell. **41**(11), 2599–2613 (2018)
9. Cheng, X., Wang, P., Yang, R.: Learning depth with convolutional spatial propagation network. In: ECCV (2018)
10. Liu, S., Mello, S.D., Gu, J., Zhong, G., Yang, M.H., Kautz, J.: Learning affinity via spatial propagation networks (2017)
11. Cheng, X., Wang, P., Guan, C., Yang, R.: CSPN++: learning context and resource aware convolutional spatial propagation networks for depth completion (2019)
12. Park, J., Joo, K., Hu, Z., Liu, C.-K., So Kweon, I.: Non-local spatial propagation network for depth completion. In: Vedaldi, A., Bischof, H., Brox, T., Frahm, J.-M. (eds.) ECCV 2020. LNCS, vol. 12358, pp. 120–136. Springer, Cham (2020). https://doi.org/10.1007/978-3-030-58601-0_8
13. Xu, Z., Wang, Y., Yao, J.: Deformable spatial propagation network for depth completion (2020)

14. Song, X., et al.: ApolloCar3D: a large 3D car instance understanding benchmark for autonomous driving. IEEE (2018)
15. Liao, Y., Huang, L., Yue, W., Kodagoda, S., Yong, L.: Parse geometry from a line: monocular depth estimation with partial laser observation. IEEE (2017)
16. Armbruester, C., Wolter, M., Kuhlen, T., Spijkers, W., Fimm, B.: Depth perception in virtual reality: distance estimations in peri-and extrapersonal space. CyberPsychology **11**(1), 9–15 (2008)
17. Uhrig, J., Schneider, N., Schneider, L., Franke, U., Brox, T., Geiger, A.: Sparsity invariant CNNs, pp. 11–20. IEEE Computer Society (2017)
18. Chodosh, N., Wang, C., Lucey, S.: Deep convolutional compressed sensing for LiDAR depth completion. In: Jawahar, C.V., Li, H., Mori, G., Schindler, K. (eds.) ACCV 2018. LNCS, vol. 11361, pp. 499–513. Springer, Cham (2019). https://doi.org/10.1007/978-3-030-20887-5_31
19. Ma, F., Cavalheiro, G.V., Karaman, S.: Self-supervised sparse-to-dense: self-supervised depth completion from LiDAR and monocular camera (2018)
20. Chen, Y., Yang, B., Liang, M., Urtasun, R.: Learning joint 2D–3D representations for depth completion. In: 2019 IEEE/CVF International Conference on Computer Vision (ICCV) (2020)
21. Qiu, J., Cui, Z., Zhang, Y., Zhang, X., Pollefeys, M.: DeepLiDAR: deep surface normal guided depth prediction for outdoor scene from sparse lidar data and single color image. In: 2019 IEEE/CVF Conference on Computer Vision and Pattern Recognition (CVPR) (2019)
22. Zhao, S., Gong, M., Fu, H., Tao, D.: Adaptive context-aware multi-modal network for depth completion. IEEE Trans. Image Process. **30**, 5264–5276 (2021)
23. Liu, L., Song, X., Lyu, X., Diao, J., Zhang, L.: FCFR-Net: feature fusion based coarse-to-fine residual learning for monocular depth completion (2020)
24. Tang, J., Tian, F.P., Feng, W., Li, J., Tan, P.: Learning guided convolutional network for depth completion. IEEE Trans. Image Process. **30**, 1116–1129 (2021)
25. Lin, Y., Cheng, T., Zhong, Q., Zhou, W., Yang, H.: Dynamic spatial propagation network for depth completion (2022)
26. Van Gansbeke, W., Neven, D., Brabandere, B.D., Van Gool, L.: Sparse and noisy LiDAR completion with RGB guidance and uncertainty. In: 2019 16th International Conference on Machine Vision Applications (MVA) (2019)
27. Paszke, A., Gross, S., Massa, F., Lerer, A., Chintala, S.: PyTorch: an imperative style, high-performance deep learning library (2019)
28. Yan, Z., et al.: RigNet: repetitive image guided network for depth completion. In: Avidan, S., Brostow, G., Cissé, M., Farinella, G.M., Hassner, T. (eds.) ECCV 2022. LNCS, vol. 13687, pp. 214–230. Springer, Cham (2021). https://doi.org/10.1007/978-3-031-19812-0_13
29. Nazir, D., Liwicki, M., Stricker, D., Afzal, M.Z.: SemAttNet: towards attention-based semantic aware guided depth completion. arXiv e-prints (2022)

W19 - Challenge on People Analysis: From Face, Body and Fashion to 3D Virtual Avatars

W19 - Challenge on People Analysis: From Face, Body and Fashion to 3D Virtual Avatars

In the workshop and challenge on people analysis we address human-centered data analysis. These data are extremely widespread and have been intensely investigated by researchers belonging to very different fields, including Computer Vision, Machine Learning, and Artificial Intelligence. These research efforts are motivated by the several highly-informative aspects of humans that can be investigated, ranging from corporal elements (e.g. bodies, faces, hands, anthropometric measurements) to emotions and outward appearance (e.g. human garments and accessories). The huge amount and the extreme variety of this kind of data make the analysis and the use of learning approaches extremely challenging. The workshop provided a forum for novel research in the area of human understanding and the winners of the 3D human body and 3D face reconstruction challenge were announced. It also featured two invited talks by experts in the field.

October 2022

Alberto Bimbo
Mohamed Daoudi
Roberto Vezzani
Xavier Alameda-Pineda
Marcella Cornia
Guido Borghi
Claudio Ferrari
Federico Becattini
Andrea Pilzer
Zhiwen Chen
Xiangyu Zhu
Ye Pan
Xiaoming Liu

PSUMNet: Unified Modality Part Streams Are All You Need for Efficient Pose-Based Action Recognition

Neel Trivedi[✉] and Ravi Kiran Sarvadevabhatla

Centre for Visual Information Technology, IIIT Hyderabad, Hyderabad 500032, India
neel.trivedi@research.iiit.ac.in, ravi.kiran@iiit.ac.in

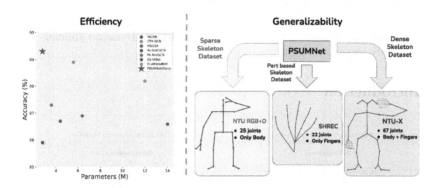

Fig. 1. The plot on left shows accuracy against # parameters for our proposed architecture PSUMNet (⋆) and existing approaches for the large-scale NTURGB+D 120 human actions dataset (cross subject). PSUMNet achieves state of the art performance while competing recent methods use 100%–400% more parameters. The diagram on right illustrates that PSUMNet scales to sparse pose (SHREC [6]) and dense pose (NTU-X [26]) configurations in addition to the popular NTURGB+D [15] configuration.

Abstract. Pose-based action recognition is predominantly tackled by approaches which treat the input skeleton in a monolithic fashion, i.e. joints in the pose tree are processed as a whole. However, such approaches ignore the fact that action categories are often characterized by localized action dynamics involving only small subsets of part joint groups involving hands (e.g. 'Thumbs up') or legs (e.g. 'Kicking'). Although part-grouping based approaches exist, each part group is not considered within the global pose frame, causing such methods to fall short. Further, conventional approaches employ independent modality streams (e.g. joint, bone, joint velocity, bone velocity) and train their network multiple times on these streams, which massively increases the number of training parameters. To address these issues, we introduce PSUMNet, a novel approach for scalable and efficient pose-based action recognition. At the representation level, we propose a global frame based part

Supplementary Information The online version contains supplementary material available at https://doi.org/10.1007/978-3-031-25072-9_14.

stream approach as opposed to conventional modality based streams. Within each part stream, the associated data from multiple modalities is unified and consumed by the processing pipeline. Experimentally, PSUMNet achieves state of the art performance on the widely used NTURGB+D 60/120 dataset and dense joint skeleton dataset NTU 60-X/120-X. PSUMNet is highly efficient and outperforms competing methods which use 100%–400% more parameters. PSUMNet also generalizes to the SHREC hand gesture dataset with competitive performance. Overall, PSUMNet's scalability, performance and efficiency makes it an attractive choice for action recognition and for deployment on compute-restricted embedded and edge devices. Code and pretrained models can be accessed at https://github.com/skelemoa/psumnet.

Keywords: Human action recognition · Skeleton · Dataset · Human activity recognition · Part

1   Introduction

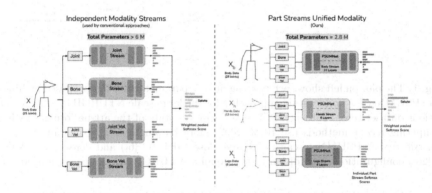

Fig. 2. Comparison between conventional training procedure used in most of the previous approaches (left) and our approach (right). Conventional methods [2,16] use dedicated independent streams and train separate instances of the same network for each of the four modalities, i.e joint, bone, joint velocity and bone velocity. This method increases the number of total parameters by a huge margin and involves a monolithic representation. Our method processes the modalities in a unified manner and creates part group based independent stream with a superior performance compared to existing methods which use 100%–400% more parameters - see Fig. 3 for architectural details of PSUMNet.

Skeleton based human action recognition at scale has gained a lot of focus recently, especially with the release of large scale skeleton action datasets such as NTURGB+D [19] and NTURGB+D 120 [15]. A plethora of RNN [8,9], CNN [10, 30] and GCN [16,28] based approaches have been proposed to tackle this important problem. The success of approaches such as ST-GCN [28] which modeled spatio-temporal joint dynamics using GCN has given much prominence to GCN-based approaches. Furthermore, approaches such as RA-GCN [23] and 2s-AGCN [21] built upon this success and demonstrated additional gains by introducing multi modal (bone and velocity) streams – see Fig. 2 (left). This multi stream approach has been adopted as convention by state of the art approaches.

However, the conventional setup has three major drawbacks. *First,* each modality stream is trained independently and the results are combined using late (decision) fusion. This deprives the processing pipeline from taking advantage of correlations across modalities. *Second,* with addition of each new modality, the number of parameters increase by a significant margin since a separate network with the same model architecture is trained for each modality. *Third,* the skeleton is considered in a monolithic fashion. In other words, the entire input pose tree at each time step is treated as a whole and at once. This is counter intuitive to the fact that a lot of action categories often involve only a subset of the available joints. For example, action categories such as "Cutting paper" or "Writing" can be easily identified using only hand joints whereas action categories such as "Walking" or "Kicking" can be easily identified using only leg joints. Additionally, monolithic processing increases compute requirements when the number of joints in the pose representation increases [26]. Non-monolithic approaches which decompose the pose tree into disjoint part groups do exist [24,25]. However, each part group is not considered within the global pose frame, causing such methods to fall short.

Our proposed approach tackles all of the aforementioned drawbacks - see Fig. 2 (right). Our contributions are the following:

– We propose a unified modality processing approach as opposed to conventional independent modality approaches. This enables a significant reduction in the number of parameters (Sect. 3.2).
– We propose a part based stream processing approach which enables richer and dedicated representations for actions involving a subset of joints (Sect. 3.1). The part stream approach also enables efficient generalization to dense joint (NTU-X [26]) and small joint (SHREC [6]) datasets.
– Our architecture, dubbed Part Stream Unified Modality Network (PSUM-Net) achieves SOTA performance on NTU 60-X/120-X, and NTURGB+D 60/120 datasets compared to existing competing methods which use 100%–400% more parameters. PSUMNet also generalizes to SHREC hand gestures dataset with competitive performance (Sect. 4.3).
– We perform extensive experiments and ablations to analyze and demonstrate the superiority of PSUMNet (Sect. 4).

The high accuracy provided by PSUMNet, coupled with its efficiency in terms of compute (number of parameters and floating-point operations) makes our approach an attractive choice for real world deployment on compute restricted embedded and edge devices - see Fig. 1. Code and pretrained models can be accessed at https://github.com/skelemoa/psumnet.

2   Related Work

Skeleton Action Recognition: Since the release of large scale skeleton based datasets [15,19] various CNN [10,14,30], RNN [8,9,30,31] and recently GCN based methods have been proposed for skeleton action recognition. ST-GCN [28] was the first successful approach to model the spatio-temporal relationships for skeleton actions at scale. Many state of the art approaches [2,5,16,22] have adopted and modified this approach to achieve superior results. However, these approaches predominantly process the skeleton joints in a monolithic manner, i.e these approaches process the entire input skeleton at once which can create a bottleneck when the input skeleton becomes denser, e.g. NTU-X [26].

Part Based Approaches: The idea of grouping skeleton joints into different groups has few precedents. Du et al. [8] propose a RNN-based hierarchical grouping of part group representations. Thakkar et al. [25] propose a GCN based approach which applies modified graph convolutions to different part groups. Huang et al. [12] propose a GCN-based approach in which they utilize the higher order part level graph for better pooling and aligning of nodes in the main skeleton graph. More recently, Song et al. [24] propose a part-aware GCN method which utilizes part factorization to aid an attention mechanism to find the most informative part. Some previous part based approaches segment the limbs based on left and right orientation as well (left/right arm, left/right leg etc.) [24,25]. Such segmentation leads to disjoint part groups which contain very small number of joints and are unable to convey useful information. In contrast, our part stream approach creates overlapping part groups with sufficient number of joints to model useful relationships. Also, each individual part group in our setup is registered to the global frame unlike the per-group coordinate system setup in existing approaches. In addition, we employ a combination of part group and coarse version of the full skeleton instead of part-group only approach seen in existing approaches. Our part stream approach allows each part based sub-skeleton to contribute towards the final prediction via decision fusion. To the best of our knowledge, such globally registered independent part stream approach has never been used before.

Multi Stream Training: Earlier approaches [5,16,21] and more recent approaches [2,27] create multiple modalities termed joint, bone and velocity from the raw input skeleton data. The conventional method is to train the given architecture multiple times using different modality data followed by decision fusion. However, this conventional approach with multiple versions of the base architecture greatly increases the total number of parameters. Song et al. [24] attempt a unified modality pipeline wherein early fusion of different modality streams is used to achieve a unified modality representation. However, before the fusion, each modality is processed via multiple independent networks which again increases the count of trainable parameters.

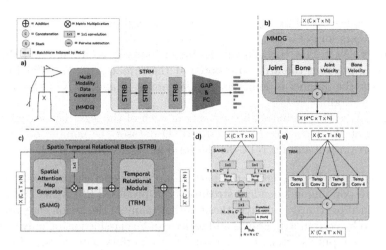

Fig. 3. (a) Overall Architecture of one stream of the proposed architecture. The input skeleton is passed through Multi modality data generator (MMDG), which generates joint, bone, joint velocity and bone velocity data from input and concatenates each modality data into channel dimension as shown in (b). This multi-modal data is processed via Spatio Temporal Relational Module (STRM) followed by global average pooling and FC. (c) Spatio Temporal Relational Block (STRB), where input data is passed through Spatial Attention Map Generator (SAMG) for spatial relation modeling, followed by Temporal Relational Module. As shown in (a) multiple STRB stacked together make the STRM. (d) Spatial Attention Map Generator (SAMG), dynamically models adjacency matrix (A_{hyb}) to model spatial relations between joints. Predefined adjacency matrix (A) is used for regularization. (e) Temporal Relational Module (TRM) consists of multiple temporal convolution blocks in parallel. Output of each temporal convolution block is concatenated to generate final features.

3   Methodology

We first describe our approach for factorizing the input skeleton into part groups and a coarser version of the skeleton (Sect. 3.1). Subsequently, we provide the architectural details of our deep network PSUMNet which processes these part streams (Sect. 3.2).

3.1   Part Stream Factorization

Let X ($\in \mathbb{R}^{3 \times T \times N}$) represent the T-frames, N-joint skeleton configuration of a 3D skeleton pose sequence for an action. We factorize X into following three part groups – see Fig. 2 (right):

1. **Coarse body** (X_b): This is comprised of all joints in the original skeleton for NTURGB+D skeleton topology, 25 joints in total. For the 67-joint dense skeleton topology of NTU-X [26], this stream comprises of all the body joints but without the intermediate joints of each finger for each hand. Specifically,

only 6 joints out of 21 finger joints are considered per hand resulting in total of 37 joints for NTU-X.

2. **Hands** (X_h): This contains all the finger joints in each hand and the arm joints. Note that the arms are rooted at the throat joint. For NTURGB+D dataset, the number of joints for this stream is 13 and for NTU-X, the total number of joints is 48.

3. **Legs** (X_l): This includes all the joints in each of the legs. The leg joints are rooted at the hip joint. For NTURGB+D dataset the number of joints for this stream is 9 and for NTU-X, the total number of joints is 13.

As shown in Fig. 2 (right), the part group sub skeletons are used to train three corresponding independent streams of our proposed PSUMNet (Sect. 3.2). As explained previously, our hypothesis is that many of the action categories are dominated by certain part groups and hence can be classified using only a subset of the entire input skeleton. To leverage this, we perform late decision fusion by performing a weighted average of the prediction scores from each of the part streams to obtain the final classification. Crucially, we change the number of layers in each of the streams in proportion to number of input joints. We use 10, 6 and 4 layers respectively for body, hands and legs streams. This helps restrict the total number of parameters used for the entire protocol.

In contrast with other part based approaches, the part groups in our setting are not completely disjoint. More crucially, the part groups are defined with respect to a shared global coordinate space. Though seemingly redundant due to multiple common joints across part groups, this design choice actually enables global motion information to propagate to the corresponding groups. Another significant advantage of such part stream approach is the better scalability to much denser skeleton datasets such as NTU-X [26] and to sparser datasets such as SHREC [6].

3.2   PSUMNet

In what follows, we explain the architecture of a single part stream of PSUMNet (e.g. $X = X_b$) since the architecture is shared across the part streams. An overview of PSUMNet's single stream architecture can be seen in Fig. 3(a). First, the input skeleton X is passed through Multi Modality Data Generator (MMDG) to create a rich modality aware representation. This feature representation is processed by Spatio-Temporal Relation Module (STRM). Global average pooling (GAP) of the processed result is transformed via fully connected layers (FC) to obtain the per-layer prediction for the single part stream.

Next, we provide details for various modules included in our architecture.

3.3   Multi Modality Data Generator (MMDG)

As shown in Fig. 3(b), this module processes the raw skeleton data and generates the corresponding multi modality data, i.e. joint, bone, joint-velocity and bone-velocity. The joint modality is the raw skeleton data represented by $X = \{x \in$

$\mathbb{R}^{C \times T \times N}\}$, where C, T and N are channels, time steps and joints. The bone modality data is obtained using the following equation:

$$X_{bone} = \{x[:, :, i] - x[:, :, i_{nei}] \,|\, i = 1, 2, ..., N\} \tag{1}$$

where i_{nei} denotes neighboring joint of i based on predefined adjacency matrix. Next we create joint-velocity and bone-velocity modality data using following equations:

$$X_{joint-vel} = \{x[:, t + 1, :] - x[:, t, :] \,|\, t = 1, 2, ..., T, \; x \in X_{joint}\} \tag{2}$$

$$X_{bone-vel} = \{x[:, t + 1, :] - x[:, t, :] \,|\, t = 1, 2, ..., T, \; x \in X_{bone}\} \tag{3}$$

Finally, we concatenate all these four modality data into channel dimension to generate $X \in \mathbb{R}^{4C \times T \times N}$ which is fed as input to the network. Concatenating the modality data helps model the inter-modality relations in a more direct manner.

3.4   Spatio Temporal Relational Module (STRM)

The modality aware representation obtained from MMDG is processed by the Spatial Temporal Relational Module (STRM) as shown in Fig. 3(a). STRM consists of multiple Spatio Temporal Relational Blocks (STRB) stacked one after another. The architecture of a single STRB is shown in Fig. 3(c). Each STRB block contains a Spatial Attention Map Generator (SAMG) to dynamically model different spatial relations between joints followed by Temporal Relational Module (TRM) to model temporal relations between joints.

Spatial Attention Map Generator (SAMG): We dynamically model an Spatial Attention Map for the spatial graph convolutions [2,20]. As shown in Fig. 3(d), we pass the input skeleton through two parallel branches, each consisting a 1×1 convolution and a temporal pooling block. We pair-wise subtract outputs from the parallel branches to model the Attention Map. We add the predefined adjacency matrix A as a regularization to the Attention Map to generate the final hybrid adjacency matrix A_{hyb}, i.e.

$$A_{hyb} = \alpha M(X_{in}) + A \tag{4}$$

where α is a learnable parameter and A is the predefined adjacency matrix. M is defined as:

$$M(X_i) = \sigma(TP(\phi(X_{in})) - TP(\psi(X_{in}))) \tag{5}$$

where σ, ϕ and ψ are 1×1 convolutions, TP is temporal pooling.

Once we obtain this adjacency matrix A_{hyb}, we pass the original input through a 1×1 convolution and multiply the results with the dynamic adjacency matrix to characterize the spatial relations between the joints as follows:

$$X_{out} = A_{hyb} \bigotimes (\theta(X_{in})) \tag{6}$$

where θ is 1×1 convolution block. $\otimes$ is matrix multiplication operation.

Temporal Relation Module (TRM): We use multiple parallel convolution blocks to model the temporal relation between the joints of the input skeleton as shown in Fig. 3(e). Each temporal convolution block is a standard 2D convolution with varying kernel sizes in temporal dimension and with dilation. This helps model temporal relations at multiple scales. The outputs from each of the temporal convolution blocks are concatenated. The result is processed by GAP and FC layers and mapped to a prediction (softmax) layer as mentioned previously.

Since each part group (body, hands, legs) contains significantly different number of joints, we adjust the number of STRBs and depth of the network for each stream accordingly as shown in Fig. 2 (Right). This design choice provides two advantages. *First,* it reduces the total number of parameters by 50%–80%. *Second,* adjusting the depth of the network in proportion to the joint count enables richer dedicated representations for actions whose dynamics are confined to the corresponding part groups, resulting in better performance overall.

4   Experiments

4.1   Datasets

NTURGB+D [19] is a large scale skeleton action recognition dataset with 60 different actions performed by 40 different subjects. The dataset contains 25 joints human skeleton captured using Microsoft Kinect V2 cameras. There are a total of 56,880 action sequences. There are two evaluation protocols for this dataset - First, Cross Subject (XSub) split where action performed by 20 subjects falls into training set and rest into the test set. Second, Cross View (XView) protocol where actions captured via camera ID 2 and 3 are used as training set and actions captured via camera ID 1 are used as test set.

NTURGB+D 120 [15] is an extension of NTURGB+D dataset with additional 60 action categories and a total of 113,945 action samples. The actions are performed by a total of 106 subjects. There are two evaluation protocols for this dataset - First, Cross Subject (XSub) split where action performed by 53 subjects falls into training set and rest into the test set. Second, Cross Setup (XSet) protocol where actions even setup IDs are used as training set and rest as test set.

NTU60-X [26] is a RGB derived skeleton dataset for the sequences of the original NTURGB+D dataset. The skeleton provided in this dataset is much denser and contains 67 joints. There are total of 56,148 action samples and the evaluation protocols are same as the NTURGB+D dataset.

NTU120-X [26] is the extension of NTU60-x dataset and corresponds to the action sequences provided by NTURGB+D 120 dataset. There are total of 113,821 samples in this dataset and the evaluation protocols are same as the NTURGB+D 120 dataset. Following [26], we evaluate our model on only Cross Subject protocol of NTU60-X and NTU120-X datasets.

SHREC [6] is a 3d skeleton based hand gesture recognition dataset. There are a total of 2800 samples with 1960 samples in train set and 840 samples in test set. Each samples has 20–50 frames and gestures are performed by 28 participants ones using only one finger and ones using the whole hand. There are 14 gestures and 28 gestures splits provided by the creators and we report results on both of these splits.

4.2 Implementation and Optimization Details

As shown in Fig. 2 (right), the input skeleton to each of the part stream contains different number of joints. For NTURGB+D dataset, the body stream has input skeleton with a total of 25 joints, hands stream has the input skeleton with a total of 13 joints and legs stream with a total of 9 joints. Within the PSUMNet architecture, we use 10 STRBs for the body stream, 6 STRBs for the hands stream and 4 STRBs to process the legs stream.

We implement PSUMNet using the Pytorch deep learning framework. We use SGD optimizer with 0.1 as the base learning rate and a weight decay of 0.0005. All the models are trained on 4 1080Ti 12 GB GPU systems. For training of 25 joints datasets-NTU60 and NTU120, we use a batch size of 200. For 67 joints datasets-NTU60-X and NTU120-X, due to much denser skeleton, smaller batch size of 65 is used.

4.3 Results

Table 1 compares the performance of proposed PSUMNet with other approaches on Cross Subject (XSub) and Cross View (XView) splits of NTURGB+D dataset [19] and Cross subject (XSub) and Cross Setup (Xset) splits of the NTURGB+D 120 dataset [15]. As can be seen from the Params. column in Table 1, PSUMNet uses the least number of parameters compared to other methods and achieves better or very comparable results across different splits of the datasets. For the harder Cross Subject split of both NTURGB+D and NTURGB+D 120, PSUMNet achieves state of the art performance compared to other approaches which use 100%–400% more parameters. This shows the superiority of PSUMNet both in terms of performance and efficiency - also see Fig. 1.

We also compare the performance of only body stream of PSUMNet with single stream (i.e. only joint, only bone) performance of other approaches in Table 2 for Xsub split of NTURGB+D and NTURGB+D 120 datasets. As can be seen, PSUMNet outperforms other approaches by a margin of 1–2% for NTURGB+D and by 2–3% for NTURGB+D 120 using almost the same or lesser number of parameters. This also supports our hypothesis that part stream based unified modality approach is much more efficient compared to conventional independent modality streams approach.

Trivedi et al. [26] introduced NTU60-X and NTU120-X, extensions of existing NTURGB+D and NTURGB+D 120 datasets with 67 joint dense skeletons containing fine-grained finger joints within the full body pose tree. Handling such large number of joints while keeping the total parameters of the model in bounds

Table 1. Comparison with state of the art approaches for NTURGB+D and NTURGB+D 120 dataset. Model parameters are in millions ($\times 10^6$) and FLOPs are in billions ($\times 10^9$). *: These numbers are cumulative over all the streams used by respective models as per their training protocol.

Type	Model	Params. (M) *	FLOPs (G) *	NTU60		NTU120	
				XSub	XView	Xsub	XSet
CNN based	VA-Fusion [30]	24.6	-	89.4	95.0	-	-
	TaCNN+ [27]	4.4	1.0	90.7	95.1	86.7	87.3
GCN based	ST-GCN [28]	3.1	16.3	81.5	88.3	70.7	73.2
	RA-GCN [23]	6.2	32.8	87.3	93.6	81.1	82.7
	2s-AGCN [21]	6.9	37.3	88.5	95.1	82.9	84.9
	PA-ResGCN [24]	3.6	18.5	90.9	96.0	87.3	88.3
	DDGCN [13]	-	-	91.1	97.1	-	-
	DGNN [20]	26.2	-	89.9	95.0	-	-
	MS-G3D [16]	6.4	48.8	91.5	96.2	86.9	88.4
	4s-ShiftGCN [5]	2.8	10.0	90.7	96.5	85.9	87.6
	DC-GCN+ADG [4]	4.9	25.7	90.8	96.6	86.5	88.1
	DualHead-Net [1]	12.0	-	92.0	96.6	88.2	89.3
	CTR-GCN [2]	5.6	7.6	92.4	96.8	88.9	90.6
Attention based	DSTA-Net [22]	14.0	64.7	91.5	96.4	86.6	89.0
	ST-TR [18]	12.1	259.4	89.9	96.1	82.7	84.7
	4s-MST-GCN [3]	12.0	-	91.5	96.6	87.5	88.8
	PSUMNet (ours)	2.8	2.7	**92.9**	96.7	**89.4**	90.6

Table 2. Comparison of only body stream of PSUMNet with the best performing modality (i.e. only joint, only bone) of state of the art approaches for NTURGB+D 60 and 120 dataset on Cross Subject protocol.

Model	Params. (M)	NTU60	NTU120
PA-ResGCN [24]	3.6	90.9	87.3
MS-G3D (joint) [16]	3.2	89.4	84.5
1s-ShiftGCN (joint) [5]	0.8	87.8	80.9
DSTA-Net (bone) [22]	3.5	88.4	84.4
DualHead-Net (bone) [1]	3.0	90.7	86.7
CTR-GCN (bone) [2]	1.4	90.6	85.7
TaCNN+ (joint) [27]	1.1	89.6	82.6
MST-GCN (bone) [3]	3.0	89.5	84.8
PSUMNet (ours) (body)	1.4	**91.9**	**88.1**

is a difficult task. However, as shown in Table 3, PSUMNet achieves state of the art performance for both NTU60-X and NTU120-X datasets. Total parameters increase by a small margin for PSUMNet to handle the additional joints, yet it is worth noting that other competing approaches use 100%–400% more parameters

Table 3. Comparison with state of the art approaches for dense skeleton datasets NTU60-X and NTU120-X.

Model	Params. (M)	NTU60-X	NTU120-X
PA-ResGCN [24]	3.6	91.6	86.4
MS-G3D [16]	6.4	91.8	87.1
4s-ShiftGCN [5]	2.8	91.8	86.2
DSTA-Net [22]	14.0	93.5	87.8
CTR-GCN [2]	5.6	93.9	88.3
PSUMNet (ours)	3.2	**94.7**	**89.1**

Table 4. Comparison with state of the art approaches for SHREC skeleton hand gesture recognition dataset.

Model	Params. (M)	14 gestures	28 gestures
Key-Frame CNN [6]	7.9	82.9	71.9
CNN+LSTM [17]	8.0	89.8	86.3
Parallel CNN [7]	13.8	91.3	84.4
STA-Res TCN [11]	6.0	93.6	90.7
DDNet [29]	1.8	94.6	91.9
DSTANet [22]	14.0	**97.0**	**93.9**
PSUMNet (ours)	**0.9**	95.5	93.1

as compared to PSUMNet. This shows the benefit of using part based streams approach for dense skeleton representation as well.

To further explore the generalization capability of our proposed method, we evaluate performance of PSUMNet for skeleton based hand gestures recognition dataset, SHREC [6]. Taking advantage of part based stream approach, we train only the hands stream of PSUMNet. As shown in Table 4, PSUMNet achieves comparable results to existing state of the art method (DSTANet [22]) which uses 1400% more parameters. PSUMNet outperforms the second best approach (DDNet [29]) which uses 100% more parameters.

Overall, Tables 1, 3 and 4 comprehensively show that proposed PSUMNet achieves state of the art performance, generalizes easily across a range of action datasets and uses a significantly smaller number of parameters compared to other methods.

4.4 Analysis

As explained in Sect. 3.1, we train PSUMNet using three part streams namely body, hands and legs streams and report the ensembled results from all the three streams. To understand the advantage of the proposed part stream approach, we compare stream wise per class accuracy for NTU120-X and NTU60-X of

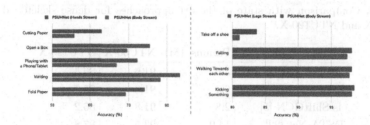

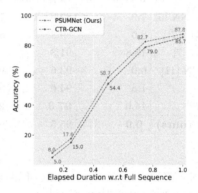

Fig. 4. Comparing per class accuracy after training PSUMNet using only Hands stream and only body stream for NTU120-X dataset (Left) and only Legs stream with only body stream for NTU60-X dataset (Right). On observing the class labels we can see that all the actions in the left plot are dominated by hand joints movements and all the actions in the right plot are dominated by leg joints movement and hence streams corresponding to these parts are able to classify these classes better which is in line with our hypothesis

Fig. 5. Comparing PSUMNet with current state of the art method, CTR-GCN on partially observed sequences for NTURGB+D 120 (XSub) dataset. Annotated numbers for each line plot denote accuracy of both models on partial sequences.

PSUMNet. Figure 4 (left) depicts the per class comparison setting for per class accuracy comparison between the 'only hands stream' and 'only body stream' setting of PSUMNet for NTU120-X dataset. The classes shown correspond to those with largest (positive) gain in per class accuracy while using only hand stream. Upon observing the action labels of these classes, ("Cutting Paper", "Writing", "Folding Paper"), it is evident that these classes are dominated by hand joints movements and hence are better classified using only a subset of input skeleton which has dedicated representations for hand joints as opposed to using entire skeleton in a monolithic fashion.

Similarly, we also compare the per class accuracy while using only legs stream against only body stream of PSUMNet for NTU60-X dataset as shown in Fig. 4 (right). In this case too, the class labels with highest positive gain while using only legs stream are dominated correspond to expected classes such as "Walking", "Kicking".

The above results can also be appreciated better by studying the number of parameters in each of the part based stream. The body stream in PSUMNet has $1.4M$ parameters, Hands stream has $0.9M$ and legs stream has $0.5M$ parameters. Hence, hands stream while using only 65% of the total parameters of the body stream and legs stream while using only 35% of the body stream parameters can identify those classes better which are dominated by joints corresponding to each part stream.

Early Action Recognition: In the experiments so far, evaluation was conducted on the full action sequence. In other words, the predicted label is known only after all the frames are provided to the network. However, there can be anticipatory scenarios where we may wish to know the predicted action label without waiting for the entire action to finish. To examine the performance in such scenarios, we create test sequences whose length is determined in terms of a fraction of the total sequence length. We study the trends in accuracy as the % of total sequence length is steadily increased. For comparison, we study PSUMNet with the state of the art network, CTR-GCN [2]. As can be seen in Fig. 5, PSUMNet consistently outperforms CTR-GCN for partially observed sequences, indicating its suitability for early action recognition.

4.5   Ablations

To understand the contribution of each part stream in PSUMNet, we provide individual stream wise performance of PSUMNet on NTU60 and NTU120 datasets Cross Subject splits as ablations in Table 5.

At a single stream level, the body stream achieves higher accuracy compared to hands and legs stream. This is expected since the body stream includes a coarse version of all the joints. However, as mentioned previously (Sect. 4.4), hands and legs streams classify actions dominated by respective joints better. Therefore, accuracies of Body+Hands (row 4 in Table 5) and Body+Legs (row 5) variants are higher than only the body stream. Legs stream achieves lower accuracy as compared to body and hands stream because there are only a small subset of action categories which are dominated by leg joints movements. However, as with hands stream, legs stream benefits classes which involve significant leg joints movements.

Our proposed part groups factorization registers each group's sub-skeleton in a global frame of reference (see Fig. 2). Further, all the part groups are not disjoint and have overlapping joints to better propagate global motion information through the network (Sect. 3.1). To justify our choice of globally registered part groups, we perform an ablation with a different part grouping strategy, with each part group being disjoint and in a local frame of reference. Specifically, the ablation setup for body stream includes on 9 torso joints (including shoulders and hips joints), hands stream includes only 12 joints and legs stream includes only 8 joints. It is important to notice here that unlike our original strategy, both legs

Table 5. Individual streams performance on NTURGB+D and NTURGB+D 120 Cross Subject dataset.

Type	Stream	Params. (M)	NTU60	NTU120
Part streams	Body	1.4	91.9	88.1
	Hands	0.9	90.3	85.8
	Legs	0.5	60.4	50.6
	Body + Hands	2.3	92.4	89.0
	Body + Legs	1.9	92.1	87.9
	Hands + Legs	1.4	90.9	86.5
Disjoint parts	Body + Hands + Legs	2.8	89.6	86.1
Modalities in PSUMNet	Joint	2.8	90.3	86.1
	Bone	2.8	90.1	87.6
	Joint-Vel	2.8	88.5	82.7
	Bone-Vel	2.8	87.6	83.2
	Joint + Bone	2.8	91.4	88.6
PSUMNet		2.8	**92.9**	**89.4**

and hands in corresponding part stream are not connected. As expected, such strategy fails to capture global motion information unlike our proposed method (c.f. 'Disjoint parts' row and last row in Table 5).

To further investigate contribution of each data modality in our proposed unified modality method, we provide ablation studies with PSUMNet trained on single and two modalities instead of four (c.f. 'Modalities in PSUMNet' rows and last row in Table 5). We notice that PSUMNet benefits most by joint and bone modalities compared to velocity modalities. However, the best performance is obtained by utilizing all the modalities.

5   Conclusion

In this work, we present Part Streams Unified Modality Network PSUMNet to efficiently tackle the challenging task of scalable pose-based action recognition. PSUMNet uses part based streams and avoids treating the input skeleton in monolithic fashion as done by contemporary approaches. This choice enables richer and dedicated representations especially for actions dominated by a small subset of localized joints (hands, legs). The unified modality approach introduced in this work enables efficient utilization of the inter-modality correlations. Overall, the design choices provide two key benefits – (1) they help attain state of the art performance using significantly smaller number of parameters compared to existing methods (2) they allow PSUMNet to easily scale to both sparse and dense skeleton action datasets in distinct domains (full body, hands) while maintaining high performance. PSUMNet is an attractive choice for pose-based action recognition especially in real world deployment scenarios involving compute restricted embedded and edge devices.

References

1. Chen, T., et al.: Learning multi-granular spatio-temporal graph network for skeleton-based action recognition. In: Proceedings of the 29th ACM International Conference on Multimedia, pp. 4334–4342 (2021)
2. Chen, Y., Zhang, Z., Yuan, C., Li, B., Deng, Y., Hu, W.: Channel-wise topology refinement graph convolution for skeleton-based action recognition. In: Proceedings of the IEEE/CVF International Conference on Computer Vision, pp. 13359–13368 (2021)
3. Chen, Z., Li, S., Yang, B., Li, Q., Liu, H.: Multi-scale spatial temporal graph convolutional network for skeleton-based action recognition. In: Proceedings of the AAAI Conference on Artificial Intelligence, vol. 35, pp. 1113–1122 (2021)
4. Cheng, K., Zhang, Y., Cao, C., Shi, L., Cheng, J., Lu, H.: Decoupling GCN with DropGraph module for skeleton-based action recognition. In: Vedaldi, A., Bischof, H., Brox, T., Frahm, J.-M. (eds.) ECCV 2020. LNCS, vol. 12369, pp. 536–553. Springer, Cham (2020). https://doi.org/10.1007/978-3-030-58586-0_32
5. Cheng, K., Zhang, Y., He, X., Chen, W., Cheng, J., Lu, H.: Skeleton-based action recognition with shift graph convolutional network. In: Proceedings of the IEEE Conference on Computer Vision and Pattern Recognition (CVPR) (2020)
6. De Smedt, Q., Wannous, H., Vandeborre, J.P., Guerry, J., Saux, B.L., Filliat, D.: 3D hand gesture recognition using a depth and skeletal dataset: SHREC'17 track. In: Proceedings of the Workshop on 3D Object Retrieval, pp. 33–38 (2017)
7. Devineau, G., Xi, W., Moutarde, F., Yang, J.: Convolutional neural networks for multivariate time series classification using both inter-and intra-channel parallel convolutions. In: Reconnaissance des Formes, Image, Apprentissage et Perception (RFIAP 2018) (2018)
8. Du, Y., Wang, W., Wang, L.: Hierarchical recurrent neural network for skeleton based action recognition. In: 2015 IEEE Conference on Computer Vision and Pattern Recognition (CVPR), pp. 1110–1118 (2015). https://doi.org/10.1109/CVPR.2015.7298714
9. Han, F., Reily, B., Hoff, W., Zhang, H.: Space-time representation of people based on 3D skeletal data: a review. Comput. Vis. Image Underst. **158**, 85–105 (2017)
10. Hernandez Ruiz, A., Porzi, L., Rota Bulò, S., Moreno-Noguer, F.: 3D CNNs on distance matrices for human action recognition. In: Proceedings of the 2017 ACM on Multimedia Conference, MM 2017, pp. 1087–1095. ACM, New York (2017). https://doi.org/10.1145/3123266.3123299
11. Hou, J., Wang, G., Chen, X., Xue, J.-H., Zhu, R., Yang, H.: Spatial-temporal attention res-TCN for skeleton-based dynamic hand gesture recognition. In: Leal-Taixé, L., Roth, S. (eds.) ECCV 2018. LNCS, vol. 11134, pp. 273–286. Springer, Cham (2019). https://doi.org/10.1007/978-3-030-11024-6_18
12. Huang, L., Huang, Y., Ouyang, W., Wang, L.: Part-level graph convolutional network for skeleton-based action recognition. In: Proceedings of the AAAI Conference on Artificial Intelligence, vol. 34, no. 07, pp. 11045–11052, April 2020
13. Korban, M., Li, X.: DDGCN: a dynamic directed graph convolutional network for action recognition. In: Vedaldi, A., Bischof, H., Brox, T., Frahm, J.-M. (eds.) ECCV 2020. LNCS, vol. 12365, pp. 761–776. Springer, Cham (2020). https://doi.org/10.1007/978-3-030-58565-5_45

14. Li, C., Zhong, Q., Xie, D., Pu, S.: Co-occurrence feature learning from skeleton data for action recognition and detection with hierarchical aggregation. In: Proceedings of the 27th International Joint Conference on Artificial Intelligence, IJCAI 2018, pp. 786–792. AAAI Press (2018)
15. Liu, J., Shahroudy, A., Perez, M., Wang, G., Duan, L.Y., Kot, A.C.: NTU RGB+D 120: a large-scale benchmark for 3D human activity understanding. IEEE Trans. Pattern Anal. Mach. Intell. (2019). https://doi.org/10.1109/TPAMI.2019.2916873
16. Liu, Z., Zhang, H., Chen, Z., Wang, Z., Ouyang, W.: Disentangling and unifying graph convolutions for skeleton-based action recognition. In: The IEEE Conference on Computer Vision and Pattern Recognition (CVPR), June 2020
17. Nunez, J.C., Cabido, R., Pantrigo, J.J., Montemayor, A.S., Velez, J.F.: Convolutional neural networks and long short-term memory for skeleton-based human activity and hand gesture recognition. Pattern Recogn. **76**, 80–94 (2018)
18. Plizzari, C., Cannici, M., Matteucci, M.: Skeleton-based action recognition via spatial and temporal transformer networks. Comput. Vis. Image Underst. **208**, 103219 (2021)
19. Shahroudy, A., Liu, J., Ng, T.T., Wang, G.: NTU RGB+D: a large scale dataset for 3D human activity analysis. In: The IEEE Conference on Computer Vision and Pattern Recognition (CVPR), June 2016
20. Shi, L., Zhang, Y., Cheng, J., Lu, H.: Skeleton-based action recognition with directed graph neural networks. In: Proceedings of the IEEE/CVF Conference on Computer Vision and Pattern Recognition, pp. 7912–7921 (2019)
21. Shi, L., Zhang, Y., Cheng, J., Lu, H.: Two-stream adaptive graph convolutional networks for skeleton-based action recognition. In: Proceedings of the IEEE/CVF Conference on Computer Vision and Pattern Recognition, pp. 12026–12035 (2019)
22. Shi, L., Zhang, Y., Cheng, J., Lu, H.: Decoupled spatial-temporal attention network for skeleton-based action-gesture recognition. In: ACCV (2020)
23. Song, Y.F., Zhang, Z., Shan, C., Wang, L.: Richly activated graph convolutional network for robust skeleton-based action recognition. IEEE Trans. Circ. Syst. Video Technol. **31**(5), 1915–1925 (2020)
24. Song, Y.F., Zhang, Z., Shan, C., Wang, L.: Stronger, faster and more explainable: a graph convolutional baseline for skeleton-based action recognition. In: Proceedings of the 28th ACM International Conference on Multimedia (ACMMM), pp. 1625–1633. Association for Computing Machinery, New York (2020). https://doi.org/10.1145/3394171.3413802
25. Thakkar, K.C., Narayanan, P.J.: Part-based graph convolutional network for action recognition. In: British Machine Vision Conference 2018, BMVC 2018, Newcastle, UK, 3–6 September 2018, p. 270. BMVA Press (2018). bmvc2018.org/contents/papers/1003.pdf
26. Trivedi, N., Thatipelli, A., Sarvadevabhatla, R.K.: NTU-X: an enhanced large-scale dataset for improving pose-based recognition of subtle human actions. In: Proceedings of the Twelfth Indian Conference on Computer Vision, Graphics and Image Processing, pp. 1–9 (2021)
27. Xu, K., Ye, F., Zhong, Q., Xie, D.: Topology-aware convolutional neural network for efficient skeleton-based action recognition. In: Proceedings of the AAAI Conference on Artificial Intelligence, vol. 36 (2022)
28. Yan, S., Xiong, Y., Lin, D.: Spatial temporal graph convolutional networks for skeleton-based action recognition. In: AAAI (2018)
29. Yang, F., Wu, Y., Sakti, S., Nakamura, S.: Make skeleton-based action recognition model smaller, faster and better. In: Proceedings of the ACM Multimedia Asia, pp. 1–6 (2019)

30. Zhang, P., Lan, C., Xing, J., Zeng, W., Xue, J., Zheng, N.: View adaptive neural networks for high performance skeleton-based human action recognition. IEEE Trans. Pattern Anal. Mach. Intell. **41**(8), 1963–1978 (2019)
31. Zhao, R., Wang, K., Su, H., Ji, Q.: Bayesian graph convolution LSTM for skeleton based action recognition. In: The IEEE International Conference on Computer Vision (ICCV), October 2019

YOLO5Face: Why Reinventing a Face Detector

Delong Qi[1], Weijun Tan[1,2,3]([✉]) [iD], Qi Yao[2], and Jingfeng Liu[2]

[1] Shenzhen Deepcam, Shenzhen, China
{delong.qi,weijun.tan}@deepcam.com
[2] Jovision-Deepcam Research Institute, Shenzhen, China
{sz.twj,sz.yaoqi,sz.ljf}@jovision.com, {qi.yao,jingfeng.liu}@deepcam.com
[3] LinkSprite Technology, Longmont, CO 80503, USA
weijun.tan@linksprite.com

Abstract. Tremendous progress has been made on face detection in recent years using convolutional neural networks. While many face detectors use designs designated for the detection of face, we treat face detection as a general object detection task. We implement a face detector based on YOLOv5 object detector and call it YOLO5Face. We add a five-point landmark regression head into it and use the Wing loss function. We design detectors with different model sizes, from a large model to achieve the best performance to a super small model for real-time detection on an embedded or mobile device. Experiment results on the WiderFace dataset show that our face detectors can achieve state-of-the-art performance in almost all the Easy, Medium, and Hard subsets, exceeding the more complex designated face detectors. The code is available at https://github.com/deepcam-cn/yolov5-face.

Keywords: Face detection · Convolutional neural network · YOLO · Real-time · Embedded device · Object detection

1 Introduction

Face detection is a very important computer vision task. Tremendous progress has been made since deep learning, particularly convolutional neural network (CNN), has been used in this task. As the first step of many tasks, including face recognition, verification, tracking, alignment, expression analysis, face detection attracts a lot of research and developments in academia and industry. And the performance of face detection has improved significantly over the years. For a survey of the face detection, please refer to the benchmark results [39,40]. There are many methods in this field from different perspectives. Research directions include the design of CNN networks, loss functions, data augmentations, and training strategies. For example, in the YOLOv4 paper, the authors explore all these research directions and propose the YOLOV4 object detector based on optimizations of network architecture, selection of bags of freebies, and selection of bags of specials [1].

L. Karlinsky et al. (Eds.): ECCV 2022 Workshops, LNCS 13805, pp. 228–244, 2023.
https://doi.org/10.1007/978-3-031-25072-9_15

In our approach, we treat face detection as a general object detection task. We have the same intuition as the TinaFace [55]. Intuitively, a face is an object. As discussed in the TinaFace [55], from the perspective of data, the properties that a face has, like pose, scale, occlusion, illumination, blur, etc., also exist in other objects. The unique properties of faces, like expression and makeup, can also correspond to distortion and color in objects. Landmarks are special to face, but they are not unique either. They are just key points of an object. For example, in license plate detection, landmarks are also used. And adding landmark regression in the object prediction head is straightforward. Then from the perspective of challenges encountered by face detection like multi-scale, small faces, and dense scenes, they all exist in generic object detection. Thus, face detection is just a sub-task of general object detection.

In this paper, we follow this intuition and design a face detector based on the YOLOv5 object detector [42]. We modify the design for face detection considering large faces, small faces, and landmark supervision for different complexities and applications. Our goal is to provide a portfolio of models for different applications, from very complex ones to get the best performance to very simple ones to get the best trade-off of performance and speed on embedded or mobile devices.

Our main contributions are summarized as following,

- We redesign the YOLOV5 object detector [42] as a face detector and call it YOLO5Face. We implement key modifications to the network to improve the performance in terms of mean average precision (mAP) and speed. The details of these modifications will be presented in Sect. 3.
- We design a series of models of different model sizes, from large models to medium models, to super small models, for needs in different applications. In addition to the backbone used in YOLOv5 [42], we implement a backbone based on ShuffleNetV2 [26], which gives the state-of-the-art (SOTA) performance and fast speed for a mobile device.
- We evaluate our models on the WiderFace [40] dataset. On VGA resolution images, almost all our models achieve the SOTA performance and fast speed. This proves our goal; as the tile of this paper claims, we do not need to reinvent a face detector since the YOLO5Face can accomplish it.

Please note that our work contributes not to the novelty of new ideas but the engineering community of face detection. Since open source, our code has been used widely in many projects.

2   Related Work

2.1   Object Detection

General object detection aims at locating and classifying the pre-defined objects in a given image. Before deep CNN is used, traditional face detection uses hand-crafted features, like HAAR, HOG, LBP, SIFT, DPM, ACF, etc. The seminal work by Viola and Jones [51] introduces integral image to compute HAAR-like features. For a survey of face detection using hand-crafted features, please refer to [38,52].

Since deep CNN shows its power in many machine learning tasks, face detection is dominated by deep CNN methods. There are two-stage and one-stage object detectors. Typical two-stage methods are the RCNN family, including RCNN [12], fast-RCNN [11], faster-RCNN [31], mask-RCNN [15], Cascade-RCNN [2].

The two-stage object detectors have very good performance but suffer from long latency and slow speed. To overcome this problem, one-stage object detectors are studied. Typical one-stage networks include SSD [23], YOLO [1,28–30,42].

Other object detection networks include FPN [21], MMDetection [4], EfficientDet [33], transformer (DETR) [3], Centernet [8,54], and so on.

2.2 Face Detection

The research for face detection follows general object detection. After the most popular and challenging face detection benchmark WiderFace dataset [40] is released, face detection develops rapidly, focusing on the extreme and real variation problems including scale, pose, occlusion, expression, makeup, illumination, blur, etc.

A lot of methods are proposed to deal with these problems, particularly the scale, context, and anchor in order to detect small faces. These methods include MTCNN [44], FaceBox [48], S3FD [47], DSFD [19], RetinaFace [7], RefineFace [45], and the most recent ASFD [43], MaskFace [41], TinaFace [55], MogFace [24], and SCRFD [13]. For a list of popular face detectors, the readers are referred to the WiderFace website [39].

It is worth noting that some of these face detectors explore unique characteristics of a human face; the others are just general object detectors adopted and modified for face detection. Use RetinaFace [7] as an example. It uses landmark (2D and 3D) regression to help the supervision of face detection, while TinaFace [55] is simply a general object detector.

2.3 YOLO

YOLO first appeared in 2015 [29] as a different approach than popular two-stage approaches. It treats object detection as a regression problem rather than a classification problem. It performs all the essential stages to detect an object using a single neural network. As a result, it achieves not only very good detection performance but also achieves real-time speed. Furthermore, it has excellent generalization capability, and can be easily trained to detect different objects.

Over the next five years, the YOLO algorithm has been upgraded to five versions with many innovative ideas from the object detection community. The first three versions - YOLOv1 [29], YOLOv2 [30], YOLOv3 [28] are developed by the author of the original YOLO algorithm. Out of these three versions, the YOLOv3 [28] is a milestone with big improvements in performance and speed by introducing multi-scale features (FPN) [21], better backbone network (Darknet53), and replacing the Softmax classification loss with the binary cross-entropy loss.

In early 2020, after the original YOLO authors withdrew from the research field, YOLOv4 [1] was released by a different research team. The team explores a lot of options in almost all aspects of the YOLOv3 [28] algorithm, including the backbone and what they call bags of freebies and bags of specials. It achieves 43.5% AP (65.7% AP50) for the MS COCO dataset at a real-time speed of 65 FPS on Tesla V100.

One month later, the YOLOv5 [42] was released by another different research team. From the algorithm perspective, the YOLOv5 [42] does not have many innovations. And the team does not publish a paper. These bring quite some controversies about if it should be called YOLOv5. However, due to its significantly reduced model size, faster speed, and similar performance as YOLOv4 [1], and full implementation in Python (Pytorch), it is welcome by the object detection community.

3   YOLO5Face Face Detector

Summary of Key Modifications. In this section, we present the key modifications we make in YOLOv5 and make it a face detector - YOLO5Face. 1) We add a landmark regression head to the YOLOv5 network. 2) We replace the Focus layer of YOLOv5 [42] with a Stem block structure [37]. 3) We change the SPP block [16] and use a smaller kernel. 4) We add a P6 output block with a stride of 64. 5) We optimize the data augmentation methods for face detection. 6) We design two super light-weight models based on ShuffleNetV2 [26].

3.1   Network Architecture

We use the YOLOv5 object detector [42] as our baseline and optimize it for face detection. The network architecture of our YOLO5Face face detector is depicted in Fig. 1. It consists of the backbone, neck, and head. In YOLOv5, a newly designed backbone called CSPNet [42] is used. In the neck, an SPP [16] and a PAN [22] are used to aggregate the features. In the head, regression and classification are both used.

In Fig. 1(a), the overall network architecture is depicted, where a newly added P6 block is highlighted with a box. In Fig. 1(b), a key block called CBS is defined, which consists of a Conv layer, BN layer, and a SILU [9] activation function. This CBS block is used in many other blocks. In Fig. 1(c), an output label for the head is shown, which includes the bounding box (bbox), confidence (conf),

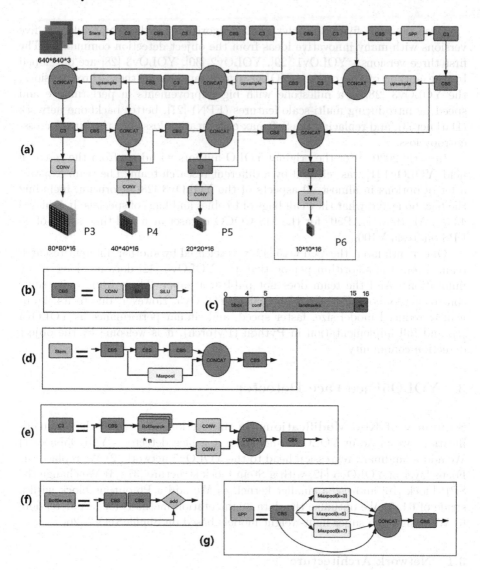

Fig. 1. The proposed YOLO5Face network architecture.

classification (cls), and five-point landmarks. The landmarks are our addition to the YOLOv5 to make it a face detector with landmark output. If without the landmark, the last dimension 16 should be 6. Please note that the output dimensions 80 * 80 * 16 in P3, 40 * 40 * 16 in P4, 20 * 20 * 16 in P5, and 10 * 10 * 16 in optional P6 are for every anchor. The real dimension should be multiplied by the number of anchors.

In Fig. 1(d), a Stem structure [37] is shown, which is used to replace the original Focus layer in YOLOv5. The introduction of the Stem block into YOLOv5 for face detection is one of our innovations.

In Fig. 1(e), a CSP block (C3) is shown. This block is inspired by the DenseNet [18]. However, instead of adding the full input and the output after some CNN layers, the input is separated into two halves. One half is passed through a CBS block, a number of Bottleneck blocks, which is shown in Fig. 1(f), the another half is sent to a Conv layer. The outputs are then concatenated, followed by another CBS block.

Figure 1(g), an SPP block [16] is shown. In this block, the three kernel sizes 13×13, 9×9, and 5×5 in YOLOv5 are revised to 7×7, 5×5, and 3×3 in our face detector. This has been shown as one of the innovations that improve face detection performance.

Note that we only consider VGA-resolution input images. The longer edge of the input image is scaled to 640, and the shorter edge is scaled accordingly. The shorter edge is also adjusted to be a multiple of the largest stride of the SPP block. For example, when P6 is not used, the shorter edge needs to be a multiple of 32; when P6 is used, the shorter edge needs to be a multiple of 64.

3.2 Landmark Regression

Landmarks are important characteristics of a human face. They can be used to do face alignment, face recognition, face express analysis, age analysis, etc. Traditional landmarks consist of 68 points. They are simplified to 5 points in MTCNN [44] Since then, the five-point landmarks have been used widely in face recognition. The quality of landmarks affects the quality of face alignment and face recognition.

The general object detector does not include landmarks. It is straightforward to add it as a regression head. Therefore, we add it to our YOLO5Face. The landmark outputs will be used in aligning face images before they are sent to the face recognition network.

General loss functions for landmark regression are L2, L1, or smooth-L1. The MTCNN [44] uses the L2 loss function. However, it has been found these loss functions are not sensitive to small errors. To overcome this problem, the Wing-loss is proposed [10],

$$wing(x) = \begin{cases} w \cdot ln(1 + |x|/e), & \text{if } x < w \\ |x| - C, & \text{otherwise} \end{cases} \quad (1)$$

where x is the error signal. The non-negative w sets the range of the non-linear part to $(-w, w)$, e limits the curvature of the nonlinear region, and $C = w - w ln(1 + w/e)$ is a constant that smoothly links the piecewise-defined linear and nonlinear parts. Plotted in Fig. 2 is this Wing loss function with different parameters w and e. It can be seen that the response at a small error area near zero is boosted compared to the L2, L1, or smooth-L1 functions.

The loss functions for landmark point vector $s = \{s_i\}$, and its ground truth $s' = \{s_i\}$, where $i = 1, 2, ..., 10$, is defined as,

$$loss_L(s) = \sum_i wing(s_i - s_i') \tag{2}$$

Let the general object detection loss function of YOLOv5 be $loss_O(bounding\text{-}box, class, probability)$, then the new total loss function is,

$$loss(s) = loss_O + \lambda_L \cdot loss_L \tag{3}$$

where the λ_L is a weighting factor for the landmark regression loss function.

3.3   Stem Block Structure

We use a stem block similar to [37]. The stem block is shown in Fig. 1(d). With this stem block, we implement a stride = 2 in the first spatial down-sampling on the input image and increase the number of channels. With this stem block, the computation complexity only increases marginally, while a strong representation capability is ensured.

3.4   SPP with Smaller Kernels

Before forwarding to the feature aggregation block in the neck, the output feature maps of the YOLO5 backbone are sent to an additional SPP block [16] to increase the receptive field and separate out the most important features. Instead of many CNN models containing fully connected layers which only accept input images of specific dimensions, SPP is proposed to aim at generating a fixed-size output irrespective of the input size. In addition, SPP also helps to extract important features by pooling multi-scale versions of itself.

In YOLO5, three kernel sizes 13×13, 9×9, 5×5 are used [42]. We revise them to use smaller size kernels 7×7, 5×5, and 3×3. These smaller kernels help to detect small faces more efficiently, and increase the overall face detection performance.

3.5   P6 Output Block

The backbone of the YOLO object detector has many layers. As the feature becomes more and more abstract as the layers go deeper, the spatial resolution of feature maps decreases due to downsampling, which leads to a loss of spatial information as well as fine-grained features. In order to preserve these fine-grained features, the FPN [21] is introduced to YOLOv3 [28].

In FPN [21], the fine-grained features take a long path traveling from low-level to high-level layers. To overcome this problem, the PAN is proposed to add a bottom-up augmentation path along the top-down path used in FPN. In addition, in the connection of the feature maps to the lateral architecture, the element-wise addition operation is replaced with concatenation. In FPN, object predictions are made independently on different scale levels, which do not utilize information from other feature maps and may produce duplicated predictions. In PAN [22], the output feature maps of the bottom-up augmentation pyramid are fused by using (Region of Interest) ROI align and fully connected layers with the element-wise max operation.

In YOLOv5, there are three output blocks in the PAN output feature maps, called P3, P4, and P5, corresponding to $80 \times 80 \times 16$, $40 \times 40 \times 16$, $20 \times 20 \times 16$, with strides 8, 16, 32, respectively. In our YOLO5Face, we add an extra P6 output block, whose feature map is $10 \times 10 \times 16$ with stride 64. This modification particularly helps the detection of large faces. While almost all face detectors focus on improving the detection of small faces, the detection of large faces can be easily overlooked. We fill this hole by adding the P6 output block.

3.6 ShuffleNetV2 as Backbone

The ShuffleNet [49] is an extremely efficient CNN for a mobile device. The key block is called the ShuffleNet block. It utilizes two new operations, pointwise group convolution and channel shuffle, to greatly reduce computation cost while maintaining accuracy.

The ShuffleNetv2 [49] is an improved version of ShuffleNet. It borrows the shortcut network architecture similar to the DenseNet [18], and element-wise addition is changed to concatenation, similar to the change in PAN [22] in YOLOv5 [42]. But different from DenseNet, ShuffleNetV2 does not densely concatenate, and after the concatenation, channel shuffling is used to mix the features. This makes the ShuffleNetV2 a super-fast network.

We use the ShuffleNetV2 as the backbone in YOLOv5 and implement super small face detectors YOLOv5n-Face, and YOLOv5n0.5-Face.

4 Experiments

4.1 Dataset

The WiderFace dataset [40] is the largest face detection dataset, which contains 32,203 images and 393,703 faces. For its large variety of scale, pose, occlusion, expression, illumination, and event, it is close to reality and is very challenging.

The whole dataset is divided into train/validation/test sets by ratio 50%/10%/40% within each event class. Furthermore, each subset is defined into three levels of difficulty: Easy, Medium, and Hard. As its name indicates, the Hard subset is the most challenging. So the performance on the Hard subset reflects best the effectiveness of a face detector.

Unless specified otherwise, the WiderFace dataset [40] is used in this work. In the face recognition with YOLO5Face landmark and alignment, the Webface dataset [56] is used. The FDDB dataset [36] is used in testing to demonstrate our model's performance on cross-domain datasets.

Table 1. Detail of implemented YOLO5Face models, where (D, W) are the depth and width multiples of the YOLOv5 CSPNet [42]. The number of parameters and Flops are listed in Table 4.

Model	Backbone	(D, W)	With P6?
YOLOv5s	YOLO5-CSPNet [42]	(0.33, 0.50)	No
YOLOv5s6	YOLO5-CSPNet	(0.33, 0.50)	Yes
YOLOv5m	YOLO5-CSPNet	(0.50, 0.75)	No
YOLOv5m6	YOLO5-CSPNet	(0.50, 0.75)	Yes
YOLOv5l	YOLO5-CSPNet	(1.0, 1.0)	No
YOLOv5l6	YOLO5-CSPNet	(1.0, 1.0)	Yes
YOLOv5n	ShuffleNetv2 [26]	-	No
YOLOv5n-0.5	ShuffleNetv2-0.5 [26]	-	No

4.2   Implementation Details

We use the YOLOv5-4.0 codebase [42] as our starting point and implement all the modifications we describe earlier in PyTorch.

The SGD optimizer is used. The initial learning rate is 1E−2, the final learning rate is 1E−5, and the weight decay is 5E−3. A momentum of 0.8 is used in the first three warming-up epochs. After that, a momentum is changed to 0.937. The training runs 250 epochs with a batch size of 64. The $\lambda_L = 0.5$ is optimized by exhaust search.

Implemented Models. We implement a series of face detector models, as listed in Table 1. We implement eight relatively large models, including extra large-size models (YOLOv5x, YOLOv5x6), large-size models (YOLOv5l, YOLOv5l6), medium-size models (YOLOv5m, YOLOv5m6), and small-size models (YOLOv5s, YOLOv5s6). In the name of the model, the last postfix 6 means it has the P6 output block in the SPP. These models all use the YOLOv4 CSPNet as the backbone with different depth and width multiples, denoted as D and W in Table 1.

Furthermore, we implement two super small-size models, YOLOv5n and YOLOv5n0.5, which use the ShuffleNetv2 and ShuffleNetv2-0.5 [26] as the backbone. Except for the backbone, all other main blocks, including the stem block, SPP, and PAN, are the same as in the larger models.

The number of parameters and number of flops of all these models is listed in Table 4 for comparison with existing methods.

4.3   Ablation Study

In this subsection, we present the effects of the modifications we have in our YOLO5Face. In this study, we use the YOLO5s model. We use the WiderFace [40] validation dataset and use the mAP as the performance metric. The results are presented in Table 2. Please note we do the experiments incrementally, where we add a modification at a time.

Table 2. Ablation study results on the WiderFace validation dataset.

Modification	Easy	Medium	Hard
Baseline (focus block)	94.70	92.90	83.27
+ Stem block	94.46	92.72	83.57
+ Landmark	94.47	92.79	83.21
+ SPP (3, 5, 7)	94.66	92.83	83.33
+ Ignore small faces	94.71	93.01	83.33
+ P6 block	95.29	93.61	83.27

Stem Block. We use the standard YOLOv5 [42] as a baseline which has a focus layer. By replacing the focus layer with our stem block, the mAP on the hard subset increased by 0.30%, which is non-trivial. The mAPs on the easy and medium subsets degrade slightly.

Landmark. Landmarks are used in many applications, including face recognition, where the landmarks are used to align the detected face image. Using the stem block as a baseline, adding the landmark achieves about the same performance on the easy and medium subsets while causing a little performance loss on the hard dataset. We will show in the next subsection the benefits of our landmarks to face recognition.

SPP with Smaller Size Kernels. The stem block and landmark are used in this test. The SPP kernel sizes $(13 \times 13, 9 \times 9, 5 \times 5)$ are changed to $(7 \times 7, 5 \times 5, 3 \times 3)$. mAPs are improved in all easy, medium, and hard subsets. The mAPs are improved by 0.1–0.2%.

Data Augmentation. A few data augmentation methods are studied. We find that ignoring small faces combined with random crop and Mosaic [1] helps the mAPs. It improves the mAPs on the easy and medium subset by 0.1–0.2%.

P6 Output Block. The P6 block brings significant mAP improvements, about 0.5–0.6%, on the easy and medium subsets. Perhaps motivated by our findings, the P6 block is added to YOLOv5 [42] after us.

Please note that since we do not use exactly the same parameters in this ablation study as in the final benchmark tests, the results in Table 2 may be slightly different from that in Table 4.

4.4   YOLO5Face for Face Recognition

Landmark is critical for face recognition accuracy. In RetinaFace [7], the accuracy of the landmark is evaluated with the MSE between estimated landmark coordinates and their ground truth and with the face recognition accuracy. The results show that the RetinaFace has better landmarks than the older MTCNN [44].

In this work, we use the face recognition framework in [27] to evaluate the accuracy of landmarks of the YOLO5Face. We use the Webface test dataset, which

Table 3. Evaluation of YOLO5Face landmark on Webface test dataset [56]. Two YOLO5Face models, the small model YOLOv5s and the medium model YOLOv5m, are used.

Backbone	FaceDetect	Training dataset	FNMR
R124	RetinaFace [7]	WiderFace [40]	0.1065
R124	YOLOv5s	WiderFace	0.1060
R124	YOLOv5s	+Multi-task facial [53]	0.1058
R124	YOLOv5m	WiderFace	0.1056
R124	YOLOv5m	+Multi-task facial	0.1051
R124‖R152	YOLOv5m	All above	0.1021
R152‖R200	YOLOv5m	All above	**0.0923**

is the largest face dataset with noisy 4M identities/260M faces, and cleaned 2M identities/42M faces [56]. This dataset is used in the ICCV2021 Masked Face Recognition (MFR) challenge [57]. In this challenge, both masked face images and standard face images are included, and a metric False Non-Match Rate (FNMR) at False Match Rate (FMR) = 1E−5 is used. The FNMR*0.25 for MFR plus FNMR*0.75 for standard face recognition are combined as the final metric.

By default, the RetinaFace [17] is used as the face detector on the dataset. We compare our YOLO5Face with the RetinaFace on this dataset. We use Arc-Face [6] framework with Resnet124 or larger backbones [14] as backbone. We also explore using concatenated features from two parallel models to get better performance. We replace the RetinaFace with our YOLO5Face. We test two models, a small model YOLOv5s and a medium model YOLOv5m. More details can be found in [27].

The results are listed in Table 3. From the results, we see that both our small and medium models outperform the RetinaFace [7]. In addition, we notice that there are very few large face images in the WiderFace dataset, so we add some large face images from the Multi-task-facial dataset [53] into the YOLO5Face training dataset. We find that this technique improves face recognition performance shown in Fig. 3 are some detected Webface [56] faces and landmarks using the RetinaFace [7] and our YOLOv5m. On the faces of a large pose, we can visually observe that our landmarks are more accurate, which has been proved in our face recognition results shown in Table 3.

4.5 YOLO5Face on WiderFace Dataset

We compare our YOLO5Face with many existing face detectors on the Wider-Face dataset. The results are listed in Table 4, where the previous SOTA results and our best results are both highlighted.

We first look at the performance of relatively large models whose number of parameters is larger than 3M and the number of flops is larger than 5G. All existing methods achieve mAP in 94.27–96.06% on the Easy subset,

91.9–94.92% on the Medium subset, and 71.39–85.29% on the Hard subset. The most recently released SCRFD [13] achieves the best performance in all subsets. Our YOLO5Face (YOLOv5x6) achieves 96.67%, 95.08%, and 86.55% on the three subsets, respectively. We achieve the SOTA performance on all the Easy, Medium, and Hard subsets.

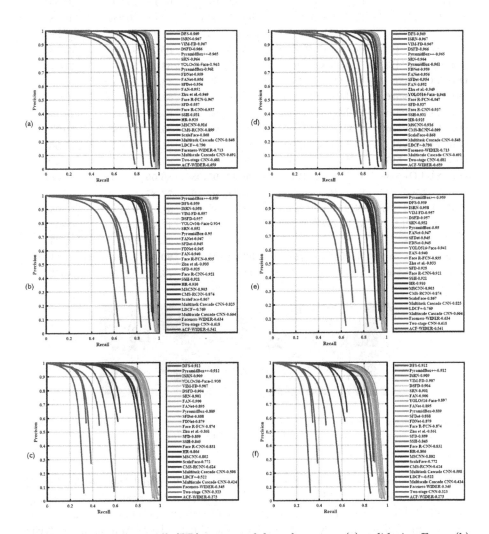

Fig. 2. The precision-recall (PR) curves of face detectors, (a) validation-Easy, (b) validation-Medium, (c) validation-Hard, (d) test-Easy, (e) test-Medium, (f) test-Hard.

Table 4. Comparison of our YOLO5Face and existing face detectors on the WiderFace validation dataset [40].

Detector	Backbone	Easy	Medium	Hard	Params (M)	Flops (G)
DSFD [19]	ResNet152 [14]	94.29	91.47	71.39	120.06	259.55
RetinaFace [7]	ResNet50 [14]	94.92	91.90	64.17	29.50	37.59
HAMBox [25]	ResNet50 [14]	95.27	93.76	76.75	30.24	43.28
TinaFace [55]	ResNet50 [14]	95.61	94.25	81.43	37.98	172.95
SCRFD-34GF [13]	Bottleneck ResNet	**96.06**	**94.92**	**85.29**	9.80	34.13
SCRFD-10GF [13]	Basic ResNet [14]	95.16	93.87	83.05	3.86	9.98
Our YOLOv5s	YOLOv5-CSPNet [42]	94.33	92.61	83.15	7.075	5.751
Our YOLOv5s6	YOLOv5-CSPNet	95.48	93.66	82.8	12.386	6.280
Our YOLOv5m	YOLOv5-CSPNet	95.30	93.76	85.28	21.063	18.146
Our YOLOv5m6	YOLOv5-CSPNet	95.66	94.1	85.2	35.485	19.773
Our YOLOv5l	YOLOv5-CSPNet	95.9	94.4	84.5	46.627	41.607
Our YOLOv5l6	YOLOv5-CSPNet	96.38	94.90	85.88	76.674	45.279
Our YOLOv5x6	YOLOv5-CSPNet	**96.67**	**95.08**	**86.55**	141.158	88.665
SCRFD-2.5GF [13]	Basic Resnet	**93.78**	**92.16**	**77.87**	0.67	2.53
SCRFD-0.5GF [13]	Depth-wise Conv	90.57	88.12	68.51	0.57	0.508
RetinaFace [7]	MobileNet0.25 [32]	87.78	81.16	47.32	0.44	0.802
FaceBoxes [48]	-	76.17	57.17	24.18	1.01	0.275
Our YOLOv5n	ShuffleNetv2 [26]	**93.61**	**91.54**	**80.53**	1.726	2.111
Our YOLOv5n0.5	ShuffleNetv2-0.5 [26]	90.76	88.12	73.82	0.447	0.571

Fig. 3. Some examples of detected face and landmarks, where the first row is from RetinaFace [7], and the second row is from our YOLOv5m.

Next, we look at the performance of super small models whose number of parameters is less than 2M and the number of flops is less than 3G. All existing methods achieve mAP in 76.17–93.78% on the Easy subset, 57.17–92.16% on the Medium subset, and 24.18–77.87% on the Hard subset. Again, the SCRFD [13] achieves the best performance in all subsets. Our YOLO5Face (YOLOv5n) achieves 93.61%, 91.54%, and 80.53% on the three subsets, respectively. Our face detector has a little bit worse performance than the SCRFD [13] on the Easy and Medium subsets. However, on the Hard subset, our face detector is leading by 2.66%. Furthermore, our smallest model, YOLOv5n0.5, has good performance, even though its model size is much smaller.

Table 5. Evaluation of YOLO5Face on the FDDB dataset [36].

Method	MAP
ASFD [43]	**0.9911**
RefineFace [45]	**0.9911**
PyramidBox [34]	0.9869
FaceBoxes [48]	0.9598
Our YOLOv5s	0.9843
Our YOLOv5m	0.9849
Our YOLOv5l	0.9867
Our YOLOv5l6	0.9880

The precision-recall (PR) curves of our YOLO5Face face detector, along with the competitors, are shown in Fig. 2. The leading competitors include DFS [35], ISRN [46], VIM-FD [50], DSFD [19], PyramidBox++ [20], SRN [5], Pyramid-Box [34] and more. For a full list of the competitors and their results on the WiderFace [40] validation and test datasets, please refer to [39]. In the results on the validation dataset, our YOLOv5x6-Face detector achieves 96.9%, 96.0%, 91.6% mAP on the Easy, Medium, and Hard subset, respectively, exceeding the previous SOTA by 0.0%, 0.1%, 0.4%. In the results on the test dataset, our YOLOv5x6-Face detector achieves 95.8%, 94.9%, 90.5% mAP on the Easy, Medium, and Hard subset, respectively with 1.1%, 1.0%, 0.7% gap to the previous SOTA. Please note that, in these evaluations, we only use multiple scales and left-right flipping without using other test-time augmentation (TTA) methods. Our focus is more on the VGA input images, where we achieve the SOTA in almost all conditions.

4.6   YOLO5Face on FDDB Dataset

FDDB dataset [36] is a small dataset with 5171 faces annotated in 2845 images. To demonstrate our YOLO5Face's performance on the cross-domain dataset, we test it on the FDDB dataset without retraining on it. The performances of the true positive rate (TPR) when the number of false-positive is 1000 are listed in Table 5. Please note that it is pointed out in RefineFace [45] that the annotation of FDDB misses many faces. In order to achieve their performance of 0.9911, the RefineFace modifies the FDDB annotation. In our evaluation, we use the original FDDB annotation without modifications. The RetinaFace [7] is not evaluated on the FDDB dataset.

5   Conclusion

In this paper, we present our YOLO5Face based on the YOLOv5 object detector [42]. We implement eight models. Both the largest model YOLOv5l6 and the super small model YOLOv5n achieve close to or exceeding SOTA performance on the WiderFace [40] validation Easy, Medium, and Hard subsets. This proves the effectiveness of our YOLO5Face in not only achieving the best performance but also running fast. Since we open-source the code, a lot of applications and mobile apps have been developed based on our design and achieved impressive performance.

References

1. Bochkovskiy, A., Wang, C.Y., Liao, H.Y.M.: YOLOv4: optimal speed and accuracy of object detection. arXiv preprint arXiv:2004.10934 (2020)
2. Cai, Z., Vasconcelos, N.: Cascade R-CNN: delving into high quality object detection. In: CVPR (2018)
3. Carion, N., Massa, F., Synnaeve, G., Usunier, N., Kirillov, A., Zagoruyko, S.: End-to-end object detection with transformers. In: Vedaldi, A., Bischof, H., Brox, T., Frahm, J.-M. (eds.) ECCV 2020. LNCS, vol. 12346, pp. 213–229. Springer, Cham (2020). https://doi.org/10.1007/978-3-030-58452-8_13
4. Chen, K., et al.: MMDetection: open MMLab detection toolbox and benchmark. In: ECCV (2020)
5. Chi, C., Zhang, S., Xing, J., Lei, Z., Li, S.Z.: SRN - selective refinement network for high performance face detection. arXiv preprint arXiv:1809.02693 (2018)
6. Deng, J., Guo, J., Xue, N., Zafeiriou, S.: ArcFace: additive angular margin loss for deep face recognition. In: CVPR, June 2019
7. Deng, J., Guo, J., Zhou, Y., Yu, J., Kotsia, I., Zafeiriou, S.: RetinaFace: single-stage dense face localisation in the wild. In: CVPR (2020)
8. Duan, K., Bai, S., Xie, L., Qi, H., Huang, Q., Tian, Q.: CenterNet: keypoint triplets for object detection. In: ICCV (2019)
9. Elfwinga, S., Uchibea, E., Doyab, K.: Sigmoid-weighted linear units for neural network function approximation in reinforcement learning. arXiv preprint arXiv:1702.03118 (2017)
10. Feng, Z., Kittler, J., Awais, M., Huber, P., Wu, X.: Wing loss for robust facial landmark localisation with convolutional neural networks. In: CVPR (2018)
11. Girshick, R.: Fast R-CNN. In: Proceedings of the International Conference on Computer Vision (ICCV) (2015)
12. Girshick, R., Donahue, J., Darrell, T., Malik, J.: Rich feature hierarchies for accurate object detection and semantic segmentation. In: Proceedings of the IEEE Conference on Computer Vision and Pattern Recognition (CVPR) (2014)
13. Guo, J., Deng, J., Lattas, A., Zafeiriou, S.: Sample and computation redistribution for efficient face detection. arXiv preprint arXiv:2105.04714 (2021)
14. He, K., Zhang, X., Ren, S., Sun, J.: Deep residual learning for image recognition. In: CVPR (2016)
15. He, K., Gkioxari, G., Dollár, P., Girshick, R.: Mask R-CNN. In: Proceedings of the International Conference on Computer Vision (ICCV) (2017)
16. He, K., Zhang, X., Ren, S., Sun, J.: Sigmoid-weighted linear units for neural network function approximation in reinforcement learning. TPAMI (2015)
17. He, T., Zhang, Z., Zhang, H., Zhang, Z., Xie, J., Li, M.: Bag of tricks for image classification with convolutional neural networks. In: CVPR (2019)
18. Huang, G., Liu, Z., Maaten, L., Weinberger, K.: Densely connected convolutional networks. In: CVPR (2017)

19. Li, J., et al.: DSFD: dual shot face detector. arXiv preprint arXiv:1810.10220 (2018)
20. Li, Z., Tang, X., Han, J., Liu, J., He, Z.: PyramidBox++: high performance detector for finding tiny face. arXiv preprint arXiv:1904.00386 (2019)
21. Lin, T., Dollár, P., Girshick, R., He, K., Hariharan, B., Belongie, S.: Feature pyramid networks for object detection. In: CVPR (2017)
22. Liu, S., Qi, L., Qin, H., Shi, J., Jia, J.: Path aggregation network for instance segmentation. arXiv preprint arXiv:1803.01534 (2018)
23. Liu, W., et al.: YOLOv3: an incremental improvement. In: ECCV (2016)
24. Liu, Y., Wang, F., Sun, B., Li, H.: MogFace: rethinking scale augmentation on the face detector. arXiv preprint arXiv:2103.11139 (2021)
25. Liu, Y., Tang, X., Wu, X., Han, J., Liu, J., Ding, E.: HAMBox: delving into online high-quality anchors mining for detecting outer faces. In: CVPR (2020)
26. Ma, M., Zhang, X., Zheng, H., Sun, J.: ShuffleNet V2: practical guidelines for efficient CNN architecture design. arXiv preprint ArXiv:1807.11164 (2018)
27. Qi, D., Hu, K., Tan, W., Yao, Q., Liu, J.: Balanced masked and standard face recognition. In: ICCV Workshops (2021)
28. Redmon, J., Farhadi, A.: YOLOv3: an incremental improvement. arXiv preprint arXiv:1804.02767 (2015)
29. Redmon, J., Divvala, S., Girshick, R., Farhadi, A.: You only look once: unified, real-time object detection. In: CVPR (2016)
30. Redmon, J., Farhadi, A.: YOLO9000: better, faster, stronger. In: CVPR (2017)
31. Ren, S., He, K., Girshick, R., Sun, J.: Faster R-CNN: towards real-time object detection with region proposal networks. IEEE Trans. Pattern Anal. Mach. Intell. (2016)
32. Sandler, M., Howard, A., Zhu, W., Zhmoginov, A., Chen, L.: MobileNetV2: inverted residuals and linear bottlenecks. In: CVPR (2018)
33. Tan, M., Pang, R., Le, Q.: EfficientDet: scalable and efficient object detection. In: CVPR (2020)
34. Tang, X., Du, D.K., He, Z., Liu, J.: PyramidBox: a context-assisted single shot face detector. arXiv preprint ArXiv:1803.07737 (2018)
35. Tian, W., Wang, Z., Shen, H., Deng, W., Chen, B., Zhang, X.: Learning better features for face detection with feature fusion and segmentation supervision. arXiv preprint arXiv:1811.08557 (2018)
36. Jain, V., Learned-Miller, E.: FDDB: a benchmark for face detection in unconstrained settings. University of Massachusetts Report (UM-CS-2010-009) (2010)
37. Wang, R.J., Li, X., Ling, C.X.: Pelee: a real-time object detection system on mobile devices. In: NeurIPS (2018)
38. Yang, M., Kriegman, D., Ahuja, N.: Detecting faces in images: a survey. TPAMI 24(1), 34–58 (2002)
39. Yang, S., Luo, P., Loy, C.C., Tang, X.: Wider face: a face detection benchmark. shuoyang1213.me/WIDERFACE/index.html
40. Yang, S., Luo, P., Loy, C.C., Tang, X.: Wider face: a face detection benchmark. In: CVPR (2016)
41. Yashunin, D., Baydasov, T., Vlasov, R.: MaskFace: multi-task face and landmark detector. arXiv preprint arXiv:2005.09412 (2020)
42. YOLOv5. github.com/ultralytics/yolov5
43. Zhang, B., et al.: Automatic and scalable face detector. arXiv preprint arXiv:2003.11228 (2020)
44. Zhang, K., Zhang, Z., Li, Z., Qiao, Y.: Joint face detection and alignment using multitask cascaded convolutional networks. IEEE Sig. Process. Lett. 23(10), 1499–1503 (2016)

45. Zhang, S., Chi, C., Lei, Z., Li, S.: RefineFace: refinement neural network for high performance face detection. arXiv preprint arXiv:1909.04376 (2019)
46. Zhang, S., et al.: ISRN - improved selective refinement network for face detection. arXiv preprint arXiv:1901.06651 (2019)
47. Zhang, S., Zhu, X., Lei, Z., Shi, H., Wang, X., Li, S.Z.: S^3FD: single shot scale-invariant face detector. In: ICCV (2017)
48. Zhang, S., Zhu, X., Lei, Z., Shi, H., Wang, X., Li, S.Z.: FaceBoxes: a CPU real-time face detector with high accuracy. In: IJCB (2017)
49. Zhang, X., Zhou, X., Lin, M., Sun, J.: ShuffleNet: an extremely efficient convolutional neural network for mobile devices. arXiv preprint arXiv:1707.01083 (2017)
50. Zhang, Y., Xu, X., Liu, X.: Robust and high performance face detector. arXiv preprint arXiv:1901.02350 (2019)
51. Zhang, C., Zhang, Z.: Robust real-time face detection. IJCV **57**, 137–154 (2004). https://doi.org/10.1023/B:VISI.0000013087.49260.fb
52. Zhang, C., Zhang, Z.: A survey of recent advances in face detection. Technical report, Microsoft Research (2010)
53. Zhao, R., Liu, T., Xiao, J., Lun, D.P.K., Lam, K.M.: Deep multi-task learning for facial expression recognition and synthesis based on selective feature sharing. In: ICPR (2020)
54. Zhou, X., Wang, D., Philipp, K.: Objects as points. arXiv preprint arXiv:1904.07850 (2019)
55. Zhu, Y., Cai, H., Zhang, S., Wang, C., Xiong, W.: TinaFace: strong but simple baseline for face detection. arXiv preprint arXiv:2011.13183 (2020)
56. Zhu, Z., et al.: WebFace260M: a benchmark unveiling the power of million-scale deep face recognition. In: CVPR (2021)
57. Zhu, Z., et al.: Masked face recognition challenge: the WebFace260M track report. In: ICCV Workshops (2021)

Counterfactual Fairness for Facial Expression Recognition

Jiaee Cheong[1,2]([✉]), Sinan Kalkan[3], and Hatice Gunes[1]

[1] University of Cambridge, Cambridge, UK
hatice.gunes@cl.cam.ac.uk
[2] Alan Turing Institute, London, UK
jc2208@cam.ac.uk
[3] Middle East Technical University, Ankara, Turkey
skalkan@metu.edu.tr

Abstract. Given the increasing prevalence of facial analysis technology, the problem of bias in these tools is becoming an even greater source of concern. Causality has been proposed as a method to address the problem of bias, giving rise to the popularity of using counterfactuals as a bias mitigation tool. In this paper, we undertake a systematic investigation of the usage of counterfactuals to achieve both statistical and causal-based fairness in facial expression recognition. We explore bias mitigation strategies with counterfactual data augmentation at the pre-processing, in-processing, and post-processing stages as well as a stacked approach that combines all three methods. At the in-processing stage, we propose using Siamese Networks to suppress the differences between the predictions on the original and the counterfactual images. Our experimental results on RAF-DB with counterfactuals added show that: (1) The in-processing method outperforms at the pre-processing and post-processing stages, in terms of accuracy, F1 score, statistical fairness and counterfactual fairness, and (2) stacking the pre-processing, in-processing and post-processing stages provides the best performance.

Keywords: Bias mitigation · Counterfactual fairness · Facial expression recognition

1 Introduction

Given the increasing prevalence and stakes involved in machine learning applications, the problem of bias in such applications is now becoming an even greater source of concern. The same is true for facial affect analysis technologies [9]. Several studies have highlighted the pervasiveness of such discrimination [4,19,24] and a number of works have sought to address the problem by proposing solutions for mitigation [5,8,49]. In order to assess whether bias has been mitigated, a reliable measure of bias and fairness is sorely needed. A significant number of fairness definitions have been proposed [1,18,37,46]. The more prevalent and long-standing definitions are often based upon statistical measures. However,

© The Author(s), under exclusive license to Springer Nature Switzerland AG 2023
L. Karlinsky et al. (Eds.): ECCV 2022 Workshops, LNCS 13805, pp. 245–261, 2023.
https://doi.org/10.1007/978-3-031-25072-9_16

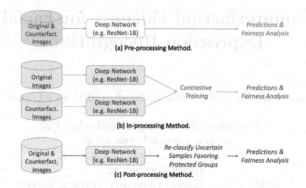

Fig. 1. Three methods for bias mitigation with counterfactual images.

statistical measures of fairness are mired with gaps. The definitions can often end up being mutually exclusive [7] and fail to distinguish spurious correlations between a sensitive attribute and the predicted outcome [15,29,39].

Causal reasoning has been proposed as a potential instrument to address such gaps [29,34,39] and counterfactuals, one of the key ideas within causal reasoning, are increasingly used as a tool to achieve such goals. One such use case is to rely on counterfactuals as a *data augmentation* strategy. Such an approach has proved promising for several use cases within the field of natural language processing [11, 12,14,35,36], recommendation systems [47] as well as visual question answering systems [41]. However, this approach has yet to be explored within the field of facial expression analysis. Our first contribution thus involves exploring the use of counterfactuals as a data augmentation strategy for the task of facial expression recognition. Second, as bias mitigation can be performed at the pre-processing, in-processing or post-processing stage [5], we will make use of counterfactuals at all three stages to mitigate bias as illustrated in Fig. 1. To the best of our knowledge, no existing works describe a comprehensive system for mitigating bias at all three stages. Our key contributions are summarised as follows:

1. We make use of counterfactuals at the pre-processing, in-processing and post-processing stage for the first time in the literature in order to mitigate for bias in facial expression recognition.
2. We do an in-depth analysis of bias at the pre-processing, in-processing and post-processing stages using both statistical and causal measures of fairness. We show that different forms of bias can exist at different levels and are captured by the different measures used.

2   Literature Review

2.1   Fairness in Machine Learning

Fairness is now recognised as a significantly important component of Machine Learning (ML) given how the problem of bias can result in significant impact on human lives. The general Machine Learning (ML) approaches tackle bias typically at the pre-processing, in-processing and post-processing stages [5,17,44]. The pre-processing methods address the problem of bias at the data-level [44]. This typically involves some form of data augmentation or modification to the input data. The in-processing methods involve making modification to the learning algorithm itself [8]. The post-processing methods occur at the end of the learning and prediction process [28]. This usually involves altering the output to achieve fairer predictions. The pre- and post-processing methods are usually model agnostic and can be applied without having to re-train the model. In contrast, the in-processing method will involve making changes in the model or the training method.

Different fairness definitions result in very different quantitative measures which can result in very different algorithmic outcomes [1,18]. To exacerbate matters, the different definitions can even sometimes be at odds with each other and improving the score on one fairness metric may very well involve a trade-off on another [2]. Hence, selecting the right definition and metric is a highly challenging and important task. A more thorough examination of these issues can be found in the following papers [1,2,18,37].

There are two main groups of fairness measures. Statistical notion of fairness is based on statistical measures of the variables of interest, e.g. accuracy, true positive rate. As statistical measures are only able to capture correlation and not causation, this form of fairness is sometimes also referred to as *associational* fairness. Some well-known examples of such forms of fairness include demographic parity [50] and equality of opportunity [22]. As highlighted in several recent research, there are many gaps stemming from statistical fairness. As a result, causal notion of fairness has been proposed to address these gaps [29,31,44]. Causal fairness assumes the existence of an underlying causal model. Some examples of causal fairness include counterfactual fairness [31] and proxy fairness [29]. In this research, we will build upon the definition of *equality of opportunity* and *counterfactual fairness* to conduct a comparison between both types of fairness.

2.2 Facial Affect Fairness

Facial affect recognition involves automatically analysing and predicting facial affect [45]. The most common method is the discrete category method which assumes six basic emotion categories of facial expressions recognized universally (i.e., happiness, surprise, fear, disgust, anger and sadness) [16]. Another method is to rely on the Facial Action Coding System (FACS), a taxonomy of facial expressions in the form of Action Units (AUs), where the emotions can then be defined according to the combination of AUs activated [16]. The six basic categories have been criticized for being unrealistic and limited at representing a full spectrum of emotions using only a handful of categories [20]. Another alternative is to use a dimensional description of facial affect which views any affective state as being represented as a bipolar entity existing on a continuum. [43]. Depending on the method of distinguishing facial affect, this can be achieved by either training an algorithm to classify the facial expressions of emotion [49], predict the valence and the arousal value of the displayed facial expression or detect the activated facial

action units [8]. To date, investigating bias and fairness in facial affect recognition
is still very much a understudied problem [5,49]. Only a handful of studies have
been done to highlight the bias and propose fairer solutions for facial affect analysis [8,25,40,49]. In this paper, we focus on the task of classifying facial expressions and attempt to investigate a solution which addresses bias at the pre-, in-
and post-processing stages with the use of counterfactuals.

2.3   Counterfactuals and Bias

A counterfactual is the result of an intervention after modifying the state of a set
of variables X to some value $X = x$ and observing the effect on some output Y.
Using Pearl's notation [42], this intervention is captured by the $do(\cdot)$ operator
and the resulting computation then becomes $P(Y|do(X = x))$.Several existing
frameworks offer methods for countering bias with the use of counterfactuals.
Existing methods typically rely on using counterfactuals as a data augmentation
strategy at the pre-processing stage to mitigate for bias. This method has proven
to be successful within the field of natural language processing [11,12,14,35,36].
Its use case includes hate speech detection [11,12], machine translation [35,48]
and dialogue generation [14]. Experiments done on recommendation systems [47]
and more recently, in Visual Question Answering (VQA) systems [41] indicated
that such an approach is promising.

Such an approach has yet to be explored for facial *expression* analysis. In the
domain of facial analysis, counterfactuals have been used to identify [13,27] and
mitigate for bias [10]. Our research resembles that of [13] and [27] in that we used
a generative adversarial network (GAN), STGAN [33], to generate adversarial
counterfactual facial images to assess for counterfactual fairness. Though alike in
spirit, our paper differs as follow. First, the above studies focused on investigating
different methods for counterfactual generation [10,13,27]. In our case, we do not
propose an alternative method to generate adversarial or counterfactual images.
Instead, we use a pre-trained GAN [33] to generate images which would then be
used to augment the original dataset.

Second, [13] and [27] focused on using the generated counterfactual images
to measure the bias present in either a publicly available dataset [13] or existing
black-box image analysis APIs [27] but did not propose any method to mitigate bias. Though we do conduct a bias analysis of the model's performance
on counterfactuals, the focus is chiefly on deploying counterfactuals explicitly to
mitigate for bias which was attempted by [10] as well. However, the goal in [10]
focuses on "attractiveness" prediction rather than facial expression recognition.
In addition, we focus on investigating methods to mitigate bias at all three stages
which is distinctly different from all the previous research mentioned above.

3   Methodology

As highlighted in [5], we can intervene to mitigate bias at either the pre-
processing, in-processing or post-processing stage. To investigate our bias

mitigation proposal for the task of facial expression recognition, we conduct a comparative study using counterfactuals at the three different stages. The first method is the pre-processing method which involves augmenting the training set using counterfactual images. Subsequently, we implemented an in-processing approach using a Siamese Network [3] to investigate further downstream methods for mitigating bias with the use of counterfactuals. Finally, we explore the use of a post-processing method, the Reject-Option Classification proposed by Kamiran et al. to mitigate bias [28].

3.1 Notation and Problem Definition

We adopt a machine learning approach where we have a dataset $D = \{(\mathbf{x}_i, y_i)\}_i$ such that $\mathbf{x}_i \in X$ is a tensor representing information (e.g., facial image, health record, legal history) about an individual I and $y_i \in Y$ is an outcome (e.g., identity, age, emotion, facial action unit labels) that we wish to predict. In other words, we assume that we have a classification problem and we are interested in finding a parametric predictor/mapping f with $f : X \to Y$. We use $\hat{y}_i$ to denote the predicted outcome for input $\mathbf{x}_i$ and $p(y_i|\mathbf{x}_i)$ is the predicted probability for $\mathbf{x}_i$ to be assigned to the correct class y_i. Each input $\mathbf{x}_i$ is associated (through an individual I) with a set of sensitive attributes $\{s_{j \in a}\}_a \subset S$ where a is e.g. *race* and $j \in \{$Caucasian, African-American, Asian$\}$. The minority group are those with sensitive attributes which are fewer in numbers (e.g. African-American) compared to the main group (e.g. Caucasian). Note that there are other attributes $\{z_{j \in a}\}_a \subset Z$ that are not sensitive. In bias mitigation, we are interested in diminishing the discrepancy between $p(\hat{y}_i = c|\mathbf{x}_i)$ and $p(\hat{y}_j = c|\mathbf{x}_j)$ if $\mathbf{x}_i$ and $\mathbf{x}_j$ are facial images for different individuals and their sensitive attributes $s^{\mathbf{x}_i}$ and $s^{\mathbf{x}_j}$ are different.

3.2 Counterfactual Fairness

Causality-based fairness reasoning [29,31,44] assumes that there exists a cause-and-effect relationship between the attribute variables and the outcome. We follow the counterfactual fairness notation used by Kusner et al. [31]. Given a predictive problem with fairness considerations, where S, X and Y represent the sensitive attributes, the input, and the output of interest respectively, let us assume that we are given a causal model (U, V, F), where U is a set of latent background variables, which are factors not caused by any variable in the set V of observable variables and F is a set of functions over $V_i \in V$. In our problem setup, $V \equiv S \cup X$. Kusner et al. [31] postulate the following criterion for predictors of Y: Predictor $f(\cdot\,; \theta)$ is counterfactually fair if, under any context $X = \mathbf{x}$ and $S = s$, the following is true:

$$p(f(\mathbf{x}; \theta, U) = y \mid \mathbf{x}, s) = p(f(\mathbf{x}^{S \leftarrow s'}; \theta, U) = y \mid \mathbf{x}, s), \tag{1}$$

where $\mathbf{x}^{S \leftarrow s'}$ denotes the counterfactual input (image) obtained by changing the sensitive attribute to s'. To simplify notation, we will use $\mathbf{x}'$ to denote the counterfactual image, $\mathbf{x}^{S \leftarrow s'}$, in the rest of the paper. With reference to Eq. 1, we would like to highlight that this definition is an individual-level fairness definition. For the set of experiments that we will be doing in this paper, we will be aggregating the counterfactual *counts* dissected according to sensitive attributes and class in order to facilitate measurement and comparison.

3.3 Counterfactual Image Generation

We used a state-of-the-art GAN, a pre-trained STGAN model [33] trained on the CelebA dataset to generate a set of counterfactual images for RAF-DB. In our experiments, the attribute that we have chosen to manipulate is that of skin tone – see Fig. 2 for samples. This is because manipulating skin tone produces more consistent results than manipulating other sensitive attributes such as age or gender. The adversarial counterfactual images modified across the other sensitive attributes are less stable and more likely to be corrupted. Our counterfactuals involve lightening but not darkening skin tone as GANs are currently still incapable of effectively doing so [26]. We are not conflating skin tone with race. We recognise that they are separate entities with some overlaps. Our analysis is focused on mitigating bias stemming from difference in skin tone which aligns with the approach taken in other bias investigation research [4,13]. Moreover, though the images generated may not be completely satisfactory, this is due to the limitations of GANs. This is a contextual challenge and the solutions proposed here can still be deployed when GANs have been further improved.

Fig. 2. Sample counterfactual images obtained by changing the skin tone of the original images (without changing facial expression) using a pre-trained STGAN model [33].

3.4 Baseline Approach

We take prior of Tian et al. [49] as baseline. For this, we use a 18-layer Residual Network (ResNet) [23] architecture and train it from scratch with a Cross Entropy loss to predict a single expression label y_i for each $\mathbf{x}_i$:

$$\mathcal{L}_{CE}(\mathbf{x}_i, y_i) = -\sum_{y \in Y} \mathbb{1}[y_i = \hat{y}_i] \log p(y|\mathbf{x}_i), \qquad (2)$$

where $p(y|\mathbf{x}_i)$ denotes the predicted probability for $\mathbf{x}_i$ being assigned to class $y_i \in Y$ and $\mathbb{1}[\cdot]$ denotes the indicator function. The baseline model is trained on the original images.

3.5 Pre-processing: Data Augmentation with Counterfactuals

Similar to the works done in the field of natural language processing (discussed in Sect. 2.3), we make use of counterfactuals as a data augmentation strategy. In this approach, we generate a counterfactual for each image and feed them as input to train a new network (Fig. 1). Hence, instead of having N image samples in D, we now have $2N$ number of training samples. The network is trained with these $2N$ images using a Cross Entropy loss defined in Eq. 2.

3.6 In-processing: Contrastive Counterfactual Fairness

Counterfactual fairness is defined with respect to the discrepancy between the predictions on an image $\mathbf{x}_i$ and its counterfactual version $\mathbf{x}'_i$. To be specific, as discussed in Sect. 3.2, counterfactual fairness requires the gap between $p(y_i|\mathbf{x}_i)$ and $p(y_i|\mathbf{x}'_i)$ to be minimal. An in-processing solution that fits very well to this requirement is contrastive learning.

In general, the goal of contrastive learning [6] is to learn an embedding space which minimises the embedding distance between a pair of images which are of the same class (positive pair) and maximizes the distance between a pair of "unrelated" images (negative pair). However, in our setting, we seek to minimise the difference between the prediction probabilities on an image and its counterfactual version. We realize contrastive counterfactual fairness by feeding $\mathbf{x}_i$ and its counterfactual version $\mathbf{x}'_i$ through a Siamese network (Fig. 3). The following contrastive loss then seeks to minimise the discrepancy (bias) between the predictions of the two branches:

$$\mathcal{L}_{con}(\mathbf{x}_i, \mathbf{x}'_i) = -\sum_{y \in Y} \mathbb{1}[f(\mathbf{x}_i; \theta) = \hat{y}_i] \log p(y_i|\mathbf{x}'_i), \tag{3}$$

where we penalize the Siamese network if the counterfactual prediction is not consistent with respect to the predicted label $f(\mathbf{x}_i; \theta)$ of the original image $\mathbf{x}_i$.

Each branch of the Siamese network has its own Cross Entropy Loss so that each branch can predict the correct label independently. The overall loss is then defined as:

$$\mathcal{L}_i = \alpha(\mathcal{L}_{CE}(\mathbf{x}_i, y_i) + \mathcal{L}_{CE}(\mathbf{x}'_i, y_i)) + \mathcal{L}_{con}(\mathbf{x}_i, \mathbf{x}'_i), \tag{4}$$

where $\mathcal{L}_{CE}$ is as defined in Eq. 2, and α is a hyper-parameter which we tuned as 1.5. By jointly minimizing the two Cross Entropy objectives and $\mathcal{L}_{con}$, the network learns a representation that minimises the difference between the predictions on the original and counterfactual images as well as the individual prediction errors for both the original and counterfactual images. The authors of [10] leveraged on a similar idea though both research were done independently. It is similar in that, to enforce fairness, they added a regularizer between the logits of the image and its counterfactual which is similar to the functionality of $\mathcal{L}_{con}$. The slight difference is that the overall loss function only accounts for the CE loss of the original image whereas ours attempt to account for both the original and counterfactual loss via $\mathcal{L}_{CE}(\mathbf{x}_i, y_i)$ and $\mathcal{L}_{CE}(\mathbf{x}'_i, y_i)$ respectively.

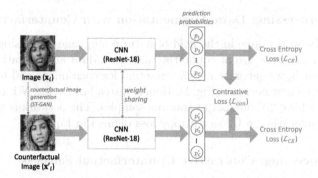

Fig. 3. An overview of the in-processing method: Original image $\mathbf{x}_i$ and its counterfactual $\mathbf{x}'_i$ are fed through a Siamese network, which is trained to minimize the discrepancy between the branches is penalized with a contrastive loss ($\mathcal{L}_{con}$) and to maximize the prediction probabilities of both branches using Cross Entropy loss ($\mathcal{L}_{CE}$).

3.7   Post-processing: Reject Option Classification

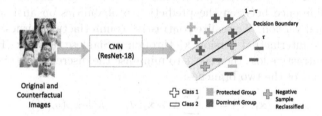

Fig. 4. Post-processing method by Kamiran et al. [28] reclassifies samples around the decision boundary in favor of protected groups.

Post-processing approaches take the output of the model and modify the output in a manner to achieve greater fairness. Here, we employ the Reject Option Classification suggested by Kamiran et al. [28]. This approach re-classifies the outputs where the model is less certain about, i.e., the predictions that fall in a region around the decision boundary parameterised by τ (see Fig. 4). For instance, given a typical classification probability threshold of 0.5, if the model prediction is 0.85 or 0.15, this means that the model is highly certain of its prediction. If the values range around 0.53 or 0.44, this means that the input falls very close to the decision boundary and the model is less certain about its prediction. In order to improve fairness of the prediction, we reclassify the predicted output if it belongs to that of the minority group. More formally, if a sample $\mathbf{x}_i$ that falls in the "critical" region $1 - \tau \leq p(y|\mathbf{x}_i) \leq \tau$ where $0.5 \leq \tau \leq 1$, we reclassify $\mathbf{x}_i$ as y if $\mathbf{x}_i$ belongs to a minority group. Otherwise, i.e. when $p(y|\mathbf{x}_i) > \tau$, we accept the predicted output class y. In our experiments, we set $\tau = 0.6$ as suggested by Kamiran et al. [28]. This method captures the innate human intuition to give the benefit of the doubt to samples from the minority group which they are unsure of.

4   Experimental Setup

4.1   Dataset

We chiefly conducted our experiments on the RAF-DB [32] dataset. RAF-DB contains labels in terms of facial expressions of emotions (Surprise, Fear, Disgust, Happy, Sad, Anger and Neutral) and sensitive attribute labels along gender, race and age. We excluded images labelled as "Unsure" for gender. In addition, the age binning system is likely to cause greater variation noise. For instance, an individual who is age 4 is likely to look very different from someone who is age 19 but yet they are categorised in the same category. As such, we have chosen to restrict our analysis to the sensitive attributes gender and race. We utilised a subset of the dataset consisting of 14,388 images, with 11,512 samples used for training and 2,876 samples used for testing within our experiments. This training and testing split has been pre-defined according to the instructions in the original dataset [32].

4.2   Implementation and Training Details

We first generated a set of counterfactual images using the method delineated in Sect. 3.3. This is done for both the training and testing images within RAF-DB. Our task subsequently focuses on categorising the seven categories of facial expressions of emotion. We then reclassified the counterfactual RAF-DB images (Fig. 2) to evaluate for counterfactual biases as shown in Table 2.

Training Details. ResNet-18 [23] is used as the baseline model as well as for the mitigation models as illustrated in Fig. 1. For all models, we take the PyTorch implementation of ResNet and train it from scratch with the Adam optimizer [30], with a mini-batch size of 64, and an initial learning rate of 0.001 (except for the in-processing method for which 0.0005 worked slightly better). The learning rate decays linearly by a factor of 0.1 every 40 epochs. The maximum training epochs is 100, but early stopping is applied if the accuracy does not increase after 30 epochs. For the pre-processing method, we train a network with both the original and counterfactual images (Fig. 1). For the in-processing method, we have two Siamese branches with shared weights (Fig. 3). For the post processing approach, we train a network with both the original and counterfactual images (Fig. 4) but take the output predictions and reclassify them according to the methodology delineated in Sect. 3.7.

Image Pre-processing and Augmentation. All images are cropped to ensure faces appear in approximately similar positions. The images are subsequently normalized to a size of 128×128 pixels which are then fed into the networks as input. During the training step, we apply the following commonly used augmentation methods: Randomly cropping the images to a slightly smaller size (i.e. 96×96); rotating them with a small angle (i.e. range from −15° to 15°); and horizontally mirroring them in a randomized manner.

4.3   Evaluation Measures

In this paper, we use two measures: accuracy and F1-score, to evaluate prediction quality and two measures: equal opportunity and causal fairness, to evaluate fairness. *Equal opportunity* ($\mathcal{M}_{EO}$), a group-based metric, is used to compare the group fairness between models [22]. This can be understood as the largest accuracy gap among all demographic groups:

$$\mathcal{M}_{EO} = \frac{\min_{s \in S} \mathcal{M}_{ACC}(s)}{\max_{s \in S} \mathcal{M}_{ACC}(s)}, \tag{5}$$

where $\mathcal{M}_{ACC}(s)$ is the accuracy for a certain demographic group s. We also include a causality-based fairness metric *Counterfactual fairness* ($\mathcal{M}_{CF}$) because it has often been noted that commonly-used fairness metrics based on classification evaluation metrics such as accuracy, precision, recall and TP rate are insufficient to capture the bias present [29,44]. $\mathcal{M}_{CF}$ is defined as:

$$\mathcal{M}_{CF} = \frac{1}{N} \sum_{i \in N} \mathbb{1}[f(\mathbf{x}_i; \theta) = f(\mathbf{x}'_i; \theta)], \tag{6}$$

where we compare the labels predicted by $f(\cdot; \theta)$. This is not a newly defined metric but a prevalent one based on an *aggregated* form of Counterfactual Fairness defined accordingly in Sect. 3.2 and [31].

Table 1. RAF-DB Test Set Distribution (Cauc: Caucasian, AA: African-American).

Emotion	Gender		Race			
	Male	Female	Cauc	AA	Asian	Percent
Surprise	138	159	260	16	21	10.3%
Fear	43	36	61	5	13	2.7%
Disgust	69	89	125	6	27	5.5%
Happy	429	712	855	98	188	39.7%
Sad	147	239	291	30	65	13.4%
Angry	119	45	144	10	10	5.7%
Neutral	312	339	489	39	123	22.6%
Percent.	43.7%	56.3%	77.4%	7.1%	16.4%	

5   Results

5.1   An Analysis of Dataset Bias and Counterfactual Bias

Dataset Bias Analysis. First, we conduct a preliminary bias analysis by attempting to highlight the different biases present. As highlighted in Table 1, there is a slight dataset bias across gender. 56.3% of the subjects are female, while 43.7% are male. There is a greater bias across race. 77.4% of the subjects are Caucasian, 15.5% are Asian, and only 7.1% are African-American.

Table 2. Proportion of samples that stayed consistent with the original classification after counterfactual manipulation (skin tone change). The values are the $\mathcal{M}_{CF}$ values. Classification is performed using the baseline model.

Emotion	Gender		Race		
	Male	Female	Cauc	AA	Asian
Surprise	0.34	0.31	0.33	0.38	0.24
Fear	0.16	0.25	0.16	0.20	0.38
Disgust	0.20	0.13	0.15	0.17	0.22
Happy	0.44	0.56	0.52	0.53	0.48
Sad	0.22	0.27	0.26	0.20	0.25
Angry	0.29	0.27	0.28	0.20	0.30
Neutral	0.33	0.36	0.36	0.28	0.33
$\mathcal{M}_{CF}$	0.34	0.41	0.39	0.39	0.37

Counterfactual Bias Analysis. This involves calculating the proportion of the baseline model's predictions that remained the same between the original and counterfactual images. The specific formulation is captured by Eq. 6. Though simple in nature, it forms a crucial cornerstone in evaluating Counterfactual Fairness. With reference to Table 2, we see that, for a majority of samples, the predicted labels did not remain the same for the counterfactual images. Across the sensitive attribute Gender, performance accuracy is slightly more consistent for Females. Across the sensitive attribute Race, performance accuracy is comparatively more consistent for Caucasians and Asians. This phenomena may be correlated with class size numbers as evidenced in Table 1. A similar trend is true across emotions. We see that the emotion "Happy" has the highest consistency. This may be due to the larger sample size and the fact that "Happy" is considered an emotion that is relatively easier to recognise and label. On the other hand, we see that the "Fear" class has the lowest consistency. This might be due to the fact that the emotion fear has lesser samples and is more ambiguous, variable and perhaps harder to identify and label. This hints that bias may not only vary across sensitive attribute but across emotion categories as well. We would like to highlight that this counterfactual analysis only analyses whether

Table 3. Accuracy and fairness scores from the fine-tuned standalone models trained on the combined training and tested on the combined test set which includes both the original and counterfactual images.

Emotion	1. Pre-processing					2. In-processing					3. Post-processing				
	Gender		Race			Gender		Race			Gender		Race		
	M	F	Cau	AA	A	M	F	Cau	AA	A	M	F	Cau	AA	A
Surprise	0.58	0.59	0.58	0.59	0.60	0.59	0.63	0.63	0.50	0.52	0.55	0.67	0.61	0.72	0.60
Fear	0.35	0.26	0.34	0.30	0.19	0.31	0.28	0.33	0.30	0.15	0.37	0.22	0.34	0.40	0.12
Disgust	0.29	0.26	0.26	0.33	0.31	0.25	0.31	0.30	0.33	0.24	0.24	0.35	0.28	0.25	0.39
Happy	0.75	0.76	0.72	0.87	0.84	0.89	0.92	0.90	0.90	0.91	0.90	0.93	0.92	0.88	0.94
Sad	0.44	0.48	0.47	0.35	0.51	0.45	0.49	0.47	0.42	0.50	0.46	0.47	0.46	0.37	0.52
Angry	0.52	0.47	0.52	0.45	0.30	0.51	0.43	0.50	0.50	0.25	0.48	0.37	0.45	0.45	0.40
Neutral	0.66	0.63	0.65	0.51	0.66	0.63	0.68	0.67	0.58	0.64	0.61	0.62	0.62	0.58	0.62
$\mathcal{M}_{ACC}$	0.61	0.63	0.61	0.65	0.67	0.65	0.72	0.69	0.69	0.68	0.65	0.71	0.68	0.68	0.70
$\mathcal{M}_{EO}$	**0.97**		0.91			0.91		**0.99**			0.91		0.96		
$\mathcal{M}_{CF}$	0.53	0.57	0.54	0.56	0.61	0.58	0.63	0.59	0.63	0.65	0.39	0.44	0.41	0.44	0.39
$\mathcal{M}_{CF}$ (Avg.)	0.55		0.57			**0.60**		**0.62**			0.42		0.41		

predictions remained consistent between the original and counterfactual images and has no bearing on whether the initial prediction was correct.

5.2 Bias Mitigation Results with Counterfactual Images

Next, we evaluate to what extent we are able to mitigate bias via the methods proposed in the Methodology section: At the pre-processing, in-processing and post-processing stage. With reference to Table 3, in terms of accuracy, there does not seem to be a difference between the pre-processing, in-processing and post-processing methods. All were able to improve accuracy prediction to largely the same effect. However, we witness a difference in outcome across $\mathcal{M}_{EO}$. According to $\mathcal{M}_{EO}$, the pre-processing method is best for achieving fairness across gender whilst the in-processing method is best for achieving fairness across race. Though this measure highlight different effectiveness, all three methods seem comparable in terms of their ability to improve $\mathcal{M}_{EO}$ across board.

It is only in terms of $\mathcal{M}_{CF}$ where we manage to observe a wider difference in performance disparity. The pre-processing and in-processing methods were able to improve $\mathcal{M}_{CF}$ to a greater extent compared to the post-processing method. Out of the first two, it is evident that the in-processing method manages to outperform the other two across both sensitive attributes. It gives the highest $\mathcal{M}_{CF}$ score of 0.60 and 0.62 across gender and race respectively compared to 0.55 and 0.57 for the pre-processing method. On the other hand, we see that the post-processing method only gives 0.42 and 0.41 across gender and race respectively. This result is noteworthy in several ways. First, this highlights the importance of using different metrics for bias evaluation as this methodological gap would not have been picked up by the other two metrics ($\mathcal{M}_{ACC}$ and $\mathcal{M}_{EO}$). Second,

Table 4. Stacked model combines the pre-, in- and post-processing stages.

Emotion	Gender		Race		
	Male	Female	Caucasian	AA	Asian
Surprise	0.68	0.74	0.73	0.72	0.55
Fear	0.47	0.44	0.47	0.50	0.38
Disgust	0.37	0.42	0.42	0.33	0.30
Happy	0.95	0.98	0.98	0.91	0.93
Sad	0.59	0.62	0.60	0.63	0.61
Angry	0.67	0.50	0.64	0.60	0.50
Neutral	0.75	0.84	0.80	0.72	0.78
$\mathcal{M}_{ACC}$	0.75	0.81	0.79	0.78	0.76
$\mathcal{M}_{EO}$	0.93		0.96		
$\mathcal{M}_{CF}$	0.59	0.65	0.62	0.61	0.63
$\hookrightarrow \mathcal{M}_{CF}(Avg.)$	0.62		0.62		

this underlines the need to use a variety of methods to tackle the problem of bias as a standalone post-processing method might be inadequate. Finally, with reference to Table 4, we see that the stacked approach improved scores across all evaluation metrics. This model comprises of a combination of the fine-tuned pre, in and post-processing methods. In Table 5, we see that the combined approach is comparatively better than all the standalone methods across most metrics. Out of the standalone methods, the in-processing method seems to be best in terms of achieving both $\mathcal{M}_{EO}$ and $\mathcal{M}_{CF}$ fairness.

Table 5. Results summary showing that a combined stacked approach supersedes all other standalone models on most metrics.

Model	$\mathcal{M}_{ACC}$	$\mathcal{M}_{F1}$	$\mathcal{M}_{EO}$	$\mathcal{M}_{CF}$
Original	0.65	0.54	**0.97**	0.30
Pre-processing	0.63	0.67	0.94	0.56
In-processing	0.69	0.68	0.95	0.61
Post-processing	0.68	0.65	0.94	0.42
Combined	**0.78**	**0.71**	0.95	**0.62**

6   Conclusion and Discussion

Overall, the stacked approach supersedes the rest across most measures: accuracy ($\mathcal{M}_{ACC}$), F1-score ($\mathcal{M}_{F1}$), and Counterfactual Fairness ($\mathcal{M}_{CF}$). A significant point is that our work agrees with the findings in [17] in many ways. One of the key findings was how pre-processing methods can have a huge effect on disparity in prediction outcomes. Second, their evaluation showed that many of the measures of bias and fairness strongly correlate with each other. This is evident in our findings too as we see that the equal opportunity fairness correlates with class-wise performance accuracy across board. Hence, we argue the importance of using an orthogonal measure to capture the bias which would otherwise go unnoticed. In our experiments, the Counterfactual Fairness measure $\mathcal{M}_{CF}$ fulfills this criteria. Indeed, we see that it captures the efficacy difference in achieving Counterfactual Fairness across the different bias mitigation strategies. This provides empirical evidence for the gaps highlighted in Sect. 2.1.

Further, though the overall evaluation metrics have improved, we still observe bias when conducting a disaggregated analysis partitioned across the different sensitive attributes or emotion categories. For instance, the mitigation algorithms consistently performed poorly for certain emotion categories, e.g. "Disgust". This aligns with the findings in [25,49] as such expressions are hypothesised to be less well-recognised than other prevalent emotions e.g. "Happy".

A key limitation of our work is that the methods that we propose are limited by dataset availability. We have used the annotations as provided by the original dataset owners [32] which were crowd-sourced and labelled by humans. However, this approach of treating race as an attribute fails to take into account the multi-dimensionality of race [21] which represents a research area that future researchers can look into. In addition, we have solely relied upon the original training-test split provided by the data repository. As highlighted in [17], algorithms are highly sensitive to variation in dataset composition and changes in training-test splits resulted in great variability in the accuracy and fairness measure performance of all algorithms. Another limitation is that of robustness and further research on adversarial attacks on fairness [38] should be investigated.

We recognise that face recognition and by its extension, facial affect recognition has received criticism for its misuse and parallelism to facial phrenology. First, we would like to underscore that the ideas in this paper are meant to address the existing problem of bias and our intention is for it to be used for good. Second, we concur that many of the concerns raised are valid and we view this as an opportunity for future work extensions. Third, we hope this piece of work will encourage researchers and companies to shape solutions that ensure that the technology and applications developed are fair and ethical for all.

Data Access Statement: This study involved secondary analyses of pre-existing datasets. All datasets are described in the text and cited accordingly. Licensing restrictions prevent sharing of the datasets. The authors thank Shan Li, Prof Weihong Deng and JunPing Du from the Beijing University of Posts and Telecommunications (China) for providing access to RAF-DB.

Acknowledgement. J. Cheong is supported by the Alan Turing Institute doctoral studentship and the Cambridge Commonwealth Trust. H. Gunes' work is supported by the EPSRC under grant ref. EP/R030782/1.

References

1. Barocas, S., Hardt, M., Narayanan, A.: Fairness in machine learning. NIPS Tutor. **1**, 2 (2017)
2. Binns, R.: Fairness in machine learning: Lessons from political philosophy. In: Conference on Fairness, Accountability and Transparency (2018)
3. Bromley, J., et al.: Signature verification using a "Siamese" time delay neural network. Int. J. Pattern Recogn. Artif. Intell. **7**(04), 669–688 (1993)
4. Buolamwini, J., Gebru, T.: Gender shades: intersectional accuracy disparities in commercial gender classification. In: Conference on Fairness, Accountability and Transparency, pp. 77–91. PMLR (2018)
5. Cheong, J., Kalkan, S., Gunes, H.: The hitchhiker's guide to bias and fairness in facial affective signal processing: overview and techniques. IEEE Signal Process. Mag. **38**(6), 39–49 (2021)
6. Chopra, S., Hadsell, R., LeCun, Y.: Learning a similarity metric discriminatively, with application to face verification. In: CVPR (2005)
7. Chouldechova, A.: Fair prediction with disparate impact: a study of bias in recidivism prediction instruments. Big Data **5**(2), 153–163 (2017)
8. Churamani, N., Kara, O., Gunes, H.: Domain-incremental continual learning for mitigating bias in facial expression and action unit recognition. arXiv preprint arXiv:2103.08637 (2021)
9. Crawford, K.: Time to regulate AI that interprets human emotions. Nature **592**(7853), 167–167 (2021)
10. Dash, S., Balasubramanian, V.N., Sharma, A.: Evaluating and mitigating bias in image classifiers: a causal perspective using counterfactuals. In: WACV (2022)
11. Davani, A.M., Omrani, A., Kennedy, B., Atari, M., Ren, X., Dehghani, M.: Fair hate speech detection through evaluation of social group counterfactuals. arXiv preprint arXiv:2010.12779 (2020)
12. Davani, A.M., Omrani, A., Kennedy, B., Atari, M., Ren, X., Dehghani, M.: Improving counterfactual generation for fair hate speech detection. In: Workshop on Online Abuse and Harms (WOAH) (2021)
13. Denton, E., Hutchinson, B., Mitchell, M., Gebru, T.: Detecting bias with generative counterfactual face attribute augmentation. arXiv e-prints, pp. arXiv-1906 (2019)
14. Dinan, E., Fan, A., Williams, A., Urbanek, J., Kiela, D., Weston, J.: Queens are powerful too: mitigating gender bias in dialogue generation. In: EMNLP (2020)
15. Dwork, C., Hardt, M., Pitassi, T., Reingold, O., Zemel, R.: Fairness through awareness. In: Proceedings of the 3rd Innovations in Theoretical Computer Science Conference, pp. 214–226 (2012)
16. Ekman, R.: What the Face Reveals: Basic and Applied Studies of Spontaneous Expression Using the Facial Action Coding System (FACS). Oxford University Press, USA (1997)
17. Friedler, S.A., Scheidegger, C., Venkatasubramanian, S., Choudhary, S., Hamilton, E.P., Roth, D.: A comparative study of fairness-enhancing interventions in machine learning. In: Conference on Fairness, Accountability, and Transparency (2019)
18. Gajane, P., Pechenizkiy, M.: On formalizing fairness in prediction with machine learning. arXiv preprint arXiv:1710.03184 (2017)

19. Garcia, R., Wandzik, L., Grabner, L., Krueger, J.: The harms of demographic bias in deep face recognition research. In: Proceedings of International Conference on Biometrics (ICB), pp. 1–6 (2019)

20. Gunes, H., Schuller, B.: Categorical and dimensional affect analysis in continuous input: current trends and future directions. Image Vis. Comput. **31**(2), 120–136 (2013)

21. Hanna, A., Denton, E., Smart, A., Smith-Loud, J.: Towards a critical race methodology in algorithmic fairness. In: Proceedings of the 2020 Conference on Fairness, Accountability, and Transparency, pp. 501–512 (2020)

22. Hardt, M., Price, E., Srebro, N.: Equality of opportunity in supervised learning. In: NIPS (2016)

23. He, K., Zhang, X., Ren, S., Sun, J.: Deep residual learning for image recognition. In: CVPR (2016)

24. Hoffman, A.: Where fairness fails: data, algorithms and the limits of antidiscrimination discourse. J. Inf. Commun. Soc. **22**, 900–915 (2019)

25. Howard, A., Zhang, C., Horvitz, E.: Addressing bias in machine learning algorithms: A pilot study on emotion recognition for intelligent systems. In: Proceedings of Advanced Robotics Social Impacts (ARSO) (2017)

26. Jain, N., Olmo, A., Sengupta, S., Manikonda, L., Kambhampati, S.: Imperfect imaganation: Implications of GANs exacerbating biases on facial data augmentation and snapchat face lenses. Artif. Intell. **304**, 103652 (2022)

27. Joo, J., Kärkkäinen, K.: Gender slopes: Counterfactual fairness for computer vision models by attribute manipulation. In: Workshop on Fairness, Accountability, Transparency and Ethics in Multimedia (2020)

28. Kamiran, F., Karim, A., Zhang, X.: Decision theory for discrimination-aware classification. In: International Conference on Data Mining (2012)

29. Kilbertus, N., Rojas-Carulla, M., Parascandolo, G., Hardt, M., Janzing, D., Schölkopf, B.: Avoiding discrimination through causal reasoning. In: NIPS, pp. 656–666 (2017)

30. Kingma, D.P., Ba, J.: Adam: A method for stochastic optimization. arXiv preprint arXiv:1412.6980 (2014)

31. Kusner, M., Loftus, J., Russell, C., Silva, R.: Counterfactual fairness. In: NIPS (2017)

32. Li, S., Deng, W., Du, J.: Reliable crowdsourcing and deep locality-preserving learning for expression recognition in the wild. In: CVPR (2017)

33. Liu, M., et al.: StGAN: a unified selective transfer network for arbitrary image attribute editing. In: CVPR (2019)

34. Loftus, J.R., Russell, C., Kusner, M.J., Silva, R.: Causal reasoning for algorithmic fairness. arXiv preprint arXiv:1805.05859 (2018)

35. Lu, K., Mardziel, P., Wu, F., Amancharla, P., Datta, A.: Gender bias in neural natural language processing. In: Nigam, V., et al. (eds.) Logic, Language, and Security. LNCS, vol. 12300, pp. 189–202. Springer, Cham (2020). https://doi.org/10.1007/978-3-030-62077-6_14

36. Maudslay, R.H., Gonen, H., Cotterell, R., Teufel, S.: It's all in the name: mitigating gender bias with name-based counterfactual data substitution. In: EMNLP-IJCNLP (2019)

37. Mehrabi, N., Morstatter, F., Saxena, N., Lerman, K., Galstyan, A.: A survey on bias and fairness in machine learning. ACM Comput. Surv. (CSUR) **54**(6), 1–35 (2019)

38. Mehrabi, N., Naveed, M., Morstatter, F., Galstyan, A.: Exacerbating algorithmic bias through fairness attacks. In: AAAI (2021)

39. Nabi, R., Shpitser, I.: Fair inference on outcomes. In: AAAI (2018)
40. Ngxande, M., Tapamo, J., Burke, M.: Bias remediation in driver drowsiness detection systems using generative adversarial networks. IEEE Access **8**, 55592–55601 (2020). https://doi.org/10.1109/ACCESS.2020.2981912
41. Niu, Y., Tang, K., Zhang, H., Lu, Z., Hua, X.S., Wen, J.R.: Counterfactual VQA: a cause-effect look at language bias. In: CVPR (2021)
42. Pearl, J.: Causality. Cambridge University Press, Cambridge (2009)
43. Russell, J.A.: A circumplex model of affect. J. Pers. Soc. Psychol. **39**(6), 1161 (1980)
44. Salimi, B., Rodriguez, L., Howe, B., Suciu, D.: Interventional fairness: causal database repair for algorithmic fairness. In: International Conference on Management of Data (2019)
45. Sariyanidi, E., Gunes, H., Cavallaro, A.: Automatic analysis of facial affect: a survey of registration, representation, and recognition. IEEE TPAMI **37**(6), 1113–1133 (2014)
46. Verma, S., Rubin, J.: Fairness definitions explained. In: International Workshop on Software Fairness (Fairware), pp. 1–7. IEEE (2018)
47. Wang, W., Feng, F., He, X., Zhang, H., Chua, T.S.: Clicks can be cheating: counterfactual recommendation for mitigating clickbait issue. In: ACM SIGIR Conference on Research and Development in Information Retrieval (2021)
48. Wong, A.: Mitigating gender bias in neural machine translation using counterfactual data. M.A. thesis, City University of New York (2020)
49. Xu, T., White, J., Kalkan, S., Gunes, H.: Investigating bias and fairness in facial expression recognition. In: Bartoli, A., Fusiello, A. (eds.) ECCV 2020. LNCS, vol. 12540, pp. 506–523. Springer, Cham (2020). https://doi.org/10.1007/978-3-030-65414-6_35
50. Zafar, M.B., Valera, I., Rogriguez, M.G., Gummadi, K.P.: Fairness constraints: mechanisms for fair classification. In: Artificial Intelligence and Statistics, pp. 962–970. PMLR (2017)

Improved Cross-Dataset Facial Expression Recognition by Handling Data Imbalance and Feature Confusion

Manogna Sreenivas[1]([✉]), Sawa Takamuku[2], Soma Biswas[1], Aditya Chepuri[3],
Balasubramanian Vengatesan[3], and Naotake Natori[2]

[1] Indian Institute of Science, Bangalore, India
manognas@iisc.ac.in
[2] Aisin Corporation, Tokyo, Japan
[3] Aisin Automotive Haryana Pvt Ltd., Bangalore, India

Abstract. Facial Expression Recognition (FER) models trained on one dataset (source) usually do not perform well on a different dataset (target) due to the implicit domain shift between different datasets. In addition, FER data is naturally highly imbalanced, with a majority of the samples belonging to few expressions like neutral, happy and relatively fewer samples coming from expressions like disgust, fear, etc., which makes the FER task even more challenging. This class imbalance of the source and target data (which may be different), along with other factors like similarity of few expressions, etc., can result in unsatisfactory target classification performance due to confusion between the different classes. In this work, we propose an integrated module, termed **DIFC**, which can not only handle the source **D**ata **I**mbalance, but also the **F**eature **C**onfusion of the target data for improved classification of the target expressions. We integrate this DIFC module with an existing Unsupervised Domain Adaptation (UDA) approach to handle the domain shift and show that the proposed simple yet effective module can result in significant performance improvement on four benchmark datasets for Cross-Dataset FER (CD-FER) task. We also show that the proposed module works across different architectures and can be used with other UDA baselines to further boost their performance.

Keywords: Facial expression recognition · Unsupervised domain adaptation · Class imbalance

1 Introduction

The need for accurate facial expression recognition models is ever increasing, considering its applications in driver assistance systems, understanding social behaviour in different environments [8,28], Human Computer Interaction [10], security applications [1], etc. Though deep learning has proven to work remarkably well for generic object classification tasks, classifying human expressions still remains challenging due to the subjective nature of the task [3,16]. Since data annotation is itself an expensive task, it is important that FER models trained on one dataset can be adapted seamlessly to other datasets. The large domain shift between the real-world datasets [7,13,21,25,35] makes this problem extremely challenging. This problem is addressed in the UDA setting, where the objective is to transfer discriminative knowledge learnt from a labelled source domain to a target domain with different data distribution, from which only unlabelled samples are accessible. Most of the current UDA approaches [11,27,31] try to address this domain shift by aligning the source and target features.

In this work, we aim to address complementary challenges which can adversely affect the FER performance. One such challenge is the huge source data imbalance that is usually present in FER datasets since the number of labelled training examples for a few expressions like neutral and happy is usually significantly more compared to that of the less frequently occurring expressions like disgust, fear, etc. However, the final goal is to recognize the expressions in the target dataset, which may not have the same amount of imbalance as the source. In addition, there may be other factors, like the inherent similarity between two expressions etc. that may result in confusion between different target classes.

Here, we propose a simple, yet effective module termed as **DIFC** (**D**ata **I**mbalance and **F**eature **C**onfusion), which addresses both these challenges simultaneously. Specifically, we modify the state-of-the-art technique LDAM [4] which is designed to handle class imbalance for the supervised classification task, such that it can also handle target feature confusion in an unsupervised domain adaptation scenario. This module can be seamlessly integrated with several UDA methods [11,27,31]. In this work, we integrate it with an existing UDA technique, Maximum Classifier Discrepancy (MCD) [27] and evaluate its effectiveness for the CD-FER task on four benchmark datasets, namely JAFFE [25], SFEW2.0 [7], FER 2013 [13] and ExpW [35]. Thus, the contributions of this work can be summarized as:

1. We propose a simple yet effective DIFC module for the CD-FER task.
2. The module can not only handle source data imbalance, but also target feature confusion in an integrated manner.
3. The proposed module can be used with several existing UDA approaches for handling these challenges.
4. Experiments on four real-world benchmark datasets show the effectiveness of the proposed DIFC module.

2   Related Work

Unsupervised Domain Adaptation: The objective of UDA is to utilize labelled source domain samples along with unlabelled target domain samples to learn a classifier that can perform well on the target domain. A wide spectrum of UDA methods follow adversarial training mechanism inspired by Generative Adversarial Networks [12]. Domain Adversarial Neural Networks [11] uses a two-player game, where the feature extractor aims to align the source and target features, while the domain discriminator aims to distinguish between the domains. This results in a domain invariant feature space and hence, translating the source classifier to the target. Conditional Adversarial Domain Adaptation [23] leverages the discriminative information provided by the classifier to aid domain alignment. Another line of works [15,26] perform distribution alignment in the pixel space instead of feature space. Maximum Classifier Discrepancy (MCD) [27] uses two classifiers to identify misaligned target features and further aligns them in an adversarial fashion. In MDD [34], a new distribution divergence measure called Margin Disparity Discrepancy is proposed, which has a strong generalization bound that can ease minimax optimization. In SWD [17], the authors use a Wasserstein distance metric to measure the discrepancy between classifiers. SAFN [31] is a simple non-adversarial UDA method where the features are learnt progressively with large norms as they are better transferable to the target domain.

Cross-Dataset FER: Domain shifts in facial expression recognition can be attributed to several factors like pose, occlusion, race, lighting conditions etc. Also, inconsistencies in data due to noisy source labels, high class imbalance, etc., add to the challenges. We briefly discuss prior works [5,20,29,32,36] that address this problem. In [32], unsupervised domain adaptive dictionary learning is used to minimize the discrepancy between the source and target domains. A transductive transfer subspace learning method that combines the labelled source domain with an auxiliary set of unlabelled samples from the target domain to jointly learn a discriminative subspace was proposed in [36]. In [29], after pretraining the network using source samples, a Generative Adversarial Network is trained to generate image samples from target distribution, which is then used along with the source samples to finetune the network. Contrary to these works where only the marginal distribution between source and target are aligned, [20] proposes to minimize the discrepancy between the class conditional distributions across FER datasets as well. In [37], they propose to preserve class-level semantics while adversarially aligning source and target features, along with an auxiliary uncertainty regularization to mitigate the effect of uncertain images in CD-FER task. The recent state-of-the-art AGRA framework [5,30] proposes to integrate graph representation adaptation with adversarial learning to adapt global and local features across domains effectively.

Class Imbalance Methods: The problem of class imbalance has been quite extensively studied in the supervised image classification setting. The traditional ways of mitigating imbalance are re-sampling and re-weighting. The majority classes are under-sampled while minority classes are over-sampled to balance the classes in a re-sampling strategy [2]. However, this duplicates the minority classes, leading to overfitting, while undersampling the majority classes could leave out important samples. On the other hand, re-weighting refers to adaptively weighing a sample's loss based on its class. Weighting by inverse class frequency is a commonly used strategy. [6] propose to use a re-weighting scheme by estimating the effective number of samples per class. In [14], a hybrid loss is proposed that jointly performs classification and clustering. A max-margin constraint is used to

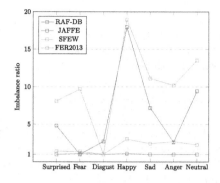

Fig. 1. Imbalance ratio for different classes vary across FER datasets.

achieve equispaced classification boundaries by simultaneously minimizing intra-class variance and maximizing inter-class variance. In LDAM [4], class imbalance is handled by enforcing a larger margin for the minority classes. They show that enforcing class-dependent margins along with a deferred re-weighting strategy effectively handles class imbalance in the supervised classification problem. We propose to modify this LDAM loss for the UDA setting to simultaneously handle source data imbalance and target confusion.

3   Problem Definition and Notations

We assume that we have access to a labelled source dataset $\mathcal{D}_s = \{(x_s^i, y_s^i)\}_{i=1}^{n_s}$, where $y_s^i \in \{1, \ldots, K\}$ and K denotes the number of classes. The objective is to correctly classify the samples from the target dataset $\mathcal{D}_t = \{x_t^i\}_{i=1}^{n_t}$, which does not have the class labels, but shares the same label space as $\mathcal{D}_s$. Here, n_s and n_t denote the number of source and target samples respectively. Now, we discuss the proposed approach in detail.

4   Proposed Approach

Most of the existing UDA approaches [11,23,27,34] aim to align the source and target features to mitigate the domain shift between them. In this work, we address two complementary challenges, namely source data imbalance and target feature confusion, which have been relatively less explored in the context of CD-FER task. These two challenges are handled in an integrated manner in the proposed DIFC module, which we describe next.

4.1   Handling Data Imbalance in DIFC

As already mentioned, FER datasets are naturally very imbalanced, with few classes like happy and neutral having a large number of samples, compared to classes like fear, disgust, etc. This can be seen from Fig. 1, where we plot the imbalance ratio of all the classes for the four datasets used in this work.

For a given dataset, if we denote the number of samples available for class k by N_k, its imbalance ratio is calculated as

$$Imbalance\_ratio_k = \frac{N_k}{\min_j(N_j)}, \qquad j, k \in \{1, \ldots, K\} \tag{1}$$

As mentioned earlier, K denotes the number of classes. We observe that the amount of imbalance varies across datasets, and the majority and minority classes may also differ when conditioned on the dataset. We now describe how we handle this imbalance for the FER task.

The goal of UDA is to perform well on the annotated source data and adapt to the target data simultaneously. Many UDA approaches [11,27,34] utilize two losses for this task, namely (i) classification loss on the annotated source samples and (ii) adaptation loss to bridge the domain gap between the source and target. Usually, the supervision from the labelled source domain $\mathcal{D}_s$ is used to learn the final classifier by minimizing the Cross-Entropy(CE) loss. Specifically, given a sample $(x_s, y_s) \in \mathcal{D}_s$, let the corresponding output logits from the model be denoted as $\mathbf{z_s} = [z_1, z_2, \ldots, z_K]^\top$. The CE loss is then computed as

$$\mathcal{L}_{\text{CE}}(\mathbf{z_s}, y_s) = -\log \left(\frac{\exp(z_{y_s})}{\sum_{k=1}^K \exp(z_k)} \right) \tag{2}$$

In general, if the training data is highly imbalanced, the standard cross-entropy loss may lead to poor generalization for minority classes because of the smaller margin between these classes. This problem due to data imbalance has been extensively explored in the supervised classification setting [4,6,14]. Here, we adopt the very successful Label Distribution Aware Margin (LDAM) approach [4] and modify it suitably for our task. The LDAM loss addresses the problem of class imbalance in classification by enforcing a class-dependent margin, with the margin being larger for the minority classes relative to that of the majority classes. Given the number of samples in each class $N_k, k \in \{1, .., K\}$, the class dependent margins $\boldsymbol{\Delta} \in \mathcal{R}^K$ and the LDAM loss (termed here as Data Imbalance or DI loss) for a sample (x_s, y_s) is computed as

$$\mathcal{L}_{\text{DI}}(\mathbf{z_s}, y_s; \boldsymbol{\Delta}) = -\log \frac{e^{z_{y_s} - \Delta_{y_s}}}{e^{z_{y_s} - \Delta_{y_s}} + \sum_{k \neq y_s} e^{z_k}}$$
$$\text{where } \Delta_k = \frac{\gamma}{N_k^{1/4}} \text{ for } k \in \{1, \ldots, K\} \tag{3}$$

where γ is a hyperparameter. In [4], this loss, along with a deferred re-weighting scheme complementing each other, is proven to address the imbalance issue effectively for long-tail label distributed classification tasks without trading off the performance on the majority classes. Following this scheme, we weight this loss using class-specific weights after a few epochs. We use the subscript DI to indicate that this is used for mitigating the effect of source data imbalance in the proposed DIFC module. Although this is very successful in handling data imbalance for supervised classification tasks, our goal is to correctly classify the target data. Figure 1 shows that amount of imbalance varies across datasets, suggesting that the margins computed using the source label distribution as in Eq. (3) may not be optimal for the target data. We now discuss how we address the target confusion along with the source data imbalance seamlessly in the proposed DIFC module.

4.2   Handling Feature Confusion in DIFC

The final goal is to improve the classification accuracy of the target features. Usually, samples from the minority classes tend to be confused with its neighbouring classes in the feature space. The target features can be confused because of the source as well as target data imbalance. But the target data imbalance may not be identical to that of the source. In addition, there may be other factors, e.g., few expressions are inherently close to one another compared to the others, which may result in increased confusion between certain classes. These unknown, difficult to quantify factors result in target feature confusion, which cannot be handled by LDAM or by aligning source and target distributions using UDA techniques. We now explain how this challenge is addressed in the proposed DIFC module.

The goal is to quantitatively measure and reduce the confusion between the target classes at any stage of the adaptation process. As the target data is unlabeled, one commonly used technique is to minimize the entropy of the unlabeled target data to enforce confident predictions, which in turn reduces the confusion among target classes. For this task, we empirically found that this decreased the target accuracy. Upon further analysis, we observed that several target examples were being pushed towards the wrong classes, which resulted in this performance decrease. This might be because classifying subtle facial expressions is, in general, very challenging and subjective. Thus, in the DIFC module, we try to find the confusion in a class-wise manner, hence reducing the influence of wrong class predictions of the individual target samples.

Here, we propose to use the softmax scores $\mathbf{p_t} = softmax(\mathbf{z_t})$ predicted by the model to estimate a confusion score for each class. We first group the target samples $x_t \in \mathcal{D}_t$ based on the predictions made by the model into sets $S_k, k \in \{1, \ldots, K\}$, where S_k contains all the target features classified into class k by the current model, i.e.,

$$S_k = \{x_t | \hat{y}_t = k\}; \quad \text{where} \quad \hat{y}_t = argmax(\mathbf{p_t}) \tag{4}$$

Using these predicted labels $\hat{y}_t$, we aim to quantify the confusion of each class in the target dataset. In general, samples from a particular class are confused with its neighbouring classes. But there are few classes which are more confusing than the others. Conditioned on the target dataset, this trend of confusion can differ from that of source. For a sample $x_t \in S_k$, if p_t^1 and p_t^2 are the top two softmax scores in the prediction vector $\mathbf{p_t} \in \mathcal{R}^K$, we formulate the confusion score of class k as

$$\text{conf}_k = \frac{1}{\mathbf{E}_{x_t \in S_k}|p_t^1 - p_t^2|} \tag{5}$$

As $x_t \in S_k$, p_t^1 corresponds to its confidence score for class k, however the second score p_t^2 can correspond to any class $j \neq k$. This implies that, for all the target samples predicted to belong to class k, we compute the mean differences between the highest and second-highest softmax scores. If the mean difference is large, it implies that the model is not confusing its predicted class k with any other class, and the computed confusion score for class k is less. On the other hand, if for majority of the samples predicted to belong to a class, the average difference between the top-2 scores is small, it implies that this class is more confusing and thus gets a high confusion score. We emphasize that this confusion is not computed in a sample-wise manner as it can be noisy due to wrong predictions.

We propose to modify the margins in Eq. (3) based on these estimated class confusion scores. Towards this goal, the class indices are sorted in decreasing order of confusion to get the ordered class indices C, such that $C[1]$ corresponds to the most confusing class and $C[K]$ corresponds to the least confusing class. We propose to use exponentially decreasing margin updates that emphasize on increasing the margin for the top few confusing classes without affecting the decision boundaries of the other classes. If $\Delta_{C[j]}$ denotes the margin for class $C[j]$, the class margins and DIFC loss are computed as follows:

$$\Delta'_{C[j]} = \Delta_{C[j]} + \frac{\epsilon}{2^{j-1}}$$
$$\mathcal{L}_{DIFC} = \mathcal{L}_{DI}(x_t, y_t; \mathbf{\Delta}') \tag{6}$$

where ϵ is a hyperparameter and refers to the increase in margin for the most confusing class i.e., $C[1]$. Using this formulation, we not only learn good decision boundaries taking into account the imbalanced source samples, but also reduce the confusion among the target domain samples.

4.3   Choosing Baseline UDA Approach

There has been a plethora of work in UDA for image classification, and it is not clear which one is better suited for this task. To choose a baseline UDA approach, we perform an empirical study of a few UDA approaches on SFEW2.0 [7] dataset, using RAF-DB [21] as the source data. The five approaches chosen are: (i) DANN [11]: classical UDA technique which forms the backbone for SOTA methods like AGRA [5], (ii) CADA [23]: approach which utilizes classifier outputs in a DANN-like framework, (iii) MCD [27]: approach which uses multiple classifiers to perform adversarial domain alignment, (iv) STAR [24]: very recent approach which extends the concept of multiple classifiers in MCD using a stochastic classifier, (v) SAFN [31]: a non-adversarial UDA approach that encourages learning features with larger norm as they are more transferable.

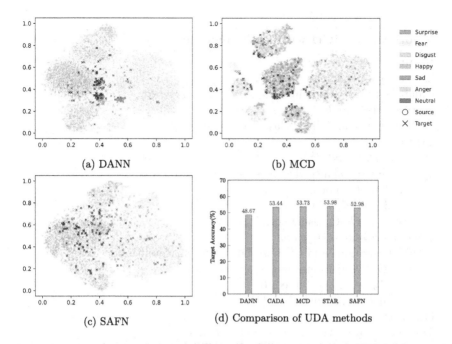

Fig. 2. We observe from the t-SNE plots for DANN (a), MCD (b) and SAFN (c) that the classes are better clustered and separated in MCD. (d) empirically shows the effectiveness of MCD when compared to others.

We observe from Fig. 2d that CADA [23], MCD [27] and STAR [24] achieve very similar accuracy, and all of them perform significantly better than DANN and SAFN. This implies that incorporating class information or class discrimination improves the classification performance of the target features. CD-FER task is characterized by large intra-class variations, small inter-class variations and domain shift. In such a fine-grained classification task, class-agnostic alignment based on domain confusion may not be sufficient. Confusing target samples often lie between two source clusters and could get aligned to the wrong class in this case. The effectiveness of using class discrimination is also evident from the t-SNE plots using DANN, SAFN and MCD (similar to CADA and STAR) in Fig. 2. Finally, in this work, we decided to use MCD [27] as the baseline UDA approach for the two following reasons: 1. It performs similar to the other approaches like CADA [23], STAR [24]. 2. Several recent works like SWD [17], MDD [34], STAR [24] build upon the principle of MCD. However, we emphasize that the proposed DIFC module can be seamlessly integrated with other UDA methods (based on completely different techniques) as an alternative for CE loss, which we demonstrate in Sect. 5.4.

Algorithm 1: MCD with DIFC module

Input:
Labelled source domain data $\mathcal{D}_s = \{(x_s^i, y_s^i), i = 1, ., n_s\}$
Unlabelled target domain data $\mathcal{D}_t = \{x_t^i, i = 1, ., n_t\}$.
Feature extractor F
Two randomly initialized classifiers C_1 and C_2.
Training:
Handling Data Imbalance
   Use Data Imbalance (DI) loss as classification loss.
   for i=1 to T_1 epochs
      i.e $\mathcal{L}_{cls} \leftarrow \mathcal{L}_{DI}$ in eq. (3)
      Train F, C_1, C_2 using eq. (7).
Handling Feature Confusion
   Update margins and then use DIFC loss as classification loss.
   for i=$T_1 + 1$ to T_2 epochs
      i.e. $\mathcal{L}_{cls} \leftarrow \mathcal{L}_{DIFC}$ in eq. (6)
      Train F, C_1, C_2 using eq. (7).
Output:
Trained feature extractor F and classifiers C_1 and C_2.

4.4 Integrating DIFC with Baseline UDA

Here, we briefly describe Maximum Classifier Discrepancy (MCD) [27] and how the DIFC module is integrated with it. MCD consists of three modules, a feature extractor F and two classifiers C_1 and C_2. Here, the feature extractor acts as the generator, and the classifiers C_1 and C_2 play the role of the discriminator. The generator and classifiers are trained to learn accurate decision boundaries by minimizing cross entropy loss on source domain samples. In order to align the source and target domain features, the classifiers C_1 and C_2 are leveraged to identify target samples beyond the support of source by maximizing the prediction discrepancy between the classifiers. The generator F is then trained to minimize the discrepancy so that the target features align with that of source. Given $(x_s, y_s) \in \mathcal{D}_s$ and $x_t \in \mathcal{D}_t$, the generator F and classifiers C_1, C_2 are optimized in three steps as follows

$$\min_{F,C_1,C_2} \mathcal{L}_{cls}\left(C_1\left(F\left(x_s\right)\right), y_s\right) + \mathcal{L}_{cls}\left(C_2\left(F\left(x_s\right)\right), y_s\right)$$

$$\max_{C_1,C_2} \|C_1\left(F\left(x_t\right)\right) - C_2\left(F\left(x_t\right)\right)\|_1 \qquad (7)$$

$$\min_{F} \|C_1\left(F\left(x_t\right)\right) - C_2\left(F\left(x_t\right)\right)\|_1$$

where $\mathcal{L}_{cls}$ refers to the standard CE loss in MCD.

First, we propose to address data imbalance using $\mathcal{L}_{DI}$ as the classification loss. Once the source samples are well separated, feature confusion is addressed using the final DIFC module. As two classifiers are used in MCD, the class prediction is based on the average of the two scores given by

$$\mathbf{p_{t,i}} = softmax(C_i(F(x_t))), \quad i = 1, 2$$
$$\mathbf{p_t} = \frac{1}{2}(\mathbf{p_{t,1}} + \mathbf{p_{t,2}}) \tag{8}$$
$$\hat{y}_t = argmax_k(\mathbf{p_t})$$

The DIFC loss is then computed as described in Eq. (6). During testing, given a target sample x_t, the class prediction is made using Eq. (8). The final integrated algorithm (MCD with DIFC module) is described in Algorithm 1.

5 Experiments

5.1 Datasets Used

We demonstrate the effectiveness of the proposed method on several publicly available datasets spanning different cultures, pose, illumination, occlusions, etc. For fair comparison, we follow the benchmark [5] and use RAF-DB [21] as the source dataset. We use other FER datasets, namely JAFFE [25], SFEW2.0 [7], FER2013 [13] and ExpW [35] as the target data. For all the datasets, the images are labelled with one of the seven expressions, namely surprise, fear, disgust, happy, anger, sad and neutral.

5.2 Implementation Details

Model Architecture. We use the same backbone as that used in [5] for fair comparison. A ResNet-50 backbone pretrained on MS-Celeb-1M dataset is used to extract local and global features. Firstly, MT-CNN [33] network is used to obtain the face bounding boxes as well as five landmarks corresponding to the two eyes, the nose and the two ends of the lips. Following [5], the face crop is resized to $112 \times 112 \times 3$ image resolution and fed to the ResNet-50 backbone, from which a feature map of size $7 \times 7 \times 512$ is obtained at the end of the fourth ResNet block. The global features are obtained by convolving this with a $7 \times 7 \times 64$ filter followed by Global Average Pooling, resulting in a 64-dimensional feature vector. The feature maps from the second ResNet block of dimension $28 \times 28 \times 128$ are used to extract the local features. This feature map is used to obtain five crops of $7 \times 7 \times 128$ surrounding each corresponding landmark location. Each such feature map is further convolved with a $7 \times 7 \times 64$ filter followed by Global Average Pooling, resulting in a 64-dimensional local feature descriptor for each landmark. The global and local features are then concatenated to obtain a 384 $(64 \times (1+5))$ dimensional feature. An additional fully connected layer enables global and local feature connections to get the final feature vector of 384 dimensions. We use two classifiers with different random initializations for MCD, each being a fully connected layer that does a 7-way classification of the 384-dimensional feature vectors.

Table 1. Target classification accuracy (%) of the proposed framework compared with state-of-the-art methods.

Method	JAFFE	SFEW2.0	FER2013	ExpW	Avg
CADA [23]	52.11	53.44	57.61	63.15	56.58
SAFN [31]	61.03	52.98	55.64	64.91	58.64
SWD [17]	54.93	52.06	55.84	68.35	57.79
LPL [22]	53.05	48.85	55.89	66.90	56.17
DETN [19]	55.89	49.40	52.29	47.58	51.29
ECAN [20]	57.28	52.29	56.46	47.37	53.35
JUMBOT [9]	54.13	51.97	53.56	63.69	55.84
ETD [18]	51.19	52.77	50.41	67.82	55.55
AGRA [5]	61.50	56.43	**58.95**	68.50	61.34
Proposed DIFC	**68.54**	**56.87**	58.06	**71.20**	**63.67**

Training Details. We train the backbone and the two classifiers using stochastic gradient descent (SGD) with an initial learning rate of 0.001, momentum of 0.9, and a weight decay of 0.0005. Initially, during the domain adaptation, only the DI loss is used for the initial epochs which addresses the source imbalance. The complete DIFC loss is used after few epochs to ensure that the target class predictions are more reliable and can be used to compute the feature confusion. In all our experiments, after training the model with only the DI loss for the initial 20 epochs, the learning rate is reduced to 0.0001 and the model is further fine-tuned with the complete DIFC loss for another 20 epochs. We tune γ to normalise the class margins $\Delta_j, j \in \{1, \ldots, K\}$, so that the largest enforced margin is 0.3. We set ϵ to 0.02 for all the experiments. The same protocol without any modification is followed for all the target datasets. We use Pytorch framework and perform all experiments on a single NVIDIA GTX 2080Ti GPU using a batch size of 32.

5.3   Results on Benchmark Datasets

We report the results of the proposed framework on four datasets, i.e. JAFFE [25], SFEW2.0 [7], FER2013 [13] and ExpW [35] as the target and RAF-DB as the source in Table 1. Comparison with the state-of-the-art approaches is also provided. The results of the other approaches are taken directly from [5]. We observe that the proposed framework performs significantly better than the state-of-the-art for almost all datasets, except for SFEW2.0, where it is second to [5]. But on an average, it outperforms all the other approaches.

5.4   Additional Analysis

Here, we perform additional experiments to study the effect of different components of the loss function, integrate it with existing UDA methods. These show that our module is effective in widely varying experimental settings.

Table 2. Importance of both components of the proposed DIFC loss.

Method	JAFFE	SFEW2.0	FER2013	ExpW	Avg
CE	64.79	53.73	55.80	69.10	60.85
DI	67.14	55.90	57.30	70.70	62.76
DIFC	**68.54**	**56.87**	**58.06**	**71.20**	**63.67**

Table 3. Order of confusion for different target datasets

Dataset	Surprise	Fear	Disgust	Happy	Sad	Anger	Neutral
JAFFE	5	1	2	7	3	6	4
SFEW2.0	4	2	5	7	3	6	1
FER2013	6	2	1	4	3	7	5
ExpW	3	1	2	7	4	5	6

Ablation Study: To understand the impact of different components of the proposed DIFC loss, we perform experiments using MCD as baseline UDA method with CE (Eq. 2) loss, DI or LDAM loss (Eq. 3) and the proposed DIFC loss (Eq. 6). The results in Table 2 show that using the DI loss improves target accuracy when compared to using the standard CE loss. Specifically, addressing source data imbalance improves the target accuracy by about 2% on average across all four target datasets. However, as the imbalance exists not only in the source but also in the target datasets, which may be different as shown in Fig. 1, the margins derived using source label distribution may not be optimal for the target dataset. The DIFC module adapts these margins based on the target feature confusion, further improving the target accuracy.

Analysis of Confusion Scores: In order to analyze the metric proposed to measure confusion (Eq. 5), we compute the order of confusion obtained for each dataset after addressing the source data imbalance in Table 3. We observe that the confusing classes vary across datasets and also differ from the minority classes (Fig. 1). Based on this order, we divide the classes into two sets, *confusing* (with order 1, 2, 3) and *non-confusing* (with order 4,5,6,7). Since the total accuracy as reported in Table 1 can be biased towards the majority classes, here we analyze the average class accuracy for each of these two sets in Table 4 after handling data imbalance and further target confusion, which we refer as **DI** and **DIFC** respectively. We observe that (1) the average accuracy of the non-confusing classes is, in general, significantly more than that of the confusing classes, except for SFEW2.0, where they are very similar; (2) the DIFC module significantly improves the average accuracy for confusing classes without compromising the performance on the other classes. Table 6 shows a few model predictions from different target datasets for DI loss and the final DIFC loss. These results show that incorporating DIFC loss can correct several samples which were misclassified even after addressing source imbalance.

Table 4. Average accuracy for confusing vs other classes.

Dataset	Confusing classes		Non confusing classes	
	DI	DIFC	DI	DIFC
JAFFE	48.01	**51.22**	80.88	**80.88**
SFEW2.0	50.70	**51.89**	46.88	**48.02**
FER2013	34.57	**38.09**	**67.58**	67.41
ExpW	30.78	**31.20**	66.25	**66.75**

Table 5. DIFC with baseline UDA methods. Backbone: ResNet-18, Source: RAF-DB, Target: SFEW2.0

Method	DANN	SAFN	MCD	Avg
CE	48.67	50.46	52.75	50.63
DI	50.12	51.81	53.25	51.73
DIFC	**51.33**	**53.01**	**55.66**	**53.33**

DIFC Module Integrated with Other UDA Approaches: Most UDA methods [5,11,27,31] use CE loss to learn the classifier from the labelled source dataset and an adaptation loss driving the alignment between source and target features. However, the decision boundaries learnt using CE loss are not very effective in the presence of data imbalance. Here, we show that the DIFC module can be integrated with other UDA methods as an alternative for CE loss. We select three UDA approaches whose adaptation mechanisms are principally different from each other. DANN [11] and MCD [27] are adversarial methods that use domain discriminator and multiple classifiers respectively. As discussed, MCD incorporates class-discriminative information in the adaptation process unlike DANN. On the other hand, SAFN [31] is a non-adversarial UDA method. We use ResNet-18 backbone, RAF-DB as source and SFEW2.0 as target dataset in these experiments. Incorporating DIFC with each of these improves the target accuracy on SFEW2.0 dataset by about 2.5% when compared to using CE loss, as shown in Table 5.

Complexity Analysis. The base feature extractor being common for AGRA [5] and our method, we report the additional number of parameters in the two methods. AGRA has a total of 318,464 extra parameters accounting for two intra-domain GCNs $(2 \times 64 \times 18)$, inter-domain GCN (64×64), Classifier (384×7) and Domain discriminator $(2 \times 384 \times 384 + 384 \times 1)$. On the other hand, the proposed framework has only 152,832 additional parameters which is due to an extra FC layer (384×384) and two classifiers (384×7) used in MCD. The proposed method outperforms current state-of-the-art results using only about half the additional parameters when compared to AGRA.

Table 6. Comparison of model predictions for DI (blue) and DIFC (orange) loss. The correct and incorrect predictions are marked with ✓and ✗respectively.

	Surprise	Fear	Disgust	Happy	Sad	Anger	Neutral
JAFFE	✓✓	✗✓	✗✓	✓✓	✗✓	✗✗	✓✓
SFEW2.0	✗✓	✗✓	✗✗	✓✓	✓✗	✗✓	✗✓
FER2013	✓✓	✗✓	✗✓	✓✓	✓✓	✗✓	✓✓
ExpW	✗✓	✓✓	✗✓	✓✗	✓✓	✗✓	✗✓

6   Conclusion

In this work, we propose a novel Data Imbalance and Feature Confusion (DIFC) module for the Cross-Dataset FER task. Firstly, the proposed module effectively mitigates the effect of source data imbalance and hence learns better decision boundaries. But this can be insufficient due to the shift in label distribution of the target data compared to the source data. To handle this and other unknown subjective factors that might be present, we devise the DIFC module to mitigate such confusion among the target classes. Specifically, the proposed DIFC module incorporates confusion in target data into the supervised classification loss of the baseline UDA framework to learn improved decision boundaries. Extensive experiments in varied settings demonstrate the effectiveness of the DIFC module for the CD-FER task. Additionally, the DIFC module can be seamlessly integrated with several existing UDA methods as an alternative for the standard CE loss, thereby further improving their performance.

Acknowledgements. This work is partly supported through a research grant from AISIN.

References

1. Al-Modwahi, A.A.M., Sebetela, O., Batleng, L.N., Parhizkar, B., Lashkari, A.H.: Facial expression recognition intelligent security system for real time surveillance. In: Proceedings of World Congress in Computer Science, Computer Engineering, and Applied Computing (2012)
2. Barandela, R., Rangel, E., Sánchez, J.S., Ferri, F.J.: Restricted decontamination for the imbalanced training sample problem. In: Sanfeliu, A., Ruiz-Shulcloper, J. (eds.) CIARP 2003. LNCS, vol. 2905, pp. 424–431. Springer, Heidelberg (2003). https://doi.org/10.1007/978-3-540-24586-5_52
3. Brooks, J.A., Chikazoe, J., Sadato, N., Freeman, J.B.: The neural representation of facial-emotion categories reflects conceptual structure. Proc. Natl. Acad. Sci. **116**(32), 15861–15870 (2019)
4. Cao, K., Wei, C., Gaidon, A., Arechiga, N., Ma, T.: Learning imbalanced datasets with label-distribution-aware margin loss. In: NeurIPS, vol. 32 (2019)
5. Chen, T., Pu, T., Wu, H., Xie, Y., Liu, L., Lin, L.: Cross-domain facial expression recognition: A unified evaluation benchmark and adversarial graph learning. IEEE Trans. Pattern Anal. Mach. Intell. (2021)
6. Cui, Y., Jia, M., Lin, T.Y., Song, Y., Belongie, S.: Class-balanced loss based on effective number of samples. In: CVPR, pp. 9268–9277 (2019)
7. Dhall, A., Goecke, R., Lucey, S., Gedeon, T.: Static facial expression analysis in tough conditions: data, evaluation protocol and benchmark. In: ICCV Workshops, pp. 2106–2112 (2011)
8. Edwards, J., Jackson, H., Pattison, P.: Erratum to "emotion recognition via facial expression and affective prosody in schizophrenia: a methodological review" [clinical psychology review 22 (2002) 789–832]. Clin. Psychol. Rev. **22**, 1267–1285 (2002)
9. Fatras, K., Sejourne, T., Flamary, R., Courty, N.: Unbalanced minibatch optimal transport; applications to domain adaptation. In: Meila, M., Zhang, T. (eds.) ICML, pp. 3186–3197 (2021)
10. Fragopanagos, N., Taylor, J.: Emotion recognition in human-computer interaction. Neural Netw. **18**(4), 389–405 (2005)
11. Ganin, Y., et al.: Domain-adversarial training of neural networks. J. Mach. Learn. Res. **17**(59), 1–35 (2016)
12. Goodfellow, I., et al.: Generative adversarial nets. In: NeurIPS, vol. 27 (2014)
13. Goodfellow, I.J., et al.: Challenges in representation learning: a report on three machine learning contests. Neural Netw. **64**, 59–63 (2015)
14. Hayat, M., Khan, S., Zamir, S.W., Shen, J., Shao, L.: Gaussian affinity for max-margin class imbalanced learning. In: ICCV (2019)
15. Hoffman, J., et al.: CyCADA: cycle-consistent adversarial domain adaptation. In: ICML, vol. 80, pp. 1989–1998 (2018)
16. Jack, R.E., Garrod, O.G.B., Yu, H., Caldara, R., Schyns, P.G.: Facial expressions of emotion are not culturally universal. Proc. Natl. Acad. Sci. **109**(19), 7241–7244 (2012)
17. Lee, C.Y., Batra, T., Baig, M.H., Ulbricht, D.: Sliced wasserstein discrepancy for unsupervised domain adaptation. In: CVPR, pp. 10277–10287 (2019)
18. Li, M., Zhai, Y.M., Luo, Y.W., Ge, P.F., Ren, C.X.: Enhanced transport distance for unsupervised domain adaptation. In: CVPR (2020)
19. Li, S., Deng, W.: Deep emotion transfer network for cross-database facial expression recognition. In: ICPR, pp. 3092–3099 (2018)

20. Li, S., Deng, W.: A deeper look at facial expression dataset bias. IEEE Trans. Affect. Comput. (2020)
21. Li, S., Deng, W., Du, J.: Reliable crowdsourcing and deep locality-preserving learning for expression recognition in the wild. In: CVPR, pp. 2584–2593. IEEE (2017)
22. Li, S., Deng, W., Du, J.: Reliable crowdsourcing and deep locality-preserving learning for expression recognition in the wild. In: CVPR, pp. 2584–2593 (2017)
23. Long, M., CAO, Z., Wang, J., Jordan, M.I.: Conditional adversarial domain adaptation. In: NeurIPS, vol. 31 (2018)
24. Lu, Z., Yang, Y., Zhu, X., Liu, C., Song, Y.Z., Xiang, T.: Stochastic classifiers for unsupervised domain adaptation. In: CVPR, pp. 9108–9117 (2020)
25. Lyons, M., Akamatsu, S., Kamachi, M., Gyoba, J.: Coding facial expressions with gabor wavelets. In: Proceedings Third IEEE International Conference on Automatic Face and Gesture Recognition, pp. 200–205 (1998)
26. Murez, Z., Kolouri, S., Kriegman, D., Ramamoorthi, R., Kim, K.: Image to image translation for domain adaptation. In: CVPR, pp. 4500–4509 (2018)
27. Saito, K., Watanabe, K., Ushiku, Y., Harada, T.: Maximum classifier discrepancy for unsupervised domain adaptation. In: CVPR (2018)
28. Sajjad, M., Zahir, S., Ullah, A., Akhtar, Z., Muhammad, K.: Human behavior understanding in big multimedia data using CNN based facial expression recognition. Mob. Netw. Appl. **25**(4), 1611–1621 (2020)
29. Wang, X., Wang, X., Ni, Y.: Unsupervised domain adaptation for facial expression recognition using generative adversarial networks. Comput. Intell. Neurosci. (2018)
30. Xie, Y., Chen, T., Pu, T., Wu, H., Lin, L.: Adversarial graph representation adaptation for cross-domain facial expression recognition. In: Proceedings of the 28th ACM International conference on Multimedia (2020)
31. Xu, R., Li, G., Yang, J., Lin, L.: Larger norm more transferable: an adaptive feature norm approach for unsupervised domain adaptation. In: ICCV (2019)
32. Yan, K., Zheng, W., Cui, Z., Zong, Y.: Cross-database facial expression recognition via unsupervised domain adaptive dictionary learning. In: NeurIPS, pp. 427–434 (2016)
33. Zhang, K., Zhang, Z., Li, Z., Qiao, Y.: Joint face detection and alignment using multitask cascaded convolutional networks. IEEE Signal Process. Lett. **23**(10), 1499–1503 (2016)
34. Zhang, Y., Liu, T., Long, M., Jordan, M.: Bridging theory and algorithm for domain adaptation. In: ICML, pp. 7404–7413 (2019)
35. Zhang, Z., Luo, P., Loy, C.C., Tang, X.: Learning social relation traits from face images. In: ICCV, pp. 3631–3639 (2015)
36. Zheng, W., Zong, Y., Zhou, X., Xin, M.: Cross-domain color facial expression recognition using transductive transfer subspace learning. IEEE Trans. Affect. Comput. **9**(1), 21–37 (2016)
37. Zhou, L., Fan, X., Ma, Y., Tjahjadi, T., Ye, Q.: Uncertainty-aware cross-dataset facial expression recognition via regularized conditional alignment. In: Proceedings of the 28th ACM International Conference on Multimedia, pp. 2964–2972 (2020)

Video-Based Gait Analysis for Spinal Deformity

Himanshu Kumar Suman[1] and Tanmay Tulsidas Verlekar[1,2]([ENVELOPE]) [ORCID]

[1] Department of CSIS, BITS Pilani, K K Birla, Goa Campus,
Sancoale 403726, Goa, India
{h20210066,tanmayv}@goa.bits-pilani.ac.in
[2] APPCAIR, BITS Pilani, K K Birla, Goa Campus, Sancoale 403726, Goa, India
https://www.bits-pilani.ac.in/goa/tanmayv/profile

Abstract. In this paper, we explore the area of classifying spinal deformities unintrusively using machine learning and RGB cameras. We postulate that any changes to posture due to spinal deformity can induce specific changes in people's gait. These changes are not limited to the spine's bending but manifest in the movement of the entire body, including the feet. Thus, spinal deformities such as Kyphosis and Lordosis can be classified much more effectively by observing people's gait. To test our claim, we present a bidirectional long short-term memory (BiLSTM) based neural network that operates using the key points on the body to classify the deformity. To evaluate the system, we captured a dataset containing 29 people simulating Kyphosis, Lordosis and their normal gait under the supervision of an orthopaedic surgeon using an RGB camera. Results suggest that gait is a better indicator of spinal deformity than spine angle.

Keywords: Gait analysis · LSTM · Spinal deformity · Dataset · Pattern recognition

1 Introduction

Gait can be defined as the manner of walking that results from coordinated interplay between the nervous, musculoskeletal and cardiorespiratory systems [7]. It consists of repeating a sequence of movements which comprise a gait cycle. Analysis of the gait cycle can reveal a wide range of information about people's identity [11], gender, age group [8], body weight, feelings and even emotions [21]. The gait cycle can be altered by factors such as neurological or systemic disorders, diseases, genetics, injuries, or ageing. Thus, medical diagnosis of such factors often involves capturing patients' gait to detect patterns unique to those factors. Most medical professionals capture people's gait through the use of sensor-based systems, such as optoelectronics, force plates or electromyographs [13]. The output of these systems can be used to estimate various gait-related features that medical professionals inspect to perform their diagnoses. With the advent of machine learning (ML), systems can now be developed to learn the patterns and perform gait analysis automatically.

L. Karlinsky et al. (Eds.): ECCV 2022 Workshops, LNCS 13805, pp. 278–288, 2023.
https://doi.org/10.1007/978-3-031-25072-9_18

A limitation of most sensor-based systems is that they are expensive. It limits their availability in most general clinics. They are also challenging to operate, requiring significant training and calibrations to use effectively. The limitations have motivated the use of RGB cameras to capture people's gait [14]. Their presence on even the most inexpensive cell phones and surveillance cameras makes gait analysis for medical diagnoses affordable and accessible to people in the comfort of their own homes. This paper explores using ML and RGB cameras to develop a system that can classify spinal deformities through gait analysis.

1.1 State-of-the-Art Review

In the field of medicine, sensor-based systems are extensively used to capture gait for medical diagnoses. For instance, in [25], posture and movement-related features are captured with ultrasonic systems attached to an athlete's heels and analysed to prevent injury. Recovery from surgical procedures, such as knee arthroplasty, is often analysed using joint motion, estimated from the output of optoelectronic motion capture systems [27]. Some daily tasks, such as walking up the stairs, can be good indicators of people's health. Works such as [26] captured older adults' gait during such tasks using inertial sensors. They then used ML techniques to predict the likelihood of a fatal fall. ML can also be used to learn specific foot patterns, such as spastic gait, which manifests in people with cerebral palsy. In [20] ground reaction forces captured using force plate sensors are used to detect the presence of spastic gait. Since some diseases become more prominent with time, the changes in regular gait patterns can be learnt to analyse the severity of certain diseases. It is demonstrated in [6], where the stage of Parkinson's disease is identified using the step length and back angle, estimated from the output of a marker-based optoelectronic system.

As discussed in Sect. 1, the use of RGB cameras and ML can make gait analysis possible without the need for medical professionals. The framework for such remote gait analysis is presented in [2], where the system hosted on the server processes uploaded video recordings. It then emails the diagnosis of people classifying their gait across five different types. It should be noted that most systems that operate using RGB cameras extract the silhouettes from the video sequences during pre-processing. Most aggregate them to create gait representation, such as the gait energy image (GEI). The GEIs can then be used by ML techniques such as support vector machines (SVM) [15] or convolutional neural networks (CNN) such as the finetuned VGG 16 [24] to perform classification. Apart from GEI, other gait representations, such as the skeleton energy image, are also used by CNNs to improve their accuracy [10]. To capture relevant temporal features, some systems have also used recurrent neural networks (RNN) to process the silhouettes directly [3].

While most methods focus on classification, some RGB camera-based systems focus on extracting biomechanical features that medical professionals can directly use in their diagnoses [23]. These systems emphasise detecting features, such as the time of heel strike and toe-off, with accuracies equivalent to the gold standard in gait analysis, the optoelectric motion capture system [22]. However, the operation of such systems is currently limited to observations in the sagittal plane. Some systems use these features to learn changes in gait due to restrictions in movement [16] or the severity of diseases such as Parkinson's [17].

1.2 Motivation and Contribution

The state-of-the-art review suggests there is scope for using RGB cameras to perform gait analysis in the field of medicine. Existing systems based on RGB cameras have demonstrated their effectiveness by operating equivalently to most sensor-based systems. Additionally, they offer advantages such as being unintrusive, affordable, and operable without medical professionals. The advantages make such systems of great interest to research and the medical community. Currently, most of these RGB camera-based systems focus on the movement of feet and legs to perform gait analysis. Apart from feet and legs, gait also contains information about people's posture. Thus, this paper explores a novel problem of using gait analysis to classify spinal deformity captured using RGB camera. We present:

- A gait classification system that can classify Kyphosis, Lordosis and normal gait;
- A video dataset containing 29 people simulating Kyphosis and Lordosis and normal gait (captured under the supervision of an orthopaedic surgeon);
- An evaluation of the proposed system and a comparison with spinal analysis (a benchmark suggested by the orthopaedic surgeon).

2 Proposed System

This paper considers two spinal deformities, Kyphosis [1] and Lordosis, that affect a large amount of the elderly population [12]. They are usually diagnosed by observing spinal curvatures from X-Rays images. Recently, ML has also been explored to develop deep learning-based systems for automatic analysis of X-Ray images [9]. Even with such advancements, elderly people are still required to self-assess their discomfort and approach a medical professional to obtain a diagnosis.

We propose a 2D camera-based gait analysis system that can classify captured gait as Kyphosis, Lordosis or normal. It analyses the recording of people captured using a cell phone or surveillance cameras to offer a preliminary diagnosis at home or elderly care facilities. An advantage of such analysis is that it can occur before people experience any significant discomfort. The proposed system consists of two blocks, as illustrated in Fig. 1:

- Pose estimation: to obtain a 2D pose from an input image;
- BiLSTM-based network: to learn the temporal patterns available in the 2D poses across a video sequence and classify the gait as Kyphosis, Lordosis or normal.

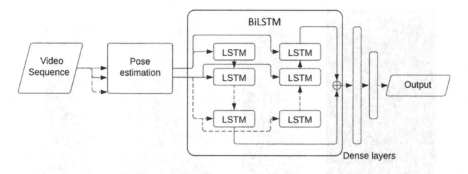

Fig. 1. Block diagram of the proposed system

2.1 Pose Estimation

The pose estimation is performed using the OpenPose [4], a deep CNN capable of identifying and classifying 135 key points in the human body. Given an RGB image, OpenPose returns a set of (X, Y) coordinates along with the mapping between them to create a skeleton-like representation of the body. The Open-Pose network uses VGG-19 [19] as its backbone. Its output branches into two additional CNNs. The first network builds a confidence map for each key point. The second network estimates the degree of association between two key points to create a mapping. This process is repeated in stages, where the network successively refines its output.

Out of the available 135 key points, we select 25 key points, as illustrated in Fig. 2 (a), to create a 50-dimensional vector. The keypoint vectors belonging to a video sequence are concatenated to create the input to the BiLSTM-based network. The dataset discussed in Sect. 3 contains video sequences with a large number of frames capturing multiple gait cycles. Thus, we sample 128 frames at a time from the video sequence, such that the frames captures at least one gait cycle. The frames at the beginning and the end capture a slightly different view of the people when compared to frames captured when people are perpendicular to the camera -see Fig. 2 (c).

2.2 BiLSTM-Based Network

The vectorised output of the OpenPose is processed by a BiLSTM to capture the temporal information available between frames. It is composed of two LSTM [5] networks arranged to process information in two directions, beginning to end and end to the beginning [18]. Since a gait cycle consists of a sequence of events that follow one another, an LSTM can effectively capture the context of the events at the beginning of the gait cycle. Feeding the input in the reverse direction to an LSTM will allow it to hold the context of the events at the end of the gait cycle. Combining the two through a BiLSTM allows preservation of context in both directions at any point in time.

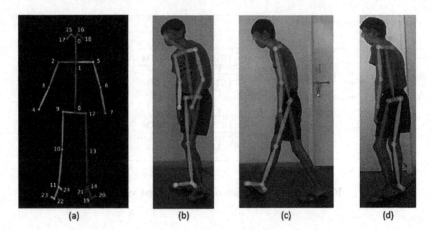

Fig. 2. Output of OpenPose: (a) selected key points, (b) pose at the beginning of the sequence, (c) pose when people are perpendicular to the camera, (d) pose at the end of the sequence.

Structurally, LSTM is a special type of recurrent neural network capable of handling long-term dependency in sequential data. It is achieved by introducing a memory cell, which is capable of retaining additional information. The access to the memory cell is controlled by a number of neural networks called gates. The first gate, called the output gate, is used to read the entries from the memory cell. The second gate, referred to as the input gate, decides when to read the data into the memory cell. Lastly, the forget gait governs the mechanism to reset the content of the cell. The LSTM processes the input sequentially and passes the content of only the hidden state as its output. To offer the LSTM a look-ahead ability, a second LSTM is introduced in a BiLSTM, which processes the input in the reverse direction. The hidden states are shared across the two LSTMs to process the input more flexibly. The outputs of the two networks are concatenated to create the final output of the BiLSTM network. The proposed system uses a single layer of BiLSTM to generate a 100-dimensional output vector containing the relevant temporal information. It can then use the dense layers to perform the final classification task.

The proposed network contains two fully connected dense layers with a dropout of 0.2 between them. The first layer contains 64 neurons with a sigmoid activation function. The second layer has an output size of three with a softmax activation function. It outputs the probabilities of an input gait being Kyphosis, Lordosis and normal.

The proposed system is trained using categorical cross-entropy as the loss function and the Adam optimiser with a learning rate of 0.0001. The epoch is set to 300 with early stopping set to monitor loss with patience of 20. The batch size for training is set to 20.

3   Spinal Deformity Dataset

The use of RGB cameras and ML for the classification of spinal deformity through gait analysis is a novel problem to our best knowledge. No publicly available datasets currently exist to evaluate such systems. Hence, we captured a dataset called the spinal deformity dataset with 29 people simulating Kyphosis, Lordosis and their normal gait. The decision to simulate was made due to the ethics and privacy concerns of the patients. Most patients hesitate to be recorded with a camera. Access to elderly people is also restricted due to the COVID-19 pandemic. However, to maintain quality control, an orthopaedic surgeon was consulted. The surgeon advised on the movements and posture to be maintained during simulations and later validated all the captured sequences. The sequences were captured using GoPro Hero 9, a 4K RGB video camera that operated at 30fps. People walked along a line perpendicular to the camera, approximately 3 m away. The camera was mounted on a tripod stand, raised 1.25 m from the ground. The 29 people aged 20 to 27 years participated in the construction of the dataset. They were captured on three separate days, where each participant simulated three different types of gait. Each participant was captured twice per simulation, walking from right to left. Each sequence was between 7 to 10 s, capturing multiple (3–4) gait cycles.

Before capturing the sequences, the participants were instructed on how to simulate Kyphosis and Lordosis and underwent practise runs. Kyphosis is caused by forwarding bending of the dorsal spine and appears as a hump or curve to the upper back, as illustrated in Fig. 3 (a). The participants simulated it by leaning forward their heads compared to the rest of the body. Kyphosis also causes fatigue in the back and legs simulated by smaller step lengths. Lordosis is caused by backward curving of the lumbar spine, appearing swayback, with their buttocks being more pronounced. The participants simulated it by adopting a backwards-leaning posture while walking - see Fig. 3 (b). Also, back pain and discomfort in moving can be simulated with restricted leg movement. No instructions were offered during the recording of the normal gait. However, adhering to the medical protocol, arm movement was restricted in all three cases -see Fig. 3.

It should be noted that the participants for the simulation are in their 20 s, while the demographics for the proposal are aged people. The purpose of the dataset, in its current form, is to highlight the ability of the proposed system to classify spatial deformity from bad posture. We intend to improve the dataset by capturing actual patients in the future.

4   Results

To evaluate the proposed system, we split spinal deformity dataset into a training and a test set. The training set is composed of 48 Kyphosis, 42 Lordosis and 45 normal sequences captured from 24 people captured on the first and the second day. The test set comprises 5 Kyphosis, 5 Lordosis and 5 normal sequences captured from 5 people captured on the third day. The evaluation performed

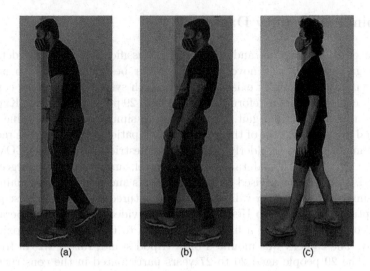

Fig. 3. Simulation of (a) Kyphosis, (b) Lordosis, (c) normal gait.

such that the training and the test set are mutually exclusive with respect to the participants. It should be noted that although 29 people participated in the dataset acquisition process, several sequences were discarded by the orthopaedic surgeon due to poor or exaggerated simulations.

Following the orthopaedic surgeon's guidance, a benchmark is created to compare the performance of the proposed system. The benchmark analysis focuses on the bending of the spine. The bend is indicated by the angle value computed using the key point 0 and 8 (as illustrated in Fig 2 (a)) from the OpenPose and the vertical axis. K-nearest neighbour is then used to train the benchmark system using all the training sequences and evaluated using the test set, similar to the proposed system. The benchmark system is designed with the assumption that even when using an RGB camera, the captured spine contains sufficient information to classify the spinal deformity. The results of the evaluation are reported in Table 1.

Table 1. Performance of the systems on the spinal deformity dataset.

System	Accuracy(%)
Benchmark	82.23
Proposed system	**91.97**

The results suggest that gait is far more effective than just the spine angle data. Using the benchmark, the mean angles associated with Kyphosis, Lordosis and normal training data are $-9.1°$, $10.8°$ and $-1.2°$ respectively. However, each of them has a standard deviation of approximately ± 6. Thus, as many as 35% of normal sequences get classified as Kyphosis - see Table 2. The problem can become severe if all the frames in the beginning and end of the video sequence are considered for processing. As the position of the people in the field of view of the camera changes, the skeleton-like structure of the key points may appear skewed - see Fig. 2. The proposed system addresses this issue by learning the change in view of the people captured by the RGB camera.

Table 2. Confusion matrix for the benchmark system.

		Prediction		
Class		Kyphosis	Lordosis	Normal
Ground truth	Kyphosis	91.25	0.31	8.44
	Lordosis	0.63	94.53	4.84
	Normal	35.31	3.75	60.94

Some misclassifications can also occur because some people have a slouching posture while walking normally. In such cases, the spine angle might not be a reliable feature. The proposed system is an improvement over the benchmark system as it does not limit its focus to just the spine angle. Kyphosis and Lordosis cause the centre of gravity to shift forward or backwards due to the bending of the spine, which limits the body's overall movement while walking. A similar effect can also be seen in the legs, where the discomfort caused by the spinal deformities restricts their movement. The proposed system uses the body and leg movement information to improve its classification accuracy. It should also be noted that the benchmark system performs misclassification across Kyphosis, Lordosis and the normal gait. In contrast, the proposed system has a precision of 100% in classifying Lordosis - see Table 3. Thus, it can be concluded that the use of gait by the proposed system allows it to obtain better results.

Table 3. Confusion matrix for the proposed system.

		Prediction		
Class		Kyphosis	Lordosis	Normal
Ground truth	Kyphosis	85.16	0	14.84
	Lordosis	0	97.81	2.19
	Normal	7.03	0	92.97

5   Conclusions

The use of an RGB camera for gait analysis has gained significant interest in recent years. Several systems have been proposed that use ML to classify different types of gait. Most of these systems focus on the movement of the feet and legs regions. Since gait also captures posture information, we propose a BiLSTM-based system to classify gait as Kyphosis, Lordosis or normal. The system processes a video sequence using OpenPose to obtain important key points in people's bodies. It then trains the system using the key points to perform classification. The proposed system addresses a novel problem not yet discussed by the research community to our best knowledge. Thus, to test the proposed system, we capture a dataset containing 29 people simulating Kyphosis, Lordosis and normal gait under the guidance of an orthopaedic surgeon. We also create a benchmark system using the orthopaedic surgeon's inputs. The results highlight the advantage of using gait for classification over the spine angle.

This paper can be considered a pilot study. Possible future directions can include improving the system to assess the severity of the spinal deformity. The domain of explainable artificial intelligence can also be explored so that the proposed system can offer preliminary diagnoses to people and useful insights to medical professionals. Finally, the dataset can be improved by enrolling patients suffering from spinal deformities.

Acknowledgement. We thank the orthopaedic surgeon Dr. Hemant Patil for helping us construct the spinal deformity dataset.

References

1. Ailon, T., Shaffrey, C.I., Lenke, L.G., Harrop, J.S., Smith, J.S.: Progressive spinal kyphosis in the aging population. Neurosurgery. **77**(suppl_1), S164–S172 (2015)
2. Albuquerque, P., Machado, J.P., Verlekar, T.T., Correia, P.L., Soares, L.D.: Remote gait type classification system using markerless 2d video. Diagnostics **11**(10), 1824 (2021)
3. Albuquerque, P., Verlekar, T.T., Correia, P.L., Soares, L.D.: A spatiotemporal deep learning approach for automatic pathological gait classification. Sensors **21**(18), 6202 (2021)
4. Cao, Z., Simon, T., Wei, S.E., Sheikh, Y.: Realtime multi-person 2d pose estimation using part affinity fields. In: Proceedings of the IEEE Conference on Computer Vision and Pattern Recognition, pp. 7291–7299 (2017)
5. Hochreiter, S., Schmidhuber, J.: Long short-term memory. Neural Comput. **9**(8), 1735–1780 (1997)
6. Jellish, J., et al.: A system for real-time feedback to improve gait and posture in Parkinson's disease. IEEE J. Biomed. Health Inform. **19**(6), 1809–1819 (2015)
7. Kerrigan, C.K.: Gait analysis in The Science of Rehabilitation, vol. 2. Diane Publishing, Collingdale (2000)
8. Kozlowski, L.T., Cutting, J.E.: Recognizing the sex of a walker from a dynamic point-light display. Percept. Psychophys. **21**(6), 575–580 (1977)

9. Lee, H.M., Kim, Y.J., Cho, J.B., Jeon, J.Y., Kim, K.G.: Computer-aided diagnosis for determining sagittal spinal curvatures using deep learning and radiography. J. Digital Imaging. **35**, 1–14 (2022)

10. Loureiro, J., Correia, P.L.: Using a skeleton gait energy image for pathological gait classification. In: 2020 15th IEEE International Conference on Automatic Face and Gesture Recognition (FG 2020), pp. 503–507. IEEE (2020)

11. Makihara, Y., Nixon, M.S., Yagi, Y.: Gait recognition: databases, representations, and applications. Computer Vision: A Reference Guide, pp. 1–13. Springer, Cham (2020). https://doi.org/10.1007/978-3-030-03243-2_883-1

12. Milne, J., Lauder, I.: Age effects in kyphosis and lordosis in adults. Ann. Hum. Biol. **1**(3), 327–337 (1974)

13. Muro-De-La-Herran, A., Garcia-Zapirain, B., Mendez-Zorrilla, A.: Gait analysis methods: an overview of wearable and non-wearable systems, highlighting clinical applications. Sensors **14**(2), 3362–3394 (2014)

14. Nieto-Hidalgo, M., Ferrández-Pastor, F.J., Valdivieso-Sarabia, R.J., Mora-Pascual, J., García-Chamizo, J.M.: Vision based gait analysis for frontal view gait sequences using RGB camera. In: García, C.R., Caballero-Gil, P., Burmester, M., Quesada-Arencibia, A. (eds.) UCAmI 2016. LNCS, vol. 10069, pp. 26–37. Springer, Cham (2016). https://doi.org/10.1007/978-3-319-48746-5_3

15. Nieto-Hidalgo, M., García-Chamizo, J.M.: Classification of pathologies using a vision based feature extraction. In: Ochoa, S.F., Singh, P., Bravo, J. (eds.) UCAmI 2017. LNCS, vol. 10586, pp. 265–274. Springer, Cham (2017). https://doi.org/10.1007/978-3-319-67585-5_28

16. Ortells, J., Herrero-Ezquerro, M.T., Mollineda, R.A.: Vision-based gait impairment analysis for aided diagnosis. Med. Biol. Eng. Comput. **56**(9), 1553–1564 (2018). https://doi.org/10.1007/s11517-018-1795-2

17. Rupprechter, S., et al.: A clinically interpretable computer-vision based method for quantifying gait in Parkinson's disease. Sensors **21**(16), 5437 (2021)

18. Schuster, M., Paliwal, K.K.: Bidirectional recurrent neural networks. IEEE Trans. Signal Process. **45**(11), 2673–2681 (1997)

19. Simonyan, K., Zisserman, A.: Very deep convolutional networks for large-scale image recognition. arXiv preprint arXiv:1409.1556 (2014)

20. Slijepcevic, D., et al.: Automatic classification of functional gait disorders. IEEE J. Biomed. Health Inform. **22**(5), 1653–1661 (2017)

21. Troje, N.F.: Decomposing biological motion: a framework for analysis and synthesis of human gait patterns. J. Vis. **2**(5), 2 (2002)

22. Verlekar, T.T., De Vroey, H., Claeys, K., Hallez, H., Soares, L.D., Correia, P.L.: Estimation and validation of temporal gait features using a markerless 2d video system. Comput. Methods Programs Biomed. **175**, 45–51 (2019)

23. Verlekar, T.T., Soares, L.D., Correia, P.L.: Automatic classification of gait impairments using a markerless 2d video-based system. Sensors **18**(9), 2743 (2018)

24. Verlekar, T.T., Correia, P.L., Soares, L.D.: Using transfer learning for classification of gait pathologies. In: 2018 IEEE International Conference on Bioinformatics and Biomedicine (BIBM), pp. 2376–2381. IEEE (2018)

25. Wahab, Y., Bakar, N.A.: Gait analysis measurement for sport application based on ultrasonic system. In: 2011 IEEE 15th International Symposium on Consumer Electronics (ISCE), pp. 20–24. IEEE (2011)

26. Wang, K., et al.: Differences between gait on stairs and flat surfaces in relation to fall risk and future falls. IEEE J. Biomed. Health Inform. **21**(6), 1479–1486 (2017)
27. Wang, W., Ackland, D.C., McClelland, J.A., Webster, K.E., Halgamuge, S.: Assessment of gait characteristics in total knee arthroplasty patients using a hierarchical partial least squares method. IEEE J. Biomed. Health Inform. **22**(1), 205–214 (2017)

TSCom-Net: Coarse-to-Fine 3D Textured Shape Completion Network

Ahmet Serdar Karadeniz, Sk Aziz Ali$^{(\boxtimes)}$, Anis Kacem, Elona Dupont, and Djamila Aouada

SnT, University of Luxembourg, 1511 Kirchberg, Luxembourg
{AhmetSerdar.Karadeniz,SkAziz.Ali,Anis.Kacem,Elona.Dupont,
Djamila.Aouada}@uni.lu

Abstract. Reconstructing 3D human body shapes from 3D partial textured scans remains a fundamental task for many computer vision and graphics applications – *e.g.*, body animation, and virtual dressing. We propose a new neural network architecture for 3D body shape and high-resolution texture completion – TSCom-Net – that can reconstruct the full geometry from mid-level to high-level partial input scans. We decompose the overall reconstruction task into two stages – first, a joint implicit learning network (SCom-Net and TCom-Net) that takes a voxelized scan and its occupancy grid as input to reconstruct the full body shape and predict vertex textures. Second, a high-resolution texture completion network, that utilizes the predicted coarse vertex textures to inpaint the missing parts of the partial 'texture atlas'. A Thorough experimental evaluation on 3DBodyTex.V2 dataset shows that our method achieves competitive results with respect to the state-of-the-art while generalizing to different types and levels of partial shapes. The proposed method has also ranked second in the track1 of SHApe Recovery from Partial textured 3D scans (SHARP [2,37]) 2022 (https://cvi2.uni.lu/sharp2022/) challenge.

Keywords: 3D Reconstruction · Shape completion · Texture-inpainting · Implicit function · Signed distance function

1 Introduction

3D textured body shape completion and reconstruction from visual sensors plays a key role in a wide range of applications such as gaming, fashion, and virtual reality [30,31]. The challenges of this task and the methods to tackle it varies depending on the used sensors to capture the human shape and texture. Some existing methods focus mainly on 3D textured body shape reconstruction from a single monocular image [17,19,22,42,49]. For instance, under the body shape symmetry assumption, methods like Human Mesh Recovery (HMR) [18], use a statistical body model [25] to reconstruct the body shape and pose. Despite the impressive advances in this line of work [22,42,49], the reconstructed shapes still cannot capture high-level geometrical details, such as wrinkles, ears. This is due to the lack of original geometrical information and the projective distortions in

© The Author(s), under exclusive license to Springer Nature Switzerland AG 2023
L. Karlinsky et al. (Eds.): ECCV 2022 Workshops, LNCS 13805, pp. 289–306, 2023.
https://doi.org/10.1007/978-3-031-25072-9_19

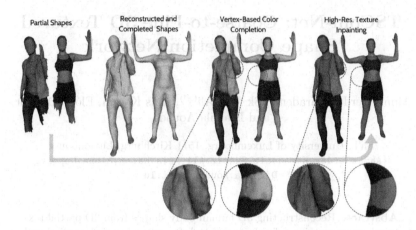

Fig. 1. 3D Partial Textured Body Shape Completion. The proposed deep learning-based method reconstructs and completes the surface geometry of a dressed or minimally-clothed partial body scan, and inpaints the missing regions with high-resolution texture. TSCom-Net has the flexibility to either apply dense texture mapping or vertex-based color estimation as per desired application scenarios.

2D images. On the other hand, 3D scanners and RGB-D sensors can provide richer information about the geometry of body shapes [41]. Nevertheless, they are often subject to partial acquisitions due to self occlusions, restricted sensing range, and other limitations of scanning systems [8,46]. While most of existing works focus on completing the geometry of body shapes from 3D partial scans [8,38,46], less interest has been dedicated to complete both the texture and the geometry at the same time. Nonetheless, this problem remains critical in real-world applications where complete and realistic human reconstructions are usually required. Aware of this need, recent competitions, such as SHApe Recovery from Partial Textured 3D Scans (SHARP) challenges [1,3,37], emerged in the research community to foster the development of joint shape and texture completion techniques from partial 3D acquisitions. The results of these competitions showed promising techniques [9,36] with a room for improvements.

The problem of textured 3D body shape completion from 3D partial scans, as shown in Fig. 1, can naturally be decomposed into two challenging subtopics – (i) partial shape surface completion and (ii) partial texture completion of the reconstructed shape. In this setup, especially when no fixed UV-parametrization of shape is available, methods need to rely on inherent shape structure and texture correlation cues from the available body parts. Taking this direction, IFNet-Texture [9] uses implicit functions to perform high-quality shape completion and vertex-based texture predictions. Despite the impressive results in SHARP challenges [1,37], the method [9] does not output high-resolution texture and often over-smooths vertex colors over the whole shape. On the other hand, 3DBooSTeR method [36] decouples the problems of shape and texture completions and solves them in a sequential model. The shape completion of [36] is based on body models [12], while the texture completion is tackled as an image based *texture-atlas* inpainting. [36] can complete high-resolution partial texture, but suffers from large shape and texture modelling artifacts.

Our method overcomes the weaknesses of both [9] and [36]. The shape completion part of our TSCom-Net is designed to learn a continuous *implicit function* representation [28, 29, 48] as also followed by Chibane *et al.* [8]. Furthermore, we identify that the IFNet-Texture method [9], an extension of [8], uses a vertex-based inference-time texture estimation process that ignores the shape and color feature correlation cues at the learning phase. For this reason, we employ an end-to-end trainable vertex-based shape and texture completion networks – SCom-Net and TCom-Net – to boost their performance with joint-learning. Next, we propose to refine the predicted vertex-based texture completions directly in the texture-atlas of 3D partial scans. This is achieved using an image inpainting network [23, 36], reusing the predicted vertex textures, and yielding high-resolution texture. It is notable that, unlike [20, 36], TSCom-Net does not require a fixed UV-parametrization which makes the considered 'texture charts' random, discontinuous, and more complex in nature across the samples. Overall, our **contributions** can be summarized to:

1. An end-to-end trainable joint implicit function network for 3D partial shape and vertex texture completion. We propose an early fusion scheme of shape and texture features within this network.
2. A high-resolution texture-atlas inpainting network, which uses partial convolutions [23] to refine the predicted vertex textures. At the same time, this module is flexible and can be plugged with any other 3D shape and vertex texture reconstruction module.
3. An extensive experimental evaluation of TSCom-Net showing its multi-stage improvements, its comparison *w.r.t.* the participants in SHARP 2022 challenge [3], and its generalization capabilities to different types of input data partiality.

The rest of the paper is organized as follows. After summarizing the methods related to our line of work in Sect. 2, we present the core components of TSCom-Net in Sect. 3. The experiments and evaluation parts are reported and discussed in Sect. 4. Finally, Sect. 5 concludes the paper and draws some future works.

2   Related Works

Deep Implicit Function and 3D Shape Representation Learning. Supervised learning methods for 3D shape processing require an efficient input data representation. Apart from common learning representations of 3D shapes – such as regular voxel grids [24, 49], point clouds [32], Oc-trees [44], Barnes-Hut 2^D-tree [5], depth-maps [27], and meshes [16], the popularity of implicit representation [28, 29, 48] has recently increased. These representations [29, 48] serve as a continuous representation on the volumetric occupancy grid [28] to encode the iso-surface of a shape. Therefore, given any query 3D point, it returns its binary occupancy in the encoded iso-surface. This continuous representation is extremely useful when the input shapes are partial and the spatial locations can be queried on missing regions based on the available regions.

Learning-based 3D Shape Completion. 3D reconstruction and completion of shapes, especially for human 3D body scans, is tackled in different ways [13–15, 39, 40, 43, 46, 47]. Some methods deal with partial point clouds as input [13, 15, 46, 47]. A subset of these methods do not consider textured scans and focus on different types of shape quantization – e.g., voxel grids [15], octrees, and sparse vectors [46] with the aim of capturing only fine geometric details. Furthermore, many of these models [7, 13, 46] cannot always predict a single shape corresponding to the partial input shape. Another traditional way for body shape completion from a set of partial 3D scans is by non-rigidly fitting a template mesh [4, 6, 11]. However, improper scaling [4], computational speed [6], and articulated body pose matching [11] is a major problem for these methods.

More closely related works to ours are [8, 9, 36]. Saint *et al.* [36] and Chibane *et al.* [9] solve the same 3D textured shape completion, where the former uses 3D-Coded [12] and non-rigid refinement to reconstruct the shape and the latter uses deep implicit functions [8]. While [36] provides a fixed UV-parametrization useful for the texture completion, it cannot recover extreme partial body scans due to its restriction to a template body model [25]. On the other hand, [8] can recover corase-to-fine geometrical details using multiscale implicit representations. However, the implicit representations cannot preserve the UV-parametrization of the input partial shape, which restricts the texture completion to a vertex-based solution [9]. TSCom-Net builds on top the implicit shape representation of [8,9] and extends it to produce higher-resolution texture.

Learning-Based 3D Textured Shape Completion. Completing the texture of 3D body shapes is a challenging task due to many factors – e.g., varying shades, unknown clothing boundaries, and complex styles and stripes. This task can be either tackled by completing the RGB texture for each vertex of the completed mesh [9], or directly completing the texture-atlas [10, 36]. While the former takes advantage of the shape structure of the vertices to predict textures, its resolution depends on the resolution of the shape, which is often limited due to memory constraints. On the other hand, the texture-atlas completion allows for high-resolution texture generation, but does not take advantage of the structure of the shape when no fixed UV-parametrization is given [20, 36], or multiple charts are provided [10, 36]. To address these limitations, TSCom-Net uses both paradigms. Firstly, it completes the vertex-based textures, then refines them in the texture-atlas image space using a dedicated inpainting module.

3   Proposed Approach – TSCom-Net

Given a 3D partial body scan $\mathbf{X} = (\mathcal{V}, \mathcal{T}, \mathcal{E})$, tuple defining set of vertices $\mathcal{V}$, triangles $\mathcal{T}$, and edges $\mathcal{E}$ along with its corresponding partial texture $\mathcal{A}$, our aim is to recover the 3D complete body scan $\mathbf{Y}$ with high-resolution texture.

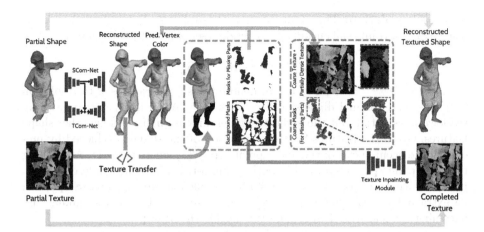

Fig. 2. Overview of TSCom-Net. TSCom-Net has two main components – (I) an end-to-end trainable vertex-based shape and color completion network (*on the left*) that returns low resolution and often over-smoothed vertex colors. The underlying architecture of this modules is based on IFNet [8] and IFNet-Texture [9] along with newly added modifications to boost both the shape and color completion jointly. (II) A coarse-to-fine texture inpainting module (using 2D partial convolutions [23]) that takes partially textured and fully reconstructed body shape, background coarse masks, and coarse-texture and its masks for the missing regions as input, and outputs a high-resolution, complete 'texture atlas'.

To achieve that, we propose a deep learning-based framework TSCom-Net with two intermediate stages (see Fig. 2). First, we reconstruct the shape and coarse vertex texture of the given partial scan. Then, we further inpaint the partial texture atlas with another network in a coarse-to-fine manner.

3.1   Joint Shape and Texture Completion

Let $\mathcal{X}_s = \mathbb{R}^{N \times N \times N}$ be the discretized voxel grid of the input partial scan **X** with resolution N. Each occupied voxel has the value of 1 (0 otherwise). Both the ground-truth and partial scans are normalized. Similarly, another voxel grid, $\mathcal{X}_c$, is constructed with the texture of the input partial scan **X**. Each occupied voxel of $\mathcal{X}_c$ has an RGB texture value in $[0, 1]^3$ obtained from the partial texture ($\langle -1, -1, -1 \rangle$ otherwise).

TSCom-Net consists of two implicit feature networks SCom-Net and TCom-Net that learn to complete the shape and texture of a partial surface, respectively. Different from [9], the two sub-networks are trained jointly to enable the interaction between shape and texture learning. An early fusion technique is utilized to improve the color accuracy by incorporating the shape features in addition to the color features of the partial scan. Consequently, the texture network TCom-Net learns to use the shape information extracted by SCom-Net to predict a more accurate texture. In what follows, we start by describing the shape completion, then present the texture completion.

Shape Completion: The input discretized voxel grid $\mathcal{X}_s$ is encoded as in [8] using a 3D convolutional encoder g_s to obtain a set of n multiscale shape features,

$$g_s(\mathcal{X}_s) := \mathbf{S}_1, \mathbf{S}_2, \dots, \mathbf{S}_n, \tag{1}$$

where $\mathbf{S}_i \in \mathbb{R}^{d_i \times K \times K \times K}$, $1 \leq i \leq n$, denote the shape features with resolution $K = \frac{N}{2^{i-1}}$ and channel dimension $d_i = d_1 \times 2^{i-1}$. Accordingly, $\mathbf{S}_n$ would have the lowest resolution but the highest channel dimensionality among other shape features. Given the features $\mathbf{S}_i$, the completion of the shape is achieved by predicting the occupancies of the query points $\{\mathbf{p}_j\}_{j=1}^M$, where $\mathbf{p}_j \in \mathbb{R}^3$. At training time, the points $\mathbf{p}_j$ are sampled from the ground-truth $\mathbf{Y}$. During inference time, $\mathbf{p}_j$ are the centroids of all voxels in $\mathcal{X}_s$. Multiscale features $\mathbf{S}_{i,j}^p := \phi(\mathbf{S}_i, \mathbf{p}_j)$ can be extracted for each point $\mathbf{p}_j$ using a grid sampling function ϕ, taking and flattening the features of $\mathbf{S}_i$ around the neighborhood of $\mathbf{p}_j$.

Finally, $\mathbf{p}_j$ and $\mathbf{S}_{i,j}^p$ are decoded with f_s consisting of sequential 1D convolutional layers to predict the occupancy value s_j of $\mathbf{p}_j$,

$$f_s(\mathbf{p}_j, \mathbf{S}_{1,j}^p, \mathbf{S}_{2,j}^p, \dots, \mathbf{S}_{n,j}^p) := s_j. \tag{2}$$

The completed mesh structure $\hat{\mathbf{X}}$ is obtained by applying marching cubes [26] on the voxel grid $\mathcal{X}_s$ with the predicted occupancy values s_j.

Texture Completion: For the texture completion, the colored voxel grid $\mathcal{X}_c$ is encoded using a 3D convolutional encoder g_c to obtain a set of m multiscale texture features,

$$g_c(\mathcal{X}_c) := \mathbf{C}_1, \mathbf{C}_2, \dots, \mathbf{C}_m , \tag{3}$$

where $\mathbf{C}_i \in \mathbb{R}^{r_i \times L \times L \times L}$, $1 \leq i \leq m$, denote the texture features with resolution $L = \frac{N}{2^{i-1}}$ and channel dimension $r_i = r_1 \times 2^{i-1}$. Texture completion is carried out by predicting the RGB values of the query points $\{\mathbf{q}_j\}_{j=1}^M$, where $\mathbf{q}_j \in \mathbb{R}^3$. The points $\mathbf{q}_j$ are sampled from the ground-truth $\mathbf{Y}$ during training. At the inference time, $\mathbf{q}_j$ are the vertices of the reconstructed mesh $\hat{\mathbf{X}}$. Similar to the shape completion, the grid sampling function $\phi(\mathbf{C}_i, \mathbf{q}_j)$ allows extracting multiscale texture features $\mathbf{C}_{i,j}^q$ for each point $\mathbf{q}_j$.

The shape and texture completions are fused at the texture decoder level. This is achieved by concatenating the multiscale shape and texture features before feeding them to a decoder f_c consisting of sequential 1D convolutional layers. In particular, the RGB values $\mathbf{c}_j$ for each point $\mathbf{q}_j$ are given by,

$$f_c(\mathbf{q}_j, \mathbf{S}_{1,j}^q, \mathbf{S}_{2,j}^q, \dots, \mathbf{S}_{n,j}^q, \mathbf{C}_{1,j}^q, \mathbf{C}_{2,j}^q, \dots, \mathbf{C}_{m,j}^q) := \mathbf{c}_j. \tag{4}$$

Finally, vertex textures are obtained by attaching the predicted $\mathbf{c}_j$ to $\hat{\mathbf{X}}$.

3.2   Texture Refinement

The proposed joint shape and texture networks are able to predict the vertex textures of a completed mesh. However, the predicted vertex textures have two limitations: (1) they are predicted for the full body, including the existing regions of the partial body. Consequently, high-level texture details of input partial body might be lost in the predicted textured body mesh; (2) the resolution of the vertex textures depends on the resolution of the reconstructed shape. This implies that high-resolution vertex textures come at a cost of high-resolution predicted shape, which is not straightforward to obtain due to memory constraints.

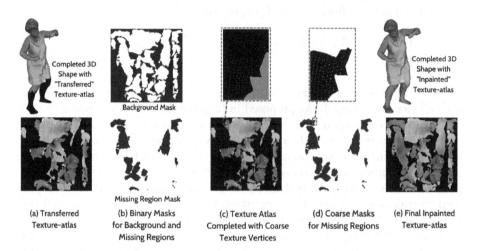

Fig. 3. Texture Refinement. (a) the texture-atlas is transferred to the completed 3D shape, (b) the masks for the missing regions and background are identified, (c) the vertex textures are projected into the transferred texture-atlas, (d) the masks for missing regions are updated by unmasking the regions of the projected vertex textures, (e) the final inpainted texture-atlas.

To overcome these issues, we use a *texture-atlas* [21,36] based refinement of the predicted vertex textures. This refinement reuses the coarse vertex textures predicted by the joint implicit shape and texture network and refines them in the 2D image space, while preserving the original texture from the partial input scan. In particular, the original texture of the partial scan is transferred to the completed shape by a ray-casting algorithm, as in [36]. This allows the creation of a UV map and a texture atlas $\mathcal{A}$ for the completed mesh as depicted in Fig. 3(a). Following [36], the missing regions and the background regions are identified in the transferred texture atlas to create the two binary masks M and M_b shown in Fig. 3(b). Using the UV map created by the texture transfer, the vertex textures are projected to obtain a coarsely completed texture-atlas $\mathcal{A}_c$ as sketched in Fig. 3(c). The mask for missing regions M is then updated with the projected vertex textures, yielding a coarse mask M_c as displayed in Fig. 3(d).

Given the coarsely completed texture-atlas $\mathcal{A}_c$, the coarse mask of missing regions M_c, and the background mask M_b, the problem of texture refinement is formulated as an image inpainting one. Specifically, we opt for the texture-atlas inpainting method proposed in [36], which adapts partial convolution inpainting [23] to the context of texture-atlas. Partial convolutions in [23,36] extend standard convolutions to convolve the information from unmasked regions (*i.e.* white regions in the binary masks). Formally, let us consider a convolution filter defined by the weights w, the bias b and the feature values $\mathcal{A}_c^w$ of the texture-atlas $\mathcal{A}_c$ for the current sliding window. Given the coarse mask of missing regions M_c and the corresponding background mask M_b, the partial convolution at every location, similarly defined in [36], is expressed as,

$$
a_c = \begin{cases} w^T(\mathcal{A}_c^w \odot M_c \odot M_b) \cdot \frac{\mathrm{sum}(\mathbf{1})}{\mathrm{sum}(M_c \odot M_b)} + b & \text{if } \mathrm{sum}(M_c \odot M_b) > 0 \\ 0 & \text{otherwise} \end{cases}, \quad (5)
$$

where $\odot$ denotes element-wise multiplication, and $\mathbf{1}$ has same shape as M_c but with all elements being 1. As proposed in [36], the masks M_c are updated after every partial convolution, while the background masks are passed to all partial convolutions layers without being updated by applying *do-nothing* convolution kernels. The partial convolutional layers are employed in a UNet architecture [33] instead of standard convolutions. At training time, the vertex textures are sampled from the ground-truth texture-atlas. The same loss functions in [23,36] are used to train the network. It is important to highlight that the proposed texture refinement is different from [36] as it reuses the predicted vertex textures instead of inpainting the texture-atlas from scratch. We show in Sect. 4 that the proposed refinement outperforms the inpainting from scratch and the vertex texture based completion. Furthermore, we reveal that such refinement can be used to improve other vertex based texture completion.

4   Experimental Results

Dataset and Evaluation Metrics. Our method was trained on the 3DBody-Tex.v2 dataset [34,35,37] which has been recently used as a benchmark for the SHARP challenge [3]. The dataset contains a large variety of human poses and different clothing types from 500 different subjects. Each subject is captured with 3 different poses and different clothing types, such as close-fitting or arbitrary casual clothing. The number of ground truth scans in the training set is 2094. A number of 15904 partial scans for training and validation were generated using the routines provided by the SHARP challenge organizers [2]. The number of unseen (during training) scans in the evaluation set is 451.

As considered in SHARP challenges, the evaluation is conducted in terms of shape scores S_s, texture scores S_t, and the area scores S_a. Shape and texture scores are calculated via measuring surface-to-surface distances by sampling points from the ground truth and the reconstructed meshes. The area score estimates the similarity between the triangle areas of the meshes. The final score S_r is calculated as $S_r = \frac{1}{2}S_a(S_s + S_t)$. Details about these metrics can be found in $[2, 37]^1$. The evaluation of the completed meshes is performed via the Codalab system provided by the SHARP challenge[2].

4.1   Network Training Details

We trained the joint implicit networks using the Adam optimizer with a learning rate of 10^{-4} for 54 epochs. The model was trained on an NVIDIA RTX A6000 GPU using the Pytorch library. The query points for the training are obtained from the ground truth surfaces by sampling 100000 points. During training, we sub-sample 50000 of these points at each iteration. Gaussian random noise $\mathcal{N}(\sigma)$ is added to each point to move the sampled point near or far from the surface, depending on the σ value. Similar to [9], the σ is chosen as 0.01 for half of the points and 0.1 for the other half. We did not add noise to the query points of the texture network, as its goal is to predict the color value of the points sampled from the surface. Partial scans are voxelized by sampling 100000 points from the partial surface and setting the occupancy value in the nearest voxel grid to 1. Similarly for the colored voxelization, the value of the nearest voxel is set to the RGB value of the sampled point obtained from the corresponding texture-atlas. The input voxel resolution is 128 and the resolution for the final retrieval is 256.

We follow a similar naming convention as in [45] for our architecture details. Let *c3-k* denote Conv3D-ReLU-BatchNorm block and *d3-k* denote two Conv3D-ReLU blocks followed by one BatchNorm layer where k is the number of filters. The kernel size for all 3D convolutional layers is 3×3. Let *gs-i* represent the grid feature sampling of the query points from the output of the previous layer, and *mp* denote 3D max pooling layer. Let *c1-k* denote the Conv1D layer with 1×1 kernel where k is the number of output features and ReLU activation except the output layer where the activation is linear. The encoder architecture for shape and texture networks is composed of the following layers: *gs-0, c3-16, gs-1, mp, d3-32, d3-32, gs-2, mp, d3-64, d3-64, gs-3, mp, d3-128, gs-4, mp, d3-128, d3-128, gs-5.*[3] The decoder architecture for the shape network contains: *c1-512, c1-256, c1-256, c1-1*. The decoder architecture for the texture network consists of *c1-512, c1-256, c1-256, c1-3*. The partial convolutional network is trained with Adam optimizer and learning rate of 10^{-4} for 330000 iterations. The original texture size of 2048×2048 was used for training the network with a batch size of 1.

4.2   Results and Evaluation

In this section, a qualitative and quantitative analysis of the results of TSCom-Net against the results of the SHARP challenge participants are presented. Other

---

[1] The details about the metrics can be accessed via: https://gitlab.uni.lu/cvi2/cvpr2022-sharp-workshop/-/blob/master/doc/evaluation.md.

[2] The leaderboard on Codalab can be accessed via: https://codalab.lisn.upsaclay.fr/competitions/4604#results.

[3] Sampling features *gs-1, gs-2, ..., gs-5* are flattened and concatenated. Shape features are also added here for the texture encoding.

Table 1. Quantitative Results for SHARP 2022. The best and second best scores are denoted in **bold-black** and bold-gray colors respectively.

Method	Shape Score(%)	Area Score(%)	Texture Score(%)	Final Score(%)
IFNet-Texture [9]	85.44 ± 2.93	96.26 ± 6.35	81.25 ± 7.61	83.34 ± 6.86
Method Raywit	85.91 ± 7.14	93.96 ± 3.96	83.45 ± 8.43	84.68 ± 7.63
Method Rayee	86.13 ± 7.32	96.26 ± 3.61	83.23 ± 8.31	84.68 ± 7.74
Method Janaldo	**89.76 ± 4.97**	**96.76 ± 2.28**	**87.10 ± 6.33**	**88.43 ± 5.56**
TSCom-Net (**Ours**)	85.75 ± 6.15	96.68± 2.89	83.72 ± 6.95	84.73 ± 6.5

participants of the challenge employ implicit networks for the shape reconstruction with vertex-based texture completion [3]. We also compare our method to the state-of-the-art IFNet-Texture [9] method, which won the previous editions of SHARP 2021 and 2020 [37]. The shape network of [9] is re-trained with the generated partial data of SHARP 2022. For the vertex texture predictions, the available pretrained model provided by the authors was used.

Quantitative Evaluation: In Table 1, we illustrate the quantitative results using the metrics of SHARP challenge [37]. Overall, our method is ranked second in the leaderboard with a final score of 84.73%, obtaining a better texture score than Method-Rayee, Method-Raywit and IFNet-Texture [9]. It should be noted that the texture score depends on the shape score as well, since it is not possible to get a correct texture score for an incorrectly predicted shape.

Table 2. Effectiveness of our texture refinement.

Method	Texture score (%)
Method-Janaldo	87.10± 6.33
Method-Janaldo + Our texture refinement	**87.54 ± 6.19**

Furthermore, we demonstrate the effectiveness of our texture refinement method by applying it to the predictions of Method-Janaldo. The results in Table 2 show that our inpainting network introduced in Sect. 3.2 improves the texture score by 0.44%. This shows that the proposed texture refinement can be used as an additional component to improve other shape and texture completion methods that predict low-resolution vertex textures.

Finally, Fig. 4 show the distributions of shape, texture, and final scores for the predictions of TSCom-Net, Method-Janaldo, Method-Janaldo with our texture refinement, and IFNet-Texture [9]. The first row of this figure illustrates the overall distribution of the scores and highlights the superiority of our texture scores *w.r.t.* IFNet-Texture [9]. While Method-Janaldo outperforms our results, it can be observed that if we endow it with our texture refinement scheme, the distribution of texture scores becomes slightly higher. In the second and third

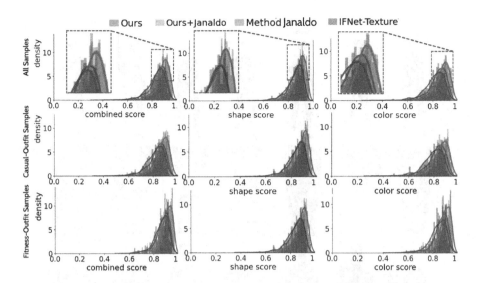

Fig. 4. Score Distribution. Distribution of shape, color and combined final scores ($\frac{score}{100}$ along x-axes of the plots). The first row reports the distribution of the scores on all the scans. The second row focuses on casual-outfit scans while the third one reports the distributions on fitness-outfit scans.

row of Fig. 4, we report the distribution of the scores on scans with casual-outfit and fitness-outfit. Unsurprisingly, casual-outfit scans were more challenging than fitness-outfit ones for all methods. Nevertheless, our approaches recorded more improvements *w.r.t.* Method-Janaldo and IFNet-Texture on the casual-outfit scans than the fitness-outfit ones.

It is important to mention that the reported scores are the result of mapping distance values (shape or texture) to scores in percentage using a parametric function. This mapping might make the scores from different methods close to each other, depending on the chosen parameters. We note that the original parameters of the SHARP challenge 2022 metrics are used. Thus, we further conducted a qualitative evaluation of our method.

Qualitative Evaluation: Figure 5 shows a qualitative comparison of our approach to other competing methods on models with casual outfit and fitness outfit. Considering the coat missing region of the top row model, it can be noted that IFNet-Texture [9] and Method-Raywit are unable to predict the correct colors. On the other hand, Method-Rayee and Method-Janaldo produce over-smoothed textures, creating blurry artifacts. Neither of these effects can be observed on the TSCom-Net predictions. In the second row, the bottom legs and feet appear to be the most difficult regions to recover. Method-Rayee and Method-Raywit fail to produce the correct shape for these missing regions. While IFNet-Texture [9] and Method-Janaldo are able to recover the correct shape, both generate white color artifacts on the jeans. Our method is the only one to produce more reasonable texture predictions, showing a sharper change in color between the jeans and

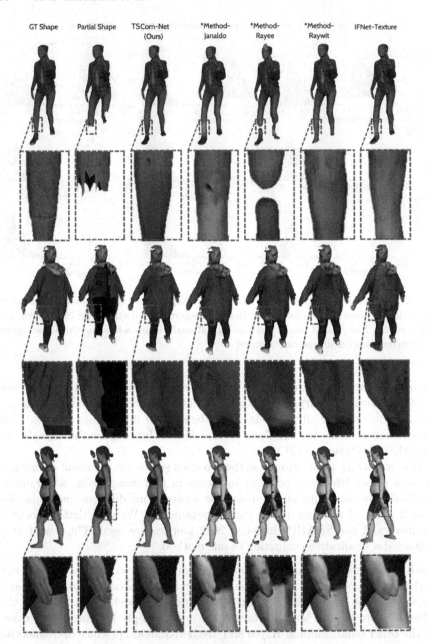

Fig. 5. Qualitative results. Visual comparisons of textured body shape completion results by different competing methods (as per Table 1). The *first two rows* depict results on samples with casual outfits, *i.e.,* more variation on garment texture pattern. The *last two rows* depict results on samples with fitness outfits.

the feet. Similar to the second model, it is apparent that our results on the third model are sharper for the regions with a color change from skin to black when compared to the other results. The visual comparisons illustrate that our results are of higher resolution with better texture representation than the competing approaches despite the close quantitative scores.

4.3   Ablation Study

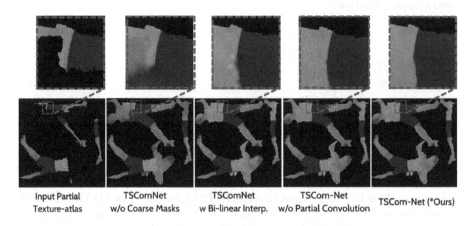

| Input Partial Texture-atlas | TSComNet w/o Coarse Masks | TSComNet w Bi-linear Interp. | TSCom-Net w/o Partial Convolution | TSCom-Net (*Ours) |

Fig. 6. Multi-stage improvements of texture-atlas inpainting in TSCom-Net.

Table 3. Quantitative scores of TSCom-Net with multi-stage improvements.

Method	Shape Score(%)	Area Score(%)	Texture Score(%)	Final Score(%)
IFNet-Texture [9]	85.44 ± 2.93	96.26 ± 6.35	81.25 ± 7.61	83.34 ± 6.86
Texture-transfer Baseline	85.75± 6.15	96.68 ± 2.89	56.51 ± 18.98	71.13 ± 11.11
Ours w/o Coarse Masks	85.75± 6.15	96.68 ± 2.89	81.04 ± 7.92	83.39± 6.87
Ours w/o Tex. Refine.	85.75± 6.15	96.68 ± 2.89	83.27 ± 7.08	84.53 ± 6.54
Ours w/ Bilinear Interp.	85.75 ± 6.15	96.68 ± 2.89	83.66 ± 6.95	84.71 ± 6.50
Ours w/o Partial Conv.	85.75 ± 6.15	96.68 ± 2.89	83.68 ± 6.96	84.71 ± 6.50
TSCom-Net (Ours)	**85.75 ± 6.15**	**96.68± 2.89**	**83.72 ± 6.95**	**84.73± 6.50**

Our joint implicit function learning for shape and texture (TSCom-Net w/o Texture Refinement) gives a 1.19% (cf. Table 3) increase in the final score with respect to [9] demonstrating the effect of early fusion between SCom-Net and TCom-Net. Only transferring the partial texture to the reconstructed shape, where the color values for the missing regions are all black, gives us a baseline texture score of 56.51%. Training an inpainting network directly on the partial textures (TSCom-Net w/o Coarse Masks) increase texture score to 81.04% which is 0.21% lower than [9] and 24.53% higher than the baseline. Conducting bilinear interpolation of the vertex colors in missing regions is giving a texture score of 83.66%. All-in-all partial convolutions, instead of standard convolutions, gives stable and more sharper texture inpainting results. Figure 6 depicts how the different types artifacts or lack of sharpness appear when other options of inpainting are tested. Finally, TSCom-Net consisting of all the components is giving the highest final score of 84.73%.

Table 4. Generalizability of TSCom-Net when trained and tested on samples with different types of partiality.

Method	Partiality type		Shape score (%)	Texture score (%)	Final score (%)
	Training	Testing			
IFNet-Texture [9]	T2	T1	87.99 ± 4.65	84.32 ± 5.73	86.15 ± 5.09
IFNet-Texture [9]	T1	T1	86.49 ± 3.96	**89.28 ± 2.89**	**87.88 ± 3.32**
3DBooster [36]	T1	T1	58.81 ± 14.99	72.31 ± 6.79	65.57 ± 3.32
TSCom-Net (**Ours**)	T2	T1	**88.05 ± 4.84**	85.46 ± 5.56	86.75 ± 5.14

Generalization to Other Types of Partiality: In this section, we evaluate the generalization capability of our model to other types of partiality. In particular, we consider the *view-based* partiality (T2) introduced in SHARP 2022 [3] to train the models and use the *hole-based* partiality (T1) introduced in previous editions of SHARP [37] for inference. This implies that the networks trained on (T2) have never seen hole-based partial scans (T1). Table 4 demonstrates a comparison of the generalization capability of our method to IFNet-Texture [9] that is also trained on (T2) and tested on (T1) subset. Following these settings, we obtain an increase of 1.14% for the texture score compared to [9]. We also compare our model trained on (T2) and tested on (T1) to IFNet-Texture [9] and 3DBooster [36] both trained on (T1) and tested on (T1). In this case, our approach significantly outperforms [36] while achieving comparable results to [9] although both were trained on (T1).

5   Conclusion

This paper presents a method for completing the shape and texture of 3D partial scans. Joint implicit feature networks are proposed for learning to complete the shape and textures. Moreover, a new coarse-to-fine texture refinement network was introduced. It generates high-resolution texture from the predicted coarse vertex texture and the available partial texture. Experimental evaluations show that our method gives visually more appealing results than the state-of-the-art and is positioned second in the SHARP 2022 challenge. In future, we plan to make the entire TSCom-Net modules end-to-end trainable for completing 3D scans and refining the texture. At the same time, we will investigate neural implicit radiance field for texture completion (with editable 2D UV texture map) and 3D surface reconstruction.

Acknowledgement. This work is partially funded by the National Research Fund (FNR), Luxembourg under the project references BRIDGES21/IS/16849599/FREE-3D, CPPP17/IS/11643091/IDform, and industrial project grant IF/17052459/CAS CADES by Artec 3D.

References

1. SHARP 2021, the 2nd shape recovery from partial textured 3d scans. https://cvi2.uni.lu/sharp2021/. Accessed 23 July 2022
2. SHARP 2022 Repository, the repository of the 3rd shape recovery from partial textured 3d scans. https://gitlab.uni.lu/cvi2/cvpr2022-sharp-workshop. Accessed 23 July 2022
3. SHARP 2022, the 3rd shape recovery from partial textured 3d scans. https://cvi2.uni.lu/sharp2022/. Accessed 23 July 2022
4. Ali, S.A., Golyanik, V., Stricker, D.: NRGA: gravitational approach for non-rigid point set registration. In: International Conference on 3D Vision (3DV) (2018)
5. Ali, S.A., Kahraman, K., Reis, G., Stricker, D.: RPSRNet: end-to-end trainable rigid point set registration network using Barnes-hut 2d-tree representation. In: IEEE Conference on Computer Vision and Pattern Recognition (CVPR) (2021)
6. Ali, S.A., Yan, S., Dornisch, W., Stricker, D.: FoldMatch: accurate and high fidelity garment fitting onto 3d scans. In: IEEE International Conference on Image Processing (ICIP) (2020)
7. Arora, H., Mishra, S., Peng, S., Li, K., Mahdavi-Amiri, A.: Multimodal shape completion via implicit maximum likelihood estimation. In: IEEE Conference on Computer Vision and Pattern Recognition (CVPR) (2022)
8. Chibane, J., Alldieck, T., Pons-Moll, G.: Implicit functions in feature space for 3d shape reconstruction and completion. In: IEEE Conference on Computer Vision and Pattern Recognition (CVPR) (2020)
9. Chibane, J., Pons-Moll, G.: Implicit feature networks for texture completion from partial 3D Data. In: Bartoli, A., Fusiello, A. (eds.) ECCV 2020. LNCS, vol. 12536, pp. 717–725. Springer, Cham (2020). https://doi.org/10.1007/978-3-030-66096-3_48
10. Deng, J., Cheng, S., Xue, N., Zhou, Y., Zafeiriou, S.: UV-GAN: adversarial facial UV map completion for pose-invariant face recognition. In: Proceedings of the IEEE Conference on Computer Vision and Pattern Recognition, pp. 7093–7102 (2018)
11. Golyanik, V., Reis, G., Taetz, B., Strieker, D.: A framework for an accurate point cloud based registration of full 3d human body scans. In: International Conference on Machine Vision and Applications (ICMVA) (2017)
12. Groueix, T., Fisher, M., Kim, V.G., Russell, B.C., Aubry, M.: 3D-CODED: 3D correspondences by deep deformation. In: Ferrari, V., Hebert, M., Sminchisescu, C., Weiss, Y. (eds.) ECCV 2018. LNCS, vol. 11206, pp. 235–251. Springer, Cham (2018). https://doi.org/10.1007/978-3-030-01216-8_15
13. Gurumurthy, S., Agrawal, S.: High fidelity semantic shape completion for point clouds using latent optimization. In: IEEE Winter Conference on Applications of Computer Vision (WACV) (2019)
14. Han, X.F., Laga, H., Bennamoun, M.: Image-based 3d object reconstruction: state-of-the-art and trends in the deep learning era. IEEE Trans. Pattern Anal. Mach. Intell. **43**(5), 1578–1604 (2019)

15. Han, X., Li, Z., Huang, H., Kalogerakis, E., Yu, Y.: High-resolution shape completion using deep neural networks for global structure and local geometry inference. In: IEEE International Conference on Computer Vision (ICCV) (2017)
16. Hanocka, R., Hertz, A., Fish, N., Giryes, R., Fleishman, S., Cohen-Or, D.: MeshCNN: a network with an edge. ACM Trans. Graph. (TOG) **38**(4), 1–12 (2019)
17. Hasler, N., Ackermann, H., Rosenhahn, B., Thormählen, T., Seidel, H.P.: Multilinear pose and body shape estimation of dressed subjects from image sets. In: IEEE Computer Society Conference on Computer Vision and Pattern Recognition (2010)
18. Kanazawa, A., Black, M.J., Jacobs, D.W., Malik, J.: End-to-end recovery of human shape and pose. In: IEEE Conference on Computer Vision and Pattern Recognition (CVPR) (2018)
19. Kanazawa, A., Tulsiani, S., Efros, A.A., Malik, J.: Learning category-specific mesh reconstruction from image collections. In: Ferrari, V., Hebert, M., Sminchisescu, C., Weiss, Y. (eds.) ECCV 2018. LNCS, vol. 11219, pp. 386–402. Springer, Cham (2018). https://doi.org/10.1007/978-3-030-01267-0_23
20. Lazova, V., Insafutdinov, E., Pons-Moll, G.: 360-degree textures of people in clothing from a single image. In: International Conference on 3D Vision (3DV) (2019)
21. Lévy, B., Petitjean, S., Ray, N., Maillot, J.: Least squares conformal maps for automatic texture atlas generation. ACM Trans. Graph. (TOG) **21**(3), 362–371 (2002)
22. Li, X., et al.: Self-supervised single-view 3D reconstruction via semantic consistency. In: Vedaldi, A., Bischof, H., Brox, T., Frahm, J.-M. (eds.) ECCV 2020. LNCS, vol. 12359, pp. 677–693. Springer, Cham (2020). https://doi.org/10.1007/978-3-030-58568-6_40
23. Liu, G., et al.: Image inpainting for irregular holes using partial convolutions. In: Ferrari, V., Hebert, M., Sminchisescu, C., Weiss, Y. (eds.) ECCV 2018. LNCS, vol. 11215, pp. 89–105. Springer, Cham (2018). https://doi.org/10.1007/978-3-030-01252-6_6
24. Liu, Z., Tang, H., Lin, Y., Han, S.: Point-voxel CNN for efficient 3d deep learning. In: Advances in Neural Information Processing Systems, vol. 32 (2019)
25. Loper, M., Mahmood, N., Romero, J., Pons-Moll, G., Black, M.J.: SMPL: a skinned multi-person linear model. ACM Trans. Graph. (TOG) **34**(6), 1–16 (2015)
26. Lorensen, W.E., Cline, H.E.: Marching cubes: a high resolution 3d surface construction algorithm. ACM SIGGRAPH Comput. Graph. **21**(4), 163–169 (1987)
27. Malik, J., et al.: HandVoxNet: deep voxel-based network for 3d hand shape and pose estimation from a single depth map. In: IEEE Conference on Computer Vision and Pattern Recognition (CVPR) (2020)
28. Mescheder, L., Oechsle, M., Niemeyer, M., Nowozin, S., Geiger, A.: Occupancy networks: learning 3d reconstruction in function space. In: IEEE Conference on Computer Vision and Pattern Recognition (CVPR) (2019)
29. Park, J.J., Florence, P., Straub, J., Newcombe, R., Lovegrove, S.: DeepSDF: learning continuous signed distance functions for shape representation. In: IEEE Conference on Computer Vision and Pattern Recognition (CVPR) (2019)
30. Patel, P., Huang, C.H.P., Tesch, J., Hoffmann, D.T., Tripathi, S., Black, M.J.: Agora: avatars in geography optimized for regression analysis. In: Proceedings of the IEEE/CVF Conference on Computer Vision and Pattern Recognition, pp. 13468–13478 (2021)
31. Prokudin, S., Black, M.J., Romero, J.: SMPLpix: neural avatars from 3d human models. In: Proceedings of the IEEE/CVF Winter Conference on Applications of Computer Vision, pp. 1810–1819 (2021)

32. Qi, C.R., Su, H., Mo, K., Guibas, L.J.: PointNet: deep learning on point sets for 3d classification and segmentation. In: IEEE Conference on Computer Vision and Pattern Recognition (CVPR) (2017)

33. Ronneberger, O., Fischer, P., Brox, T.: U-net: convolutional networks for biomedical image segmentation. In: International Conference on Medical Image Computing and Computer-assisted Intervention (2015)

34. Saint, A., et al.: 3dbodytex: textured 3d body dataset. In: International Conference on 3D Vision (3DV) (2018)

35. Saint, A., Cherenkova, K., Gusev, G., Aouada, D., Ottersten, B., et al.: BODY-FITR: robust automatic 3d human body fitting. In: IEEE International Conference on Image Processing (ICIP) (2019)

36. Saint, A., Kacem, A., Cherenkova, K., Aouada, D.: 3DBooSTeR: 3D body shape and texture recovery. In: Bartoli, A., Fusiello, A. (eds.) ECCV 2020. LNCS, vol. 12536, pp. 726–740. Springer, Cham (2020). https://doi.org/10.1007/978-3-030-66096-3_49

37. Saint, A., et al.: SHARP 2020: the 1st shape recovery from partial textured 3d scans challenge results. In: Bartoli, A., Fusiello, A. (eds.) ECCV 2020. LNCS, vol. 12536, pp. 741–755. Springer, Cham (2020). https://doi.org/10.1007/978-3-030-66096-3_50

38. Sarkar, K., Varanasi, K., Stricker, D.: Learning quadrangulated patches for 3d shape parameterization and completion. In: International Conference on 3D Vision (3DV) (2017)

39. Sinha, A., Unmesh, A., Huang, Q., Ramani, K.: SurfNet: generating 3d shape surfaces using deep residual networks. In: IEEE Conference on Computer Vision and Pattern Recognition (CVPR) (2017)

40. Tatarchenko, M., Dosovitskiy, A., Brox, T.: Octree generating networks: efficient convolutional architectures for high-resolution 3d outputs. In: IEEE International Conference on Computer Vision (ICCV) (2017)

41. Tian, Y., Zhang, H., Liu, Y., Wang, L.: Recovering 3d human mesh from monocular images: a survey. arXiv preprint arXiv:2203.01923 (2022)

42. Wang, J., Zhong, Y., Li, Y., Zhang, C., Wei, Y.: Re-identification supervised texture generation. In: IEEE Computer Vision and Pattern Recognition (CVPR) (2019)

43. Wang, N., et al.: Pixel2Mesh: generating 3d mesh models from single RGB images. In: Ferrari, V., Hebert, M., Sminchisescu, C., Weiss, Y. (eds.) ECCV 2018. LNCS, vol. 11215, pp. 55–71. Springer, Cham (2018). https://doi.org/10.1007/978-3-030-01252-6_4

44. Wang, P.S., Liu, Y., Guo, Y.X., Sun, C.Y., Tong, X.: O-CNN: octree-based convolutional neural networks for 3d shape analysis. ACM Trans. Graph. (TOG) **36**(4), 1–11 (2017)

45. Wang, T.C., Liu, M.Y., Zhu, J.Y., Tao, A., Kautz, J., Catanzaro, B.: High-resolution image synthesis and semantic manipulation with conditional GANs. In: IEEE Conference on Computer Vision and Pattern Recognition (CVPR) (2018)

46. Yan, X., Lin, L., Mitra, N.J., Lischinski, D., Cohen-Or, D., Huang, H.: Shapeformer: transformer-based shape completion via sparse representation. In: IEEE Conference on Computer Vision and Pattern Recognition (CVPR) (2022)

47. Yuan, W., Khot, T., Held, D., Mertz, C., Hebert, M.: PCN: point completion network. In: International Conference on 3D Vision (3DV) (2018)

48. Zheng, Z., Yu, T., Dai, Q., Liu, Y.: Deep implicit templates for 3d shape representation. In: IEEE Conference on Computer Vision and Pattern Recognition (CVPR) (2021)
49. Zheng, Z., Yu, T., Wei, Y., Dai, Q., Liu, Y.: DeepHuman: 3d human reconstruction from a single image. In: IEEE International Conference on Computer Vision (ICCV) (2019)

Deep Learning-Based Assessment of Facial Periodic Affect in Work-Like Settings

Siyang Song[1(✉)], Yiming Luo[2], Vincenzo Ronca[3], Gianluca Borghini[3], Hesam Sagha[4], Vera Rick[5], Alexander Mertens[5], and Hatice Gunes[1]

[1] AFAR Lab, University of Cambridge, Cambridge, UK
ss2796@cam.ac.uk
[2] Tsinghua University, Beijing, China
[3] BrainSigns, Rome, Italy
[4] audEERING, Gilching, Germany
[5] RWTH Aachen University, Aachen, Germany

Abstract. Facial behaviour forms an important cue for understanding human affect. While a large number of existing approaches successfully recognize affect from facial behaviours occurring in daily life, facial behaviours displayed in work settings have not been investigated to a great extent. This paper presents the first study that systematically investigates the influence of spatial and temporal facial behaviours on human affect recognition in work-like settings. We first introduce a new multi-site data collection protocol for acquiring human behavioural data under various simulated working conditions. Then, we propose a deep learning-based framework that leverages both spatio-temporal facial behavioural cues and background information for workers' affect recognition. We conduct extensive experiments to evaluate the impact of spatial, temporal and contextual information for models that learn to recognize affect in work-like settings. Our experimental results show that (i) workers' affective states can be inferred from their facial behaviours; (ii) models pre-trained on naturalistic datasets prove useful for predicting affect from facial behaviours in work-like settings; and (iii) task type and task setting influence the affect recognition performance.

1 Introduction

Human affect is a key indicator of various human internal states including mental well-being [3,17] and personality [12,14], as well as people's working behaviours [11,13,33]. Therefore, accurately understanding employees' affect in their working environment would enable managers identify risks for workers' safety, collect a history of risk exposures and monitor individuals' health status, which would further help employers in shaping organizational attitudes and decisions.

Since facial behaviours are a reliable source for affect recognition and can be easily recorded in an non-invasive way, a large number of existing approaches have been devoted to inferring affect from the human face. These approaches

---

S. Song and Y. Luo—Equal contribution.

frequently claim that static facial displays, spatio-temporal facial behaviours and even the background in a face image can provide useful cues for affect recognition. Subsequently, existing approaches can be categorized as: static face-based solutions that infer affect from each static facial display [7,10,24], full frame-based solutions that utilize not only facial display but also background cues [1], and spatio-temporal solutions that consider facial temporal evolution of the target face [5,15,22,28].

Although some of these approaches can accurately identify human affect in both in-lab and naturalistic conditions, none of these works have focused on analysing affect in working environments. This is largely due to the fact that there is neither a well-developed protocol for worker facial behaviour data collection nor a publicly available worker facial behaviour dataset for developing affect recognition systems in work settings (**Problem 1**). Since different working conditions would lead workers to express behaviours in different manners [6,18, 29], existing approaches that are developed using naturalistic or in-lab facial data may fail to accurately recognize affect from facial expressions expressed in work environments (**Problem 2**). Another key issue is that these approaches mainly provide affect prediction for a single facial image or video frame. However, in real-world working conditions, a large number of workers may need to be recorded and assessed many times a day. Subsequently, the limited computing and disk resources that are typical for SMEs (small-to-medium scale businesses) would not always allow making frame-level affect predictions in real time, and storing these for a high number of workers (**Problem 3**). More specifically, a common and realistic requirement for a real-world worker affect recognition system that could be adopted by SMEs is to predict each worker's affective state regularly but for a certain period of time, i.e., providing periodical affect predictions.

In this paper, we aim to address the three problems described above. Firstly, we introduce a new data collection protocol for acquiring naturalistic human audio-visual and physiological behavioural signals stimulated under different work-like conditions, tasks and stress levels. Using this protocol, we acquired the first human working facial behaviour database called WorkingAge DB, which is collected in four different sites with participants of different backgrounds (**addressing Problem 1**). Then, we benchmark several standard deep learning-based solutions on the collected WorkingAge DB, providing a set of baseline results for worker's periodical facial affect recognition (**addressing Problem 2 and Problem 3**). We further investigate the influence of the frame-rate on different models' periodical facial affect recognition performance with the assumption that SMEs where such a system is to be deployed typically have limited computational and disk resources, as well as the influence of task type, recording site, gender, and various feature representations. In summary, the main contributions of this paper are listed as follows:

- We propose an new protocol to acquire human facial behavioural data under various simulated working conditions together with the self-reported emotion/affect/workload state for each condition.
- Based on the proposed protocol, we collect a facial behaviour dataset in work-like settings across multiple sites with participants from different cultural

and language backgrounds. To the best of our knowledge, this is the first cross-cultural human facial behaviour dataset that is collected under various simulated working conditions.

- We benchmark several standard deep learning-based video-level behaviour approaches for worker's periodical dimensional affect recognition, and specifically investigate the influence of influence of task type, recording site, gender, and feature representations on models' performance. This provides a set of baselines for future studies that will focus on facial behaviour understanding in work settings. **Code access:** our code for the experiments is made available at https://github.com/takuyara/Working-Age-Baselines.

2 A Protocol for Human Facial Behaviour Data Acquisition in Work-Like Settings

Although many existing face datasets [15,16,20,21,23,32] have been annotated with dimensional affect labels, to the best of our knowledge, none of them has been collected in a working environment. Moreover, these datasets only provided static face/frame-level labels without periodical affect labels (labels that reflect the subject's affective state for a certain period of time (each clip in this paper)). As a result, none of them is suitable to be used for the purpose of investigating the relationship between human facial behaviours and affective states in working environments. To bridge this research gap, we propose a new protocol for acquiring human facial behavioural data displayed in various simulated working conditions, and annotated with self-reported affect labels that reflect workers' periodical affective states. The details of this protocol is described below.

Sensors Setup: The sensor setup of the protocol is illustrated in Fig. 1. During the recording, the participant sits at a table, where a laptop is placed to display slides that guide the participant to undertake a number of tasks based on a predefined order. To record visual information (including facial behaviours), a Logit web camera is placed in front of the participant. Additionally, a GoPro camera is also placed on the keyboard of the laptop to record facial behaviours during the Operations Task Game when the participant lowers the head on a panel to pick up items. Specifically, we build our model based on facial behaviours recorded by both cameras, i.e., only facial behaviours triggered by the operation task (explained in following paragraphs) are recorded by the GoPro camera.

Work-like Tasks. To simulate several working conditions, the proposed protocol consists of three work-like tasks including the N-back tasks, the video conference tasks and the operation game (i.e., the Doctor game). Additionally, we set an Eyes Open and Close task as the first task to acquire the baseline behaviours of each participant. The details of the four tasks are listed below:

- **Eyes Open and Close:** this task contains two sub-tasks, **(i) Eyes Open:** keeping the eyes open for 1 min; and **(ii) Eyes Closed:** keeping the eyes closed for 1 min. This task aims to acquire the baseline behaviours of each participant and help them get familiar with the task and the data acquisition environment.

310   S. Song et al.

Fig. 1. Hardware settings and example recordings, where the bottom-left figure displays the participant conducting a sub-task under the 'stressful condition' and is recorded by the logit camera. The bottom-right figure displays the participant conducting an operation sub-task and is recorded by the GoPro camera.

- **N-back task:** this task is a continuous performance task that is commonly employed to simulate different working memory capacity [9]. In our protocol, it simulates an office-related activity that does not need intensive physical work but causes mental strain. This task contains six sub-tasks in the following order: **(i) Baseline (NBB):** looking at the N-back interface for 1 min without reacting; **(ii) Easy game 1 (NBE01):** playing 0-back game for 2 min (typing the number that has just been shown on the screen); **(iii) Hard game 1 (NBH01):** playing 2-back game for 2 min (typing the number that has just been shown on the screen two turns ago); **(iv) Hard game 2 (NBH02):** playing 2-back game for 2 min; **(v) Easy game 2 (NBE02):** playing 0-back game for 2 min; and **(vi) Stressful hard game (NBS):** playing 2-back game for 2 min with 85 dB background noise while a human experimenter is seated in the same room as the participant.
- **Video conference task:** this task simulates a teleworking scenario, in which employees are frequently requested to interact and coordinate with colleagues who are not physically present. It contains three sub-tasks: **(i) Baseline (WEB):** looking at Microsoft Teams screen without reacting; **(ii) Positive emotions (WEP):** describing the happiest memory in one's life to the human experimenter for 2 min via Microsoft Teams video conferencing application (aiming to stimulate positive emotions); and **(iii) Negative emotions (WEN):** describing the most negative/sad memory in one's life to the

human experimenter for 2 min via Microsoft Teams video conferencing application (aiming to stimulate negative emotions). During both tasks, if needed, the human experimenter would ask several neutral and factual questions to keep the participant talking for about 2 min.
- **Operation task (Doctor Game):** this task requires the participant to use tweezers to pick up objects from a panel, simulating an assembly line scenario. It contains six sub-tasks: **(i) Baseline (DB):** looking at the doctor game panel without reacting; **(ii) Easy game 1 (DE01):** picking up and removing 5 objects from the panel within 2 min; **(iii) Hard game 1 (DH01):** picking up and removing as many objects as possible from the panel within 3 min; **(iv) Easy game 2 (DE02):** picking up and removing 5 objects from the panel within 2 min; **(v) Hard game 2 (DH02):** picking up and removing as many objects as possible from the panel within 3 min; and **(vi) Stressful hard game (DS):** removing as many objects as possible from the panel within 3 min with 85 dB background noise while a human experimenter is seated in the same room as the participant.

After finishing each sub-task, the participant is asked to fill in two questionnaires.

Self-Reported Questionnaires. In our protocol, we propose to obtain periodical emotion and affect annotations from each participant, i.e., sub-task-level emotion and affect annotations. Specifically, two questionnaires are employed, namely, Geneva Emotion Wheel (GEW) [25] that measures the intensities of several categorical emotions of the participant along 5 scales (from low to high), and Self-Assessment Manikin (SAM) [4] that measures the intensities of three affect dimensions (arousal, valence and power) from unhappy/calm/controlled (1) to happy/excited/in-control (9) using a scale of 9.

Data Acquisition. The study was approved by the relevant Departmental Ethics Committee. Additional COVID-19 related measures were also put in place prior to the study. Prior to data acquisition, each participant is provided with an information sheet and is asked to read and sign a consent form. Following these procedures, the experimenter explains all the details (e.g., purpose, tasks, questionnaires, etc.) of the study to the participant. Then, the participant is asked to enter the room and sit in front of the laptop. The experimenter then leaves the recording the room and goes to the operations room next door to remotely start both cameras. The instructions related to each work-like task are displayed to the participant on the laptop screen. After each main task (N-back, video conference and operation tasks), the participant is instructed to relax for 4 minutes by listening a calming music. This aims to help the participant to get back to their baseline/neutral affective and cognitive state, to prevent the positive or negative emotions caused by the previous tasks from impacting the future tasks.

3   The WorkingAge Facial Behaviour Dataset

Based on the proposed protocol, we acquired a human facial behaviour dataset in work-like settings in four different sites located in three countries: the audEERING and RWTH Aachen University in Germany, the University of Cambridge in United Kingdom, and the BrainSigns in Italy (i.e., they are individually referred as AUD, RWTH, UCAM and BS in this paper). The collected dataset contains data from a total of 55 participants, with 7 participants coming from the AUD site, 16 participants coming from the BS site, 20 participants coming from the RWTH site, and 12 participants coming from the UCAM site. It contains 935 clips (each clip corresponds to a sub-task, and 17 sub-tasks were recorded for each participant), where 605 of them are annotated using the proposed protocol (i.e., participants were not asked to provide self report annotations for NBE01, NBH01, DB01 and DE01 tasks to reduce their annotation burden. As a result, these clips were not used for experiments.). Specifically, the videos collected by Logit camera and GoPro Camera are set as 24 and 30 fps during the recording, with the frame resolutions of 1280×720, and 1920×1080, respectively. The mean value, standard deviation, maximum and minimum values of each annotated sub-task's duration in our dataset are listed in Table 1. It is clear that the participants need longer periods of time to undertake the WEP and WEN tasks, while DB, DE02 and DS tasks take shorter time to complete.

Table 1. Statistics of clips' duration (seconds) in the WorkingAge dataset.

Sub-task	Mean	Standard deviation	Maximum	Minimum
NBB	73.1	15.9	111.4	25.9
NBE02	91.4	29.8	131.2	26.1
NBH02	93.4	29.8	135.3	29.1
NBS	97.1	27.0	129.9	36.5
WEB	57.8	13.5	81.2	20.8
WEP	136.6	40.2	206.2	49.0
WEN	134.5	35.0	224.0	54.1
DB	61.0	7.2	77.5	38.1
DE02	60.6	41.2	204.2	10.5
DH02	100.3	41.8	185.3	32.4
DS	61.6	39.3	212.1	14.2

In addition, the distributions of the self-reported arousal and valence labels (based on the SAM questionnaire) are illustrated in Fig. 2. In this paper, we further quantize the valence/arousal labels into three classes: positive (scores 7, 8, or 9), neutral (scores 4, 5, or 6) and negative (scores 1, 2, or 3). As we can see, during the three baseline sub-tasks, participants reported a relatively high valence (72, 88 and 11 subjects have positive, neutral and negative valence status) and low arousal value (6, 46 and 119 subjects have positive, neutral and negative arousal). However, it can be seen that the increasing mental workload requirement clearly leads participants to report lower valence (with mean valence scores of 5.9, 5.8 and 5.7 for DE02, DH02 and DS as well as 6.0, 5.5 and 5.3 for NBE02, NBH02 and NBS) and higher arousal (with mean arousal scores of 4.4, 4.8 and 5.0 for DE02, DH02 and DS as well as 3.4, 4.0 and 5.2 for NBE02, NBH02 and NBS). We also see that the valence label distributions are quite different for the three video conference sub-tasks, i.e., most participants reported a relatively neutral valence status during the baseline condition, while positive and negative memory-based conversations caused most participants to report corresponding positive and negative valence values, with mean valence scores of 5.8, 7.1 and 4.0 for WEB, WEP and WEN, respectively.

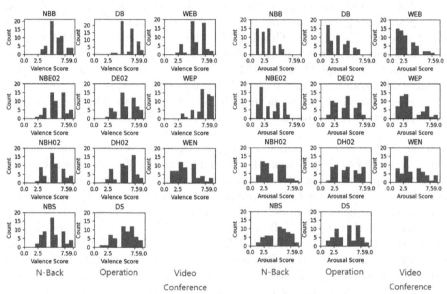

a). The distribution of self-reported valence intensities for each subtask

b). The distribution of self-reported arousal intensities for each subtask.

Fig. 2. The distribution of the self-reported valence intensities for each subtask.

4   Periodical Facial Affect Recognition

Although facial behaviours have been proven to be informative for inferring human affect under natural or controlled conditions, none of the previous studies evaluated the feasibility of using workers' facial behaviours to infer affect in terms of valence and arousal. In this section, we implement three standard deep learning-based short video-level modelling approaches to provide a benchmark for the task of facial affect recognition in work-like settings.

Baseline 1. Given a short face video, the first baseline starts with generating frame-level affect predictions for all frames, which are then combined to output periodical affect prediction. In particular, we individually employed two frame-level facial analysis models, i.e., a ResNet-50 [8] that is pre-trained for facial expression recognition (i.e., pre-trained on the FER 2013 dataset) and a GraphAU model [19] that is pre-trained for facial action units (AUs) recognition (i.e., pre-trained on the BP4D dataset [31]). Specifically, we individually use the latent feature output by the second-last fully connected layer of the ResNet-50, as well as the 12 AU predictions generated by the GraphAU model, as the frame-level facial features. Then, we individually apply a multi-layer perceptron (MLP) on each of them to provide frame-level valence and arousal predictions, which is trained by re-using the clip-level self-reported valence/arousal scores as the frame-level label. To obtain periodical (clip-level) affect predictions, we combine all frame-level predictions of the target clip with the following widely-used strategies: (i) using the mode prediction of all frame-level predictions as the periodical affect predictions (i.e., GraphAU(P)-MODE and ResNet(P)-MODE); (ii) applying a Long-short-term-memory Network (LSTM) to combine all frame-level predictions (i.e., GraphAU(P)-LSTM and ResNet(P)-LSTM); and (iii) applying spectral encoding algorithm [26,27] to produce a spectral heatmap from all frame-level predictions, which is then fed to a 1D-CNN to generate periodical affect predictions (i.e., GraphAU(P)-SE and ResNet(P)-SE).

Baseline 2. The second baseline also applies the same two pre-trained models used in baseline 1 to provide frame-level facial features. Differently from the baseline 1, we employ three long-term modelling strategies to combine all frame-level facial features (the latent feature vectors produced by the last FC layer of the corresponding frame-level model) of the clip as the clip-level (periodical) affect representation: (i) averaging all frame-level facial features (i.e., GraphAU(F)-AVERAGE and ResNet(F)-AVERAGE); (ii) applying LSTM to process all frame-level facial features (i.e., GraphAU(F)-LSTM and ResNet(F)-LSTM); and (iii) spectral encoding all frame-level facial features (i.e., GraphAU(F)-SE and ResNet(F)-SE). These clip-level affect representations are then fed to either an MLP (for (i)) or 1D-CNN (for (ii) and (iii)) to generate clip-level affect predictions.

Baseline 3. The third baseline applies a spatio-temporal CNN (Temporal Pyramid Network (TPN) [30]) to process the facial sequence. In particular, we first divide each clip into several segments, where each consists of 160 frames, and down-sample each segment to 32 frames. We then feed the cropped face sequence (32 frames) to TPN for affect classification. If a clip contains multiple segments, then the clip-level predictions are achieved by averaging all segment-level predictions.

5 Experiments

5.1 Experimental Setup

Data Pre-processing: We re-sampled all video clips to 24 fps to make videos recorded by different cameras to have the same frame rate. Then, we used

OpenFace 2.0 [2] to crop and align the face from each frame, where frames with failed or low-confidence face detection are treated as black images.

Training and Evaluation Protocol: We use the leave-one-site-out validation protocol for models' training and evaluation, i.e., at each time, we use all clips from three sites to train the model, and evaluate the trained model on the rest one. The final reported results are obtained by averaging validation results of four folds.

Model Settings and Training Details: There are three main settings for our baseline models: (i) frame-level feature extraction; (ii) clip-level (periodical) facial behaviour representation extraction; and (iii) classifiers. Specifically, we used the output (2048D) of the second-last layer of the ResNet-50 and the 12 predicted action units' occurrences as the frame-level facial features, respectively. In terms of the clip-level representation extraction, we used bidirectional LSTM with one hidden layer, as well as a spectral encoding algorithm with a resolution of 256 (80 lowest frequencies are selected). Finally, the MLP classifier is set as 3 layers for mode/averaging-based classification while the 1D-CNN is set to have 3 convolution blocks where each consists of a convolution layer, a ReLU activation, as well as a dropout layer, whose channels and hidden sizes varies depending the size of the input feature. We used the batch size of 512 and initial learning rate of 0.001 for all experiments.

Evaluation Metrics: Considering that the samples are unbalanced in terms of arousal and valence label distribution, in this paper we employ the Unweighted Average Recall (UAR) as the measurement to evaluate different baseline performances on facial affect recognition in work-like settings.

5.2   Baseline Results of Leave-One-Site-Out Cross-Validation

Tables 2 and 3 list the valence and arousal classification UAR results achieved by all baseline systems for all sub-tasks (the models are trained using all clips in the training set regardless of the task type). It can be seen that almost all baselines achieved over the chance-level classification UAR (33.33%), with the *GraphAU(P)-SE* system achieving the best valence UAR result (42.31%) and the *ResNet(F)-SE* system achieving the best arousal UAR result (41.20%). Meanwhile, we found that if we only use the mode prediction of all frame-level predictions, both valence and arousal classification results are clearly worse than most of other systems, i.e., the two corresponding systems only achieved less than 34% valence and arousal classification accuracy. These results indicate that: (i) according to Fig. 3, the long-term modelling for either frame-level predictions or features is a crucial step to achieve more reliable periodical arousal/valence predictions, as simply choosing the mode prediction from all frame-level predictions or averaging all frame-level features clearly provided the worst results; (ii) the frame-level facial analysis (AU recognition/facial expression recognition) models that are pre-trained using the lab-based facial datasets (AU or facial expression datasets) can still extract human affect-informative facial features from

Table 2. The UAR results achieved for worker's valence recognition, where the name of each method is formatted as *frame level facial feature-long term model*, where P and F represent the frame-level prediction and facial features, respectively. For example, *ResNet(P)-SE* denote the system that applies ResNet facial features to make frame-level affect predictions, and then using spectral encoding algorithm to summarise all frame-level valence/arousal predictions as the clip-level valence/arousal prediction.

Model	NBB	NBE02	NBH02	NBS	DB	DE02	DH02	DS	WEB	WEP	WEN	Total
GraphAU(P)-SE	0.3814	0.4279	0.4206	0.4691	0.6667	**0.4026**	0.4611	**0.4461**	0.5324	0.3280	0.2857	**0.4231**
GraphAU(P)-LSTM	0.3921	0.4254	0.4444	0.4414	0.6667	0.3898	0.4611	0.4428	0.5139	0.2899	0.3175	0.4209
GraphAU(P)-MODE	0.2918	0.2667	0.3254	0.3241	0.5417	0.2700	0.2019	0.2525	0.3287	0.2899	0.2619	0.3009
GraphAU(F)-SE	0.3584	0.4019	0.4246	**0.4784**	0.6875	0.3929	0.3963	0.3956	0.4491	0.2984	0.3810	0.3995
GraphAU(F)-LSTM	0.3921	0.4254	0.4444	0.4414	0.6667	0.3898	0.4611	0.4428	0.5139	0.3090	0.2857	0.4153
GraphAU(F)-AVERAGE	0.3685	0.3994	0.3651	0.3951	0.6250	0.3570	0.4500	0.4209	0.4722	**0.3640**	0.3333	0.3970
ResNet(P)-SE	0.3653	0.4402	0.4243	0.4484	0.7150	0.3796	0.4956	0.3881	0.5207	0.3564	0.3254	0.4226
ResNet(P)-LSTM	0.3847	**0.4438**	0.4473	0.4444	0.6558	0.3977	0.4974	0.4171	0.5207	0.3023	0.2857	0.4176
ResNet(P)-MODE	0.3153	0.3394	0.3798	0.2897	0.4558	0.2778	0.3465	0.3124	0.3622	0.3504	**0.4127**	0.3344
ResNet(F)-SE	**0.3951**	**0.4438**	**0.4473**	0.4722	0.6550	0.3750	**0.5140**	0.4211	**0.5393**	0.3023	0.2857	0.4225
ResNet(F)-LSTM	0.3847	**0.4438**	0.4358	0.4444	0.6542	0.3838	**0.5140**	0.4171	**0.5393**	0.3023	0.2857	0.4191
ResNet(F)-AVERAGE	0.3847	**0.4438**	**0.4473**	0.4444	**0.7167**	0.3801	0.4974	0.4316	0.5207	0.3023	0.2857	0.4219
TPN	0.3751	0.3740	0.3016	0.2940	0.4785	0.2686	0.2571	0.4002	0.3062	0.2915	0.3379	0.3350

Table 3. The UAR results achieved for worker's arousal recognition.

Model	NBB	NBE02	NBH02	NBS	DB	DE02	DH02	DS	WEB	WEP	WEN	Total
GraphAU(P)-SE	0.5605	0.3417	0.3506	0.4058	0.3464	**0.4842**	0.4077	**0.4118**	0.3538	0.4167	0.3801	0.3999
GraphAU(P)-LSTM	0.5474	0.3625	0.3975	0.4058	0.4071	0.3667	0.3310	0.3725	0.3547	0.3774	0.3581	0.3678
GraphAU(P)-MODE	0.5658	**0.4319**	0.3232	0.2580	0.2389	0.3741	0.3902	0.3081	0.3155	**0.4358**	0.2327	0.3427
GraphAU(F)-SE	0.5711	0.3819	**0.4149**	0.4014	0.4224	0.4201	0.3634	0.3880	0.4153	0.4100	0.3973	0.3967
GraphAU(F)-LSTM	0.5474	0.4153	0.4134	0.4058	0.4071	0.3667	0.3310	0.3725	0.3645	0.3774	0.3581	0.3759
GraphAU(F)-AVERAGE	0.5974	0.4042	0.3983	0.4203	**0.4309**	0.3667	0.3310	0.3725	0.3645	0.3774	0.3581	0.3790
ResNet(P)-SE	0.5244	0.3636	0.3837	0.3610	0.4115	0.3842	0.3860	0.3889	0.3873	0.3988	**0.4074**	0.3834
ResNet(P)-LSTM	0.5231	0.3206	0.3692	0.3900	0.4122	0.3491	0.3651	0.4074	0.3775	0.4129	0.3454	0.3671
ResNet(P)-MODE	0.4359	0.3011	0.3202	0.3320	0.3118	0.3667	0.3684	0.3519	0.3595	0.4014	0.3639	0.3386
ResNet(F)-SE	**0.6346**	0.4133	0.3775	0.4295	0.4036	0.4719	0.4106	0.3519	0.4101	0.3775	0.3406	**0.4120**
ResNet(F)-LSTM	0.5462	0.3925	0.4037	**0.4466**	0.4029	0.4649	**0.4496**	0.4021	**0.4869**	0.3837	0.3285	0.4032
ResNet(F)-AVERAGE	0.5231	0.3945	0.3996	0.3900	0.4122	0.3991	0.4002	0.4074	0.3954	0.3758	0.3639	0.3858
TPN	0.4823	0.4877	0.3543	0.3403	0.3168	0.1935	0.2391	0.2951	0.1807	0.2602	0.3498	0.3182

facial displays triggered by work-like tasks, as their features frequently provide around 40% UAR for both tasks (the chance-level UAR should be around 33% for three class classification); and (iii) directly pairing workers' facial sequences with clip-level affect labels to train spatio-temporal models does not provide superior results, which further validates that frame-level facial analysis models pre-trained on facial datasets acquired in naturalistic settings are beneficial for facial affect analysis in work-like settings.

5.3   Ablation Studies

Task Type: Since workers' facial behaviours are highly correlated with the task type and task setting, we also specifically investigate which tasks can trigger most affect-informative facial behaviours in Fig. 4, where we report the average

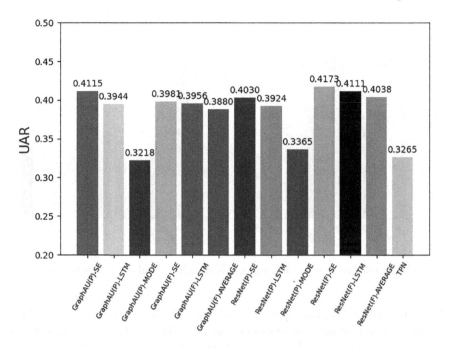

Fig. 3. The average valence and arousal UAR results of all baseline models.

UAR results achieved by all baselines for each task. The facial behaviours triggered by four sub-tasks allow the model to achieve over 40% valence recognition UAR results, which are clearly superior than the results achieved on other sub-tasks. Therefore, we hypothesize that different subjects display affect in different intensity (valence) when undertaking different sub-tasks of the same task (even though their facial behaviour may be similar). Meanwhile, facial behaviours displayed during N-back baseline sub-task is very informative for predicting subjects' arousal, i.e., the arousal UAR result achieved for the baseline sub-tasks of N-Back has more than 14.76% absolute accuracy improvements over the results of other sub-tasks. This finding suggests that human facial displays before conducting the memory task may be reliable for inferring worker's arousal state.

Table 4. The results of the four-fold cross-validation results achieved by our best models (GraphAU(P)-SE for valence, ResNet(P)-SE for arousal).

Validation set	AUD	BS	RWTH	UCAM
Valence	**0.3715**	0.3319	0.3438	0.3337
Arousal	**0.3621**	0.3385	0.3405	0.3523

318   S. Song et al.

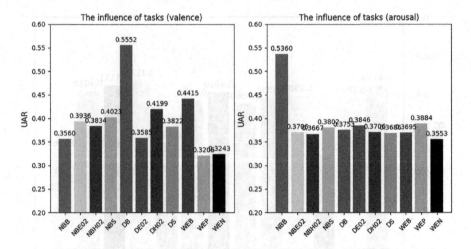

Fig. 4. The influence of different tasks on valence and arousal prediction.

Recording Site: We also explore the differences in affect classification for different sites. Table 4 displays the leave-one-site-out four-fold cross-validation results. It is clear that the data collected at different sites impact the valence classification results, with around 4% UAR difference between the lowest (0.3319 (BS)) and the highest (0.3715 (AUD)). These results indicate that people at different sites may display different facial behaviours when expressing valence. On the other hand, the performance variations for arousal classification are much smaller, indicating that the relationship between arousal and workers' facial behaviours are more stable as compared to valence.

Gender: We report the gender-dependent worker valence/arousal classification UAR results using our all baselines, where leave-one-site-out cross validation is applied to either male or female facial data. Both the valence prediction results (0.3912 for male, 0.3714 for female and 0.3943 for gender-independent) and arousal prediction results (0.3711 for male, 0.3588 for female and 0.3878 for gender-independent) indicate that male facial behaviours are more correlated with their affect status. Moreover, it should be noted that the gender-independent experiment achieved better performance than gender dependent experiments. This might be caused by that fact that the gender-dependent experiments have less data for model training. In addition, these results might indicate that females and males do not display large variations when expressing their affect via facial behaviours in work-like settings, i.e., such small variations can not compensate the negative impact of the reduced number of training data on worker affect prediction models (Table 5).

Table 5. The average UAR results achieved for different genders.

Gender	Male	Female	Both
Valence	0.3912	0.3741	**0.3943**
Arousal	0.3711	0.3588	**0.3878**

Feature Representation and Long-Term Modelling: We also specifically investigate the influence of different model configurations on periodical affect classification performance. As we can see from Table 6, the backbone that was pre-trained using naturalistic facial expression dataset achieved slightly better affect recognition UAR results than the backbone that was pre-trained using a facial AU dataset, both of which clearly are higher than the chance-level prediction. This means both facial expression and AU-related facial features obtained from the workers' faces are correlated with their self-reported affective state. Then, using facial features as the frame-level representation to construct clip-level facial behaviour representation is a superior way, as this setting achieved better UAR results for the recognition of both valence and arousal. We assume this is because frame-level features retain more affect-related facial cues than frame-level predictions, and thus during long-term modelling, frame-level feature-based clip-level representations can encode more affect-related temporal behavioural cues. Finally, simply choosing the mode of all frame-level predictions or the average feature of all frame-level features provided the worst results among all long-term modelling strategies, while spectral encoding achieved the best average performance for all baselines. This is because the encoded spectral representation contains multi-scale clip-level temporal dynamics.

Table 6. The average results of different baseline configurations.

	Strategy	Valence	Arousal
Backbone	GraphAU	0.3928	0.3770
	ResNet	**0.4064**	**0.3817**
Frame-level feature	Prediction	0.3866	0.3666
	Features	**0.4126**	**0.3921**
Long-term modelling	SE	0.4169	**0.3980**
	LSTM	**0.4182**	0.3785
	MODE/AVG	0.3635	0.3615

6 Conclusion

In this paper, we presented the first study that systematically investigated the face-based periodical valence and arousal analysis in work-like settings. More specifically, this paper introduced a worker facial behaviour data acquisition protocol and the first cross-cultural human facial behaviour dataset in work-like settings. We also provided a set of deep learning-based baselines for face-based worker affect recognition. The results show that facial behaviours triggered by different tasks are informative for inferring valence and arousal states, but the performance is dependant on the task type and task setting. Our future work will focus on developing more advanced domain-specific loss functions and network architectures for multi-modal worker affect recognition.

Acknowledgement. This work is funded by the European Union's Horizon 2020 research and innovation programme project WorkingAge, under grant agreement No. 82623. S. Song and H. Gunes are also partially supported by the EPSRC/UKRI under grant ref. EP/R030782/1. Y. Luo contributed to this work while undertaking a summer research study at the Department of Computer Science and Technology, University of Cambridge.

References

1. Antoniadis, P., Pikoulis, I., Filntisis, P.P., Maragos, P.: An audiovisual and contextual approach for categorical and continuous emotion recognition in-the-wild. In: Proceedings of the IEEE/CVF International Conference on Computer Vision, pp. 3645–3651 (2021)
2. Baltrusaitis, T., Zadeh, A., Lim, Y.C., Morency, L.P.: Openface 2.0: facial behavior analysis toolkit. In: 2018 13th IEEE International Conference on Automatic Face and Gesture Recognition (FG 2018), pp. 59–66. IEEE (2018)
3. Borghini, G., et al.: Stress assessment by combining neurophysiological signals and radio communications of air traffic controllers. In: International Conference of the IEEE Engineering in Medicine and Biology Society (EMBC), pp. 851–854. IEEE (2020)
4. Bradley, M.M., Lang, P.J.: Measuring emotion: the self-assessment manikin and the semantic differential. J. Behav. Ther. Exp. Psych. **25**(1), 49–59 (1994)
5. Du, Z., Wu, S., Huang, D., Li, W., Wang, Y.: Spatio-temporal encoder-decoder fully convolutional network for video-based dimensional emotion recognition. IEEE Trans. Affect. Comput. **12**, 565–572 (2019)
6. Giorgi, A., et al.: Wearable technologies for mental workload, stress, and emotional state assessment during working-like tasks: a comparison with laboratory technologies. Sensors **21**(7), 2332 (2021)
7. Guo, J., et al.: Dominant and complementary emotion recognition from still images of faces. IEEE Access **6**, 26391–26403 (2018)
8. He, K., Zhang, X., Ren, S., Sun, J.: Deep residual learning for image recognition. In: Proceedings of the IEEE Conference on Computer Vision and Pattern Recognition, pp. 770–778 (2016)
9. Herreras, E.B.: Cognitive neuroscience; the biology of the mind. Cuadernos de Neuropsicología/Panamerican J. Neuropsychol. **4**(1), 87–90 (2010)
10. Ilyas, C.M.A., Rehm, M., Nasrollahi, K., Madadi, Y., Moeslund, T.B., Seydi, V.: Deep transfer learning in human-robot interaction for cognitive and physical rehabilitation purposes. Pattern Anal. App. **25**, 1–25 (2021)
11. Ilyas, C.M.A., Song, S., Gunes, H.: Inferring user facial affect in work-like settings. arXiv preprint arXiv:2111.11862 (2021)

12. Izard, C.E.: Human Emotions. Emotions, Personality, and Psychotherapy. Plenum-Press, New York (1977)

13. Jenkins, J.M.: Self-monitoring and turnover: the impact of personality on intent to leave. J. Organ. Behav. **14**(1), 83–91 (1993)

14. Keltner, D.: Facial expressions of emotion and personality. In: Handbook of Emotion, Adult Development, and Aging, pp. 385–401. Elsevier (1996)

15. Kollias, D., et al.: Deep affect prediction in-the-wild: AFF-wild database and challenge, deep architectures, and beyond. Int. J. Comput. Vision **127**(6), 907–929 (2019)

16. Kossaifi, J., et al.: SEWA DB: a rich database for audio-visual emotion and sentiment research in the wild. IEEE Trans. Pattern Anal. Mach. Intell. **43**, 1022–1040 (2019)

17. Lerner, J.S., Li, Y., Valdesolo, P., Kassam, K.S.: Emotion and decision making. Annu. Rev. Psychol. **66**, 799–823 (2015)

18. Lohse, M., Rothuis, R., Gallego-Pérez, J., Karreman, D.E., Evers, V.: Robot gestures make difficult tasks easier: the impact of gestures on perceived workload and task performance. In: Proceedings of the SIGCHI Conference on Human Factors in Computing Systems, pp. 1459–1466 (2014)

19. Luo, C., Song, S., Xie, W., Shen, L., Gunes, H.: Learning multi-dimensional edge feature-based au relation graph for facial action unit recognition. In: Proceedings of the Thirty-First International Conference on International Joint Conferences on Artificial Intelligence (2022)

20. McKeown, G., Valstar, M., Cowie, R., Pantic, M., Schroder, M.: The semaine database: Annotated multimodal records of emotionally colored conversations between a person and a limited agent. IEEE Trans. Affect. Comput. **3**(1), 5–17 (2011)

21. Mollahosseini, A., Hasani, B., Mahoor, M.H.: AffectNet: a database for facial expression, valence, and arousal computing in the wild. IEEE Trans. Affect. Comput. **10**(1), 18–31 (2017)

22. Mou, W., Gunes, H., Patras, I.: Alone versus in-a-group: a multi-modal framework for automatic affect recognition. ACM Trans. Multimed. Comput. Commun. App. (TOMM) **15**(2), 1–23 (2019)

23. Ringeval, F., Sonderegger, A., Sauer, J., Lalanne, D.: Introducing the recola multimodal corpus of remote collaborative and affective interactions. In: 2013 10th IEEE International Conference and Workshops on Automatic Face and Gesture Recognition (FG), pp. 1–8. IEEE (2013)

24. Sariyanidi, E., Gunes, H., Cavallaro, A.: Automatic analysis of facial affect: a survey of registration, representation, and recognition. IEEE Trans. Pattern Anal. Mach. Intell. **37**(6), 1113–1133 (2014)

25. Scherer, K.R.: What are emotions? And how can they be measured? Soc. Sci. Inf. **44**(4), 695–729 (2005)

26. Song, S., Jaiswal, S., Shen, L., Valstar, M.: Spectral representation of behaviour primitives for depression analysis. IEEE Trans. Affect. Comput. **13**, 829–844 (2020)

27. Song, S., Shen, L., Valstar, M.: Human behaviour-based automatic depression analysis using hand-crafted statistics and deep learned spectral features. In: 2018 13th IEEE FG (2018), pp. 158–165. IEEE (2018)

28. Song, S., Sánchez-Lozano, E., Kumar Tellamekala, M., Shen, L., Johnston, A., Valstar, M.: Dynamic facial models for video-based dimensional affect estimation. In: Proceedings of the IEEE/CVF International Conference on Computer Vision Workshops (2019)

29. Tsai, Y.F., Viirre, E., Strychacz, C., Chase, B., Jung, T.P.: Task performance and eye activity: predicting behavior relating to cognitive workload. Aviat. Space Environ. Med. **78**(5), B176–B185 (2007)
30. Yang, C., Xu, Y., Shi, J., Dai, B., Zhou, B.: Temporal pyramid network for action recognition. In: Proceedings of the IEEE/CVF Conference on Computer Vision and Pattern Recognition, pp. 591–600 (2020)
31. Zhang, X., et al.: BP4D-spontaneous: a high-resolution spontaneous 3d dynamic facial expression database. Image Vis. Comput. **32**(10), 692–706 (2014)
32. Zhao, G., Huang, X., Taini, M., Li, S.Z., Pietikälnen, M.: Facial expression recognition from near-infrared videos. Image Vis. Comput. **29**(9), 607–619 (2011)
33. Zimmerman, R.D.: Understanding the impact of personality traits on individuals' turnover decisions: a meta-analytic path model. Pers. Psychol. **61**(2), 309–348 (2008)

Supervision by Landmarks: An Enhanced Facial De-occlusion Network for VR-Based Applications

Surabhi Gupta[(✉)], Sai Sagar Jinka, Avinash Sharma, and Anoop Namboodiri

Center for Visual Information Technology
International Institute of Information Technology, Hyderabad, India
{surabhi.gupta,jinka.sagar}@research.iiit.ac.in,
{asharma,anoop}@iiit.ac.in

Abstract. Face possesses a rich spatial structure that can provide valuable cues to guide various face-related tasks. The eyes are considered an important socio-visual cue for effective communication. They are an integral feature of facial expressions as they are an important aspect of interpersonal communication. However, virtual reality headsets occlude a significant portion of the face and restrict the visibility of certain facial features, particularly the eye region. Reproducing this region with realistic content and handling complex eye movements such as blinks is challenging. Previous facial inpainting methods are not capable enough to capture subtle eye movements. In view of this, we propose a working solution to refine the reconstructions, particularly around the eye region, by leveraging inherent eye structure. We introduce spatial supervision and a novel landmark predictor module to regularize per-frame reconstructions obtained from an existing image-based facial de-occlusion network. Experiments verify the usefulness of our approach in enhancing the quality of reconstructions to capture subtle eye movements.

Keywords: Face image inpainting · Landmark guided facial de-occlusion · HMD removal · Virtual reality · Eye consistency

1 Introduction

Social telepresence and interaction are essential for human survival. Since globalization, there has been a considerable increase in users interacting remotely, which has witnessed a tremendous surge during the Covid-19 pandemic. Traditional video conferencing platforms such as Microsoft Teams, WhatsApp, etc., gained immense popularity during the pandemic. However, they lack immersiveness and compromise realism that impacts the user's experience, which is undesirable. With the integration of virtual reality in a communication platform, the current technologies have witnessed a breakthrough in enhancing user experience with a sense of heightened social existence and

Supplementary Information The online version contains supplementary material available at https://doi.org/10.1007/978-3-031-25072-9_21.

interaction. Faces convey vital socio-visual cues that are important for effective communication. However, one of the major challenges with virtual reality, such as HMDs, is the occlusion it cause over the face when wearing these devices. These devices occlude almost 30–40 percent of the face, obscuring essential social cues, particularly the eye region, which hinders the user's experience. Several approaches have been proposed in the literature to tackle this problem, but none of them produces photorealistic results that could be integrated into hybrid telepresence systems.

Existing face image inpainting approaches often suffer from incoherency in generating smooth reconstructions when applied to video frames. This incoherency is highly noticeable in the eye region, which is undesirable. It is generally visible as jittering in eyelids in successive video frames. Specific eye movements, such as blinking, are usually involuntary act in humans that is natural and unavoidable. Thus, it is important to retain this characteristic for effective communication. Synthesizing eyes, including iris and eyelid reconstruction with appropriate eye gaze, have been attempted before using 3D model-based approaches. Nonetheless, they require high-quality data and incur expensive training costs. [6] and [17] are such examples of 3D models based approaches to HMD de-occlusion. However, they work only with frontal face images and fail in cases of extreme head-poses. Another set of works like [14, 15] use an inpainting approach to correct and animate the eye gaze of high-resolution, unconstrained portrait face images. Since these methods have not been validated on videos, they fail to generate consistent eye motions across frames.

Interestingly, one of the biggest advantages when dealing with digital face images in computer vision is its rich spatial structure. In digital images, this structure is generally represented in the form of 2D/3D coordinates, heatmaps, and edges and is provided as an auxiliary input to the network. Many works in the literature have exploited these spatial constraints for achieving better quality reconstruction in face inpainting and generation tasks. Recently, [11] proposed image-to-face video inpainting using spatio-temporal nested gan architecture. They used 3D residual blocks to capture inter-frame dependencies. The authors showed that conditioning facial inpainting on landmarks yielded stable reconstructions. Nonetheless, it is only validated with a specific type of circular mask that covers the eye, nose, and mouth. [12] is another face image inpainting method that is guided using facial landmarks. Personalized facial de-occlusion networks such as [3] have been proposed in the literature to generate plausible reconstructions. However, they are not controllable and thus cannot handle eye movements. Thus, we tackle this problem using an image generation/image synthesis approach applied to faces using additional information that is easily accessible using modern HMD devices with eye-tracking capabilities.

This work aims to generate high-quality facial reconstructions in and around the eye region with consistent eye motions in the presence of occluders such as HMDs. Our primary focus is to handle instability during eye movements that are noticeable mainly around the eye region. For this, we leverage the spatial property of faces, i.e., facial landmarks, to guide the model to synthesize the eye region with minimal artifacts that look realistic and plausible. Figure 1 presents high-quality and photo-realistic results produced by our method showing the efficacy of our approach of using spatial supervision to control complex eye motions such as eye blinks and rolling of eyeballs.

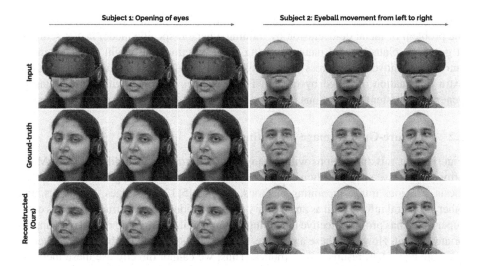

Fig. 1. Photo-realistic results generated by our proposed facial de-occlusion network, targeting complex eye motions.

To summarize, we make the following contributions:

1. We propose a potential solution to refine the reconstructions in the eye region.
2. To achieve this, we leverage the spatial constraints such as landmarks to improve upon consistency in the eye region by feeding eye landmarks heatmaps as an auxiliary input to the network along with occluded face image.
3. To further improve the fidelity of the reconstruction, we use an additional loss function to regularize the training based on the landmarks.

2   Related Work

To see what is not present in the image is one of the most exciting yet challenging tasks in computer vision. We often refer to it as image restoration/image inpainting in the digital domain. It has applications in medical image processing, watermark removal, restoring old photographs, and object removal. Inpainting has been an active research topic for many years, and several works have been proposed in the literature. Recently, this area has seen tremendous interest in image synthesis/image completion in AR/VR. This section will discuss the most relevant existing works in detail.

2.1   Facial De-occlusion and HMD Removal Methods

De-occluding face images in the presence of large occluders such as HMDs is highly an ill-posed problem. Several works such as [6, 9] have been proposed in the literature to address this issue. However, none promises to provide usable results in practice as these have only been validated for frontal face images with rectangular masks. Since they use an additional reference image of the person, they fail in cases of different pose variations between occluded and reference images. Recently, [3] presented an approach to tackle the problem of facial de-occlusion by training a person-specific model in VR settings. It generates plausible and natural-looking reconstructions but might fail to maintain smooth eye movements across consecutive frames. To address this issue, we can use extra information provided by modern HMD devices equipped with eye-tracking to generate consistent eye motions.

2.2 Structure-Guided Image Inpainting

Figuring out missing regions without any prior information is a difficult task. Many prior works have successfully used landmarks for the task of face generation and synthesis. Previous image inpainting methods, such as [5, 12] use edges, landmarks, and other structural information as an auxiliary input to guide the reconstructions. This extra supervision has proven effective in helping the model fill the missing region with appropriate content. However, these are image-based approaches and might not guarantee to generate consistent results across frames. Thus, we cannot directly use these methods to generate smooth reconstructions, particularly in the eye region.

3 Proposed Method

3.1 The Architecture

We built upon the architecture proposed in [3] and used an attention-enabled encoder-decoder architecture followed by a novel Landmark Heatmap Predictor (LHP) module that acts as a regularizer to enhance the reconstruction in and around the eye region. We train this network in an end-to-end fashion in two stages using a dedicated loss function. [3] is an existing facial de-occlusion network, and we consider it a baseline.

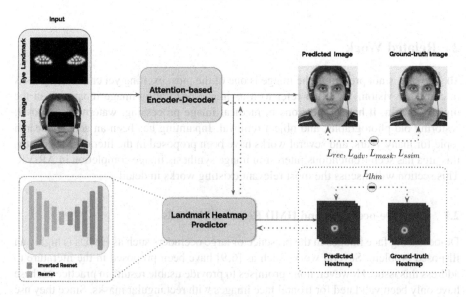

Fig. 2. Illustration of our proposed architecture.

Attention-Based Encoder-Decoder: We utilize an encoder-decoder architecture with an attention module to inpaint the missing regions of the face, particularly the eye region. The primary function of the attention module is to focus on reconstructing this region with high-frequency details such as hair, facial accessories, and appearances. It also helps the model to generalize to unseen and novel appearances, hairstyles, etc. The encoder-decoder comprises ResNet and Inverted ResNet layers, with a bottleneck layer of 99 dimensions. For Inverted ResNet, the first convolution in the ResNet block is replaced by a 4×4 deconv layer. The attention module is composed of four convolution layers: $Conv(4 * m, 3)$, $Conv(4 * m, 3)$, $Conv(8 * m, 3)$ and $Conv(2 * m, 3)$, where m denotes the base number of filters and $Conv(m, k)$ denotes a convolutional layer with output number of channels m and kernel size k. Given an occluded face image as an input X_{occ}, this network hallucinates the missing region in order to reconstruct the generated unoccluded image, X_{rec} against the ground truth unoccluded image, X_{gt}.

Landmark Heatmap Predictor Module: We employ another encoder-decoder network to refine the reconstruction around the eye region. The primary aim of this network is to predict the eye landmark heatmap of the reconstructed image, based on which we can regularize the final reconstructed image using a loss function. This landmark heatmap predictor network is composed of ResNet and Inverted ResNet layers. The input to this network is the reconstructed image, X_{recon} produced from the attention-based encoder-decoder network. The output is a 42-channeled landmark heatmap, denoted by $LHP(X_{rec})$, where each channel corresponds to one of 42 eye landmarks. Figure 2 illustrates an overview of the proposed pipeline.

3.2 Spatial Supervision Using Landmarks

Per-frame predictions from traditional image-based facial de-occlusion network such as [3] suffer from temporal discontinuity and flickering, especially in eyelids. Therefore, to stabilize the eye movements, we leverage eye landmarks as an auxiliary input to guide smooth reconstructions in the eyelids that are much more realistic and consistent. This supervision helps the model preserve the structure of the eyelids. For better and enriched representations, we prefer 2D heatmaps over 2D coordinates. Each landmark is represented by a separate heatmap, interpreted as a grayscale image. We convolve all heatmaps to a single-channel grayscale image which is then concatenated with the occluded RGB input image in the channel dimension. This is further fed to the attention-based encoder-decoder to generate plausible and stable reconstructions.

3.3 Loss Functions

The primary goal of this pipeline is to generate plausible facial inpainted reconstructions consistent with other frames in sequence while preserving the landmark structure of the eyes. To serve this purpose, we use the following loss functions:

The first loss function ensures that the generated reconstruction is in close proximity to the ground-truth unoccluded image. Thus, we formulate pixel-based $L1$ loss to penalize reconstruction errors.

$$L_{rec} = \|X_{gt} - X_{rec}\|_1 \tag{1}$$

However, using only reconstruction loss generates blurry reconstructions. Thus we adopt the architecture of the DCGAN discriminator [7] in the pipeline, denoted by D, to compute the adversarial loss that forces the encoder-decoder to reconstruct high-fidelity outputs by sharpening the blurred images.

$$L_{adv} = log(D(X_{gt})) + log(1 - D(X_{rec})) \qquad (2)$$

To further stabilize the adversarial training, we use SSIM based structural similarity loss, as defined in [1], that helps to improve the alignment of high-frequency image elements.

$$L_{ssim} = SSIM(X_{rec}, X_{gt}) \qquad (3)$$

In order to emphasize the quality of reconstruction in the masked region, i.e., invalid pixels, we use a mask-based loss function. Here, we use the binary mask image as additional supervision to the network and input image while training. This helps mitigate the blinking artifacts around the eye region for stable reconstructions.

$$L_{mask} = \|I_{mask} \odot X_{gt} - I_{mask} \odot X_{rec}\|_1 \qquad (4)$$

where, I_{mask} refers to single channel binary mask image where white pixels (1) correspond to occluded region and black pixels (0) correspond to the remaining unoccluded region and $\odot$ is element-wise multiplication.

Apart from providing a landmark heatmap along with occluded input, we also regularize the reconstructions based on landmarks using a loss function. To prevent irregularities in the eye region and preserve eyelid shape, we utilize landmark heatmap prediction loss that regularizes the inpainted reconstructions based on predicted eye landmark heatmaps, $LHP(X_{rec})$ and ground-truth eye landmark heatmaps, H. Here, for each landmark $l_i \in R^2$, H_i consists of a 2D normal distribution centered at li and a standard deviation of σ.

$$L_{lhm} = \|H - LHP(X_{rec})\|_2 \qquad (5)$$

Thus, the final training objective loss function can be written as,

$$L_{final} = \lambda_{rec} * L_{rec} + \lambda_{adv} * L_{adv} + \lambda_{ssim} * L_{ssim} + \lambda_{mask} * L_{mask} + \lambda_{lhm} * L_{lhm} \qquad (6)$$

where, $\lambda_{rec}, \lambda_{adv}, \lambda_{ssim}, \lambda_{mask}$ and λ_{lhm} are the corresponding weight parameters for each loss term.

4   Experiments and Results

4.1   Dataset and Training Settings

Dataset Preparation: We train the network on different face video sequences for multiple identities. We train a person-specific model for every identity on 4–5 sequences captured in various appearances, including apparel, hairstyle, facial accessories, and different head poses. Videos are recorded at a resolution of 1280×720 at 30 fps using a regular smart-phone and then cropped to 256×256 for training. Note that there is no overlap between the training and test set. To test the ability of our model to generalize to novel appearances, we validate it with completely unseen videos that are not seen during the training process. The dataset is available here. For the provision of spatial supervision, we use Mediapipe [2] to detect and localize 42 landmarks around the eye, including iris landmarks. As discussed in Sect. 3.2, we create a heatmap for every landmark coordinate. Since we extract pseudo landmarks directly from unoccluded ground truth that is inherently spatially aligned with the occluded face, we directly append landmark heatmaps with the occluded input image without any further processing.

Inference with Real Occlusion: The eye information might not be directly accessible during inference when wearing regular virtual reality headsets. Fortunately, modern devices allow eye tracking using IR cameras mounted inside headsets. We can extract this information from eye images captured using these cameras. Unfortunately, the images captured by these cameras are not aligned with the face image. Hence, we need to calibrate both the eye and face camera as proposed in [8,17] to align the eye images with the face image coordinate system. However, due to the unavailability of these headsets, we opt for pseudo landmarks extracted from ground-truth images to provide supervision to the model. As discussed, we extract these landmarks using the Mediapipe [2] face landmark detector. It is to be noted that these landmarks do not adhere well to an anatomically defined point across every video frame and thus have local noise in them generated due to the inaccuracy of the facial landmark detector.

Training Strategy: We follow a similar two-stage training strategy proposed in [3]. In the first stage, we only train the encoder-decoder network without an attention module on unoccluded images along with their corresponding eye landmark images of the person using the first three losses aforementioned in Sect. 3.3, each added incrementally after 400, 100, and 300 epochs, respectively. In the second stage, we fine-tune the same encoder-decoder with the attention module and the landmark prediction module on occluded images of the same person and their corresponding eye landmark images using two additional loss functions. We use the same three losses as the first stage. Apart from this, we also use a landmark heatmap prediction loss to regularize the reconstructions generated from the attention-based encoder-decoder network and a mask-based loss to minimize reconstruction errors in the masked region. We use $\lambda_{rec} = 1, \lambda_{adv} = 0.25, \lambda_{ssim} = 60, \lambda_{lhm} = 1$ and $\lambda_{mask} = 1$.

4.2 Results

In this section, we present the results of our method and discuss its superiority over existing approaches. We first compare the visual quality of the reconstruction generated by our method with popular state-of-the-art inpainting methods, followed by a quantitative analysis using standard evaluation metrics. To further validate the efficacy of our approach, we also report an ablation study conducted in the scope of this work.

Qualitative Comparison: For visual comparisons, we evaluate our method against various image inpainting methods across 20 subjects. Results highlighted in Figs. 4, 6,

7 show that the reconstructions generated using our method are visually pleasing and consistent across frames compared to other inpainting methods. Reconstructions generated using our approach, as shown in row (C) of Fig. 4 show the significance of landmark supervision and regularization loss in capturing eye movements such as blinks. However, predictions using other approaches are often incoherent across frames. As visible in row (F), DeepFillv2 [13] fails poorly to generate plausible reconstruction in the eye region. LaFIn [12] and Edge-Connect [5] generate superior reconstructions compared to DeepFillv2, however, it cannot handle eye movements. Besides, there is a noticeable discrepancy in the left and right eyes that looks unnatural. Baseline [3] produces naturally-looking reconstructions but cannot handle eye blinks. For better comparison, refer to the supplementary video. In Fig. 3, we also show the reconstruction error (l2 error) between the results generated by different image inpainting methods and the ground truth for better justification. Please refer to the supplemental video.

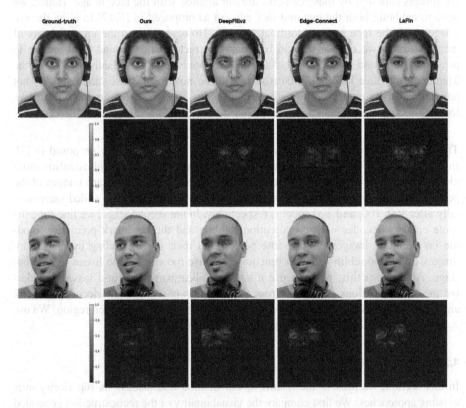

Fig. 3. Qualitative result that showing the reconstruction error (l2 error) between the results generated by different image inpainting methods and the ground truth.

Frame 1	Frame 2	Frame 3	Frame 4	Frame 5

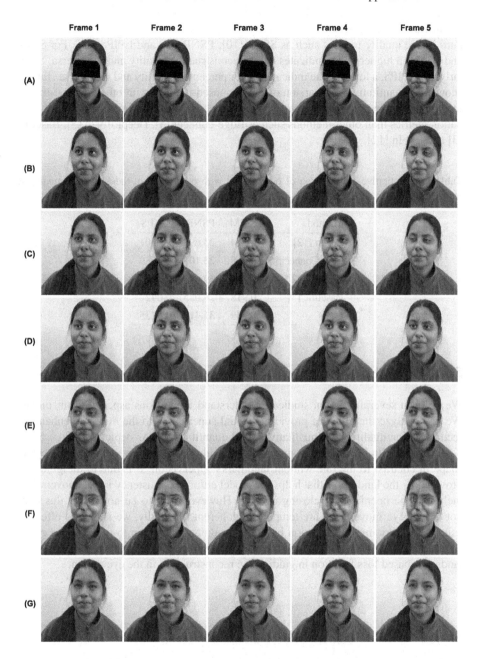

Fig. 4. Qualitative comparison with SOTA image inpainting methods. From row (A-G) are Occluded (input), Original (ground-truth), Ours, Baseline [3], Edge-connect [5], DeepFillv2 [13] and LaFIn [12] respectively. From left to right are consecutive frames of unseen testing video.

Quantitative Comparison: To quantify the quality of reconstructions, we use standard image quality metrics such as SSIM [10], PSNR [4], and LPIPS [16]. For SSIM and PSNR, a higher value indicates better reconstruction quality and vice-versa. Similarly, for LPIPS, a lower value indicates better perceptual quality and vice-versa. Table 1 shows the quantitative comparison of our proposed method with other state-of-the-art face inpainting methods. As reported, our method (**in bold**) performs better in all evaluation metrics than other methods such as Edge-connect [5], DeepFillv2 [13], Baseline [3] and LaFIn [12].

Table 1. Quantitative comparision of our method with other state-of-the-art image inpainting methods.

Method	SSIM↑	PSNR↑	LPIPS↓
LaFIn [12]	0.914	23.693	0.0601
EdgeConnect [5]	0.908	23.10	0.0689
DeepFillv2 [13]	0.845	19.693	0.117
Baseline [3]	0.918	29.025	0.042
Ours	**0.949**	**31.417**	**0.0235**

5   Ablation Studies

We perform several ablation studies to understand the various aspects of our model. We first analyze the effect of providing spatial supervision to the model in enhancing reconstruction quality, both qualitatively and quantitatively. As depicted in Figs. 5 and 8, our model with landmarks produces aesthetically pleasing eyes and preserves eyelid shape in contrast to the one without landmarks supervision. It is due to the guidance provided by the landmarks that helps the model enforce consistency in eye movements, including the opening and closing of eyes. However, it is to be noted that this does not ensure eye movements are temporally coherent. Secondly, we show the effect of using a regularizing loss function based on landmarks heatmap to penalize the errors caused by the model. Table 2 reports the positive impact of using eye landmarks and landmark-based loss function in guiding the reconstruction in the eye region.

Table 2. Ablation study showing the significance of using landmark supervision on the reconstruction quality. Here, LHM is the auxiliary landmark heatmap provided along with the input image and L_{lhm} is the regularizing loss function.

Method	SSIM↑	PSNR↑	LPIPS↓
Baseline [3]	0.918	29.025	0.042
Ours (with LHM)	0.926	29.272	0.0418
Ours (with LHM + L_{lhm})	**0.949**	**31.417**	**0.0235**

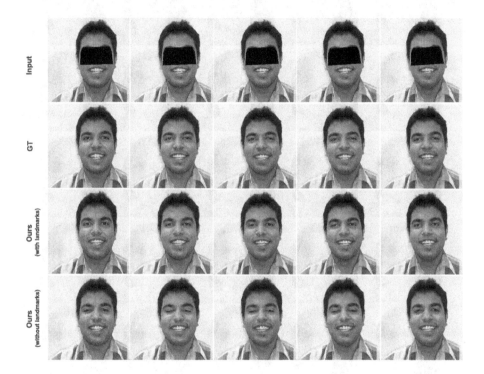

Fig. 5. Testing results showing the effect of using landmarks as auxillary input to the network. From row (1–4) are occluded (input), original (ground-truth), results with and without landmarks respectively. From left to right is temporal continuously images of original 30 fps videos.

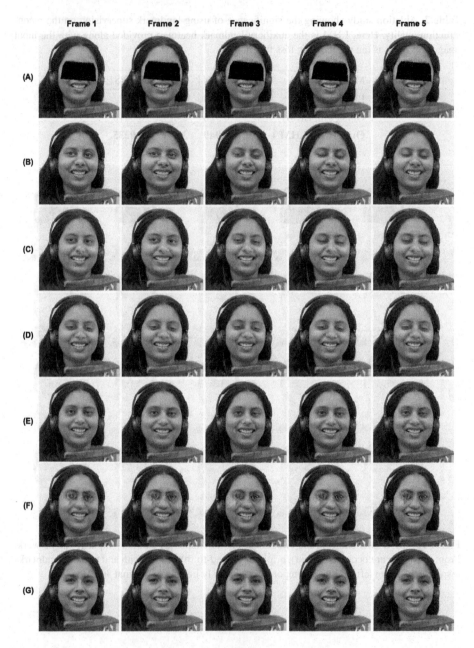

Fig. 6. Qualitative comparison with SOTA image inpainting methods. From row (A-G) are Occluded (input), Original (ground-truth), Ours, Baseline [3], Edge-connect [5], DeepFillv2 [13] and LaFIn [12] respectively. From left to right are consecutive frames of unseen testing video.

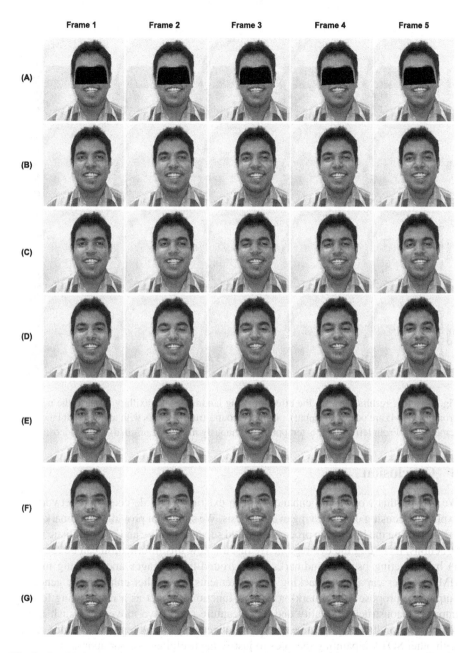

Fig. 7. Qualitative comparison with SOTA image inpainting methods. From row (A-G) are Occluded (input), Original (ground-truth), Ours, Baseline [3], Edge-connect [5], DeepFillv2 [13] and LaFIn [12] respectively. From left to right are consecutive frames of unseen testing video.

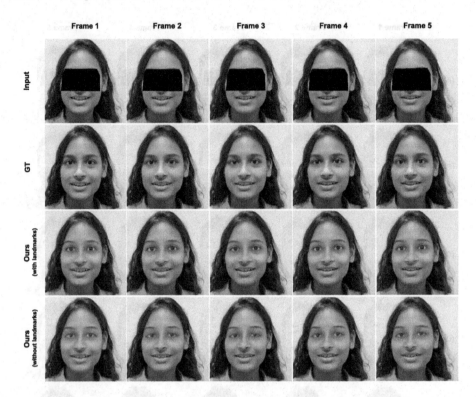

Fig. 8. Testing results showing the effect of using landmarks as auxillary input to the network. From row (1–4) are occluded (input), original (ground-truth), results with and without landmarks respectively. From left to right is temporal continuously images of original 30 fps videos.

6   Conclusion

We present this work as an enhancement in existing facial de-occlusion networks by explicitly focusing on improving eye synthesis. We show that providing landmark information during the inpainting process can yield superior quality and photorealistic reconstructions, including the eye region. We discuss how this information can be retrieved: 1) by extracting pseudo landmarks from ground-truth images and 2) using modern HMD devices capable of tracking eye movements. To further enhance the generated output, we propose a landmark-based loss function that act as a regularizing term to improve reconstruction quality and helps capture subtle eye movements such as eye blinks. We conducted qualitative and quantitative analysis and reported superior results with other SOTA inpainting methods to justify the usefulness of our approach.

References

1. Browatzki, B., Wallraven, C.: 3FabRec: fast few-shot face alignment by reconstruction. In: Proceedings of the IEEE/CVF Conference on Computer Vision and Pattern Recognition (2020)

2. Grishchenko, I., Ablavatski, A., Kartynnik, Y., Raveendran, K., Grundmann, M.: Attention mesh: high-fidelity face mesh prediction in real-time (2020)
3. Gupta, S., Shetty, A., Sharma, A.: Attention based occlusion removal for hybrid telepresence systems. In: 19th Conference on Robots and Vision (CRV) (2022)
4. Horé, A., Ziou, D.: Image quality metrics: PSNR vs. SSIM. In: ICPR (2010)
5. Nazeri, K., Ng, E., Joseph, T., Qureshi, F.Z., Ebrahimi, M.: Edgeconnect: generative image inpainting with adversarial edge learning (2019)
6. Numan, N., ter Haar, F., Cesar, P.: Generative RGB-D face completion for head-mounted display removal. In: 2021 IEEE Conference on Virtual Reality and 3D User Interfaces Abstracts and Workshops (VRW). IEEE (2021)
7. Radford, A., Metz, L., Chintala, S.: Unsupervised representation learning with deep convolutional generative adversarial networks (2016)
8. Thies, J., Zollöfer, M., Stamminger, M., Theobalt, C., Nießner, M.: FaceVR: Real-Time Facial Reenactment and Eye Gaze Control in Virtual Reality (2016)
9. Wang, M., Wen, X., Hu, S.M.: Faithful face image completion for HMD occlusion removal. In: 2019 IEEE International Symposium on Mixed and Augmented Reality Adjunct (ISMAR-Adjunct). IEEE (2019)
10. Wang, Z., Bovik, A.C., Sheikh, H.R., Simoncelli, E.P.: Image quality assessment: from error visibility to structural similarity. IEEE Trans. Image Process. **13**, 600–612 (2004)
11. Wu, Y., Singh, V., Kapoor, A.: From image to video face inpainting: spatial-temporal nested GAN (STN-GAN) for usability recovery. In: 2020 IEEE Winter Conference on Applications of Computer Vision (2020)
12. Yang, Y., Guo, X., Ma, J., Ma, L., Ling, H.: LaFIn: Generative landmark guided face inpainting (2019)
13. Yu, J., Lin, Z., Yang, J., Shen, X., Lu, X., Huang, T.S.: Generative image inpainting with contextual attention (2018)
14. Zhang, J., et al.: Unsupervised high-resolution portrait gaze correction and animation. IEEE Trans. Image Process. **31**, 1572–1586 (2022)
15. Zhang, J., et al.: Dual in-painting model for unsupervised gaze correction and animation in the wild. In: ACM MM (2020)
16. Zhang, R., Isola, P., Efros, A.A., Shechtman, E., Wang, O.: The unreasonable effectiveness of deep features as a perceptual metric. In: Proceedings of the IEEE Conference on Computer Vision and Pattern Recognition (CVPR), June 2018
17. Zhao, Y., et al.: Mask-off: synthesizing face images in the presence of head-mounted displays. In: 2019 IEEE Conference on Virtual Reality and 3D User Interfaces (VR) (2019)

Consistency-Based Self-supervised Learning for Temporal Anomaly Localization

Aniello Panariello(✉) ⓘ, Angelo Porrello ⓘ, Simone Calderara ⓘ,
and Rita Cucchiara ⓘ

AImageLab, University of Modena and Reggio Emilia, Modena, Italy
{Aniello.Panariello,Angelo.Porrello,Simone.Calderara,
Rita.Cucchiara}@unimore.it

Abstract. This work tackles Weakly Supervised Anomaly detection, in which a predictor is allowed to learn not only from normal examples but also from a few labeled anomalies made available during training. In particular, we deal with the localization of anomalous activities within the video stream: this is a very challenging scenario, as training examples come only with video-level annotations (and not frame-level). Several recent works have proposed various regularization terms to address it *i.e.* by enforcing sparsity and smoothness constraints over the weakly-learned frame-level anomaly scores. In this work, we get inspired by recent advances within the field of self-supervised learning and ask the model to yield the same scores for different augmentations of the same video sequence. We show that enforcing such an alignment improves the performance of the model on XD-Violence.

Keywords: Video anomaly detection · Temporal action localization · Weakly supervised · Self supervised learning

1 Introduction

The goal of Video Anomaly Detection is to detect and localize anomalous events occurring in a video stream. Usually, these events involve human actions such as abuse, falling, fighting, theft, etc. In the last decades, such a task has gained relevance thanks to its potential for a video surveillance pipeline. In light of the widespread presence of CCTV cameras, employing enough personnel to examine all the footage is indeed infeasible and, therefore, automatic tools must be exploited.

Traditional approaches leverage low level features [16] to deal with this task, computed either on visual cues [3,22] or object trajectories [8,35]. These techniques – which could suffer and be unreliable in some contexts [20,32] – have been surpassed by reconstruction-based approaches [19]: these methods require a set of normal data to learn a model of regularity; afterward, an example can be assessed as "anomalous" if it substantially deviates from the learned model. In this regard, the architectural design often resorts to deep unsupervised autoencoders [45,55], which may include several strategies for regularizing the latent space.

The major recent trend [42] regards the exploitation of weak supervision, which provides the learning phase also with examples from the anomalous classes. However, the annotations come in a video-level format: the learner does not know in which time steps the anomaly will show, but only if it does at least one time. For such a reason, the proposed techniques often fall into the framework of multiple instance learning [17,43,47], devising additional optimization constraints that mitigate the lack of frame-level annotations.

In this work, we attempt to complement the existing regularization techniques with an idea from the fields of self-supervised learning and consistency regularization [6,15,41,48]. In this context, data augmentation is exploited to generate positive pairs, consisting of a couple of slightly different versions of the same example. Afterward, the network is trained to output very similar representations, in a self-supervised manner (no class labels are required). Similarly, we devise a data augmentation strategy tailored to sequences and encourage the network to assign the same frame-level anomaly scores for the elements of each positive pair. Such a strategy resembles the smoothness constraint often imposed in recent works; however, while those approaches focus on adjacent time steps, our approach can randomly span over longer temporal windows.

To evaluate the merits of our proposal, we conduct experiments on the XD-Violence [47] dataset. We found that the presence of our regularization term yields remarkable improvements, although being not yet enough to reach state-of-the-art approaches.

2 Related Work

Video Anomaly Detection. Traditionally researchers and practitioners exploit object trajectories [8,34,35], low-level handcrafted features [3,22,33] (such as Histogram of Gradients and Histogram of Flows), and connected component analysis cues [2,4,5]. These traditional approaches have proven to be effective on benchmark datasets, but turn out to be still ineffective when used on a real domain. This means that these methods do not adapt well to anomalies that have never been seen before. For this reason, the task of anomaly detection has recently been approached with deep neural networks.

Most recent works lean towards unsupervised methods [1,19] usually based on deep autoencoders. These models leverage the reconstruction error to measure how much an incoming example conforms to the training set distribution. Remarkably, the authors of [29] resorted to a variation of this common paradigm: namely, they proposed an approach that learns the distribution of normal data by guessing the appearance of future frames. In this respect, the idea was to compare the actual frame and the predicted one: if their difference is high, then it is possible to assume that an anomalous event occurred.

While it is easy to label whole videos (*e.g.*, anomaly present or not), the availability of fine-grained labeled anomalous videos is often scarce. For this reason, weakly supervised methods have seen a great advance. Among these works,

Sultani *et al.* [42] introduced a new pattern for video anomaly detection termed Multiple Instance Learning (MIL), upon which most of the subsequent weakly supervised methods developed. In this scenario, a positive video (*i.e.* containing an anomaly) and a negative one are taken into account at each iteration. These videos are usually split into segments; the model assigns them an anomaly score through a sequence of convolutional layers followed by an MLP. The scores are collected into a positive bag and a negative bag: as in the former, there is at least an anomaly (while in the latter there are no anomalies) the minimum score from the positive bag has to be as far as possible from the maximum score of the negative bag. To impose this constraint, the objective function contains a MIL ranking loss as well as smoothness and sparsity constraints. Zhou *et al.* [55] introduced the attention mechanism combined with the MIL approach, to improve the localization of the anomalies.

Temporal Action Localization. The majority of the works dealing with Temporal Action Localization are either fully supervised or weakly supervised. The former ones can be broadly categorized into one stage [27,30] and two stages methods [26,49,53]. For the first type, in works such as [27], action boundaries and labels are predicted simultaneously; in [30] this concept is developed by exploiting Gaussian kernels to dynamically optimize the temporal scale of each action proposal. On the other hand, two-stage methods initially generate temporal action proposals and then classify them. A manner to produce proposals is by exploiting the anchor mechanism [13,18,50], sliding window [40] or combining confident starting and ending frames of an action [26,28].

Weakly supervised methods have been pioneered by UntrimmedNet [44], STPN [36] and AutoLoc [39], in which the action instances are localized by applying a threshold on the class activation sequence. This paradigm has been recently brought into the video anomaly detection settings in [31,46].

3   Proposed Method

In the following, we present the proposed model and its training objective. The third part is dedicated to explaining the process of proposal generation, which consists of a post-processing step grouping adjacent similar scores into contiguous discrete intervals.

3.1   Model

In our set-up, each video is split into segments of 16 frames, with no overlap between consecutive segments. To extract video-level features, some works [12, 52] opted for combining 2D-CNNs and Recurrent Neural Networks; instead, we give each segment to a pre-trained I3D network [10]. Indeed, the authors of [24] have shown that the I3D features can prove to be effective for video-anomaly detection, even when a shallow classifier as XGBoost [9,14] is employed for later classification. In our case, the I3D network is pre-trained on the Kinetics [21] dataset and not fine-tuned on our target data.

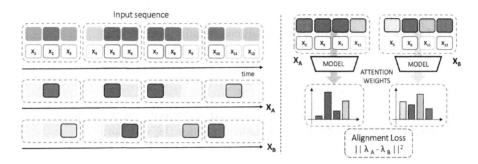

Fig. 1. Overview of the proposed framework. (left) Augmentation function sampling two slightly different sequences out of a single one. The original sequence gets split into windows and for each, we randomly sample a single feature vector. This is done twice to obtain two sequences X^A and X^B. (right) Both sequences are separately fed to the model, obtaining two sequences of attention weights λ_A and λ_B, pulled closer together by the alignment loss.

This way, each example is represented by a variable-length sequence of T feature vectors $\mathcal{X} = (x_1, x_2, \ldots, x_T)$. During training, examples come with a label y indicating whether an anomalous event appears in that sequence at least one time. Hence, given a training set of examples $\{(\mathcal{X}_j, y_j)\}_{j=1}^{\mathcal{N}}$ forged in the above-mentioned manner, we seek to train a neural network $f(\cdot; \theta)$ that solves such a task with the lowest empirical error:

$$\min_{\theta} \quad \frac{1}{\mathcal{N}} \sum_{i=1}^{N} \text{BCE}(f(\mathcal{X}_i; \theta), y_i) \qquad (1)$$

where $\text{BCE}(\cdot, \cdot)$ stands for the binary cross entropy loss function. For the architectural design of $f(\cdot; \theta)$, we took inspiration from [36]. It consists of two main parts, discussed in the following paragraphs: namely, the computation of attention coefficients and the creation of an aggregate video-level representation.

Attention Coefficients. The aim of this module is to assign a weight $\lambda_t \in [0, 1]$ to each element of the input sequence. As explained in the following, these weights will identify the most salient segments of the video *i.e.*, the likelihood of having observed an abnormal event within each segment. The module initially performs a masked temporal 1D convolution [25] to allow each feature vector to encode information from the past. Such a transformation – which does not alter the number of input feature vectors – is followed by two fully connected layers activated by ReLU functions, except for the last layer where a sigmoid function is employed.

Video-Level Representation. Once we have guessed the attention values, we exploit them to aggregate the input feature vectors. Such an operation – which resembles a temporal weighted average pooling presented in [37] – produces a single feature vector with unchanged dimensionality; formally:

$$\mathbf{x} = \sum_{t=1}^{T} \lambda_t x_t. \tag{2}$$

We finally feed it to a classifier $g(\cdot)$, composed of two fully connected layers. The final output represents the guess of the network for the value of y.

3.2   Training Objective

As mentioned before, we train our network in a weakly supervised fashion *i.e.*, only video level labels are provided to the learner. However, to provide a stronger training signal and to encourage attention coefficients to highlight salient events, we follow recent works [7,51] and use some additional regularization terms that encode a prior knowledge we retain about the dynamics of abnormality.

Often, the presence of anomalous activities is characterized by the properties of sparseness and smoothness. Namely, anomalies appear rarely (*i.e.*, normal events dominate) and transitions between the two modalities usually occur throughout multiple frames. Such a peculiarity represents a prior we would like to enforce over the scores learned by the model: the majority of them should be close to zero and have similar values for video segments. In formal terms, the first constraint can be injected by penalizing the l_1 norm [11] of the attention weights, as follows:

$$\mathcal{L}_{sp} = \|\lambda\|_1, \tag{3}$$

while the second one can be carried out by imposing adjacent coefficients to vary as little as possible:

$$\mathcal{L}_{sm} = \sum_{t=1}^{T-1} (\lambda_t - \lambda_{t+1})^2. \tag{4}$$

Alignment Loss. Our main contribution consists in adding a consistency-based regularization term to the overall objective function. Overall, the idea is to generate two slightly different sequences out of a single one and, then, to encourage the model to produce the same attention coefficients for the two inputs.

To do so, we introduce a data augmentation function, shown in Fig. 1, that allows us to forge different versions $\mathcal{X}_A$ and $\mathcal{X}_B$ from the same example $\mathcal{X}$. In more detail, we split each sequence $(x_1, x_2, \ldots, x_T)$ into fixed-size blocks, whose length L is a hyperparameter we always set to 3; afterward, we randomly choose a feature vector within each block.

Once the variants $\mathcal{X}_A$ and $\mathcal{X}_B$ have been created, we ask the network to minimize the following objective function:

$$\mathcal{L}_a = \sum_{t=1}^{T} (\lambda_t^A - \lambda_t^B)^2, \tag{5}$$

where λ^A and λ^B are respectively the attention coefficients computed by the network for $\mathcal{X}_A$ and $\mathcal{X}_B$. With this additional regularization term, we seek to enforce that not only adjacent time-steps should have the same weight, but also those lying within a wider temporal horizon.

Overall Objective. Finally, the objective function will be:

$$\mathcal{L} \equiv \mathcal{L}_{cl} + \alpha\mathcal{L}_{sp} + \beta\mathcal{L}_{sm} + \gamma\mathcal{L}_a, \tag{6}$$

where the parameters α, β, and γ give different weights to each loss component and $\mathcal{L}_{cl}$ is the binary cross entropy loss function.

3.3 Temporal Proposal

During inference, we refine the anomaly scores by applying a post-processing step. Usually, two segments considered paramount by the network are interleaved by "holes", mostly due to noisy acquisitions or poor representations. The purpose of this phase is therefore to merge temporally close detections in a single retrieved candidate. In particular, as done in [54], we initially take out from the candidate set all those time-steps whose corresponding attention scores are lower than a certain threshold (in our experiments, < 0.35). The remaining non-zero scores will be used to generate the temporal proposal.

To generate the proposals, we do not use the rough coefficients, but instead a more refined version. In particular, we compute a 1-d activation map in the temporal domain, called Temporal Class Activation Map (T-CAM) [54], which indicates the relevance of the segment t in the prediction of one of the two classes involved (*normal vs anomalous*). Each value a_t of such activation map is computed as $a_t = g(x_t)$, *i.e.*, the guess of the classifier $g(\cdot)$ (introduced in Sect. 3.1) if masking the contributions of all time-steps except the t-th one. Furthermore, we extract the Weighted T-CAM, which combines the attention weight and the T-CAM activation values, *i.e.*, $\psi_t = \lambda_t \cdot a_t$. This operation let us emphasize the most important features for generating the proposal.

The last operation involves interpolating the weighted scores in the temporal axis and taking the bounding box that covers the largest connected component [36] to generate the final proposal [54]. The anomaly score for each proposal is then computed as:

$$\sum_{t=t_{start}}^{t_{end}} = \lambda_t \cdot \frac{a_t}{t_{end} - t_{start} - 1}, \tag{7}$$

where t_{start} and t_{end} represent the beginning and the ending of a single proposal.

4   Experiments

We conduct our experiments on the **XD-Violence Dataset** [47], a multi-modal dataset that contains scenes from different sources such as movies, sports, games,

A. Panariello et al.

Table 1. For different levels and metrics, results of our model with and without the proposed align loss. There is an improvement on almost all metrics when leveraging the proposed term.

Align loss	Video level AUC%	Video level AP%	Segment level AUC%	Segment level AP%	Frame level proposal AUC%	Frame level proposal AP%	Frame level AUC%	Frame level AP%
-	**97.91**	**98.36**	84.39	66.75	85.14	68.01	84.57	65.96
✓	97.79	98.28	**85.49**	**66.87**	**90.23**	**71.68**	**85.65**	**66.05**

Table 2. Comparison with recent works. We report the frame-level AP score on XD-Violence for both unsupervised and weakly-supervised methods. All the competitors exploit the I3D network for extracting features from RGB frames.

Supervision	Method	AP%
Unsupervised	SVM baseline	50.78
	OCSVM [38]	27.25
	Hasan et al. [19]	30.77
Weakly Supervised	Ours (no align loss)	68.01
	Ours	71.68
	Sultani et al. [42]	75.68
	Wu et al. [47]	75.41
	RTFM [43]	77.81

news, and live scenes. It holds a great variety concerning the devices used for video acquisition; indeed, the examples were captured by CCTV cameras, hand-held cameras, or car driving recorders. There are 4754 videos for a total of 217 hours: among all these, 2405 are violent videos and 2349 are non-violent. The training set consists of 3954 examples; the test set, instead, features 800 ones, split into 500 violent and 300 non-violent videos. The dataset comes with multiple modalities such as RGB, optical flow, and audio; however, we restrict our analysis only to the RGB input domain.

XD-Violence also comes with segment-level ground truth labels; however, we only use video-level annotation during training, conforming to the weakly supervised setting. Differently, we exploit both the segment-level and frame-level annotations during the test phase, thus evaluating the model's capabilities for fine-grained localization.

Metrics. We used the most popular metrics in anomaly detection settings, namely the Area Under Receiver Operating Characteristic Curve (AUC) and the Area Under Precision-Recall Curve (AP). While the AUC tends to be optimistic in the case of unbalanced datasets, the AP gives a more accurate evaluation of these scenarios.

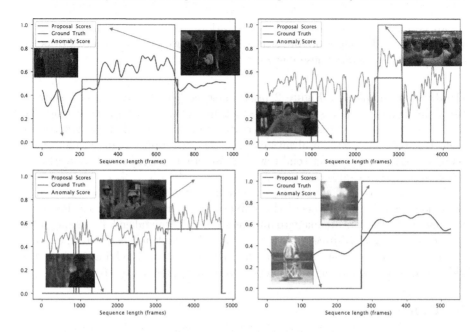

Fig. 2. Qualitative examples of the capabilities of our model to perform anomaly localization. The temporal proposal scores are indicated with a blue line, while the weighted T-CAM scores and the ground truth are shown in green and red respectively. (Color figure online)

We assess the performance of the classification head by providing these two metrics at the video level; therefore, we use the attention scores, their interpolation, as well as the temporal proposals to assess the other grains.

Training Details. We use the Adam optimizer [23] with a learning rate of 10^{-4} for the first 10 epochs and 10^{-5} for the remaining 40 epochs. The hyperparameters for the loss components are set to: $\alpha = 2 \times 10^{-8}$, $\beta = 0.002$, $\gamma = 0.5$. The threshold for taking out the low weighted T-CAM scores is set to 0.35; the batch size equals 8.

Results. Table 1 reports the comparison between the baseline approach with and without the proposed alignment objective. It can be observed that its addition leads to a remarkable improvement in most of the metrics. In particular, we have a gain of about 1% in AUC for segment and frame level metrics, while the AP remains almost the same. The greatest improvement subsists for the temporal proposal metric, where we gain 5 points in AUC and around 4 points in AP. The video level metrics remain approximately the same but still very high.

When comparing our approach with other recent works (see Table 2), it can be seen that it outperforms the unsupervised state-of-the-art methods; however, it is in turn surpassed by the weakly supervised ones. We conjecture that such a gap is mainly due to the bag representations inherent in these approaches, which could confer superior robustness; therefore, we leave to future works the extension of our idea to these methods.

Qualitative Analysis. Figure 2 presents several qualitative results, showing two fight scenes in the first row, and a riot and an explosion in the second one. We notice that the scores rightly increase when an anomalous action begins. Unfortunately, they remain always close to the uncertainty regime (assuming a score around 0.5) and never tend towards discrete decisions. In future works, we are going to address also this issue.

Looking at the original videos, we could also explain why the scores yielded by the model are noisy and subject to local fluctuations. Indeed, as sudden characters or camera' movements are likely to intervene, they are mistaken for real anomalies by the model, which is susceptible to these visual discontinuities due to the lack of segment-level annotations during training. Differently, when relying on the entire video sequence, the model can conversely recognize them correctly.

5   Conclusions

This work proposes a novel strategy to learn effective frame-level scores in weakly supervised settings when only video-level annotations are made available to the learner. Our proposal – which builds upon recent advances in the field of self-supervised learning – relies on maximizing the alignment between the attention weights of two different augmentations of the same input sequence. We show that a base network equipped also with other common regularization strategies (*e.g.* sparsity and smoothness) brings even more improvements. We found that it is not enough to achieve state-of-the-art performance; however, we leave aside for future works a comprehensive investigation of its applicability to more advanced architectures.

Acknowledgments. This work has been supported in part by the InSecTT project, funded by the Electronic Component Systems for European Leadership Joint Undertaking under grant agreement 876038. The Joint Undertaking receives support from the European Union's Horizon 2020 research and innovation programme and AU, SWE, SPA, IT, FR, POR, IRE, FIN, SLO, PO, NED and TUR. The document reflects only the author's view and the Commission is not responsible for any use that may be made of the information it contains.

References

1. Abati, D., Porrello, A., Calderara, S., Cucchiara, R.: Latent space autoregression for novelty detection. In: Proceedings of the IEEE Conference on Computer Vision and Pattern Recognition (2019)
2. Amraee, S., Vafaei, A., Jamshidi, K., Adibi, P.: Anomaly detection and localization in crowded scenes using connected component analysis. Multim. Tools Appl. **77**,14767–14782 (2018)

3. Benezeth, Y., Jodoin, P.M., Saligrama, V., Rosenberger, C.: Abnormal events detection based on spatio-temporal co-occurrences. In: Proceedings of the IEEE Conference on Computer Vision and Pattern Recognition. IEEE (2009)
4. Bolelli, F., Allegretti, S., Baraldi, L., Grana, C.: Spaghetti labeling: directed acyclic graphs for block-based connected components labeling. IEEE Trans. Image Process. **29**, 1999 –2012 (2019)
5. Bolelli, F., Allegretti, S., Grana, C.: One DAG to rule them all. IEEE Trans. Pattern Anal. Mach. Intell. (99), 1–1 (2021)
6. Boschini, M., Buzzega, P., Bonicelli, L., Porrello, A., Calderara, S.: Continual semi-supervised learning through contrastive interpolation consistency. arXiv preprint arXiv:2108.06552 (2021)
7. Cai, R., Zhang, H., Liu, W., Gao, S., Hao, Z.: Appearance-motion memory consistency network for video anomaly detection. In: Proceedings of the AAAI Conference on Artificial Intelligence (2021)
8. Calderara, S., Heinemann, U., Prati, A., Cucchiara, R., Tishby, N.: Detecting anomalies in people's trajectories using spectral graph analysis. Comput. Vis. Image Underst. **115**, 1099–1111 (2011)
9. Candeloro, L., et al.: Predicting WNV circulation in Italy using earth observation data and extreme gradient boosting model. Remote Sens **12**(18), 3064 (2020)
10. Carreira, J., Zisserman, A.: Quo vadis, action recognition? A new model and the kinetics dataset. In: Proceedings of the IEEE Conference on Computer Vision and Pattern Recognition (2017)
11. Cascianelli, S., Costante, G., Crocetti, F., Ricci, E., Valigi, P., Luca Fravolini, M.: Data-based design of robust fault detection and isolation residuals via lasso optimization and Bayesian filtering. Asian J. Control **23**, 57–71 (2021)
12. Cascianelli, S., Costante, G., Devo, A., Ciarfuglia, T.A., Valigi, P., Fravolini, M.L.: The role of the input in natural language video description. IEEE Trans. Multim. **22**, 271 –283 (2019)
13. Chao, Y.W., Vijayanarasimhan, S., Seybold, B., Ross, D.A., Deng, J., Sukthankar, R.: Rethinking the faster r-CNN architecture for temporal action localization. In: Proceedings of the IEEE Conference on Computer Vision and Pattern Recognition (2018)
14. Chen, T., Guestrin, C.: Xgboost: a scalable tree boosting system. In: Proceedings of the 22nd ACM SIGKDD International Conference on Knowledge Discovery and Data Mining (2016)
15. Chen, T., Kornblith, S., Norouzi, M., Hinton, G.: A simple framework for contrastive learning of visual representations. In: International Conference on Machine Learning (2020)
16. Dalal, N., Triggs, B.: Histograms of oriented gradients for human detection. In: Proceedings of the IEEE Conference on Computer Vision and Pattern Recognition. IEEE (2005)
17. Feng, J.C., Hong, F.T., Zheng, W.S.: MIST: multiple instance self-training framework for video anomaly detection. In: Proceedings of the IEEE Conference on Computer Vision and Pattern Recognition (2021)
18. Gao, J., Yang, Z., Chen, K., Sun, C., Nevatia, R.: Turn tap: temporal unit regression network for temporal action proposals. In: IEEE International Conference on Computer Vision (2017)
19. Hasan, M., Choi, J., Neumann, J., Roy-Chowdhury, A.K., Davis, L.S.: Learning temporal regularity in video sequences. In: Proceedings of the IEEE Conference on Computer Vision and Pattern Recognition (2016)

20. Hu, X., Hu, S., Huang, Y., Zhang, H., Wu, H.: Video anomaly detection using deep incremental slow feature analysis network. IET Comput. Vis. 10, 258–267 (2016)
21. Kay, W., et al.: The kinetics human action video dataset. arXiv preprint arXiv:1705.06950 (2017)
22. Kim, J., Grauman, K.: Observe locally, infer globally: a space-time MRF for detecting abnormal activities with incremental updates. In: Proceedings of the IEEE Conference on Computer Vision and Pattern Recognition. IEEE (2009)
23. Kingma, D.P., Ba, J.: Adam: a method for stochastic optimization. In: International Conference on Learning Representations (2014)
24. Koshti, D., Kamoji, S., Kalnad, N., Sreekumar, S., Bhujbal, S.: Video anomaly detection using inflated 3D convolution network. In: International Conference on Inventive Computation Technologies (ICICT). IEEE (2020)
25. Lea, C., Flynn, M.D., Vidal, R., Reiter, A., Hager, G.D.: Temporal convolutional networks for action segmentation and detection. In: Proceedings of the IEEE Conference on Computer Vision and Pattern Recognition (2017)
26. Lin, T., Liu, X., Li, X., Ding, E., Wen, S.: BMN: boundary-matching network for temporal action proposal generation. In: IEEE International Conference on Computer Vision (2019)
27. Lin, T., Zhao, X., Shou, Z.: Single shot temporal action detection. In: Proceedings of the 25th ACM International Conference on Multimedia (2017)
28. Lin, T., Zhao, X., Su, H., Wang, C., Yang, M.: BSN: boundary sensitive network for temporal action proposal generation. In: Ferrari, V., Hebert, M., Sminchisescu, C., Weiss, Y. (eds.) ECCV 2018. LNCS, vol. 11208, pp. 3–21. Springer, Cham (2018). https://doi.org/10.1007/978-3-030-01225-0_1
29. Liu, W., Luo, W., Lian, D., Gao, S.: Future frame prediction for anomaly detection-a new baseline. In: Proceedings of the IEEE Conference on Computer Vision and Pattern Recognition (2018)
30. Long, F., Yao, T., Qiu, Z., Tian, X., Luo, J., Mei, T.: Gaussian temporal awareness networks for action localization. In: Proceedings of the IEEE Conference on Computer Vision and Pattern Recognition (2019)
31. Lv, H., Zhou, C., Cui, Z., Xu, C., Li, Y., Yang, J.: Localizing anomalies from weakly-labeled videos. IEEE Trans. Image Process. 30, 4505–4515 (2021)
32. Medel, J.R., Savakis, A.: Anomaly detection in video using predictive convolutional long short-term memory networks. arXiv preprint arXiv:1612.00390 (2016)
33. Mehran, R., Oyama, A., Shah, M.: Abnormal crowd behavior detection using social force model. In: Proceedings of the IEEE Conference on Computer Vision and Pattern Recognition. IEEE (2009)
34. Monti, A., Porrello, A., Calderara, S., Coscia, P., Ballan, L., Cucchiara, R.: How many observations are enough? knowledge distillation for trajectory forecasting. In: Proceedings of the IEEE Conference on Computer Vision and Pattern Recognition (2022)
35. Morris, B.T., Trivedi, M.M.: Trajectory learning for activity understanding: Unsupervised, multilevel, and long-term adaptive approach. IEEE Trans. Pattern Anal. Mach. Intell. 33, 2287–2301 (2011)
36. Nguyen, P., Liu, T., Prasad, G., Han, B.: Weakly supervised action localization by sparse temporal pooling network. In: Proceedings of the IEEE Conference on Computer Vision and Pattern Recognition (2018)
37. Porrello, A., et al.: Spotting insects from satellites: modeling the presence of culicoides imicola through deep CNNs. In: 2019 15th International Conference on Signal-Image Technology & Internet-Based Systems (SITIS). IEEE (2019)

38. Schölkopf, B., Williamson, R.C., Smola, A., Shawe-Taylor, J., Platt, J.: Support vector method for novelty detection. In: Advances in Neural Information Processing Systems (1999)

39. Shou, Z., Gao, H., Zhang, L., Miyazawa, K., Chang, S.-F.: AutoLoc: weakly-supervised temporal action localization in untrimmed videos. In: Ferrari, V., Hebert, M., Sminchisescu, C., Weiss, Y. (eds.) ECCV 2018. LNCS, vol. 11220, pp. 162–179. Springer, Cham (2018). https://doi.org/10.1007/978-3-030-01270-0_10

40. Shou, Z., Wang, D., Chang, S.F.: Temporal action localization in untrimmed videos via multi-stage CNNs. In: Proceedings of the IEEE Conference on Computer Vision and Pattern Recognition (2016)

41. Sohn, K., et al.: Fixmatch: simplifying semi-supervised learning with consistency and confidence. In; Advances in Neural Information Processing Systems (2020)

42. Sultani, W., Chen, C., Shah, M.: Real-world anomaly detection in surveillance videos. In: Proceedings of the IEEE Conference on Computer Vision and Pattern Recognition (2018)

43. Tian, Y., Pang, G., Chen, Y., Singh, R., Verjans, J.W., Carneiro, G.: Weakly-supervised video anomaly detection with robust temporal feature magnitude learning. In: IEEE International Conference on Computer Vision (2021)

44. Wang, L., Xiong, Y., Lin, D., Van Gool, L.: Untrimmednets for weakly supervised action recognition and detection. In: Proceedings of the IEEE Conference on Computer Vision and Pattern Recognition (2017)

45. Wang, X., et al.: Robust unsupervised video anomaly detection by multipath frame prediction. IEEE Trans. Neural Netw. Learn. Syst. (2021)

46. Wu, J., et al.: Weakly-supervised spatio-temporal anomaly detection in surveillance video. In: International Joint Conferences on Artificial Intelligence (2021)

47. Wu, P., et al.: Not only look, but also listen: learning multimodal violence detection under weak supervision. In: Vedaldi, A., Bischof, H., Brox, T., Frahm, J.-M. (eds.) ECCV 2020. LNCS, vol. 12375, pp. 322–339. Springer, Cham (2020). https://doi.org/10.1007/978-3-030-58577-8_20

48. Xie, Q., Dai, Z., Hovy, E., Luong, T., Le, Q.: Unsupervised data augmentation for consistency training. In: Conference on Advances in Neural Information Processing Systems (2020)

49. Xu, H., Das, A., Saenko, K.: R-c3d: region convolutional 3d network for temporal activity detection. In: IEEE International Conference on Computer Vision (2017)

50. Yang, L., Peng, H., Zhang, D., Fu, J., Han, J.: Revisiting anchor mechanisms for temporal action localization. IEEE Trans. Image Process. (99), 1–1 (2020)

51. Yu, G., et al.: Cloze test helps: effective video anomaly detection via learning to complete video events. In: Proceedings of the 28th ACM International Conference on Multimedia, pp. 583–591 (2020)

52. Yue-Hei Ng, J., Hausknecht, M., Vijayanarasimhan, S., Vinyals, O., Monga, R., Toderici, G.: Beyond short snippets: deep networks for video classification. In: Proceedings of the IEEE Conference on Computer Vision and Pattern Recognition (2015)

53. Zeng, R., et al.: Graph convolutional networks for temporal action localization. In: IEEE International Conference on Computer Vision (2019)

54. Zhou, B., Khosla, A., Lapedriza, A., Oliva, A., Torralba, A.: Learning deep features for discriminative localization. In: Proceedings of the IEEE Conference on Computer Vision and Pattern Recognition (2016)

55. Zhu, Y., Newsam, S.: Motion-aware feature for improved video anomaly detection. In: British Machine Vision Conference (2019)

Perspective Reconstruction of Human Faces by Joint Mesh and Landmark Regression

Jia Guo[1], Jinke Yu[1], Alexandros Lattas[2], and Jiankang Deng[1,2(✉)]

[1] Insightface, London, UK
[2] Imperial College London, London, UK
{a.lattas,j.deng16}@imperial.ac.uk

Abstract. Even though 3D face reconstruction has achieved impressive progress, most orthogonal projection-based face reconstruction methods can not achieve accurate and consistent reconstruction results when the face is very close to the camera due to the distortion under the perspective projection. In this paper, we propose to simultaneously reconstruct 3D face mesh in the world space and predict 2D face landmarks on the image plane to address the problem of perspective 3D face reconstruction. Based on the predicted 3D vertices and 2D landmarks, the 6DoF (6 Degrees of Freedom) face pose can be easily estimated by the PnP solver to represent perspective projection. Our approach achieves 1st place on the leader-board of the ECCV 2022 WCPA challenge and our model is visually robust under different identities, expressions and poses. The training code and models are released to facilitate future research. https://github.com/deepinsight/insightface/tree/master/reconstruction/jmlr.

Keywords: Monocular 3D face reconstruction · Perspective projection · Face pose estimation

1 Introduction

Monocular 3D face reconstruction has been widely applied in many fields such as VR/AR applications (e.g., movies, sports, games), video editing, and virtual avatars. Reconstructing human faces from monocular RGB data is a well-explored field and most of the approaches can be categorized into optimization-based methods [2,3,14] or regression-based methods [7,11,13,28].

For the optimization-based methods, a prior of face shape and appearance [2,3] is used. The pioneer work 3D Morphable Model (3DMM) [2] represents the face shape and appearance in a PCA-based compact space and the fitting is then based on the principle of analysis-by-synthesis. In [3], an in-the-wild texture model is employed to greatly simplify the fitting procedure without optimization on the illumination parameters. To model textures in high fidelity, GANFit [14] harnesses Generative Adversarial Networks (GANs) to train a very powerful generator of facial texture in UV space and constrains the latent parameter by

L. Karlinsky et al. (Eds.): ECCV 2022 Workshops, LNCS 13805, pp. 350–365, 2023.
https://doi.org/10.1007/978-3-031-25072-9_23

state-of-the-art face recognition model [1,6,8,9,29,30]. However, the iterative optimization in these methods is not efficient for real-time inference.

For the regression-based methods, a series of methods are based on synthetic renderings of human faces [15,26] or 3DMM fitted data [28] to perform a supervised training of a regressor that predicts the latent representation of a prior face model (e.g., 3DMM [24], GCN [27], CNN [25]) or 3D vertices in different representation formats [7,13,17,19]. Genova et al. [15] propose a 3DMM parameter regression technique that is based on synthetic renderings and Tran et al. [24] directly regress 3DMM parameters using a CNN trained on fitted 3DMM data. Zhu et al. [28] propose 3D Dense Face Alignment (3DDFA) by taking advantage of Projected Normalized Coordinate Code (PNCC). In [27], joint shape and texture auto-encoder using direct mesh convolutions is proposed based on Graph Convolutional Network (GCN). In [25], CNN-based shape and texture decoders are trained on unwrapped UV space for non-linear 3D morphable face modelling. Jackson et al. [19] propose a model-free approach that reconstructs a voxel-based representation of the human face. DenseReg [17] regresses horizontal and vertical tessellation which is obtained by unwrapping the template shape and transferring it to the image domain. PRN [13] predicts a position map in the UV space of a template mesh. RetinaFace [7] directly predicts projected vertices on the image plane within the face detector. Besides these supervised regression methods, there are also self-supervised regression approaches. Deng et al. [11] train a 3DMM parameter regressor based on photometric reconstruction loss with skin attention masks, a perception loss based on FaceNet [23], and multi-image consistency losses. DECA [12] robustly produces a UV displacement map from a low-dimensional latent representation.

Although the above studies have achieved good face reconstruction results, existing methods mainly use orthogonal projection to simplify the projection process of the face. When the face is close to the camera, the distortion caused by perspective projection can not be ignored [20,31]. Due to the use of orthogonal projection, existing methods can not well explain the facial distortion caused by perspective projection, leading to poor performance.

To this end, we propose a joint 3D mesh and 2D landmark regression method in this paper. Monocular 3D face reconstruction usually includes two tasks: 3D face geometric reconstruction and face pose estimation. However, we avoid explicit face pose estimation as in RetinaFace [7] and DenseLandmark [26]. The insight behind this strategy is that the evaluation metric for face reconstruction is usually in the camera space and the regression error on 6DoF parameters will repeat to every vertex. Even though dense vertex or landmark regression seems redundant, the regression error of each point is individual. Most important, explicit pose parameter regression can result in drift-alignment problem [5], which will make the 2D visualization of face reconstruction unsatisfying. By contrast, the 6 Degrees of Freedom (6DoF) face pose can be easily estimated by the PnP solver based on our predicted 3D vertices and 2D landmarks.

To summarize, the main contributions of this work are:

- We propose a Joint Mesh and Landmark Regression (JMLR) method to reconstruct 3D face shape under perspective projection, and the 6DoF face pose can be further estimated by PnP.

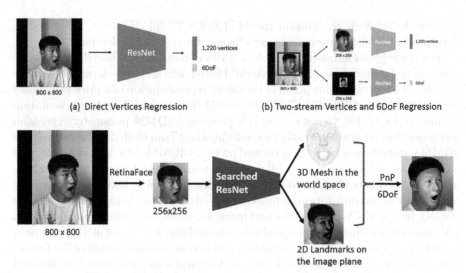

(a) Direct Vertices Regression   (b) Two-stream Vertices and 6DoF Regression

(c) Joint mesh and landmark regression for perspective face reconstruction

Fig. 1. Straightforward solutions for perspective face reconstruction. (a) direct 3D vertices regression, (b) two-stream 3D vertices regression and 6DoF prediction, and (c) our proposed Joint Mesh and Landmark Regression (JMLR).

- The proposed JMLR achieves first place on the leader-board of the ECCV 2022 WCPA challenge.
- The visualization results show that the proposed JMLR is robust under different identities, exaggerated expressions and extreme poses.

2 Our Method

2.1 3D Face Geometric Reconstruction

In Fig. 1, we show three straightforward solutions for perspective face reconstruction. The simplest solution as shown in Fig. 1(a) is to directly regress the 3D vertices and 6DoF (i.e., Euler angles and translation vector) from the original image (i.e., 800×800), however, the performance of this method is very limited as mentioned by the challenge organizer. Another improved solution as illustrated in Fig. 1(b) is that the face shape should be predicted from the local facial region while the 6DoF can be obtained from the global image. Therefore, the local facial region is cropped and resized into 256×256, and then this face patch is fed into the ResNet [18] to predict the 1,220 vertices. To predict 6DOF information, the region outside the face is blackened and the original 800×800 image is then resized into 256×256 as the input of another ResNet.

In RetinaFace [7], explicit pose estimation is avoided by direct mesh regression on the image plane as direct pose parameter regression can result in misalignment under challenging scenarios. However, RetinaFace only considered orthographic face reconstruction. In this paper, we slightly change the regression target of RetinaFace, but still employ the insight behind RetinaFace, that is, avoiding direct pose parameter regression. More specifically, we directly regress 3D facial mesh in the world space as well as projected 2D facial landmarks on the image plane as illustrated in Fig. 1(c).

For 3D facial mesh regression in the world space, we predict a fixed number of $N = 1,220$ vertices ($\mathbf{V} = [v_x^0, v_y^0, v_z^0; \cdots ; v_x^{N-1}, v_y^{N-1}, v_x^{N-1}]$) on a pre-defined topological triangle context (i.e., 2,304 triangles). These corresponding vertices share the same semantic meaning across different faces. With the fixed triangle topology, every pixel on the face can be indexed by the barycentric coordinates and the triangle index, thus there exists a pixel-wise correspondence with the 3D face. In [26], 703 dense landmarks covering the entire head, including ears, eyes, and teeth, are proved to be effective and efficient to encode facial identity and subtle expressions for 3D face reconstruction. Therefore, 1K-level dense vertices/landmarks are a good balance between accuracy and efficiency for face reconstruction.

As each 3D face is represented by concatenating its N vertex coordinates, we employ the following vertex loss to constrain the location of vertices:

$$\mathcal{L}_{vert} = \frac{1}{N} \sum_{i=1}^{N} ||v_i(x,y,z) - v_i^*(x,y,z)||_1, \tag{1}$$

where $N = 1,220$ is the number of vertices, v is the prediction of our model and v^* is the ground-truth. By taking advantage of the 3D triangulation topology, we consider the edge length loss [7]:

$$\mathcal{L}_{edge} = \frac{1}{3M} \sum_{i=1}^{M} ||e_i - e_i^*||_1, \tag{2}$$

where $M = 2,304$ is the number of triangles, e is the edge length calculated from the prediction and e^* is the edge length calculated from the ground truth. The edge graph is a fixed topology as shown in Fig. 1(c).

For projected 2D landmark regression, we also employ distance loss to constrain predicted landmarks close to the projected landmarks from ground-truth:

$$\mathcal{L}_{land} = \frac{1}{N} \sum_{i=1}^{N} ||p_i(x,y) - p_i^*(x,y)||_1, \tag{3}$$

where $N = 1,220$ is the number of vertices, p is the prediction of our model and p^* is the ground-truth generated by perspective projection.

By combining the vertex loss and the edge loss for 3D mesh regression and the landmark loss for 2D projected landmark regression, we define the following prospective face reconstruction loss:

$$\mathcal{L} = \mathcal{L}_{vert} + \lambda_0 \mathcal{L}_{edge} + \lambda_1 \mathcal{L}_{land}, \tag{4}$$

where λ_0 is set to 0.25 and λ_1 is set to 2 according to our experimental experience.

2.2   6DoF Estimation

Based on the predicted 3D vertices in the world space, projected 2D landmarks on the image plane, and the camera intrinsic parameters, we can easily employ the Perspective-n-Point (PnP) algorithm [21] to compute the 6D facial pose parameters (i.e., the rotation of roll, pitch, and yaw as well as the 3D translation of the camera with respect to the world). Even though directly regressing 6DoF pose parameters from a single image by CNN (Fig. 1(a)) is also feasible, it achieves much worse performance than our method due to the non-linearity of the rotation space.

In perspective projection, 3D face shape V_{world} is transformed from the world coordinate system to the camera coordinate system by using 6DoF face pose (i.e., the rotation matrix $R \in \mathbb{R}^{3 \times 3}$ and the translation vector $T \in \mathbb{R}^{1 \times 3}$) with known intrinsic camera parameters K,

$$V_{camera} = K(V_{world}R + T), \tag{5}$$

where K is related to (1) the coordinates of the principal point (the intersection of the optical axes with the image plane) and (2) the ratio between the focal length and the size of the pixel. The intrinsic parameters K can be easily obtained through an off-line calibration step. Knowing 2D-3D point correspondences as well as the intrinsic parameters, pose estimation is straightforward by calling *cv.solvePnP()*.

3   Experimental Results

3.1   Dataset

In the perspective face reconstruction challenge, 250 volunteers were invited to record the training and test dataset. These volunteers sit in a random environment, and the 3D acquisition equipment (i.e., iPhone 11) is fixed in front of them, with a distance ranging from about 0.3 to 0.9 m. Each subject is asked to perform 33 specific expressions with two head movements (from looking left to looking right/from looking up to looking down). The triangle mesh and head pose information of the RGB image is obtained by the built-in ARKit toolbox. Then, the original data are pre-processed to unify the image size and camera intrinsic, as well as eliminate the shifting of the principal point of the camera.

In this challenge, 200 subjects with 356, 640 instances are used as the training set, and 50 subjects with 90, 545 instances are used as the test set. Note that there is only one face in each image and the location of each face is provided by the face-alignment toolkit [4]. The ground truth of 3D vertices and pose transform matrix of the training set is provided. The unit of 3D ground-truth is in meters. The facial triangle mesh is made up of 1, 220 vertices and 2, 304 triangles. The indices of 68 landmarks [10] from 1, 220 vertices are also provided.

3.2   Evaluation Metrics

For each test image, the challenger should predict the 1,220 3D vertices in world space (i.e., $V_{world} \in \mathbb{R}^{1220 \times 3}$) and the pose transform matrix (i.e., the rotation matrix $R \in \mathbb{R}^{3 \times 3}$ and the translation vector $T \in \mathbb{R}^{1 \times 3}$) from world space to camera space.

$$V_{world} = \begin{bmatrix} v_x^0 & v_y^0 & v_z^0 \\ v_x^1 & v_y^1 & v_z^1 \\ \vdots & \vdots & \vdots \\ v_x^{1219} & v_y^{1219} & v_z^{1219} \end{bmatrix}, R = \begin{bmatrix} r_{00} & r_{01} & r_{02} \\ r_{10} & r_{11} & r_{12} \\ r_{20} & r_{21} & r_{22} \end{bmatrix}, T = \begin{bmatrix} t_x & t_y & t_z \end{bmatrix}.$$

Then, we can compute the transformed vertices in camera space by $V_{camera} = V_{world}R + T$. In this challenge, the error of 3D vertices and face pose is measured in camera space. Four transformed sets of vertices are computed as follows:

$$V_1 = V^{gt}R^{gt} + T^{gt}, V_2 = V^{pred}R^{pred} + T^{pred}, \tag{6}$$
$$V_3 = V^{gt}R^{pred} + T^{pred}, V_4 = V^{pred}R^{gt} + T^{gt}.$$

Finally, L_2 distance between pair $(V_1, V_2), (V_1, V_3), (V_1, V_4)$ are calculated and combined into the final distance error:

$$L_{error} = \|V_1 - V_2\|_2 + \|V_1 - V_3\|_2 + 10\|V_1 - V_4\|_2, \tag{7}$$

where geometry accuracy across different identities and expressions is emphasized by $\times 10$. On the challenge leader-board, the distance error is multiplied by 1,000, thus the distance error is in millimeters instead of meters.

3.3   Implementation Details

Input. We employ five facial landmarks predicted by RetinaFace [7] for normalized face cropping at the resolution of 256×256. Colour jitter and flip data augmentation are also used during our training.

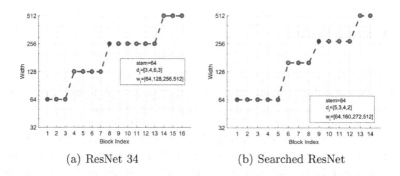

(a) ResNet 34   (b) Searched ResNet

Fig. 2. Computation redistribution for the backbone. (a) the default ResNet-34 design. (b) our searched ResNet structure for perspective face reconstruction.

Table 1. Ablation study regarding augmentation, network structure, and loss. The results are reported on 40 subjects from the training set and the rest 160 subjects are used for training.

Aug	Network	Loss	Flops	Score
Flip	ResNet34	$\mathcal{L}$	5.12G	34.31
No Flip	ResNet34	$\mathcal{L}$	5.12G	34.76
Flip	ResNet18	$\mathcal{L}$	3.42G	34.95
Flip	ResNet50	$\mathcal{L}$	5.71G	34.70
Flip	Searched ResNet	$\mathcal{L}$	5.13G	34.13
Flip	ResNet34	$\mathcal{L}_{vert}+\mathcal{L}_{land}$	5.12G	34.60
Flip	ResNet34 & Searched ResNet	$\mathcal{L}$	10.25 G	33.39

Backbone. We employ the ResNet [18] as our backbone. In addition, we refer to SCRFD [16] to optimize the computation distribution over the backbone. For the ResNet design, there are four stages (i.e., C2, C3, C4 and C5) operating at progressively reduced resolution, with each stage consisting of a sequence of identical blocks. For each stage i, the degrees of freedom include the number of blocks d_i (i.e., network depth) and the block width w_i (i.e., number of channels). Therefore, the backbone search space has 8 Degrees of freedom as there are 4 stages and each stage i has 2 parameters: the number of blocks d_i and block width w_i. Following RegNet [22], we perform uniform sampling of $d_i \leq 24$ and $w_i \leq 512$ (w_i is divisible by 8). As state-of-the-art backbones have increasing widths, we also constrain the search space, according to the principle of $w_{i+1} \geq w_i$. The searched ResNet structure for perspective face reconstruction is illustrated in Fig. 2(b).

Loss Optimization. We train the proposed joint mesh and landmark regression method for 40 epochs with the SGD optimizer. We set the learning rate as 0.1 and the batch size is set to 64×8. The training is warmed up by 1,000 steps and then the Poly learning scheduler is used.

3.4   Ablation Study

In Table 1, we run several experiments to validate the effectiveness of the proposed augmentation, network structure, and loss. We split the training data into 80% subjects for training and 20% subjects for validation. From Table 1, we have the following observations: (1) flipping augmentation can decrease the reconstruction error by 0.45 as we check experiments 1 and 2, (2) the capacity of ResNet-34 is well matched to the dataset as we compare the performance between different network structures (e.g., ResNet-18, ResNet-34, and ResNet-50), (3) searched ResNet with similar computation cost as ResNet-34 can slightly decrease the reconstruction error by 0.18, (4) without the topology regularization by the proposed edge loss, the reconstruction error increase by 0.29, confirming the effectiveness of mesh regression proposed in RetinaFace [7], (5) by the model ensemble, the reconstruction error significantly reduces by 0.92.

Table 2. Leaderboard results. The proposed JMLR achieves best overall score compared to other methods.

Rank	Team	Score	$\|V_1 - V_2\|_2$	$\|V_1 - V_3\|_2$	$10\,\|V_1 - V_4\|_2$
1st (JMLR)	EldenRing	**32.811318**	**8.596190**	**8.662822**	15.552306
2nd	faceavata	32.884849	8.833569	8.876343	**15.174936**
3rd	raccoon&bird	33.841992	8.880053	8.937624	16.024315

3.5   Benchmark Results

For the submission of the challenge, we employ all of the 200 training subjects and combine all the above-mentioned training tricks (e.g. flip augmentation, searched ResNet, joint mesh and landmark regression, and model ensemble). As shown in Table 2, we rank first on the challenge leader-board and our method outperforms the runner-up by 0.07.

3.6   Result Visualization

In this section, we display qualitative results for perspective face reconstruction on the test dataset. As we care about the accuracy of geometric face reconstruction and 6DoF estimation, we visualize the projected 3D vertices on the image space both in the formats of vertex mesh and rendered mesh. As shown in Fig. 3, the proposed method shows accurate mesh regression results across 50 different test subjects (i.e., genders and ages). In Fig. 4 and Fig. 5, we select two subjects under 33 different expressions. The proposed method demonstrates precise mesh regression results across different expressions (i.e., blinking and mouth opening). In Fig. 6, Fig. 7, and Fig. 8, we show the mesh regression results under extreme poses (e.g., large yaw and pitch variations). The proposed method can easily handle profile cases as well as large pitch angles. From these visualization results, we can see that our method is effective under different identities, expressions and large poses.

358   J. Guo et al.

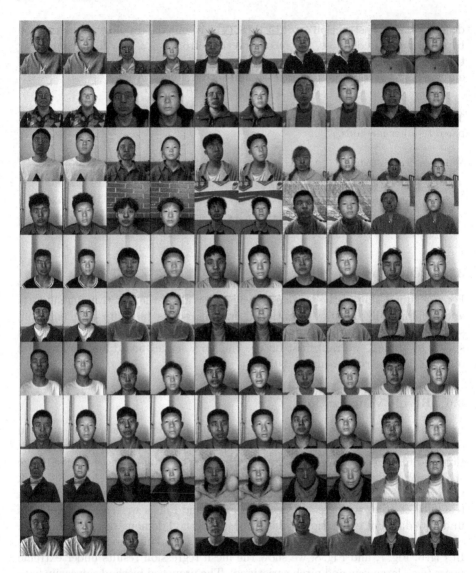

Fig. 3. Predicted meshes on 50 test subjects. 68 landmarks are also indexed for better visualization. The proposed method shows stable and accurate mesh regression results across different subjects (i.e., genders and ages).

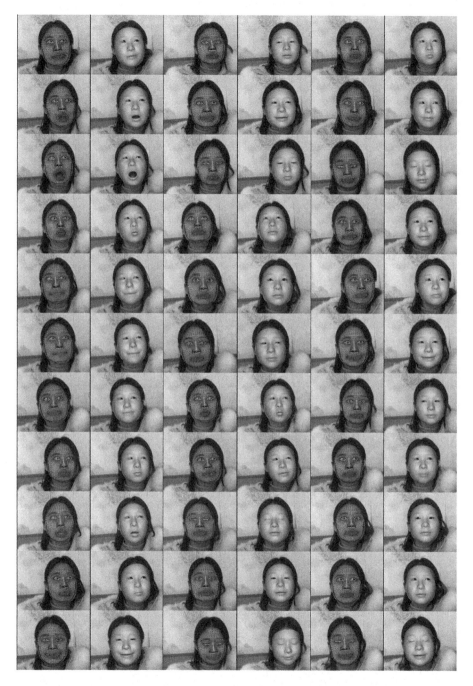

Fig. 4. Predicted meshes of ID-6159 under 33 different expressions. 68 landmarks are also indexed for better visualization. The proposed method shows stable and accurate mesh regression results across different expressions (i.e., blinking and mouth opening).

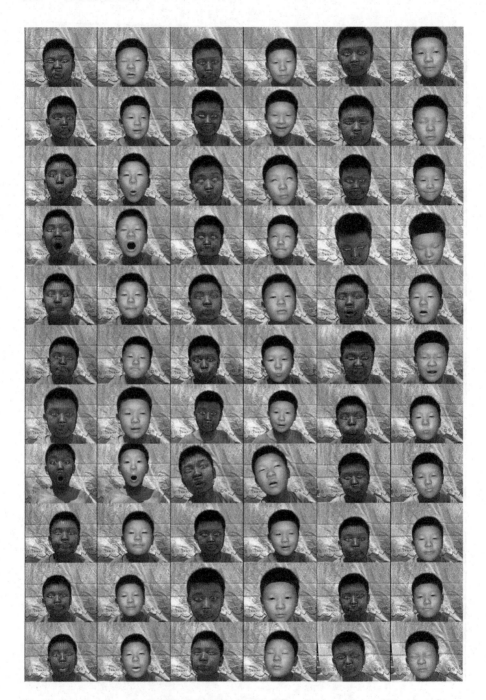

Fig. 5. Predicted meshes of ID-239749 under 33 different expressions. 68 landmarks are also indexed for better visualization. The proposed method shows stable and accurate mesh regression results across different expressions (i.e., blinking and mouth opening).

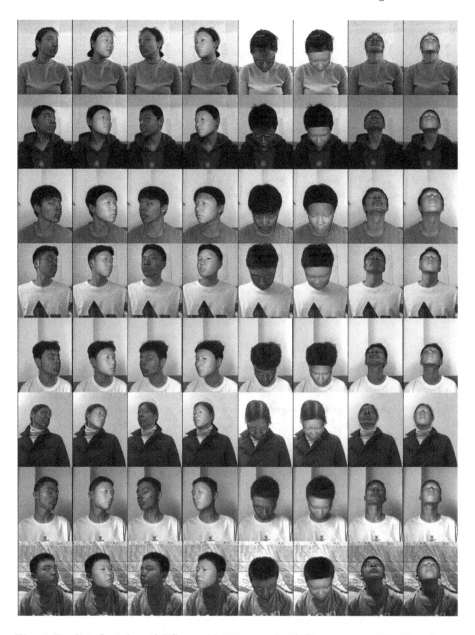

Fig. 6. Predicted meshes of different identities under different poses. 68 landmarks are also indexed for better visualization. The proposed method shows stable and accurate mesh regression results across different facial poses (i.e., yaw and pitch).

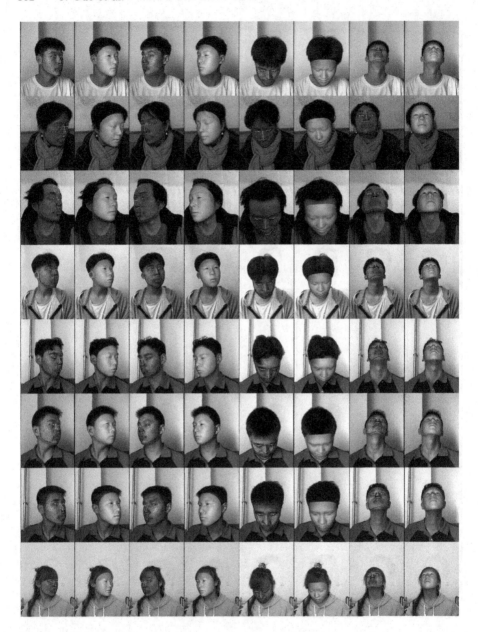

Fig. 7. Predicted meshes of different identities under different poses. 68 landmarks are also indexed for better visualization. The proposed method shows stable and accurate mesh regression results across different facial poses (i.e., yaw and pitch).

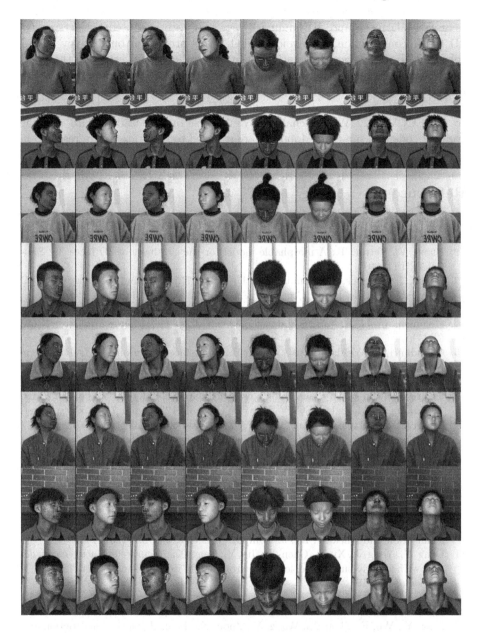

Fig. 8. Predicted meshes of different identities under different poses. 68 landmarks are also indexed for better visualization. The proposed method shows stable and accurate mesh regression results across different facial poses (i.e., yaw and pitch).

4   Conclusion

In this paper, we explore 3D face reconstruction under perspective projection from a single RGB image. We implement a straightforward algorithm, in which joint face 3D mesh and 2D landmark regression are proposed for perspective 3D face reconstruction. We avoid explicit 6DoF prediction but employ a PnP solver for 6DoF face pose estimation given the predicted 3D vertices in the world space and predicted 2D landmarks on the image space. Both quantitative and qualitative experimental results demonstrate the effectiveness of our approach for perspective 3D face reconstruction and 6DoF pose estimation. Our submission to the ECCV 2022 WCPA challenge ranks first on the leader-board and the training code and pre-trained models are released to facilitate future research in this direction.

References

1. An, X., et al.: Killing two birds with one stone: Efficient and robust training of face recognition CNNs by partial fc. In: CVPR (2022)
2. Blanz, V., Vetter, T.: A morphable model for the synthesis of 3D faces. In: SIGGRAPH (1999)
3. Booth, J., Roussos, A., Ponniah, A., Dunaway, D., Zafeiriou, S.: Large scale 3D morphable models. Int. J. Ccomput. Vis. **126**, 233–254 (2018)
4. Bulat, A., Tzimiropoulos, G.: How far are we from solving the 2D & 3D face alignment problem? (and a dataset of 230,000 3D facial landmarks). In: ICCV (2017)
5. Chaudhuri, B., Vesdapunt, N., Wang, B.: Joint face detection and facial motion retargeting for multiple faces. In: CVPR (2019)
6. Deng, J., Guo, J., Liu, T., Gong, M., Zafeiriou, S.: Sub-center ArcFace: boosting face recognition by large-scale noisy web faces. In: Vedaldi, A., Bischof, H., Brox, T., Frahm, J.-M. (eds.) ECCV 2020. LNCS, vol. 12356, pp. 741–757. Springer, Cham (2020). https://doi.org/10.1007/978-3-030-58621-8_43
7. Deng, J., Guo, J., Ververas, E., Kotsia, I., Zafeiriou, S.: Retinaface: Single-shot multi-level face localisation in the wild. In: CVPR (2020)
8. Deng, J., Guo, J., Xue, N., Zafeiriou, S.: Arcface: additive angular margin loss for deep face recognition. In: CVPR (2019)
9. Deng, J., Guo, J., Yang, J., Lattas, A., Zafeiriou, S.: Variational prototype learning for deep face recognition. In: CVPR (2021)
10. Deng, J., et al.: The Menpo benchmark for multi-pose 2D and 3D facial landmark localisation and tracking. Int. J. Comput. Vis. **127**, 599–624 (2019)
11. Deng, Y., Yang, J., Xu, S., Chen, D., Jia, Y., Tong, X.: Accurate 3D face reconstruction with weakly-supervised learning: from single image to image set. In: CVPR Workshops (2019)
12. Feng, Y., Feng, H., Black, M.J., Bolkart, T.: Learning an animatable detailed 3D face model from in-the-wild images. In: SIGGRAPH (2021)
13. Feng, Y., Wu, F., Shao, X., Wang, Y., Zhou, X.: Joint 3D face reconstruction and dense alignment with position map regression network. In: Ferrari, V., Hebert, M., Sminchisescu, C., Weiss, Y. (eds.) Computer Vision – ECCV 2018. LNCS, vol. 11218, pp. 557–574. Springer, Cham (2018). https://doi.org/10.1007/978-3-030-01264-9_33
14. Gecer, B., Ploumpis, S., Kotsia, I., Zafeiriou, S.: Ganfit: Generative adversarial network fitting for high fidelity 3d face reconstruction. In: CVPR (2019)
15. Genova, K., Cole, F., Maschinot, A., Sarna, A., Vlasic, D., Freeman, W.T.: Unsupervised training for 3D morphable model regression. In: CVPR (2018)

16. Guo, J., Deng, J., Lattas, A., Zafeiriou, S.: Sample and computation redistribution for efficient face detection. In: ICLR (2022)
17. Güler, R.A., Trigeorgis, G., Antonakos, E., Snape, P., Zafeiriou, S., Kokkinos, I.: Densereg: fully convolutional dense shape regression in-the-wild. In: CVPR (2017)
18. He, K., Zhang, X., Ren, S., Sun, J.: Deep residual learning for image recognition. In: CVPR (2016)
19. Jackson, A.S., Bulat, A., Argyriou, V., Tzimiropoulos, G.: Large pose 3D face reconstruction from a single image via direct volumetric CNN regression. In: ICCV (2017)
20. Kao, Y., et al.: Single-image 3D face reconstruction under perspective projection. arXiv:2205.04126 (2022)
21. Marchand, E., Uchiyama, H., Spindler, F.: Pose estimation for augmented reality: a hands-on survey. IEEE Trans. Visual. Comput. Graph. **22**, 2633–2651 (2015)
22. Radosavovic, I., Kosaraju, R.P., Girshick, R., He, K., Dollár, P.: Designing network design spaces. In: CVPR (2020)
23. Schroff, F., Kalenichenko, D., Philbin, J.: FaceNet: a unified embedding for face recognition and clustering. In: CVPR (2015)
24. Tran, A.T., Hassner, T., Masi, I., Medioni, G.: Regressing robust and discriminative 3D morphable models with a very deep neural network. In: CVPR (2017)
25. Tran, L., Liu, X.: Nonlinear 3D face morphable model. In: CVPR (2018)
26. Wood, E., et al.: 3D face reconstruction with dense landmarks. In: In: Avidan, S., Brostow, G., Cisse, M., Farinella, G.M., Hassner, T. (eds.) Computer Vision–ECCV 2022. ECCV 2022. LNCS, vol .13673. Springer, Cham (2021). https://doi.org/10.1007/978-3-031-19778-9_10
27. Zhou, Y., Deng, J., Kotsia, I., Zafeiriou, S.: Dense 3D face decoding over 2500fps: Joint texture & shape convolutional mesh decoders. In: CVPR (2019)
28. Zhu, X., Lei, Z., Liu, X., Shi, H., Li, S.Z.: Face alignment across large poses: a 3D solution. In: CVPR (2016)
29. Zhu, Z., et al.: Webface260m: a benchmark for million-scale deep face recognition. Trans. Pattern Anal. Mach. Intell. (2022)
30. Zhu, Z., et al.: Webface260m: a benchmark unveiling the power of million-scale deep face recognition. In: CVPR (2021)
31. Zielonka, W., Bolkart, T., Thies, J.: Towards metrical reconstruction of human faces. In: ECCV (2022). https://doi.org/10.1007/978-3-031-19778-9_15

Pixel2ISDF: Implicit Signed Distance Fields Based Human Body Model from Multi-view and Multi-pose Images

Jianchuan Chen[1], Wentao Yi[1], Tiantian Wang[2], Xing Li[3], Liqian Ma[4], Yangyu Fan[3], and Huchuan Lu[1(✉)]

[1] Dalian University of Technology, Dalian, China
{Janaldo,raywit}@mail.dlut.edu.cn, lhchuan@dlut.edu.cn
[2] University of California, Merced, USA
twang61@ucmerced.edu
[3] Northwestern Polytechnical University, Xi'an, China
fan_yangyu@nwpu.edu.cn
[4] ZMO AI Inc., Sacramento, USA

Abstract. In this report, we focus on reconstructing clothed humans in the canonical space given multiple views and poses of a human as the input. To achieve this, we utilize the geometric prior of the SMPLX model in the canonical space to learn the implicit representation for geometry reconstruction. Based on the observation that the topology between the posed mesh and the mesh in the canonical space are consistent, we propose to learn latent codes on the posed mesh by leveraging multiple input images and then assign the latent codes to the mesh in the canonical space. Specifically, we first leverage normal and geometry networks to extract the feature vector for each vertex on the SMPLX mesh. Normal maps are adopted for better generalization to unseen images compared to 2D images. Then, features for each vertex on the posed mesh from multiple images are integrated by MLPs. The integrated features acting as the latent code are anchored to the SMPLX mesh in the canonical space. Finally, latent code for each 3D point is extracted and utilized to calculate the SDF. Our work for reconstructing the human shape on canonical pose achieves 3rd performance on WCPA MVP-Human Body Challenge.

Keywords: Implicit representation · SMPLX · Signed distance function · Latent codes fusion

1 Introduction

High-fidelity human digitization has attracted a lot of interest for its application in VR/AR, image and video editing, telepresence, virtual try-on, etc. In this

---

J. Chen, W. Yi, T. Wang and X. Li—Equal contributions.

Supplementary Information The online version contains supplementary material available at https://doi.org/10.1007/978-3-031-25072-9_24.

L. Karlinsky et al. (Eds.): ECCV 2022 Workshops, LNCS 13805, pp. 366–375, 2023.
https://doi.org/10.1007/978-3-031-25072-9_24

work, we target at reconstructing the high-quality 3D clothed humans in the canonical space given multiple views and multiple poses of a human performer as the input.

Our network utilizes multiple images as the input and learns the implicit representation for the given points in the 3D space. Inspired by the advantages of implicit representation such as arbitrary topology and continuous representation, we adopt this representation to reconstruct high-fidelity clothed 3D humans. To learn geometry in the canonical space, we utilize the SMPLX mesh [15] in the canonical space as the geometric guidance. Due to the correspondence between the posed mesh and the mesh in the canonical space, we propose to first learn the latent codes in the posed mesh and then assign the latent codes to the canonical mesh based on the correspondence. By utilizing the posed mesh, image information in the 2D space can be included by projecting the posed mesh to the image space.

Given the multi-view images as input, normal and geometry networks are utilized to extract the features for the vertices on the SMPLX mesh [15]. We utilize a normal map as the intermediate prediction which helps generate sharp reconstructed geometry [21].

To integrate the features from different views or poses, we utilize a fusion network to generate a weighted summation of multiple features. Specifically, we first concatenate the features with the means and variances of features from all inputs followed by a Multi-layer Perceptron (MLP) predicting the weight and transformed features. Then, the weighted features will be integrated into the latent code through another MLP.

The latent code learned by the neural network is anchored to the SMPLX mesh in the canonical space, which serves as the geometry guidance to reconstruct the 3D geometry. Because of the sparsity of the vertices, we utilize a SparseConvNet to generate effective features for any 3D point following [19]. Finally, we use the trilinear interpolation to extract the latent code followed by an MLP to produce the SDF which models the signed distance of a point to the nearest surface.

2   Related Work

Implicit Neural Representations. Recently, the implicit representation encoded by a deep neural network has gained extensive attention, since it can represent continuous fields and can generate the details on the clothes, such as wrinkles and facial expressions. Implicit representation has been applied successfully to shape representation [16,18,26]. Here we utilize the signed distance function (SDF) as the implicit representation, which is a continuous function that outputs the point's signed distance to the closest surface, whose sign encodes whether the point is inside (negative) or outside (positive) of the watertight surface. The underlying surface is implicitly represented by the isosurface of $SDF(\cdot) = 0$.

3D Clothed Human Reconstruction. Reconstructing humans has been widely studied given the depth maps [4,22,27], images [2,10,14], or videos [11,13]

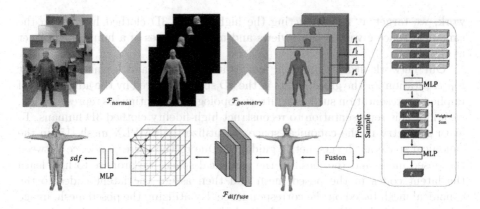

Fig. 1. Overview of our clothed human reconstruction pipeline.

as the input. Utilizing the SMPLX mesh, the existing works show promising results with the RGB image as the input. NeuralBody [19] adopts the posed SMPLX mesh to construct the latent code volume that aims to extract the implicit pose code. ICON [24] proposes to refine the SMPLX mesh and normal map iteratively. SelfRecon [9] utilizes the initial canonical SMPLX body mesh to calculate the canonical implicit SDF. Motivated by these methods, we propose to utilize the SMPLX mesh in the canonical space as a geometric prior to reconstruct the clothed humans.

3   Methodology

Given the multi-view and multi-pose RGB images of human and SMPLX parameters as the input, we aim to reconstruct the clothed 3D geometry in the canonical space. Here the input images are denoted as $\{I_k\}_{k=1}^K$, where k denotes the image index and K is the number of images. The corresponding SMPLX parameters are denoted as $\{\theta_k, \beta_k, s_k, t_k\}_{k=1}^K$, where θ_k, β_k are the pose and shape parameters of SMPLX and s_k, t_k are the camera parameters used for projection.

3.1   Multi-view and Multi-pose Image Feature Encoding

Our feature extraction networks utilizes multi-view and multi-pose images $\{I_k\}_{k=1}^K$ as input and outputs the geometric feature maps that help to predict the 3D geometry in the canonical space. Specifically, we first adopt the normal network $\mathcal{F}_{normal}$ to extract the normal maps and then utilize the geometry network $\mathcal{F}_{geo}$ to generate geometric feature maps $\{F_k\}_{k=1}^K$ that will be further utilized to extract the pixel-aligned features for the vertices on the posed mesh.

$$F_k = \mathcal{F}_{\text{geometry}}\left(\mathcal{F}_{\text{normal}}\left(I_k\right)\right) \quad k = 1, 2, \cdots, K \tag{1}$$

In particular, we adopt the pretrained image-to-image translation network from PIFuHD [21] as our normal network, and use the pretrained ResUNet34 [8] backbone as our geometry network.

3.2 Structured Latent Codes

After obtaining the geometric feature maps $\{F_k\}_{k=1}^{K}$, we extract the pixel-aligned features for each vertex $\mathbf{v}_k^i$ on the posed mesh $M(\theta_k, \beta_k)$. For each vertex $\mathbf{v}_k^i$, we first project it to the image space by utilizing the weak-perspective projection Φ according to camera parameters scale s_k and translation t_k, then adopt the bilinear interpolation operation to extract the pixel-aligned features $\boldsymbol{f}_k^i$.

$$\Phi : \hat{\mathbf{x}} = s\Pi(\mathbf{x}) + t, \quad \mathbf{x} \in \mathbb{R}^3, \hat{\mathbf{x}} \in \mathbb{R}^2 \tag{2}$$

$$\boldsymbol{f}_k^i = \text{bilinear}\left(F_k, \Phi\left(\mathbf{v}_k^i, s_k, t_k\right)\right) \tag{3}$$

where Π is the orthogonal projection, $\mathbf{x}$ is the point in 3D space and $\hat{\mathbf{x}}$ is the projected points in 2D image space.

To integrate the feature of the i-th vertex in the canonical space from multiple views/poses, we use a fusion network that takes $\{\boldsymbol{f}_k^i\}_{k=1}^{K}$ as the input and outputs the integrated feature $\boldsymbol{l}^i$, which is illustrated in Fig. 1. Specifically, the mean $\boldsymbol{\mu}^i$ and variance $\boldsymbol{\sigma}^i$ of features $\{\boldsymbol{f}_k^i\}_{k=1}^{K}$ is calculated and then concatenated with $\boldsymbol{f}_k^i$ to serve as the input of a MLP. The MLP predicts the new feature vector and weight $\{w_k^i\}_{k=1}^{K}$ for each feature, which generates a weighted sum of features from multiple inputs. The weighted feature is finally forwarded to an MLP for feature integration, $\boldsymbol{l}^i$ which serves as the structured latent code for the vertex $\mathbf{v}^i$.

Different from the latent code in NeuralBody [19] which is initialized randomly for optimizing specific humans, our latent code is the feature vector learned by a network with the normal map as the input, which can generalize to humans unseen from training ones.

The above-mentioned latent codes are generated based on the posed mesh. The posed mesh and the canonical mesh share the same latent codes because of their topology correspondence which forms the set of latent codes by $\mathcal{Q} = \{\boldsymbol{l}^1, \boldsymbol{l}^2, \cdots, \boldsymbol{l}^{N_V}\}, \boldsymbol{l}^i \in \mathbb{R}^d$. Here N_V represents the number of vertices.

3.3 Implicit Neural Shape Field

The learned latent codes are anchored to a human body model (SMPLX [15]) in the canonical space. SMPLX is parameterized by shape and pose parameters with $N_V = 10,475$ vertices and $N_J = 54$ joints. The locations of the latent codes $\mathcal{Q} = \{\boldsymbol{l}^1, \boldsymbol{l}^2, \cdots, \boldsymbol{l}^{N_V}\}$ are transformed for learning the implicit representation by forwarding the latent codes into a neural network

To query the latent code at continuous 3D locations, trilinear interpolation is adopted for each point. However, the latent codes are relatively sparse in the 3D space, and directly calculating the latent codes using trilinear interpolation will generate zero vectors for most points. To overcome this challenge, we use a SparseConvNet [19] to form a latent feature volume $\mathcal{V} \in \mathbb{R}^{H \times H \times H \times d}$ which diffuses the codes defined on the mesh surface to the nearby 3D space.

$$\mathcal{V} = \mathcal{F}_{diffuse}\left(\{l^1, l^2, \ldots, l^{N_V}\}\right) \tag{4}$$

Specifically, to obtain the latent code for each 3D point, trilinear interpolation is employed to query the code at continuous 3D locations.

Here the latent code will be forwarded into a neural network φ to predict the SDF $\mathcal{F}_{sdf}(\mathbf{x})$ for 3D point $\mathbf{x}$.

$$\mathcal{F}_{sdf}(\mathbf{x}) = \varphi(\text{ trilinear } (\mathcal{V}, \mathbf{x})) \tag{5}$$

3.4   Loss Function

During training, we sample N_S spatial points $\mathcal{X}$ surrounding the ground truth canonical mesh. To train the implicit SDF, we deploy a mixed-sampling strategy: 20% for uniform sampling on the whole space and 80% for sampling near the surface. We adopt the mixed-sampling strategy because of the following two reasons. First, sampling uniformly in the 3D space will put more weight on the points outside the mesh during network training, which results in overfitting when sampling around the iso-surface. Second, sampling points far away from the reconstructed surface contribute little to geometry reconstruction, which increases the pressure of network training.

Overall, we enforce 3D geometric loss L_{sdf} and normal constraint loss L_{normal}.

$$\mathcal{L} = \lambda_{sdf}\mathcal{L}_{sdf} + \lambda_{normal}\mathcal{L}_{normal} \tag{6}$$

3D Geometric Loss. Given a sampling point $\mathbf{x} \in \mathcal{X}$, we employ the L2 loss between the predicted $\mathcal{F}_{sdf}(\mathbf{x})$ and the ground truth $\mathcal{G}_{sdf}(\mathbf{x})$, which are truncated by a threshold δ,

$$\mathcal{L}_{sdf} = \frac{1}{N_S} \sum_{\mathbf{x} \in \mathcal{X}} \|\mathbb{C}(\mathcal{F}_{sdf}(\mathbf{x}), \delta) - \mathbb{C}(\mathcal{G}_{sdf}(\mathbf{x}), \delta)\|_2. \tag{7}$$

Here $\mathbb{C}(\cdot, \delta) = max(-\delta, min(\cdot, \delta))$.

Normal Constraint Loss. Beyond the geometric loss, to make the predicted surface smoother, we deploy Eikonal loss [6] to encourage the gradients of the sampling points to be close to 1 (Fig. 2).

$$\mathcal{L}_{normal} = \frac{1}{N_S} \sum_{\mathbf{x} \in \mathcal{X}} \left\| \|\nabla_{\mathbf{x}}\mathcal{F}_{sdf}(\mathbf{x})\|_2 - 1 \right\|_2. \tag{8}$$

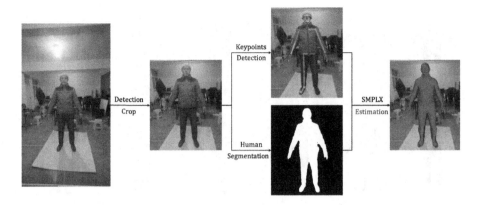

Fig. 2. The pipeline for data processing.

4   Experiments

4.1   Datasets

We use the WCPA [1] dataset as training and testing datasets, which consists of 200 subjects for training and 50 subjects for testing. Each subject contains 15 actions, and each action contains 8 RGB images from different angles (0, 45, 90, 135, 180, 225, 270, 315). Each image is an 1280×720 jpeg file. The ground truth of the canonical pose is a high-resolution 3D mesh with detailed information about clothes, faces, etc. For the training phase, we randomly select 4 RGB images with different views and poses as inputs to learn 3D human models.

4.2   Image Preprocessing

Image Cropping. For each image, we first apply VarifocalNet [28] to detect the bounding boxes to localize the humans. Next, we crop the input images with the resolution of 512×512 according to the bounding boxes. When the cropped images exceed the bounds of the input images, the cropped images will be padded with zeros.

Mask Generation. For each image, we first apply DensePose [7] to obtain part segmentations. Then we set the parts on the human as the foreground human mask and set the values of the background image pixels as zero. The masked images are served as the input of our network.

The mask is then refined using the MatteFormer [17], which can generate better boundary details. Following [23], the trimap adopted in [17] is generated based on the mask using the erosion and dilation operation.

4.3   SMPLX Estimation and Optimization

It is very challenging to get accurate SMPLX from 2D RGB images due to the inherent depth ambiguity. First, we use Openpose [3] to detect the 2D keypoints

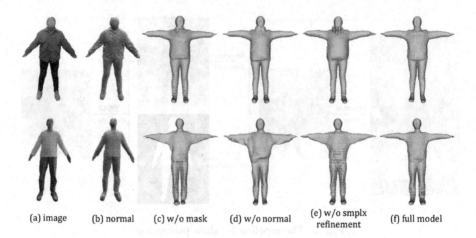

(a) image   (b) normal   (c) w/o mask   (d) w/o normal   (e) w/o smplx refinement   (f) full model

Fig. 3. The visualization results on the test dataset.

of the person in the image and then use ExPose [5] to estimate the SMPLX parameters and camera parameters. However, due to extreme illumination conditions and complex backgrounds, the SMPLX obtained in this way is not sufficiently accurate, and we need to refine the SMPLX parameters further. We utilize the 2D keypoints and masks to optimize SMPLX parameters θ, β. For the 2D keypoints loss, given the SMPLX 3D joints location $J(\theta, \beta)$, we project them to the 2D image using the weak-perspective camera parameters s, t. For the mask loss, we utilize PyTorch3D [20] to render the 2D mask given the posed mesh $M(\theta, \beta)$. Then the mask loss is calculated based on the rendered mask and the pseudo ground truth mask $\mathcal{M}_{gt}$.

$$\theta^*, \beta^* = \min_{\theta, \beta} \|\Phi(J(\theta, \beta), s, t) - \mathcal{J}_{gt}\| + \lambda \|\Psi(M(\theta, \beta), s, t) - \mathcal{M}_{gt}\| \quad (9)$$

4.4   Implementation Details

During training, we randomly choose $K = 4$ images from total of $8 * 15$ images as input to extract the 2D feature map. Taking into account the memory limitation, we set the latent feature volume $\mathcal{V} \in \mathbb{R}^{224 \times 224 \times 224 \times 64}$, where each latent code $l^i \in \mathbb{R}^{64}$. During training, we randomly sample $N_t = 10,000$ points around the complete mesh. To stably train our network, we initialize the SDF to approximate a unit sphere [25]. We adopt the Adam optimizer [12] and set the learning rate $lr = 5e - 4$, and it spends about 40 h on 2 Nvidia GeForce RTX 3090 24 GB GPUs. For inference, the surface mesh is extracted by the zero-level set of SDF by running marching cubes on a 256^3 grid.

4.5   Results

Our method achieved very good results in the challenge, demonstrating the superiority of our method for 3D human reconstruction from multi-view and multi-pose images. To further analyze the effectiveness of our method, we performed different ablation experiments. According to the results as shown in the Table 1 and the Fig. 3, removing the background allows the model to better reconstruct the human body. The two-stage approach of predicting the normal vector map as input has a stronger generalization line on unseen data compared to directly using the image as input. What is more important is that the precision of SMPLX has a significant impact on the reconstruction performance, demonstrating the necessity of SMPLX optimization.

Table 1. Quantitative metrics of different strategies on the test dataset.

	w/o mask	w/o normal map	w/o SMPLX refinement	Full model
Chamfer distance ↓	1.1277	1.0827	1.1285	**0.9985**

5   Conclusion

Modeling 3D humans accurately and robustly from challenging multi-views and multi-posed RGB images is a challenging problem, due to the varieties of body poses, viewpoints, light conditions, and other environmental factors. To this end, we have contributed a deep-learning based framework to incorporate the parametric SMPLX model and non-parametric implicit function for reconstructing a 3D human model from multi-images. Our key idea is to overcome these challenges by constructing structured latent codes as the inputs for implicit representation. The latent codes integrate the features for vertices in the canonical pose from different poses or views.

References

1. Eccv 2022 WCPA challenge (2022). https://tianchi.aliyun.com/competition/entrance/531958/information
2. Bogo, F., Kanazawa, A., Lassner, C., Gehler, P., Romero, J., Black, M.J.: Keep It SMPL: automatic estimation of 3D human pose and shape from a single image. In: Leibe, B., Matas, J., Sebe, N., Welling, M. (eds.) ECCV 2016. LNCS, vol. 9909, pp. 561–578. Springer, Cham (2016). https://doi.org/10.1007/978-3-319-46454-1_34
3. Cao, Z., Hidalgo, G., Simon, T., Wei, S., Sheikh, Y.: OpenPose: realtime multi-person 2d pose estimation using part affinity fields. IEEE Trans. Pattern Anal. Mach. Intell. **43**, 172–186 (2021)
4. Chibane, J., Alldieck, T., Pons-Moll, G.: Implicit functions in feature space for 3D shape reconstruction and completion. In: Proceedings of the IEEE/CVF Conference on Computer Vision and Pattern Recognition, pp. 6970–6981 (2020)
5. Choutas, V., Pavlakos, G., Bolkart, T., Tzionas, D., Black, M.J.: Monocular expressive body regression through body-driven attention. In: Vedaldi, A., Bischof, H., Brox, T., Frahm, J.-M. (eds.) ECCV 2020. LNCS, vol. 12355, pp. 20–40. Springer, Cham (2020). https://doi.org/10.1007/978-3-030-58607-2_2
6. Gropp, A., Yariv, L., Haim, N., Atzmon, M., Lipman, Y.: Implicit geometric regularization for learning shapes. arXiv preprint arXiv:2002.10099 (2020)

7. Güler, R.A., Neverova, N., Kokkinos, I.: DensePose: dense human pose estimation in the wild. In: Proceedings of the IEEE Conference on Computer Vision and Pattern Recognition, pp. 7297–7306 (2018)
8. He, K., Zhang, X., Ren, S., Sun, J.: Deep residual learning for image recognition. In: Proceedings of the IEEE Conference on Computer Vision and Pattern Recognition, pp. 770–778 (2016)
9. Jiang, B., Hong, Y., Bao, H., Zhang, J.: SelfRecon: self reconstruction your digital avatar from monocular video. In: Proceedings of the IEEE/CVF Conference on Computer Vision and Pattern Recognition, pp. 5605–5615 (2022)
10. Kanazawa, A., Black, M.J., Jacobs, D.W., Malik, J.: End-to-end recovery of human shape and pose. In: Proceedings of the IEEE Conference on Computer Vision and Pattern Recognition, pp. 7122–7131 (2018)
11. Kanazawa, A., Zhang, J.Y., Felsen, P., Malik, J.: Learning 3D human dynamics from video. In: Proceedings of the IEEE/CVF Conference on Computer Vision and Pattern Recognition, pp. 5614–5623 (2019)
12. Kingma, D.P., Ba, J.: Adam: a method for stochastic optimization. arXiv preprint arXiv:1412.6980 (2014)
13. Kocabas, M., Athanasiou, N., Black, M.J.: VIBE: video inference for human body pose and shape estimation. In: Proceedings of the IEEE/CVF Conference on Computer Vision and Pattern Recognition, pp. 5253–5263 (2020)
14. Kolotouros, N., Pavlakos, G., Black, M.J., Daniilidis, K.: Learning to reconstruct 3d human pose and shape via model-fitting in the loop. In: Proceedings of the IEEE/CVF International Conference on Computer Vision, pp. 2252–2261 (2019)
15. Loper, M., Mahmood, N., Romero, J., Pons-Moll, G., Black, M.J.: SMPL: a skinned multi-person linear model. ACM Trans. Graphics **34**(6), 1–16 (2015)
16. Mescheder, L., Oechsle, M., Niemeyer, M., Nowozin, S., Geiger, A.: occupancy networks: Learning 3d reconstruction in function space. In: Proceedings of the IEEE/CVF Conference on Computer Vision and Pattern Recognition, pp. 4460–4470 (2019)
17. Park, G., Son, S., Yoo, J., Kim, S., Kwak, N.: MatteFormer: transformer-based image matting via prior-tokens. In: Proceedings of the IEEE/CVF Conference on Computer Vision and Pattern Recognition (CVPR), pp. 11696–11706 (June 2022)
18. Park, J.J., Florence, P., Straub, J., Newcombe, R., Lovegrove, S.: DeepSDF: learning continuous signed distance functions for shape representation. In: Proceedings of the IEEE/CVF Conference on Computer Vision and Pattern Recognition, pp. 165–174 (2019)
19. Peng, S., et al.: Neural body: Implicit neural representations with structured latent codes for novel view synthesis of dynamic humans, In: CVPR (2021)
20. Ravi, N., et al.: Accelerating 3d deep learning with pytorch3d. arXiv preprint arXiv:2007.08501 (2020)
21. Saito, S., Simon, T., Saragih, J., Joo, H.: PIFuHD: multi-level pixel-aligned implicit function for high-resolution 3d human digitization. In: Proceedings of the IEEE/CVF Conference on Computer Vision and Pattern Recognition, pp. 84–93 (2020)
22. Wang, L., Zhao, X., Yu, T., Wang, S., Liu, Y.: NormalGAN: learning detailed 3d human from a single RGB-D image. In: Vedaldi, A., Bischof, H., Brox, T., Frahm, J.-M. (eds.) ECCV 2020. LNCS, vol. 12365, pp. 430–446. Springer, Cham (2020). https://doi.org/10.1007/978-3-030-58565-5_26
23. Wang, T., Liu, S., Tian, Y., Li, K., Yang, M.H.: Video matting via consistency-regularized graph neural networks. In: Proceedings of the IEEE/CVF International Conference on Computer Vision, pp. 4902–4911 (2021)

24. Xiu, Y., Yang, J., Tzionas, D., Black, M.J.: Icon: Implicit clothed humans obtained from normals. In: Proceedings of IEEE Conference on Computer Vision and Pattern Recognition (CVPR). vol. 2 (2022)

25. Yariv, I., et al.: Multiview neural surface reconstruction by disentangling geometry and appearance. Adv. Neural. Inf. Process. Syst. **33**, 2492–2502 (2020)

26. Yifan, W., Wu, S., Oztireli, C., Sorkine-Hornung, O.: ISO-points: optimizing neural implicit surfaces with hybrid representations. In: Proceedings of the IEEE/CVF Conference on Computer Vision and Pattern Recognition, pp. 374–383 (2021)

27. Yu, T., et al.: Doublefusion: Real-time capture of human performances with inner body shapes from a single depth sensor. In: Proceedings of the IEEE Conference on Computer Vision and Pattern Recognition, pp. 7287–7296 (2018)

28. Zhang, H., Wang, Y., Dayoub, F., Sunderhauf, N.: Varifocalnet: An IOU-aware dense object detector. In: Proceedings of the IEEE/CVF Conference on Computer Vision and Pattern Recognition, pp. 8514–8523 (2021)

UnconFuse: Avatar Reconstruction from Unconstrained Images

Han Huang[1]([✉])(iD), Liliang Chen[1,2](iD), and Xihao Wang[1,3]

[1] OPPO Research Institute, Beijing, China
huanghan@oppo.com
[2] University of Pittsburgh, Pittsburgh, USA
[3] Xidian University, Xian, China

Abstract. The report proposes an effective solution about 3D human body reconstruction from multiple unconstrained frames for ECCV 2022 WCPA Challenge: From Face, Body and Fashion to 3D Virtual avatars I (track1: Multi-View Based 3D Human Body Reconstruction). We reproduce the reconstruction method presented in MVP-Human [19] as our baseline, and make some improvements for the particularity of this challenge. We finally achieve the score 0.93 on the official testing set, getting the 1st place on the leaderboard.

Keywords: 3D human reconstruction · Avatar · Unconstrained frames

1 Method

The purpose of this challenge is to explore various methods to reconstruct a high-quality T-pose 3D human in the canonical space from multiple unconstrained frames where the character can have totally different actions. Many existing works [3,11,12,14,18] have proved that pixel-aligned image features are significant to recover details of the body surface, and spatially-aligned local features queried from multi-view images can greatly improve the robustness of the reconstructions since more regions of the body can be observed. These methods use well-calibrated camera parameters to ensure the accurate inter-view alignment and 3D-to-2D alignment, while the setup of multiple unconstrained frames makes it much difficult to achieve precision alignments. Recent years, tracking based methods are explored to reconstruct human with single-view video or unconstrained multiple frames [1,8,15,18,19]. To our knowledge, the research purpose of MVP-Human [19] is most related to our challenge, aiming to reconstructing an avatar consisting of geometry and skinning weights [5] from multiple unconstrained images. We reproduce the methods proposed in MVP-Human and make some modifications. In this section, we will deliver a introduction about the details of our pipeline and experiments settings through which we have achieved a fairly high accuracy on the challenging MVP-Human dataset [19].

In general, we query pixel-aligned image features from several images and fuse them through parametric body motion. Then, occupancy field of the canonical space are predicted from these fused features. Like MVP-Human, normals from

L. Karlinsky et al. (Eds.): ECCV 2022 Workshops, LNCS 13805, pp. 376–381, 2023.
https://doi.org/10.1007/978-3-031-25072-9_25

SMPL models are also introduced to improve the reconstruction quality. Considering the requirements of this challenge, we focus more on geometry reconstruction quality in canonical space and only use one backbone for both occupancy and skinning weights prediction. Our experiments demonstrate that our data preprocessing, data augmentation, model architecture design, feature aggregation strategy and online hard example mining (OHEM) [13] all contribute to higher score.

1.1 Data Preprocessing

Data preprocessing is the basis of the entire task, which strongly affects the reconstruction accuracy. We perform the following processes for training and testing data:

Human Detection and Image Crop. The size of the raw images is 1280 × 720 and the character in the image is lying down. We rotate the images and detect the human bounding box with off-the-shelf method HumanDet [4]. To make images more friendly to encoder networks, we crop the them to 512 × 512 and ensure the character roughly on the center of the image.

Segmentation. We extract the foreground mask with the resolution 512 × 512 through PGN [2] .

SMPL Fitting. We apply the state-of-the-art method PyMAF [16] to fit SMPL parameters for each image. Ideally, one character with different actions or viewpoint should share the same SMPL shape, but the shape parameters extracted frame by frame are slightly different. To regularize the shape consistency of the same character, we take a joint optimization strategy for SMPL fitting, achieving only one set of shape parameters for each character. The fitting error at the beginning and the ending of iteration of the joint optimization process are respectively shown in Fig. 1 and Fig. 2, we can see that the fitting error have been definitely minimized when joint optimization finished.

Sampling. Like the training strategy proposed in PIFu [11], we sample points in the 3D space and around the iso-surface. For each training iteration, we sample 14756 points close to the ground-truth body surface and 1628 points randomly distributed in a 3D bounding box containing the ground-truth body.

Normals. In order to better utilize the prior knowledge of the SMPL model, we calculate the point-wise normals of the SMPL, and use K-Nearest Neighbor (KNN) algorithm to obtain the normal information of all sampling points.

Dataset Split. Considering that the challenge doesn't provide the ground-truth for testing set, we manually split the training data to be training set and validation set. Specifically, we randomly select 190 subjects from all 200 subjects for training, and the rest 10 subjects for validation.

Data Augmentation. In order to achieve a model with better generalizability, we attempt to enhance the diversity of the training data by image augmentation which simulates various texture, material or lighting conditions at the image level. We randomly jitter the brightness, hue, saturation, and contrast where brightness factor, saturation and contrast is all chosen as 0.2, hue is set to be 0.05.

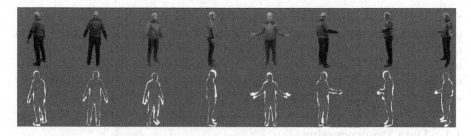

Fig. 1. Part of results at the beginning of the iteration. On the Bottom row, the white contour represents the fitting error and the thinner the contour width, the smaller the error is.

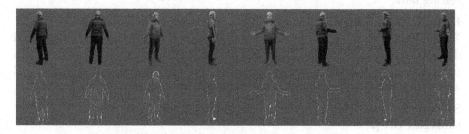

Fig. 2. Part of results at the end of the iteration. On the Bottom row, the white contour represents the fitting error and the thinner the contour width, the smaller the error is.

1.2   Model Design and Configuration

The architecture of our method is shown in Fig. 3. As mentioned above, this challenge aims to reconstruct high quality T-pose human in the canonical space with detailed garment from multiple unconstrained images. Referring to the framework of MVP-Human [19], voxels or sampling points in canonical spaces are warped with SMPL motion to align with posed images. Then, occupancy and skinning weights are predicted from pixel-aligned features through MLPs. Finally, the explicit mesh will be extracted through Marching Cubes [9]. The loss functions for training our model is quite similar to the ones used in PIFu and MVP-Human. Since this challenge care more about the geometry in canonical space, we don't assign a specialized backbone for skinning weight prediction. Instead of the stacked hourglass network used in MVP-Human, we introduce HRNet as image encoder that shows better reconstruction accuracy and also faster runtime according to MonoPort [6,7].

In addition, we perform the position encoding [10] to the 3D coordinates of queried points in the canonical space and concatenate them with corresponding local image features and noramls extracted from SMPL body surface.

In our baseline method, we simply apply average pooling to fuse features from different frames. In order to weigh the contribution of image features from different view-point to a 3D point, we introduce the self-attention mechanism

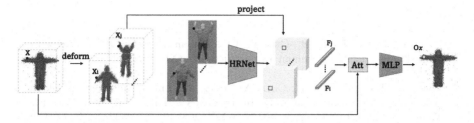

Fig. 3. The overview of our model architecture. Points in the canonical space are warped with SMPL motion to align with unconstrained multi-frame images. Then, local image features are fused by self-attention. Finally, the occupancy or skinning weights are predicted from fused features through MLPs.

proposed in DeepMultiCap [17] for feature aggregation. Our experiments proved that such feature fusion strategy leads to better performance in this task.

2 Results

As shown in Table 1, after all strategies mentioned above are used, our model finally achieve score (chamfer distance with ground-truth mesh) 0.93 on the testing set.

Table 1. Quantitative results of our methods

Our methods with different settings	Chamfer↓
Our baseline	1.13
Baseline + PE	1.09
Baseline + PE + Att	0.96
Baseline + PE + Att + DA	0.95
Our final model (baseline + PE + Att + DA + HM)	0.93

PE: position encoding for points coordinates, Att: self-attention for multi-frame feature fusion, DA: data augmentation, HM: online hard case mining.

3 Discussion

We appreciate the challenge for providing this valuable academic exchange opportunities. The novel and interesting setup contributes to the innovation of this specific research area. We are glad to explore the avatar reconstruction methods and lead the race. Though our quantitative results look good and get 1st place on the leaderboard, the meshes generated by our method are still very smooth without high-fidelity details. We will further improve our methods for better reconstruction quality under limited data.

References

1. Chen, L., Li, J., Huang, H., Guo, Y.: Crosshuman: learning cross-guidance from multi-frame images for human reconstruction. arXiv preprint arXiv:2207.09735 (2022)
2. Gong, K., Liang, X., Li, Y., Chen, Y., Yang, M., Lin, L.: Instance-level human parsing via part grouping network. In: Ferrari, V., Hebert, M., Sminchisescu, C., Weiss, Y. (eds.) ECCV 2018. LNCS, vol. 11208, pp. 805–822. Springer, Cham (2018). https://doi.org/10.1007/978-3-030-01225-0_47
3. Huang, Z., Xu, Y., Lassner, C., Li, H., Tung, T.: Arch: Animatable reconstruction of clothed humans. In: Proceedings of the IEEE/CVF Conference on Computer Vision and Pattern Recognition (CVPR) (June 2020)
4. HumanDet (2020). https://github.com/Project-Splinter/human_det
5. Kavan, L., Collins, S., Žára, J., O'Sullivan, C.: Skinning with dual quaternions. In: Proceedings of the 2007 Symposium on Interactive 3D Graphics and Games, pp. 39–46 (2007)
6. Li, R., Olszewski, K., Xiu, Y., Saito, S., Huang, Z., Li, H.: Volumetric human teleportation. In: ACM SIGGRAPH 2020 Real-Time Live, pp. 1–1 (2020)
7. Li, R., Xiu, Y., Saito, S., Huang, Z., Olszewski, K., Li, H.: Monocular real-time volumetric performance capture. In: Vedaldi, A., Bischof, H., Brox, T., Frahm, J.-M. (eds.) ECCV 2020. LNCS, vol. 12368, pp. 49–67. Springer, Cham (2020). https://doi.org/10.1007/978-3-030-58592-1_4
8. Li, Z., Yu, T., Zheng, Z., Guo, K., Liu, Y.: Posefusion: pose-guided selective fusion for single-view human volumetric capture. In: IEEE Conference on Computer Vision and Pattern Recognition (June 2021)
9. Lorensen, W.E., Cline, H.E.: Marching cubes: a high resolution 3D surface construction algorithm. ACM SIGGRAPH Comput. Graphics **21**(4), 163–169 (1987)
10. Mildenhall, B., Srinivasan, P.P., Tancik, M., Barron, J.T., Ramamoorthi, R., Ng, R.: NeRF: representing scenes as neural radiance fields for view synthesis. In: Vedaldi, A., Bischof, H., Brox, T., Frahm, J.-M. (eds.) ECCV 2020. LNCS, vol. 12346, pp. 405–421. Springer, Cham (2020). https://doi.org/10.1007/978-3-030-58452-8_24
11. Saito, S., Huang, Z., Natsume, R., Morishima, S., Kanazawa, A., Li, H.: PIFu: Pixel-aligned implicit function for high-resolution clothed human digitization. In: Proceedings of the IEEE/CVF International Conference on Computer Vision, pp. 2304–2314 (2019)
12. Saito, S., Simon, T., Saragih, J., Joo, H.: PIFuHD: multi-level pixel-aligned implicit function for high-resolution 3D human digitization. In: Proceedings of the IEEE/CVF Conference on Computer Vision and Pattern Recognition (CVPR) (June 2020)
13. Shrivastava, A., Gupta, A., Girshick, R.: Training region-based object detectors with online hard example mining. In: Proceedings of the IEEE Conference on Computer Vision and Pattern Recognition, pp. 761–769 (2016)
14. Xiu, Y., Yang, J., Tzionas, D., Black, M.J.: ICON: implicit Clothed humans Obtained from Normals. In: Proceedings of the IEEE/CVF Conference on Computer Vision and Pattern Recognition (CVPR). pp. 13296–13306 (June 2022)
15. Yu, T., et al.: DoubleFusion: real-time capture of human performances with inner body shapes from a single depth sensor. In: The IEEE International Conference on Computer Vision and Pattern Recognition(CVPR). IEEE (June 2018)

16. Zhang, H., et al.: 3D human pose and shape regression with pyramidal mesh alignment feedback loop. In: Proceedings of the IEEE/CVF International Conference on Computer Vision, pp. 11446–11456 (2021)
17. Zheng, Y., et al.: DeepMultiCap: performance capture of multiple characters using sparse multiview cameras. In: IEEE Conference on Computer Vision (ICCV 2021) (2021)
18. Zheng, Z., Yu, T., Liu, Y., Dai, Q.: Pamir: Parametric model-conditioned implicit representation for image-based human reconstruction. IEEE Trans. Pattern Anal. Mach. Intell. (99), 1–1 (2021). https://doi.org/10.1109/TPAMI.2021.3050505
19. Zhu, X., et al.: MVP-human dataset for 3D human avatar reconstruction from unconstrained frames. arXiv preprint arXiv:2204.11184 (2022)

HiFace: Hybrid Task Learning for Face Reconstruction from Single Image

Wei Xu[1,2], Zhihong Fu[2(✉)], Zhixing Chen[2], Qili Deng[2], Mingtao Fu[2],
Xijin Zhang[2], Yuan Gao[2], Daniel K. Du[2], and Min Zheng[2]

[1] Beijing University of Posts and Telecommunications, Beijing, China
xuwei2020@bupt.edu.cn
[2] ByteDance Inc., Beijing, China
{fuzhihong.2022,chenzhixing.omega,dengqili,fumingtao.fmt,zhangxijin,
gaoyuan.azeroth,dukang.daniel,zhengmin.666}@bytedance.com

Abstract. The task of 3D face reconstruction in the WCPA challenge requires a monocular image as input and outputs 3D face geometry, which has been a prevalent field for decades. Considerable works have been published, in which PerspNet significantly outperforms the other methods under perspective projection. However, as the UV coordinates distribute unevenly, the UV mapping process introduces inevitable precision degradation in dense regions of reconstructed 3D faces. Thus, we design a vertex refinement module to overcome the precision degradation. We also design a multi-task learning module to enhance 3D features. By carefully designing and organizing the vertex refinement module and the multi-task learning module, we propose a hybrid task learning based 3D face reconstruction method called HiFace. Our HiFace achieves the 2nd place in the final official ranking of the ECCV 2022 WCPA Challenge, which demonstrates the superiority of our HiFace.

Keywords: 3D face reconstruction · Perspective projection · 6DoF pose estimation

1 Introduction

3D human face modeling enables various applications such as interactive face editing, video-conferencing, extended reality, metaverse applications, and visual effects for movie post-production. However, most 3D face reconstruction methods employ orthogonal projection, which limits the application's potential. The WCPA challenge requires modeling the facial distortion caused by perspective projection. To our knowledge, PerspNet [10] achieves great success in this area and is selected as the official baseline. However, PerspNet suffers from severe precision degradation in dense regions such as eye edges and mouse corners when employing UV mapping due to the uneven distribution of 3D vertices.

To address this, we design a vertex refinement module that is capable of getting all 3D vertices more accurate by predicting extra offsets. Moreover, in order

L. Karlinsky et al. (Eds.): ECCV 2022 Workshops, LNCS 13805, pp. 382–391, 2023.
https://doi.org/10.1007/978-3-031-25072-9_26

to enhance 3D features, we propose a multi-task learning module. Specifically, it learns to regress the normal UV map, depth map, and normal map for each single image. Since the three of them each have rich 3D structure information, the networks are guided to learn more semantic 3D features. Notably, the learning module is only used in the training phase and would be removed in the inference phase, which improves the face reconstruction performance while keeping zero inference cost.

By carefully designing and organizing the vertex refinement module and the multi-task learning module, we propose a hybrid task learning based 3D face reconstruction method called HiFace. We conduct extensive experiments that indicate the effectiveness of the proposed modules and the superiority of our HiFace.

Summarily, the main contributions of our work in the WCPA challenge are three-fold.

1. We propose a vertex refinement module to overcome the precision degradation introduced by the UV mapping process.
2. We propose a multi-task learning module that guides the networks to enhance 3D features without additional inference costs.
3. Extensive experiments and the final official ranking show that our method outperforms a significant number of approaches by a large margin on the test set, demonstrating the superiority of our method.

2   Related Work

We focus on the 3D face reconstruction methods that only use single-image as the input under perspective projection.

2.1   3DMM Based

Most of the related methods [1,7,15–17] rely on 3D Morphable Model (3DMM) [2], which is a powerful reconstruction tool due to its effective control of facial space. However, the statistical 3DMM adds structure and priors to the face reconstruction process. It sets representation limitations, especially when requiring fine-grained facial details. S2F2 [5] proposes to use coarse-to-fine fashion to deliver high-fidelity face reconstruction from multi inference stages. It achieves visually appealing reconstruction. However, the reference mode also restricts the reconstructed face geometry from the S2F2.

2.2   Model-Free

Some methods [6,9,10] propose to achieve model-free reconstruction. Specifically, [9] maps the image pixels to full 3D facial geometry via volumetric CNN regression. [6] reconstructs full facial structure along with semantic meaning. [10] proposes Perspective Network (PerspNet) to recover 3D facial geometry under

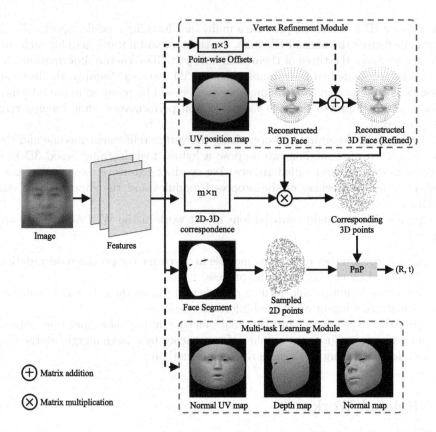

Fig. 1. The overview of our method. The pipeline is inherited from the PerspNet. The components newly employed by us are highlighted with the dotted boxes.

perspective projection by estimating the 6DoF pose. The PerspNet outperforms current state-of-the-art methods on public datasets. In this paper, We mainly explore the face reconstruction performance in a model-free framework on the WCPA challenge dataset in this report.

3 Method

3.1 Architecture

The whole architecture of our method is shown in Fig. 1. The UV position map, reconstructed 3D face, 2D-3D correspondence, corresponding 3D points, face segment, and the sampled 2D points are inherited from PerspNet. We newly propose the vertex refinement module (VRM) and the multi-task learning module. In the next, we introduce all proposed modules in detail.

3.2 Vertex Refinement Module

The proposed VRM is designed to refine all 3D vertices. Specifically, following the 3D face shape reconstruction in the PerspNet, VRM first predicts the UV position map $S \in \mathbb{R}^{H \times W \times 3}$ that records the 3D coordinates of 3D facial structure. Then, taking UV coordinates, we apply grid sampling on the predicted UV position map to obtain the 3D vertices X'. In sparse regions, the UV coordinates are distinguishable. Thus, the corresponding regions on the UV position map are accurate enough. However, since the UV coordinates distribute unevenly, there is inherent precision degradation in generating the target UV position map where the UV coordinates are dense. Therefore, in addition to predicting the UV position map, VRM also predicts extra vertex-wise offsets $\Delta \in \mathbb{R}^{n \times 3}$ for the reconstructed 3D face shape. The offsets $\Delta \in \mathbb{R}^{n \times 3}$ are added to the 3D vertices X' to obtain the refined reconstructed 3D face shape $X \in \mathbb{R}^{n \times 3}$.

In the training phase, we employ a weighted L_1 loss function and a naive L_1 loss function to supervise the training of the UV position map and the 3D vertices, respectively. They can be denoted as:

$$\mathcal{L}_S = \sum \left\| S - \hat{S} \right\| \cdot W\left(S_{coord}\right), \tag{1}$$

where $\hat{S}$ is the target UV position map. $S_{coord} \in \mathbb{R}^{2 \times n}$ represents the UV coordinates that records the 2D locations of n 3D vertices in the UV position map. $W(\cdot)$ is a weight function for S, set as 1 in the UV coordinates, 0 elsewhere. Notably, the WCPA challenge dataset does not provide target UV position maps. Hence, we generate them offline.

$$X = X' + \Delta$$
$$\mathcal{L}_X = \left\| X - \hat{X} \right\|, \tag{2}$$

where $\hat{X}$ is the target 3D vertices.

3.3 Multi-task Learning Module

PerspNet extracts 3D features to predict UV position maps. During the experiments, we find that the using heavy backbones bring overfitting. To alleviate this, we design a multi-task learning module (MTLM) to guide the networks to learn more 3D semantic information. Specifically, the multi-task learning module contains three outputs, *i.e.* normal UV maps $S_{normal\_UV} \in \mathbb{R}^{H \times W \times 3}$, depth maps $S_{depth} \in \mathbb{R}^{H \times W \times 3}$, and normal maps $S_{normal} \in \mathbb{R}^{H \times W \times 3}$. Each one contains specific 3D structure information. The module shares the same decoded 3D features with the UV position map prediction branch, guiding the 3D features to involve richer 3D information. We use the L_1 loss to supervise the training of the multi-task learning module, which can be denoted as:

$$\mathcal{L}_M = \lambda_1 \left\| S_{normal\_UV} - \hat{S}_{normal\_UV} \right\| + \lambda_2 \left\| S_{depth} - \hat{S}_{depth} \right\|$$
$$+ \lambda_3 \left\| S_{normal} - \hat{S}_{normal} \right\|, \tag{3}$$

where $\lambda_1, \lambda_2, \lambda_3$ are balance coefficients, we set them to $40, 1, 1$.

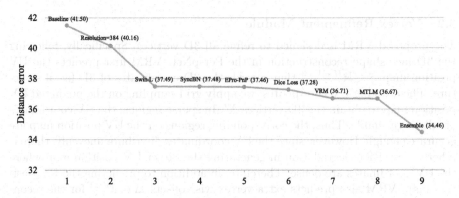

Fig. 2. We present the strategies that boost the face reconstruction performance on the WCPA challenge dataset (validation set).

4 Experiments

4.1 Dataset

The WCPA challenge dataset contains $447,185$ images captured from 250 subjects. Each subject takes 33 sets of actions throughout the frames such as turning the head from left to right with mouse closing and turning the head from up to down with eyes opening. Variations in persons, postures, and environments present challenges for the face reconstruction task. Here 50 subjects are officially divided into the test set. We divide the rest 200 subjects into training and validation sets in the ratio of 3 : 1. The following results and comparisons are evaluated on the validation set.

4.2 Implementation Details

Hyper-parameters. The number of points n in the reconstructed 3D face is $1,220$. We use AdamW optimizer to train the networks. The initial learning rate is 1e–4, decreasing to 1e–5 after the 15-th epoch. The weight decay is set as 1e–3. The resolutions of input images, segments, and ground-truth maps are 384×384. The number of sampled 2D points is $1,024$. The batch size is set as 64. We train all models for 25 epochs.

Strategies. In order to make distance error as low as possible, we utilize extensive strategies. Specifically, as shown in Fig. 2, our baseline is reproduced PerspNet that sets a distance error of 41.50 on the validation set. PerspNet resizes the input images to 192×192. We argue that the face detail information is inadequate in such settings. Thus, we resize the input images and target maps to 384×384. This setting reduces distance error by 1.34. Besides, the original backbone ResNet-18 is replaced with Swin-L pretrained on ImageNet-22k [4] to enhance features. Quantitatively, after being equipped with a deeper backbone, our model further decreased distance error by 2.67. We also change all batch normalization (BN) layers to synchronized batch normalization (SyncBN) [14] layers, which brings distance error reduction by 0.01. For the estimation of rotation and translation, we replace the non-differentiable solvePnP implemented by OpenCV with differentiable EPro-PnP [3], which reduces distance error by 0.02. Moreover, dice loss [13] is used to supervise the training of the face segment branch, getting distance error reduction by 0.18. The proposed VRM and MTLM decrease distance error by 0.61. Finally, we ensemble all results outputted by several models, which sets the distance error to 34.46.

4.3 Comparisons and Ablation Studies

Backbones. We first explore the impacts of different backbones for our baseline. As listed in Table 1, Swin-L performs better than ResNet-18 and CBNet on the validation dataset in terms of distance error. Thus, Swin-L is used to be the feature extraction network in the next experiments.

Table 1. The impacts of different backbones on the WCPA challenge dataset (validation set).

Backbone	Distance error ↓
ResNet-18 [8]	40.16
CBNet [11]	38.79
Swin-L [12]	**37.49**

VRM and MTLM. Then, we further analyze the effectiveness of our proposed VRM and our proposed MTLM on the validation set. As shown in Table 2, the two modules are both effective in reducing distance error. In particular, VRM brings a significant distance error reduction. When they work together in our method HiFace, our method results in the lowest distance error, demonstrating their effectiveness.

Table 2. The effectiveness of VRM and MTLM on the WCPA challenge dataset (validation set).

VRM	MTLM	Distance error ↓
✗	✗	37.28
✔	✗	36.71
✗	✔	37.23
✔	✔	**36.67**

Visualization. As shown in Fig. 3, we present visualizations on the WCPA challenge dataset (test set). The pure images are the inputs. The estimated

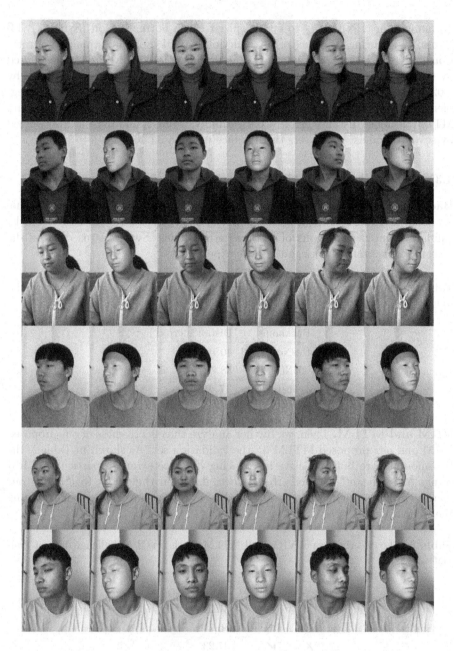

Fig. 3. Visualizations on the WCPA challenge dataset (test set). Our HiFace achieves high-quality face reconstruction with the input images containing different persons in various postures.

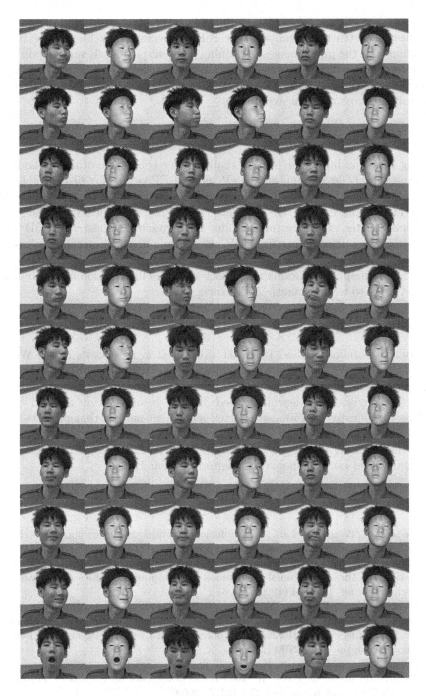

Fig. 4. We present the predicted meshes of ID-239763 in the WCPA challenge dataset (test set). There are 33 different expressions. Our HiFace shows reliable face reconstruction abilities.

6DoFs and reconstructed 3D faces are rendered on the input images to present the face reconstruction qualities. Our HiFace successfully achieves face reconstruction with the input images containing different persons in various postures. We further present the face reconstruction performance of our HiFace in 33 different expressions. As shown in Fig. 4, the reconstruction results are accurate and robust across different expressions.

4.4 Discussions

During participating in the WCPA challenge, we tried some other methods. 1) utilizing the SOTA differentiable PnP method EPro-PnP for 6DoF solving to achieve end-to-end training and inference. Experiments show that distance error reduction is quite limited. We speculate that EPro-PnP may disturb the learning phase on the structure information. 2) Regressing reconstructed 3D faces directly instead of sampling from UV position maps, which decreases distance error slightly. However, it is a sub-optimal method compared to VRM. We argue that the UV mapping introducing 3D face structure information naturally helps the networks to learn semantic 3D features.

4.5 Conclusion

The official baseline method PerspNet works well in most cases of the WCPA challenge. However, the inevitable precision degradation introduced by the UV mapping process limits further performance improvement. To address this, we proposed a hybrid task learning based 3D face reconstruction method called HiFace. Specifically, we propose a vertex refinement module that predicts additional vertex-wise offsets to compensate for the precision degradation and a multi-task training module to enhance 3D features. Extensive experiments and the final official ranking show the superiority of our HiFace.

References

1. Andrus, C., et al.: FaceLab: scalable facial performance capture for visual effects. In: The Digital Production Symposium, pp. 1–3 (2020)
2. Blanz, V., Vetter, T.: A morphable model for the synthesis of 3d faces. In: Proceedings of the 26th Annual Conference on Computer Graphics and Interactive Techniques, pp. 187–194 (1999)
3. Chen, H., Wang, P., Wang, F., Tian, W., Xiong, L., Li, H.: EPro-PnP: generalized end-to-end probabilistic perspective-n-points for monocular object pose estimation. In: Proceedings of the IEEE/CVF Conference on Computer Vision and Pattern Recognition, pp. 2781–2790 (2022)
4. Deng, J., Dong, W., Socher, R., Li, L.J., Li, K., Fei-Fei, L.: ImageNet: a large-scale hierarchical image database. In: 2009 IEEE Conference on Computer Vision and Pattern Recognition, pp. 248–255. IEEE (2009)
5. Dib, A., Ahn, J., Thebault, C., Gosselin, P.H., Chevallier, L.: S2f2: self-supervised high fidelity face reconstruction from monocular image. arXiv preprint arXiv:2203.07732 (2022)

6. Feng, Y., Wu, F., Shao, X., Wang, Y., Zhou, X.: Joint 3D face reconstruction and dense alignment with position map regression network. In: Ferrari, V., Hebert, M., Sminchisescu, C., Weiss, Y. (eds.) Computer Vision – ECCV 2018. LNCS, vol. 11218, pp. 557–574. Springer, Cham (2018). https://doi.org/10.1007/978-3-030-01264-9_33

7. Gerig, T., et al.: Morphable face models-an open framework. In: 2018 13th IEEE International Conference on Automatic Face & Gesture Recognition (FG 2018), pp. 75–82. IEEE (2018)

8. He, K., Zhang, X., Ren, S., Sun, J.: Deep residual learning for image recognition. In: Proceedings of the IEEE Conference on Computer Vision and Pattern Recognition, pp. 770–778 (2016)

9. Jackson, A.S., Bulat, A., Argyriou, V., Tzimiropoulos, G.: Large pose 3D face reconstruction from a single image via direct volumetric CNN regression. In: Proceedings of the IEEE International Conference on Computer Vision, pp. 1031–1039 (2017)

10. Kao, Y., et al.: Single-image 3D face reconstruction under perspective projection. arXiv preprint arXiv:2205.04126 (2022)

11. Liu, Y., et al.: CBNet: a novel composite backbone network architecture for object detection. In: Proceedings of the AAAI Conference on Artificial Intelligence, vol. 34, pp. 11653–11660 (2020)

12. Liu, Z., et al.: Swin transformer: hierarchical vision transformer using shifted windows. In: Proceedings of the IEEE/CVF International Conference on Computer Vision, pp. 10012–10022 (2021)

13. Milletari, F., Navab, N., Ahmadi, S.A.: V-Net: fully convolutional neural networks for volumetric medical image segmentation. In: 2016 Fourth International Conference on 3D Vision (3DV), pp. 565–571. IEEE (2016)

14. Peng, C., et al.: MegDet: a large mini-batch object detector. In: Proceedings of the IEEE Conference on Computer Vision and Pattern Recognition, pp. 6181–6189 (2018)

15. Suwajanakorn, S., Kemelmacher-Shlizerman, I., Seitz, S.M.: Total moving face reconstruction. In: Fleet, D., Pajdla, T., Schiele, B., Tuytelaars, T. (eds.) ECCV 2014. LNCS, vol. 8692, pp. 796–812. Springer, Cham (2014). https://doi.org/10.1007/978-3-319-10593-2_52

16. Thies, J., Zollhofer, M., Stamminger, M., Theobalt, C., Nießner, M.: Face2face: Real-time face capture and reenactment of RGB videos. In: Proceedings of the IEEE Conference on Computer Vision and Pattern Recognition, pp. 2387–2395 (2016)

17. Wu, C., Bradley, D., Gross, M., Beeler, T.: An anatomically-constrained local deformation model for monocular face capture. ACM Trans. Graphics **35**(4), 1–12 (2016)

Multi-view Canonical Pose 3D Human Body Reconstruction Based on Volumetric TSDF

Xi Li[✉]

University of Electronic Science and Technology of China, Chengdu, China
xi-li@foxmail.com

Abstract. In this report, we present our solution for track1, multi-view based 3D human body reconstruction, of the ECCV 2022 WCPA Challenge: From Face, Body and Fashion to 3D Virtual Avatars 1. We developed a variant network based on TetraTSDF to reconstruct detailed 3D human body models with canonical T-pose from that multi-view images of the same person with different actions and angles, which is called TetraTSDF++ in this report. This method first fuses the features of different views, then infers the volumetric truncated signed distance function (TSDF) in the preset human body shell, and finally obtains the human surface through marching cube algorithm. The best chamfer distance score of our solution is 0.9751, and our solution got the 2nd place on the leaderboard.

Keywords: Multi-view · Human body reconstruction · Canonical T-pose · TSDF

1 Introduction

3D human body reconstruction from a single image is ill-posed due to the lack of sufficient and clear visual clues. With the help of deep neural networks and parametric 3D human body model like SMPL [5], it is possible to improve the rationality of prediction results to some extent. Obviously, with the integration of more views, neural networks can rely on clear visual information to reconstruct a more refined three-dimensional model. Most of the current studies are aimed at different perspectives of the same human body maintaining the same posture [3,7]. MVPHuman [8] proposed a new problem, that is, to reconstruct 3D human avatars from multiple unconstrained frames, which is independent of the assumptions of camera calibration, capture space and constrained motion. Recently, MVPHuman has raised a new issue that has nothing to do with camera calibration, capture space, and posture, that is, rebuilding 3D human body models from multiple views that are not constrained by posture. To address this valuable new issue, this report developed TetraTSDF++, a variant network based on TetraTSDF [6], to reconstruct detailed T-pose 3D human avatars from

L. Karlinsky et al. (Eds.): ECCV 2022 Workshops, LNCS 13805, pp. 392–397, 2023.
https://doi.org/10.1007/978-3-031-25072-9_27

multi-view images of a person with different poses. Our method first used PaddleSeg to get the foreground from the pictures, then fuses the features of different views and predict the volumetric TSDF in the preset human body outer shell, finally, the human body surface is obtained by marching cube algorithm.

The reason why TSDF is used as the expression method of human model is that the fine mesh model contains millions of vertices and faces. It is still difficult to process meshes directly by neural networks such as graph convolutional networks, and the standard voxel expression will consume huge computing resources. In addition, the parametric human body model like SMPL is limited by too few patches and does not have sufficient surface details such as clothes, hair and so on. Therefore, TetraTSDF proposes to use the volumetric TSDF value in the preset human body shell to express the human body surface, which effectively reduces the computational pressure and reconstructs the human avatars with details.

2   Dataset

MVP-Human (Multi-View and multi-Pose 3D Human) is a large-scale dataset [8], which contains 400 subjects, each of which has 15 scans in different poses and 8-view images for each pose, providing 6, 000 3D scans and 48, 000 in-the-wild images in total. Each image is a 1280×720 jpeg file. Each action contains 8 RGB images from different angles [0, 45, 90, 135, 180, 225, 270, 315]. Each sample includes a number of pairs of multi-view RGB images, texture maps, canonical pose ground truth and skinning weights.

2.1   Data Analysis

After analyzing the dataset, we eliminated bad case subject 226831, so there are 199 valid samples, we set 150 samples as training sets and 49 as val sets preliminarily. In order to get better scores in the competition, we counted the data distribution of the test set and selected actions and angles online in a approximately the same distribution during the training.

2.2   Data Preprocessing

Data processing includes two aspects, one is to separate the picture foreground, and the other is to generate TSDF binary files according to the canonical pose ground truth model.

Human Segmentation. The images of MVPHuman dataset are all collected from the real world, The person is in the center of the images, but inevitably, there are many redundant background informations in the picture, which will interfere with the result prediction. Therefore, correctly segmenting the foreground can effectively ensure the reasoning ability of neural network. We used the PP-HumanSeg algorithm from the PaddleSeg [1,4] as a tool to separate the foreground, and finally placed the human body in the center of the newly generated 512×512 size image. The body length occupies about 95% of the area of the whole image, and the background is uniformly set to dark green. (Fig. 1 shows some image processing result).

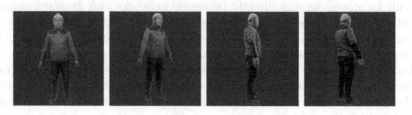

Fig. 1. Image processing results of different angles from subject 197019 Action01

Generate TSDF. On the other hand, in order to reduce the time required for training, we generate the corresponding TSDF binary file from each ground truth model offline, which is different from calculating the volumetric TSDF value after changing the posture of the human body into a star-shape by SMPL parameter in method [6]. In order to avoid the loss of accuracy caused by the posture change, we adopt T-pose to calculate the volumetric TSDF value directly, and the preset human body shell still retains the method in [6].

3 Method

3.1 Network Architecture

We developed TetraTSDF++ to reconstruct detailed 3D human avatars with T-pose from that multi-view images of the same person with different actions and angles. This method first fuses the features of different views, then infers the volumetric truncated signed distance function (TSDF) in the preset human body shell. (Fig. 2 shows the network architecture of our TetraTSDF++).

3.2 TSDF

we calculate the truncate TSDF values between the voxel summit and its nearest point according to the threshold of the Euclidean distance. (Fig. 3 shows the preset human body outer shell and canonical pose ground truth 3D mesh).

Set v as a voxel summit, $\hat{v}$ as the nearest point of v in the canonical pose GT, then the Euclidean distance calculation formula is as follows

$$\varepsilon = \|v - \hat{v}\|_2 \tag{1}$$

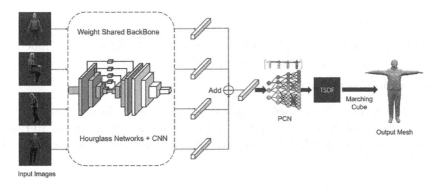

Fig. 2. Network architecture of TetraTSDF++, the input images are different actions and different angles from the same person, then extract the image features of multiple views through a weight sharing backbone network, which consists of an hourglass network and convolutional neural network. The image feature of backbone network output is one-dimensional vector, then add the image features of each view and feed them into PCN to forecast the volumetric TSDF value of input features. PCN is a coarse to fine up sampling network, and each level of PCN conforms to the preset human body model. Finally, marching cube algorithm is used to reconstruct mesh from TSDF.

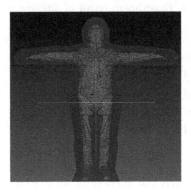

Fig. 3. Calculate volumetric TSDF value in the preset human body outer shell.

Set $\hat{n}$ is the normal vector at $\hat{v}$, then let

$$\rho = \hat{n} \cdot (v - \hat{v}) \tag{2}$$

To sum up, the calculation formula of tsdf is as follows

$$tsdf = \begin{cases} \rho/\tau, & if \ \tau \geq \varepsilon \\ 0, & if \ \tau < \varepsilon \ and \ \rho \leq 0 \\ 1, & if \ \tau < \varepsilon \ and \ \rho > 0 \end{cases} \tag{3}$$

where τ is the threshold value of 3 cm.

396   X. Li

3.3 Loss Function

We use the mean square error between predict and GT TSDF as the loss function.

3.4 Implementation Details

We build networks on tensorflow and keras frameworks. We set batch size of 4, learning rate of 4e–4, epoch of 120, and the learning rate decreases 10 times when the epoch reaches 100. We train and infer our model on a single NVIDIA Tesla V100 GPU.

4 Experiment

In this challenge, we calculate chamfer distance (CD) between GT meshes and predicted meshes as score, the smaller the CD, the better the reconstruction quality. We recorded the scores of different versions of our method, which are given in Table 1.

Table 1. Experiment result of different methods

Methods	Chamfer distance (CD)
Mean Shape	1.2187
TetraTSDF++(Xception)	1.1641
TetraTSDF++(concat)	1.0159
TetraTSDF++(add)	0.9932
TetraTSDF++(add*)	0.9751

Note that, we calculated the average of all 3D human models given in the dataset and take the 'Mean Shape' model as the inference result of each sample in the test set as a benchmark. Different versions of our method outperform this benchmark, which validates the effectiveness of our method.

'TetraTSDF++(Xception)' means that the traditional convolutional neural network Xception [2] is used as the backbone network. For 'TetraTSDF++ (concat)', the backbone network is Hourglass Network + CNN, and multi view feature fusion through concat. The experimental results show that, Hourglass Network + CNN is better than Xception in this task. Finally, 'TetraTSDF++(add)' replace concat operation with add, and for 'TetraTSDF++(add*)', we set 191 samples as training sets and 8 as val sets from all 199 valid samples, and get a best score when more training samples are divided.

Finally, a visualization result of human body reconstruction in this solution are given below Fig. 4.

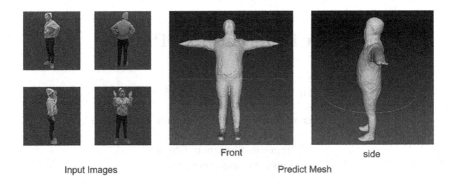

Front   side

Input Images   Predict Mesh

Fig. 4. Reconstruction result visualization.

5 Conclusion

This challenge and MVPHuman proposed a new problem, that is, to reconstruct 3D human avatars from multiple unconstrained images, which is independent of the assumptions of camera calibration, capture space and constrained motion. In this report, we developed TetraTSDF++, a variant network based on TetraTSDF, to reconstruct detailed T-pose 3D human avatars from multi-view images of one person with different poses. We utilized appropriate data processing, effective networks and some strategies and finally achieved good results in this challenge.

References

1. Authors, P.: PaddleSeg, end-to-end image segmentation kit based on paddlepaddle (2019). https://github.com/PaddlePaddle/PaddleSeg
2. Chollet, F.: Xception: deep learning with depthwise separable convolutions. In: Proceedings of the IEEE Conference on Computer Vision and Pattern Recognition. pp. 1251–1258 (2017)
3. Kolotouros, N., Pavlakos, G., Daniilidis, K.: Convolutional mesh regression for single-image human shape reconstruction. In: Proceedings of the IEEE/CVF Conference on Computer Vision and Pattern Recognition, pp. 4501–4510 (2019)
4. Liu, Y., et al.: PaddleSeg: a high-efficient development toolkit for image segmentation (2021)
5. Loper, M., Mahmood, N., Romero, J., Pons-Moll, G., Black, M.J.: SMPL: a skinned multi-person linear model. ACM Trans. Graphics **34**(6), 1–16 (2015)
6. Onizuka, H., Hayirci, Z., Thomas, D., Sugimoto, A., Uchiyama, H., Taniguchi, R.i.: Tetratsdf: 3D human reconstruction from a single image with a tetrahedral outer shell. In: Proceedings of the IEEE/CVF Conference on Computer Vision and Pattern Recognition, pp. 6011–6020 (2020)
7. Zheng, Z., Yu, T., Liu, Y., Dai, Q.: Pamir: Parametric model-conditioned implicit representation for image-based human reconstruction. IEEE Trans. Pattern Anal. Mach. Intell. **44**(6), 3170–3184 (2021)
8. Zhu, X., et al.: MVP-human dataset for 3d human avatar reconstruction from unconstrained frames. arXiv preprint arXiv:2204.11184 (2022)

End to End Face Reconstruction
via Differentiable PnP

Yiren Lu[1]([⊠])[iD] and Huawei Wei[2][iD]

[1] State University of New York at Buffalo, Buffalo, USA
yirenlu@buffalo.edu
[2] Tencent, Shenzhen, China
huaweiwei@tencent.com

Abstract. This is a challenge report of the ECCV 2022 WCPA Challenge, Face Reconstruction Track. Inside this report is a brief explanation of how we accomplish this challenge. We design a two-branch network to accomplish this task, whose roles are Face Reconstruction and Face Landmark Detection. The former outputs canonical 3D face coordinates. The latter outputs pixel coordinates, i.e. 2D mapping of 3D coordinates with head pose and perspective projection. In addition, we utilize a differentiable PnP (Perspective-n-Points) layer to finetune the outputs of the two branch. Our method achieves very competitive quantitative results on the MVP-Human dataset and wins a 3rd prize in the challenge.

Keywords: Computer vision · Face alignment · Head pose estimation · Face reconstruction · Face landmark detection

1 Introduction

Face reconstruction from a single image has a wide range of applications in entertainment and CG fields, such as make-up in AR and face animation. The current methods [1,4,5,7] generally use a simple orthogonal projection in face reconstruction. The premise of this assumption is that the face is far enough away from the camera. When the distance is very close, the face will suffer from severe perspective distortion. In this case, the reconstructed face has a large error with the real face. To solve this problem, perspective projection must be adopted as the camera model. In this paper, we propose a method to reconstruct 3D face under perspective projection. We develop a two-branch network to regress 3D facial canonical mesh and 2D face landmarks simultaneously. For 3D branch, we adopt a compact PCA representation [7] to replace the dense 3d points. It makes the network learn easier. For 2D branch, we take uncertainty learning into account. Specifically, Gaussion Negative Log Loss [11] is introduced as our loss function. It can enhance the robustness of the model, especially for

---

Y. Lu—Work done during an internship in Tencent.
Y. Lu and H. Wei—Contributed equally to this work.

L. Karlinsky et al. (Eds.): ECCV 2022 Workshops, LNCS 13805, pp. 398–406, 2023.
https://doi.org/10.1007/978-3-031-25072-9_28

faces with large poses and extreme expressions. In addition, a differentiable PnP (Perspective-n-Points) layer [2] is leveraged to finetune our network after the two branches converge. It brings interaction of the two branches, so that the two branches can be jointly trained and promote each other. Our method achieves very competitive results on the MVP-Human dataset.

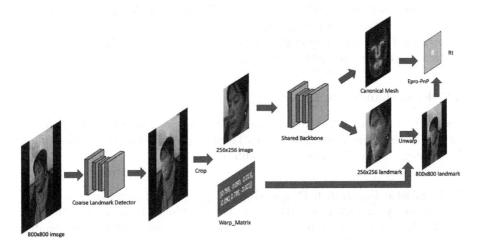

Fig. 1. This figure shows the overall pipeline of our method. First, the 800×800 image will go through a Coarse Landmark Detector to get coarse landmarks so that we can utilize them to crop the face out. Then the cropped 256×256 image will be fed into a shared backbone with different regressors. After that, a canonical mesh and a 256×256 landmark will be output. Finally, we use the unwarped landmark and the canonical mesh to find out the Rotation and Translation through PnP.

2   Methodology

We decouple the problem into two parts: head pose estimation and canonical mesh prediction. For head pose estimation, the commonly used methods are generally divided into two categories. The first category is to directly regress the 6DoF pose from the image. The second is to utilize PnP (Perspective-n-Points) to figure out the 6DoF pose with the correspondence between 2D and 3D coordinates. We find it's hard for the network to directly regress the 6DoF well, especially for the translation of the depth axis. So we choose the second category of method as our framework. Our strategy is to regress canonical 3D mesh and 2D facial landmarks separately. To this end, we develop a two-branch network, which respectively outputs canonical 3D face coordinates and the landmarks, i.e. the perspective projection of 3D coordinates. When both branches converge, we use a differentiable PnP layer to jointly optimize the two branches. Its purpose is to make the 6DoF pose calculated from 2D and 3D coordinates as close to the groundtruth pose as possible. In inference phase, we directly use PnP to compute 6DoF pose from the output of two branches. An overview of our pipeline is showed in Fig. 1.

2.1 Data Preparation

The original image size is 800 × 800, training with images of this size is very inefficient. Hence, it is necessary to crop the image to an appropriate size. To achieve this purpose, we train a network that predicts the 2D face landmarks on 800×800 image. We utilize these landmarks to frontalize the face (rotate the face to a status that the roll is 0), and crop the face out. After this process, we obtain a 256 × 256 aligned image and a warp matrix for each image, with which the predicted 256×256 landmarks can be transformed back to the corresponding positions on the original 800 × 800 image.

2.2 Facial Landmark Detection

Generally, we consider Facial Landmark Detection as a regression task to find out the corresponding position of each landmark on the image. We also follow this strategy, but with some small modifications.

Probabilistic Landmark Regression. [14] Instead of regressing the coordinates $\mathbf{x}$ and $\mathbf{y}$ of each landmark point. We introduce Gaussion Negative Log Loss(GNLL) to regress the uncertainty of each point along with the coordinates:

$$L_{gnll} = \sum_{i=1}^{|Lmk|} \left(\log(\sigma_i^2) + \frac{||\mu_i - \hat{\mu}_i||^2}{2\sigma_i^2} \right). \tag{1}$$

In which Lmk represents the set of all 1220 landmarks, and $Lmk_i = (\mu_i, \sigma_i)$. In Lmk_i, $\mu_i = (x_i, y_i)$ represents the ground truth coordinates of each landmark, and σ_i indicates the uncertainty or invisibility of each landmark. We find that, with the help of GNLL, the predicted landmarks will be more accurate than training using regular MSE loss.

CoordConv. [13] We apply CoordConv in our landmark detection branch, which plays a position encoding role. With this technique, the input of the landmark detection model would be 5 channels (RGBXY) instead of 3 channels (RGB). We find CoordConv helps to improve the accuracy of the landmark regression.

2.3 3D Face Reconstruction

In this branch, we need to predict the coordinates of each point from the canonical mesh. The most direct way is to roughly regress the coordinates. However, we find this strategy not that ideal. Our strategy is to convert the mesh into a compact PCA representation. It is verified the network training is easier and higher accuracy can be obtained with this strategy.

Construct PCA Space. We manually select a face with neutral expression from each of the 200 persons in the training set. In addition, to enhance the representation ability of PCA, 51 ARKit[1] blendshape meshes are involved (the TongueOut blendshape is excluded). We use these 251 face mesh to fit a PCA model, whose dimension of components is eventually 250. Then the PCA model is utilized to fit the coefficient for each training sample.

Loss Function. We use Vertex Distance Cost (VDC) and Weighted Parameter Distance Cost (WPDC) proposed by 3DDFA V2 [6,7] as our loss function. And the algorithms are showed below,

VDC loss:

$$L_{vdc} = ||X_{3d}^{pred} - X_{3d}^{gt}||^2, \tag{2}$$

where X_{3d}^{pred} represents the 3d mesh predicted by network and X_{3d}^{gt} represents the ground truth 3d mesh.

WPDC loss:

$$L_{wpdc} = ||W \cdot (PCA^{pred} - PCA^{gt})||^2, \tag{3}$$

where PCA^{pred} represents the predicted PCA coefficients and PCA^{gt} presents the ground truth ones, and W represents the weight of each coefficient.

2.4   PnPLoss

In order to make the two branches have stronger interaction, we try to jointly optimize the two branches by a differentiable PnP layer. Specifically, we utilize Epro-PnP [2] proposed recently on CVPR2022 to achieve this. Epro-PnP interprets pose as a probability distribution and replace the argmin function with a softargmin in the optimization procedure. Hence, it is a differentiable operation and can be plugged into neural network training. The inputs of the pnp layer is 2D landmarks and 3D mesh coordinates, then the pose will be output. By making the pose as close to the groundtruth pose as possible, the coordinates locations of 2D/3D branch will be finetuned to a more accurate status. The expression of PnPLoss is showed below,

$$P^{pred} = EproPnP(X_{2d}^{pred}, X_{3d}^{pred}), \tag{4}$$

$$L_{pnp} = ||P^{pred}X_{3d}^{gt} - P^{gt}X_{3d}^{gt}||_2 + ||P^{pred}X_{3d}^{pred} - P^{gt}X_{3d}^{gt}||_2, \tag{5}$$

where X_{2d}^{pred}, X_{3d}^{pred} represents the 3d mesh, 2d landmark predicted by network and P^{pred} stands for the predicted pose output by Epro-PnP.

Finally, the total loss function should look like this:

$$L_{total} = \lambda_1 L_{gnll} + \lambda_2 L_{vdc} + \lambda_3 L_{wpdc} + \lambda_4 L_{pnp}. \tag{6}$$

---

[1] https://developer.apple.com/documentation/arkit/arfaceanchor/blendshapelocation.

One thing need to be informed is that we do not train from scratch with this pnp layer. We first train the network for several epoches without it. Then we use Epro-PnP layer to finetune the training.

2.5 Inference Phase

In inference phase, we utilize opencv to implement the PnP module. Before start to solve PnP, we first need to unwarp the landmarks predicted on 256×256 to 800×800 by multipling the inverse matrix of the warp matrix.

3 Experiments

3.1 Experimental Details

We use the aligned 256×256 images as training samples and add color jitter and flip augmentation in training process. Resnet50 [8] (pretrained on ImageNet [3]) is utilized as our backbone. The regressor layer of 2D/3D branch is a simple FC layer. In training stage, we use Adam optimizer [12] and 1e–4 as the learning rate. We first train for 10 epochs, and then finetune the model for 5 epochs with Epro-PnP. Besides, the λ's in the total loss function are 0.01, 20, 10, 2 respectively.

3.2 Head Pose Estimation

In this experiment section we compare the performance of several methods to estimate the headpose Rt. We choose MAE_r, MAE_t, ADD as our evaluation metrics, where MAE_r is the mean absolute error of the 3 components of eular angle, yaw, pitch and roll; MAE_t is the mean absolute error of the 3 components of translation, x, y and z; ADD is the second term of challenge loss, which is shown in Eq. (7)

$$ADD = ||P^{gt}X^{gt}_{3d} - P^{pred}X^{gt}_{3d}||_2. \tag{7}$$

Direct 6DoF with CoordConv. This is our baseline method. As we are trying to regress Rt which corresponds to the 800×800 image on a 256×256 image, we need to give it some extra information. So we add a 2 channel coordmap XY to indicate where each pixel in the 256×256 image locates on the original 800×800 image. And then feed the 5 channel input to a ResNet50 backbone and a linear regressor. The output should be 9 numbers, 6 of them for Rotation and 3 for Translation. We supervise the training process with Geodesic Loss [9] and MSE Loss for Rt. From Table 1, we can see that though it cannot compete with PerspNet [10], it is much better than Direct 6DoF without coordmap.

MSELoss & PnP. In this experiment we try to use landmark detection together with PnP to calculate Rt, the choice of backbone and regressor is the same as the previous experiment. The input of network is simply 3 channel RGB image and we utilize only landmark MSE loss to supervise the training process. We can see that it outperforms PerspNet and result in a 9.79 ADD loss.

GNLL & PnP. In this experiment, we increase the input of the network from 3 channels to 5 channels, we add a 2 channel coordmap XY(the range of coordmap is [0, 255] and normalized into [−1, 1]) and concatenate them with RGB channels. Besides, we utilize GNLL to supervise the training process instead of landmark MSE loss. This experiment reaches a 9.66 ADD loss, which is slightly better than the previous one.

Table 1. Comparison with the performance of different methods for 6DoF Head Pose Estimation on validation set. The experiment result of Direct 6DoF and PerspNet is from the PerspNet [10] paper.

Method	MAE_r	MAE_t	ADD
Direct 6DoF	1.87	9.06	21.39
Direct 6DoF(CoordConv)	1.11	5.58	13.42
PerspNet	0.99	4.18	10.01
MSELoss & PnP	0.89	4.02	9.79
GNLL & PnP	**0.82**	**3.95**	**9.66**

3.3 Face Reconsturction

Direct 3D Points. In this experiment we try to directly regress each 3d coordinate of the canonical mesh. Points MSE Loss is used to supervise the training process. From Table 2, we can see that the result is not that satisfying, it only reaches a 1.82 mean error.

Regress PCA Coefficient. Instead of regressing 3D coordinates, we try to regress 250 PCA coefficients, which should be easier for the network to learn. Also, backbone and regressor remains the same as in the Direct 3D experiment. Just as the way we think, PCA results in a 1.68mm mean error and has performed better than PerspNet.

3.4 Finetuning with PnP Layer

In this experiment, we introduce the newly proposed Epro-PnP. Instead of training from scratch, we choose to finetune the network based on the previous experiment. After finetuning, we can find from Table 3 that MAE_t and ADD loss reduces obviously. We attribute the effectiveness to that the pnp layer can

Table 2. Comparison with the performance of different methods for Face Reconstruction on validation set. The experiment result of PerspNet is from the PerspNet paper.

Method	Medium (mm)	Mean (mm)
PerspNet	1.72	1.76
Direct 3D	1.79	1.82
PCA	**1.63**	**1.68**

enhance the interaction of the 2D/3D branches. It makes the two branches not only for the pursuit of lower coordinates regression error, but to make the pose estimation more accurate.

Table 3. Comparison with the performance for 6DoF Head Pose Estimation with and without Epro-PnP on validation set.

Method	MAE_r	MAE_t	ADD
GNLL & PnP	0.82	3.95	9.66
GNLL & PnP & Epro-PnP	**0.81**	**3.86**	**9.48**

3.5 Qualitative Results

We show some qualitative results in Fig. 2. It can be observed that our method can accurately reconstruct faces even in large poses and extreme expressions. In some cases, our method even outperforms the ground truth.

4 Tricks

4.1 Flip and Merge

When doing inference on test set, we flip the input image and let the flipped and unflipped go through the network at the same time. After getting two results, we merge them and calculate an average as the final result. One thing worth noting is that this strategy only works in the face reconstruction branch in our test, in the landmark detection branch it is ineffective. We think this is due to the face reconstruction task with PCA is to some extent similar to a classification problem.

Fig. 2. Visualization comparison between our method and the ground truth obtained by the structured light sensor of iphone11. We can find that in some extreme cases such as the first column and the fifth column, our method outperforms the ground truth result, which means we have learned the essence of this task.

5   Conclusions

In this competition, we design a two-branch network to solve the 3D face reconstruction task. Meanwhile, we propose to utilize a differentiable PnP layer to jointly optimize the two branches. Finally, our method achieves a competitive result and scores 33.84 on leaderboard with a 3rd rank.

References

1. Bai, Z., Cui, Z., Liu, X., Tan, P.: Riggable 3D face reconstruction via in-network optimization. In: Proceedings of the IEEE/CVF Conference on Computer Vision and Pattern Recognition (CVPR), pp. 6216–6225 (June 2021)
2. Chen, H., Wang, P., Wang, F., Tian, W., Xiong, L., Li, H.: EPro-PnP: generalized end-to-end probabilistic perspective-n-points for monocular object pose estimation. In: Proceedings of the IEEE/CVF Conference on Computer Vision and Pattern Recognition, pp. 2781–2790 (2022)
3. Deng, J., Dong, W., Socher, R., Li, L.J., Li, K., Fei-Fei, L.: ImageNet: a large-scale hierarchical image database. In: 2009 IEEE Conference on Computer Vision and Pattern Recognition, pp. 248–255 (2009). https://doi.org/10.1109/CVPR.2009.5206848
4. Feng, Y., Feng, H., Black, M.J., Bolkart, T.: Learning an animatable detailed 3D face model from in-the-wild images (2020). https://doi.org/10.48550/ARXIV.2012.04012,https://arxiv.org/abs/2012.04012
5. Feng, Y., Wu, F., Shao, X., Wang, Y., Zhou, X.: Joint 3D face reconstruction and dense alignment with position map regression network (2018). https://doi.org/10.48550/ARXIV.1803.07835.https://arxiv.org/abs/1803.07835

6. Guo, J., Zhu, X., Lei, Z.: 3ddfa. https://github.com/cleardusk/3DDFA (2018)
7. Guo, J., Zhu, X., Yang, Y., Yang, F., Lei, Z., Li, S.Z.: Towards fast, accurate and stable 3D dense face alignment. In: Vedaldi, A., Bischof, H., Brox, T., Frahm, J.-M. (eds.) ECCV 2020. LNCS, vol. 12364, pp. 152–168. Springer, Cham (2020). https://doi.org/10.1007/978-3-030-58529-7_10
8. He, K., Zhang, X., Ren, S., Sun, J.: Deep residual learning for image recognition. In: Proceedings of the IEEE Conference on Computer Vision and Pattern Recognition, pp. 770–778 (2016)
9. Hempel, T., Abdelrahman, A.A., Al-Hamadi, A.: 6D rotation representation for unconstrained head pose estimation. arXiv preprint arXiv:2202.12555 (2022)
10. Kao, Y., et al.: Single-image 3D face reconstruction under perspective projection. arXiv preprint arXiv:2205.04126 (2022)
11. Kendall, A., Gal, Y.: What uncertainties do we need in Bayesian deep learning for computer vision? In: Guyon, I.,et al. (eds.) Advances in Neural Information Processing Systems, vol. 30. Curran Associates, Inc. (2017)
12. Kingma, D.P., Ba, J.: Adam: a method for stochastic optimization. CoRR abs/1412.6980 (2015)
13. Liu, R., et al.: An intriguing failing of convolutional neural networks and the coord-conv solution. ArXiv abs/1807.03247 (2018)
14. Wood, E., et al.: 3D face reconstruction with dense landmarks (2022). https://doi.org/10.48550/ARXIV.2204.02776,https://arxiv.org/abs/2204.02776

W20 - Safe Artificial Intelligence for Automated Driving

W20 - Safe Artificial Intelligence for Automated Driving

The realization of highly automated driving relies heavily on the safety of AI. Demonstrations of current systems that are showcased on appropriate portals can give the impression that AI has already achieved sufficient performance and is safe. However, this by no means represents statistically significant evidence that AI is safe. A changed environment in which the system is deployed quickly leads to significantly reduced performance of DNNs. The occurrence of natural or adversarial perturbations to the input data has fatal consequences for the safety of DNNs. In addition, DNNs have an insufficient explainability of their behavior, which drastically complicates the detection of mispredictions as well as the proof that AI is safe. This workshop addresses all topics related to the safety of AI in the context of highly automated driving.

October 2022

Timo Saemann
Oliver Wasenmüller
Markus Enzweiler
Peter Schlicht
Joachim Sicking
Stefan Milz
Fabian Hüger
Seyed Ghobadi
Ruby Moritz
Oliver Grau
Frederik Blank
Thomas Stauner

One Ontology to Rule Them All: Corner Case Scenarios for Autonomous Driving

Daniel Bogdoll[1,2]([✉]), Stefani Guneshka[2], and J. Marius Zöllner[1,2]

[1] FZI Research Center for Information Technology, Karlsruhe, Germany
bogdoll@fzi.de
[2] Karlsruhe Institute of Technology, Karlsruhe, Germany

Abstract. The core obstacle towards a large-scale deployment of autonomous vehicles currently lies in the long tail of rare events. These are extremely challenging since they do not occur often in the utilized training data for deep neural networks. To tackle this problem, we propose the generation of additional synthetic training data, covering a wide variety of corner case scenarios. As ontologies can represent human expert knowledge while enabling computational processing, we use them to describe scenarios. Our proposed master ontology is capable to model scenarios from all common corner case categories found in the literature. From this one master ontology, arbitrary scenario-describing ontologies can be derived. In an automated fashion, these can be converted into the OpenSCENARIO format and subsequently executed in simulation. This way, also challenging test and evaluation scenarios can be generated.

Keywords: Corner cases · Ontology · Scenarios · Synthetic data · Simulation · Autonomous driving

1 Introduction

For selected Operational Design Domains (ODD), autonomous vehicles of the SAE level 4 [32] can already be seen on the roads [43]. However, it remains highly debated, how the safety of these vehicles can be shown and steadily improved in a structured way. In Germany, the first country with a federal law for level 4 autonomous driving, the safety of such vehicles needs to be demonstrated based on a catalog of test scenarios [17,18]. However, the coverage of rare, but highly relevant corner cases [26] in scenario-based descriptions poses a significant challenge [34]. Data-driven, learned scenario generation approaches currently tend to focus on adversarial scenarios with a high risk of collision [13,20,36,42], neglecting other forms of corner cases. While there exist comprehensive taxonomies on the types and categories of corner cases [8,21], there exist no generation method tailored to these most important long-tail scenes and scenarios. Based on this, the verification and validation during testing and ultimately the scalability of

---

D. Bogdoll and S. Guneshka—Contributed equally.

© The Author(s), under exclusive license to Springer Nature Switzerland AG 2023
L. Karlinsky et al. (Eds.): ECCV 2022 Workshops, LNCS 13805, pp. 409–425, 2023.
https://doi.org/10.1007/978-3-031-25072-9_29

autonomous driving systems to larger ODDs in real world deployments remain enormous challenges. To tackle these challenges, it is necessary to generate a large variety of rare corner case scenarios for the purposes of training, testing, and evaluation of autonomous vehicle systems. As shown by Tuncali et al. [40], model-, data-, and scenario-based methods can be used for this purpose. An extensive overview on these can be found in [6]. However, the authors find that, "while there are knowledge-based descriptions and taxonomies for corner cases, there is little research on machine-interpretable descriptions" [6].

To fill this gap between knowledge- and data-driven approaches for the description and generation of corner case scenarios[1], we propose the first scenario generation method which is capable of generating all corner case categories, also called levels, described by Breitenstein et al. [8] in a scalable fashion, where all types of scenarios can be derived from a single master ontology. Based on the resulting scenario-describing ontologies, synthetic data of rare corner case scenarios can be generated automatically. This newly created training data hopefully contributes to an increased robustness of deep neural networks to anomalies, helping make deep neural networks safer. For a general introduction to ontologies in the domain of information sciences, we refer to [31]. More details can be found in [19]. All code and data is available on GitHub.

The remainder of this work is structured as follows: In Sect. 2, we provide an overview of related ontology-based scene and scenario description methods and outline the identified research gap. In Sect. 3, we describe how our proposed master ontology is designed and the automated process to design and generate scenario ontologies from it. In Sect. 4, we demonstrate how different, concrete scenarios can be derived from the master ontology and how the resulting ontologies can be used to execute these in simulation. Finally, in Sect. 5, we provide a brief summary and outline next steps and future directions.

2   Related Work

While there exist many ways of describing scenarios [6], ontologies are the most powerful way of doing so, as these are not only human- and machine-readable, but also extremely scalable for the generation of scenarios, when used in the right fashion [22]. Ontologies are being widely used for the description of scenarios. In the work of Bagschik et al. [5], an ontology is presented which describes simple highway scenarios based on a set of pre-defined keywords. In a later work, Menzel et al. [29] extend the concept to generate OpenSCENARIO and OpenDRIVE scenarios, while many of the relevant details were not modelled in the ontology itself, but in post-processing steps. For the description of the surrounding environment of a vehicle, Fuchs et al. [15] especially focus on lanes and occupying traffic participants, while neglecting their actions. Li et al. [28] also create scenarios which are executed in a simulation environment, covering primarily situations, where sudden braking maneuvers are necessary. Thus, their ontology

---

[1] We follow the definitions of scene and scenario by Ulbrich et al. [41], where a scene is a snapshot, and a scenario consists of successive scenes.

is very domain-specific. They build upon their previous works [27,39,44]. Tahir and Alexander [38] propose an ontology that focuses on intersections due to their high collision rates. They show that their scenarios can be executed in simulation, while focusing on changing weather conditions. While they claim to have developed an ontology, the released code [45] only contains scripted scenarios, which might be derived from an ontology structurally. Hermann et al. [22] propose an ontology for dataset creation, with a demonstrated focus on pedestrian detection, including pedestrian occlusions. Their ontology is structurally inspired by the Pegasus model [37] and consists of 22 sub-ontologies. It is capable of describing a wide variety of scenarios and translate them into simulation. However, since the ontology itself is neither described in detail nor publicly available, it does not become clear whether each frame requires a separate ontology or whether the ontology itself is able to describe temporal scenarios. In the OpenXOntology project by ASAM [3], an ontology is being developed with the purpose to unify their different products, such as OpenSCENARIO or OpenDRIVE. Based on the large body of previous work in the field of scenario descriptions, this ontology is promising for further developments. However, at the moment, it serves the purpose of a taxonomy. Finally, Gelder et al. [16] propose an extensive framework for the development of a "full ontology of scenarios". At the moment, they have not developed the ontology itself yet, which is why their work cannot be compared to existing ontologies.

Table 1. Comparison of related ontologies and our proposed ontology for scenario descriptions in the field of autonomous driving.

Authors	Year	Temporal scenario description	Arbitrary environments	Arbitrary objects	Scenario simulation	Corner case categorization	Ontology available
Fuchs et al. [15]	2008	–	–	✓	–	–	–
Hummel [24]	2010	–	✓	–	–	–	–
Hülsen et al. [25]	2011	✓	✓	–	–	–	–
Armand et al. [1]	2014	–	–	–	–	–	–
Zhao et al. [46]	2017	–	✓	–	–	–	✓
Bagschik et al. [5]	2018	✓	–	–	–	–	–
Chen and Kloul [12]	2018	✓	–	–	–	–	–
Huang et al. [23]	2019	✓	–	–	–	–	–
Menzel et al. [29]	2019	✓	✓	–	✓	–	–
Li et al. [28]	2020	✓	✓	–	✓	–	–
Tahir and Alexander [38]	2022	✓	–	–	✓	–	–
Hermann et al. [22]	2022	–	✓	✓	✓	–	–
ASAM [3]	2022	–	–	–	–	–	–
Proposed ontology		✓	✓	✓	✓	✓	✓

Next to ontologies which are explicitly designed to describe scenarios, more exist which also focus on decision-making aspects. In this category, Hummel [24] developed an ontology capable of describing intersections to a degree, where the ontology can also be used to infer knowledge about the scenes. While this is a general attribute of ontologies, she provides a set of rules for the analysis.

Hülsen et al. [25] also describe intersections based on an ontology, focusing on the road layout, while interactions between entities cannot be modeled in detail. In [1], this issue is addressed, as Armand et al. focus on such interactions. They also propose rules to infer knowledge from their ontology. These rules are partly attributed to the decision-making of an ego vehicle, e.g., whether it should stop or continue. Due to their strong focus on actions and interactions, they struggle to describe complex scenarios in a more general way. Zhao et al. [46] developed a set of three ontologies, namely Map, Car, and Control. Based on these, they are capable of describing complex scenes for vehicles only. While the scenes do contain temporal information, such as paths for vehicles, these are only broad descriptions and not detailed enough to model complex scenarios. Huang et al. [23] present a similar work that is able to describe a wide variety of scenarios based on classes for road networks for highway and urban scenarios, the ego vehicle and its behavior, static and dynamic objects, as well as scenario types. However, it is designed to derive driving decisions from the descriptions instead of simulating these scenarios. Chen and Kloul [12] on the other hand propose an ontology that is primarily designed to describe highway scenarios, with a special focus on weather circumstances.

To model corner case scenarios, the requirements for an ontology are very complex. In general, it needs to be able to describe all types of scenes and scenarios. For the temporal context, an ontology needs to be able to *a) describe scenarios*. Furthermore, it needs to be able to *b) describe arbitrary environments* and *c) arbitrary objects*. Following an open world assumption, we define "arbitrary", in respect to environments and objects, as the possibility to include such without changing any classes or properties of the ontology. This means, e.g., referencing to external sources, such as OpenDRIVE files for environments or CAD files for objects. An ontology needs to be designed in a way that *d) the described scenarios can also be simulated*. Finally, *e) information about the corner case levels* needs to be included for details and knowledge extraction. In Table 1, we provide an overview of the previously introduced ontologies related to these attributes and also mention, whether the ontology itself is published online. While some authors, such as [15,24], released their ontologies previously, the provided links do not contain them anymore, which is why we excluded outdated sources. A trend can be observed, where recent approaches focus more on the aspect of scenario simulation. However, to the best of our knowledge, there exists no ontology to date that is able to describe and simulate long-tail corner case events. Our proposed ontology fills this gap, being able to generate ontology scenarios for all corner case levels and execute them in simulation.

3   Method

In order to generate corner case scenarios, we have developed a *Master Ontology* which is the foundation for the creation of specific scenarios and provides the structure for all elements of a scenario. Based on this, all common corner case categories found in the literature can be addressed. For the creation of scenarios, we have developed an *Ontology Generator* module, which is our interface to human *Scenario Designers* which do not need any expertise in the field on ontologies in order to design scenarios. For each designed scenario, a concrete *Scenario Ontology* is created. This is a major advantage over purely coded scenarios, as the complete scenario description is available in a human- and machine-readable form, which directly enables knowledge extraction, analysis, and further processing, such as exports into others formats or combinations of scenarios, for all created scenarios on any level of detail. Finally, our *OpenSCENARIO Conversion* module converts this ontology into an OpenSCENARIO file, which can directly be simulated in the CARLA simulator. An overview can be found in Fig. 1.

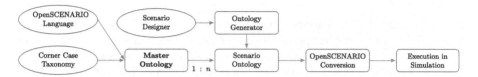

Fig. 1. Flow diagram of our proposed method for the description and generation of corner case scenarios. Based on a corner case taxonomy and the OpenSCENARIO language, a *Master Ontology* was developed, containing all necessary attributes to describe complex scenarios. In a 1 : n relation, ontologies describing individual scenarios can be derived. In an automated fashion, these scenarios are then converted into the OpenSCENARIO format, enabling the direct execution in simulation environments.

3.1 Master Ontology

At first, we describe the *Master Ontology*, which is the skeleton of every concrete scenario. With its help, different scenarios can be described by instantiating the different classes, using individuals, and setting property assertions between them. The *Master Ontology* is closely aligned to the OpenSCENARIO documentation [4], since the ontology is used for automatic generation of scenarios. Within the ontology, it is also possible to describe concrete categories of a corner cases, based on the categories introduced by Breitenstein et al. [8].

The master ontology, as shown in Fig. 2, consists of 100 classes, 53 object properties, 44 data properties, 67 individuals, and 683 axioms. The 100 classes are either classes for the description of the corner case category or derived from the OpenSCENARIO documentation [4], which means that the definitions of the different OpenSCENARIO elements can also be found there. They are used as parents for the different individuals we have created within the ontology. The 53 object properties and the 44 data properties are used to connect the different parts of a scenario, in order to embed individuals into concrete scenarios. For a better understanding and more structured explanation, the proposed *Master Ontology* can be divided into seven main groups - Scenario and Environment, Entities, Main Scenario Elements, Actions, Conditions, Weather and Time, and Corner Case Level. We will describe these in more detail in the following, with each section marked with an individual color in Fig. 2.

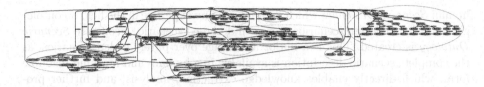

Fig. 2. Master ontology, best viewed at 1,600% and in color. The ontology is capable to describe scenarios based on the seven sections scenario and environment, entities, main scenario elements, actions, conditions, weather and time of day, and corner case level. These seven sections are further explained in Sect. 3.1. Adapted from [19]. (Color figure online)

Scenario and Environment (Red). In order to be able to describe a scenario, the *Master Ontology* provides the Scenario class, which acts as the root of the ontology. Together with the scenario class, different object and data properties are provided. Those are used as connections between the different scenario elements, such as the Entities, Towns or the Storyboard. Towns are CARLA specific environments used in the ontology. CARLA allows users to create custom and thus arbitrary environments.

Entities (Green). This group holds the different entities Vehicle, Pedestrian, Bicycle, and Misc. For arbitrary Entities, the Misc class can be utilized. If specific movement patterns are wanted, the classes Vehicle, Pedestrian and Bicycle are also already available. The individuals can be then connected to 3D assets from the CARLA blueprint library [9], which can be extended with external objects. This way, a *Scenario Designer* is able to add arbitrary assets into a scenario.

Main Scenario Elements (Yellow). The main scenario elements are used to build the core of any scenario. The highest level is the Storyboard, which includes an Init and a Story. A Story has at least one Act, which needs at least a StartTrigger and can optionally include a StopTrigger. Acts also are a container for different ManeuverGroups, that logically include Maneuvers. The Maneuvers then have to have minimum one Event, which is also activated by a StartTrigger. At last, each Event needs to include at least one Action. These are the main components of the OpenSCENARIO scenario description language and thus necessary parts of each scenario. For each of them, also a corresponding connecting property exists, i.e. *has_event, has_action, has_init_action*.

Actions (Dark Blue). To be able to describe the maneuvers of the different Entities, different Actions are represented within the Ontology. Those include, e.g., TeleportAction, which sets the position of an Entity, or Relative-LaneChangeAction, which describes a lane change of an Entity.

Conditions (Light Blue). As part of the StartTrigger and StopTrigger elements, Conditions are used to activate them. Conditions are divided into the two subclasses ByEntityCondition and ByValueCondition. In general the difference between those two is that the ByEntityCondition is always in regard to an entity, i.e., how close a vehicle is to another vehicle, while the ByValueCondition is always in regard to a value, i.e., the passed simulation time. Depending on the type of the Condition, different values must be met in order for the StartTrigger or StopTrigger to get activated. As an example, the SimulationTimeCondition can be used as a trigger with respect to the simulation time, using arithmetic rules.

Weather and Time of Day (Orange). To set the weather, the underlying CARLA town can be modified individually. This includes the weather conditions, which are subdivided into fog, precipitation, and the position of the sun. Also, the time of day can be set.

Corner Case Level (Pink). In the long tail of rare scenarios, each can be related to a specific corner case level. The first ones to propose an extensive taxonomy on these were Breitenstein et al. [7] with a focus on camera-based sensor setups. This taxonomy was extended by Heidecker et al. [21] to generalize it to a set of three top-level layers, namely Sensor Layer, Content Layer, and Temporal Layer, as shown in Fig. 3. In this work, we focus on camera-related corner cases, which is why the master ontology uses a mixed model, where the top level layers from Heidecker et al. and the underlying corner case levels from Breitenstein et al. are used, as shown in Fig. 2, making a future extension of the master ontology to further sensors effortless, as they fall into the same top level layers. Occurrences on the hardware or physical-level, such as dead pixels or overexposure, can be simulated with subsequent scripts during the simulation phase. Details on those corner cases can be placed in the individual scenario ontologies by creating specific individuals of the respective corner case classes of the *Master Ontology*.

	Sensor Layer		Content Layer			Temporal Layer
	Hardware Level	Physical Level	Domain Level	Object Level	Scene Level	Scenario Level
LiDAR-based corner cases	Laser Error • Broken mirror • Misaligned actuator	Beam-Based Corner Case • Black cars disappear • ...	Domain Shift on Single Point Cloud • Shape of Road markings	Single-Point Anomaly on Single Point Cloud • Dust cloud • ...	Contextual/Collective Anomaly on Single Point Cloud • Sweeper cleaning the sidewalk	Corner Cases on Multiple Point Clouds and Frames
Camera-based corner cases	Pixel Error • Dead pixel • Broken lense	Pixel-Based Corner Case • Dirt on lense • Overexposure	Domain Shift on Single Frame • Location (EU-U.S.A.) • ...	Single-Point Anomaly on Single Frame • Animal • ...	Contextual/Collective Anomaly on Single Frame • People on a billboard • ...	• Person breaks traffic rule • Overtaking a cyclist • Car accident
RADAR-based corner cases	Impulse Error • Low voltage • Low temperature	Impulse-Based Corner Case • Interference • ...	Domain Shift on Single Point Cloud • Weather, e.g., snow, rain, etc.	Single-Point Anomaly on Single Point Cloud • Lost objects • ...	Contextual/Collective Anomaly on Single Point Cloud • Demonstration • Tree on street	• ...

Fig. 3. Corner case categorization from Heidecker et al. [21]. The columns show different layers and levels of corner cases, while the rows are related to specific sensors. For each combination, examples are provided.

Next to those groups, additional 67 individuals exist, which are divided into Constants and Default Individuals. There are two types of constants: OpenSCE-NARIO Constants, such as arithmetic or priority rules, and CARLA Constants, such as assets. The default individuals are used to help a *Scenario Designer* to create scenarios faster and easier. These include common patterns, such as default weather conditions or a trigger, which activates when the simulation starts running. In addition, a default ego vehicle is also included in the *Master Ontology*, which has a set of cameras and a BoundingBox attached to it. As the last part of the ontology, the 683 axioms represent the connections and rules between the entities and the properties within the ontology, along with the individuals.

3.2   Scenario Ontology Generation

The manual creation of ontologies is a very time-consuming, exhausting and error-prone process, which additionally needs expertise in the general field of ontologies and related software. Thus, and to ensure, that the *OpenSCENARIO Conversion* module functions properly, we have developed the *Ontology Generator* module, which takes as input a scripted version of a scenario and creates a *Scenario Ontology* as a result. The concept behind the *Ontology Generator* is to use the *Master Ontology* as a base for a scenario description and automatically create the necessary individuals and property assertions between them. To read and write ontologies, we utilize the python library Owlready2 [33]. The *Master Ontology* is read by the *Ontology Generator*, and it uses, depending on the scenario, all classes, properties, and default individuals needed. The result is a new *Scenario Ontology*, which has the same structure as the *Master Ontology* with respect to classes and properties, but includes newly created individuals for the scenario designed by the *Scenario Designer*.

Since the *Master Ontology* is built based on the OpenSCENARIO documentation [2], which is a very powerful and flexible framework, it allows for many possible combinations. This gives a *Scenario Designer* a large flexibility with respect to the design of new scenarios. The *Ontology Generator* is well documented. This way, no prior experience with the OpenSCENARIO format is necessary. Table 2 provides an overview of the functions available to the human *Scenario Designer* to create scenarios.

With the help of the *Ontology Generator*, every part which was defined within the *Master Ontology* can be utilized. For example, the functions within the first group shown in Table 2, are used to create every scenario main element. Algorithm 1 shows, how a partly abstracted implementation, as done by a *Scenario Designer*, looks like. The full example can be found in the GitHub repository. In Sect. 4 we demonstrate an exemplary *Scenario Ontology* which was generated by the *Ontology Generator*. In this demonstration, the scenario ontology from Algorithm 1 is related to the visualization in Fig. 5.

Table 2. Overview of methods of the *Ontology Generator* module, which are available to a human *Scenario Designer*. The methods are divided in seven groups, five from which were introduced in Sect. 3.1. The groups Assets and Other are related to 3D objects and miscellaneous functions, respectively.

Main scenario elements	Entities	Actions	Environment and weather
newScenario	newEgoVehicle	newEnvironmentAction	newEnvironment
newStoryboard	newCar	newSpeedAction	newTimeOfDay
newInit	newPedestrian	newTeleportActionWithPosition	newWeather
newStory	newMisc	newTeleportActionWithRPAO	newFog
newAct		newRelativeLaneChangeAction	newPrecipitation
newManeuverGroup			newSun
newManeuver			changeWeather
newEvent			
newAction			
Conditions	Assets	Other	
newSimulationCondition	newAsset	newStopTrigger	
newRelativeDistanceCondition	getBicycleAssets	newStartTrigger	
newStoryboardESCondition	getCarAssets	newRoadCondition	
newTraveledDistanceCondition	getPedestrianAssets	newTransitionDynamics	
	getMiscAssets	setCornerCase	

Algorithm 1: Creation of a scenario ontology with the Ontology Generator, where the ego vehicle enters a foggy area (incl. abstract elements)

```
1  import OntologyGenerator as OG
2  import MasterOntology as MO
3  ego_vehicle ← MO.ego_vehicle //Default ego vehicle individual
4  weather_def ← MO.def_weather //Default weather individual
5  Initialize teleport_action(ego_vehicle), speed_action(ego_vehicle)
6  init_scenario ← OG.newInit(speed_action, teleport_action, weather_def)
   //Starting conditions for storyboard
7  Initialize traveled_distance_condition
8  Trigger ← OG.newStartTrigger(traveled_distance_condition)
   //Trigger Condition: Ego vehicle travelled defined distance
9  Initialize weather(sun, fog, precipitation)
10 Initialize time_of_day, road_condition
11 env ← OG.newEnv(time_of_day, weather, road_condition)
12 env_action ← OG.newEnvAction(env)
   //Foggy environment after trigger
13 Initialize Event, Maneuver, ManeuverGroup, Act, Story, Storyboard
   //Necessary OpenSCENARIO elements
14 Export ScenarioOntology
```

3.3 Scenario Simulation

After a scenario is described with the help of individuals within a scenario ontology, it is being read by the *OpenSCENARIO Conversion* module, as shown in Fig. 1. From these concrete scenarios, the conversion module generates

OpenSCENARIO files. These can be directly simulated without any further adjustments. Since the OpenSCENARIO files include simulator-specific details, we have focused on the CARLA[2] simulation environment [10]. In an earlier stage, we were also able to show compatibility with the esmini [14] environment.

When the *Ontology Generator* module is used to create the scenario ontologies, their structural integrity is ensured, which is a necessary requirement for the conversion module. This means that each scenario ontology is correctly provided to be processed by the conversion module. Theoretically, scenario ontologies could also be created manually to be processed by the conversion module. However, human errors are likely, preventing the correct processing by the conversion module.

While each scenario ontology is able to cover multiple corner cases, the created ontologies are fully modular. This means, given the same environment, our method is capable of combining multiple, already existing scenario ontologies into a new single scenario ontology. In such cases, where the number of scenario individuals is $n > 1$, a pre-processing stage is triggered which extends the ontology to combine all n provided scenarios into a single new scenario S_{fusion}. For this purpose, this stage creates a new scenario, storyboard, and init. Subsequently, for every included scenario, the algorithm goes through its stories, entities, and init actions and merges them in S_{fusion}. For the final creation of the OpenSCENARIO file, the conversion module utilizes the property assertions between individuals to create the according python objects, which are then used by the PYOSCX library [35] to create the OpenSCENARIO file. These files can then be read by the ScenarioRunner [11] and executed in CARLA. In the following Sect. 4, we demonstrate a set of ten simulated scenarios.

4   Evaluation

For the evaluation, we have created a diverse scenario catalog containing scenarios from all corner case levels. Following Breitenstein et al. [7], these cover different levels of complexity and thus criticalities, starting with simpler sensor layer cases and ending with highly complex temporal layer corner cases. For the qualitative evaluation we show the feasibility of the approach as shown in Fig. 1 and demonstrate it with a set of ten scenarios, that descriptions made by a human *Scenario Designer* get translated into proper *Scenario Ontologies* and correctly simulated.

For the selection of the exemplary corner case scenarios, we considered three types of sources. First, we used examples provided by the literature, such as the ones provided by Breitenstein et al. [8]. Second, various video sources, such as third-person videos, and dash-cam videos of traffic situations [30] were used for inspiration. Third, multiple brainstorming sessions took place, where personal experiences were collected. Afterwards, we narrowed the selection down to a set of eight representative scenarios. Two more were created by combining two of

---

[2] CARLA version 0.9.13 was utilized.

Table 3. Overview of scenario ontologies, which were derived from the master ontology and subsequently executed in simulation. These exemplary scenarios cover all corner case categories.

#	Corner case level	Individuals	Scenario description
(a)	Sensor Layer - Hardware Level	–	Dead Pixel: Camera sensor affected
(b)	Content Layer - Domain Level	94	Domain Shift: Sudden weather change
(c)	Content Layer - Object Level	93	Single-Point Anomaly: Unknown object on the road
(d)	Content Layer - Scene Level	164	Collective Anomaly: Multiple known objects on the road
(e)	Content Layer - Scene Level	111	Contextual Anomaly: Known non-road object on the road
(f)	Temporal Layer - Scenario Level	94	Novel Scenario: Unexpected event in another lane
(g)	Temporal Layer - Scenario Level	104	Risky Scenario: A risky maneuver
(h)	Temporal Layer - Scenario Level	95	Anomalous Scenario: Unexpected traffic participant behaviour
(i)	Combined: (d) and (f)	156	Combined: Collective and Novel Scenario
(j)	Combined: (f) and (h)	122	Combined: Novel and Anomalous Scenario

those eight scenarios. An overview of these scenarios can be found in Table 3. Visualizations of all scenarios can be found in Fig. 4.

In the *a) Dead Pixel* scenario, an arbitrary scenario can be chosen, where only the sensor itself will be affected. The *b) Domain Shift* is a sudden weather change, where we created a scenario where the ego vehicle suddenly drives into a dense fog. For the *c) Single-Point Anomaly*, we chose to simulate a falling vending machine on the road. This scenario also inspired the next scenario, *d) Contextual Anomaly*, which also has falling objects on the road, but in this case the objects are traffic signs. This can be considered, for example, in a very windy environment. As a *e) Collective Anomaly*, we chose to simulate a lot of running pedestrians in front of the ego vehicle, which can for example happen during a sports event. For the *f) Novel Scenario*, we described a scenario where a cyclist performs unexpected maneuvers in the opposite lane. The *g) Risky Scenario* which we chose is a close cut-in maneuver in front of the ego vehicle. The last corner case category is the *h) Anomalous scenario*, for which we have chosen a pedestrian who suddenly runs in front of the ego vehicle. To demonstrate the scalability of our approach, we have also combined scenarios. In the *i)* combination of *Novel Scenario* and the *Collective Scenario*, a lot of running pedestrians are in front of the ego vehicle, while a cyclist performs unexpected maneuvers in the opposite lane, next to the pedestrians. In addition, we also merged the *Novel Scenario* with the *Anomalous Scenario*, resulting in *j)* a pedestrian walking in front of the ego vehicle and the cyclist.

4.1   Scenario Ontologies

At the core of each demonstrated scenario lies a *Scenario Ontology*. In the following, we present the construction of an exemplary scenario and describe how it is represented in the *Scenario Ontology* with individuals. An example graph of the exemplary ontology can be seen in Fig. 5. The ontology has 94 individuals, which means that 27 new individuals were created, since the *Master Ontology* has 67 default individuals.

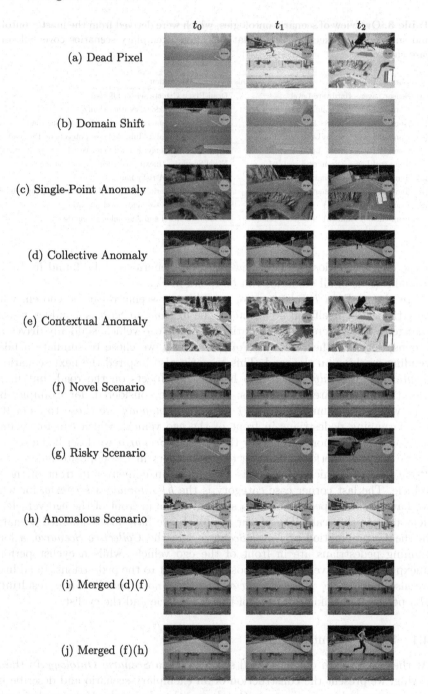

Fig. 4. Visualization of the realized corner case scenarios as listed in Table 3.

In Fig. 5, every used class, property and individual is presented. Each individual, which name starts with *"indiv_"*, is a newly created part of the *Scenario ontology*, every other individual is either a default or a constant. The graph starts from the top with the *Scenario* individual, which is connected to a CARLA town and a newly created Storyboard. As mentioned earlier, every *Storyboard* has an *Init* and a *Story*. In this particular *Init*, there are only the *Actions*, which are responsible for the position and the speed of the *Ego Vehicle*, and connections to the default *EnvironmentAction*. The most interesting part of this *Scenario* however can be found deep within the *Story* - namely the second *EnvironmentAction*, which creates a dense fog inside the scenario. This *Action* gets triggered by the *indiv_DistanceStartTrigger*, which has a *TraveledDistanceCondition* as a *Condition*. Since this type of *Condition* is an *EntityCondition*, it requires a connection to an *Entity*, in our case the *ego_vehicle*. This *StartTrigger* gets activated when the *ego_vehicle* has travelled a certain distance. After this *Event* is executed, the Scenario comes to its end.

Descriptions and visualizations of the remaining nine *Scenario Ontologies* can be found in [19], and the code for all ten scenarios, in order to recreate them, can be found in the GitHub repository.

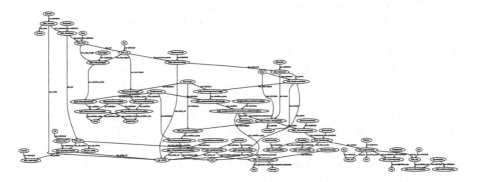

Fig. 5. Scenario ontology describing a vehicle entering a foggy area. Best viewed at 1,600 %. The ontology describes the scenario based on the seven sections scenario and environment, entities, main scenario elements, actions, conditions, weather and time of day, and corner case level. Reprinted from [19].

5   Conclusion

Our work focuses on the generation of rare corner case scenarios in order to provide additional input training data as well as test and evaluation scenarios. This way, it contributes to potentially safer deep neural networks, which might be able to become more robust to anomalies. The proposed *Master Ontology* is the core of our approach and enables the creation of specific scenarios for all of the corner case levels developed by Breitenstein et al. [8]. From the one *Master Ontology*, concrete scenario ontologies can be derived. The whole process after the design of such an ontology is fully automated, which includes the generation of ontologies itself, based on the human input for the *Ontology Generator* as well as the conversion into the OpenSCENARIO format. This allows for a direct simulation of the scenarios in the CARLA simulation environment without any further adjustments. Since ontologies are a highly complex domain, we have put an emphasis on the design of the *Ontology Generator* module, which does not require expertise in the field of ontologies for a human to create scenarios. We have demonstrated our approach with a set of ten concrete scenarios, which cover all corner case levels.

As an outlook, we would first like to discuss existing limitations of our work. At the moment, the master ontology is focused on camera-related corner cases, which is why the current implementation only includes hardware-related corner cases for camera sensors. Also, the subsequent scripts for camera-related corner cases are currently not triggered automatically, but manually. For the *Master Ontology*, we have implemented a vast body of the OpenSCENARIO standard in order to demonstrate our designed scenarios. However, for future scenarios, modifications of the master ontology might be necessary. We have provided instructions for such extensions in our GitHub repository. While the master ontology supports arbitrary objects, in our demonstrated scenarios we were only able to utilize assets which were already included within CARLA, as we ran into compilation issues with respect to the Unreal Engine and the utilized CARLA version. Our learnings as well as ideas to address this issue can also be found in the GitHub repository.

Finally, we would like to point out future directions. The extraction of corner case levels from the ontology itself could be automated, e.g., with knowledge extraction methods based on Semantic Web Rule Language (SWRL) rules. However, this is a challenging field itself. Based on the generated scenarios, which were created by human *Scenario Designers*, automated variations can be introduced to drastically increase the number of available scenarios, as shown by [22,28,29]. This way, a powerful combination of knowledge- and data-driven scenario generation can be achieved for the long tail of rare corner cases.

Acknowledgment. This work results partly from the project KI Data Tooling (19A20001J) funded by the Federal Ministry for Economic Affairs and Climate Action (BMWK).

References

1. Armand, A., Filliat, D., Ibañez-Guzman, J.: Ontology-based context awareness for driving assistance systems. In: IEEE Intelligent Vehicles Symposium Proceedings (2014)
2. ASAM: ASAM OpenSCENARIO. https://www.asam.net/standards/detail/opens cenario. Accessed 28 Feb 2022
3. ASAM: ASAM OpenXOntology. https://www.asam.net/project-detail/asam-open xontology/. Accessed 28 Feb 2022

4. ASAM: OpenSCENARIO Documentation. https://releases.asam.net/OpenSCEN ARIO/1.0.0/ASAM_OpenSCENARIO_BS-1-2_User-Guide_V1-0-0.html. Accessed 28 Jan 2022
5. Bagschik, G., Menzel, T., Maurer, M.: Ontology based scene creation for the development of automated vehicles. In: IEEE Intelligent Vehicles Symposium (IV) (2018)
6. Bogdoll, D., et al.: Description of corner cases in automated driving: goals and challenges. In: Proceedings of the IEEE/CVF International Conference on Computer Vision (ICCV) Workshops (2021)
7. Breitenstein, J., Termöhlen, J.A., Lipinski, D., Fingscheidt, T.: Systematization of corner cases for visual perception in automated driving. In: IEEE Intelligent Vehicles Symposium (IV) (2020)
8. Breitenstein, J., Termöhlen, J.A., Lipinski, D., Fingscheidt, T.: Corner cases for visual perception in automated driving: some guidance on detection approaches. arXiv:2102.05897 (2021)
9. CARLA: CARLA Blueprint Library. https://carla.readthedocs.io/en/latest/bp_ library/. Accessed 28 Feb 2022
10. CARLA: CARLA Simulator. https://carla.org/. Accessed 28 Feb 2022
11. CARLA: Scenario Runner Github. https://github.com/carla-simulator/scenario_ runner. Accessed 28 Feb 2022
12. Chen, W., Kloul, L.: An ontology-based approach to generate the advanced driver assistance use cases of highway traffic. In: Proceedings of the 10th International Joint Conference on Knowledge Discovery, Knowledge Engineering and Knowledge Management (KEOD) (2018)
13. Ding, W., Chen, B., Li, B., Eun, K.J., Zhao, D.: Multimodal safety-critical scenarios generation for decision-making algorithms evaluation. IEEE Robot. Autom. Lett. **6**, 1551–1558 (2021)
14. esmini: Esmini github. https://github.com/esmini/esmini. Accessed 28 Feb 2022
15. Fuchs, S., Rass, S., Lamprecht, B., Kyamakya, K.: A model for ontology-based scene description for context-aware driver assistance systems. In: International ICST Conference on Ambient Media and Systems (2010)
16. de Gelder, E., et al.: Towards an ontology for scenario definition for the assessment of automated vehicles: an object-oriented framework. IEEE Trans. Intell. Veh. **7**(2), 300–314 (2022)
17. (Germany), B.: Verordnung zur Regelung des Betriebs von Kraftfahrzeugen mit automatisierter und autonomer Fahrfunktion und zur Änderung straßen-verkehrsrechtlicher Vorschriften. https://dserver.bundestag.de/brd/2022/0086-22. pdf. Accessed 15 June 2022
18. (Germany), B.: Entwurf eines Gesetzes zur Änderung des Straßenverkehrsgesetzes und des Pflichtversicherungsgesetzes - Gesetz zum autonomen Fahren (2021). https://www.bmvi.de/SharedDocs/DE/Anlage/Gesetze/Gesetze-19/gesetz-aende rung-strassenverkehrsgesetz-pflichtversicherungsgesetz-autonomes-fahren.pdf?__bl ob=publicationFile. Accessed 15 June 2022
19. Guneshka, S.: Ontology-based corner case scenario simulation for autonomous driving. Bachelor thesis, Karlsruhe Institute of Technology (KIT) (2022)
20. Hanselmann, N., Renz, K., Chitta, K., Bhattacharyya, A., Geiger, A.: King: generating safety-critical driving scenarios for robust imitation via kinematics gradients. arXiv:2204.13683 (2022)
21. Heidecker, F., et al.: An application-driven conceptualization of corner cases for perception in highly automated driving. In: IEEE Intelligent Vehicles Symposium (IV) (2021)

22. Herrmann, M., et al.: Using ontologies for dataset engineering in automotive AI applications. In: Design, Automation and Test in Europe Conference and Exhibition (DATE) (2022)
23. Huang, L., Liang, H., Yu, B., Li, B., Zhu, H.: Ontology-based driving scene modeling, situation assessment and decision making for autonomous vehicles. In: Asia-Pacific Conference on Intelligent Robot Systems (ACIRS) (2019)
24. Hummel, B.: Description logic for scene understanding at the example of urban road intersections. Ph.D. thesis, Karlsruhe Institute of Technology (KIT) (2009)
25. Hülsen, M., Zöllner, J.M., Weiss, C.: Traffic intersection situation description ontology for advanced driver assistance. In: IEEE Intelligent Vehicles Symposium (IV) (2011)
26. Karpathy, A.: Tesla Autonomoy Day (2019). https://youtu.be/Ucp0TTmvqOE?t=8671. Accessed 15 June 2022
27. Klueck, F., Li, Y., Nica, M., Tao, J., Wotawa, F.: Using ontologies for test suites generation for automated and autonomous driving functions. In: IEEE International Symposium on Software Reliability Engineering Workshops (ISSREW) (2018)
28. Li, Y., Tao, J., Wotawa, F.: Ontology-based test generation for automated and autonomous driving functions. Inf. Softw. Technol. **117**, 106200 (2020)
29. Menzel, T., Bagschik, G., Isensee, L., Schomburg, A., Maurer, M.: From functional to logical scenarios: Detailing a keyword-based scenario description for execution in a simulation environment. In: IEEE Intelligent Vehicles Symposium (IV) (2019)
30. Minute, M.: Trees falling on road (2017). https://www.youtube.com/watch?v=3VsLeUtXvxk&ab_channel=Mad1Minute. Accessed 21 July 2022
31. Noy, N., Mcguinness, D.: Ontology development 101: a guide to creating your first ontology. In: Knowledge Systems Laboratory, vol. 32 (2001)
32. On-Road Automated Driving Committee: Taxonomy and Definitions for Terms Related to Driving Automation Systems for On-Road Motor Vehicles. Standard J3016-202104, SAE International (2021)
33. OWLReady2: Welcome to Owlready2's documentation! (2021). https://owlready2.readthedocs.io/en/v0.36/. Accessed 28 Feb 2022
34. Pretschner, A., Hauer, F., Schmidt, T.: Tests für automatisierte und autonome Fahrsysteme. Informatik Spektrum **44**, 214–218 (2021)
35. pyoscx: scenariogeneration (2022). https://github.com/pyoscx/scenariogeneration. Accessed 20 July 2022
36. Rempe, D., Philion, J., Guibas, L.J., Fidler, S., Litany, O.: Generating useful accident-prone driving scenarios via a learned traffic prior. In: Proceedings of the IEEE/CVF Conference on Computer Vision and Pattern Recognition (CVPR) (2022)
37. Schoener, H.P., Mazzega, J.: Introduction to Pegasus. In: China Autonomous Driving Testing Technology Innovation Conference (2018)
38. Tahir, Z., Alexander, R.: Intersection focused situation coverage-based verification and validation framework for autonomous vehicles implemented in Carla. In: Mazal, J., et al. (eds.) Modelling and Simulation for Autonomous Systems (2022)
39. Tao, J., Li, Y., Wotawa, F., Felbinger, H., Nica, M.: On the industrial application of combinatorial testing for autonomous driving functions. In: IEEE International Conference on Software Testing, Verification and Validation Workshops (ICSTW) (2019)
40. Tuncali, C.E., Fainekos, G., Prokhorov, D., Ito, H., Kapinski, J.: Requirements-driven test generation for autonomous vehicles with machine learning components. IEEE Trans. Intell. Veh. **5**(2), 265–280 (2020)

41. Ulbrich, S., Menzel, T., Reschka, A., Schuldt, F., Maurer, M.: Defining and substantiating the terms scene, situation, and scenario for automated driving. In: Proceedings of ITSC (2015)
42. Wang, J., et al.: AdvSim: generating safety-critical scenarios for self-driving vehicles. In: Proceedings of the IEEE/CVF Conference on Computer Vision and Pattern Recognition (CVPR) (2021)
43. Waymo: Waymo One (2022). https://waymo.com/waymo-one/. Accessed 15 June 2022
44. Wotawa, F., Li, Y.: From ontologies to input models for combinatorial testing. In: International Conference on Testing Software and Systems (ICTSS) (2018)
45. Zaid, T.: Intersection focused situation coverage-based verification and validation framework for autonomous vehicles implemented in CARLA (2022). https://github.com/zaidtahirbutt/Situation-Coverage-based-AV-Testing-Framework-in-CARLA. Accessed 20 July 2022
46. Zhao, L., Ichise, R., Liu, Z., Mita, S., Sasaki, Y.: Ontology-based driving decision making: a feasibility study at uncontrolled intersections. IEICE Trans. Inf. Syst. **100**(D(7)), 1425–1439 (2017)

Parametric and Multivariate Uncertainty Calibration for Regression and Object Detection

Fabian Küppers[1,2]([✉]), Jonas Schneider[1], and Anselm Haselhoff[2]

[1] e:fs TechHub GmbH, Gaimersheim, Germany
{fabian.kueppers,jonas.schneider}@efs-auto.com
[2] Ruhr West University of Applied Sciences, Bottrop, Germany
anselm.haselhoff@hs-ruhrwest.de

Abstract. Reliable spatial uncertainty evaluation of object detection models is of special interest and has been subject of recent work. In this work, we review the existing definitions for uncertainty calibration of probabilistic regression tasks. We inspect the calibration properties of common detection networks and extend state-of-the-art recalibration methods. Our methods use a Gaussian process (GP) recalibration scheme that yields parametric distributions as output (e.g. Gaussian or Cauchy). The usage of GP recalibration allows for a local (conditional) uncertainty calibration by capturing dependencies between neighboring samples. The use of parametric distributions such as Gaussian allows for a simplified adaption of calibration in subsequent processes, e.g., for Kalman filtering in the scope of object tracking.

In addition, we use the GP recalibration scheme to perform **covariance estimation** which allows for post-hoc introduction of local correlations between the output quantities, e.g., position, width, or height in object detection. To measure the joint calibration of multivariate and possibly correlated data, we introduce the *quantile calibration error* which is based on the Mahalanobis distance between the predicted distribution and the ground truth to determine whether the ground truth is within a predicted quantile.

Our experiments show that common detection models overestimate the spatial uncertainty in comparison to the observed error. We show that the simple Isotonic Regression recalibration method is sufficient to achieve a good uncertainty quantification in terms of calibrated quantiles. In contrast, if normal distributions are required for subsequent processes, our GP-Normal recalibration method yields the best results. Finally, we show that our **covariance estimation** method is able to achieve best calibration results for joint multivariate calibration. All code is open source and available at https://github.com/EFS-OpenSource/calibration-framework.

© The Author(s), under exclusive license to Springer Nature Switzerland AG 2023
L. Karlinsky et al. (Eds.): ECCV 2022 Workshops, LNCS 13805, pp. 426–442, 2023.
https://doi.org/10.1007/978-3-031-25072-9_30

Fig. 1. Qualitative example of spatial uncertainty regression for object detection tasks. (a) A probabilistic `RetinaNet` [22] outputs (independent) normal distributions for each dimension. (b) On the one hand, we can recalibrate these distributions using *GP-Beta* [31] which yields multiple (independent) distributions of arbitrary shape. (c) On the other hand, we can also use our multivariate (mv.) *GP-Normal* method which recalibrates the uncertainty of Gaussian distributions and is also able to capture correlations between different dimensions.

1 Introduction

Obtaining reliable uncertainty information is a major concern especially for safety-critical applications such as autonomous driving [7,8,35]. For camera-based environment perception, it is nowadays possible to utilize object detection algorithms that are based on neural networks [1]. Such detection methods can be used in the context of object tracking (tracking-by-detection) to track individual objects within an image sequence [2,34]. Besides the position of individual objects, common tracking algorithms such as Kalman filtering require additional uncertainty information. However, these detection models are known to produce unreliable confidence information [10,19,27]. In this context, uncertainty calibration is the task of matching the predicted model uncertainty with the observed error. For example, in the scope of classification, calibration requires that the predicted confidence of a forecaster should match the observed accuracy. In contrast, the task for probabilistic regression models is to target the true ground-truth score for an input using a probability distribution as the model output [13,15]. An object detection model can also be trained to output probabilistic estimates for the position information using Gaussian distributions [7,11–13]. We refer to this as the probabilistic regression branch of a detection model.

While there is a large consent about the metrics and definitions in the scope of classification calibration [10,17,18,24,25], research for regression calibration still differs in definitions as well as in the used evaluation metrics [16,20,21,31]. Recent work has proposed several methods to apply post-hoc calibration for regression tasks [16,20,21,31]. A flexible and input-dependent recalibration method is the GP-Beta provided by [31] that utilizes a Gaussian Process (GP) for recalibration parameter estimation. However, this method provides non-parametric distributions as calibration output which might be disadvantageous for subsequent applications such as Kalman filtering since these applications commonly require parametric distributions, e.g., Gaussians [2]. In contrast, the authors in [20,21] propose a temperature scaling [10] for the variance of a Gaussian distribution which, however, is not sensitive to a specific input. Furthermore, most of the current research only focuses on the calibration of 1D regression problems. However, especially for object detection, it might be necessary to jointly inspect the calibration properties in all dimensions of the probabilistic regression branch. For example, a large object might require a different uncertainty calibration compared to smaller objects. A representative uncertainty quantification is of special interest especially for safety-critical applications such as autonomous driving or tracking tasks, e.g., Kalman filtering. However, a joint recalibration of multiple dimensions is not possible using the current existing regression calibration methods. Therefore, we seek for a multivariate regression recalibration method that offers the flexibility in the parameter estimation of methods such as GP-Beta [31] but also provides parametric probability distributions.

Contributions. In this work, we focus on the safety-relevant task of object detection and provide a brief overview over the most important definitions for regression calibration. We adapt the idea of Gaussian process (GP) recalibration [31] by using parametric probability distributions (Gaussian or Cauchy) as calibration output. Furthermore, we investigate a method for a joint regression calibration of multiple dimensions. The effect of joint uncertainty calibration is qualitatively demonstrated in Fig. 1. On the one hand, this method is able to capture possible (conditional) correlations within the data to subsequently learn covariances. On the other hand, our method is also able to recalibrate multivariate Gaussian probability distributions with full covariance matrices. The task of joint multivariate recalibration for regression problems hasn't been addressed so far to best of our knowledge.

2 Definitions for Regression Calibration and Related Work

In this section, we give an overview over the related work regarding uncertainty calibration for regression. Furthermore, we summarize the most important definitions for regression calibration. In this way, we can examine calibration in a unified context that allows us to better compare the calibration definitions as well as the associated calibration methods.

In the scope of uncertainty calibration, we work with models that output an uncertainty quantification for each forecast. Let $X \in \mathcal{X}$ denote the input for a regression model $h(X)$ that predicts the aleatoric (data) uncertainty as a probability density distribution $f_{Y|X} \in \mathcal{S}$ where $\mathcal{S}$ denotes the set of all possible probability distributions, so that $S_Y := f_{Y|X}(y|x) = h(x)$, $S_Y \in \mathcal{S}$. The predicted probability distribution targets the ground-truth $Y \in \mathcal{Y} = \mathbb{R}$ in the output space. Let further denote $F_{Y|X} : \mathcal{Y} \to (0,1)$ the respective cumulative density function (CDF), and $F_{Y|X}^{-1} : (0,1) \to \mathcal{Y}$ the (inverse) quantile function.

Quantile Calibration. We start with the first definition for regression calibration in terms of quantiles provided by [16]. The authors argue that a probabilistic forecaster is well calibrated if an estimated prediction interval for a certain quantile level $\tau \in (0,1)$ covers the ground-truth Y in $100\tau\%$ cases, i.e., if the predicted and the empirical CDF match given sufficient data. More formally, a probabilistic forecaster is *quantile calibrated*, if

$$\mathbb{P}\left(Y \leq F_{Y|X}^{-1}(\tau)\right) = \tau, \quad \forall \tau \in (0,1), \tag{1}$$

is fulfilled. This also holds for two-sided quantiles [16]. Besides their definition of *quantile calibration*, the authors in [16] propose to use the isotonic regression calibration method [37] from classification calibration to recalibrate the cumulative distribution function predicted by a probabilistic forecaster. We adapt this method and use it as a reference in our experiments.

Distribution Calibration. In classification calibration, the miscalibration of a classifier is conditioned on the predicted confidence [10], i.e., we inspect the miscalibration of samples with equal confidence. In contrast, the definition of *quantile calibration* in (1) only faces the marginal probability for all distributional moments [20,21,31]. Therefore, the authors in [31] extend the definition for regression calibration to *distribution calibration*. This definition requires that an estimated probability distribution should match the observed error distribution given sufficient data. Therefore, a model is *distribution calibrated*, if

$$f(Y = y | S_Y = s) = s(y) \tag{2}$$

holds for all continuous distributions $s \in \mathcal{S}$ that target Y, and $y \in \mathcal{Y}$ [31]. Following Theorem 1 in [31], a *distribution calibrated* probabilistic forecaster is also *quantile calibrated* (but not vice-versa). In addition, the authors in [31] propose the *GP-Beta* recalibration method, an approach that uses beta calibration [17] in conjunction with GP parameter estimation to recalibrate the cumulative distribution and to achieve *distribution calibration*. The *Isotonic Regression* as well as the *GP-Beta* approaches exploit the fact that the CDF is defined in the $[0, 1]$ interval which allows for an application of recalibration methods from the scope of classification calibration. While the former method [16] seeks for the optimal recalibration parameters globally, the latter approach [31] utilizes GP parameter estimation to find the optimal recalibration parameters for a single distribution using all samples whose distribution moments are close to the target sample. This allows for a local uncertainty recalibration.

Variance Calibration. Besides the previous definitions, the authors in [21] and [20] consider normal distributions over the output space with mean $\mu_Y(X)$ and variance $\sigma_Y^2(X)$ that are implicit functions of the base network with X as input. The authors define the term of *variance calibration*. This definition requires that the predicted variance of a forecaster should match the observed variance *for* *a certain variance level.* For example, given a certain set of predictions by an unbiased probabilistic forecaster *with equal variance*, the mean squared error (which is equivalent to the observed variance in this case) should match the predicted variance. This must hold for all possible variances. Thus, *variance calibration* is defined as

$$\mathbb{E}_{X,Y}\left[\left(Y - \mu_Y(X)\right)^2 \Big| \sigma_Y^2(X) = \sigma^2\right] = \sigma^2, \tag{3}$$

for all $\sigma^2 \in \mathbb{R}_{>0}$ [20,21]. Note that any *variance calibrated* forecaster is *quantile calibrated*, if (a) the forecaster is unbiased, i.e., $\mathbb{E}_{Y|X}[Y] = \mu = \mu_Y(X)$ for all X, and (b) the ground-truth data is normally distributed around the predicted mean. In this case, the predicted normal distribution $\mathcal{N}(\mu_Y(X), \sigma_Y^2(X))$ matches the observed data distribution $\mathcal{N}(\mu, \sigma^2)$ for any $\sigma^2 \in \mathbb{R}_{>0}$, and thus $F_{Y|X}^{-1}(\tau) = F_{\mathrm{obs}}^{-1}(\tau)$ for any $\tau \in (0,1)$. If the observed variance does not depend on the mean, i.e., $\mathrm{Cov}(\mu, \sigma^2) = 0$, and conditions (a) and (b) are fulfilled, then a *variance calibrated* forecaster is also *distribution calibrated*, as the condition in (2) is met because $\mathcal{S}$ is restricted to normal distributions and μ is no influential factor. For uncertainty calibration, the authors in [21] and [20] use temperature scaling [10] to rescale the variance of a normal distribution. This method optimizes the negative log likelihood (NLL) to find the optimal rescaling parameter.

In conclusion, although the definition of *quantile calibration* [16] is straightforward and intuitive, only the marginal probabilities are considered which are independent of the predicted distribution moments. In contrast, the definition for *distribution calibration* [31] is more restrictive as it requires a forecaster to match the observed distribution conditioned on the forecaster's output. If a forecaster is distribution calibrated and unbiased, and if the ground-truth data follows a normal distribution, then the definition for *variance calibration* [20,21] is met which is the most restrictive definition for regression calibration but also very useful when working with normal distributions.

Further Research. A different branch of yielding prediction uncertainties is quantile regression where a forecaster outputs estimates for certain quantile levels [4,6]. The advantage of quantile regression is that no parametric assumptions about the output distribution is necessary. However, these methods haven't been used for object detection so far, thus, we focus on detection models that output a parametric estimate for the object position [7,11–13]. As opposed to post-hoc calibration, there are also methods that aim to achieve regression calibration during model training, e.g., calibration loss [8], maximum mean discrepancy (MMD) [5] or *f*-Cal [3]. The former approach adds a loss term to match the predicted variances with the observed squared error, while the latter methods introduce a second loss term to perform distribution matching during model training. The advantage of these approaches is that an additional calibration data set is not needed. However, it is necessary to retrain the whole network architecture. Thus, we focus on post-hoc calibration methods in this work.

3   Joint Parametric Regression Calibration

Several tasks such as object detection are inference problems with multiple dimensions, i.e., the joint estimation of position, width, and height of objects is necessary. Therefore, the output space extends to $\mathcal{Y} = \mathbb{R}^K$, where K denotes the number of dimensions. In the scope of object detection, most probabilistic models predict independent distributions for each quantity [7,11,13].

We start with the GP-Beta framework by [31]. Given a training set $\mathcal{D}$, the goal is to build a calibration mapping from uncalibrated distributions to calibrated ones. Since the CDF of a distribution is bound to the $[0, 1]$ interval, the authors in [31] derive a *beta link function* from the beta calibration method [17] to rescale the CDF of a distribution with a total amount of 3 rescaling parameters. These rescaling parameters $\mathbf{w} \in \mathcal{W} = \mathbb{R}^3$ are implicitly obtained by a GP model $\mathrm{gp}(0, k, \mathbf{B})$ with zero mean, where k is the covariance kernel function of the GP, and $\mathbf{B}$ is the coregionalization matrix that captures correlations between the parameters [31]. The authors argue that the usage of a GP model allows for finding the optimal recalibration parameters w.r.t. distributions that are close to the actual prediction. During model training, the parameters $\mathbf{w}$ for the beta link function are trained using the training data distribution given by

$$f(Y|\mathcal{D}) = \int_{\mathcal{W}} f(Y|\mathbf{w}, \mathcal{D}) f(\mathbf{w}|\mathcal{D}) \partial \mathbf{w}, \tag{4}$$

with the likelihood $f(Y|\mathbf{w}, \mathcal{D})$ obtained by the beta link function and $f(\mathbf{w}|\mathcal{D})$ as the posterior obtained by the GP model. However, the integral in (4) is analytically intractable due to the non-linearity within the link function in the likelihood [31]. Thus, the authors in [31] use a Monte-Carlo sampling approach for model training and inference. Furthermore, the authors use an approximate GP because the GP covariance matrix has the complexity $\mathcal{O}(N^2)$. Thus, during optimization, N^* pseudo inducing points are learnt to represent the training set [14,30]. This guarantees a stable and fast computation of new calibrated distributions during inference. The inducing points, the coregionalization matrix, and the kernel parameters are optimized using the evidence lower bound (ELBO). We refer the reader to [31] for a detailed derivation of the beta link likelihood.

Parametric Recalibration. A major advantage of this approach is that it captures dependencies between neighboring input samples, i.e., it reflects a conditional recalibration of the estimated probability distributions. However, the resulting distribution has no parametric form as it is represented by a CDF of arbitrary shape. For this reason, we propose a novel recalibration method **GP-Normal** which can be seen as a temperature scaling that rescales the variance σ^2 by a scaling factor w [20,21]. In contrast to the baseline approaches in [20,21], we use the GP scheme by [31] to estimate a weight $w \in \mathbb{R}_{>0}$ for each distribution individually, so that

$$\log(w) \sim \mathrm{gp}(0, k) \tag{5}$$

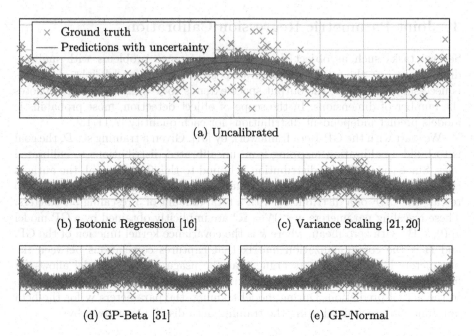

(a) Uncalibrated

(b) Isotonic Regression [16]   (c) Variance Scaling [21, 20]

(d) GP-Beta [31]   (e) GP-Normal

Fig. 2. Regression example using a cosine function with noise to generate ground-truth samples (green). The noise increases towards the function's maximum. (a) shows an unbiased estimator with equally sampled variance (blue). (b) *Isotonic Regression* accounts for marginal uncertainty across all samples, while (c) *Variance Scaling* only has a dependency on the predicted variance. In contrast, (d) *GP-Beta* and (e) *GP-Normal* capture information from the whole predicted distribution and thus are able to properly recalibrate the uncertainty. (Color figure online)

can be used to rescale the variance $\hat{\sigma}^2 = w \cdot \sigma^2$ with the same training likelihood given in (4). Note that a coregionalization matrix $\mathbf{B}$ is not required as only a single parameter is estimated. We use the exponential function to guarantee positive estimates for w. The advantage of this approach compared to the standard *GP-Beta* is that the output distribution has a closed-form representation. Furthermore, the uncertainty recalibration is sensitive to neighboring samples, i.e., it allows for a conditional uncertainty recalibration of the input distributions, compared to the standard *Isotonic Regression* [16] or *Variance Scaling* [20,21]. A comparison of common calibration methods and our *GP-Normal* method is given in Fig. 2 for an artificial example.

In general, we can use any parametric continuous probability distribution to derive a calibrated distribution. For example, if we inspect the distribution of the error obtained by a Faster R-CNN [26] on the MS COCO dataset [23] (shown in Fig. 3), we can observe that the error rather follows a Cauchy distribution $Cauchy(x_0, \gamma)$ with location parameter $x_0 \in \mathbb{R}$ and scale $\gamma \in \mathbb{R}_{>0}$. Therefore, we further propose the **GP-Cauchy** method that is similar to *GP-Normal* but utilizes a Cauchy distribution as the likelihood within the GP framework. This

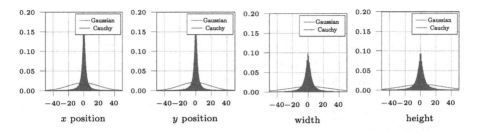

Fig. 3. Prediction error of the mean estimate $\mu_Y(X)$ of a `Faster R-CNN` on MS COCO. We can only get a poor quality of fit using a Gaussian distribution in this case. In contrast, the Cauchy distribution achieves a good approximation of the error distribution as this distribution allows for heavier tails.

allows for deriving a scale weight w (cf. (5)) so that $\hat{\gamma} = w \cdot \sigma$ leads to a calibrated probability distribution $Cauchy(x_0, \hat{\gamma})$ using $\mu_Y(X) = x_0$ as an approximation.

Joint Recalibration. In many applications, such as object detection, a probabilistic regression model needs to jointly infer the output distribution for multiple dimensions. Usually, the output of such a model is parameterized using independent Gaussian distributions for each dimension [11,13,15]. We can adapt the *GP-Normal*, *GP-Cauchy*, as well as the *GP-Beta* methods [31] to jointly recalibrate these independent distributions. Since we assert normal distributions as calibration input, we use the same kernel function as [31] and developed by [32], that is defined by

$$k\big((\mu_i, \Sigma_i), (\mu_j, \Sigma_j)\big) = \theta^K |\Sigma_{ij}|^{-\frac{1}{2}} \exp\Big(-\frac{1}{2}(\mu_i - \mu_j)^\top \Sigma_{ij}^{-1}(\mu_i - \mu_j)\Big), \quad (6)$$

where $\theta \in \mathbb{R}$ is the kernel lengthscale parameter and $\Sigma_{ij} = \Sigma_i + \Sigma_j + \theta^2 \mathbf{I}$ defines the covariance, with $\mathbf{I}$ as the identity matrix. In the *GP-Beta* framework, the coregionalization matrix $\mathbf{B}$ is used to introduce correlations in the output data. We extend this framework to *jointly* recalibrate probability distributions within a multidimensional output space $\mathbf{Y} \in \mathcal{Y} = \mathbb{R}^K$. We extend the coregionalization matrix $\mathbf{B}$ by additional entries for each dimension. Therefore, the parameter estimation in (5) for the *GP-Normal* and the *GP-Cauchy* methods extends to

$$\log(\mathbf{w}) \sim gp(0, k, \mathbf{B}), \quad (7)$$

where $\mathbf{B} \in \mathbb{R}^{K \times K}$ and $\mathbf{w} \in \mathbb{R}^K_{>0}$. The likelihood of GP calibration methods can be interpreted as the product of K multiple independent dimensions.

Covariance Estimation. Since the *GP-Normal* method yields a parametric normal distribution, it is also possible to capture correlations between all output quantities in $\mathcal{Y}$ by introducing a **covariance estimation** scheme. First, we capture the *marginal* correlation coefficients $\rho_{ij} \in [-1, 1]$ between dimensions i

and j. This allows for the computation of a covariance matrix Σ for all samples in the training set $\mathcal{D}$. Second, we use the $\mathbf{LDL}^\top$ factorization of Σ and perform a rescaling of the lower triangular $\mathbf{L}$ and the diagonal matrix $\mathbf{D}$ by the weights $\mathbf{w_L} \in \mathbb{R}^{J-K}$ and $\mathbf{w_D} \in \mathbb{R}_{>0}^K$, respectively, with the total number of parameters $J = \frac{K}{2}(K+1) + K$, where

$$\log(\mathbf{w_D}), \mathbf{w_L} \sim \mathrm{gp}(0, k, \mathbf{B}), \tag{8}$$

using $\mathbf{B} \in \mathbb{R}^{J \times J}$. After covariance reconstruction, the likelihood of the model is obtained by the multivariate normal distribution using the mean $\boldsymbol{\mu}$ and the rescaled covariance matrix $\hat{\boldsymbol{\Sigma}} = \hat{\mathbf{L}}\hat{\mathbf{D}}\hat{\mathbf{L}}^\top$, where $\hat{\mathbf{L}} = \mathbf{w_L} \odot \mathbf{L}$ and $\hat{\mathbf{D}} = \mathbf{w_D} \odot \mathbf{D}$ using the broadcasted weights ($\odot$ denotes the element-wise multiplication). On the one hand, it is now possible to introduce conditional correlations between independent Gaussian distributions using the *GP-Normal* method. On the other hand, we can also use this method for **covariance recalibration** if covariance estimates are already given in the input data set $\mathcal{D}$ [12]. In this case, we can directly use the $\mathbf{LDL}^\top$ factorization on the given covariance matrices to perform a covariance recalibration. Both approaches, *covariance estimation* and *covariance recalibration*, might be used to further improve the uncertainty quantification. This is beneficial if these uncertainties are incorporated in subsequent processes.

4   Measuring Miscalibration

Besides proper scoring rules such as negative log likelihood (NLL), there is actually no common consent about how to measure regression miscalibration. A common metric to measure the quality of predicted quantiles is the *Pinball loss* $\mathcal{L}_{\mathrm{Pin}}(\tau)$ [33] for a certain quantile level τ. For regression calibration evaluation, a mean Pinball loss $\overline{\mathcal{L}_{\mathrm{Pin}}} = \mathbb{E}_\tau[\mathcal{L}_{\mathrm{Pin}}(\tau)]$ for multiple quantile levels is commonly used to get an overall measure. Recently, the authors in [20] defined the *uncertainty calibration error* (UCE) in the scope of *variance calibration* that measures the difference between predicted uncertainty and the actual error. Similar to the *expected calibration error* (ECE) within classification calibration, the UCE uses a binning scheme with M bins over the variance to estimate the calibration error w.r.t. the predicted variance. More formally, the UCE is defined by

$$\mathrm{UCE} := \sum_{M=1}^{M} \frac{N_M}{N} |\mathrm{MSE}(M) - \mathrm{MV}(M)|, \tag{9}$$

with $\mathrm{MSE}(M)$ and $\mathrm{MV}(M)$ as the mean squared error and the mean variance within each bin, respectively, and N_M as the number of samples within bin M [20]. Another similar metric, the *expected normalized calibration error* (ENCE), has been proposed by [21] and is defined by

$$\mathrm{ENCE} := \frac{1}{M} \sum_{M=1}^{M} \frac{|\mathrm{RMSE}(M) - \mathrm{RMV}(M)|}{\mathrm{RMV}(M)}, \tag{10}$$

where RMSE(M) and RMV(M) denote the root mean squared error and the root mean variance within each bin, respectively. The advantage of the ENCE is that the miscalibration can be expressed as the percentage of the RMV. Thus, the error is independent of the output space.

Measuring Multivariate Miscalibration. Besides NLL, none of these metrics is able to jointly measure the miscalibration of multiple dimensions or to capture the influence of correlations between multiple dimensions. Therefore, we derive a new metric, the **quantile calibration error** (QCE). This metric is based on the well-known *normalized estimation error squared* (NEES) that is used for Kalman filter consistency evaluation [2, p. 232], [34, p. 292]. Given an unbiased base estimator that outputs a mean $\boldsymbol{\mu_Y}(X)$ and covariance $\boldsymbol{\Sigma}_Y(X)$ estimate with K dimensions for an input X, the NEES ϵ is the squared Mahalanobis distance between the predicted distribution f_Y and the ground-truth $\mathbf{Y}$. The hypothesis, that the predicted estimation errors are consistent with the observed estimation errors, is evaluated using a χ^2-test. This test is accepted if the mean NEES is below a certain quantile threshold a_τ for a target quantile τ, i.e., $\epsilon_{\mu,\Sigma} \leq a_\tau$. The threshold $a_\tau \in \mathbb{R}_{>0}$ is obtained by a χ^2_K distribution with K degrees of freedom, so that $a_\tau = \chi^2_K(\tau)$. The advantage of the NEES is that it measures a *normalized* estimation error and can also capture correlations between multiple dimensions.

For regression calibration evaluation, we adapt this idea and measure the difference between the fraction of samples that fall into the acceptance interval for a certain quantile τ and the quantile level τ itself. However, this would only reflect the *marginal* calibration error. Therefore, we would like to measure the error conditioned on distributions with similar properties. For instance, the dispersion of a distribution can be captured using the *standardized generalized variance* (SGV) which is defined by $\sigma_G^2 = \det(\boldsymbol{\Sigma})^{\frac{1}{k}}$ [28,29]. In the univariate case, the SGV is equal to the variance σ^2. Using the SGV as a property of the distributions, we define the QCE as

$$\mathrm{QCE}(\tau) := \mathbb{E}_{X,\sigma_G}\Big[\big|\mathbb{P}(\epsilon(X) \leq a_\tau|\sigma_G) - \tau\big|\Big]. \qquad (11)$$

Similar to the ECE, UCE, and ENCE, we use a binning scheme but rather over the square root σ_G of the SGV to achieve a better data distribution for binning. The QCE measures the fraction of samples whose NEES falls into the acceptance interval for each bin, separately. Therefore, the QCE for a certain quantile τ is approximated by

$$\mathrm{QCE}(\tau) \approx \sum_{M=1}^{M} \frac{N_M}{N}\Big|\mathrm{freq}(M) - \tau\Big|, \qquad (12)$$

where $\mathrm{freq}(M) = \frac{1}{N_M}\sum_{I=1}^{N_M} \mathbb{1}\big(\epsilon(x_I) \leq a_\tau\big)$ using $\mathbb{1}(\cdot)$ as the indicator function. Similar to the Pinball loss, we can approximate a mean $\overline{\mathrm{QCE}} = \mathbb{E}_\tau[\mathrm{QCE}(\tau)]$ for multiple quantile levels to get an overall quality measure. Furthermore, we can use this metric to measure the quality of multivariate uncertainty estimates in the following experiments. Note that we can also use the QCE to measure non-Gaussian distributed univariate random variables. The derivation of the QCE for multivariate, non-parametric distributions is subject of future work.

5 Experiments

We evaluate our methods using probabilistic object detection models that are trained on the Berkeley DeepDrive 100k [36] and the MS COCO 2017 [23] data sets. We use a `RetinaNet R50-FPN` [22] and a `Faster R-CNN R101-FPN` [26] as base networks. These networks are trained using a probabilistic regression output where a mean and a variance score is inferred for each bounding box quantity [11,13,15]. For uncertainty calibration training and evaluation, these networks are used to infer objects on the validation sets of the Berkeley DeepDrive 100k and COCO 2017 data sets. Unfortunately, we cannot use the respective test sets as we need the ground-truth information for both, calibration training and evaluation. Therefore, we split both validation datasets and use the first half for the training of the calibration methods, while the second half is used for calibration evaluation. Finally, it is necessary to match all bounding box predictions with the ground-truth objects to determine the estimation error. For all experiments, we use an intersection over union (IoU) score of 0.5 for object matching.

For calibration training and evaluation, we adapt the methods *Isotonic Regression* [16], *Variance Scaling* [20,21], and *GP-Beta* [31] and provide an optimized reimplementation in our publicly available repository[1]. Furthermore, this repository also contains the new *GP-Normal* method and its multivariate extension for covariance estimation and covariance recalibration, as well as the *GP-Cauchy* method. The GP methods are based on GPyTorch [9] that provides an efficient implementation for GP optimization. For calibration evaluation, we use the negative log likelihood (NLL), Pinball loss, UCE [20], ENCE [21], and our proposed QCE. The UCE, ENCE, and QCE use binning schemes with $M = 20$ bins, respectively, while the Pinball loss and the QCE use confidence levels from $\tau = 0.05$ up to $\tau = 0.95$ within steps of 0.05 to obtain an average calibration score over different quantile levels. Note that we cannot report the UCE and ENCE for the *GP-Cauchy* method as the Cauchy distribution has no variance defined. Furthermore, we only report the multivariate QCE for Gaussian distributions in our experiments since this metric is based on the squared Mahalanobis distance for Gaussian distributions in the multivariate case. On the one hand, we examine the calibration properties of each dimension independently and denote the average calibration metrics over all dimensions. On the other hand, we further compute the multivariate NLL and QCE to measure calibration for multivariate distributions. The results are given in Table 1.

Observations. The experiments show that the simple *Isotonic Regression* [16] achieves the best results regarding *quantile calibration* (cf. metrics QCE and

---
[1] https://github.com/EFS-OpenSource/calibration-framework

Table 1. Calibration results for `Faster R-CNN R101-FPN` and `RetinaNet R50-FPN` evaluated on Berkeley DeepDrive and MS COCO evaluation datasets before and after uncertainty calibration with different calibration methods.

Setup			Univariate					Multivariate	
	DB	Method	NLL	$\overline{\text{QCE}}$	$\mathcal{L}_{\text{Pin}}$	UCE	ENCE	NLL	$\overline{\text{QCE}}$
`Faster R-CNN`	BDD	Uncalibrated	3.053	0.040	1.079	19.683	0.454	12.210	**0.071**
		Isotonic Reg. [16]	**2.895**	**0.017**	**1.059**	39.157	0.303	**11.579**	–
		GP-Beta [31]	2.941	0.057	1.077	3.635	0.199	11.764	–
		Var. Scaling [20,21]	2.962	0.061	1.086	3.361	**0.175**	11.848	0.131
		GP-Normal	2.962	0.059	1.084	3.289	0.188	11.848	0.128
		GP-Normal (mv.)	2.968	0.054	1.150	**3.234**	0.191	11.584	0.133
		GP-Cauchy	3.011	0.050	1.189	–	–	12.045	–
	COCO	Uncalibrated	3.561	0.154	3.055	32.899	0.096	14.245	0.256
		Isotonic Reg. [16]	**3.340**	**0.020**	**2.715**	42.440	0.121	**13.360**	–
		GP-Beta [31]	3.412	0.074	2.750	51.455	0.140	13.649	–
		Var. Scaling [20,21]	3.554	0.131	2.952	33.155	0.093	14.216	0.222
		GP-Normal	3.554	0.130	2.949	48.167	0.132	14.235	**0.200**
		GP-Normal (mv.)	3.562	0.121	3.298	**28.382**	**0.087**	13.955	0.216
		GP-Cauchy	3.406	0.039	2.897	–	–	13.624	–
`RetinaNet`	BDD	Uncalibrated	4.052	0.130	1.847	39.933	0.491	16.208	0.234
		Isotonic Reg. [16]	**3.224**	**0.068**	**1.814**	39.498	0.252	**12.898**	–
		GP-Beta [31]	3.419	0.091	2.169	**14.892**	**0.173**	13.677	–
		Var. Scaling [20,21]	3.392	0.095	2.203	15.833	0.180	13.568	**0.131**
		GP-Normal	3.392	0.095	2.205	15.820	0.180	13.568	**0.131**
		GP-Normal (mv.)	3.434	0.097	3.356	15.342	0.200	13.353	0.133
		GP-Cauchy	3.316	0.097	1.944	–	–	13.262	–
	COCO	Uncalibrated	4.694	0.115	4.999	553.244	0.530	18.778	**0.153**
		Isotonic Reg. [16]	**3.886**	**0.029**	**4.534**	105.343	0.113	**15.544**	–
		GP-Beta [31]	4.169	0.142	5.093	**77.017**	**0.071**	16.677	–
		Var. Scaling [20,21]	4.204	0.167	5.606	83.653	0.072	16.815	0.207
		GP-Normal	4.204	0.167	5.606	84.367	0.072	16.815	0.207
		GP-Normal (mv.)	4.236	0.158	6.233	82.074	0.087	16.455	0.194
		GP-Cauchy	3.936	0.043	4.716	–	–	15.745	–

Pinball loss) as well as for the NLL. In contrast, the calibration methods *Variance Scaling* and *GP-Normal* only lead to slight or no improvements w.r.t. to *quantile calibration* but achieve better results for UCE and ENCE. These metrics measure the miscalibration by asserting normal distributions which leads to the assumption that these methods are beneficial if normal distributions are required for subsequent processes. To get a better insight, we further inspect the error distribution as well as the reliability diagrams for quantile regression in Figs. 3 and 4, respectively. On the one hand, our experiments show that the estimation error of the base networks is centered around 0 but also has heavier tails compared to a representation by a normal distribution. In these cases, the error distribution can be better expressed in terms of a Cauchy distribution (*GP-Cauchy*). On

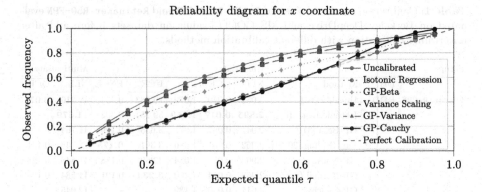

Fig. 4. Reliability diagram which shows the fraction of ground-truth samples that are covered by the predicted quantile for a certain τ. The `Faster R-CNN` underestimates the prediction uncertainty. The best calibration performance w.r.t. *quantile calibration* is reached by *Isotonic Regression* [16] and *GP-Cauchy*.

the other hand, we cannot find a strong connection between the error and the bounding box position and/or shape. This can be seen by comparing the calibration performance of *GP-Normal* and the more simple *Variance Scaling* method that does not take additional distribution information into account. Therefore, marginal recalibration in terms of the basic *quantile calibration* is sufficient in this case which explains the success of the simple *Isotonic Regression* calibration method. Although the *GP-Beta* is also not bounded to any parametric output distribution and should also lead to *quantile calibration*, it is restricted to the beta calibration family of functions which limits the representation power of this method. In contrast, *Isotonic Regression* calibration is more flexible in its shape for the output distribution. However, applications such as Kalman filtering require a normal distribution as input. Although *Isotonic Regression* achieves good results with respect to *quantile calibration*, it only yields poor results in terms of *variance calibration* (cf. UCE and ENCE). In contrast, *Variance Scaling* and especially the multivariate *GP-Normal* achieve the best results since these methods are designed to represent the error distribution as a Gaussian. In a nutshell, if the application requires to extract statistical moments (e.g. mean and variance) for subsequent processing, *Isotonic Regression* is not well suited since the calibration properties are aligned w.r.t. the quantiles and not a parametric representation. In comparison, a direct calibration with parametric models (e.g. *GP-Normal*) leads to better properties regarding the statistical moments.

Therefore, we conclude that the error distribution is represented best by the simple *Isotonic Regression* when working with quantiles, and by *Variance Scaling* and multivariate *GP-Normal* when working with normal distributions. In our experiments, the error distribution rather follows a Cauchy distribution than a normal distribution. Furthermore, we could not find a strong connection between position information and the error distribution. Thus, marginal recalibration is sufficient in this case. Therefore, the choice of the calibration method depends on the use case: if the quantiles are of interest, we advise the use of *Isotonic Regression* or *GP-Beta*. If a representation in terms of a continuous parametric distribution is required, the *GP-Cauchy* or *GP-Normal* achieve reasonable results. If a normal distribution is required, we strongly suggest the use of the multivariate *GP-Normal* recalibration method as it yields the best calibration results for the statistical moments and is well suited for a Gaussian representation of the uncertainty.

6   Conclusion

For environment perception in the context of autonomous driving, calibrated spatial uncertainty information is of major importance to reliably process the detected objects in subsequent applications such as Kalman filtering. In this work, we give a brief overview over the existing definitions for regression calibration. Furthermore, we extend the GP recalibration framework by [31] and use parametric probability distributions (Gaussian, Cauchy) as calibration output on the one hand. This is an advantage when subsequent processes require a certain type of distribution, i.e., a Gaussian distribution for Kalman filtering. On the other hand, we use this framework to perform *covariance estimation* of independently learnt distribution for multivariate regression tasks. This enables a post-hoc introduction of (conditional) dependencies between multiple dimensions and further improves the uncertainty representation.

We adapt the most relevant recalibration techniques and compare them to our parametric GP methods to recalibrate the spatial uncertainty of probabilistic object detection models [11–13]. Our examinations show that common probabilistic object detection models are too underconfident in their spatial uncertainty estimates. Furthermore, the distribution of the prediction error is more similar to a Cauchy distribution than to a normal distribution (cf. Fig. 3). We cannot find significant correlations between prediction error and position information. Therefore, the definition for *quantile calibration* is sufficient to represent the uncertainty of the examined detection models. This explains the superior performance of the simple *Isotonic Regression* method which performs a mean- and variance-agnostic (marginal) recalibration of the cumulative distribution functions. In contrast, we can show that our multivariate *GP-Normal* method achieves the best results in terms of *variance calibration* which is advantageous when normal distributions are required as output.

Acknowledgement. The authors gratefully acknowledge support of this work by Elektronische Fahrwerksysteme GmbH, Gaimersheim, Germany. The research leading to the results presented above are funded by the German Federal Ministry for Economic Affairs and Energy within the project "KI Absicherung - Safe AI for automated driving".

References

1. Banerjee, K., Notz, D., Windelen, J., Gavarraju, S., He, M.: Online camera lidar fusion and object detection on hybrid data for autonomous driving. In: 2018 IEEE Intelligent Vehicles Symposium (IV), pp. 1632–1638. IEEE (2018)
2. Bar-Shalom, Y., Li, X.R., Kirubarajan, T.: Estimation with Applications to Tracking and Navigation: Theory Algorithms and Software. Wiley, New York (2004)
3. Bhatt, D., Mani, K., Bansal, D., Murthy, K., Lee, H., Paull, L.: f-Cal: calibrated aleatoric uncertainty estimation from neural networks for robot perception. arXiv preprint arXiv:2109.13913 (2021)
4. Chung, Y., Neiswanger, W., Char, I., Schneider, J.: Beyond pinball loss: quantile methods for calibrated uncertainty quantification. In: Advances in Neural Information Processing Systems, vol. 34 (2021)
5. Cui, P., Hu, W., Zhu, J.: Calibrated reliable regression using maximum mean discrepancy. In: Advances in Neural Information Processing Systems, vol. 33 (2020)
6. Fasiolo, M., Wood, S.N., Zaffran, M., Nedellec, R., Goude, Y.: Fast calibrated additive quantile regression. J. Am. Stat. Assoc. **116**, 1402–1412 (2020). https://doi.org/10.1080/01621459.2020.1725521
7. Feng, D., Harakeh, A., Waslander, S.L., Dietmayer, K.: A review and comparative study on probabilistic object detection in autonomous driving. IEEE Trans. Intell. Transp. Syst. (2021)
8. Feng, D., Rosenbaum, L., Glaeser, C., Timm, F., Dietmayer, K.: Can we trust you? On calibration of a probabilistic object detector for autonomous driving. arXiv preprint (2019)
9. Gardner, J.R., Pleiss, G., Bindel, D., Weinberger, K.Q., Wilson, A.G.: GPyTorch: blackbox matrix-matrix Gaussian process inference with GPU acceleration. In: Advances in Neural Information Processing Systems (2018)
10. Guo, C., Pleiss, G., Sun, Y., Weinberger, K.Q.: On calibration of modern neural networks. In: Proceedings of the 34th International Conference on Machine Learning, Proceedings of Machine Learning Research, vol. 70, pp. 1321–1330. PMLR, August 2017
11. Hall, D., et al.: Probabilistic object detection: definition and evaluation. In: The IEEE Winter Conference on Applications of Computer Vision, pp. 1031–1040 (2020)
12. Harakeh, A., Smart, M., Waslander, S.L.: BayesOD: a Bayesian approach for uncertainty estimation in deep object detectors. In: 2020 IEEE International Conference on Robotics and Automation (ICRA), pp. 87–93. IEEE (2020)
13. He, Y., Zhu, C., Wang, J., Savvides, M., Zhang, X.: Bounding box regression with uncertainty for accurate object detection. In: Proceedings of the IEEE Conference on Computer Vision and Pattern Recognition, pp. 2888–2897 (2019)
14. Hensman, J., Matthews, A., Ghahramani, Z.: Scalable variational gaussian process classification. In: Artificial Intelligence and Statistics, pp. 351–360. PMLR (2015)
15. Kendall, A., Gal, Y.: What uncertainties do we need in Bayesian deep learning for computer vision? In: Advances in Neural Information Processing Systems (NIPS), pp. 5574–5584 (2017)
16. Kuleshov, V., Fenner, N., Ermon, S.: Accurate uncertainties for deep learning using calibrated regression. In: International Conference on Machine Learning (ICML), pp. 2801–2809 (2018)
17. Kull, M., Silva Filho, T., Flach, P.: Beta calibration: a well-founded and easily implemented improvement on logistic calibration for binary classifiers. In: Artificial Intelligence and Statistics, pp. 623–631 (2017)

18. Kumar, A., Liang, P.S., Ma, T.: Verified uncertainty calibration. In: Advances in Neural Information Processing Systems, vol. 32, pp. 3792–3803. Curran Associates, Inc. (2019). http://papers.nips.cc/paper/8635-verified-uncertainty-calibration.pdf

19. Küppers, F., Kronenberger, J., Shantia, A., Haselhoff, A.: Multivariate confidence calibration for object detection. In: Proceedings of the IEEE/CVF Conference on Computer Vision and Pattern Recognition Workshops, pp. 326–327 (2020)

20. Laves, M.H., Ihler, S., Fast, J.F., Kahrs, L.A., Ortmaier, T.: Well-calibrated regression uncertainty in medical imaging with deep learning. In: Medical Imaging with Deep Learning, pp. 393–412. PMLR (2020)

21. Levi, D., Gispan, L., Giladi, N., Fetaya, E.: Evaluating and calibrating uncertainty prediction in regression tasks. arXiv preprint abs/1905.11659 (2019)

22. Lin, T.Y., Goyal, P., Girshick, R., He, K., Dollár, P.: Focal loss for dense object detection. In: Proceedings of the IEEE International Conference on Computer Vision (ICCV), pp. 2980–2988 (2017)

23. Lin, T.-Y., et al.: Microsoft COCO: common objects in context. In: Fleet, D., Pajdla, T., Schiele, B., Tuytelaars, T. (eds.) ECCV 2014. LNCS, vol. 8693, pp. 740–755. Springer, Cham (2014). https://doi.org/10.1007/978-3-319-10602-1_48

24. Naeini, M.P., Cooper, G.F., Hauskrecht, M.: Binary classifier calibration using a Bayesian non-parametric approach. In: Proceedings of the 2015 SIAM International Conference on Data Mining, pp. 208–216 (2015). https://doi.org/10.1137/1.9781611974010.24

25. Naeini, M., Cooper, G., Hauskrecht, M.: Obtaining well calibrated probabilities using Bayesian binning. In: Proceedings of the 29th AAAI Conference on Artificial Intelligence, pp. 2901–2907 (2015)

26. Ren, S., He, K., Girshick, R., Sun, J.: Faster R-CNN: towards real-time object detection with region proposal networks. In: Advances in Neural Information Processing Systems (NIPS), pp. 91–99 (2015)

27. Schwaiger, F., Henne, M., Küppers, F., Schmoeller Roza, F., Roscher, K., Haselhoff, A.: From black-box to white-box: examining confidence calibration under different conditions. In: Proceedings of the Workshop on Artificial Intelligence Safety 2021 (SafeAI 2021) Co-located with the Thirty-Fifth AAAI Conference on Artificial Intelligence (AAAI 2021) (2021)

28. SenGupta, A.: Tests for standardized generalized variances of multivariate normal populations of possibly different dimensions. J. Multivariate Anal. **23**(2), 209–219 (1987)

29. SenGupta, A.: Generalized variance. Encycl. Stat. Sci. **6053** (2004)

30. Snelson, E., Ghahramani, Z.: Sparse Gaussian processes using pseudo-inputs. Adv. Neural Inf. Process. Syst. **18**, 1257 (2006)

31. Song, H., Diethe, T., Kull, M., Flach, P.: Distribution calibration for regression. In: Proceedings of the 36th International Conference on Machine Learning. Proceedings of Machine Learning Research, vol. 97, pp. 5897–5906. PMLR, Long Beach, California, USA, 9–15 June 2019

32. Song, L., Zhang, X., Smola, A., Gretton, A., Schölkopf, B.: Tailoring density estimation via reproducing Kernel moment matching. In: Proceedings of the 25th International Conference on Machine Learning, pp. 992–999 (2008)

33. Steinwart, I., Christmann, A.: Estimating conditional quantiles with the help of the pinball loss. Bernoulli **17**(1), 211–225 (2011)

34. Van Der Heijden, F., Duin, R.P., De Ridder, D., Tax, D.M.: Classification, Parameter Estimation and State Estimation: An Engineering Approach Using MATLAB. Wiley, New York (2005)

35. Yang, Z., Li, J., Li, H.: Real-time pedestrian and vehicle detection for autonomous driving. In: 2018 IEEE Intelligent Vehicles Symposium (IV), pp. 179–184. IEEE (2018)
36. Yu, F., et al.: BDD100K: a diverse driving video database with scalable annotation tooling. CoRR (2018)
37. Zadrozny, B., Elkan, C.: Transforming classifier scores into accurate multiclass probability estimates. In: Proceedings of the Eighth International Conference on Knowledge Discovery and Data Mining, 23–26 July 2002, Edmonton, Alberta, Canada, pp. 694–699 (2002). https://doi.org/10.1145/775047.775151

Reliable Multimodal Trajectory Prediction via Error Aligned Uncertainty Optimization

Neslihan Kose[1(✉)] , Ranganath Krishnan[2] , Akash Dhamasia[1] ,
Omesh Tickoo[2] , and Michael Paulitsch[1]

[1] Intel Labs, Munich, Germany
{neslihan.kose.cihangir,akash.dhamasia,michael.paulitsch}@intel.com
[2] Intel Labs, Hillsboro, USA
{ranganath.krishnan,omesh.tickoo}@intel.com

Abstract. Reliable uncertainty quantification in deep neural networks
is very crucial in safety-critical applications such as automated driving for trustworthy and informed decision-making. Assessing the quality
of uncertainty estimates is challenging as ground truth for uncertainty
estimates is not available. Ideally, in a well-calibrated model, uncertainty
estimates should perfectly correlate with model error. We propose a novel
error aligned uncertainty optimization method and introduce a trainable
loss function to guide the models to yield good quality uncertainty estimates aligning with the model error. Our approach targets continuous
structured prediction and regression tasks, and is evaluated on multiple
datasets including a large-scale vehicle motion prediction task involving real-world distributional shifts. We demonstrate that our method
improves average displacement error by 1.69% and 4.69%, and the uncertainty correlation with model error by 17.22% and 19.13% as quantified
by Pearson correlation coefficient on two state-of-the-art baselines.

Keywords: Reliable uncertainty quantification · Robustness ·
Multimodal trajectory prediction · Error aligned uncertainty
calibration · Safety-critical applications · Informed decision-making ·
Safe artificial intelligence · Automated driving · Real-world
distributional shift

1 Introduction

Conventional deep learning models often tend to make unreliable predictions [16,30] and do not provide a measure of uncertainty in regression tasks.
Incorporating uncertainty estimation to deep neural networks enables informed
decision-making as it indicates how certain the model is with its' prediction.
In safety-critical applications such as automated driving (AD), in addition to
making accurate predictions, it is important to quantify uncertainty associated
with predictions [15] in order to establish trust in the models.

© The Author(s), under exclusive license to Springer Nature Switzerland AG 2023
L. Karlinsky et al. (Eds.): ECCV 2022 Workshops, LNCS 13805, pp. 443–458, 2023.
https://doi.org/10.1007/978-3-031-25072-9_31

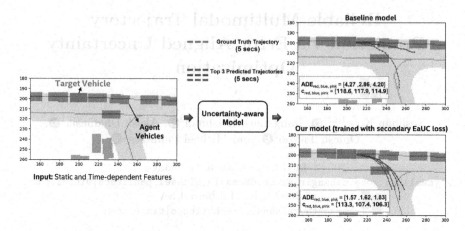

Fig. 1. Figure shows how our approach improves model calibration when it is incorporated to state-of-the-art uncertainty-aware vehicle trajectory prediction baseline (Deep Imitative Model [34]) on Shift Dataset [29]. Here, c and ADE denote log-likelihood score (certainty measure) and average displacement error (robustness measure), respectively. In well-calibrated models, we expect higher c value for samples with lower ADE and lower c value for samples with higher ADE.

A well-calibrated model should yield low uncertainty when the prediction is accurate, and indicate high uncertainty for inaccurate predictions. Since the ground truths for uncertainty estimates are not available, uncertainty calibration is challenging, and has been explored mostly for classification tasks or post-hoc fine-tuning so far. In this paper, we propose a novel **error aligned uncertainty optimization** technique introducing a differentiable secondary loss function to obtain well-calibrated models for regression-based tasks including the challenging continuous structured prediction task such as vehicle motion prediction.

Figure 1 shows how our approach improves model calibration when it is incorporated to state-of-the art (SoTA) uncertainty-aware vehicle trajectory prediction model, which involves predicting the next trajectories (i.e., possible future states) of agents around the automated vehicle. Vehicle trajectory prediction is a crucial component of AD. This area has recently received considerable attention from both industry and academia [5,14,18,26]. Reliable uncertainty quantification for prediction module is of critical importance for the planning module of autonomous system to make the right decision leading to safer decision-making. Related safety standards like ISO/PAS 21448 [20] suggest uncertainty is an important safety aspect. In addition to being a safety-critical application, trajectory prediction is also a challenging task for the uncertainty estimation domain due to the structured and continuous nature of predictions.

In this paper, our contributions are as follows:

- We propose a novel error aligned uncertainty (EaU) optimization method and introduce EaU calibration loss to guide the models to provide reliable uncertainty estimates correlating with the model error. To the best of our knowledge, this is the first work to introduce a trainable loss function to obtain reliable uncertainty estimates for continuous structured prediction (e.g., trajectory prediction) and regression tasks.

- We evaluate the proposed method with the large-scale real-world Shifts dataset and benchmark [29] for vehicle trajectory prediction task using two different baselines. We also use UCI datasets [9] as an additional evaluation of our approach considering multiple regression tasks. The results from extensive experiments demonstrate our method improves SoTA model performance on these tasks and yields well-calibrated models in addition to improved robustness even under real-world distributional shifts.

In the rest of the paper, our approach is explained considering vehicle motion prediction as it is a safety-critical application of AD, and challenging application of uncertainty quantification due to the continuous, structured and multimodal nature of predictions. Here, multimodality refers to multiple trajectory predictions that are plausible for a target vehicle in a certain context. Our method is applicable to any regression task, which we later show on UCI benchmark.

2 Related Work

This paper focuses on reliable uncertainty quantification for continuous structured prediction tasks such as trajectory prediction in addition to any regression-based tasks introducing a novel differentiable loss function that correlates uncertainty estimates with model error.

There are various approaches to estimate uncertainty in neural network predictions including Bayesian [4,10,13,36,37] and non-Bayesian [24,25,35] methods. Our proposed error aligned uncertainty optimization is orthogonal to and can complement existing uncertainty estimation methods as it can be applied as a secondary loss function together with any of these methods to get calibrated good quality uncertainty estimates in addition to improved robustness.

The existing solutions to get well-calibrated uncertainties are mainly developed for classification tasks or based on post-hoc calibration so far.

In [21], the authors propose an optimization method that is based on the relationship between accuracy and uncertainty. For this purpose, they introduced differentiable accuracy versus uncertainty calibration (AvUC) loss which enables the model to provide well-calibrated uncertainties in addition to improved accuracy. The paper shows promising results, however it is applicable only for classification tasks. We follow the insights from AvUC to build a loss-calibrated inference method for time-series based prediction and regression tasks.

Post-hoc methods [22,23,27] are based on recalibrating a pretrained model with a sample of independent and identically distributed (i.i.d.) data, and have been devised mainly for classification or small-scale regression problems. [22] proposed a post-hoc method to calibrate the output of any 1D regression algorithm based on Platt scaling [32]. In [27], a truth discovery framework is proposed integrating ensemble-based and post-hoc calibration methods based on an accuracy preserving truth estimator. The authors show that post-hoc methods can be enhanced by truth discovery-regularized optimization. To the best of our knowledge, there is no existing post-hoc calibration solution for the time-series based multi-variate regression tasks such as trajectory prediction as they bring additional challenges when designing these methods. [30] has also shown that post-hoc methods fail to provide well-calibrated uncertainties under distributional shifts in the real-world.

Recently, Shifts dataset and benchmark [29] were introduced for uncertainty and robustness to distributional shift analysis on multimodal vehicle trajectory prediction. Multimodal techniques in this context output a distribution over multiple trajectories [6,8,17,28,31]. The Shifts dataset is currently the largest publicly available vehicle motion prediction dataset with 600,000 scenes including real-world distributional shifts as well. The Shifts benchmark incorporates uncertainties to vehicle motion prediction pipeline and proposes metrics and solutions on how to jointly assess robustness and uncertainties in this task. In [29], the authors emphasize that in continuous structured prediction tasks such as vehicle trajectory prediction, there is still much potential for further development of informative measures of uncertainty and well-calibrated models.

To the best of our knowledge, this is the first work proposing an orthogonal solution based on a new trainable secondary loss function that can complement existing uncertainty estimation methods for continuous structured prediction tasks to improve the quality of their uncertainty measures.

3   Setup for Vehicle Trajectory Prediction

Vehicle trajectory prediction aims to predict the future states of agents in the scene. Here, the training set is denoted as $\mathcal{D}_{train} = (x_i, y_i)_{i=1}^{N}$. y and x denote the ground truth trajectories and high-dimensional features representing scene context, respectively. Each $y = (s_1, ..., s_T)$ denotes the future trajectory of a given vehicle observed by the automated vehicle perception stack. Each state s_t corresponds to the d_x- and d_y-displacement of the vehicle at timestep t, where $y \in R^{T \times 2}$ (Fig. 1).

Each scene has M seconds duration. It is divided into K seconds of context features and L seconds of ground truth targets for prediction that is separated by the time $T = 0$. The goal is to predict the future trajectory of vehicles at time $T \in (0, L]$ based on the information available for time $T \in [-K, 0]$.

In [29], the authors propose a solution on how to incorporate uncertainty to motion prediction, and introduce two types of uncertainty quantification metrics:

- *Per-trajectory confidence-aware metric:* For a given input x, the stochastic model accompanies its D top trajectory predictions with scalar per-trajectory confidence scores $(c^{(i)}|i \in 1, .., D)$.
- *Per-prediction request uncertainty metric:* U is computed by aggregating the D top per-trajectory confidence scores to a single uncertainty score, which represents model's uncertainty in making any prediction for the target agent.

In our paper, per-trajectory metric c represents our certainty measure based on the log-likelihood (Fig. 1). Per-prediction request uncertainty metric U is computed with negating the mean average of D top per-trajectory certainty scores.

As shown in Fig. 1, the input of the model is a single scene context x consisting of static (map of the environment that can be augmented with extra information such as crosswalk occupancy, lane availability, direction, and speed limit) and time-dependent input features (e.g., occupancy, velocity, acceleration and yaw for vehicles and pedestrians in the scene). The output of the model is D top trajectory predictions $(y^{(d)}|d \in 1, .., D)$ for the future movements of the target vehicle together with their corresponding certainty $(c^{(d)}|d \in 1, .., D)$ as in Fig. 1 or uncertainty $(u^{(d)}|d \in 1, .., D)$ scores as well as a single per-prediction request uncertainty score U. In the paper, we use c and u as our uncertainty measure interchangeably according to the context. Higher c indicates lower u.

In the existence of unfamiliar or high-risk scene context, an automated vehicle associates a high per-prediction request uncertainty [29]. However, since uncertainty estimates do not have ground truth, it is challenging to assess their quality. In our paper, we jointly assess the quality of uncertainty measures and robustness to distributional shift according to the following problem dimensions:

Robustness to distributional shift is mainly assessed via metrics such as Average Displacement Error (ADE) or Mean Square Error (MSE) in case of continuous structured prediction and regression tasks, respectively. ADE is the standard performance metric for time-series data and measures the quality of a prediction y with respect to the ground truth y^* as given in Eq. 1, where $y = (s_1, ..., s_T)$.

$$ADE(\mathbf{y}) := \frac{1}{T} \sum_{t=1}^{T} ||s_t - s_t^*||_2. \tag{1}$$

The analysis is done with two types of evaluation datasets, which are the in-distribution and shifted datasets. Models are considered more robust in case smaller degradation in performance is observed on the shifted data.

Quality of uncertainty estimates should be jointly assessed with robustness measure as there could be cases such that the model performs well on shifted data (e.g., for certain examples of distributional shift) and poorly on in-distribution data (e.g., on data from a not well-represented part of the training set) [29]. Joint assessment enables to understand whether measures of uncertainty correlate well with the presence of an incorrect prediction or high error.

Our goal is to have reliable uncertainty estimates which correlate well with the corresponding error measure. For this purpose, we propose an error aligned uncertainty (EaU) optimization technique to get well-calibrated models in addition to improved robustness.

4   Error Aligned Uncertainty Optimization

In order to develop well-calibrated models for continuous structured prediction and regression tasks, we propose a novel **Error aligned Uncertainty Calibration (EaUC)** loss, which can be used as a secondary loss utility function relying on theoretically sound Bayesian decision theory [3]. A task specific utility function is used for optimal model learning in Bayesian decision theory. EaUC loss is intended to be used as a secondary task specific utility function to achieve the objective of training the model to provide well-calibrated uncertainty estimates.

We build our method with the insights from AvUC loss [21] that was introduced for classification tasks, and extend previous work on classification to continuous structured prediction and regression tasks.

In regression and continuous structured prediction tasks, robustness is measured typically in terms of Mean Square Error (MSE) and Average Displacement Error (ADE), respectively, instead of accuracy score. Therefore our robustness measure is ADE here as we explain our technique on motion prediction task. Lower MSE and ADE indicate more accurate results.

		Uncertainty	
		certain	uncertain
ADE	low	LC	LU
	high	HC	HU

$$EaU = \frac{n_{LC} + n_{HU}}{n_{LC} + n_{LU} + n_{HC} + n_{HU}}. \tag{2}$$

In the rest of this paper, we use the notations in Eq. 2 to represent the number of accurate & certain samples (n_{LC}), inaccurate & certain samples (n_{HC}), accurate & uncertain samples (n_{LU}) and inaccurate & uncertain samples (n_{HU}). Here, L and H refer to low and high error, respectively, and C and U refer to certain and uncertain sample, respectively.

Ideally, we expect the model to be certain about its prediction when it is accurate and provide high uncertainty when making inaccurate predictions (Fig. 1). So, our aim is to have more certain samples when the predictions are accurate (LC), and uncertain samples when they are inaccurate (HU) compared to having uncertain samples when they are accurate (LU), and certain samples when they are inaccurate (HC). For this purpose, we propose the Error aligned Uncertainty (EaU) measure shown in Eq. 2. A reliable and well-calibrated model provides higher EaU measure ($EAU \in [0, 1]$).

Equation 3 shows how we assign each sample to the corresponding class of samples.

$$
\begin{aligned}
n_{LU} &:= \sum_i \mathbb{1}(ade_i \leq ade_{th} \quad \text{and} \quad c_i \leq c_{th}); \\
n_{HC} &:= \sum_i \mathbb{1}(ade_i > ade_{th} \quad \text{and} \quad c_i > c_{th}); \\
n_{LC} &:= \sum_i \mathbb{1}(ade_i \leq ade_{th} \quad \text{and} \quad c_i > c_{th}); \\
n_{HU} &:= \sum_i \mathbb{1}(ade_i > ade_{th} \quad \text{and} \quad c_i \leq c_{th});
\end{aligned}
\tag{3}
$$

In Eq. 3, we use average displacement error, ade_i, as our robustness measure to classify the sample as accurate or inaccurate comparing it with a task-dependant threshold ade_{th}. The samples are classified as certain or uncertain according to their certainty score c_i that is based on log-likelihood in our motion prediction use case. Similarly, log-likelihood of each sample, which is our certainty measure, is compared with a task-dependant threshold c_{th}.

As the equations in Eq. 3 are not differentiable, we propose differentiable approximations to these functions and introduce a trainable uncertainty calibration loss (L_{EaUC}) in Eq. 4. This loss serves as the utility-dependent penalty term within the loss-calibrated approximate inference framework for regression and continuous structured prediction tasks.

$$L_{EaUC} = -\log\left(\frac{n_{LC} + n_{HU}}{n_{LC} + n_{LU} + n_{HC} + n_{HU}}\right). \tag{4}$$

where;

$$n_{LU} = \sum_{i \in \left\{ \substack{ade_i \leq ade_{th} \\ \text{and} \quad c_i \leq c_{th}} \right\}} (1 - tanh(ade_i))(1 - c_i);$$

$$n_{LC} = \sum_{i \in \left\{ \substack{ade_i \leq ade_{th} \\ \text{and} \quad c_i > c_{th}} \right\}} (1 - tanh(ade_i))(c_i);$$

$$n_{HC} = \sum_{i \in \left\{ \substack{ade_i > ade_{th} \\ \text{and} \quad c_i > c_{th}} \right\}} tanh(ade_i)(c_i);$$

$$n_{HU} = \sum_{i \in \left\{ \substack{ade_i > ade_{th} \\ \text{and} \quad c_i \leq c_{th}} \right\}} tanh(ade_i)(1 - c_i);$$

We utilize hyperbolic tangent function as bounding function to be able to scale the error and/or certainty measures to the range $[0, 1]$. The intuition behind the approximate functions is that the bounded error $tanh(ade) \to 0$ when the predictions are accurate and $tanh(ade) \to 1$ when inaccurate.

In our approach, the proposed EaUC loss is used as secondary loss. Equation 5 shows the final loss function used for continuous structured prediction task:

$$L_{Final} = L_{NLL} + (\beta \times L_{EaUC}). \tag{5}$$

β is a hyperparameter for relative weighting of EaUC loss with respect to primary loss (e.g. Negative Log Likelihood (NLL) in Eq. 5).

Our approach consists of three hyperparameters, which are ade_{th}, c_{th} and β. The details on how to set these hyperparameters are provided in Sect. 5.3.

Under ideal conditions, the proxy functions in Eq. 4 are equivalent to indicator functions defined in Eq. 3. This calibration loss enables the model to learn to provide well-calibrated uncertainties in addition to improved robustness.

Algorithm 1 shows the implementation of our method with Behavioral Cloning (BC) model [7], which is the baseline model of Shifts benchmark. BC consists of the encoder and decoder stages. The encoder is a Convolutional Neural Network (CNN) which captures scene context, and decoder consists of Gated Recurrent Unit (GRU) cells applied at each time step to capture the predictive distribution for future trajectories. In our approach, we add the proposed L_{EaUC} loss as secondary loss to this pipeline.

5   Experiments and Results

In our experiments, we incorporated the proposed calibration loss, L_{EaUC}, to two stochastic models of the recently introduced Shifts benchmark for uncertainty quantification of multimodal trajectory prediction and one Bayesian model for uncertainty quantification of regression tasks as additional evaluation of our approach. The results are provided incorporating the new loss to state-of-the art and diverse baseline architectures.

5.1   Multimodal Vehicle Trajectory Prediction

We use the real-world Shifts vehicle motion prediction dataset and benchmark [29]. In this task, distributional shift is prevalent involving real 'in-the-wild' distributional shifts, which brings challenges for robustness and uncertainty estimation.

Shifts dataset has data from six geographical locations, three seasons, three times of day, and four weather conditions to evaluate the quality of uncertainty under distributional shift. Currently, it is the largest publicly available vehicle motion prediction dataset containing $600,000$ scenes, which consists of both in-distribution and shifted partitions.

In Shifts benchmark, optimization is done based on NLL objective, and results are reported for two baseline architectures, which are stochastic Behavioral Cloning (BC) Model [7] and Deep Imitative Model (DIM) [34]. We report our results incorporating the '**Error Aligned Uncertainty Calibration**' loss L_{EaUC} as secondary loss to Shifts pipeline as shown in Algorithm 1.

Algorithm 1. EaUC loss incorporated to BC model for trajectory prediction

---

1: Given dataset $\mathcal{D} = \{X, Y\}$
2: Given K seconds of context features (x) and L seconds of ground truth targets (y) separated by $T = 0$ for each scene
3: let parameters of MultiVariate Normal Distribution (MVN) (μ, Σ) ▷ approx MVN distribution $\mathbf{Y} = \mathcal{N}(\mu, \Sigma)$
4: define learning rate schedule α
5: **repeat**
6:      Sample B index set of training samples; $\mathcal{D}_m = \{(x_i, y_i)\}_{i=1}^{B}$ ▷ batch-size
7:      **for** $i \in B$ **do**
8:          $z_i = encoder(x_i)$ ▷ encodes visual input x_i
9:          **for** $t \in L$ **do**   ▷ decode a trajectory
10:             $z_i = decoder(\hat{y}^t, z_i)$ ▷ unrolls the GRU
11:             $\mu^t, \Sigma^t = linear(z_i)$ ▷ Predicts the location and scale of the MVN distribution.
12:             Sample $\epsilon \sim \mathcal{N}(\mu^t, \Sigma^t)$
13:             $\hat{y}_i^{\,t} = \hat{y}_i^{\,t} + \epsilon$
14:         **end for**
15:         Calculate Average Displacement Error (ADE)
16:         $ade_i := \frac{1}{L}\sum_{t=1}^{L} ||\hat{y}_i^t - y_i^t||_2$.   ▷ ADE wrt to ground truth y
17:         Calculate Certainty Score
18:         $c_i = \log(\Pi_{t=1}^{L}\mathcal{N}\left(\hat{y}_i^{\,t}|\mu(y_i^t, x_i; \theta), \Sigma(y_i^t, x_i; \theta)\right))$ ▷ log-likelihood of a trajectory $\hat{y}_i$ in
19:         context x_i to come from an expert y_i (teacher-forcing)
20:     **end for**
21:     Compute $\mathbf{n_{LC}, n_{LU}, n_{HC}, n_{HU}}$ ▷ components of EaUC loss penalty
22:     Compute loss-calibrated objective (total loss): $\mathcal{L} = \mathcal{L}_{\text{NLL}} + \beta\mathcal{L}_{\text{EaUC}}$
23:     Compute the gradients of loss function w.r.t to weights w, $\Delta\mathcal{L}_w$
24:     Update the parameters w: $w \leftarrow w - \alpha\Delta\mathcal{L}_w$
25: **until** $\mathcal{L}$ has converged, or when stopped

Our aim is to learn predictive distributions that capture reliable uncertainty. This allows us to quantify uncertainties in the predictions during inference through sampling method and predict multiple trajectories of a target vehicle for the next 5 *sec* using data collected with 5 *Hz* sampling rate.

During training, for each BC and DIM models, density estimator (likelihood model) is generated by teacher-forcing (i.e. from the distribution of ground truth trajectories). The same settings of the Shifts benchmark is used: The model is trained with AdamW [33] optimizer with a learning rate (LR) of 0.0001, using a cosine annealing LR schedule with 1 epoch warmup, and gradient clipping at 1. Training is stopped after 100 epochs in each experiment.

During inference, Robust Imitative Planning [12] is applied. Sampling is applied on the likelihood model considering a predetermined number of predictions $G = 10$. The top $D = 5$ predictions of the model (or multiple models when using ensembles) are selected according to their log-likelihood based certainty scores.

In this paper, we demonstrate the predictive performance of our model using the *weightedADE* metric:

$$weightedADE_D(q) := \sum_{d \in D} \tilde{c}^{(d)} \cdot ADE(y^{(d)}) \qquad (6)$$

In Eq. 6, $\tilde{c}$ represents the confidence score for each prediction computed by applying softmax to log-likelihood scores as in Shifts benchmark.

The joint assessment of uncertainty quality and model robustness is analyzed following the same experimental setup and evaluation methodology as in Shifts benchmark, which applies error and F1 retention curves for this purpose. Lower *R-AUC* (area under error retention curve) and higher *F1-AUC* (area under F1 retention curve) indicate better calibration performances. In these retention curves, *weightedADE* is the error and the retention fraction is based on per-prediction uncertainty score U. Mean averaging is applied while computing U, which is based on the per-plan log-likelihoods, as well as for the aggregation of ensemble results. The improvement with area under these curves can be achieved either with a better model providing lower overall error, or providing better uncertainty estimates such that more errorful predictions are rejected earlier [29].

The secondary loss incentivizes the model to align the uncertainty with average displacement error (ADE) while training the model. Our experimental results are conducted by setting β (see Eq. 5) as 200, ade_{th} and u_{th} (see Eq. 4) as 0.8 and 0.6, respectively, for both BC and DIM models. In order to scale the robustness (ade_i) and certainty (c_i) measures to a proper range for the bounding function $tanh$ or to be used directly, we apply post-processing. We scale the ade_i values with 0.5 so that we can assign samples with ADE below 1.6 as an accurate sample. Our analysis on the Shifts dataset shows that the log-likelihood based certainty measures, c_i, of most samples are in between $[0, 100]$ range so we applied a clipping of values for the values below 0 and above 100 by setting these values to 0 and 100, respectively, and then these certainty measures are normalized to $[0, 1]$ range for direct use in Eq. 4. The applied post-processing steps and hyperparameter selection should be adapted according to each performed task considering initial training epochs (see Sect. 5.3 for details).

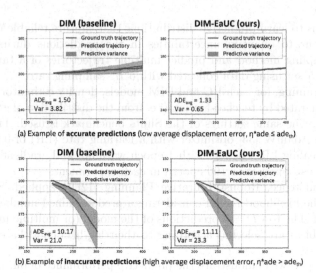

(a) Example of **accurate predictions** (low average displacement error, η*ade ≤ ade$_{th}$)

(b) Example of **inaccurate predictions** (high average displacement error, η*ade > ade$_{th}$)

Fig. 2. Figure shows correlation of accuracy with variance of the top-3 predicted trajectories based on certainty score with/without our loss for DIM baseline on Shift Dataset. With EaUC loss, we observe lower variance for accurate predictions and higher variance for inaccurate predictions.

Correlation Analysis: Table 1 shows the correlation between uncertainty measure and prediction error using Pearson correlation coefficient (r) [2] and the classification of samples as accurate and inaccurate using AUROC metric.

- We observe 17.22% and 19.13% improvement in correlation between uncertainty and error as quantified by Pearson's r when L_{EaUC} loss is incorporated to BC and DIM models, respectively (Table 1).
- Setting accurate prediction threshold as 1.6, we observe 6.55% and 8.02% improvement in AUROC when L_{EaUC} is incorporated to BC and DIM models, respectively, for detecting samples as accurate versus inaccurate based on the uncertainty measures (Table 1).

Fig. 2 provides example cases for both accurate and inaccurate trajectory predictions for DIM baseline, which shows EaUC loss improves model calibration.

Quality Assessment of Uncertainty and Model Performance Using Retention Curves: Table 2 shows the results for the joint quality assessment of uncertainty and robustness using $R\text{-}AUC$, $F1\text{-}AUC$, $F1@95\%$ metrics. Predictive performance is computed with $weightedADE$ metric, which is also the error metric of retention plots. Figure 3 shows error retention plots and F1-weightedADE retention plots for both BC and DIM baselines with/without our calibration loss. The results in Table 2 and Fig. 3 show that:

- There is an improvement of 1.69% and 4.69% for $weightedADE$, and 20.67% and 21.41% for $R\text{-}AUC$ for BC an DIM models, respectively, using the full

Table 1. Evaluating the correlation between uncertainty estimation and prediction error with Pearson correlation coefficient r and AUROC metric for binary classification of samples as accurate and inaccurate.

Model (MA, K = 1)	Pearson corr. co-eff. r ↑			AUROC ↑		
	In	Shifted	Full	In	Shifted	Full
BC	0.476	0.529	0.482	0.756	0.817	0.763
BC-EaUC*	**0.557**	**0.624**	**0.565**	**0.811**	**0.833**	**0.813**
DIM	0.475	0.522	0.481	0.754	0.816	0.761
DIM-EaUC*	**0.567**	**0.620**	**0.573**	**0.819**	**0.838**	**0.822**

Table 2. *Predictive robustness (weightedADE)* and *joint assessment of uncertainty and robustness performance (F1-AUC, F1@95%, R-AUC)* of the two baselines with/without EaUC loss. Lower is better for *weightedADE* and *R-AUC*, and higher is better for *F1-AUC, F1@95%*.

Model (MA, K = 1)	WeightedADE ↓			R-AUC ↓			F1-AUC ↑			F1@95% ↑		
	In	Shifted	Full	In	Shifted	Full	In	Shifted	Full	In	Shifted	Full
BC	1.475	1.523	1.481	0.423	0.357	0.416	0.631	0.646	0.633	0.848	0.884	0.852
BC-EaUC*	**1.450**	**1.521**	**1.456**	**0.336**	**0.290**	**0.330**	**0.652**	**0.666**	**0.654**	**0.875**	0.884	**0.876**
DIM	1.465	1.523	1.472	0.418	0.358	0.411	0.632	0.647	0.634	0.851	0.884	0.855
DIM-EaUC*	**1.388**	**1.514**	**1.403**	**0.328**	**0.290**	**0.323**	**0.651**	**0.664**	**0.653**	**0.881**	**0.888**	**0.882**

Table 3. Impact of assigning higher weights to LC class in L_{EaUC} loss on robustness and calibration performance.

Model (MA, K = 1)	WeightedADE ↓	R-AUC ↓
BC-EaUC	1.880	0.355
BC-EaUC*	**1.456**	**0.330**
DIM-EaUC	1.697	0.344
DIM-EaUC*	**1.403**	**0.323**

dataset. We outperform the results on two baselines providing well-calibrated uncertainties in addition to improved robustness.

- *R-AUC, F1-AUC and F1@95%* improve for both models using all Full, In, and Shifted datasets with our L_{EaUC} loss, which indicate better calibration performance using all three metrics.
- *weightedADE* is observed to be higher for Shifted dataset compared to In dataset, which proves that error is higher for out-of-distribution data.

Impact of Weight Assignment to the Class of Accurate&Certain Samples (LC) in the EaUC Loss: In safety-critical scenarios, it is very important to be certain about our prediction when the prediction is accurate in addition to having a greater number of accurate samples.

To incentivize the model during training more towards accurate predictions and be certain, we experimented by giving higher weightage to the class of LC

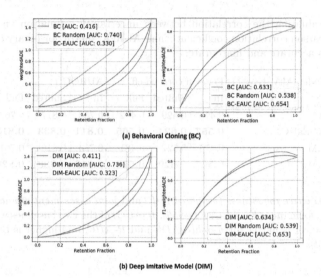

Fig. 3. Error and F1-weightedADE as a function of amount of retained samples based on uncertainty scores. The retention plots are presented for robust imitative planning method applied to two baselines (BC and DIM) with/without EaUC loss using full dataset. Retention fraction is based on per-prediction request uncertainty metric U, which is computed with Mean Averaging (MA) of per-trajectory log-likelihood scores. Lower weightedADE and higher F1-weightedADE are better.

samples while computing Eq. 4. Equation 7 shows how we assign high weights to these samples in our loss where $\gamma > 1$. We force the algorithm to better learn the samples of this class and obtain both improved calibration (well-calibrated) and robustness (lower weightedADE) with this approach as shown in Table 3.

$$L_{EaUC} = -\log\left(\frac{(\gamma \cdot n_{LC}) + n_{HU}}{(\gamma \cdot n_{LC}) + n_{LU} + n_{HC} + n_{HU}}\right). \tag{7}$$

BC-EaUC/DIM-EaUC and BC-EaUC*/DIM-EaUC* denote the results according to Eq. 4 and Eq. 7, respectively. BC-EaUC* and DIM-EaUC* provide better performance in terms of robustness (*weightedADE*) and model calibration (R-AUC) compared to BC-EaUC and DIM-EaUC, hence all the experiments in this section are reported applying our loss according to Eq. 7, where $\gamma = 3$, for both models. Our analysis showed that increasing γ value above some threshold causes degradation in performance. Table 3 shows the results on Full dataset.

Another observation in this experiment is that even though BC-EaUC and DIM-EaUC provide slightly higher *weightedADE* compared to their corresponding baselines (BC and DIM in Table 2), they provide better quality uncertainty estimates as quantified by R-AUC.

Study with Ensembles: Table 4 shows that using ensembles, better performances are achieved for both predictive model performance and calibration performances with the joint assessment of uncertainty and robustness using both BC and BC-EaUC*. EaUC together with Ensembles provides the best results.

Table 4. Impact analysis of L_{EaUC} loss on ensembles using the aforementioned metrics for robustness and joint assessment

Model (MA)	WeightedADE ↓			R-AUC ↓			F1-AUC ↑			F1@95% ↑		
	In	Shifted	Full	In	Shifted	Full	In	Shifted	Full	In	Shifted	Full
BC (K = 1)	1.475	1.523	1.481	0.423	0.357	0.416	0.631	0.646	0.633	0.848	0.884	0.852
BC (K = 3)	**1.420**	**1.433**	**1.421**	**0.392**	**0.323**	**0.384**	**0.639**	**0.651**	**0.640**	**0.853**	**0.887**	**0.857**
BC-EaUC* (K = 1)	1.450	1.521	1.456	0.336	0.290	0.330	0.652	0.666	0.654	0.875	0.884	0.876
BC-EaUC* (K = 3)	**1.290**	**1.373**	**1.300**	**0.312**	**0.271**	**0.308**	**0.647**	**0.659**	**0.648**	**0.890**	**0.898**	**0.891**

5.2 UCI Regression

We evaluate our method on UCI [9] datasets following [19] and [13]. We use the Bayesian neural network (BNN) with Monte Carlo dropout [13] approximate Bayesian inference. In this setup, we use neural network with two hidden layers fully-connected with 100 neurons and a ReLU activation. A dropout layer with probability of 0.5 is used after each hidden layer, with 20 Monte Carlo samples for approximate Bayesian inference. Following [1], we find the optimal hyperparameters for each dataset using Bayesian optimization with HyperBand [11] and train the models with SGD optimizer and batch size of 128. The predictive variance from Monte Carlo forward passes is used as the uncertainty measure in the EaUC loss function. BNN trained with a secondary EaUC loss yields lower predictive negative log-likelihood and lower RMSE on multiple UCI datasets as shown in Table 5. Figure 4 shows the model error retention plot based on the model uncertainty estimates for UCI boston housing regression dataset. BNN trained with loss-calibrated approximate inference framework using EaUC loss yields lower model error as most certain samples are retained with 15.5% lower AUC compared to baseline BNN and 31.5% lower AUC compared to Random retention (without considering uncertainty estimation). These results demonstrate that EaUC enables to obtain informative and well-calibrated uncertainty estimates from neural networks for reliable decision making.

Table 5. Avg. predictive negative log-likelihood (NLL)(lower is better) and test root mean squared error (RMSE)(lower is better) on the UCI regression benchmark.

UCI dataset	Test NLL ↓		Test RMSE ↓	
	BNN	BNN (EaUC)	BNN	BNN (EaUC)
boston	2.424	**2.203**	2.694	**2.317**
concrete	3.085	**3.016**	5.310	**4.580**
energy	1.675	**1.643**	0.973	**0.671**
yacht	2.084	**1.684**	2.024	**0.787**
power	2.885	**2.829**	4.355	**4.018**
naval	**−4.729**	−4.294	0.001	0.001
kin8nm	−0.930	**−0.967**	0.086	**0.074**
protein	2.888	**2.851**	4.351	**4.181**

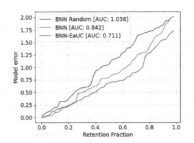

Fig. 4. Error retention plot for BNN on UCI boston housing dataset to evaluate the quality of model uncertainty (lower AUC is better).

5.3   Hyperparameter Selection

Our solution requires three hyperparameters, which are the thresholds ade_{th} and c_{th} to assign the sample to appropriate category of prediction (certain & accurate; certain & inaccurate; uncertain & inaccurate; uncertain & accurate) as well as the β parameter to assign relative weighting to our secondary loss compared to primary loss. Initially we train the model without secondary calibration loss for few epochs, then we analyze both ADE and log-likelihood values to set the thresholds required for calibration loss following the same strategy as [21]. We perform grid search in order to tune the threshold parameters. Post-processing is applied on certainty measures to set the values between 0 and 1 range for direct use of these measures in our loss function. β is chosen such that the introduced secondary loss contributes significantly to the final loss, we select this value such that the secondary loss is at least half of the primary loss. For the regression experiments, we found the optimal hyperparameters for each dataset using Bayesian optimization with HyperBand [11].

6   Conclusions

Reliable uncertainty estimation is crucial for safety-critical systems towards safe decision making. In this paper, we proposed a novel error aligned uncertainty (EaU) optimization method and introduced differentiable EaU calibration loss that can be used as a secondary penalty loss to incentivize the models to yield well-calibrated uncertainties in addition to improved robustness. To the best of our knowledge, this is the first work to introduce a trainable uncertainty calibration loss function to obtain reliable uncertainty estimates for time-series based multi-variate regression tasks such as trajectory prediction. We evaluated our method on real-world vehicle motion prediction task with large-scale Shifts dataset involving distributional shifts. We also showed that the proposed method can be applied to any regression task by evaluating on UCI regression benchmark. The experimental results shows the proposed method can achieve well-calibrated models yielding reliable uncertainty quantification in addition to improved robustness even under real-world distributional shifts. The EaU loss can also be utilized for post-hoc calibration of pretrained models, which we will explore in future work. We hope our method can be used along with well-established robust baselines to advance the state-of-the-art in uncertainty estimation benefiting various safety-critical real-world AI applications.

Acknowledgements. This project has received funding from the European Union's Horizon 2020 research and innovation programme under grant agreement No 956123. This research has also received funding from the Federal Ministry of Transport and Digital Infrastructure of Germany in the project Providentia++ (01MM19008F).

References

1. Antorán, J., Allingham, J., Hernández-Lobato, J.M.: Depth uncertainty in neural networks. Adv. Neural. Inf. Process. Syst. **33**, 10620–10634 (2020)

2. Benesty, J., Chen, J., Huang, Y., Cohen, I.: Pearson correlation coefficient, pp. 37–40 (2009)
3. Berger, J.O.: Statistical Decision Theory and Bayesian Analysis. Springer, New York (1985). https://doi.org/10.1007/978-1-4757-4286-2
4. Blundell, C., Cornebise, J., Kavukcuoglu, K., Wierstra, D.: Weight uncertainty in neural network. In: International Conference on Machine Learning, pp. 1613–1622 (2015)
5. Casas, S., Gulino, C., Suo, S., Luo, K., Liao, R., Urtasun, R.: Implicit latent variable model for scene-consistent motion forecasting. In: Vedaldi, A., Bischof, H., Brox, T., Frahm, J.-M. (eds.) ECCV 2020. LNCS, vol. 12368, pp. 624–641. Springer, Cham (2020). https://doi.org/10.1007/978-3-030-58592-1_37
6. Chai, Y., Sapp, B., Bansal, M., Anguelov, D.: Multipath: multiple probabilistic anchor trajectory hypotheses for behavior prediction. Conference on Robot Learning (CoRL) (2019)
7. Codevilla, F., Müller, M., López, A., Koltun, V., Dosovitskiy, A.: End-to-end driving via conditional imitation learning. In: IEEE International Conference on Robotics and Automation (ICRA), pp. 4693–4700 (2018)
8. Cui, H., et al.: Multimodal trajectory predictions for autonomous driving using deep convolutional networks. In: International Conference on Robotics and Automation (ICRA) (2019)
9. Dua, D., Graff, C.: UCI machine learning repository (2017). http://archive.ics.uci.edu/ml
10. Dusenberry, M.W., et al.: Efficient and scalable Bayesian neural nets with rank-1 factors (2020)
11. Falkner, S., Klein, A., Hutter, F.: BOHB: robust and efficient hyperparameter optimization at scale. In: International Conference on Machine Learning, pp. 1437–1446. PMLR (2018)
12. Filos, A., Tigas, P., McAllister, R., Rhinehart, N., Levine, S., Gal, Y.: Can autonomous vehicles identify, recover from, and adapt to distribution shifts? In: International Conference on Machine Learning (2020)
13. Gal, Y., Ghahramani, Z.: Dropout as a Bayesian approximation: representing model uncertainty in deep learning. In: International Conference on Machine Learning, pp. 1050–1059. PMLR (2016)
14. Gao, J., et al.: VectorNet: encoding HD maps and agent dynamics from vectorized representation. In: 2020 IEEE/CVF Conference on Computer Vision and Pattern Recognition (CVPR), pp. 11522–11530 (2020)
15. Ghahramani, Z.: Probabilistic machine learning and artificial intelligence. Nature 521(7553), 452–459 (2015)
16. Guo, C., Pleiss, G., Sun, Y., Weinberger, K.Q.: On calibration of modern neural networks. In: Proceedings of the 34th International Conference on Machine Learning, vol. 70, pp. 1321–1330. JMLR.org (2017)
17. Hong, J., Sapp, B., Philbin, J.: Rules of the road: predicting driving behavior with a convolutional model of semantic interactions. In: IEEE/CVF Conference on Computer Vision and Pattern Recognition (CVPR), pp. 8454–8462 (2019)
18. Huang, X., McGill, S.G., Williams, B.C., Fletcher, L., Rosman, G.: Uncertainty-aware driver trajectory prediction at urban intersections. In: 2019 International Conference on Robotics and Automation (ICRA), pp. 9718–9724. IEEE (2019)
19. Immer, A., Bauer, M., Fortuin, V., Rätsch, G., Emtiyaz, K.M.: Scalable marginal likelihood estimation for model selection in deep learning. In: International Conference on Machine Learning, pp. 4563–4573. PMLR (2021)

20. ISO: Road vehicles - safety of the intended functionality. ISO/PAS 21448: 2019(EN), (Standard) (2019)
21. Krishnan, R., Tickoo, O.: Improving model calibration with accuracy versus uncertainty optimization. Adv. Neural. Inf. Process. Syst. **33**, 18237–18248 (2020)
22. Kuleshov, V., Fenner, N., Ermon, S.: Accurate uncertainties for deep learning using calibrated regression. In: Proceedings of the 35th International Conference on Machine Learning, vol. 80, pp. 2796–2804 (2018)
23. Kumar, A., Liang, P.S., Ma, T.: Verified uncertainty calibration. In: Advances in Neural Information Processing Systems, vol. 32 (2019)
24. Lakshminarayanan, B., Pritzel, A., Blundell, C.: Simple and scalable predictive uncertainty estimation using deep ensembles. In: Advances in Neural Information Processing Systems, pp. 6402–6413 (2017)
25. Liu, J.Z., Lin, Z., Padhy, S., Tran, D., Bedrax-Weiss, T., Lakshminarayanan, B.: Simple and principled uncertainty estimation with deterministic deep learning via distance awareness. arXiv preprint arXiv:2006.10108 (2020)
26. Liu, Y., Zhang, J., Fang, L., Jiang, Q., Zhou, B.: Multimodal motion prediction with stacked transformers. In: 2021 IEEE/CVF Conference on Computer Vision and Pattern Recognition (CVPR), pp. 7573–7582 (2021)
27. Ma, C., Huang, Z., Xian, J., Gao, M., Xu, J.: Improving uncertainty calibration of deep neural networks via truth discovery and geometric optimization. In: Uncertainty in Artificial Intelligence (UAI), pp. 75–85 (2021)
28. Makansi, O., Ilg, E., Cicek, O., Brox, T.: Overcoming limitations of mixture density networks: a sampling and fitting framework for multimodal future prediction. In: IEEE/CVF Conference on Computer Vision and Pattern Recognition (CVPR), pp. 7144–7153 (2019)
29. Malinin, A., et al.: Shifts: a dataset of real distributional shift across multiple large-scale tasks. In: Thirty-Fifth Conference on Neural Information Processing Systems Datasets and Benchmarks Track (2021)
30. Ovadia, Y., et al.: Can you trust your model's uncertainty? Evaluating predictive uncertainty under dataset shift. In: Advances in Neural Information Processing Systems, vol. 32 (2019)
31. Phan-Minh, T., Grigore, E.C., Boulton, F.A., Beijbom, O., Wolff, E.M.: Covernet: multimodal behavior prediction using trajectory sets. In: IEEE/CVF Conference on Computer Vision and Pattern Recognition (CVPR), pp. 14062–14071 (2020)
32. Platt, J.C.: Probabilistic outputs for support vector machines and comparisons to regularized likelihood methods. In: Advances in Large Margin Classifiers, pp. 61–74 (1999)
33. Pytorch: Weighted adam optimizer (2021). https://pytorch.org/docs/stable/generated/torch.optim.AdamW.html
34. Rhinehart, N., McAllister, R., Levine, S.: Deep imitative models for flexible inference, planning, and control. CoRR (2018)
35. Van Amersfoort, J., Smith, L., Teh, Y.W., Gal, Y.: Uncertainty estimation using a single deep deterministic neural network, vol. 119, pp. 9690–9700 (2020)
36. Welling, M., Teh, Y.W.: Bayesian learning via stochastic gradient langevin dynamics. In: Proceedings of the 28th International Conference on Machine Learning (ICML 2011), pp. 681–688 (2011)
37. Zhang, R., Li, C., Zhang, J., Chen, C., Wilson, A.G.: Cyclical stochastic gradient MCMC for Bayesian deep learning. In: International Conference on Learning Representations (2020)

PAI3D: Painting Adaptive Instance-Prior for 3D Object Detection

Hao Liu$^{(\boxtimes)}$, Zhuoran Xu, Dan Wang, Baofeng Zhang, Guan Wang, Bo Dong, Xin Wen, and Xinyu Xu

Autonomous Driving Division of X Research Department, JD Logistics, Beijing, China
{liuhao163,xuzhuoran,wangdan257,zhangbaofeng13,wangguan151,dongbo14, wenxin16,xinyu.xu}@jd.com

Abstract. 3D object detection is a critical task in autonomous driving. Recently multi-modal fusion-based 3D object detection methods, which combine the complementary advantages of LiDAR and camera, have shown great performance improvements over mono-modal methods. However, so far, no methods have attempted to utilize the instance-level contextual image semantics to guide the 3D object detection. In this paper, we propose a simple and effective Painting Adaptive Instance-prior for 3D object detection (PAI3D) to fuse instance-level image semantics flexibly with point cloud features. PAI3D is a multi-modal sequential instance-level fusion framework. It first extracts instance-level semantic information from images, the extracted information, including objects categorical label, point-to-object membership and object position, are then used to augment each LiDAR point in the subsequent 3D detection network to guide and improve detection performance. PAI3D outperforms the state-of-the-art with a large margin on the nuScenes dataset, achieving 71.4 in mAP and 74.2 in NDS on the test split. Our comprehensive experiments show that instance-level image semantics contribute the most to the performance gain, and PAI3D works well with any good-quality instance segmentation models and any modern point cloud 3D encoders, making it a strong candidate for deployment on autonomous vehicles.

Keywords: 3D object detection · Instance-level multi-modal fusion

1 Introduction

3D object detection is critical for autonomous driving. LiDAR, camera and Radar are the typical sensors equipped on autonomous driving cars to facilitate 3D object detection. Up to date point cloud alone based 3D object detection methods [22,33,39,40,44,47] achieve significantly better performance than camera alone monocular 3D object detection methods [8,26,37] in almost all

---

The authors share equal contributions.

© The Author(s), under exclusive license to Springer Nature Switzerland AG 2023
L. Karlinsky et al. (Eds.): ECCV 2022 Workshops, LNCS 13805, pp. 459–475, 2023.
https://doi.org/10.1007/978-3-031-25072-9_32

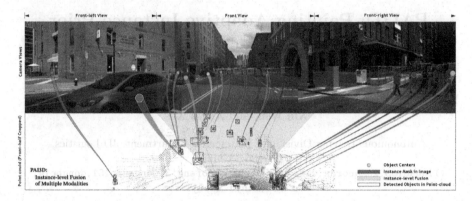

Fig. 1. PAI3D fuses point cloud and images at instance-level for 3D object detection. We utilize image instance segmentation to obtain 3D instance priors, including instance-level semantic label and 3D instance center, to augment the raw LiDAR points. We select one frame from the nuScenes dataset and show the front half point cloud and the three frontal image views. Best viewed in color.

large-scale Autonomous Driving Dataset such as nuScenes [1] and Waymo Open Dataset [32], making LiDAR the primary, first-choice sensor due to its accurate and direct depth sensing ability. However, point cloud data may be insufficient in some scenarios, e.g., where objects are far away with very few points reflected, or objects are nearby but are very small or thin. Recently more and more works employ both point cloud and images for 3D object detection [7,10,21,28,35,38,42,45,46], aiming to utilize the complementary characteristics of 3D and 2D sensors to achieve multi-modal fusion, and these methods have demonstrated great performance boost over mono-modal approaches.

Existing multi-modal fusion methods either predict semantic labels from images, which are used as semantic priors to indicate foreground points in 3D point cloud [35,38], or incorporate implicit pixel features learned from image encoder into the 3D detection backbone [7,21]. Since the objective of 3D object detection is to identify each individual object instance, most of these methods, however, fail to reveal the instance-level semantics from 2D images, and hence hardly provide clear guidance to distinguish instance-level features in 3D, especially for the objects in the same category. Therefore, it is necessary to go beyond the existing category-level fusion and explore deep into the instance-level fusion for more explicit detection guidance.

Along the line of extracting instance-level information to improve 3D object detection, previous studies [11,44] have shown that representing and detecting objects as points is effective. These methods focus on using centers to facilitate ground-truth matching during training for only mono-modality, either point cloud or images, and the pros and cons of utilizing instance center from an additional modality and sharing it throughout the detection pipeline hasn't been studied yet. Our intuition is that extracting instance centers from images as instance priors and then feeding these instance priors into a 3D object detection network for a deep fusion with the raw LiDAR points will be highly beneficial for the 3D object detection task. However, a naive, straight-forward implementation of such idea may yield degraded performance due to erroneous semantic labels caused by point-to-pixel projection errors. Such errors can occur easily in an autonomous driving system equipped with multi-modal sensors.

To address the above challenges, in this paper, we propose a novel sequential multi-modal fusion method named PAI3D, to utilize instance priors extracted from image instance segmentation to achieve improved 3D object detection. The term "PAI" stands for Painting Adaptive Instance-prior from 2D images into 3D point cloud, where "Adaptive" refers to adaptively dealing with erroneous semantic labels caused by point-to-pixel projection errors. Figure 1 schematically illustrates the instance-level fusion between point cloud and images.

As shown by the PAI3D architecture in Fig. 2, PAI3D includes three key learnable modules. Instance Painter predicts instance-level semantic labels from images, it then augments the raw point cloud with these instance-level semantic labels by point-to-pixel projection. Since the augmented point cloud sometimes contains erroneous instance labels caused by projection mismatch introduced by inaccurate sensor calibration, object occlusions or LiDAR-camera synchronization errors, we design Adaptive Projection Refiner to correct the erroneous instance labels and output a 3D center as instance prior for each augmented LiDAR point. Moreover, road scene objects can exhibit large-scale variations, ranging from long buses or trucks to small pedestrians. Although good performance can still be achieved by detecting objects of similar scale from a single feature map associated with one detection head [48], it is generally difficult to accurately detect objects with large-scale variations from a single feature map associated with one detection head. To overcome this difficulty, we propose Fine-granular Detection Head with multiple feature maps of varying receptive fields to localize specific-scaled objects from specific-scaled feature maps, which greatly improves detection accuracy for objects with large-scale variations.

To summarize, the main contributions of this paper include the following.

(1) We propose a novel instance-level fusion method for 3D object detection, PAI3D, where instance-level semantics, including semantic label and 3D instance center, are utilized as priors to augment the raw LiDAR points, resulting in significant performance improvement.
(2) Adaptive Projection Refiner is proposed to adaptively correct the erroneous instance-level semantics, which makes the proposed method more robust and tolerant to imperfect sensor calibration, objects occlusions and LiDAR-camera synchronization errors.
(3) We design Fine-granular Detection Head of varying receptive fields to detect objects with varying scales. This module creates fine-granular match between corresponding object scale and receptive field in multi-resolution feature maps, enabling more accurate detection of objects with large-scale variations.

We first evaluate PAI3D on the well-known large-scale nuScenes dataset [1]. PAI3D outperforms the state-of-the-art (SOTA) with a large margin, achieving

71.4 mAP and 74.2 NDS on the test split (no external data used). We then
present several ablation results to offer insights into the effect of each module,
different 3D backbone encoders and instance segmentation quality. Finally, we
apply PAI3D to a proprietary autonomous driving dataset. The presented experimental results demonstrate that PAI3D improves 3D detection at far ranges and
for small and thin objects.

2 Related Work

3D object detection has been thoroughly studied in recent years. Besides the
large body of works that solely employ 3D point cloud, many researches emerge
to adopt multi-modal sensors, primarily LiDAR and camera, to do 3D object
detection (also called fusion-based methods [7,10,21,28,35,38,42,45,46]), which
utilize image semantics for improved 3D detection accuracy. Our PAI3D method
belongs to this category of multi-modal fusion-based 3D object detection.

2.1 LiDAR Based 3D Object Detection

LiDAR-based 3D detection methods employ point cloud as the sole input, which
can be roughly categorized into three categories of view-based, point-based, and
voxel-based. View-based methods project points onto either Bird's-eye View
(BEV) (PointPillar [22], PIXOR [40], Complex-YOLO [31], Frustum-PointPillars
[27]) or range view (RSN [33], RangeDet [14], RangeIoUDet [23]) to form a 2D
feature map, 2D CNNs are then performed in this feature map to detect 3D
objects. Point-based methods (PointRCNN [30], STD [43], 3DSSD [41]) directly
predict 3D objects from the extracted point cloud features [29]. Such methods typically employ down sampling [41] or segmentation [30,43] strategies to
efficiently search for the objects in the point cloud. Voxel-based methods utilize 3D convolution (VoxelNet [47]) or 3D sparse convolution [15] (SECOND
[39], SASSD [16], CenterPoint [44]) to encode 3D voxelized features, and most
succeeding efforts focus on exploring auxiliary information [16,44] to enhance
detections from voxelized data.

2.2 Multi-modal Fusion Based 3D Object Detection

Recently, more and more researches have emerged to combine LiDAR and camera
sensing for 3D object detection, where rich visual semantics from images are fused
with point cloud 3D geometry to improve performance. We group these works into
two categories based on how fusion is performed: One-shot fusion methods [7,21,
46] exploit implicit information embedded in image features and adopt an end-to-
end deep NN to fuse two modalities and predict 3D bounding boxes. Sequential
fusion methods [10,28,35,38,42,45] first predict semantic labels from images, and
then augment LiDAR points with these labels by point-to-pixel projection, visual
semantics are explicitly exploited in these methods and 3D detection are done in
point cloud space once semantic labels are acquired at the first phase.

Among one-shot fusion methods, MV3D [7] builds proposal-level fusion by generating proposals from LiDAR BEV followed by multi-view feature map fusion. AVOD [21] implements feature-level fusion where two identical but separate feature extractors are used to generate feature maps from both the LiDAR BEV map and the image followed by multi-modal feature fusion with a region proposal network to generate 3D proposals. Note that the misalignment crossing multiple views, i.e. LiDAR BEV and front view, and camera perspective view, can pose great challenges to MV3D and AVOD where the inconsistency in the fused features will degrade final detection performance. 3D-CVF [46] improves the cross-view misalignment issue by representing point cloud and images in the BEV feature maps so that the two modalities can be well aligned. Nevertheless, the lack of depth in images makes the conversion from camera perspective view to LiDAR BEV less accurate.

Among sequential fusion methods, IPOD [42] utilizes image semantic segmentation to label the foreground and background in LiDAR points, then it detects objects from only the foreground points. PointPainting [35] further attaches categorical semantic labels to each LiDAR point, which improves mAP significantly. HorizonLiDAR3D [10] adopts similar idea to PointPainting, but it employs object detection instead of image semantic segmentation to label the points encompassed by the detection bounding boxes. Inspired by PointPainting, PointAugmenting [36] replaces the hard-categorical labels with softening features extracted from the backbone, leading to further improved detection performance. FusionPainting [38] applies semantic segmentation to both image and point cloud, and an attention mechanism is then introduced to fuse the two modalities. MVP [45] utilizes image instance segmentation to generate virtual points to make object point cloud denser and hence easier to be detected. MTC-RCNN [28] uses PointPainting in a cascaded manner that leverages 3D box proposals to improve 2D segmentation quality, which are then used to further refine the 3D boxes.

Most existing fusion-based approaches concentrate on feature-level or pixel-to-point-level fusion from image to point cloud. The proposed PAI3D is a sequential fusion method that utilizes instance-level image semantics as 3D instance priors to guide 3D object detection, these 3D instance priors include object category label, the point-to-object membership and object position. Our ablation experiment results show that these instance priors are able to greatly boost 3D object detection performance.

3 Painting Adaptive Instance for Multi-modal 3D Object Detection

In this section, we present the proposed PAI3D method, its architecture overview is shown in Fig. 2. Let $\ell(x, y, z, r)$ be a LiDAR point, where x, y, z are the 3D coordinates and r is the reflectance. Given a point cloud $L = \{\ell_i\}_{i=1}^{N}$, and M views of images, $\{I_j\}_{j=1}^{M}$, that cover the 360-degree surrounding view, our goal is to utilize $\{I_j\}_{j=1}^{M}$ to obtain each point's semantic label s, the 3D instance C such

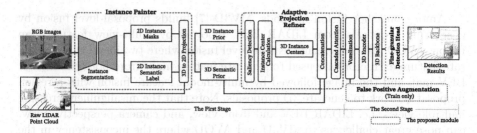

Fig. 2. PAI3D architecture overview. PAI3D consists of three main components: Instance Painter extracts 3D instance and semantic priors from image instance segmentation result, Adaptive Projection Refiner corrects the erroneous semantic labels caused by projection mismatch and extract a center for a 3D instance, and Fine-granular Detection Head is capable of detecting objects with large-scale variations. False Positive Augmentation (FPA) serves as an additional training strategy to eliminate false positive proposals by online augmentation.

that $\ell_i \in C$, and the 3D instance center $\hat{C} = (C_x, C_y, C_z)$. We augment each raw point with semantic label and 3D instance center, that is ℓ_i is represented by $(x, y, z, r, s, C_x, C_y, C_z)$, and then we detect 3D objects from the set of augmented LiDAR points. To achieve this goal, we design Instance Painter and Adaptive Projection Refiner to obtain (C, s) and $\hat{C}$ respectively. Moreover, we propose Fine-granular Detection Head with multiple feature maps of varying receptive fields to detect objects with varying scales.

3.1   Instance Painter

The goal of Instance Painter is to obtain the 3D object instances and their semantic labels (e.g. Car, Cyclist) as instance priors from point cloud utilizing image as "assistive" input (shown in Fig. 3). There are three steps to achieve this goal as described in the following.

First, image instance segmentation is performed to extract the 2D instance masks and semantic labels from images. Comparing with image semantic segmentation, instance segmentation can provide not only instance-aware semantic labels but also clear object instance delineation, allowing us to more accurately aggregate points belonging to the same 2D instance into one 3D instance in the subsequent instance aggregation step. In addition, the detailed visual texture, appearance, and color information in images can yield much higher instance segmentation performance than instance segmentation from point cloud.

Second, Instance Painter performs a frustum-based 3D to 2D projection [35, 45] to project LiDAR points to images. The transformation from LiDAR coordinate to image coordinate is defined as $KT_{(cam \leftarrow ego)} T_{(ego^{t1} \leftarrow ego^{t2})} T_{(ego \leftarrow lidar)}$, where $T_{(ego \leftarrow lidar)}$, $T_{(ego^{t1} \leftarrow ego^{t2})}$ and $T_{(cam \leftarrow ego)}$ are the transformations from LiDAR frame to ego frame, from ego frame at time t_2 to that at t_1, and from ego frame to the target camera frame, respectively. K denotes the target camera's intrinsic matrix.

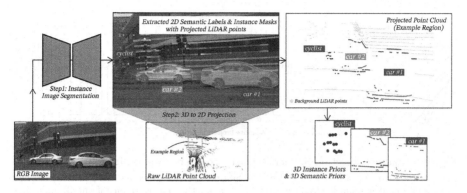

Fig. 3. Instance Painter workflow. Best viewed in color. It aims to obtain the 3D object instances and their semantic labels to facilitate the 3D object detection task in the second stage. First, it uses image instance segmentation model to extract the 2D instance masks and semantic labels. Then it projects the point cloud onto the 2D instance masks to obtain 3D instance priors with semantic labels. However, the aggregated 3D instance may contain points with erroneous semantic labels, e.g. the points at the top of car #1 actually are building points. This problem is solved by Adaptive Projection Refiner.

Finally, Instance Painter outputs a 3D instance prior by aggregating the LiDAR points which are projected into the same 2D object instance. Note that, two special cases need to be taken care of: first, the front and rear portion of an object can appear in two images of adjacent camera views due to object-at-image-border truncation, in this case we merge the two point clusters projected into the two images of the same object into one complete 3D instance; the second case is that, an object might be projected into two images of adjacent camera views if it occurs in the overlapped field of view of two adjacent cameras, and hence the 2D instance label and instance mask predicted from the two images might conflict with each other, in this case the instance label and instance mask having higher segmentation score is chosen for 3D instance aggregation.

3.2   Adaptive Projection Refiner

In a real-world autonomous car, many factors may inevitably lead to erroneous semantic labels in a 3D instance prior, these factors include imprecise 2D instance segmentation (especially at object boundary), object occlusions, point-to-pixel projection errors caused by inaccurate sensor calibration and/or multi-sensor synchronization offset. In addition, the projection path from point cloud to image is a frustum-shaped 3D subspace, implying that all the points in this subspace will be projected to the same instance mask (Fig. 4 (a)), which makes it prone to merge a 3D instance with points from a distant object (Fig. 4 (b)). Therefore, if a mean center is directly calculated from the projected points, this mean center will deviate too much from the true 3D instance center (Fig. 4 (b))).

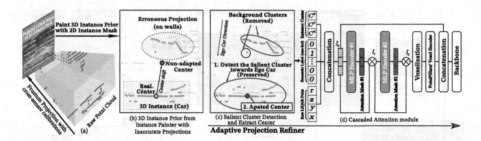

Fig. 4. Adaptive Projection Refiner. 3D instance priors obtained from Instance Painter may not accurately capture object's true shape and scale due to many inevitable factors. Mean center usually deviates too much from the real position of the 3D instance center (a) and (b). Adaptive Projection Refiner refines the shape of the 3D instance by detecting the salient cluster closest to the ego car, computing cluster medoid as the 3D instance center (c), and employing a cascaded attention module to adaptively fuse the features from all modalities (d).

To solve these problems, we design Adaptive Projection Refiner to refine the 3D instance prior that may contain erroneous semantic labels and compute the medoid as the center vector to represent the 3D instance. A cascaded attention module is then proposed to adaptively fuse the raw LiDAR points, semantic labels, and instance centers.

Salient Cluster Detection. The discovered 3D instances and the background LiDAR points with erroneous labels can manifest in arbitrary shapes and scales. We propose to utilize density-based spatial clustering, i.e., DBSCAN [12] or its efficient variants [18,24], to refine the 3D instance shape by detecting the salient cluster towards the ego car and removing the background cluster. The salient cluster here denotes the cluster with the largest number of LiDAR points and closest to the ego car. Moreover, salient cluster medoid is computed to represent the 3D instance center since medoid is more robust to outliers compared with mean center (Fig. 4 (c)).

Cascaded Attention. We design a cascaded channel attention mechanism to learn how to adaptively fuse the raw LiDAR points, semantic labels, and the 3D instance centers, which are fed to this module as channels. Specifically, each attention component uses a multilayer perceptron (MLP) to obtain a weight mask on all channels and then it produces features as weighted sum of the channels. These learned features are then encoded by a PointPillars [22] or VoxelNet [47] encoder. To preserve point-level features, we use skip connection to concatenate the voxelized features with the raw LiDAR point (Fig. 4 (d)).

Discussion. The 3D instance center plays two important roles in 3D object detection. First it represents the membership of a point to a potential object, allowing the detector more easily distinguish the boundary of different objects. Second it encodes object position and provides significant clue to the location regression task, leading to increased position prediction accuracy in 3D object detection.

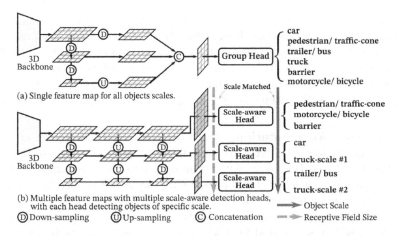

Fig. 5. A popular detection head structure (a) uses a single aggregated feature map to detect objects of all scales. Our proposed Fine-granular Detection Head (b) picks up a feature map of appropriate receptive field from pyramid feature maps to detect objects with specific scale. The size of the receptive field of a feature map is positively correlated with the object scale.

Instance Painter and Adaptive Projection Refiner form the first stage of PAI3D, it produces tremendous guidance information to the 3D detection network in the second stage, including object semantic label, point-to-object membership, and object position, before 3D detection network even begins to detect 3D objects. Experimental results in Sect. 4 prove that the two modules in the first stage bring significant 3D detection improvement.

3.3   Additional Enhancements for 3D Object Detection

Fine-Granular Detection Head. Detecting objects with various scales is crucial for a modern 3D detector. A learning strategy proposed recently, Group Head [48] divides object categories into serval groups based on object shapes or sizes, a detection head is then assigned to capture the intra-scale variations among objects for each group. However, common Group Head utilizes a single feature map as input, indicating the features it receives from the backbone are on a single-level scale. Most recently proposed detectors, e.g., CenterPoint [44] and MVP [45], aggregate three scale levels into one feature map to alleviate this problem. However, they still use one feature map for all object scales (Fig. 5 (a)). A recent study (YOLOF [6]) shows that passing multiple scaled feature maps to multiple detecting heads is essential in improving the detection performance.

We propose Fine-granular Detection Head that detects specific-scaled objects with specific-scaled feature maps (Fig. 5 (b)). In details, an FPN-like [20] (i.e. PANet [25], BiFPN [34]) detection neck is used to generate a series of feature maps with different scales of receptive-fields, and then a Group Head is attached to each generated feature map so that each detection head becomes scale aware.

Consequently, we can then dispatch different scaled objects of different categories into different scale-aware heads. Under this design, a scale-aware head naturally picks up a feature map with an appropriately scaled receptive field to match with the object scale. Likewise, this module also improves detection performance of the objects with large intra-scale (same category) variations by simultaneously learning the intra-scale variance of one category with multiple detection heads.

False Positive Augmentation. Typically, sparse LiDAR points are more prone to generate false positive detections, hence we propose False Positive Augmentation to address this issue. After training from scratch is complete, a fine-tuning is invoked, at when we build up an offline database to collect false positive proposals by running inference on the whole training dataset. Then we randomly sample false positive examples from the database and paste them into the raw point cloud, which is repeated for each successive fine-tuning step. The pasted false positive examples do not need extra annotations as they are recognized as background LiDAR points. Our ablation Experiments prove the effectiveness of this strategy in boosting detection performance.

4   Experiments

We evaluate PAI3D against existing 3D object detection methods on the nuScenes [1] dataset and present a series of ablation results to offer insights into the effect of each module, different 3D encoders, and instance segmentation quality.

4.1   The NuScenes Dataset

The nuScenes dataset [1] is one of the most well-known public large-scale datasets for autonomous driving. It is widely used by researchers to evaluate common computer vision tasks such as 3D object detection and tracking. The data was collected by a 32-beam LiDAR with a scanning frequency 20 Hz and six RGB cameras with a capture frequency 12 Hz. It provides 3D annotations for ten classes 2 Hz in a 360-degree field of view. The dataset contains 40K complete annotated frames, further split into 700 training scenes, 150 validation scenes, and 150 testing scenes. The official evaluation metrics includes mean Average Precision (mAP) [13] and NuScenes Detection Score (NDS) [1]. No external data are used for the trainings and evaluations on the nuScenes dataset.

4.2   Training and Evaluation Settings

We train and evaluate our method on four distributed NVIDIA P40 GPUs which are configured to run synchronized batch normalization with batch size 4 per GPU. Following [38], we use HTCNet[1] [3] for both image instance segmentation in our PAI3D and 2D semantic segmentation in PointPainting [35].

---

[1] We directly use the checkpoint (trained on nuImage [1]) provided by MMDetection3D [9] without making any modifications.

Table 1. The 3D object detection results on the nuScenes *test set*. Object categories include Car, truck (Tru), Bus, trailer (Tra), construction vehicle (CV), pedestrian (Ped), motorcycle (MC), bicycle(Bic), traffic cone (TC), and barrier (Bar). "L" and "C" in Modality refer to "LiDAR" and "Camera" respectively. The proposed PAI3D outperforms all existing 3D detection methods, including LiDAR-camera fusion-based and LiDAR alone methods, with a large margin on all object categories (except traffic cone) and with both mAP and NDS metrics.

Method	Modality	mAP↑	NDS↑	Car	Tru	Bus	Tra	CV	Ped	MC	Bic	TC	Bar
PointPillars [22]	L	30.5	45.3	68.4	23.0	28.2	23.4	4.1	59.7	27.4	1.1	30.8	38.9
WYSIWYG [19]	L	35.0	41.9	79.1	30.4	46.6	40.1	7.1	65.0	18.2	0.1	28.8	34.7
PointPainting [35]	L & C	46.4	58.1	77.9	35.8	36.1	37.3	15.8	73.3	41.5	24.1	62.4	60.2
CBGS [48]	L	52.8	63.3	81.1	48.5	54.9	42.9	10.5	80.1	51.5	22.3	70.9	65.7
CVCNET [4]	L	55.8	64.2	82.6	49.5	59.4	51.1	16.2	83.0	61.8	38.8	69.7	69.7
CenterPoint [44]	L	58.0	65.5	84.6	51.0	60.2	53.2	17.5	83.4	53.7	28.7	76.7	70.9
HotSpotNet [5]	L	59.3	66.0	83.1	50.9	56.4	53.3	23.0	81.3	63.5	36.6	73.0	71.6
MVP [45]	L & C	66.4	70.5	86.8	58.5	67.4	57.3	26.1	89.1	70.0	49.3	85.0	74.8
CenterPointV2 [44]	L & C	67.1	71.4	87.0	573	69.3	60.4	28.8	90.4	71.3	49.0	**86.8**	71.0
FusionPainting [38]	L & C	68.1	71.6	87.1	60.8	68.5	61.7	30.0	88.3	74.7	53.5	85.0	71.8
PAI3D (Ours)	L & C	**71.4**	**74.2**	**88.4**	**62.7**	**71.3**	**65.8**	**37.8**	**90.3**	**80.8**	**58.2**	83.2	**75.5**

For the 3D object detection task, following [1,38,44,45] we stack ten LiDAR sweeps into the keyframe sample to obtain a denser point cloud. We evaluate PAI3D's performance with two different 3D encoders, VoxelNet [47] and PointPillars [22]. For VoxelNet, the detection range is set to $[-54, 54] \times [-54, 54] \times [-5, 3]$ in meters for X, Y, Z axes respectively, and the voxel size is $(0.075, 0.075, 0.075)$ in meters. For PointPillars, the detection range is set to $[-51.2, 51.2] \times [-51.2, 51.2] \times [-5, 3]$ in meters with voxel size being $(0.2, 0.2, 8)$ in meters.

4.3 nuScenes Test Set Results

Table 1 shows the results of PAI3D in comparison to the recent SOTA approaches on the nuScenes *test* set. Following [38,44], we perform test time augmentation (TTA) operations including flip and rotation, and we apply NMS to merge the results from five models with different voxel sizes. As seen from Table 1, PAI3D achieves 71.4 mAP and 74.2 NDS, outperforming all existing 3D detection methods, including LiDAR-camera fusion based and LiDAR alone methods, with a large margin on all object categories (except traffic cone) and with both mAP and NDS metrics. Note that latest methods such as CenterPointV2 (CenterPoint [44] with PointPainting [35]) and FusionPainting [38] are already very strong, nevertheless PAI3D outperforms these two strong baselines by 4.3 in mAP and 2.8 in NDS.

Qualitative results of PAI3D in comparison to strong SOTA CenterPoint [44] (reproduced[2] + PointPainting (with HTCNet [3]) are presented in Fig. 7 (a). The

---

[2] Reproduced with the latest code from the CenterPoint's official GitHub repository: https://github.com/tianweiy/CenterPoint).

Table 2. Ablation results for contribution of each module in PAI3D on the nuScenes *validation* set. FDH: Fine-granular Detection Head, IP: Instance Painter, SL: Semantic Label, Cen: Instance Center, APR: Adaptive Projection Refiner, FPA: False Positives Augmentation. The baseline is the reproduced CenterPoint with PointPillars encoder.

Method	FDH	IP		APR	FPA	mAP↑	NDS↑
		SL	Cen				
CenterPoint [44]						50.3	60.2
CenterPoint (reproduced baseline)						52.1	61.4
PAI3D (ours)	✓					52.6 (0.5↑)	62.3 (0.9↑)
	✓	✓				62.2 (9.6↑)	66.9 (4.6↑)
	✓	✓	✓			64.1 (1.9↑)	67.7 (0.8↑)
	✓	✓	✓	✓		64.5 (0.4↑)	68.2 (0.5↑)
	✓	✓	✓	✓	✓	64.7 (0.2↑)	68.4 (0.2↑)

results show that instance-level semantics fused with point geometry in PAI3D leads to fewer false positive detections and higher recall for the sparse objects

4.4 Ablation Study

In this section, we design a series of ablation studies using the nuScenes *validation* set to analyze the contribution of each module in the proposed method.

Contribution of Each Module. Incremental experiments are performed to analyze the effectiveness of each module by benchmarking against reproduced CenterPoint [44] with PointPillars [22] encoder as the baseline. As seen from Table 2, Fine-granular Detection Head boosts over the single Group Head [48] used in the baseline by 0.5 mAP and 0.9 NDS. Notably the highest performance boost comes from the Instance Painter (IP), which increases mAP by 11.5 and NDS by 5.4 over baseline with FDH, suggesting that instance-level image semantics are tremendously helpful. Within IP, semantic label brings 9.6 mAP and 4.6 NDS boost and using instance-center further improves performance by 1.9 mAP and 0.8 NDS. Adaptive Projection Refiner yields slight increase of 0.4 mAP and 0.5 NDS, indicating that our PAI3D, even without refining the 3D instances, is still quite robust to the erroneous semantic labels. False Positives Augmentation contributes an additional 0.2 mAP and 0.2 NDS. In total, our method outperforms the baseline by 12.6 mAP and 7.0 NDS, enabling PAI3D to be the new STOA method in fusion-based 3D object detection.

3D Encoders. As shown in PAI3D architecture in Fig. 2, 3D encoders fuse the instance-level image semantics with point geometry features, and we are interested to see how sensitive PAI3D is to the choice of different 3D encoders. To that end, we compare PAI3D with strong fusion-based SOTA methods, including Center-Point [44] (reproduced) + PointPainting [35] (HTCNet), CenterPoint + virtual-point [45] and FusionPainting [38], and we select two widely used 3D encoders,

Table 3. Ablation results for 3D encoders on the nuScenes *validation* set. We implement PAI3D with two widely used 3D encoders, VoxNet and PointPillars, and benchmark its performances on nuScenes validation set with SOTA fusion-based methods. "L" and "C" in Modality refer to "LiDAR" and "Camera" respectively.

Method	Modality	3D encoders	mAP↑	NDS↑
CenterPoint [44]	L	VoxNet	56.4	64.8
CenterPoint (reproduced)	L	VoxNet	58.9	66.5
CenterPoint (reproduced) + PointPainting(HTCNet)	L & C	VoxNet	64.9	69.3
CenterPoint + virtual-point [45]	L & C	VoxNet	65.9	69.6
FusionPainting(2D Painting) [38]	L & C	VoxNet	63.8	68.5
FusionPainting(2D+3D Painting) [38]	L & C	VoxNet	66.5	70.7
PAI3D (ours)	L & C	VoxNet	**67.6**	**71.1**
CenterPoint [44]	L	PointPillars	50.3	60.2
CenterPoint (reproduced)	L	PointPillars	52.1	61.4
CenterPoint (reproduced) + PointPainting(HTCNet)	L & C	PointPillars	61.8	66.2
PAI3D (ours)	L & C	PointPillars	**64.7**	**68.4**

VoxelNet [47] and PointPillars [22], as the 3D encoders for these methods. For the sake of fair comparison, we do not use any two-stage refinement of the detected 3D proposals or dynamic voxelizations. As seen from Table 3, PAI3D consistently surpasses all existing fusion based SOTA methods for both 3D encoders with a large margin, indicating that the strength of PAI3D mainly arise from non-3D-encoder modules such as Instance Painter and its performance is less influenced by the 3D encoders. Notice that PAI3D, even only using semantics from image, still outperforms FusionPainting (2D+3D Painting) [38], which uses semantics from both image and point cloud segmentation, for the VoxelNet encoder.

Instance Segmentation Quality. [35] shows that image segmentation quality can dramatically impact the final 3D object detection performance. To examine this issue, we conduct ablation experiments to plot the correlation curve between instance segmentation quality (measured by mask AP) and 3D detection

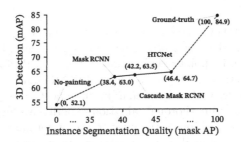

Fig. 6. Ablation results for instance segmentation quality. Instance segmentation quality and 3D detection performance are positively correlated, meaning that better 2D instance segmentation leads to better 3D object detection performance.

472   H. Liu et al.

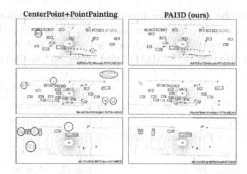

Fig. 7. Qualitative results of PAI3D compared with CenterPoint (reproduced) + Point-Painting applied to the nuScenes *validation* set. Predicted boxes are shown in red, ground truth boxes in green, and detection differences in black circles. (Color figure online)

performance (measured by mAP). We select several instance segmentation models with increasing quality, ranging from Mask R-CNN [17] to Cascade Mask R-CNN [2] to HTCNet [3] from MMDetection3D [9], and implement them into PAI3D, we then add the lower bound of no semantic painting and the upper bound of painting with ground truth label. All the instance segmentation models are trained on the nuImage [1] dataset, and all these 3D detection methods are trained with nuScenes training dataset. Finally, we obtain the 3D object detection mAP of these methods on the nuScenes *validation* set. As shown in Fig. 6, instance segmentation quality and 3D detection performance are positively correlated, meaning that better 2D instance segmentation leads to better 3D object detection performance. In addition, we observe that model-based painting yields much worse 3D detection performance than ground-truth-based painting, we point out that this performance gap can be narrowed by improving instance segmentation quality and reducing point-to-pixel projection errors. We believe more in-depth research along this direction is worthwhile to pursue in our future work.

5   Conclusions

This paper presents PAI3D, a simple and effective sequential instance-level fusion method for 3D object detection, which realizes instance-level fusion between 2D image semantics and 3D point cloud geometry features. PAI3D outperforms all the SOTA 3D detection methods, including LiDAR-camera fusion based and LiDAR alone methods, with a large margin on almost all object categories on the nuScenes dataset. Our thorough ablation experiments show that instance-level image semantics contributes the most to the performance gain, and PAI3D works well with any good-quality instance segmentation models and any modern point cloud 3D encoder networks. Qualitative results on a real-world autonomous driving dataset demonstrate that PAI3D yields fewer false positive detections and higher recall for the distant sparse objects and hard to see objects.

References

1. Caesar, H., et al.: nuScenes: a multimodal dataset for autonomous driving. In: Proceedings of the IEEE/CVF Conference on Computer Vision and Pattern Recognition, pp. 11621–11631 (2020)
2. Cai, Z., Vasconcelos, N.: Cascade R-CNN: high quality object detection and instance segmentation. IEEE Trans. Pattern Anal. Mach. Intell. **43**(5), 1483–1498 (2019)
3. Chen, K., et al.: Hybrid task cascade for instance segmentation. In: IEEE Conference on Computer Vision and Pattern Recognition (2019)
4. Chen, Q., Sun, L., Cheung, E., Yuille, A.L.: Every view counts: cross-view consistency in 3D object detection with hybrid-cylindrical-spherical voxelization. In: Advances in Neural Information Processing Systems (2020)
5. Chen, Q., Sun, L., Wang, Z., Jia, K., Yuille, A.: Object as hotspots: an anchor-free 3D object detection approach via firing of hotspots. In: Vedaldi, A., Bischof, H., Brox, T., Frahm, J.-M. (eds.) ECCV 2020. LNCS, vol. 12366, pp. 68–84. Springer, Cham (2020). https://doi.org/10.1007/978-3-030-58589-1_5
6. Chen, Q., Wang, Y., Yang, T., Zhang, X., Cheng, J., Sun, J.: You only look one-level feature. In: Proceedings of the IEEE/CVF Conference on Computer Vision and Pattern Recognition, pp. 13039–13048 (2021)
7. Chen, X., Ma, H., Wan, J., Li, B., Xia, T.: Multi-view 3D object detection network for autonomous driving. In: Proceedings of the IEEE Conference on Computer Vision and Pattern Recognition, pp. 1907–1915 (2017)
8. Chen, Y., Tai, L., Sun, K., Li, M.: MonoPair: monocular 3D object detection using pairwise spatial relationships. In: Proceedings of the IEEE/CVF Conference on Computer Vision and Pattern Recognition, pp. 12093–12102 (2020)
9. Contributors, M.: MMDetection3D: OpenMMLab next-generation platform for general 3D object detection (2020). https://github.com/open-mmlab/mmdetection3d
10. Ding, Z., et al.: 1st place solution for Waymo open dataset challenge-3D detection and domain adaptation. arXiv preprint arXiv:2006.15505 (2020)
11. Duan, K., Bai, S., Xie, L., Qi, H., Huang, Q., Tian, Q.: CenterNet: keypoint triplets for object detection. In: Proceedings of the IEEE/CVF International Conference on Computer Vision, pp. 6569–6578 (2019)
12. Ester, M., Kriegel, H.P., Sander, J., Xu, X., et al.: A density-based algorithm for discovering clusters in large spatial databases with noise. In: KDD, vol. 96, pp. 226–231 (1996)
13. Everingham, M., Van Gool, L., Williams, C.K., Winn, J., Zisserman, A.: The pascal visual object classes (VOC) challenge. Int. J. Comput. Vision **88**(2), 303–338 (2010)
14. Fan, L., Xiong, X., Wang, F., Wang, N., Zhang, Z.: RangeDet: in defense of range view for lidar-based 3D object detection. In: Proceedings of the IEEE/CVF International Conference on Computer Vision, pp. 2918–2927 (2021)
15. Graham, B., Engelcke, M., Van Der Maaten, L.: 3D semantic segmentation with submanifold sparse convolutional networks. In: Proceedings of the IEEE Conference on Computer Vision and Pattern Recognition, pp. 9224–9232 (2018)
16. He, C., Zeng, H., Huang, J., Hua, X.S., Zhang, L.: Structure aware single-stage 3D object detection from point cloud. In: Proceedings of the IEEE/CVF Conference on Computer Vision and Pattern Recognition, pp. 11873–11882 (2020)
17. He, K., Gkioxari, G., Dollár, P., Girshick, R.: Mask R-CNN. In: Proceedings of the IEEE International Conference on Computer Vision, pp. 2961–2969 (2017)

18. He, Y., et al.: MR-DBSCAN: an efficient parallel density-based clustering algorithm using MapReduce. In: 2011 IEEE 17th International Conference on Parallel and Distributed Systems, pp. 473–480. IEEE (2011)
19. Hu, P., Ziglar, J., Held, D., Ramanan, D.: What you see is what you get: exploiting visibility for 3D object detection. In: Proceedings of the IEEE/CVF Conference on Computer Vision and Pattern Recognition, pp. 11001–11009 (2020)
20. Kirillov, A., Girshick, R., He, K., Dollár, P.: Panoptic feature pyramid networks. In: Proceedings of the IEEE/CVF Conference on Computer Vision and Pattern Recognition, pp. 6399–6408 (2019)
21. Ku, J., Mozifian, M., Lee, J., Harakeh, A., Waslander, S.L.: Joint 3D proposal generation and object detection from view aggregation. In: 2018 IEEE/RSJ International Conference on Intelligent Robots and Systems (IROS), pp. 1–8. IEEE (2018)
22. Lang, A.H., Vora, S., Caesar, H., Zhou, L., Yang, J., Beijbom, O.: PointPillars: fast encoders for object detection from point clouds. In: Proceedings of the IEEE/CVF Conference on Computer Vision and Pattern Recognition, pp. 12697–12705 (2019)
23. Liang, Z., Zhang, Z., Zhang, M., Zhao, X., Pu, S.: RangeIoUDet: range image based real-time 3D object detector optimized by intersection over union. In: Proceedings of the IEEE/CVF Conference on Computer Vision and Pattern Recognition, pp. 7140–7149 (2021)
24. Liu, H., Oyama, S., Kurihara, M., Sato, H.: Landmark FN-DBSCAN: an efficient density-based clustering algorithm with fuzzy neighborhood. J. Adv. Comput. Intell. **17**(1), 60–73 (2013)
25. Liu, S., Qi, L., Qin, H., Shi, J., Jia, J.: Path aggregation network for instance segmentation. In: Proceedings of the IEEE Conference on Computer Vision and Pattern Recognition, pp. 8759–8768 (2018)
26. Liu, Z., Wu, Z., Tóth, R.: Smoke: single-stage monocular 3D object detection via keypoint estimation. In: Proceedings of the IEEE/CVF Conference on Computer Vision and Pattern Recognition Workshops, pp. 996–997 (2020)
27. Paigwar, A., Sierra-Gonzalez, D., Erkent, Ö., Laugier, C.: Frustum-PointPillars: a multi-stage approach for 3D object detection using RGB camera and lidar. In: Proceedings of the IEEE/CVF International Conference on Computer Vision, pp. 2926–2933 (2021)
28. Park, J., Weng, X., Man, Y., Kitani, K.: Multi-modality task cascade for 3D object detection. arXiv preprint arXiv:2107.04013 (2021)
29. Qi, C.R., Su, H., Mo, K., Guibas, L.J.: PointNet: deep learning on point sets for 3D classification and segmentation. In: Proceedings of the IEEE Conference on Computer Vision and Pattern Recognition, pp. 652–660 (2017)
30. Shi, S., Wang, X., Li, H.: PointRCNN: 3D object proposal generation and detection from point cloud. In: Proceedings of the IEEE/CVF Conference on Computer Vision and Pattern Recognition, pp. 770–779 (2019)
31. Simon, M., Milz, S., Amende, K., Gross, H.-M.: Complex-YOLO: an Euler-region-proposal for real-time 3D object detection on point clouds. In: Leal-Taixé, L., Roth, S. (eds.) ECCV 2018. LNCS, vol. 11129, pp. 197–209. Springer, Cham (2019). https://doi.org/10.1007/978-3-030-11009-3_11
32. Sun, P., et al.: Scalability in perception for autonomous driving: Waymo open dataset. In: Proceedings of the IEEE/CVF Conference on Computer Vision and Pattern Recognition, pp. 2446–2454 (2020)
33. Sun, P., et al.: RSN: range sparse net for efficient, accurate lidar 3D object detection. In: Proceedings of the IEEE/CVF Conference on Computer Vision and Pattern Recognition, pp. 5725–5734 (2021)

34. Tan, M., Pang, R., Le, Q.V.: EfficientDet: scalable and efficient object detection. In: Proceedings of the IEEE/CVF Conference on Computer Vision and Pattern Recognition, pp. 10781–10790 (2020)
35. Vora, S., Lang, A.H., Helou, B., Beijbom, O.: PointPainting: sequential fusion for 3D object detection. In: Proceedings of the IEEE/CVF Conference on Computer Vision and Pattern Recognition, pp. 4604–4612 (2020)
36. Wang, C., Ma, C., Zhu, M., Yang, X.: PointAugmenting: cross-modal augmentation for 3D object detection. In: Proceedings of the IEEE/CVF Conference on Computer Vision and Pattern Recognition, pp. 11794–11803 (2021)
37. Wang, T., Zhu, X., Pang, J., Lin, D.: FCOS3D: fully convolutional one-stage monocular 3D object detection. In: Proceedings of the IEEE/CVF International Conference on Computer Vision, pp. 913–922 (2021)
38. Xu, S., Zhou, D., Fang, J., Yin, J., Bin, Z., Zhang, L.: FusionPainting: multimodal fusion with adaptive attention for 3D object detection. In: 2021 IEEE International Intelligent Transportation Systems Conference (ITSC), pp. 3047–3054. IEEE (2021)
39. Yan, Y., Mao, Y., Li, B.: Second: sparsely embedded convolutional detection. Sensors **18**(10), 3337 (2018)
40. Yang, B., Luo, W., Urtasun, R.: Pixor: real-time 3D object detection from point clouds. In: Proceedings of the IEEE Conference on Computer Vision and Pattern Recognition, pp. 7652–7660 (2018)
41. Yang, Z., Sun, Y., Liu, S., Jia, J.: 3DSSD: point-based 3D single stage object detector. In: Proceedings of the IEEE/CVF Conference on Computer Vision and Pattern Recognition, pp. 11040–11048 (2020)
42. Yang, Z., Sun, Y., Liu, S., Shen, X., Jia, J.: IPOD: intensive point-based object detector for point cloud. arXiv preprint arXiv:1812.05276 (2018)
43. Yang, Z., Sun, Y., Liu, S., Shen, X., Jia, J.: STD: sparse-to-dense 3D object detector for point cloud. In: Proceedings of the IEEE/CVF International Conference on Computer Vision, pp. 1951–1960 (2019)
44. Yin, T., Zhou, X., Krahenbuhl, P.: Center-based 3D object detection and tracking. In: Proceedings of the IEEE/CVF Conference on Computer Vision and Pattern Recognition, pp. 11784–11793 (2021)
45. Yin, T., Zhou, X., Krähenbühl, P.: Multimodal virtual point 3D detection. In: Advances in Neural Information Processing Systems, vol. 34 (2021)
46. Yoo, J.H., Kim, Y., Kim, J., Choi, J.W.: 3D-CVF: generating joint camera and LiDAR features using cross-view spatial feature fusion for 3D object detection. In: Vedaldi, A., Bischof, H., Brox, T., Frahm, J.-M. (eds.) ECCV 2020. LNCS, vol. 12372, pp. 720–736. Springer, Cham (2020). https://doi.org/10.1007/978-3-030-58583-9_43
47. Zhou, Y., Tuzel, O.: VoxelNet: end-to-end learning for point cloud based 3D object detection. In: Proceedings of the IEEE Conference on Computer Vision and Pattern Recognition, pp. 4490–4499 (2018)
48. Zhu, B., Jiang, Z., Zhou, X., Li, Z., Yu, G.: Class-balanced grouping and sampling for point cloud 3D object detection. arXiv preprint arXiv:1908.09492 (2019)

Validation of Pedestrian Detectors by Classification of Visual Detection Impairing Factors

Korbinian Hagn(✉) ⓘ and Oliver Grau ⓘ

Intel Deutschland GmbH, Lilienthalstraße 15, 85579 Neubiberg, Bayern, Germany
{korbinian.hagn,oliver.grau}@intel.com

Abstract. Validation of AI based perception functions is a key cornerstone of safe automated driving. Building on the use of richly annotated synthetic data, a novel pedestrian detector validation approach is presented, enabling the detection of training data biases, like missing poses, ethnicities, geolocations, gender or age. We define a range of visual impairment factors, e.g. occlusion or contrast, which are deemed to be influential on the detection of a pedestrian object. A classifier is trained to distinguish a pedestrian object only by these visual detection impairment factors which enables to find pedestrians that should be detectable but are missed by the detector under test due to underlying training data biases. Experiments demonstrate that our method detects pose, ethnicity and geolocation data biases on the CityPersons and the EuroCity Persons datasets. Further, we evaluate the overall influence of these impairment factors on the detection performance of a pedestrian detector.

1 Introduction

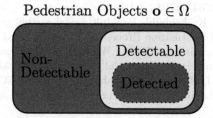

Fig. 1. A classifier can be trained to distinguish detectable from non-detectable pedestrians according to a set of visual impairment factors. Objects in the detectable set which have not been detected indicate a data bias of the pedestrian detector.

Modern deep learning-based object detectors are capable of detecting objects with unprecedented accuracy. The detection boundary of a vision task is often

---

Supplementary Information The online version contains supplementary material available at https://doi.org/10.1007/978-3-031-25072-9_33.

compared to human detection capabilities. While closing in on reaching this detection limit it is necessary to understand what properties of an object in an image does make it difficult for a human to detect this very object and how we are able to exploit this detection boundary for validation. In Fig. 1 we argue that for a set of pedestrian objects in a dataset, there is (i) a non-detectable subset, defined as persons that are not detectable due to their visual impairment factors that we will introduce in the course of this paper, (ii) the detectable subset, which is the set of pedestrian objects that are detectable by a human observer or state-of-the-art object detector trained on a global super set without any data biases, and (iii) a subset consisting of objects that are detected by a state-of-the-art detector trained on a Geo specific domain with likely data biases.

The difference between detectable and detected subsets is then mainly defined due to differences in the training data of the detector. The boundary of the detectable subset itself is defined by visual detection impairment factors, like occlusion and contrast. If for example the contrast is close to 0 or the occlusion is close to 100% then the object, while still being present in the image, simply cannot be detected even by a human observer.

Our validation approach makes use of all these three subsets by identifying pedestrians that should be detectable according to their visual impairment factors, i.e., classifying if a pedestrian object is in the non-detectable or the detectable set. We then evaluate the difference between the detectable subset and the actually detected subset which we receive from predictions of a pedestrian detector. We demonstrate by investigation of this difference that we can reveal data biases in the pedestrian detector under test, like missing poses, ethnicity or geolocations originating from the used training dataset.

This validation strategy is majorly enabled by the recent progress in synthetic data generation giving new possibilities to produce high-realism data for computer vision tasks. Especially for understanding the prediction decision-making of a detector, the pixel-accurate ground-truth and the rich meta-data that can be extracted from the data generation process are highly beneficial to find relations of a visual factor, e.g., contrast, of an object and its detection.

The contributions of this paper are as follows:

- We identify factors that are impairing the visual detection capability of a pedestrian object and investigate each factor's importance to the detectability.
- We demonstrate the usage of these factors by training a classifier to distinguish detectable from non-detectable pedestrian objects and utilizing this classifier to validate a pedestrian detector for biases of its underlying training dataset.

2   Related Works

Validation Strategies. A validation strategy of a perception function which is part of the automotive driving stack is of high importance to guarantee a safe driving function. Our method hereby detects pedestrian detection faults, i.e. a functional insufficiency as defined by [12], addressing the lack of generalization as defined by [22]. To validate for a functional insufficiency, the detection, creation and testing with corner cases has been addressed by several previous works [1,2]. Bolte et al. [3] define a corner case as a *"non-predictable relevant object in relevant location"*. By this definition, we propose a new method validating a pedestrian detector through the detection of corner cases by learning to classify the *"non-predictable"* property from visual detection impairment factors.

Validation with Synthetic Data. Several synthetic datasets for automotive perception functions have been proposed [19,21,25]. While these datasets are very valuable for benchmarking, methods for validation should allow to directly control the generation of synthetic data, especially for automotive tasks like pedestrian detection, and here the most notably to mention are Carla [10] and the LGSVL Simulator [20]. But, instead of a whole simulation of an automotive driving scenario also variational automotive scene creation approaches have been introduced for detector evaluation [11,18,23]. These works could already show validation results with detecting training and validation dataset domain shifts. Our method not only detects data biases with semantic relevance, i.e. geolocation, ethnicity, etc., but also missing pose information in the training data by utilizing a classifier to distinguish detectable from non-detectable pedestrians through analyzing the proposed visual detection impairment factors.

Visual Detection Impairing Factors. Besides label noise and sensor noise, detection impairing factors such as occlusion rate and contrast were analyzed for their effect on object detection performance. While there are factors that are believed to have no influence on the detection, for example contrast [26], we found evidence that contrast is still relevant to the object detection performance. Distance and occlusion on the other side is acknowledged to be one of the most influential factors and has already been addressed by several works [7,8] and incorporated into a safety relevance distance metric [17]. We not only address these factors and their influence on the detection performance, but add several new factors relevant to the detection performance and investigate their influence.

Pedestrian Detection Models. While there is a recent popularity increase of models based on the Transformer architecture [9], most automotive detection benchmarks[1,2] are still dominated by convolutional neural network (CNN) based methods. In this work we are using the Cascade R-CNN [5] pedestrian detector, a development from the R-CNN [14] and Faster R-CNN [13] model. The HRNet [24] feature extraction backbone is pre-trained on ImageNet [6].

---

[1] CityPersons: https://github.com/cvgroup-njust/CityPersons.
[2] EuroCity Persons: https://eurocity-dataset.tudelft.nl/eval/benchmarks/detection.

3   Methodology

Our method is described in several sub steps: First, the generation of synthetic training and validation data. Next, the definition of the visual detection impairment factors and their extraction from synthetic data. Last, the definition and training of a detectability classifier and validation of a pedestrian detector for data biases.

3.1   Synthetic Data Generation

The synthetic data used in this contribution is generated by a data synthesis pipeline and includes special modules to compute meta- or ground-truth data which is hard or impossible to observe and measure in real data. One example is the pixel-accurate occlusion rate of an object, by differencing the mask of the un-occluded object, computed in a separate rendering pass from the occluded object in the complete scene. To achieve a representative calibration of the detector, the synthetic data should have similar characteristics than real scenes. This is also described as domain gap between the real and synthetic data. We achieve highly realistic synthetic data by three levels: i) An automated scene generator produces scenes with similar complexity as those in real data, ii) we use similar 3D objects from an asset database and iii) a realistic sensor simulation building on the work described in [15]. The rendering process delivers realistic scene illumination in linear color space with floating accuracy based on the Blender Cycles path-tracing rendering engine[3], followed by a sensor simulation that includes simulation of effects like sensor noise, lens distortions, and chromatic aberrations and a tone mapping to integer sRGB color space. The parameters of the sensor simulation are tuned to match the characteristics of the Cityscapes data similar to [16].

For the purpose of this paper, we synthesize a dataset that contains complex urban street scenes with a variation of objects (about 300), such as different houses, vehicles, street elements, and about 150 different human characters automatically placed from an asset database. Figure 2 depicts some example frames from that data set. The synthetic data generation pipeline also computes various metadata and ground-truth, including semantic and instance segmentation, the distance of objects to the camera, occlusion rates, 2D + 3D bounding boxes, radiometric object features including contrast measures as introduced above. The dataset is split into a training set to calibrate the detectability classifier and a validation set to detect data biases of the pedestrian detector under test.

One of the advantages of synthetic data in our method is the precision and deterministic nature of the label and bounding box meta-data, which is free from noise, as all the generated labeling data is pixel accurate.

For evaluations with this dataset, the pedestrians in the images are pre-filtered to guarantee that only the considered impairment factors are influential on a pedestrian detection. This means that pedestrians that are too close to

---

[3] blender.org.

Fig. 2. Our fully parameterizable generation pipeline allows rendering pedestrians at any size, occlusion, time of day, and distance to the camera.

one another cannot be detected due to non-maximum suppression (NMS) and are therefore ignored. The resulting synthetic dataset $\mathcal{D}_{synth}$ consists of 17012 images of pedestrians for training and 26745 images of pedestrians for evaluation.

3.2 Visual Detection Impairing Factors

The major factors of a pedestrian object we consider to be influential of the detection capability of a detector are visualized in Fig. 3. Beginning with the placement of a pedestrian in an image. This information is extracted from the bounding box coordinates of the ground-truth. The coordinates are defined by the center o_{cx}, o_{cy} coordinate and the width o_w and height o_h of the box in pixels.

Next, distance to the observer o_d, i.e., camera, in $[m]$ and the number of visible pixels o_{vp} of the object. The distance to the observer is extracted from the 3D placement in the rendered scene. The information about the number of visible pixels is extracted from the instance segmentation label, by counting the pixels that belong to the person.

Determining the occlusion rate o_{ocl}, i.e., the ratio of visible pixels to the whole pixels of an object, is done by extracting the number of occluded and non-occluded pixels from the instance segmentation ground-truth pixels with, and without occluding objects.

Last, we define different contrast measures as visual impairment factors. The first contrast measure o_{cfull} is defined as the euclidean distance of the mean object RGB color to its surrounding background. This is done by dilating the instance segmentation mask of the person and subtract the undiluted segmentation mask so that we get the surrounding 5 pixel border of the object. The contrast is now calculated by the euclidean distance of the mean object color to the mean surrounding background color. Another contrast measure o_{cmean} is defined by segmenting the object into 12 smaller segments and calculate the

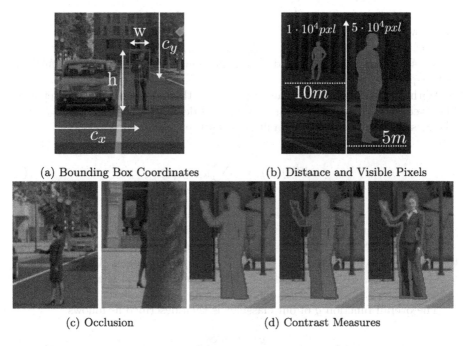

(a) Bounding Box Coordinates   (b) Distance and Visible Pixels

(c) Occlusion   (d) Contrast Measures

Fig. 3. The potential detection performance impairing factors we consider in this work: (a) bounding box coordinates $(o_{cx}, o_{cy}, o_h, o_w)$, (b) distance and number of visible pixels of a pedestrian (o_d, o_{vp}), (c) rate of occlusion (o_{ocl}), (d) contrast of a pedestrian (red) to its background (blue) calculated by the full pedestrian silhouette (o_{cfull}), segment wise (o_{cmean}) and edge wise (o_{cedge}).

mean RGB color of a segment and the euclidean distance to its neighboring mean background color. The resulting contrast measure is then derived by averaging over all segments. The last considered contrast measure o_{cedge} is calculated with only the 5 pixels on the edge of the instance label, then calculating the mean color and the euclidean distance to the mean of the surrounding background.

3.3   Classification of Detectable Pedestrians

We define the classification loss to train a classifier to distinguish persons being detectable or not. This classifier is then used to analyze a pedestrian detector for data biases due to missing data in the training data.

Classification Loss. We begin with a synthetic image dataset of pedestrian objects $\mathbf{o} \in \Omega = \{\mathbf{o}_1, \ldots, \mathbf{o}_O\}$ where O is the number of pedestrians in a set of all pedestrian objects defined as Ω. Each pedestrian was either detected (1) or missed (0) by the pedestrian detector under test. In the accumulation step each pedestrian object is enriched with the previously defined detection impairment factors, and thus we obtain the sample vector of the objects metadata $\mathbf{o}$ defined as follows:

$\mathbf{o} = (o_{cx}, o_{cy}, o_h, o_w, o_d, o_{vp}, o_{ocl}, o_{cfull}, o_{cmean}, o_{cedge})$, where every entry in this vector is normalized to $\mu = 0$ and $\sigma^2 = 1$ and equals to the detection impairment factors previously described.

With the enriched pedestrian objects and their corresponding target class (detected, missed) we then train a supervised deep neural network as classifier. The classification loss is the default cross entropy loss defined as,

$$J^{cls}(p, u) = -\sum_{i \in \mathcal{S}} u_i log(p_i). \tag{1}$$

The classification output is a discrete probability distribution $p = (p_0, \ldots, p_s)$, i.e. confidence, computed by a softmax with s being the predicted class $s \in \mathcal{S} = \{0, \ldots, S-1\}$, Here, $S = 2$, i.e., the pedestrian is detectable ($s = 1$) or the pedestrian is non-detectable ($s = 0$). The respective target class is defined as $u = (u_0, \ldots, u_s)$, with the detectable class defined as 1 and the non-detectable target class as 0.

The overall function g of our classifier is then described as follows,

$$g : \Omega \to \mathcal{S}. \tag{2}$$

The classifier learns a mapping of pedestrian objects Ω to a corresponding detectability class $\mathcal{S}$.

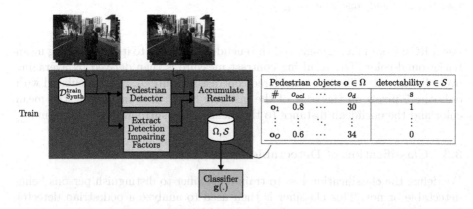

Fig. 4. Training a classifier to distinguish between detectable and non-detectable pedestrian objects.

Training the Classifier. Our approach to train a classifier to distinguish between detectable and non-detectable pedestrian objects is sketched in Fig. 4. We use the synthesized training dataset $\mathcal{D}^{train}_{Synth}$ as described in Sect. 3.2. These images are inferred by a state-of-the-art Cascade R-CNN [5] pedestrian detector with a HRNet [24] backbone pretrained on ImageNet [6]. This detector was further trained on a sub-set of our synthetic dataset. In parallel to the inference process, we extract our defined visual detection impairment factors on a per pedestrian object basis from each synthetic data frame. Next, the inference results of the detector and the extracted impairment factors are accumulated to the pedestrian object set Ω and target class set $\mathcal{S}$. Ω is the classification input with each row being a pedestrian object with its visual impairment factors $\mathbf{o}$ vectorized and a corresponding detectability class s in $\mathcal{S}$ as target class. These inputs are then used to train a fully connected deep neural network with 5 hidden layers to learn the mapping $\mathbf{g}(.)$, i.e. the detectability classification. Training this classifier results in a high classification F1 score of 0.93. The F1 score is the harmonic mean of precision and recall and regularly used to evaluate binary classification tasks.

Detection of Validation Samples. The trained detectability classifier is used to validate another pedestrian detector as described in Fig. 5. The detectors under test in our validation experiments are Cascade R-CNN detectors trained on the CityPersons [27] (CP) dataset and another trained on the EuroCity Persons [4] (ECP) dataset. For validation, we use the previously rendered synthetic validation dataset $\mathcal{D}_{Synth}^{val}$. In this dataset we integrated our validation samples, i.e., if we want to validate a pedestrian detector for biases due to geolocation of the training data we have to add different pedestrian assets from various other geolocations. Similar we can add different posed pedestrians, gender, age-group or ethnicity assets to the validation images. The Pedestrian detector under test generates inference results on the validation set images. In parallel the visual impairment factors per pedestrian object are extracted from each image. The detections and the impairment factors are accumulated into the sets Ω and $\mathcal{S}$. The previously trained classifier $\mathbf{g}(.)$ classifies the input objects into detectable or non-detectable. The results, i.e. confidence score, per object prediction and target detectability class, are stored in the validation result dataset $\mathcal{D}_{Synth}^{val\_cl}$. The validation results are then enhanced by the per pedestrian meta information from $\mathcal{D}_{Meta}^{val}$ which we obtain from the image synthesis process. This meta information consists of the asset ID, the geolocation, pose, gender, age-group and ethnicity and is useful to draw conclusions of underlying data biases in the training data.

The validation result $\mathcal{D}_{Synth}^{val\_cl}$ is now used to find detection biases of the pedestrian detector under test. Therefore, we simply filter the dataset for pedestrian objects that are classified as detectable (prediction=1) but were missed by the pedestrian detector under test (detectability=0). From this filtered validation result the data biases are extracted by statistical analysis of occurrences as described in section Results & Discussion.

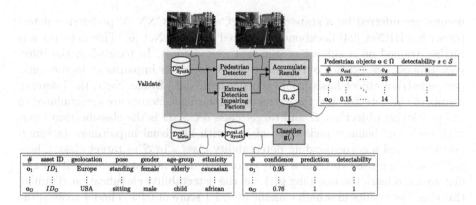

Fig. 5. Generation of validation data to detect data biases in the pedestrian detector.

4   Results and Discussion

We present two different results: First, we analyze the validation results from pedestrian detectors trained on the CityPersons (CP) and on the EuroCity Persons (ECP) dataset. This step is separated into the detection of geolocation, gender, age-group and ethnicity data biases and into the detection of data biases due to missing poses.

Last, we analyze the influence of each individual visual detection impairment factor on the detection performance, i.e. miss rate.

4.1   Evaluation of Pedestrian Detection Data Biases

Found data biases in our approach can mainly be categorized in two categories. The first category are data biases attributed to missing semantic characteristics in the training dataset and therefore resulting data bias. These errors can occur due to differences in pedestrians from different geolocations, missing gender, ethnicity or age-group in train and validation set. The second category of data biases are missing poses, which may not only originate from missing training data, but also due to incapability of the detector, i.e. missing aspect ratios of anchor boxes.

Validation of Data Bias Characteristics. To generate meaningful results from the validation result set $\mathcal{D}_{Synth}^{val\_cl}$ we filter all pedestrians according to their frequency. We deem a pedestrian to be meaningful if the ratio of frequency in the result set to overall frequency in the synthetic validation data is above 10%. Following the pre-filtering of the validation results we can plot the frequency of each pedestrian for the (a) CP and the (b) ECP dataset, as depicted in Fig. 6.

For the CP dataset 23 persons cause a data bias error comparing to only 15 persons for the ECP dataset. In the CP dataset we found that pedestrians from non-European geolocation and with a different ethnicity than Caucasian

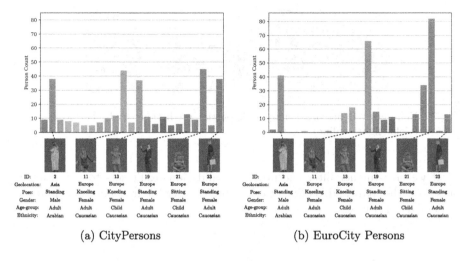

(a) CityPersons   (b) EuroCity Persons

Fig. 6. Distribution of found data biases.

will have a significant influence to miss-detections due to the data bias in the training data. This bias is very prominent if the person wears clothing typical for its geolocation but uncommon in Europe as can be seen with ID 2, a male veiled in traditional Arabian clothing. A similar observation can also be made with the ECP dataset. While many occurrences of persons with female gender can be observed, the originating cause is often in combination with an uncommon pose as in ID 19 with the waving hand or in ID 23 carrying an additional trolley. Sitting or kneeling poses, especially in combination with the child age-group, are the most common sources of data bias found in both datasets as can bee seen with the IDs 11, 13 and 21. For the child age-group as well as for the kneeling and sitting poses the overall height of the person is smaller than usual. This strongly suggests there are too few persons in sitting or kneeling pose included in both datasets. Comparing the observations from the CP and the ECP datasets we see the CP dataset to have a marginally smaller data bias on sitting and kneeling children but a higher data bias on uncommon poses of standing persons.

Validation of Pose Data Bias. To further filter the observations of found data biases we can evaluate the persons in the previous result set if they have at least a bounding box prediction with a ground-truth intersection over union (IoU) of > 0.25. Because a detection is only counted as a valid detection with an IoU threshold of > 0.5 we can lower this threshold in the validation to inspect if there was at leas at partial detection of a person. Exemplary predictions where the person was not counted as detected, but a partial detection was predicted nonetheless can be seen in Fig. 7.

Fig. 7. Non-detected persons (blue) with partial predictions (red), i.e. 50% > IoU > 25%. (Color figure online)

If we now further filter the previous validation result sets to only include persons with a partial detection of IoU > 0.25 we get the frequency distributions as depicted in Fig. 8.

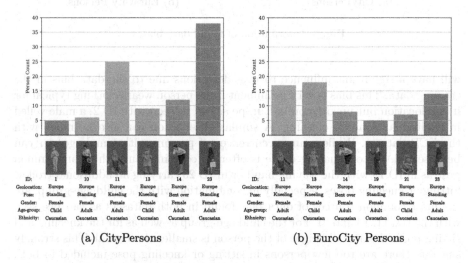

(a) CityPersons   (b) EuroCity Persons

Fig. 8. Distribution of found data biases due to missing pose data.

Again, the IDs 11, 13, 14 and 23 can be found in both datasets, suggesting that the characteristics, i.e. geolocation, age-group, gender and ethnicity, are known by the detector but the pose information is missing. The ID 19 and 21 were partially detected with the ECP dataset, but they are missing entirely now in the CP dataset, i.e. while the ECP dataset suffers from missing pose information on these persons, the CP dataset suffers from a data bias because not even a partial prediction was made. Similarly, the ID 1 and 10 are partially detected with the CP dataset but not with the ECP dataset.

Additionally, with the ID 23 person we found that the CP dataset did miss the detections because the luggage trolley is defined to be part of the bounding box of this person. While the ECP prediction would, in most cases, predict a loose bounding box for the person that is sufficient to count as detected, the

CP bounding box prediction was tighter on the person. Careful evaluation and definition of the ground-truth is therefore a key to produce meaningful validation results without any ambiguity.

Summarizing, the CP dataset is missing more pedestrian characteristics than the ECP dataset. Both datasets have a strong bias to European persons and both datasets are missing pose information on kneeling, sitting or bent over persons especially if they are children.

4.2 Detection Impact of Visual Impairment Factors

To understand if the visual impairment factors are meaningful for a detectable or non-detectable classification we have to investigate the relation of detection performance and each individual factor. Therefore, the predictions on the synthetic training set $\mathcal{D}_{Synth}^{train}$ are evaluated along with each individual impairment factor visualized as a histogram. Figure 9 shows the count of detected (blue) and non-detected (red) persons per bin. Additionally, the miss rate of each factor is calculated per bin and plotted into each histogram. The miss rate is defined as the number of missed detections (detected = 0) over the number of all persons, i.e., detected and non-detected.

A major influence factor is obviously the size of an object and therefore the number of visible pixels of a pedestrian in the image which is evident in several interrelated impairment factors. Small height, small width, high distance and a low number of visible pixels lead to a high miss rate. Increasing the occlusion rate leads to a reduced number of visible pixels and increased miss rate as well. The contrast measures show a decreasing tendency for increased values of contrast but not as pronounced as the other size related factors and with occasional outliers at higher values. Calculating the Spearman correlation of miss rate and the individual impairment factors emphasizes these observations as shown in Table 1. The size related factors (o_h, o_w, o_d, o_{vp}, o_{ocl}) show high positive or negative correlations, where the width o_w has the lowest absolute correlation to the miss rate. For the contrast measures we see a high negative correlation on the o_{cfull} contrast measure and lower negative correlations for the other two contrast measures.

Table 1. Spearman correlation of the visual impairment factors and miss rate.

	o_h	o_w	o_d	o_{vp}	o_{ocl}	o_{cfull}	o_{cmean}	o_{cedge}	o_{cy}	o_{cx}
ρ_s	-0.886	-0.744	0.995	-0.995	0.993	-0.953	-0.451	-0.343	0.093	0.017

Last, we visualize the remaining factors o_{cy} and o_{cx}, i.e. the horizontal and vertical placement of a person in the image, on a heatmap with their individual marginal distributions and the corresponding miss rate plots in Fig. 10.

488   K. Hagn and O. Grau

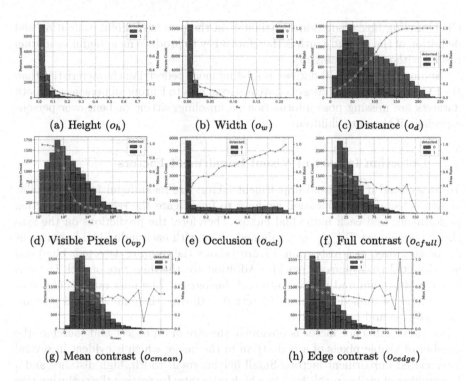

(a) Height (o_h)   (b) Width (o_w)   (c) Distance (o_d)

(d) Visible Pixels (o_{vp})   (e) Occlusion (o_{ocl})   (f) Full contrast (o_{cfull})

(g) Mean contrast (o_{cmean})   (h) Edge contrast (o_{cedge})

Fig. 9. Influence of visual impairing factors of a pedestrian on the detection performance, i.e. miss rate (gray).

The horizontal placement has no influence on the miss rate, which is confirmed by the Spearman correlation of 0.017. The vertical placement with a similar low correlation however shows several high miss rate values closer to the bottom of the image which we attribute to outliers. Additionally, a strong increase of the miss rate at the center of the image can be observed. This is due to persons at great distances which are hard to detect aligning at a vertical position around the horizon. Similarly, the sharp distribution of o_{cy} at around 0.42 stems from the fixed vertical camera angle in our synthetic data. Finally, due to missing data at lower o_{cy} values we cannot conclude that this factor has no influence.

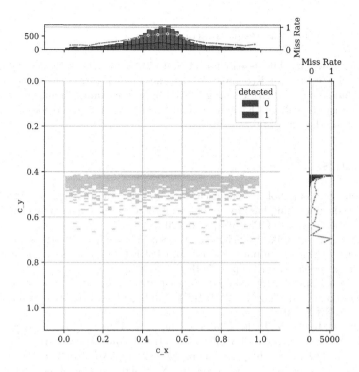

Fig. 10. Influence of person center point position in the image on detection performance, i.e. miss rate (gray).

5 Conclusions

Validation of AI based perception functions is an essential building block for safe automated driving. A failed pedestrian detection due to an underlying data bias in the training data can lead to grave consequences. In this work we presented a method to find data biases of a detector and its underlying real-world dataset by classifying a person according to several visual detection impairment factors into a detectable and non-detectable class through the use of rich annotated synthetic data. We demonstrated the effectiveness of this approach for the real-world datasets CityPersons (CP) and EuroCity Person (ECP) by detecting 23 ethnicity and geolocation based data biases in the CP and 15 in the ECP dataset. Further, 6 missing pedestrian poses were identified in both datasets. Here we remarked that the ground-truth of the validation data has to be carefully defined to generate meaningful data bias findings. Last, we investigated the influence of each visual impairment factor on the overall pedestrian detection performance and came to the conclusion that the visible pixels and distance to the observer have the highest influence with very high Spearman correlations ($|\rho_s| = 0.995$) to the miss rate. The contrast measure o_{cfull} showed a very good correlation ($|\rho_s| = 0.953$) while the horizontal position in the image has no effect ($|\rho_s| =$

0.017). The influence of the vertical position indicates no influence as well, but due to a lack of data the result is inconclusive.

Acknowledgement. The research leading to these results is funded by the German Federal Ministry for Economic Affairs and Energy within the project "Methoden und Maßnahmen zur Absicherung von KI basierten Wahrnehmungsfunktionen für das automatisierte Fahren (KI-Absicherung)". The authors would like to thank the consortium for the successful cooperation.

References

1. Abrecht, S., Gauerhof, L., Gladisch, C., Groh, K., Heinzemann, C., Woehrle, M.: Testing deep learning-based visual perception for automated driving. ACM Trans. Cyber-Phys. Syst. (TCPS) **5**(4), 1–28 (2021)
2. Bernhard, J., Schulik, T., Schutera, M., Sax, E.: Adaptive test case selection for DNN-based perception functions. In: 2021 IEEE International Symposium on Systems Engineering (ISSE), pp. 1–7. IEEE (2021)
3. Bolte, J.A., Bar, A., Lipinski, D., Fingscheidt, T.: Towards corner case detection for autonomous driving. In: 2019 IEEE Intelligent vehicles symposium (IV), pp. 438–445. IEEE (2019)
4. Braun, M., Krebs, S., Flohr, F.B., Gavrila, D.M.: Eurocity persons: a novel benchmark for person detection in traffic scenes. IEEE Trans. Pattern Anal. Mach. Intell. **41**, 1844–1861 (2019). https://doi.org/10.1109/TPAMI.2019.2897684
5. Cai, Z., Vasconcelos, N.: Cascade R-CNN: high quality object detection and instance segmentation. IEEE Trans. Pattern Anal. Mach. Intell. **43**, 1483–1498 (2019)
6. Deng, J., Dong, W., Socher, R., Li, L.J., Li, K., Fei-Fei, L.: Imagenet: a large-scale hierarchical image database. In: CVPR 2009 (2009)
7. Dollar, P., Wojek, C., Schiele, B., Perona, P.: Pedestrian detection: a benchmark. In: 2009 IEEE Conference on Computer Vision and Pattern Recognition, pp. 304–311 (2009). https://doi.org/10.1109/CVPR.2009.5206631
8. Dollar, P., Wojek, C., Schiele, B., Perona, P.: Pedestrian detection: an evaluation of the state of the art. IEEE Trans. Pattern Anal. Mach. Intell. **34**(4), 743–761 (2011)
9. Dosovitskiy, A., et al.: An image is worth 16 × 16 words: transformers for image recognition at scale. In: 9th International Conference on Learning Representations, ICLR 2021, Virtual Event, Austria, 3–7 May 2021. OpenReview.net (2021). https://openreview.net/forum?id=YicbFdNTTy
10. Dosovitskiy, A., Ros, G., Codevilla, F., Lopez, A., Koltun, V.: Carla: an open urban driving simulator. In: Conference on Robot Learning, pp. 1–16. PMLR (2017)
11. Gannamaneni, S., Houben, S., Akila, M.: Semantic concept testing in autonomous driving by extraction of object-level annotations from Carla. In: Proceedings of the IEEE/CVF International Conference on Computer Vision, pp. 1006–1014 (2021)
12. Gauerhof, L., Munk, P., Burton, S.: Structuring validation targets of a machine learning function applied to automated driving. In: Gallina, B., Skavhaug, A., Bitsch, F. (eds.) SAFECOMP 2018. LNCS, vol. 11093, pp. 45–58. Springer, Cham (2018). https://doi.org/10.1007/978-3-319-99130-6_4
13. Girshick, R.: Fast R-CNN. In: Proceedings of the IEEE International Conference on Computer Vision, pp. 1440–1448 (2015)

14. Girshick, R., Donahue, J., Darrell, T., Malik, J.: Rich feature hierarchies for accurate object detection and semantic segmentation. In: Proceedings of the IEEE Conference on Computer Vision And Pattern Recognition, pp. 580–587 (2014)
15. Grau, O., Hagn, K., Sha, Q.S.: A Variational synthesis approach for deep validation. In: Fingscheidt, T., Gottschalk, H., Houben, S. (eds.) Deep Neural Networks and Data for Automated Driving, pp. 359–381. Springer, Cham (2022). https://doi.org/10.1007/978-3-031-01233-4_13
16. Hagn, K., Grau, O.: Improved sensor model for realistic synthetic data generation. In: Computer Science in Cars Symposium. CSCS 2021, Association for Computing Machinery, New York, NY, USA (2021). https://doi.org/10.1145/3488904.3493383
17. Lyssenko, M., Gladisch, C., Heinzemann, C., Woehrle, M., Triebel, R.: From evaluation to verification: towards task-oriented relevance metrics for pedestrian detection in safety-critical domains. In: Proceedings of the IEEE/CVF Conference on Computer Vision and Pattern Recognition, pp. 38–45 (2021)
18. Lyssenko, M., Gladisch, C., Heinzemann, C., Woehrle, M., Triebel, R.: Instance segmentation in Carla: methodology and analysis for pedestrian-oriented synthetic data generation in crowded scenes. In: Proceedings of the IEEE/CVF International Conference on Computer Vision, pp. 988–996 (2021)
19. Richter, S.R., Vineet, V., Roth, S., Koltun, V.: Playing for data: ground truth from computer games. In: Leibe, B., Matas, J., Sebe, N., Welling, M. (eds.) ECCV 2016. LNCS, vol. 9906, pp. 102–118. Springer, Cham (2016). https://doi.org/10.1007/978-3-319-46475-6_7
20. Rong, G., et al.: LGSVL simulator: a high fidelity simulator for autonomous driving. In: 2020 IEEE 23rd International Conference on Intelligent Transportation Systems (ITSC), pp. 1–6. IEEE (2020)
21. Ros, G., Sellart, L., Materzynska, J., Vazquez, D., Lopez, A.M.: The Synthia dataset: a large collection of synthetic images for semantic segmentation of urban scenes. In: Proceedings of the IEEE Conference on Computer Vision and Pattern Recognition, pp. 3234–3243 (2016)
22. Sämann, T., Schlicht, P., Hüger, F.: Strategy to increase the safety of a DNN-based perception for had systems. arXiv preprint arXiv:2002.08935 (2020)
23. Syed Sha, Q., Grau, O., Hagn, K.: DNN analysis through synthetic data variation. In: Computer Science in Cars Symposium. CSCS 2020, Association for Computing Machinery, New York, NY, USA (2020). https://doi.org/10.1145/3385958.3430479
24. Wang, J., et al.: Deep high-resolution representation learning for visual recognition. TPAMI **43**, 3349–3364 (2019)
25. Wrenninge, M., Unger, J.: Synscapes: A photorealistic synthetic dataset for street scene parsing (2018)
26. Zhang, S., Benenson, R., Omran, M., Hosang, J., Schiele, B.: How far are we from solving pedestrian detection? In: Proceedings of the IEEE Conference on Computer Vision and Pattern Recognition (CVPR), June 2016
27. Zhang, S., Benenson, R., Schiele, B.: Citypersons: a diverse dataset for pedestrian detection. In: CVPR (2017)

Probing Contextual Diversity for Dense Out-of-Distribution Detection

Silvio Galesso[✉], Maria Alejandra Bravo, Mehdi Naouar, and Thomas Brox

University of Freiburg, Freiburg, Germany
galessos@cs.uni-freiburg.de

Abstract. Detection of out-of-distribution (OoD) samples in the context of image classification has recently become an area of interest and active study, along with the topic of uncertainty estimation, to which it is closely related. In this paper we explore the task of OoD segmentation, which has been studied less than its classification counterpart and presents additional challenges. Segmentation is a dense prediction task for which the model's outcome for each pixel depends on its surroundings. The receptive field and the reliance on context play a role for distinguishing different classes and, correspondingly, for spotting OoD entities. We introduce MOoSe, an efficient strategy to leverage the various levels of context represented within semantic segmentation models and show that even a simple aggregation of multi-scale representations has consistently positive effects on OoD detection and uncertainty estimation.

Keywords: Out-of-distribution detection · Semantic segmentation

1 Introduction

Imagine you see a pattern, an object, or a scene configuration you do not know. You will identify it as novel and it will attract your attention. This ability to deal with an open world and to identify novel patterns at all semantic levels is one of the many ways how human perception differs from contemporary machine learning. Most deep learning setups assume a closed world with a fixed set of known classes to choose from. However, many real-world tasks do not match this assumption. Very often, maximum deviations from the training samples are the most interesting data points.

Accordingly, novelty/anomaly/out-of-distribution detection has attracted more and more interest recently. Outside of data regimes with limited variation, such as in industrial inspection [14,50], the common approaches to identify unseen patterns derive uncertainty estimates from an existing classification model and mark samples with large uncertainty as novel or out-of-distribution [3,27,31,41,56]. This approach comes with a conflict between the classifier focusing on features that help discriminate between the known classes

---

Supplementary Information The online version contains supplementary material available at https://doi.org/10.1007/978-3-031-25072-9_34.

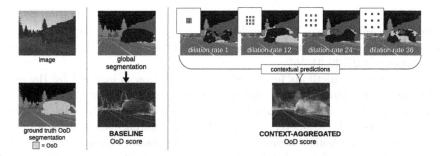

Fig. 1. Our method obtains predictions based on diverse contextual information from different dilated convolutions, exploiting the hierarchical structure of semantic segmentation architectures. On the anomalous pixels (cyan in the ground truth) the contextual predictions diverge, allowing us to improve upon the global model's uncertainty score, which is overconfident in classifying the object as a car. From improved uncertainty we get better out-of-distribution detection. More details in Fig. 2 and Sect. 3.2.

and the need for rich and diverse features that can identify out-of-distribution patterns. This is especially true for semantic segmentation, where a pixel's class is not only defined by its own appearance, but also by the context it appears in. Based on context information only and ignoring appearance, a segmentation model could assume that a large animal (or an oversized telephone, like in the example in Fig. 1) in the middle of the road is a vehicle, while based on local appearance only it could believe that pictures on a billboard are, for example, actual people in the flesh.

In order to combine context and local appearance, modern segmentation networks feature modules with different receptive fields and resolutions, designed to extract diverse representations including different amounts of contextual cues. While for known objects the different cues mostly align with the model's notion of a semantic class, in the case of novel objects the representations at multiple context levels tend to disagree. This can be used as an indicator for uncertainty. Indeed, our approach develops on this idea by having multiple heads as probes for comprehensive multi-scale cues, and obtains an aggregated uncertainty estimate for out-of-distribution (OoD) segmentation. We show that this strategy improves uncertainty estimates over using a single global prediction and often even over regular ensembles, while being substantially more efficient than the latter. It also sets up the bar on the common benchmarks for OoD segmentation. We call our model MOoSe, for **M**ulti-head **OoD Se**gmentation. Source code available at https://github.com/MOoSe-ECCV22/moose_eccv2022.

2   Related Work

Out of Distribution Detection. Out-of-distribution detection is closely related to uncertainty estimation. Under the assumption that a model should be uncertain about samples far from its training distribution, model uncertainty can be used as a proxy score for detecting outliers [29,37]. Several methods for OoD detection, including ours, rely on an existing model trained on a semantic task on the in-distribution data, such as image classification or segmentation [28,29]. Different techniques have been developed to improve outlier detection by means of uncertainty scores, either at inference time [28,41] or while learning the representations [3,27,30,31,56].

Another set of methods for uncertainty estimation is inspired by Bayesian neural networks, which produce probabilistic predictions [6]. For example Monte-Carlo Dropout [21,34] approximates the predictive distribution by sampling a finite number of parameter configurations at inference time using random dropout masks. Although arguably not strictly Bayesian, ensembles [37,55] also approximate the predictive distribution by polling a set of independently trained models fixed at inference time. Attributes of the predictive distribution, such as its entropy, can be used as a measure for uncertainty [1,53].

Alternatively, one can directly model the distribution of the training data, and the likelihood estimated by the resulting model can be used to detect outliers [35,45,51,62]. Other approaches rather rely on learning pretext tasks on the in-distribution data as a proxy for density estimation. Examples of such tasks are reconstruction [17,36,42,47,57] and classification of geometric image transformations [22].

Dense OoD Detection. Methods that are effective at recognizing outlier images do not always scale well to dense OoD detection, where individual pixels in each image need to be classified as in-distribution or anomalous. A recent work [28] has found that advanced methods like generative models [2,25,52] and Monte-Carlo dropout [21] are outperformed by metrics derived from the predictions of a pre-trained semantic segmentation network, such as the values of the segmentation logits. Several recent works [4,7,9,23] focus on the improvement of such segmentation by-products. In particular, the practice of Outlier Exposure [30], originally developed for recognition, has recently gained popularity in dense anomaly detection: several approaches revolve around using outlier data during training, either from a real data [4,9,54] or sampled from a generative model [23,36]. While our method does not need outlier exposure to work, we show that it can be beneficially combined with it.

As mentioned above, deep ensembles [37] are a versatile tool and a gold standard for uncertainty estimation, making them a popular choice for anomaly detection [38,55]. Their relative scalability and effectiveness made them a viable option for uncertainty estimation in dense contexts, including anomaly segmentation [19,20]: ensembles are a simple and almost infallible way of improving the quality of uncertainty scores of neural networks.

At the core of ensemble techniques is diversity between models, which is often provided by random weight initialization and data bootstrapping [32,37], sometimes by architectural differences [61]. While these sources of diversity are of proven efficacy and versatility, they are generic and ignore the requirements of the task at hand, introducing significant computational costs. Multi-headed ensembles mitigate this drawback by sharing the largest part of the network and drawing their diversity only from independent, lightweight heads [39,40,46].

Even though it is related to multi-headed ensembles, our method exploits an additional source of diversity: leveraging a widespread architectural design specific to semantic segmentation, MOoSe captures the variety of contextual information and receptive field within the same model. This allows for performance improvements that are equal or superior to those of bootstrapped ensembles, at a fraction of the computational cost.

3   Multi-head Context Networks

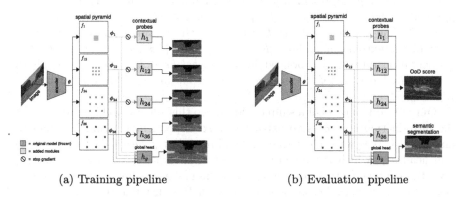

(a) Training pipeline   (b) Evaluation pipeline

Fig. 2. MOoSe: illustration of the multi-head architecture based on the DeepLabV3 semantic segmentation model. During training (a) all the probes learn standard semantic segmentation, while the rest of the network (pre-trained) is unaffected. At test time (b) the uncertainties of all heads (contextual and global) are pooled together into an improved scoremap for OoD detection.

3.1   Contextual Diversity in Semantic Segmentation Networks

Consider a semantic segmentation decoder based on the popular spatial pyramid architecture [10,11,13,63], i.e. containing either a Spatial Pyramid Pooling (SPP [59]) or Atrous Spatial Pyramid Pooling (ASPP [10]) structure. Such structures include a series of pooling or dilated convolutional operations applied in parallel to a set of feature embeddings. We refer to these operations as *spatial pyramid modules*. Each spatial pyramid module has a unique pooling scale or receptive field, allowing for scale invariance and providing the decoder with various degrees of context: the larger the receptive field the more global context at the cost of a loss in detail. The outputs of the spatial pyramid modules are concatenated and fed to a segmentation head that produces a single prediction.

In this section we propose a method for extracting the contextual diversity between the representations produced by the different spatial pyramid modules.

By exploiting said diversity we are able to improve OoD segmentation performance and network calibration without the need of expensive ensembling of multiple models. Our method is non-invasive with regard to the main semantic segmentation task and lightweight in terms of computational cost, making it suitable for real-world applications without affecting segmentation performance. In addition, since our approach improves the uncertainty scores of an existing model, it can be easily combined with other state of the art approaches.

3.2   Probing Contextual Diversity

We start from a generic semantic segmentation architecture featuring K *spatial pyramid modules* $\mathcal{F} = \{f_{c_1}, ..., f_{c_K}\}$, each with a different context size c_k, as in Fig. 2. The encoder features $\boldsymbol{\theta}$ are fed to the pyramid modules producing a set of contextual embeddings $\Phi = \{\boldsymbol{\phi}_{c_1}, ..., \boldsymbol{\phi}_{c_K}\}$, where $\boldsymbol{\phi}_{c_k} = f_{c_k}(\boldsymbol{\theta})$. The segmentation output is produced by the *global head* h_g, which takes all the concatenated features Φ. We denote the output logits of the global head as $\mathbf{z}_g := h_g(\Phi)$.

Our method consists of a simple addition to this generic architecture, namely the introduction of *probes* to extract context-dependent information from the main model. We pair each spatial pyramid module f_{c_k} with a contextual prediction head h_{c_k}, which is trained to produce segmentation logits $\mathbf{z}_{c_k} = h_{c_k}(\boldsymbol{\phi}_{c_k})$ from the context features of size c_k (and the global pooling features, see [11]). Given the input image $\mathbf{x}$, we denote the prediction distributions as:

$$p(\hat{\mathbf{y}}|\mathbf{x}, h_g) = \mathrm{softmax}(\mathbf{z}_h) \quad \text{and} \quad p(\hat{\mathbf{y}}|\mathbf{x}, h_{c_k}) = \mathrm{softmax}(\mathbf{z}_{c_k}) \qquad (1)$$

for the global and contextual heads respectively.

We train all the heads using a standard Cross Entropy loss given N classes and the ground truth segmentation $\mathbf{y}$ (we drop the mean operation over the spatial dimensions for simplicity):

$$\mathcal{L}_{\mathrm{CE}}(\mathbf{x}, \mathbf{y}) = -\sum_{k=1}^{K}\sum_{i=1}^{N} \mathbf{y}_i \cdot \log p(\hat{\mathbf{y}}_i|\mathbf{x}, h_{c_k}), \qquad (2)$$

although any semantic segmentation objective could be used.

The contextual heads are designed to act as probes and extract information from the context-specific representations. For this reason, we do not backpropagate the gradients coming from the contextual heads to the rest of the network, and only update the weights of the heads themselves. The spatial pyramid modules are distinct operations, each with its specific scope depending on its context size. By stopping the gradients before the spatial pyramid modules we force each head to solve the same segmentation task but using different features, preserving prediction diversity. As a byproduct, our architectural modifications do not interfere with the rest of the network and the main segmentation task.

Head Architecture. The architecture of the contextual heads is based on the global head of the base segmentation model. For example, for DeepLabV3 it consists of a projection block to bring the number of channels down to 256, followed by a sequence of d prediction blocks (3×3 convolution, batch normalization, ReLU), plus a final 1×1 convolution for prediction. The head depth d can be tuned according to the predictive power necessary to process contextual information, depending on the difficulty of the dataset.

3.3 Out-of-Distribution Detection with MOoSe

A model for dense OoD detection should assign to each location in the input image an anomaly score. To obtain per-pixel OoD scores we test three scoring functions, applied to the outputs of the segmentation heads: maximum softmax probability (MSP) [29], prediction entropy (H) [53] and maximum logit (ML) [28]. We adapt each scoring function to work with predictions from multiple heads. The maximum softmax probability is computed on the average predicted distribution over all the heads, including the global head:

$$S_{\text{MSP}} = - \max_{i \in [1,N]} \left[\frac{1}{K+1} \left(p(\hat{\mathbf{y}}_i | \mathbf{x}, h_g) + \sum_{k=1}^{K} p(\hat{\mathbf{y}}_i | \mathbf{x}, h_{c_k}) \right) \right], \tag{3}$$

Similarly, for the entropy we compute the entropy of the expected output distribution:

$$S_{\text{H}} = \mathcal{H} \left[\frac{1}{K+1} \left(p(\hat{\mathbf{y}} | \mathbf{x}, h_g) + \sum_{k=1}^{K} p(\hat{\mathbf{y}} | \mathbf{x}, h_{c_k}) \right) \right], \tag{4}$$

where $\mathcal{H}$ denotes the information entropy. For maximum logit we average the logits over the different heads and compute their negated maximum:

$$S_{\text{ML}} = - \max_{i \in [1,N]} \left[\frac{1}{K+1} \left(\mathbf{z}_{g,i} + \sum_{k=1}^{K} \mathbf{z}_{c_k,i} \right) \right]. \tag{5}$$

All scores should be directly proportional to the model's belief of a pixel belonging to an anomalous object, therefore for MSP and ML the negatives are taken.

4 Experiments

In this section we evaluate our approach on out-of-distribution detection, comparing it to ensembles (Sect. 4.4) and to the state of the art (Sect. 4.5).

4.1 Datasets and Benchmarks

StreetHazards [28] is a synthetic dataset for semantic segmentation and OoD detection. It features street scenes in diverse settings, created with the CARLA simulation environment [18]. The 1500 test samples feature instances from 250 different anomalous objects, diverse in appearance, location, and size.

The **BDD-Anomaly** [28] dataset is derived from the BDD100K [60] semantic segmentation dataset by removing the samples containing instances of the motorcycle, bicycle, and train classes, and using them as a test set for OoD segmentation, yielding a 6280/910/810 training/validation/test split. BDD-Anomaly and StreetHazards constitute the CAOS benchmark [28].

Fishyscapes - LostAndFound [5,49] is a dataset for road obstacle detection, designed to be used in combination with the Cityscapes [15] driving dataset. Its test split contains 1203 images of real street scenes featuring road obstacles, whose presence is marked in the segmentation ground truth.

RoadAnomaly [42] consists of 60 real world images of diverse anomalous objects in driving environments, collected from the internet. The images come with pixel-wise annotations of the anomalous objects, making them suitable for testing models trained on driving datasets.

Results for the anomaly track of the SegmentMeIfYouCan [8] benchmark can be found in the supplementary material.

4.2   Evaluation Metrics

We evaluate OoD detection performance using the area under the precision-recall curve (AUPR), and the false positive rate at 95% true positive rate (FPR_{95}). As is customary, anomalous pixels are considered positives. Other works include results for the area under the ROC curve (AUROC), however the AUPR is to be preferred to this metric in the presence of heavy class imbalance, which is the case for anomaly segmentation [16].

Table 1. CAOS benchmark. Comparison between single model/global head (Global), multi-head ensembles (MH-Ens), standard deep ensembles (DeepEns) and MOoSe on dense out-of-distribution detection. The results are for DeepLabV3 and PSPNet models with ResNet50 backbones. All three scoring functions (maximum softmax probability (MSP), entropy (H), maximum logit (ML)) are considered. All results are percentages, best results are shown in **bold**

| | | StreetHazards | | | | BDD-Anomaly | | | |
| | | DeepLabV3 | | PSPNet | | DeepLabV3 | | PSPNet | |
Score fn.	Method	AUPR↑	FPR_{95} ↓	AUPR↑	FPR_{95} ↓	AUPR↑	FPR_{95} ↓	AUPR↑	FPR_{95} ↓
MSP	Global	9.11	22.37	9.65	22.04	7.01	22.47	6.75	23.63
	MH-Ens	9.69	21.40	9.84	22.49	7.55	25.50	8.07	23.41
	DeepEns	10.22	21.09	10.61	**20.75**	7.64	**21.53**	**8.52**	**21.31**
	MOoSe	**12.53**	**21.05**	**11.28**	21.94	**8.66**	22.49	8.11	24.09
H	Global	11.89	22.07	12.28	21.77	10.23	20.64	9.89	21.69
	MH-Ens	12.59	21.10	12.45	22.29	10.62	23.51	11.73	20.76
	DeepEns	13.43	20.62	13.39	**20.35**	11.39	19.31	12.32	**18.83**
	MOoSe	**15.43**	**19.89**	**14.52**	21.20	**12.59**	**19.27**	**12.35**	20.98
ML	Global	13.57	23.27	13.43	27.71	10.69	15.60	10.68	16.79
	MH-Ens	13.99	21.86	13.64	28.30	10.69	20.19	12.40	15.08
	DeepEns	14.57	21.79	14.14	25.82	11.40	14.66	12.26	13.96
	MOoSe	**15.22**	**17.55**	**15.29**	**20.46**	**12.52**	**13.86**	**12.88**	**13.94**

4.3   Experimental Setup

MOoSe relies on semantic segmentation models to perform dense OoD detection. For the experiments on StreetHazards and BDD-Anomaly we report results for two convolutional architectures, (DeepLabV3 [10] and PSPNet [63], each with ResNet50 and ResNet101 backbones) and one transformer based (Lawin [58], see supplementary material). For the experiments on LostAndFound and Road-Anomaly we use DeepLabV3+ [12] with a ResNet101 backbone, trained on Cityscapes[1] or BDD100k [60] respectively.

Training. We build MOoSe on top of fully trained semantic segmentation networks, by adding the prediction heads and training them jointly for segmentation on the respective dataset, using a standard pixel-wise cross-entropy loss. Although nothing prevents from training the whole model together, for fairness of comparison we only apply the loss to the probes. In order to prevent any alteration to the main model while training the heads, we stop gradient propagation through the rest of the network and make sure that the normalization layers would not update their statistics during forward propagation. The heads are trained for 80 epochs, or until saturation of segmentation performance (mIoU).

MOoSe introduces two hyperparameters: learning rate and depth d of the contextual heads. By default we use $d = 1$ for the models trained on StreetHazards and $d = 3$ otherwise. While the performance gains depend on these, we find that our method is robust to configuration changes, as we show in an ablation study in the supplementary material.

4.4   Comparison with Ensembles

In this section we compare MOoSe with the single prediction baseline (global head) and with two types of ensembles. Deep ensembles [37] (DeepEns) consist of sets of independent segmentation networks, each trained on a different random subset of 67% of the original data, starting from a different random parameter initialization [32]. Similarly, multi-head ensembles (MH-Ens) are trained on random data subsets, but share the same encoder and only feature diverse prediction heads, for increased efficiency.

We compare to ensembles with 5 members/heads to match the number of heads in our method. Additionally, we pick the ensemble member with the median AUPR performance to serve both as the single model baseline and as initialization for MOoSe. The shared backbone of the multi-head ensembles also comes from the same model.

Table 1 shows results for the CAOS benchmark (StreetHazards and BDD-Anomaly) using DeepLabV3 and PSPNet as base architectures, with ResNet50 [26] backbones. We report results for the three OoD scoring functions described in Sect. 3.3; results for MOoSe are averaged over 3 runs, standard deviations are available in the supplementary material. For all datasets,

---

[1] Parameters available at:
https://github.com/NVIDIA/semantic-segmentation.

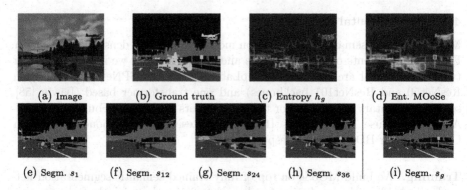

Fig. 3. OoD segmentation & head contribution. Test image from the StreetHaz-ards dataset: a street scene containing anomalous objects (indicated in cyan in the ground truth (b)). The contextual predictions (e-h) diverge on the outliers, improving the entropy score map (c, d). The example shows an interesting failure case: the street sign (in-distribution) on the right also sparks disagreement between the heads, resulting in increased entropy and thus a false positive.

architectures, scoring functions, and metrics, MOoSe consistently outperforms its respective global head. Similarly, MOoSe outperforms multi-head ensembles, as well as deep ensemble in most cases, while having a smaller computational cost than both.

In accordance with what observed in other works [28], the maximum-logit scoring function tends to outperform entropy, most notably in terms of FPR_{95} and on the BDD-Anomaly dataset. Both scoring functions consistently outperform maximum-softmax probability. Moreover, maximum-logit appears to combine well with MOoSe by effectively reducing false positives. Results for models using the ResNet101 backbone are available in the supplementary material.

Figure 3 shows an example for OoD segmentation on a driving scene. The top row compares the entropy obtained using the global head (3c) and our multi-head approach (3d). The probes of MOoSe disagree on the nature of the anomalous objects in the image, and its aggregated entropy score is able to outline the anomalous objects more clearly than the global head. However, prediction disagreement also produces false positives for smaller inlier objects, such as the street sign on the right, highlighting a possible failure mode of our approach.

Computational Costs. In Table 2 we compare our method against ensembles in terms of computational costs, reporting the number of parameters of each model and the estimated runtime of a forward pass. We consider DeepLabV3 and PSPNet with ResNet50 and MOoSe head depth 1. Deep ensembles have the highest parameter count and runtime, 5 times that of a single network. MOoSe compares favorably to both ensembles on all architectures. The larger size and runtime of PSPNet compared to DeepLab is due to its higher dimensional representations, which can be reduced with projection layers before the probes.

Table 2. Computational costs. Estimated computational costs of MOoSe in comparison with ensembles. We report the number of parameters (in millions) and the estimated forward pass runtime on StreetHazards (in milliseconds), estimated on a single Nvidia RTX2080Ti GPU using the PyTorch [48] benchmarking utilities

Architecture		Single	MH-Ens	DeepEns	MOoSe
DeepLabV3	Parameters (M)	40	104	198	43
	Runtime (ms)	113	286	583	121
PSPNet	Parameters (M)	47	139	233	94
	Runtime (ms)	107	246	542	183

4.5   Comparison with the State of the Art

Here we compare MOoSe with the best approaches for dense OoD detection that do not require negative training data (see Sect. 4.5).

The CAOS Benchmark. On StreetHazards and BDD-Anomaly we compare with TRADI [20], SynthCP [57], OVNNI [19], Deep Metric Learning (DML) [7], and the approach by Grcic et al. [23] that uses outlier exposure with generated samples. TRADI and OVNNI require multiple forward passes per sample, increasing the evaluation run-time (or memory requirements) considerably. Table 3(a) shows that MOoSe compares favorably to existing works on both datasets and on all metrics. We note that, given its non-invasive nature, MOoSe is compatible with other approaches, and can for example be combined with the loss of DML.

Table 3. State-of-the-art comparison. Left - CAOS benchmark: MOoSe in combination with the max-logit scoring function, outperforms all other methods on StreetHazards, except for DML in terms of FPR_{95}. On BDD-Anomaly MOoSe performs the best in both metrics. **Right:** MOoSe yields improvements on both **Fishyscapes LostAndFound (FS - LaF)** and **RoadAnomaly**, but on the former benchmark is outperformed by Standardized Max-Logits (Std.ML).

	Street Hazards		BDD Anomaly	
	AUPR	FPR_{95}	AUPR	FPR_{95}
TRADI[20]	7.2	25.3	5.6	26.9
SynthCP[57]	9.3	28.4	–	–
OVNNI[19]	12.6	22.2	6.7	25.0
Grcic[23]	12.7	25.2	–	–
DML[7]	14.7	**17.3**	–	–
MOoSe ML	**15.22**	17.55	**12.52**	**13.86**

	Method	FS - LaF		RoadAnomaly	
		AUPR	FPR_{95}	AUPR	FPR_{95}
MSP	Global	3.06	37.46	23.76	51.32
	MOoSe	7.13	33.72	31.53	43.41
H	Global	6.23	37.34	32.00	49.14
	MOoSe	12.08	32.58	_41.48_	_36.78_
ML	Global	10.25	37.45	37.86	39.03
	MOoSe	_13.64_	_32.32_	**43.59**	**32.12**
Resynth. [42]		5.70	48.05	–	–
DML [7]		–	–	37	37
Std.ML [33]		**31.05**	**21.52**	25.82	49.74

502   S. Galesso et al.

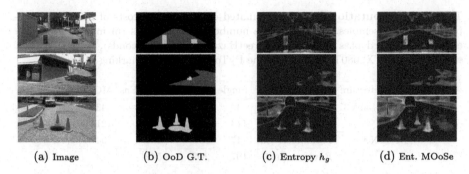

(a) Image   (b) OoD G.T.   (c) Entropy h_g (d) Ent. MOoSe

Fig. 4. OoD segmentation. Examples on LostAndFound and RoadAnomaly, anomalous objects shows in cyan in the second column. The first row shows an example in which our model is able to recognize anomalous objects in their entirety, where the global head fails. The second row shows a failure case, where our model only marks the borders of the obstacle on the road and produces false positives, performing worse than the global head. The example in the last row is from RoadAnomaly: MOoSe detects the traffic cones better than the global head, but introduces noise in the background and still fails to detect the manhole.

LostAndFound, RoadAnomaly. We extend our evaluation of MOoSe to other real world benchmarks, Fishyscapes LostAndFound and RoadAnomaly, using models trained on Cityscapes and BDD100k [60] respectively, as described in Sect. 4.3. We report our results in Table 3(b) and include results for the comparable (not needing negative training data) state-of-the-art methods DML, Standardized Max-Logits [33] and Image Resynthesis [42]. Similarly to the CAOS benchmark, we can observe that the adoption of MOoSe improves OoD detection performance on both benchmarks, regardless of the chosen scoring function. In the supplementary material we include results for the SegmentMeIfYouCan benchmark (similar to RoadAnomaly), where Image Resynthesis is currently SOTA. On the other hand, Table 3(b) shows that Image Resynthesis does not perform well on LostAndFound. Standardized Max-Logits has remarkable results on LostAndFound but not on RoadAnomaly, where MOoSe works best.

Figure 4 shows examples of OoD detection on LostAndFound and RoadAnomaly. In the first example MOoSe improves over the entropy heatmap of the global head. In the second example, however, it can be seen that our method still fails to detect the obstacles and produces more false positives. Increased false positives are visible in the background of the third example too, although here MOoSe also improves the detection of some anomalous objects.

Outlier Exposure. Several methods for anomaly segmentation rely on negative training data from a separate source. While this technique introduces some drawbacks, such as a reliance on the choice of the negative data and a potential negative impact on segmentation, it has been shown to improve OoD detection on the common benchmarks. Following the procedure described in [9] as "entropy training", we investigate whether our method can also benefit from outlier exposure. Indeed, results show that outlier exposure boosts MOoSe+ML to 53.19

AUPR (+22%) and 24.38 FPR$_{95}$ (−24%) on RoadAnomaly. Full results on all scoring functions are available in the supplementary material.

5 Analysis

Our approach relies on a collection of different predictions to improve OoD detection. Previous literature on network ensembles puts the spotlight on diversity [39,43], emphasizing that multiple estimators can be helpful only if their predictions are diverse and each contributes with useful information for the cumulative decision. In this section we address some points to better understand the working principle of the method and verify its underlying hypotheses. Specifically, we investigate: 1) the effect of MOoSe on prediction diversity, 2) whether contextual aggregation can be responsible for prediction diversity, and 3) how this translates into better OoD detection.

5.1 Quantifying Diversity: Variance and Mutual Information

We are interested in comparing MOoSe to the closely related ensembles in terms of prediction diversity, of which a simple metric is variance. We compare the average variance of the output distributions of MOoSe and ensembles on Street-Hazards and BDD-Anomaly validation, as reported in Table 4 (left). On both datasets our method's predictions have higher variance than both ensembles.

Variance, however, gives us no insights on what the predictions disagree upon, and is therefore of limited interest. From the literature on Bayesian networks we can borrow a more informative metric: the mutual information (MI) between the model distribution and the output distribution [44]. Consider an ensemble of K networks, or a multi-head model with K heads. Each model or head produces a prediction $p(\hat{y}|x, k)$. We can compute the mutual information between the distribution of the models k and the distribution of their predictions as:

$$\mathrm{MI}(\hat{y}, k|x) = \mathcal{H}\left[\frac{1}{K}\sum_{k=1}^{K} p(\hat{y}|x, k)\right] - \frac{1}{K}\sum_{k=1}^{K}\mathcal{H}\left[p(\hat{y}|x, k)\right], \qquad (6)$$

which is the entropy of the expected output distribution minus the average entropy of the output distributions. MI is high for a sample x if the predictions are *individually confident but also in disagreement with each other*. This tells us how much additional information the diversity brings to the overall model: if all the predictions are equally uncertain about the same samples they disagree on, then aggregating them will not affect the aggregated uncertainty estimate.

In Table 4 (left) we report the average MI on StreetHazards and BDD-Anomaly validation, comparing again MOoSe and ensembles. Similarly to variance, our method's predictions have higher MI than both ensemble types, indicating that contextual probing not only produces more diversity in absolute terms, but also that this diversity adds more information to the model's predictive distribution.

Finally, in Table 4 (left) we report the Expected Calibration Error [24] of all methods, to show that even if ensembles are better calibrated than the baseline, it is MOoSe that performs the best at uncertainty estimation overall.

Table 4. Left - variance and Mutual Information (MI) show higher diversity for MOoSe than for ensembles, while lower ECE shows that our approach yields better calibrated predictions. All metrics are computed on DeepLabV3-ResNet50. **Right - single-dilation:** OoD detection results (AUPR) for single dilation models (SD) with different dilation rates (1, 12, 24, 36), compared to standard multi-dilation MOoSe. First row shows results for the global head, second row adds the probes, bottom two rows show the absolute and relative improvement

Method	StreetHazards			BDD-Anomaly		
	Var.↑	MI↑	ECE↓	Var.↑	MI↑	ECE↓
Global	–	–	.038	–	–	.123
MH-Ens	0.20	.004	.039	0.30	.022	.104
DeepEns	0.54	.012	.032	1.19	.054	.103
MOoSe	1.05	.034	.031	1.34	.062	.093

	SD1	SD12	SD24	SD36	MOoSe
Var.	0.41	1.05	0.93	0.83	1.34
Global	8.2	8.1	9.1	8.6	10.2
+probes	10.1	10.1	10.6	10.5	13.1
Chng.	1.9	2.0	1.5	1.9	2.9
Chng. %	22.7	24.9	16.4	22.3	27.9

5.2   Context as a Source of Diversity

In the previous section we showed that our approach produces highly diverse predictions. In this section we investigate the source of this diversity: our hypothesis is that each head relies differently on contextual information depending on the dilation rate of their respective spatial pyramid module, resulting in diverse predictive behaviors.

We test this hypothesis by evaluating the ability of each head to perform semantic segmentation when *only* contextual information is available. We corrupt the pixels of the foreground classes in BDD-Anomaly[2] with random uniform noise while leaving the background pixels unchanged, then we evaluate how well each head can still classify the corrupted foreground pixels by relying on the context. An example of the process can be seen in Fig. 5 (left). Figure 5 (right) shows the mIoU on the noisy foreground as a percentage of the foreground mIoU on the original clean image. We can observe that dilation rate and robustness to foreground corruption are proportional to each other at multiple noise levels, as further illustrated by the qualitative example in the figure. The different result quality for different dilation rates confirm the validity of contextual aggregation as a source of prediction diversity, as anticipated by the comparison with regular (non-contextual) ensembles on variance and mutual information in Sect. 5.1.

5.3   Effect of Contextual Diversity on OoD Detection

The results presented in Sect. 4 already show that contextual probing improves performance on the task. Moreover, results obtained from the application of MOoSe to transformer-based models (7.3% average AUPR increase on StreetHazards across scoring functions), which are available in the supplementary material, indicate that the principle is applicable across architectures and its gains are not an artifact of CNNs.

---

[2] Pole, traffic light, traffic sign, person, car, truck, bus.

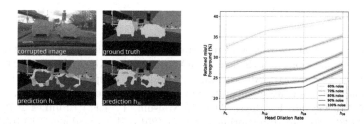

Fig. 5. Left - example of foreground corruption: the cars are corrupted with noise and the head with the largest dilation rate (h_{36}) can still largely segment them, unlike the no-dilation head h_1. **Right - corruption robustness:** We evaluate semantic segmentation of each probe on the corrupted foreground objects. The retained mIoU (mIoU$_{corrupt}$/mIoU$_{clean}$) increases with dilation rate, indicating more reliance on context. Results for BDD-Anomaly on DeepLabV3.

The last point to address is the contribution of contextual diversity to out-of-distribution detection. To quantify this contribution, we performed an ablation study removing receptive field diversity from DeepLabV3 by using the same dilation rate for all the convolutions in the spatial pyramid module. We train several versions of this single-dilation (SD) MOoSe, each with a different dilation rate, and present the comparison with standard MOoSe in Table 4 (right). Firstly, all single dilation models have lower prediction variance than regular MOoSe. Secondly, although the single-dilation models still outperform their global head, MOoSe yields larger gains than all SD models, both in absolute and relative terms. While these results confirm that contextual diversity is crucial for the success of our method, they also show that there are more contributing factors, compatibly with the known benefits of ensembles.

6   Conclusion

In this work we proposed a simple and effective approach for improving dense out-of-distribution detection by leveraging the properties of segmentation decoders to obtain a set of diverse predictions. Our experiments showed that MOoSe yields consistent gains on a variety of datasets and model architectures, and that it compares favorably with computationally much more expensive ensembles. We showed that our approach also outperforms other state-of-the-art approaches, and that due to its simplicity it could be easily combined with them. Even though we tested our method on various architectures, and despite the versatility of the main idea, one current limitation of MOoSe is its reliance on a specific architectural paradigm: the spatial pyramid. We also identified false positives among small objects to be an inherent failure mode of our approach, which potentially could be mitigated by combining MOoSe with alternative concepts that act at a single contextual scale.

References

1. Abdar, M., et al.: A review of uncertainty quantification in deep learning: Techniques, applications and challenges. arXiv preprint arXiv:2011.06225 (2020)
2. Baur, C., Wiestler, B., Albarqouni, S., Navab, N.: Deep autoencoding models for unsupervised anomaly segmentation in brain MR images. In: Crimi, A., Bakas, S., Kuijf, H., Keyvan, F., Reyes, M., van Walsum, T. (eds.) BrainLes 2018. LNCS, vol. 11383, pp. 161–169. Springer, Cham (2019). https://doi.org/10.1007/978-3-030-11723-8_16
3. Bergman, L., Hoshen, Y.: Classification-based anomaly detection for general data. In: International Conference on Learning Representations (2019)
4. Bevandić, P., Krešo, I., Oršić, M., Šegvić, S.: Simultaneous semantic segmentation and outlier detection in presence of domain shift. In: Fink, G.A., Frintrop, S., Jiang, X. (eds.) DAGM GCPR 2019. LNCS, vol. 11824, pp. 33–47. Springer, Cham (2019). https://doi.org/10.1007/978-3-030-33676-9_3
5. Blum, H., Sarlin, P.E., Nieto, J., Siegwart, R., Cadena, C.: The fishyscapes benchmark: Measuring blind spots in semantic segmentation. arXiv preprint arXiv:1904.03215 (2019)
6. Blundell, C., Cornebise, J., Kavukcuoglu, K., Wierstra, D.: Weight uncertainty in neural network. In: International Conference on Machine Learning, pp. 1613–1622. PMLR (2015)
7. Cen, J., Yun, P., Cai, J., Wang, M.Y., Liu, M.: Deep metric learning for open world semantic segmentation. In: Proceedings of the IEEE/CVF International Conference on Computer Vision, pp. 15333–15342 (2021)
8. Chan, R., et al.: SegMentMeifYouCan: A benchmark for anomaly segmentation (2021)
9. Chan, R., Rottmann, M., Gottschalk, H.: Entropy maximization and meta classification for out-of-distribution detection in semantic segmentation. CoRR abs/2012.06575 (2020). https://arxiv.org/abs/2012.06575
10. Chen, L.C., Papandreou, G., Kokkinos, I., Murphy, K., Yuille, A.L.: DeepLab: semantic image segmentation with deep convolutional nets, atrous convolution, and fully connected CRFs. IEEE Trans. Pattern Anal. Mach. Intell. 40(4), 834–848 (2017)
11. Chen, L.C., Papandreou, G., Schroff, F., Adam, H.: Rethinking atrous convolution for semantic image segmentation. arXiv preprint arXiv:1706.05587 (2017)
12. Chen, L.C., Zhu, Y., Papandreou, G., Schroff, F., Adam, H.: Encoder-decoder with atrous separable convolution for semantic image segmentation. In: European Conference on Computer Vision (ECCV) (2018)
13. Cheng, B., et al.: Panoptic-DeepLab: a simple, strong, and fast baseline for bottom-up panoptic segmentation. In: Proceedings of the IEEE/CVF Conference on Computer Vision and Pattern Recognition, pp. 12475–12485 (2020)
14. Cohen, N., Hoshen, Y.: Sub-image anomaly detection with deep pyramid correspondences. arXiv preprint arXiv:2005.02357 (2020)
15. Cordts, M., et al.: The cityscapes dataset for semantic urban scene understanding. In: Proceedings of the IEEE Conference on Computer Vision and Pattern Recognition (CVPR) (2016)
16. Davis, J., Goadrich, M.: The relationship between precision-recall and roc curves. In: Proceedings of the 23rd International Conference on Machine Learning, pp. 233–240 (2006)

17. Di Biase, G., Blum, H., Siegwart, R., Cadena, C.: Pixel-wise anomaly detection in complex driving scenes. In: Proceedings of the IEEE/CVF Conference on Computer Vision and Pattern Recognition, pp. 16918–16927 (2021)
18. Dosovitskiy, A., Ros, G., Codevilla, F., Lopez, A., Koltun, V.: CARLA: an open urban driving simulator. In: Conference on Robot Learning, pp. 1–16. PMLR (2017)
19. Franchi, G., Bursuc, A., Aldea, E., Dubuisson, S., Bloch, I.: One versus all for deep neural network incertitude (OVNNI) quantification. CoRR abs/2006.00954 (2020). https://arxiv.org/abs/2006.00954
20. Franchi, G., Bursuc, A., Aldea, E., Dubuisson, S., Bloch, I.: TRADI: tracking deep neural network weight distributions. In: Vedaldi, A., Bischof, H., Brox, T., Frahm, J.-M. (eds.) ECCV 2020. LNCS, vol. 12362, pp. 105–121. Springer, Cham (2020). https://doi.org/10.1007/978-3-030-58520-4_7
21. Gal, Y., Ghahramani, Z.: Dropout as a bayesian approximation: representing model uncertainty in deep learning. In: International Conference on Machine Learning, pp. 1050–1059. PMLR (2016)
22. Golan, I., El-Yaniv, R.: Deep anomaly detection using geometric transformations. In: Neural Information Processing Systems (NeurIPS) (2018)
23. Grcić, M., Bevandić, P., Šegvić, S.: Dense open-set recognition with synthetic outliers generated by real NVP. arXiv preprint arXiv:2011.11094 (2020)
24. Guo, C., Pleiss, G., Sun, Y., Weinberger, K.Q.: On calibration of modern neural networks. In: International Conference on Machine Learning, pp. 1321–1330. PMLR (2017)
25. Haselmann, M., Gruber, D.P., Tabatabai, P.: Anomaly detection using deep learning based image completion. In: 2018 17th IEEE International Conference on Machine Learning and Applications (ICMLA), pp. 1237–1242. IEEE (2018)
26. He, K., Zhang, X., Ren, S., Sun, J.: Deep residual learning for image recognition. In: Proceedings of the IEEE Conference on Computer Vision and Pattern Recognition, pp. 770–778 (2016)
27. Hein, M., Andriushchenko, M., Bitterwolf, J.: Why relu networks yield high-confidence predictions far away from the training data and how to mitigate the problem. In: Proceedings of the IEEE/CVF Conference on Computer Vision and Pattern Recognition, pp. 41–50 (2019)
28. Hendrycks, D., Basart, S., Mazeika, M., Mostajabi, M., Steinhardt, J., Song, D.: Scaling out-of-distribution detection for real-world settings. arXiv preprint arXiv:1911.11132 (2019)
29. Hendrycks, D., Gimpel, K.: A baseline for detecting misclassified and out-of-distribution examples in neural networks. In: Proceedings of International Conference on Learning Representations (2017)
30. Hendrycks, D., Mazeika, M., Dietterich, T.: Deep anomaly detection with outlier exposure. In: International Conference on Learning Representations (2018)
31. Hendrycks, D., Mazeika, M., Kadavath, S., Song, D.: Using self-supervised learning can improve model robustness and uncertainty. In: Advances in Neural Information Processing Systems (NeurIPS) (2019)
32. Ilg, E., et al.: Uncertainty estimates and multi-hypotheses networks for optical flow. In: Proceedings of the European Conference on Computer Vision (ECCV), pp. 652–667 (2018)
33. Jung, S., Lee, J., Gwak, D., Choi, S., Choo, J.: Standardized max logits: a simple yet effective approach for identifying unexpected road obstacles in urban-scene segmentation. In: Proceedings of the IEEE/CVF International Conference on Computer Vision (ICCV), pp. 15425–15434 (October 2021)

34. Kendall, A., Badrinarayanan, V., Cipolla, R.: Bayesian SegNet: model uncertainty in deep convolutional encoder-decoder architectures for scene understanding. In: British Machine Vision Conference 2017. BMVC 2017 (2017)

35. Kirichenko, P., Izmailov, P., Wilson, A.G.: Why normalizing flows fail to detect out-of-distribution data. arXiv preprint arXiv:2006.08545 (2020)

36. Kong, S., Ramanan, D.: OpenGAN: open-set recognition via open data generation. arXiv preprint arXiv:2104.02939 (2021)

37. Lakshminarayanan, B., Pritzel, A., Blundell, C.: Simple and scalable predictive uncertainty estimation using deep ensembles. In: Neural Information Processing Systems (NeurIPS) (2017)

38. Lee, K., Lee, K., Lee, H., Shin, J.: A simple unified framework for detecting out-of-distribution samples and adversarial attacks. In: Advances in Neural Information Processing Systems. 31 (2018)

39. Lee, S., Purushwalkam, S., Cogswell, M., Crandall, D., Batra, D.: Why m heads are better than one: training a diverse ensemble of deep networks. arXiv preprint arXiv:1511.06314 (2015)

40. Li, H., Ng, J.Y.H., Natsev, P.: EnsembleNet: end-to-end optimization of multi-headed models. arXiv preprint arXiv:1905.09979 (2019)

41. Liang, S., Li, Y., Srikant, R.: Enhancing the reliability of out-of-distribution image detection in neural networks. In: International Conference on Learning Representations (2018)

42. Lis, K.M., Nakka, K.K., Fua, P., Salzmann, M.: Detecting the unexpected via image resynthesis. In: International Conference On Computer Vision (ICCV), pp. 2152–2161 (2019). https://doi.org/10.1109/ICCV.2019.00224, http://infoscience.epfl.ch/record/269093

43. Liu, L., et al.: Deep neural network ensembles against deception: Ensemble diversity, accuracy and robustness. In: 2019 IEEE 16th International Conference on Mobile Ad Hoc and Sensor Systems (MASS), pp. 274–282. IEEE (2019)

44. Malinin, A., Mlodozeniec, B., Gales, M.: Ensemble distribution distillation. arXiv preprint arXiv:1905.00076 (2019)

45. Nalisnick, E., Matsukawa, A., Teh, Y.W., Gorur, D., Lakshminarayanan, B.: Do deep generative models know what they don't know? arXiv preprint arXiv:1810.09136 (2018)

46. Narayanan, A.R., Zela, A., Saikia, T., Brox, T., Hutter, F.: Multi-headed neural ensemble search. In: Workshop on Uncertainty and Robustness in Deep Learning (UDL@ICML2021) (2021)

47. Nguyen, D.T., Lou, Z., Klar, M., Brox, T.: Anomaly detection with multiple-hypotheses predictions. In: International Conference on Machine Learning, pp. 4800–4809. PMLR (2019)

48. Paszke, A., et al.: Automatic differentiation in pytorch. In: NIPS 2017 Workshop on Autodiff (NIPS-W) (2017)

49. Pinggera, P., Ramos, S., Gehrig, S., Franke, U., Rother, C., Mester, R.: Lost and found: detecting small road hazards for self-driving vehicles. In: 2016 IEEE/RSJ International Conference on Intelligent Robots and Systems (IROS), pp. 1099–1106. IEEE (2016)

50. Rudolph, M., Wandt, B., Rosenhahn, B.: Same Same but DifferNet: semi-supervised defect detection with normalizing flows. In: Proceedings of the IEEE/CVF Winter Conference on Applications of Computer Vision, pp. 1907–1916 (2021)

51. Schirrmeister, R., Zhou, Y., Ball, T., Zhang, D.: Understanding anomaly detection with deep invertible networks through hierarchies of distributions and features. In: Advances in Neural Information Processing Systems. **33** (2020)
52. Schlegl, T., Seeböck, P., Waldstein, S.M., Schmidt-Erfurth, U., Langs, G.: Unsupervised anomaly detection with generative adversarial networks to guide marker discovery. In: Niethammer, M. (ed.) IPMI 2017. LNCS, vol. 10265, pp. 146–157. Springer, Cham (2017). https://doi.org/10.1007/978-3-319-59050-9_12
53. Smith, L., Gal, Y.: Understanding measures of uncertainty for adversarial example detection. arXiv preprint arXiv:1803.08533 (2018)
54. Vojir, T., Sipka, T., Aljundi, R., Chumerin, N., Reino, D.O., Matas, J.: Road anomaly detection by partial image reconstruction with segmentation coupling. In: Proceedings of the IEEE/CVF International Conference on Computer Vision, pp. 15651–15660 (2021)
55. Vyas, A., Jammalamadaka, N., Zhu, X., Das, D., Kaul, B., Willke, T.L.: Out-of-distribution detection using an ensemble of self supervised leave-out classifiers. In: Proceedings of the European Conference on Computer Vision (ECCV) (September 2018)
56. Winkens, J., et al.: Contrastive training for improved out-of-distribution detection. arXiv preprint arXiv:2007.05566 (2020)
57. Xia, Y., Zhang, Y., Liu, F., Shen, W., Yuille, A.L.: Synthesize then compare: detecting failures and anomalies for semantic segmentation. In: Vedaldi, A., Bischof, H., Brox, T., Frahm, J.-M. (eds.) ECCV 2020. LNCS, vol. 12346, pp. 145–161. Springer, Cham (2020). https://doi.org/10.1007/978-3-030-58452-8_9
58. Yan, H., Zhang, C., Wu, M.: Lawin transformer: Improving semantic segmentation transformer with multi-scale representations via large window attention. CoRR abs/2201.01615 (2022), https://arxiv.org/abs/2201.01615
59. Yoo, D., Park, S., Lee, J.Y., So Kweon, I.: Multi-scale pyramid pooling for deep convolutional representation. In: Proceedings of the IEEE Conference on Computer Vision and Pattern Recognition Workshops, pp. 71–80 (2015)
60. Yu, F., Xian, W., Chen, Y., Liu, F., Liao, M., Madhavan, V., Darrell, T.: BDD100K: a diverse driving video database with scalable annotation tooling. arXiv preprint arXiv:1805.04687 2(5), 6 (2018)
61. Zaidi, S., Zela, A., Elsken, T., Holmes, C., Hutter, F., Teh, Y.W.: Neural ensemble search for performant and calibrated predictions. In: Workshop on Uncertainty and Robustness in Deep Learning (UDL@ICML2020) (2020)
62. Zhang, H., Li, A., Guo, J., Guo, Y.: Hybrid models for open set recognition. In: Vedaldi, A., Bischof, H., Brox, T., Frahm, J.-M. (eds.) ECCV 2020. LNCS, vol. 12348, pp. 102–117. Springer, Cham (2020). https://doi.org/10.1007/978-3-030-58580-8_7
63. Zhao, H., Shi, J., Qi, X., Wang, X., Jia, J.: Pyramid scene parsing network. In: Proceedings of the IEEE Conference on Computer Vision and Pattern Recognition, pp. 2881–2890 (2017)

Adversarial Vulnerability of Temporal Feature Networks for Object Detection

Svetlana Pavlitskaya[1]([✉]), Nikolai Polley[2], Michael Weber[1],
and J. Marius Zöllner[1,2]

[1] FZI Research Center for Information Technology, 76131 Karlsruhe, Germany
pavlitskaya@fzi.de
[2] Karlsruhe Institute of Technology (KIT), 76131 Karlsruhe, Germany

Abstract. Taking into account information across the temporal domain helps to improve environment perception in autonomous driving. However, it has not been studied so far whether temporally fused neural networks are vulnerable to deliberately generated perturbations, i.e. adversarial attacks, or whether temporal history is an inherent defense against them. In this work, we study whether temporal feature networks for object detection are vulnerable to universal adversarial attacks. We evaluate attacks of two types: imperceptible noise for the whole image and locally-bound adversarial patch. In both cases, perturbations are generated in a white-box manner using PGD. Our experiments confirm, that attacking even a portion of a temporal input suffices to fool the network. We visually assess generated perturbations to gain insights into the functioning of attacks. To enhance the robustness, we apply adversarial training using 5-PGD. Our experiments on KITTI and nuScenes datasets demonstrate, that a model robustified via K-PGD is able to withstand the studied attacks while keeping the mAP-based performance comparable to that of an unattacked model.

Keywords: Adversarial attacks · Temporal fusion · Object detection

1 Introduction

Deep neural networks (DNNs) have become an indispensable component of environment perception in autonomous driving systems. The inherent vulnerability of DNNs to adversarial attacks [25] makes adversarial robustness one of the crucial requirements before their wide adoption in autonomous vehicles is possible. Recent studies [3,6,11] demonstrate that adversarial attacks can be performed in the real world and thus present a significant threat to self-driving cars.

Previous works have already investigated adversarial vulnerability of DNNs for specific tasks like object detection [11] or semantic segmentation [14,17], and also of DNNs with specific architectures like sensor fusion [29] or multi-task learning [13]. In this work, we focus on temporal feature networks as a model under attack. Although the emphasis of recent studies was mostly on LiDAR

L. Karlinsky et al. (Eds.): ECCV 2022 Workshops, LNCS 13805, pp. 510–525, 2023.
https://doi.org/10.1007/978-3-031-25072-9_35

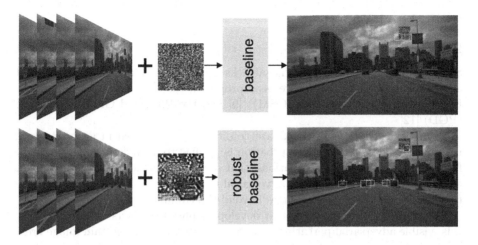

Fig. 1. We evaluate universal attacks on temporal feature networks for object detection (here – with a universal patch). Adversarial patch suppresses all detections of an undefended baseline and creates fake detections instead (top). Detection of ground truth objects by a robust model is no longer restrained by an adversarial patch (bottom). Note, that the patch against robust models has a more complex structure

data used in conjunction with camera images, the temporal fusion is a further approach to increase the object detection accuracy, which deserves attention. There has been very little research into fusing multiple images in the temporal dimension for object detection. Object detectors with temporal fusion, however, receive more context data from previous images and outperform single-image object detectors [30]. In particular, the prediction of obstructed objects that are visible in previous images might be a possible advantage.

To the best of our knowledge, we are the first to explore attacks and defenses for this type of DNNs. As an exemplary model under attack, we consider the late slow fusion architecture for object detection proposed by Weber et al. [30], which is in turn inspired by DSOD [23] and expanded to incorporate several input images. The model proposed by Weber et al. outperformed its non-temporal counterpart, thus demonstrating the importance of the temporal history for the object detection accuracy. The main goal of our work is to study, whether CNNs with temporal fusion are prone to adversarial attacks and what is the impact of the temporal history on the adversarial vulnerability of these models.

For this, we evaluate two variants of universal white-box attacks against this model: with an adversarial patch (see Fig. 1) and with adversarial noise. In our setting, a single instance of malicious input is generated to attack all possible data. This way, if printed out, the generated patch can be used for an attack in the real world. The latter was already demonstrated for state-of-the-art object detectors in previous works [3,24,26]. To increase the adversarial robustness of the model under attack, we consider adversarial training using K-PGD [12].

2   Related Work

2.1   Adversarial Attacks

Since the discovery of adversarial attacks by Szegedy et al. [25], a number of algorithms to attack DNNs have been proposed, the most prominent being the Fast Gradient Sign Method (FGSM) [9] and the Projected Gradient Descent (PGD) [12].

While the mentioned attacks are traditionally performed in a per-instance manner, i.e. an adversarial perturbation is applied to a single image in order to fool a model, universal perturbations, that are able to attack multiple instances, are also possible [15]. Universal adversarial inputs pose a special threat, because they are also able to attack images beyond the training data.

Another line of research aims at developing physical adversarial attacks. For this, visible adversarial perturbations are generated within a certain image area. The resulting *adversarial patch* can then be printed out to perform an attack in the real world. After their introduction by Brown et al. in [3], adversarial patches have been shown to successfully fool various deep learning models, including object detectors [11,18], semantic segmentation networks [17] and end-to-end driving models [19]. A combination of adversarial patches with universal attacks is especially interesting. Taking into consideration the transferability of adversarial examples across DNNs, such attacks might also be performed in a black-box manner in the real world [26,32].

Although no previous work on adversarial vulnerability of temporal fusion networks is known, a certain effort was already made in the community to develop adversarial attacks against sensor fusion models [29,33]. In particular, the work by Yu et al. has revealed, that late fusion is more robust against attacks than early fusion [33]. The evaluated dataset, however, is relatively small with only 306 training and 132 validation samples from the KITTI dataset.

2.2   Adversarial Training

Adversarial training (AT) is currently one of the few defenses that are able to combat even strong attacks [2]. It consists in training a DNN while adding adversarial inputs to each minibatch of training data. It has recently been shown, that adversarial training not only increases the robustness of neural networks to adversarial attacks but also leads to better interpretability [28].

While the idea originates from [9], the first strong defense was demonstrated with a multi-step PGD algorithm (the K-PGD adversarial training) [12]. For each minibatch, a forward/backward step is first executed k times to generate an adversarial input and then a single forward/backward step follows, which aims to update the model parameters. The PGD loop thus drastically increases the overall training time. For this reason, K-PGD adversarial training is intractable for large datasets.

One of the recently proposed strategies to speed up adversarial training is the so-called AT for free [21], which reuses gradient information during training. Instead of performing separate gradient calculations to generate adversarial examples during training, adversarial perturbations and model parameters are updated simultaneously in a single backward pass. This way, multiple FGSM steps are performed on a single minibatch to simulate the PGD algorithm while concurrently training the model. The authors were the first to apply adversarial training to the large-scale ImageNet classification task. The robustified models demonstrate resistance to attacks, comparable to that of K-PGD, while being 7 to 30 times faster. The approach, however, is still more time-consuming than standard training.

A similar method to accelerate adversarial training named YOPO (You Only Propagate Once) is proposed by Zhang et al. [34]. In YOPO, the gradients of the early network layers are frozen and reused to generate an adversarial input. YOPO is four to five times faster than K-PGD, although the results are only provided for relatively small datasets. YOPO reaches a performance similar to the free adversarial training but is less computationally expensive.

Most recently, a further approach to accelerate adversarial training was proposed by Wong et al. [31]. Starting with the assumption, that iterative attacks like K-PGD do not necessarily lead to more robust defenses, the authors propose to use R-FGSM AT instead. R-FGSM applies FGSM after a random initialization. This adversarial training method is claimed to be as effective as K-PGD.

Enhancing adversarial training to increase robustness against universal attacks was first addressed by Shafahi et al. [22]. Each training step uses FGSM to update a universal adversarial perturbation, which is then simultaneously used to update the model parameters. The proposed extension to adversarial training introduces almost no additional computational cost, which makes adversarial training on large datasets possible.

A further approach to harden DNNs against universal attacks is the shared adversarial training, proposed in [16]. To generate a shared perturbation, each batch is split into heaps, which are then attacked with single perturbations. These perturbations are aggregated and shared across heaps and further used for standard adversarial training.

3   Adversarial Attacks

3.1   Threat Model

We consider two types of white-box attacks in this work: adversarial patch and adversarial noise. All attacks are performed in a universal manner, i.e. a single perturbation is used to to attack all images [15]. We use a slightly modified version of the PGD algorithm [12] for the attack. In order to apply PGD in a universal manner, an empty mask is introduced, which is added to each input image. We then only update this mask in contrast to the original PGD, which updates the input images directly.

We use Adam [10] for faster PGD convergence as suggested in [5]. We also do not take a sign of the gradients but use actual gradient values instead. In an unsigned case, pixels, that strongly affect the prediction, can be modified to a much larger extent than less important pixels. Noise initialization strategy is a further setting, influencing training speed. For the FGSM attack, the advantage of initialization with random values has already been shown [27]. We perform patch-based attacks with randomly initialized patch values and noise-based attacks initialized with Xavier [8].

3.2   Adversarial Training

Adversarial training expands the training dataset with adversarial examples, created on the fly during training. Training on this dataset should enable the model to predict the correct label even in the presence of an attack.

We consider the established K-PGD AT with patch/noise generated for each input sequence. We create a single adversarial example (with either patch or noise) for all images in an input sequence. Thus, the generated adversarial perturbation is not universal and is only intended to fool the data from the current batch. The usage of this approach is motivated by a recent observation, that defending against non-universal attacks also protects against universal attacks [16].

Although Free [21] and YOPO [34] approaches offer a considerable speed up in training, they are not applicable in our case. Both algorithms require the same loss function to train a model and create adversarial examples. This is only possible for untargeted attacks, where the goal is to maximize the loss for the correct class. Instead, we want to perform object vanishing attacks, which require different loss functions than training the model. This way, reusing the gradients, already computed for the parameter update step, as foreseen in these AT algorithms, cannot be performed in our case.

4   Experimental Setup

4.1   Dataset

For the evaluation, we consider the late slow fusion architecture for object detection, proposed by Weber et al. [30]. While the original work by Weber et al. has focused on the KITTI Object Detection dataset [7], we additionally run our experiments on the nuScenes data [4]. We focus on the models with four input frames with equal temporal distance.

KITTI. Images from the object detection benchmark are resized to 1224×370. We only select images, that have the corresponding three temporally preceding frames, delivered in the dataset, resulting in 3689 sequences of length four for training and 3754 sequences of length four for validation.

nuScenes. The dataset contains 850 annotated scenes, whereas each scene contains about 40 keyframes per camera, taken at a frequency 2 Hz. We generate input sequences with a length of four from the keyframes as follows: if keyframes

Fig. 2. Architecture of the temporal feature network used for baselines. bs denotes the bottleneck size. Inception block uses $k \times \times 1 \times 1$ convolutions for depth reduction

$\{a, b, c, d, e\}$ belong to the same scene, the two sequences $\{a, b, c, d\}$ and $\{b, c, d, e\}$ are created. For the training dataset, we use front-facing and backward-facing images. For the latter, we have inverted the order of the sequence. Additionally, we augment the training dataset by horizontal mirroring of images. Since the nuScenes dataset contains images from a left-driving country, the introduced images remain within the domain. We use the 700/150 scene split for training and validation, as recommended by nuScenes, so that the training subset contains 52060 and the validation subset – 5569 sequences of length four. Images are resized from an initial 1600×900 to 1024×576 pixels and normalized to a range of [0,1]. For attack, we do not apply horizontally mirrored images, so that the training dataset contains 26030 images.

Since we focus on the detection of cars and pedestrians, the nuScenes classes *adult, child, construction worker, police officer, stroller* and *wheelchair* are consolidated into one new class *pedestrian*. As a second class we define *car*, which corresponds to the nuScenes class *vehicle.car*. We do not include other vehicles available in the nuScenes labels (e.g. busses, motorcycles, trucks, and trailers) to minimize intra-class variance.

4.2 Baselines

The baselines follow the late slow fusion architecture (see Fig. 2) proposed by Weber et al. [30]. The input is a sequence of images with temporal distance Delta t = 500 ms, whereas prediction is learned for the last input frame. Each input image is processed by a separate 2D convolutional stem, being identical to the 2D DSOD detector [23]. Stems are followed by a series of 3D dense blocks, intervened with transition blocks. The subsequent Inception redux blocks aim at reducing the temporal depth by half. Finally, a detector subnet similar to that of YOLOv2 [20] follows.

For each dataset, we deliberately defined new anchor boxes to have high IoU scores with the bounding boxes in the training data. For this, we analyzed all objects in the training dataset and clustered them using k-means with IoU as a distance metric. As a result, we obtained eight anchor boxes, which comprise various box forms to detect both pedestrians (upright vertical rectangles) and cars (horizontal rectangles).

We train the baselines on an NVIDIA RTX 2080 Ti GPU. KITTI models are trained for 50 epochs, whereas nuScenes models are trained for 100 epochs. For validation, the PASCAL VOC implementation of mAP is used. The mAP is always calculated for a confidence threshold of 0.01 and nms threshold of 0.5.

Table 1. AP_{car} in % of the 1-class KITTI baseline and AT models

Model / Attack	No attack	Universal patch	Universal noise $\epsilon = 5/255$	Universal noise, $\epsilon = 10/255$
Baseline	74.82	13.62	34.73	2.08
5-PGD AT with patch	66.47	52.18	24.46	23.07
5-PGD AT with noise, $\epsilon = 5/255$	62.77	42.82	61.52	61.75
5-PGD AT with noise, $\epsilon = 10/255$	53.24	32.15	53.24	53.07

Table 2. mAP in % of the 2-class KITTI baseline and AT models

Model / Attack	No attack	Universal patch	Universal noise $\epsilon = 5/255$	Universal noise, $\epsilon = 10/255$
Baseline	50.93	4.36	34.49	4.43
5-PGD AT with patch	50.86	44.18	23.88	14.83
5-PGD AT with noise, $\epsilon = 5/255$	41.95	27.90	41.95	41.35
5-PGD AT with noise, $\epsilon = 10/255$	38.50	25.76	38.49	38.49

We train models to detect either only objects of the class *car* (1-class models) or objects of the classes *car* and *pedestrian* (2-class model). Tables 1 and 2 show performance of the 1-class and 2-class KITTI baselines. Whereas average precision for the class *car* is comparable for both models (AP_{car} reaches 74.82% for the 1-class model vs. 73.21% for the 2-class), the worse performance for the underrepresented class *pedestrian* (28.65%) explains the overall worse mAP of the 2-class model.

Tables 3 and 4 summarize results on the nuScenes baselines. Due to class imbalance (108K annotated cars vs. 48K annotated pedestrians), the results for the 2-class baseline for the underrepresented class *pedestrian* are again significantly worse.

The results achieved on the KITTI and nuScenes baselines, described below, are comparable if the opposite is not stated.

4.3 Adversarial Noise Attack

For a universal noise attack, we have initially applied the proposed universal PGD with the changes motivated above. However, it turned out, that this attack leads to the detection of nonexistent new objects (see Fig. 3a). The loss maximization goal apparently favors creating new features that resemble objects rather than preventing the detection of existing objects. We have therefore adapted the attack algorithm by replacing the gradient ascent with the gradient descent on an empty label. Figure 3b shows, that adversarial noise generated using targeted PGD on an empty label can successfully suppress all objects

Table 3. AP_{car} in % of the 1-class nuScenes baseline and AT models

Model / Attack	No attack	Universal patch	Universal noise $\epsilon = 5/255$	Universal noise, $\epsilon = 10/255$
Baseline	73.20	6.20	9.03	0.15
5-PGD AT with patch	74.04	61.89	27.71	27.65
5-PGD AT with noise, $\epsilon = 5/255$	72.24	3.51	71.85	68.19

Table 4. mAP in % of the 2-class nuScenes baseline and AT models. For this model, AT with reused patches and R-FGSM AT were additionally evaluated

Model / Attack	No attack	Universal patch	Universal noise $\epsilon = 5/255$	Universal noise, $\epsilon = 10/255$
Baseline	27.55	0.98	0.74	0.16
5-PGD AT with patch	28.69	17.95	7.02	0.53
5-PGD AT with noise, $\epsilon = 5/255$	27.10	12.83	25.78	20.69
AT with reused patches	27.24	9.13	6.01	0.48
R-FGSM AT with noise, $\epsilon = 5/255$	27.51	2.72	0.24	0.02

present in the input. All perturbations are trained using the Adam optimizer for 100 epochs.

4.4 Adversarial Patch Attack

To generate a universal patch, we apply PGD with unsigned gradients using the Adam optimizer for 100 epochs. We evaluated patches of size 71×71, 51×51 and 31×31. As expected, the largest patch led to a stronger attack and was used in the following experiments. Note, that a 71×71 patch still takes only about 1% of the image area both in the case of KITTI and nuScenes images.

To evaluate the impact of patch position, we evaluated a total of 33 patch positions. Average precision for different positions fluctuated only within few percentage points, so patch position apparently has only a minor impact on its attack strength.

4.5 Adversarial Training

We apply adversarial training as a method to improve the robustness of the studied models. We consider K-PGD AT and two additional AT strategies: (1) reusing the patches, already generated for the baseline, during the training and (2) R-FGSM.

(a) Untargeted attack with gradient ascent

(b) Targeted attack for an empty label with gradient descent

Fig. 3. Predictions of the 2-class nuScenes baseline on images attacked with universal noise, $\epsilon = 5/255$

(a) Untargeted attack with gradient ascent on the 2-class model

(b) Untargeted attack with gradient ascent on the 1-class model

Fig. 4. Predictions of the nuScenes baselines on images attacked with universal patch

In the case of K-PGD AT, creating an adversarial example iteratively with k steps increases the number of forward and backward propagations by a factor of k. We have therefore used a small $k = 5$ per adversarial attack during training and drastically increased the learning rate of the Adam optimizer to make the attacks possible. Training 5-PGD AT both on nuScenes and KITTI thus took five times longer than the corresponding baseline, both for the adversarial noise and the adversarial patch.

In the case of the pre-generated patches for adversarial training, we first generated a pool of patches against the baseline as described above. We then trained a model, while adding a randomly chosen patch at each training step with a 50% probability. The AT with reused patches involved no generation of new patches, therefore its duration is comparable to regular model training.

Finally, R-FGSM AT was trained with adversarial noise with $\epsilon = 5/255$.

5   Evaluation

5.1   Attacks on 1-Class Vs. on 2-Class Baselines

Visual assessment of the adversarial noise patterns helps to understand, how the attack functions. We observed different behavior of attacks on 1-class and 2-class models. In particular, universal noise, generated to attack 1-class baselines evidently contains structures resembling cars, whereas noise attacking 2-class models exhibits no such patterns (see Fig. 5). Apparently, the attack aims at mimicking existing objects, if they all belong to one class.

Adversarial patches, generated for both types of models, however, look similar. Patch-based attacks on nuScenes baselines tend to detect non-existing cars in a patch (see Fig. 4), whereas attacks against the KITTI 2-class baseline rather find pedestrians in a patch. This might be explained by a different portion of pedestrians in the corresponding datasets. Patches against nuScenes 2-class model never mimicked pedestrians, because they are highly underrepresented in the training data.

5.2   Impact of Temporal Horizon

Temporal fusion models have better performance due to the incorporation of the temporal history. To assess which portion of the history is enough to attack the model, we perform the evaluation exemplary with the $\epsilon = 10/255$ adversarial noise attack on the 1-class KITTI baseline. We have attacked single frames using adversarial noise, which was initially generated for the whole input sequence of length four and with adversarial noise, generated for the corresponding portion of the input (see Table 5). We observed, that perturbations, deliberately generated for specific frames, work better when attacking them, than those generated for the whole input sequence. For both cases, the attack works the best, when frames, immediately preceding the current frame, are attacked. On contrary, attacking only the oldest frame leads to the worst results. Also, perturbing only the frame for which the prediction is done and not attacking the temporal history at all leads to a significantly weaker attack. Finally, the more preceding temporal history frames are attacked, the better the results.

These results confirm, that the later images in the input sequence are more important for the prediction. Furthermore, single attacked images that appear later in the input sequence, cause larger error than those which appear earlier.

Furthermore, we evaluate the impact of the temporal horizon. In addition to the already evaluated models with four input images, we also evaluated models with a smaller sequence length. In particular, for the 1-class nuScenes model, we observe about 10% for each reduction of the number of input images: 73.20% mAP for the temporal history of length four, 63.12% for the length three and 52.18% for the length two. The attack strength also decreases correspondingly. We thus conclude that a larger temporal horizon helps to enhance not only the performance on the clean data, but also the adversarial robustness.

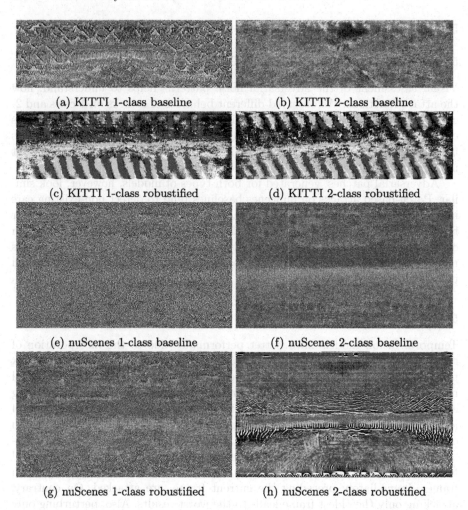

(a) KITTI 1-class baseline   (b) KITTI 2-class baseline

(c) KITTI 1-class robustified   (d) KITTI 2-class robustified

(e) nuScenes 1-class baseline   (f) nuScenes 2-class baseline

(g) nuScenes 1-class robustified   (h) nuScenes 2-class robustified

Fig. 5. Universal noise with $\epsilon = 5/255$, generated to attack baselines and models, robustified via 5-PGD AT with noise, $\epsilon = 5/255$. Original pixel value range [–5,5] mapped to [0,255] for better visibility

5.3   Robustness of the Adversarially Trained Models

To evaluate the robustness of the robustified models, we attack them again with newly generated universal patches and noises. Tables 1-4 demonstrate the results.

All models robustified via AT demonstrate performance similar to the baseline on non-attacked data and almost no accuracy drop on the malicious data for the nuScenes models and a small drop for the KITTI models. For the KITTI dataset, we have additionally evaluated AT with adversarial noise with $\epsilon = 5/255$ and $\epsilon = 10/255$ (see Tables 1 and 2). The larger epsilon leads to worse performance on clean data due to larger perturbation.

Table 5. Attacking a part of the sequence. The attacked frames are highlighted red, prediction is performed for the frame t. mAP in % is reported for the 1-class KITTI baseline, attacked with $\epsilon = 10/255$ adversarial noise, generated either for all four inputs or for each evaluated case

Attacked sequence part t_{-3} t_{-2} t_{-1} t	Noise generated for all four inputs	Noise generated for each evaluated case
	74.82	74.82
	70.68	58.46
	67.02	45.43
	53.14	36.21
	23.91	5.77
	50.84	22.14
	32.45	10.13
	6.19	2.96
	3.80	2.74
	2.08	2.08

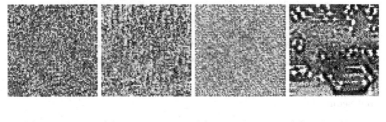

(a) Baseline   (b) AT with reused patches   (c) 5-PGD AT with patch   (d) 5-PGD AT with noise

Fig. 6. Universal patches to attack different nuScenes 2-class models

Moreover, we have evaluated AT with reused patches and R-FGSM AT on the 2-class nuScenes model (see Table 4). As expected, the defended model with reused patches is less robust to attacks than the one which was trained with 5-PGD. Surprisingly, the R-FGSM AT method has completely failed to defend against attacks. We explain this behavior with the *catastrophic overfitting* phenomenon, mentioned in the original work by Wong et al. [31] and in a more recent study by Andriushchenko et al. [1], which challenges the original claim that using randomized initialization prevents this overfitting.

Figure 6 compares patches, generated for the adversarially trained models with the patch generated against the nuScenes 2-class baseline. In the case of patch reuse, the patch contains more green and yellow pixels than the original patch. In the case of K-PGD adversarial training, the patch is brighter and contains more white pixels. Interestingly, the patch generated for a model, which

| (a) Baseline | (b) 5-PGD AT with noise |

Fig. 7. Per-instance adversarial noise with $\epsilon = 5/255$ generated to attack nuScenes 2-class model on a single input sequence. Original pixel value range [−5,5] mapped to [0,255] for better visibility

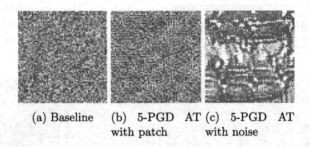

(a) Baseline   (b) 5-PGD AT with patch   (c) 5-PGD AT with noise

Fig. 8. Per-instance adversarial patch generated to attack nuScenes 2-class model on single input sequence

was adversarially trained with adversarial noise, is the only one that contains structures, resembling a car.

Figure 5 compares universal noises, generated to attack the KITTI and nuScenes baselines and the corresponding model defended via 5-PGD AT with $\epsilon = 5/255$ noise. Both contain a streak of color at the horizon line and wave-like patterns at the bottom. We again conclude, that perturbation attacking robustified models exhibit more complex structure.

5.4 Robustness of the AT-Trained Models Against Per-instance Attacks

Finally, we examine whether adversarially trained models also become robust against per-instance attacks. For this, we take an exemplary input sequence and generate adversarial perturbations against it. Each attack is trained for 1000 steps.

Per-instance noise attack (see Fig. 7) manage to completely suppress all detections of the corresponding model. Analogously to universal attacks, non-universal noise attacks against the hardened model look much more complex. Similarly, per-instance patches (see Fig. 8) against the robustified model show complex structures resembling cars. Interestingly, this patch is also unable to efficiently attack the model, several cars are still correctly detected after applying this patch.

Overall, while models hardened with the evaluated adversarial training strategies are very successful in resisting universal attacks while preserving high accuracy, they are still unprotected against per-instance attacks. Universal attacks are, however, much more feasible with regard to real-life settings.

6 Conclusion

In this work, we have studied the adversarial vulnerability of temporal feature networks for object detection. The architecture proposed by Weber et al. [30] was used as an exemplary model under attack.

Our experiments on KITTI and nuScenes datasets have demonstrated that the studied temporal fusion model is susceptible to both universal patch and noise attacks. Furthermore, we have explored different adversarial training strategies as a defense measure. Out of the three evaluated methods, the 5-PGD approach with a per-instance adversarial noise has proven to be the most powerful. The R-FGSM strategy, however, has failed to defend against the studied attacks. 5-PGD adversarial training was able to withstand newly created universal attacks. The robustified networks have also demonstrated only a slight drop in performance on clean data.

Our experiments with attacking a portion of the temporal history have demonstrated, that the frames, immediately preceding the current frame, have a greater impact on the model decision and thus lead to stronger attacks when manipulated. We have further observed, that reducing the temporal horizon leads to worse performance and adversarial robustness of the model.

We have compared the universal and per-instance perturbations generated to attack the baseline and the robustified models. In all cases, we observed that in order to attack a hardened neural network, the adversarial perturbation has to exhibit a much more complex structure. In particular, a universal patch against the most robust 5-PGD with noise contains a pattern resembling a car.

Our adversarially trained models, however, remain vulnerable to non-universal attacks like per-instance-generated noise or patch. This stresses the need for further research in this area.

Since the computation time for adversarial training is still a bottleneck, adapting gradient re-usage strategies like [21] or [34] for models, which use different loss functions to learn an adversarial perturbation and to update the model weights, might be a promising line of research for future.

Acknowledgement. The research leading to these results is funded by the German Federal Ministry for Economic Affairs and Climate Action within the project "KI Absicherung" (grant 19A19005W) and by KASTEL Security Research Labs. The authors would like to thank the consortium for the successful cooperation.

References

1. Andriushchenko, M., Flammarion, N.: Understanding and improving fast adversarial training. Adv. Neural Inf. Proc. Syst. (NIPS) **33**, 16048–16059 (2020)
2. Athalye, A., Carlini, N., Wagner, D.: Obfuscated gradients give a false sense of security: circumventing defenses to adversarial examples. In: International Conference on Machine Learning (ICML). PMLR (2018)
3. Brown, T.B., Mané, D., Roy, A., Abadi, M., Gilmer, J.: Adversarial patch. In: Advances in Neural Information Processing Systems (NIPS) - Workshops (2017)
4. Caesar, H., Bankiti, V., Lang, A.H., Vora, S., Liong, V.E., et al.: nuScenes: a multimodal dataset for autonomous driving. In: Conference on Computer Vision and Pattern Recognition (CVPR) (2020)
5. Carlini, N., Wagner, D.A.: Towards evaluating the robustness of neural networks. In: Symposium on Security and Privacy (SP). IEEE (2017)
6. Eykholt, K., et al.: Robust physical-world attacks on deep learning visual classification. In: Conference on Computer Vision and Pattern Recognition (CVPR). IEEE (2018)
7. Geiger, A., Lenz, P., Urtasun, R.: Are we ready for autonomous driving? The KITTI vision benchmark suite. In: Conference on Computer Vision and Pattern Recognition (CVPR) (2012). https://doi.org/10.1109/CVPR.2012.6248074
8. Glorot, X., Bengio, Y.: Understanding the difficulty of training deep feedforward neural networks. In: International Conference on Artificial Intelligence and Statistics (AISTATS) (2010)
9. Goodfellow, I.J., Shlens, J., Szegedy, C.: Explaining and harnessing adversarial examples. In: International Conference on Learning Representations (ICLR) (2015)
10. Kingma, D.P., Ba, J.: Adam: a method for stochastic optimization. In: International Conference on Learning Representations (ICLR) (2015)
11. Lee, M., Kolter, Z.: On physical adversarial patches for object detection. CoRR abs/1906.11897 (2019)
12. Madry, A., Makelov, A., Schmidt, L., Tsipras, D., Vladu, A.: Towards deep learning models resistant to adversarial attacks. In: International Conference on Learning Representations (ICLR) (2018)
13. Mao, C., et al.: Multitask learning strengthens adversarial robustness. In: Vedaldi, A., Bischof, H., Brox, T., Frahm, J.-M. (eds.) ECCV 2020. LNCS, vol. 12347, pp. 158–174. Springer, Cham (2020). https://doi.org/10.1007/978-3-030-58536-5_10
14. Metzen, J.H., Kumar, M.C., Brox, T., Fischer, V.: Universal adversarial perturbations against semantic image segmentation. In: International Conference on Computer Vision (ICCV). IEEE Computer Society (2017)
15. Moosavi-Dezfooli, S.M., Fawzi, A., Fawzi, O., Frossard, P.: Universal adversarial perturbations. In: Conference on Computer Vision and Pattern Recognition (CVPR). IEEE (2017)
16. Mummadi, C.K., Brox, T., Metzen, J.H.: Defending against universal perturbations with shared adversarial training. In: International Conference on Computer Vision (ICCV) (2019)
17. Nesti, F., Rossolini, G., Nair, S., Biondi, A., Buttazzo, G.C.: Evaluating the robustness of semantic segmentation for autonomous driving against real-world adversarial patch attacks. In: Proceedings of the Winter Conference on Applications of Computer Vision (WACV). IEEE (2022)
18. Pavlitskaya, S., Codau, B., Zöllner, J.M.: Feasibility of inconspicuous GAN-generated adversarial patches against object detection. In: International Joint Conference on Artificial Intelligence (IJCAI) - Workshops (2022)

19. Pavlitskaya, S., Ünver, S., Zöllner, J.M.: Feasibility and suppression of adversarial patch attacks on end-to-end vehicle control. In: International Conference on Intelligent Transportation Systems (ITSC). IEEE (2020)
20. Redmon, J., Farhadi, A.: YOLO9000: better, faster, stronger. In: Conference on Computer Vision and Pattern Recognition (CVPR). IEEE (2017)
21. Shafahi, A., at al.: Adversarial training for free! In: Advances in Neural Information Processing Systems (NIPS) (2019)
22. Shafahi, A., Najibi, M., Xu, Z., Dickerson, J., Davis, L.S., Goldstein, T.: Universal adversarial training. In: AAAI Conference on Artificial Intelligence (2020)
23. Shen, Z., Liu, Z., Li, J., Jiang, Y.G., Chen, Y., Xue, X.: DSOD: learning deeply supervised object detectors from scratch. In: International Conference on Computer Vision (ICCV). IEEE (2017)
24. Song, D., et al.: Physical adversarial examples for object detectors. In: USENIX workshop on offensive technologies, WOOT 2. USENIX association (2018)
25. Szegedy, C., et al.: Intriguing properties of neural networks. In: International Conference on Learning Representations (ICLR) (2014)
26. Thys, S., Ranst, W.V., Goedemé, T.: Fooling automated surveillance cameras: adversarial patches to attack person detection. In: Conference on Computer Vision and Pattern Recognition (CVPR) - Workshops. Computer Vision Foundation. IEEE (2019)
27. Tramèr, F., Kurakin, A., Papernot, N., Goodfellow, I., Boneh, D., McDaniel, P.: ensemble adversarial training: attacks and defenses. In: International Conference on Learning Representations (ICLR) (2018)
28. Tsipras, D., Santurkar, S., Engstrom, L., Turner, A., Madry, A.: robustness may be at odds with accuracy. In: International Conference on Learning Representations (ICLR) (2019)
29. Tu, J., et al.: Exploring adversarial robustness of multi-sensor perception systems in self driving. (2021) CoRR abs/2101.06784
30. Weber, M., Wald, T., Zöllner, J.M.: Temporal feature networks for CNN based object detection. In: Intelligent Vehicles Symposium (IV). IEEE (2021)
31. Wong, E., Rice, L., Kolter, J.Z.: Fast is better than free: revisiting adversarial training. International Conference on Learning Representations (ICLR) (2020)
32. Xu, K., et al.: Adversarial T-Shirt! Evading Person Detectors in a Physical World. In: Vedaldi, A., Bischof, H., Brox, T., Frahm, J.-M. (eds.) ECCV 2020. LNCS, vol. 12350, pp. 665–681. Springer, Cham (2020). https://doi.org/10.1007/978-3-030-58558-7_39
33. Yu, Y., Lee, H.J., Kim, B.C., Kim, J.U., Ro, Y.M.: Investigating vulnerability to adversarial examples on multimodal data fusion in deep learning. (2020) CoRR abs/2005.10987
34. Zhang, D., Zhang, T., Lu, Y., Zhu, Z., Dong, B.: You only propagate once: accelerating adversarial training via maximal principle. In: Advances in Neural Information Processing Systems (NIPS) (2019)

Towards Improved Intermediate Layer Variational Inference for Uncertainty Estimation

Ahmed Hammam[1,2(✉)], Frank Bonarens[1], Seyed Eghbal Ghobadi[3], and Christoph Stiller[2]

[1] Stellantis, Opel Automobile GmbH, Rüsselsheim am Main, Germany
`ahmedmostafa.hammam@external.stellantis.com`
[2] Institute of Measurement and Control Systems, Karlsruhe Institute of Technology, Karlsruhe, Germany
[3] Technische Hochschule Mittelhessen, Giessen, Germany

Abstract. DNNs have been excelling on many computer vision tasks, achieving many milestones and are continuing to prove their validity. It is important for DNNs to express their uncertainties to be taken into account for the whole application output, whilst maintaining their performance. The intermediate layer variational inference has been a promising approach to estimate uncertainty in real-time beating state-of-the-art approaches. In this work, we propose an enhancement of the intermediate layer variational inference by adding Dirichlet distribution to the network architecture. This improves, on one hand, the uncertainty estimation reliability and on the other hand can detect out-of-distribution samples. Results show that with the addition of Dirichlet distributions the DNN is able to maintain its segmentation performance whilst boosting its uncertainty estimation capabilities.

Keywords: Deep neural networks · Uncertainty estimation · Trustworthiness · Out-of-distribution detection

1 Introduction

Deep learning has been a breakthrough in the field of computer vision, being the go-to approach for several applications such as medical imaging [1,2] and Highly Automated Driving (HAD) [3,4]. For HAD systems, deep neural networks (DNNs) have been the most dominant technique for several applications such as sensor fusion [5,6] and path planning [7] contributing to several tasks such as object detection [8,9] and image semantic segmentation [10,11] and many more.

Even though deep learning techniques have set new benchmarks for several tasks, they are very complex and hardly interpretable [12] due to the massive amount of parameters in the DNN. Furthermore, brittleness is a main concern regarding DNNs as they are very susceptible to failure due to operational domain shift, or to minor perturbations to the input [13,14]. Moreover, DNNs show a

© The Author(s), under exclusive license to Springer Nature Switzerland AG 2023
L. Karlinsky et al. (Eds.): ECCV 2022 Workshops, LNCS 13805, pp. 526–542, 2023.
https://doi.org/10.1007/978-3-031-25072-9_36

tendency to usually be overconfident which causes great concern as this results in unreliable high confidence in incorrect predictions. These drawbacks become serious limitations for HAD functions, as they could lead to false detections in terms of misclassifications and mislocalizations which can cause problems for the whole application process [12]. Hence, this has become the main concern for safety-critical applications needing high reliability and calibrated confidence scores.

Another critical issue is the DNN's out-of-distribution (OoD) detection capability which is essential to ensure the reliability and safety of machine learning systems [15,16]. For HAD systems, it is always desired that the driving system would alert users when driving in unusual scenes or when an unknown object has been encountered presuming that it won't be able to make a safe decision whilst driving.

In recent years a common solution to these drawbacks was to empower DNNs with the ability to express their uncertainty about their output predictions [17,18]. This paradigm towards expressing uncertainty has opened the door for researchers to investigate and implement various approaches and techniques to quantify the uncertainty estimations [19,20]. Such estimations will not only allow the DNN to express its uncertainty, but can also be used as a way to find anomalies in the network's output, and can be utilized to express out-of-distribution classes apparent in the input.

Intermediate layer variational inference (IVLI) [21] has been one of the approaches to estimate epistemic uncertainty by seeking the variational inference technique with the addition of an intermediate multivariate layer. It has proved comparable uncertainty estimation results to the state-of-art approaches whilst running in real-time, making it applicable to HAD functions. In this paper, we propose an enhancement to the ILVI approach by modeling the DNN outputs using Dirichlet distributions. This improvement boosts the DNN's uncertainty estimation and out-of-distribution detection capability. Furthermore, this improves the calibration, uncertainty distributional separation efficiency and provides a more reliable uncertainty estimation whilst maintaining the segmentation performance.

The remainder of the paper is organized as follows: Sect. 2 presents related work to Dirichlet modeling and its uses in uncertainty estimation, whilst the approach proposed is discussed and explained in Sect. 3. The experiments conducted to test our approach are displayed in Sect. 4 and a conclusion and overview of the work are discussed in Sect. 5.

2 Related Work

2.1 State-of-the-Art Uncertainty Estimation Approaches

To understand and model the uncertainty in DNNs, Bayesian deep learning has been the fundamental approach used to lay the foundations for all approaches [22–24]. However, this has been a demanding and challenging task mainly due to the vast amounts of parameters in the DNN that hinders the classical standard Bayesian inference approaches yielding intractable models. Many works have been proposed recently to solve the intractability problem, but only a few have been applicable to large-scale applications [25, 26]. The two prominent approaches are Monte-Carlo Dropout (MC Dropout) [27] and Deep Ensembles [28]. The MC dropout approach adds extra dropout layers within the DNN, and at inference time the input image is passed more than once to the DNN to estimate the uncertainty. Deep ensembles follow a similar idea by training more than one DNN with different initial weights or on different splits of the training dataset. The input image will then be passed to the DNN ensemble members and all outputs will be used to estimate the uncertainty.

Even though both methods are the current state-of-the-art approaches for DNN uncertainty estimation, they both suffer from the inability to run in real-time. Furthermore, MC Dropout often decreases DNN performance. Especially in the case of HAD functions, this critical issue causes a main concern when it comes to implementation and deployment as this would affect its applicability.

2.2 Dirichlet Distributions for Uncertainty Estimation

A recently growing line of work focuses on using Dirichlet distributions in training DNNs, being the natural prior of categorical distributions [29, 30, 30, 31]. Dirichlet models have proved to provide more reliable uncertainty estimates whilst improving uncertainty estimations. Furthermore, Dirichlet distributions have also been used in several works for OoD showing prominent results [32, 33].

Previous work [29, 30] have introduced prior networks to model uncertainty using the concentration parameters from the Dirichlet model. According to Fig. 1, a Dirichlet distribution will have its distribution concentration focused on one class indicating correct prediction with low uncertainty, whereas a flat distribution over all classes indicating high uncertainty for an indecisive prediction. Authors of [29] proposed to train the DNN to mimic a Dirac distribution for in-domain samples and flat distribution for out-of-domain samples. To do so they propose to learn parameters of a Dirichlet distribution by training the DNN using Kullbeck-Leibler (KL) divergence $Loss = \mathbb{E}_{p_{in}(x)}[KL[\text{Dir}(\theta_{in}|\hat{\alpha})||\text{p}(\theta_{in}|\alpha_{in})] + \mathbb{E}_{p_{out}(x)}[KL[\text{Dir}(\theta_{out}|\tilde{\alpha})||\text{p}(\theta_{out}|\alpha_{out})]$, where α_{in} and α_{out} represent the DNN's predictions for in-distribution and out-of-distribution targets respectively and θ_{in} and θ_{out} the ground truth probability distribution for in and out of distribution targets respectively. $\hat{\alpha}$ and $\tilde{\alpha}$ resemble the in-distribution and out-of-distribution desired concentration parameters respectively, where $\tilde{\alpha}$ is a flat concentration by setting all $\tilde{\alpha}_c = 1$ and $\hat{\alpha}$ is a one-hot target. This idea has been further improved in [32] by using the reverse KL which has proven to be more reliable in uncertainty representation and improved robustness against adversarial attacks.

Similar to prior networks, [30] proposed to combine prior networks and the likelihood then maximize the whole posterior. Based on the works of the Dempster-Shafer Theory of Evidence, the Dirichlet model concentrations parameters are trained using the expected mean square error. To further penalize the Dirichlet model towards having increased uncertainty on incorrect detections, they add the second term of the KL divergence from [29]. With that, it is ensured that the network is training both on the correct classes and also getting proper uncertainty representation on incorrect predictions.

2.3   Out-of-distribution Detection

Machine learning models are mostly trained based on closed-world assumption, where the test data is assumed to be drawn from the same distribution of the training data [34,35]. However, realistically this is not always the case as when models are deployed in an open-set world, test samples may not be always the same as training data. For that, it is crucial for the system to give an alert whenever an OoD has been detected.

This task is divided into 4 subcategories: anomaly detection, novelty detection, open set recognition and out-of-distribution detection [36]. In this work we are concerned with the latter case, where we would like to detect out-of-distribution and show high uncertainty on that. Several works have been focusing on such approach due to its need for HAD functions [37–39]

To test the DNN's capabilities for OoD detection, an question arises: what data sources shall be used to test OoD detection? To resolve this issue, there are several solutions. One approach is the generation of OoD samples with generative adversarial networks [40,41]. This method showed promising results on simple images such as handwritten characters, however its complexity increases with generating natural image datasets due to the difficulty in generating high-quality images in high-dimensional space. Another line of work proposes to use external datasets to train the DNN on OoD samples. This helps the DNN to explore much more data and therefore improves the DNN's OoD detection capabilities [42]. Moreover, a more practical approach for OoD data is using the training dataset itself. This approach, named *leave-out class*, is applied by removing one class from the train dataset and testing the DNN afterwards on its OoD capabilities using this class. This has proved to be a very practical and an efficient approach towards OoD analysis [43,44]

Motivated by the previous works [29–31,33], we seek to use Dirichlet models as a method to produce a more reliable uncertainty estimation and also show an improvement on the out-of-distribution detection task. Formulating the loss function using KL divergence is usually challenging to train and can cause a poor error surface to optimize the DNN due to harsh conditions to mimic certain concentration distribution [29]. In this work we choose to formulate our loss function directly by maximizing the likelihood of the Dirichlet concentration parameters rather than using KL divergence.

3   Methodology

3.1   Dirichlet Models

A supervised network aims to predict the target value $y \in \mathcal{Y}$ for an input $x \in \mathcal{X}$, where the input space $\mathcal{X}$ corresponds to the space of images. Accordingly, a supervised machine learning problem with the task of semantic segmentation has a target $\mathcal{Y}$ consisting of a finite set of c classes where the task for the network is to predict the class of each pixel out of the set of classes K. For our purpose, a DNN is defined as a function $f_w : \mathcal{X} \rightarrow \mathcal{Y}$, parameterized by $w \in \mathbb{R}$, which maps an input $x \in \mathcal{X}$ to an output $f_w(x) \in \mathcal{Y}$.

Given the probability simplex as $\mathcal{S} = \{(\theta_1, \ldots, \theta_k) : \theta_i \geq 0, \sum_i \theta_i = 1\}$, the Dirichlet distribution is a probability density function on vectors $\theta \in \mathcal{S}$ and categorized by concentration parameters, $\alpha = \{\alpha_1, \ldots, \alpha_K\}$ as:

$$\mathrm{Dir}(\theta; \alpha) = \log \frac{1}{B(\alpha)} \prod_{i=1}^{K} \theta_i^{\alpha_i - 1} \tag{1}$$

where the normalizing constant $\frac{1}{B(\alpha)}$ denotes the multivariate Beta function $B(\alpha) = \frac{\prod_{i=1}^{K} \Gamma(\alpha_i)}{\Gamma(\alpha_0)}$, $\alpha_0 = \sum_{i=1}^{K} \alpha_i$ and $\Gamma(x)$ denotes the Gamma function and is defined to be $\int_0^\infty t^{x-1} e^{-t} dt$.

To model the Dirichlet distribution, the concentration parameters α correspond to each class output from the DNN as follows: $\alpha = f_w(x)$, where α changes with each input x, and θ denotes the ground truth probability distribution.

3.2 Proposed Loss Function

Our new loss function introduced in this paper is based upon maximizing the likelihood of the Dirichlet-based DNN model. To maximize the likelihood of the Dirichlet distribution, we minimize the negative log-likelihood as follows:

$$\begin{aligned} F(\alpha; \theta) &= \log \prod \mathrm{Dir}(\theta; \alpha) = \log \frac{1}{B(\alpha)} \prod_{i=1}^{K} \theta_i^{\alpha_i - 1} \\ &= \log \Gamma \left(\sum_{i=1}^{K} \alpha_j \right) - \sum_{i=1}^{K} \log \Gamma(\alpha_i) + \sum_{i=1}^{K} (\alpha_i - 1) \log \theta_i \end{aligned} \tag{2}$$

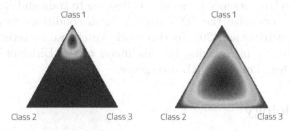

(a) Certain prediction.   (b) Uncertain prediction.

Fig. 1. Dirichlet plots

where θ represents the desired probability distribution to be maximized. The loss function aims to produce reliable uncertainty estimation by handling the DNN's correct predictions and incorrect predictions separately, as shown in Fig. 1. Our main goal is to produce correct predictions with low uncertainties, similar to Fig. 1(a). To do so the DNN should be able to produce a high concentration towards the correct class, and this can be achieved by maximizing the likelihood using the ground truth label probability, using a *one-hot vector*. For the incorrect predictions, we would like to have high uncertainty. To achieve this, equal probability for all classes would be required to yield high uncertainty, therefore the maximum likelihood should be maximized using an equal probability vector, yielding a result similar to Fig. 1(b).

Accordingly, we propose two loss functions to be minimized, very similar to each other, one for the task of out-of-distribution detection and one for full DNN training including all classes:

$$Loss = H(p,q) + Loss_{ILVI} + F(\alpha_{correct}; \theta_{correct}) + F(\alpha_{incorrect}; \theta_{incorrect}), \quad (3)$$

$$Loss_{OOD} = H(p,q) + Loss_{ILVI} + F(\alpha_{in}; \theta_{in}) + F(\alpha_{out}; \theta_{out}), \quad (4)$$

where $\alpha_{correct}$ and $\alpha_{incorrect}$ in Eq. 3 are the network's concentration parameters representing the correct and incorrect DNN predictions respectively, and α_{in} and α_{out} are the network's concentration parameters resembling the in-distribution and the out-of-distribution classes respectively and $H(p,q)$ resembles the cross-entropy loss between p the DNN's softmax output and q ground-truth labels. Accordingly, $\theta_{correct}$ and θ_{in} represent the ground truth probability distribution for the correct classes and in-distribution classes respectively, and $\theta_{incorrect}$ and θ_{out} represent the equal probability vector to yield high uncertainty. Building upon our previous ILVI work, an intermediate latent layer is added within the network, at the decoder level, to be in the form of a multivariate Gaussian distribution with mean and variance instead of point estimates. This substitution has been previously proven to enable the DNN a much faster uncertainty estimation and improve the whole segmentation performance of the network [21]. Accordingly, the first two terms in both loss functions resemble the evidence lower bound (ELBO) term for the intermediate layer of the variational inference network:

$$ELBO = \mathbb{E}_{q_\phi}[\log p(y|\theta, x)] - KL(q(\phi)||p(\theta|X, Y))$$

$$\text{and} \quad \mathbb{E}_{q_\phi}[\log p(y|\theta, x)] = -\sum_c p(y_c)\log p(\hat{y}_c) \quad (5)$$

where the KL-divergence is between the true and approximate models. In this case, we assume a prior to be Gaussian and optimize the new variational approximations using KL divergence.

4   Experiments and Results

In this section, a comparison between a DNN with ILVI only *(Baseline Loss)* and a DNN with ILVI and Dirichlet distributions *(Dirichlet Loss)* will be presented for both OoD detection and full DNN training. Furthermore, both tasks tested will be shown with their relevant metrics. For both DNNs, a modified DeepLab V3+ [10] is used to accommodate the newly added intermediate layer for the ILVI approach.

In this work, the DeepLab V3+ DNN, by Chen et. al [10], has been used as the base model to test our proposed approach. DeepLab V3+ is one of the prominent semantic segmentation DNNs performing well on several widely known benchmarking datasets.

Experiments have been performed on the task of street-scene semantic segmentation. The DNN's task is to predict the class of each pixel out of the domain of classes (car, pedestrian, etc.) from a front camera module. The Cityscapes dataset [45] is used in all experiments, where both the images and the ground-truth labels are available. In the experiments, 3475 finely annotated images are used (2975 for training and 500 for validation images which were used for evaluation), collected from various German cities.

In this paper, the uncertainty metric used is the predictive entropy, which is given by:

$$\hat{\mathbb{H}}\left[y|x\right] = - \sum_c \left(\frac{1}{T} \sum_t p(y = c|x, w_t) \right) \log \left(\frac{1}{T} \sum_t p(y = c|x, w_t) \right) \quad (6)$$

where y is the output variable, c ranges over all the classes, T is the number of stochastic forward pass samples, $p(y = c|x, w_t)$ is the softmax probability of the input x being in class c, and w_t are the model parameters for each of the tth sample.

4.1   OoD Methodology

To evaluate the efficiency of out-of-distribution detection, it is necessary to keep classes out of the training set for proper OoD detection. We follow the leave-out class approach by leaving out a class from the training set and then evaluating the OoD performance using it [43]. The Cityscapes dataset itself also defines some void labels, however, it is not consistent and not sufficient. For that, following [43], we choose the class *pedestrian* and *rider* from the Cityscapes dataset as the leave-out classes added to the void labels predefined by the Cityscapes initially.

Distributional Separational Efficiency
We aim for high certainty on correct predictions and low certainty on incorrect predictions. For that, we assess how well the DNN is able to separate in-distribution from out-of-distribution classes solely depending on the uncertainty estimation. A perfect scenario would be to have a peak at the low certainty for out-of-distribution, whilst having a peak at the high certainty for the in-distribution classes. Accordingly, both distributions are plotted with respect to

certainty, and then compared using the Wasserstein distance metric. A high value indicates dissimilar distinctive distributions and vice versa for a low Wasserstein distance value.

Figure 2 (a) shows the separation for the DNN trained with baseline loss separation both distributions are overlapping and having a peak at the high certainty. This indicates the inefficiency of the network towards distinguishing between in-distribution and out-distribution classes. By applying the Dirichlet loss addition to the baseline loss, results in Fig. 2 (b) shows where both distributions have separate non-overlapping peaks.

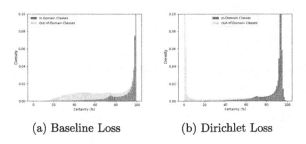

(a) Baseline Loss   (b) Dirichlet Loss

Fig. 2. OoD separation plots

Table 1. Distributional separation efficiency metrics: Wasserstein distance and average uncertainty estimation

	Wasserstein Distance (↑)	Average Uncertainty in %		
		In-Domain (↓)	Out-of-Domain(↑)	Difference(↑)
Baseline Loss	24.4	**5.1**	23.2	22.1
Dirichlet Loss	**67.3**	10.3	**82.8**	**72.4**

Table 1 shows the Wasserstein distance between both distributions. The Dirichlet loss shows great improvement by significantly more than double reflecting the separation efficiency. Furthermore, to better quantify the separation efficiency, the average uncertainty estimation for in-distribution, out-of-distribution and the difference between them have been calculated in Table 1. Moreover, the greater difference between them, the better the separation performance. The separation efficiency in Fig. 2 is also reflected in Table 1 where it can be observed through the difference between average uncertainty for in and out-of-domain. A large difference between both domains indicates efficient separation and reliable uncertainty estimation. Results show that even though the baseline loss has a lower uncertainty on in-distribution classes, the difference between both in-distribution and out-of-distribution is not as high and representative as for the Dirichlet loss.

Segmentation Performance
As seen in Table 2, both DNNs have a nearly similar performance of mean intersection over union (IoU). Moreover, Fig. 5 shows samples of the DNN output representing the effectiveness of the Dirichlet loss. As previously explained, the DNN is trained with having *ignore* labels on the pedestrian and rider pixels. This should push the DNN to an unexpected behavior with high uncertainty on the pedestrian and rider pixels whilst having high segmentation noise which indicates the inability to conclusively segment this area of the image. Figure 5 shows that the baseline DNN is unable to achieve the expected behavior, but interpolates the pedestrian and rider regions and segments them as part of the background. Additionally, there is no uncertainty on the previously assigned ignore regions indicating full confidence of the DNN towards the incorrect interpolated segmentation.

On the contrary, the hypothesis is verified with the Dirichlet loss, as in this case the DNN does generate noisy segmentation on the ignore regions indicating the inability to decisively classify the pixels which is also reflected in high uncertainty. This is because of the DNN's low confidence in its segmentation for this area. Additional black labels in the uncertainty resemble the other predefined void parts in the scene such as the car hood, as seen in the ground-truth segmentation.

Table 2. Mean intersection over union for out-of-domain detection

	Mean IoU % (↑)
Baseline Loss	**69.86**
Dirichlet Loss	69.53

4.2   Full Training with All Classes

Distributional Separation Efficiency
Separation efficiency plots from Fig. 4, show a much better separation between both distributions, where the mean difference between both distributions has increased. Furthermore, the incorrect predictions' distribution shows a much wider spread with a shifted peak more towards the low certainty. Table 3 shows a higher value for the Wasserstein distance between both distributions for the new loss indicating better separation performance. Moreover, the table also shows the average uncertainty for difference for the new loss is also much higher than the baseline loss.

Table 3. Distributional separation efficiency metrics: Wasserstein distance and average uncertainty estimation

	Wasserstein Distance ($\uparrow$)	Average Uncertainty in %		
		In-Domain ($\downarrow$)	Out-of-Domain($\uparrow$)	Difference($\uparrow$)
Baseline loss	23.7	**3.1**	28.8	25.6
Dirichlet Loss	**31.5**	9.4	**72.1**	**62.7**

Accuracy vs. Uncertainty Metrics

An important factor for deep neural networks is not only to be accurate about their predictions but also to be certain about them. Proposed by [25], ratios between accuracy and certainty are defined and quantified to compare between DNNs' uncertainty performances.

After attaining the DNN predictions and their respective uncertainty estimations, they are compared with the ground truth. Afterwards four main components are then calculated: accurate and certain (n_{ac}), accurate and uncertain (n_{au}), inaccurate and certain (n_{ic}), and inaccurate and uncertain (n_{iu}). These metrics are all pixel-based metrics, comparing the ground-truth with the segmentation class and its corresponding estimated uncertainty value. Certainty and uncertainty in these components are based upon varying thresholds, hence their value can be changed to define four different components for each uncertainty threshold. These four components are then used to calculate the three conditional probabilities needed for this evaluation test: $p(\text{accurate}|\text{certain}) = \frac{n_{ac}}{n_{ac}+n_{ic}}$, $p(\text{uncertain}|\text{inaccurate}) = \frac{n_{iu}}{n_{ic}+n_{iu}}$ and $\text{AvU} = \frac{n_{ac}+n_{iu}}{n_{ac}+n_{au}+n_{ic}+n_{iu}}$. AvU stands for accuracy vs. uncertainty, describing the probability of getting a good prediction out of the network either accurate and certain or inaccurate and uncertain. Higher values of each metric for each threshold correspond to better performance.

With the use of varying uncertainty thresholds, a new set of the four fundamental components are being generated altering the values for each metric. Figure 3 shows the three ratios for varying thresholds for both loss functions.

Table 4. Accuracy and Certainty Metrics

	P(Accurate | Certain) in % ($\uparrow$)	P(Uncertain | Inaccurate) in % ($\uparrow$)	A vs. U in % ($\uparrow$)
Baseline Loss	86.8	28.1	81.9
Dirichlet	**92.7**	**65.4**	**84.9**

Table 4 summarizes the plots of Fig. 3 by representing the area under the curve. It can be seen that the Dirichlet loss surpasses the baseline loss in all three metrics. Taking a closer look, we can see that we have a high improvement in the P(Uncertain | Inaccurate) whilst still maintaining high performance on the

P(Accurate | Certain) P(Uncertain | Inaccurate)   AvU

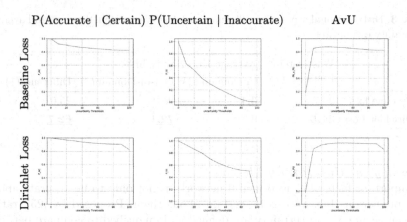

Fig. 3. Accuracy vs Certainty Plot

other two metrics. This is not usually observed as any method trying to improve uncertainty representation would come to a cost of reduced performance on the other two metrics. Relating also to the separation plots in Fig. 4, it can be stated that the Dirichlet loss presents a reliable and representative uncertainty reflecting low uncertainty on accurate and high uncertainty on inaccurate.

Looking more towards safety analysis, we plot here the remaining accuracy rate (RAR = $\frac{AC}{AC+IC+AU+IU}$) vs. remaining error rate (RER= $\frac{IC}{AC+IC+AU+IU}$) [46] in Fig. 4. Plotting both metrics together with varying thresholds provides the ratio between accurate and inaccurate results yielded at each certainty value. This helps in understanding the expected error at any certainty threshold which can be further used in safety-relevant tasks. The best case here would be to have the highest accuracy whilst maintaining the lowest possible error. The red dot in both subfigures in Fig. 4 indicates the best operating certainty threshold to yield the highest accuracy. Table 5 shows that the Dirichlet loss surpasses the baseline loss by providing a higher accuracy value by around 5% whilst simultaneously decreasing the error by almost 4%.

Receiver Operating Characteristic Curve Metrics

The Receiver Operating Characteristic (ROC) is commonly used to evaluate the performance of uncertainty estimation in recent literature [REFs]. The ROC is initially developed to relate between the true positive rate (TPR) = $\frac{TP}{TP+FN}$ and false positive rate (FPR) = $\frac{FP}{FP+TN}$ of binary classifiers. For the case in hand we label the correct as positive and the incorrect as negative [47]. Two common metrics are used to summarize the ROC performance of the classifier: area under ROC (AUROC) and the false positive rate at 95% true positive rate (FPR@0.95TPR). The latter metric shows the probability that a sample will be a negative sample when the true positive rate is as high as 95% [35]. Table 5 shows that Dirichlet loss has a much higher AUROC value than the baseline, whilst also having almost half the FPR@0.95TPR value than the baseline. These metrics reflect the reliability of the uncertainty estimation. This shows that the

Table 5. Performance comparison between both baseline and Dirichlet loss for RAR and RER, AUROC metrics, calibration and segmentation performance.

	RAR & RER		AUROC Metrics		Calibration Metrics		Segmentation Performance
	RAR % ($\uparrow$)	RER % ($\downarrow$)	AUROC % ($\uparrow$)	FPR @ 0.95 TPR % ($\downarrow$)	ECE % ($\downarrow$)	MCE % ($\downarrow$)	Mean IoU % ($\uparrow$)
Baseline Loss	85.72	12.91	87.69	51.33	6.74	17.69	**67.12**
Dirichlet Loss	**90.46**	**8.98**	**96.11**	**20.91**	**2.71**	**13.21**	66.93

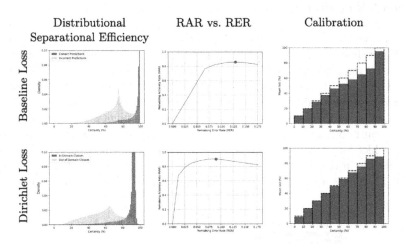

Fig. 4. Distributional separational efficiency, RER vs. RAR and calibration results for the fully trained DNN with all classes.

Dirichlet loss amends the baseline loss function by enabling the DNN to represent its fallacies through the uncertainty estimation.

Calibration Metrics

DNN calibration is an essential aspect for the safety of the ADS. Accordingly, the DNN should be able to provide a calibrated uncertainty measure for its prediction. Calibration metrics used in this work to assess the performance of the DNN are maximum calibration error (MCE) and expected calibration error (ECE) [48,49].

All DNN predictions and uncertainty should be first sorted into fixed-sized bins β_i, $i = 1, \dots, B$ according to their certainty. For each bin β_i, accuracy (acc) and certainty (cert) are computed as: $\mathrm{acc}_i = \frac{\mathrm{TP}_i}{|\beta_i|}$ and $\mathrm{cert}_i = \frac{1}{|\beta_i|} \sum_{j=1}^{|\beta_i|} \hat{c}_i$, where $|\beta_i|$ indicates the number of samples in β_i and $\hat{c}_i$ is the respective certainty. TP_i denotes the number of correctly classified in β_i. The ECE and MCE are then calculated accordingly: $MCE = max|\mathrm{acc}_i - \mathrm{cert}_i|$ and $ACE = \frac{1}{B} \sum_{i=1}^{B} |\mathrm{acc}_i - \mathrm{cert}_i|$.

Figure 4 shows the calibration plots for both losses whilst Table 5 presents the ECE and MCE values. The figure and table reflect the dominance of the

Dirichlet loss over the baseline loss where it has lower ECE and MCE values. Furthermore, it can be seen from the figure that the calibration error up to 90% is almost identical to the target calibration, which is why the Dirichlet loss has a much lower ECE than the baseline loss. Furthermore, it can be seen that Dirichlet loss has a low calibration performance for the last category of accuracy (over 90%). This can also be seen as a reflection of having maximum certainty of about 95% as represented in Fig. 4, where the Dirichlet pushes the DNN to be prudent about its detections, not allowing 100% certainty levels.

Segmentation Performance

Table 5 quantifies the segmentation performance in terms of mean IoU. It can be seen that both loss functions have almost the same mean IoU on the Cityscapes test. Figure 6 shows samples for the segmentation and uncertainty estimations. Both losses resemble almost the same segmentation performance when compared visually however it can be clearly seen that the new Dirichlet loss can represent higher uncertainty values on areas where the DNN was not able to segment the classes correctly. Furthermore, ignore regions in the semantic segmentation labels (such as car hood, and background scenery) can be seen as of high uncertainty in the Dirichlet loss with noisy segmentation, whereas on the contrary, such result is not visible in the baseline loss results. Overall this represents the reliability of the Dirichlet loss approach in better representing areas of incorrect predictions.

Input Image   Baseline   Dirichlet Loss

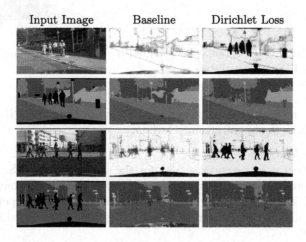

Fig. 5. Out-of-domain detection segmentation samples. Two sample images are being shown: the first column shows both input image and ground truth semantic segmentation, the middle column shows uncertainty estimation and semantic segmentation of baseline loss, and the last column shows uncertainty estimation and semantic segmentation of the Dirichlet loss.

Input Image   Baseline   Dirichlet Loss

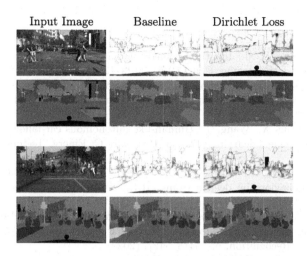

Fig. 6. Full DNN training segmentation and uncertainty samples. Two sample images are being shown: the first column shows both input image and ground truth semantic segmentation, the middle column shows uncertainty estimation and semantic segmentation of baseline loss, and the last column shows uncertainty estimation and semantic segmentation of the Dirichlet loss.

5   Conclusion and Future Work

We have presented in this work an enhanced uncertainty estimation approach by extending the ILVI approach with Dirichlet distributions and tested it on both OoD and general uncertainty estimation of the network. The new proposed approach showed that it resembles the same semantic segmentation performance as the baseline ILVI approach, but yields much more reliable uncertainty estimation and a better calibrated network with an efficient distribution separation efficiency. For future work, it is recommended the use of Dirichlet models solely without the need for cross-entropy, as this possibly would further improve the calibration and the reliability of the DNN's uncertainty estimation. Furthermore, it is recommended to verify the effectiveness of this approach on other tasks such as object detection and other datasets.

Acknowledgments. The research leading to these results is partly funded by the German Federal Ministry for Economic Affairs and Climate Action within the project "Methoden und Maßnahmen zur Absicherung von KI basierten Wahrnehmungsfunktionen für das automatisierte Fahren (KI-Absicherung)". The authors would like to thank the consortium for the successful cooperation.

References

1. Kermany, D.S., et al.: Identifying medical diagnoses and treatable diseases by image-based deep learning. Cell **172**(5), 1122–1131 (2018)
2. Alexander Selvikvåg Lundervold and Arvid Lundervold: An overview of deep learning in medical imaging focusing on MRI. Zeitschrift für Medizinische Physik **29**(2), 102–127 (2019)
3. Zhu, M., Wang, X., Wang, Y.: Human-like autonomous car-following model with deep reinforcement learning. Transp. Res. part C: Emerg. Technol. **97**, 348–368 (2018)
4. Wang, X., Jiang, R., Li, L., Lin, Y., Zheng, X., Wang, F.-Y.: Capturing car-following behaviors by deep learning. IEEE Trans. Intell. Transp. Syst. **19**(3), 910–920 (2017)
5. Wu, X., et al.: Sparse fuse dense: Towards high quality 3D detection with depth completion. In: Proceedings of the IEEE/CVF Conference on Computer Vision and Pattern Recognition, pp. 5418–5427, (2022)
6. Wang, C.-H., Chen, H.-W., Fu, L-C.: VPFNet: voxel-pixel fusion network for multiclass 3D object detection. arXiv preprint arXiv:2111.00966 (2021)
7. Aradi, S.: Survey of deep reinforcement learning for motion planning of autonomous vehicles. IEEE Trans. Intell. Transp. Syst. **23**, 740–759 (2020)
8. Liu, W., et al.: SSD: single shot MultiBox detector. In: Leibe, B., Matas, J., Sebe, N., Welling, M. (eds.) ECCV 2016. LNCS, vol. 9905, pp. 21–37. Springer, Cham (2016). https://doi.org/10.1007/978-3-319-46448-0_2
9. Redmon, J., Farhadi, A.: YOLOv3: an incremental improvement. arXiv preprint arXiv:1804.02767. (2018)
10. Chen, L.-C., Zhu, Y., Papandreou, G., Schroff, F., Adam, H.: Encoder-decoder with atrous separable convolution for semantic image segmentation. In: Proceedings of the European Conference on Computer Vision (ECCV), pp. 801–818 (2018)
11. Wang, J., et al.: Deep high-resolution representation learning for visual recognition. IEEE Trans. Pattern Anal. Mach. Intell. **43**(10), 3349–3364 (2020)
12. Willers, O., Sudholt, S., Raafatnia, S., Abrecht, S.: Safety concerns and mitigation approaches regarding the use of deep learning in safety-critical perception tasks. In: Casimiro, A., Ortmeier, F., Schoitsch, E., Bitsch, F., Ferreira, P. (eds.) SAFECOMP 2020. LNCS, vol. 12235, pp. 336–350. Springer, Cham (2020). https://doi.org/10.1007/978-3-030-55583-2_25
13. Akhtar, N., Mian, A.: Threat of adversarial attacks on deep learning in computer vision: a survey. IEEE Access **6**, 14410–14430 (2018)
14. Ozdag, M.: Adversarial attacks and defenses against deep neural networks: a survey. Procedia Comput. Sci. **140**, 152–161 (2018)
15. Smuha, N.A.: The EU approach to ethics guidelines for trustworthy artificial intelligence. Comput. Law Rev. Int. **20**(4), 97–106 (2019)
16. Dietterich, T.G.: Steps toward robust artificial intelligence. Ai Mag. **38**(3), 3–24 (2017)
17. Kendall, A.: Geometry and uncertainty in deep learning for computer vision. PhD thesis, University of Cambridge, UK (2019)
18. Gal, Y.: Uncertainty in deep learning (2016)
19. Abdar, M., et al.: A review of uncertainty quantification in deep learning: techniques, applications and challenges. Inf. Fusion **76**, 243–297 (2021)
20. Gawlikowski, J., et al.: A survey of uncertainty in deep neural networks. arXiv preprint arXiv:2107.03342 (2021)

21. Hammam, A., Ghobadi, S.E., Bonarens, F., Stiller, C.: Real-time uncertainty estimation based on intermediate layer variational inference. In: Computer Science in Cars Symposium, pp. 1–9 (2021)
22. MacKay, D.J.C.: A practical bayesian framework for backpropagation networks. Neural Comput. **4**(3), 448–472 (1992)
23. Graves, A.: Practical variational inference for neural networks. In: Advances in Neural Information Processing Systems. **24** (2011)
24. Blundell, C., Cornebise, J., Kavukcuoglu, K., Wierstra, D.: Weight uncertainty in neural network. In: International Conference on Machine Learning, pp. 1613–1622. PMLR (2015)
25. Mukhoti, J., Gal, Y.: Evaluating bayesian deep learning methods for semantic segmentation. arXiv preprint arXiv:1811.12709 (2018)
26. Gustafsson, F.K., Danelljan, M., Schon, T.B.: Evaluating scalable bayesian deep learning methods for robust computer vision. In: Proceedings of the IEEE/CVF Conference on Computer Vision and Pattern Recognition Workshops, pp. 318–319 (2020)
27. Gal, Y., Ghahramani, Z.: Dropout as a bayesian approximation: representing model uncertainty in deep learning. In: International Conference on Machine Learning, pp. 1050–1059. PMLR (2016)
28. Lakshminarayanan, B., Pritzel, A., Blundell, C.: Simple and scalable predictive uncertainty estimation using deep ensembles. arXiv preprint arXiv:1612.01474 (2016)
29. Malinin, A., Gales, M.: Predictive uncertainty estimation via prior networks. In: Advances in Neural Information Processing Systems. **31** (2018)
30. Sensoy, M., Kaplan, L., Kandemir, M.: Evidential deep learning to quantify classification uncertainty. In: Advances in Neural Information Processing Systems **31** (2018)
31. Nandy, J., Hsu, W., Lee, M.L.: Towards maximizing the representation gap between in-domain & out-of-distribution examples. Adv. Neural Inf. Process. Syst. **33**, 9239–9250 (2020)
32. Malinin, A., Gales, M.: Reverse KL-divergence training of prior networks: Improved uncertainty and adversarial robustness. In: Advances in Neural Information Processing Systems. **32** (2019)
33. Charpentier, B., Zügner, D., Günnemann, S.: Posterior network: uncertainty estimation without OOD samples via density-based pseudo-counts. Adv. Neural Inf. Process. Syst. **33**, 1356–1367 (2020)
34. Drummond, N., Shearer, R.: The open world assumption. In eSI Workshop: The Closed World of Databases meets the Open World of the Semantic Web, vol. 15, pp. 1 (2006)
35. Hendrycks, D., Gimpel, K.: A baseline for detecting misclassified and out-of-distribution examples in neural networks. arXiv preprint arXiv:1610.02136 (2016)
36. Yang, J., Zhou, K., Li, Y., Liu, Z.: Generalized out-of-distribution detection: a survey. arXiv preprint arXiv:2110.11334 (2021)
37. Nitsch, J., et al.: Out-of-distribution detection for automotive perception. In: 2021 IEEE International Intelligent Transportation Systems Conference (ITSC), pp. 2938–2943. IEEE (2021)
38. Filos, A., Tigkas, P., McAllister, R., Rhinehart, N., Levine, S., Gal, Y.: Can autonomous vehicles identify, recover from, and adapt to distribution shifts? In: International Conference on Machine Learning, pp. 3145–3153. PMLR (2020)
39. Chan, R., et al.: SegmentMeifYouCan: a benchmark for anomaly segmentation. arXiv preprint arXiv:2104.14812 (2021)

40. Vernekar, S., Gaurav, A., Abdelzad, V., Denouden, T., Salay, R., Czarnecki, K.: Out-of-distribution detection in classifiers via generation. arXiv preprint arXiv:1910.04241 (2019)
41. Zhou, D.-W., Ye, H.-J., Zhan, D.-C.: Learning placeholders for open-set recognition. In: Proceedings of the IEEE/CVF Conference on Computer Vision and Pattern Recognition, pp. 4401–4410 (2021)
42. Hendrycks, D., Mazeika, M., Dietterich, T.: Deep anomaly detection with outlier exposure. arXiv preprint arXiv:1812.04606 (2018)
43. Jourdan, N., Rehder, E., Franke, U.: Identification of uncertainty in artificial neural networks. In: Proceedings of the 13th Uni-DAS eV Workshop Fahrerassistenz und automatisiertes Fahren. vol. 2, pp. 12 (2020)
44. Vyas, A., Jammalamadaka, N., Zhu, X., Das, D., Kaul, B., Willke, T.L.: Out-of-distribution detection using an ensemble of self supervised leave-out classifiers. In: Proceedings of the European Conference on Computer Vision (ECCV), pp. 550–564 (2018)
45. Cordts, M., et al.: The cityscapes dataset for semantic urban scene understanding. In: Proceedings of the IEEE Conference on Computer Vision and Pattern Recognition, pp. 3213–3223 (2016)
46. Henne, M., Schwaiger, A., Roscher, K., Weiss, G.: Benchmarking uncertainty estimation methods for deep learning with safety-related metrics. In: SafeAI@ AAAI, pp. 83–90 (2020)
47. Liang, S., Li, Y., Srikant, R.: Enhancing the reliability of out-of-distribution image detection in neural networks. arXiv preprint arXiv:1706.02690 (2017)
48. Neumann, L., Zisserman, A., Vedaldi.: Relaxed softmax: efficient confidence auto-calibration for safe pedestrian detection (2018)
49. Naeini, M.P., Cooper, G., Hauskrecht, M.: Obtaining well calibrated probabilities using bayesian binning. In: Twenty-Ninth AAAI Conference on Artificial Intelligence (2015)

Explainable Sparse Attention for Memory-Based Trajectory Predictors

Francesco Marchetti⬤, Federico Becattini$^{(\boxtimes)}$⬤, Lorenzo Seidenari⬤,
and Alberto Del Bimbo⬤

Università degli Studi di Firenze, Florence, Italy
{francesco.marchetti,federico.becattini,lorenzo.seidenari,
alberto.bimbo}@unifi.it

Abstract. In this paper we address the problem of trajectory prediction, focusing on memory-based models. Such methods are trained to collect a set of useful samples that can be retrieved and used at test time to condition predictions. We propose Explainable Sparse Attention (ESA), a module that can be seamlessly plugged-in into several existing memory-based state of the art predictors. ESA generates a sparse attention in memory, thus selecting a small subset of memory entries that are relevant for the observed trajectory. This enables an explanation of the model's predictions with reference to previously observed training samples. Furthermore, we demonstrate significant improvements on three trajectory prediction datasets.

Keywords: Trajectory prediction · Memory networks · Explainability · Attention

1 Introduction

Decision-making in autonomous vehicles is often hard to explain and interpret due to the black-box nature of deep learning models, which are commonly deployed on self-driving cars. This makes it hard to assess responsibilities in case of accidents or anomalous behaviors and thus poses an obstacle in ensuring safety.

In this paper we focus on the task of trajectory prediction, which is a core component dedicated to safety in an autonomous driving system. In particular, we study trajectory prediction methods based on Memory Augmented Neural Networks (MANN) [32–34,54], which recently have obtained remarkable results. What makes these models interesting is the capacity to offer a certain degree of explainability about the predictions. Memory-based trajectory predictors leverage a storage of past trajectory representations to obtain cues about previously observed futures [32,33], likely endpoint goals [54] or information about the social context [34]. Methods such as MANTRA [32,33] and MemoNet [54] create a persistent memory with relevant observations corresponding to training samples. How memory is accessed plays a pivotal role in the effectiveness of the model and implies a form of attention to select relevant memory cells. Traditionally, MANNs generate a read key which is compared with all the elements stored in memory. In such a way, individual samples can be retrieved at inference time to condition predictions based on known motion patterns. This realizes an effective access-by-content mechanism but does not allow multiple elements to jointly concur to produce an output.

At the same time, each output can be explained by attributing a certain relevance to memory samples. Thus, selecting a small subset of samples from memory provides a way of explaining a decision, since the exact training samples that lead to a certain prediction can be identified. In this paper we propose an improved memory controller that we used to read relevant samples. Our controller is based on sparse attention between the read key and stored samples. Differently from prior work, this allows the model to combine cues from different samples, considerably improving the effectiveness of the predictor. By forcing the attention to be sparse, we can inspect the model decisions with reference to a restricted number of sample, thus making the model explainable. For this reason we refer to our proposed method as ESA (Explainable Sparse Attention).

The main contributions of this paper are the following:

- We present ESA, a novel addressing mechanism to enhance memory-based trajectory predictors using sparse attentions. This enables a global reasoning involving potentially every sample in memory yet focusing only on relevant instances. The advantages are twofold: at training time it reduces redundancies in memory and at test time it yields significant improvements in trajectory prediction benchmarks.
- To address the multimodal nature of the task, we predict multiple futures using a multi-head controller that attends in different ways to the samples in memory.
- To leverage information from all memory entries and at the same time exploit cues from a limited set of training samples, we enhance the memory controller with a sparsemax activation function [35] instead of a softmax. This allows the model to condition predictions on a linear combination of a restricted subset of stored samples.
- Memory Augmented Neural Networks offer explainability by design. We explore how future predictions can be explained with reference to stored samples and how sparsemax further improves the quality of the explanations.

2 Related Works

2.1 Trajectory Prediction

The task of trajectory prediction can be formalized as the problem of calculating the future state of an object, given the observable past states of that same object plus additional observable variables such as the surrounding map and other agents' past states. The main source of knowledge resides in past agents' motion [1,15]. Moreover, an accurate representation of the environment is often sought to provide physical constraints and obtain physically correct predictions [5,6,25,47,49]. Finally, a lot of effort has gone into modelling the interaction between moving agents, addressing so-called social dynamics [1,18,20,21,25,30,44,55].

Trajectory prediction may address pedestrian-only scenarios as well as auto-motive settings. For the latter a model of the road, such as lane configuration, is critical [2,5,6,10]. Indeed, road layouts physically constrain the motion of vehicles. Especially when dealing with pedestrian motion, a correct model of social interaction allows to improve future trajectory accuracy. When looking at the autonomous driving scenario, for which most of the trajectories to be forecasted are from vehicles, there is no clear evidence that social interaction modelling is beneficial [6,25]. Indeed, a vast number of approaches focused on modelling pedestrian interactions [1,18,20,21,28,38,45]. An effective and natu-ral approach to social interaction modelling is applying Graph Neural Networks (GNN) [22,36,48], modelling agents as nodes.

Recently, a different take to trajectory forecasting has gained traction. Instead of regressing a sequence of future steps, goal based methods estimate intentions, i.e. the spatial goal the agent seeks [9,19,31,54,56]. These approaches proved a high effectiveness attaining state-of-the-art results in many benchmarks.

2.2 Memory and Attention

Traditional Neural Networks have been recently augmented with Memory Mod-ules yielding a new class of methods: Memory Augmented Neural Networks (MANN) [16,53]. By augmenting a NN with a memory we enable the capa-bility to retain a state, similarly to Recurrent Neural Networks (RNN) but with more flexibility. Differently from RNNs a MANN will rely on an external address-able memory instead of exploiting a latent state. This is beneficial in terms of explainability since it is easier to establish a correspondence between memorized features and inputs. Moreover, external memories can retain information during the whole training, making it possible to learn rare samples and deal with long-tail phenomena. The Neural Turing Machine (NTM) is the first known instance of a MANN, demonstrating the capability to retain information and perform reasoning on knowledge stored in memory. For this tasks NTMs are superior to RNNs. NTMS have been recently extended and improved [17,46,50,53]. As we will show in this work MANNs are extremely flexible and can address a large vari-ety of problems: person re-identification [39], online learning [40], visual question answering [24,29] and garment recommendation [8,13].

MANNs have been also proposed to perform trajectory forecasting. MANTRA is a MANN specifically developed to perform multiple trajectory prediction [32,33]. Trajectories are encoded with recurrent units and the exter-nal memory is populated during training. At inference time the key-value store of the memory is elegantly exploited to obtain multiple predictions out of a sin-gle input past, simply by addressing multiple features in memory. Recently, an approach from the same authors, exploited an external differentiable memory module to learn social rules and perform joint predictions [34]. In this case the memory is emptied at each episode and controllers are trained to store relevant features for each agent, allowing the model to exploit the social interactions of observed agents.

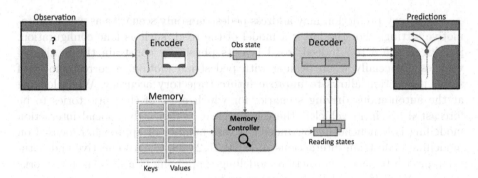

Fig. 1. General architecture of memory-based trajectory predictors.

Transformer based architectures have shown remarkable performance on sequence to sequence problems [12,52]. Attention based approaches are especially interesting since they allow to easily provide explainable outputs. In trajectory forecasting transformers were applied successfully using both their original and bidirectional variant [15]. Obtaining explainable outputs is critical, especially for autonomous driving systems. Recently, using discrete choice models has shown potential to derive interpretable predictions for trajectory forecasting [23].

3 Memory-Based Trajectory Predictors

Memory-based trajectory predictors are a class of trajectory forecasting models based on Memory-Augmented Neural Networks (MANN). The main idea is that, during a training phase, the network learns to build a knowledge base from the training samples. To this end, a controller is trained to store relevant samples in an external memory. The memory can then be accessed at inference time to retrieve relevant information to forecast the future of a given observation. The success of this approach lies in the fact that the model can match a past trajectory with multiple memory entries and obtain cues about possible outcomes of previously observed similar patterns. The actual prediction of the model can be conditioned on this recalled information to leverage additional knowledge rather than the observed sample alone.

Several variants of memory-based trajectory predictors have been proposed in literature [32,33,54]. All these methods share the same structure: (i) first, an encoder learns meaningful representations of the input data; (ii) a writing controller is trained to store relevant and non redundant samples in memory; (iii) a reading controller retrieves relevant information; (iv) finally, a decoder translates the encoded data into a future trajectory. In order to read meaningful cues to inform the decoding process, memory banks are always treated as an associative storage divided into keys and values. Existing methods [32,33,54] mainly differ in the kind of data that memory keys and values represent. Keys, however, must share a common feature space with the data fed as input to

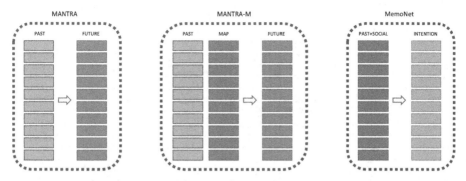

Fig. 2. Memory content for MANTRA [32], MANTRA-M [33] and MemoNet [54]. Memories are divided into keys (left) and values (right).

the model. In fact, at inference time, the current observations is encoded into a latent state which is used to access memory via a similarity function. To comply with the multimodal nature of the trajectory forecasting task, memory-based trajectory predictors can generate multiple futures by retrieving K diverse observations from memory to condition the decoder in different ways. A general scheme of a memory-based trajectory predictor is shown in Fig. 1.

In this work we consider three different models:

- MANTRA [32] was the first to propose a memory-based trajectory predictor. Memory is populated using a writing controller which decides whether or not to store individual samples based on their usefulness for the prediction task. Memory keys are encodings of past trajectories, while memory values are encodings of the respective futures.
- The MANTRA model has been improved in [33] to include contextual information from the surrounding environment. The model now stores as memory keys both encodings of the past and of the semantic segmentation of the surrounding map. We refer to this model as MANTRA-M.
- MemoNet [54] adds a social encoder to produce the latent features, thus allowing joint predictions for multiple agents and relies on a different memory structure. Memory keys are still represented by past trajectories but memory values represent intentions, i.e. final goals where the agent may be directed to. In addition, MemoNet uses a trainable reading controller and a clustering-based decoding process to improve the diversity of the predictions.

Overall, whereas the three models share the same structure, they differ in the information that they store in memory. Most importantly MANTRA and MANTRA-M use the memory bank to inform the decoder with whole future trajectories while MemoNet informs the decoder with intentions. Figure 2 summarizes what kind of information is stored in memory for each model.

All the models access memory through a similarity function: cosine similarity for MANTRA and MANTRA-M and a learnable addresser for MemoNet. It is important to underline that all the models use their similarity function to retrieve

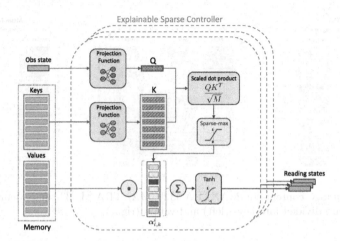

Fig. 3. Explainable Sparse Attention. Each head of the controller attends to memory samples in different ways to generate a prediction.

the top-K samples, which are used to individually generate different futures. In this paper, we propose to remove the top-K mechanism in favor of a multi-head sparse attention that combines information from all memory samples at once.

4   Method

In this paper we propose a novel reading controller for memory-based trajectory predictors named Explainable Sparse Attention (ESA). ESA is divided into multiple read heads. Heads are dedicated to extracting different information from memory, which will then be decoded in parallel into multiple diverse futures. Each head is fed with the encoding of the current observation. A projection layer maps the input into a latent query vector $Q \in \mathcal{R}^M$. Similarly, the same projection is applied to each memory key to obtain a key matrix $K \in \mathcal{R}^{N \times M}$, where N is the number of samples in memory and M the dimension of the projection space. The query Q and the key matrix K are compared using a scaled dot product, followed by a sparsemax activation function [35] to obtain attention weights over memory locations:

$$\alpha_i = \text{sparsemax}\left(\frac{Q_i K_i^T}{\sqrt{M}}\right) \tag{1}$$

The final reading state is a weighted sum of the memory values V with the attention scores $\alpha = \{\alpha_0, ..., \alpha_{N-1}\}$ followed by a tanh activation to regularize the output reading state r_s:

$$r_s = \tanh\left(\sum_i \alpha_i V_i\right) \tag{2}$$

Each head of the ESA controller learns different projections of queries and keys, thus learning different ways to attend to the samples in memory. An overview of the ESA controller is shown in Fig. 3.

Differently from prior work, the ESA controller outputs a reading state that can potentially depend on the whole memory content. This is an advantage since it allows the model to achieve better generalization capabilities by leveraging future information from multiple samples instead of just one. This makes the predictions also more robust to outliers or corrupted samples that might poison the memory bank and condition the output towards wrong predictions.

4.1 Sparsemax vs Softmax

The ESA controller uses a sparsemax activation function instead of the more common softmax. Sparsemax is an activation function that is equivalent to the Euclidean projection of the input vector onto the probability simplex, i.e., the space of points representing a probability distribution between mutually exclusive categories:

$$sparsemax(\mathbf{z}) = \underset{\mathbf{p} \in \Delta^{K-1}}{\mathrm{argmin}} \|\mathbf{p} - \mathbf{z}\|^2, \tag{3}$$

where $\mathbf{z} \in \mathbb{R}^K$ and $\Delta^{K-1} := \{\mathbf{p} \in \mathbb{R}^K | \mathbf{1}^\top \mathbf{p} = 1, \mathbf{p} \geq 0\}$ is a $(K-1)$-dimensional simplex.

It has been shown that, when the input is projected, it is likely to hit the boundary of the simplex, making the output sparse [35]. In addition to producing sparse activations, sparsemax shares most properties of softmax: by definition, the output is a probability distribution; relative ordering between input elements is maintained; the output is invariant to constant addition; differentiability.

For memory-based trajectory predictors that perform a top-K sample retrieval, softmax or sparsemax can be exchanged without affecting the output, since only ordering is important rather than the actual attention values. The usage of sparsemax in the attention mechanism of ESA, however, affects the model by changing the importance of individual samples in the decoding process. There are two important considerations to be made. First, the decoder will receive a reading state with less noise. Being r_s a linear combination of all memory samples, zeroing out the coefficients of most samples, will allow the decoder to focus only on elements that are indeed relevant to the current prediction. Second, attending only to a small subset of elements enables a better model explainability. Attended samples can in fact be used to interpret why the predictor produces its outputs.

5 Experiments

We demonstrate the effectiveness of the ESA controller by experimenting with MANTRA [32], MANTRA-M [33] and MemoNet [54]. We compare the original models against our improved version with the ESA controller. We use the original experimental setting for each model, testing the models on different trajectory prediction datasets. We first provide an overview of the experimental setting, including datasets and metrics, and we then evaluate the model both quantitatively and qualitatively.

5.1   Evaluation Metrics and Datasets

We report results using two common metrics for vehicle trajectory prediction: *Average Displacement Error* (ADE) and *Final Displacement Error* (FDE). ADE is the average L2 error between all future timesteps and FDE is the error at the final timestep. ADE indicates the overall correctness of a prediction, while FDE quantifies the quality of a prediction at a given future horizon. Following recent literature [25,34,54], we take the best out of K predictions to account for the intrinsic multimodality of the task. To evaluate our models we use the following datasets:

KITTI [14] The dataset consists of hours of navigation in real-world road traffic scenarios. Object bounding boxes, tracks, calibrations, depths and IMU data were acquired through Velodyne laser scanner, GPS localization system and stereo camera Rig. From these data the trajectories of the vehicles were extracted and divided into scenarios of fixed length. For the evaluation phase, we considered the split used in [25,32]. Each example has a total duration of 6 s where the past trajectory is 2 s long and the future trajectory 4 s. The train dataset contains 8613 examples while the test dataset 2907.

Argoverse [6] This dataset is composed by 325k vehicle trajectories acquired in an area of $1000\,km^2$ in the cities of Pittsburgh and Miami. In addition to the trajectories, HD maps containing lane centerlines, traffic direction, ground height and drivable areas are available. Each example has a duration of 5 s, 2 s for the past and 3 s for the future. The dataset is split into train, validation and test. We report results on the validation set v1.1, for which ground truth data is publicly available.

SDD [42] The Stanford Drone Dataset is composed of pedestrians and bicycles trajectories acquired by a bird's eye view drone at 2.5 Hz on a university campus. The split commonly adopted by other state-of-the-art methods (Trajnet challenge [43]) was used for the experiments. The dataset size is 14k scenarios where each trajectory is expressed in pixels. Each example is divided into past trajectories of 3.2 s and future trajectories of 4.8.

5.2   Results

Table 1, Table 2 and Table 3 show the results obtained on the KITTI [14], Argoverse [6] and SDD [42] datasets respectively. For KITTI, we added the ESA controller on MANTRA [32]. The same procedure was done for the Argoverse dataset, where we have used the MANTRA-M model [33], which leverages both trajectory and map information. We have used MemoNet [54] for demonstrating the capabilities of the ESA controller in the SDD dataset. We refer to the three enhanced models as MANTRA+ESA, MANTRA-M+ESA and MemoNet+ESA.

Table 1. Results on the KITTI dataset. ESA leads to considerable improvements against the standard version of MANTRA [32] varying the number of predictions K.

	Method	ADE				FDE			
		1 s	2 s	3 s	4 s	1 s	2 s	3 s	4 s
K = 1	Kalman [32]	0.51	1.14	1.99	3.03	0.97	2.54	4.71	7.41
	Linear [32]	0.20	0.49	0.96	1.64	0.40	1.18	2.56	4.73
	MLP [32]	**0.20**	**0.49**	0.93	1.53	**0.40**	1.17	2.39	4.12
	DESIRE [25]	-	-	-	-	0.51	1.44	2.76	4.45
	MANTRA [32]	0.24	0.57	1.08	1.78	0.44	1.34	2.79	4.83
	MANTRA+ESA	0.24	0.50	**0.91**	**1.48**	0.41	**1.13**	**2.30**	**4.01**
K = 5	SynthTraj [3]	0.22	0.38	0.59	0.89	0.35	0.73	1.29	2.27
	DESIRE [25]	-	-	-	-	**0.28**	0.67	1.22	**2.06**
	MANTRA [32]	**0.17**	0.36	0.61	0.94	0.30	0.75	1.43	2.48
	MANTRA+ESA	0.21	**0.35**	**0.55**	**0.83**	0.31	**0.66**	**1.20**	2.11
K = 20	DESIRE [25]	-	-	-	-	-	-	-	2.04
	MANTRA [32]	**0.16**	0.27	0.40	0.59	0.25	0.49	0.83	1.49
	MANTRA+ESA	0.17	0.27	**0.38**	**0.56**	**0.24**	**0.47**	**0.76**	**1.43**

Table 2. Results on Argoverse varying the number of predictions K. Errors in meters.

	Method	ADE		FDE		Off-road (%)	Memory size
		1 s	3 s	1 s	3 s		
K = 1	MANTRA-M [33]	0.72	2.36	1.25	5.31	**1.62%**	75,424
	MANTRA-M+ESA	**0.58**	**1.76**	**0.96**	**3.95**	1.84%	**9,701**
K = 6	MANTRA-M [33]	0.56	1.22	0.84	2.30	3.27%	12,467
	MANTRA-M+ESA	**0.47**	**0.93**	**0.68**	**1.57**	**2.32%**	**2,337**
K = 10	MANTRA-M [33]	0.53	1.00	0.77	1.69	4.17%	6,566
	MANTRA-M+ESA	**0.44**	**0.80**	**0.63**	**1.20**	**2.98%**	**1,799**
K = 20	MANTRA-M [33]	0.52	0.84	0.73	1.16	7.93%	2,921
	MANTRA-M+ESA	**0.45**	**0.73**	**0.65**	**0.88**	**3.14%**	**1,085**

In Table 1 we can observe that MANTRA+ESA significantly lowers the prediction error compared to its MANTRA counterpart, especially for a small number of predictions and for long term prediction horizons (4 s). Indeed, for the single prediction case (K = 1), FDE at 4s decreases from 4.83 m to 4.01 m, with an improvement of 0.82 m (17.18%). With a higher number of predictions (K = 5), the FDE error drops from 2.48 m to 2.11 m with a reduction of 0.37 m (14.91%). In almost all metrics, MANTRA+ESA achieves state-of-the-art results. Similarly, we observe significant improvements for MANTRA-M+ESA on the Argoverse dataset, compared to its original formulation [33] (Table 2). We report gains up to 1.36 m in the K = 1 FDE error at 4 s (25.61% improvement). An even larger relative error decrement is reported for K = 6 with an improvement of 31.74%. Similar considerations can be drawn for the other evaluation settings.

Moreover, we show that the amount of information that the network saves in memory with the ESA controller is drastically reduced. In Table 2, we can observe that memory size of MANTRA-M+ESA is significantly smaller than that of the original model, with a difference of 81.25% (10,130 elements). The

552   F. Marchetti et al.

Table 3. Results on SDD varying the number of predictions K. Errors in pixels.

K = 20						
Method	ADE	FDE	Method	ADE	FDE	
Social-STGCNN [36]	20.60	33.10	SimAug [27]	10.27	19.71	
Trajectron++ [45]	19.30	32.70	MANTRA [32]	8.96	17.76	
SoPhie [44]	16.27	29.38	PCCSNet [51]	8.62	16.16	
NMMP [36]	14.67	26.72	PECNet [31]	9.96	15.88	
EvolveGraph [26]	13.90	22.90	LB-EBM [37]	8.87	15.61	
EvolveGraph [26]	13.90	22.90	Expert-Goals [19]	**7.69**	14.38	
CF-VAE [4]	12.60	22.30	SMEMO [34]	8.11	13.06	
Goal-GAN [9]	12.20	22.10	MemoNet [54]	8.56	**12.66**	
P2TIRL [11]	12.58	22.07	MemoNet+ESA	8.02	12.97	

K = 5		
Method	ADE	FDE
DESIRE [25]	19.25	34.05
Ridel et al. [41]	14.92	27.97
MANTRA [32]	13.51	27.34
PECNet [31]	12.79	25.98
PCCSNet [51]	12.54	-
TNT [56]	12.23	21.16
SMEMO [34]	**11.64**	**21.12**
MemoNet [54]	13.92	27.18
MemoNet+ESA	12.21	23.03

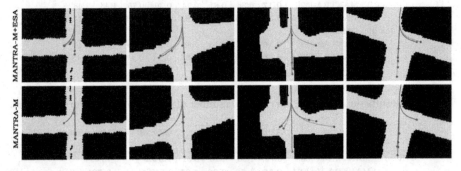

Fig. 4. Comparison between MANTRA-M+ESA and MANTRA-M [33] on Argoverse. Past trajectories are in blue, ground-truth in green and predictions in red. The gray region represents the drivable area of the map. (Color figure online)

ESA controller, thanks to its sparse attention and learned weighted combination of future states read from memory, is able to further reduce redundancy and space occupancy, as well as guaranteeing better performance. In addition, we also have a relevant improvement in the number of generated predictions that do not go off-road. Interestingly, with 20 predictions, we manage to reduce the number trajectories that go astray by half (7.93% vs 3.14%). As we can see from the qualitative examples in Fig. 4 and Fig. 5, the ESA controller allows to generate trajectories with a better multi-modality than the classic MANTRA versions. This yields a lower error and demonstrates the importance of having a model that is able to explore all the plausible directions and speeds, given a certain past movement and context.

In the experiment with the SDD dataset (Table 3), we can observe large gains for K = 5. For K = 20 the results generated by Memonet+ESA are similar to the ones obtained with the original MemoNet formulation, with an improvement for ADE and a slight drop for FDE when predicting 20 different futures. Nonetheless, using the ESA controller, we are able to reduce significantly the memory footprint. Indeed, in Table 4, we can observe that with only 7104 elements in

Fig. 5. Comparison between MANTRA+ESA and MANTRA on KITTI. Past trajectory is in blue, ground-truth in green and predictions in red. (Color figure online)

Table 4. Comparison of the results as the ratio of the Memonet thresholds varies with Memonet+ESA.

MemoNet	$\theta_{past}/\theta_{int}$	ADE/FDE	Memory size
w/o ESA	0	8.65/12.84	17970 (100.0%)
	0.5	8.59/12.70	15442 (85.9%)
	1	8.56/12.66	14652 (81.5%)
	5	9.22/14.29	10698 (59.5%)
	10	9.64/15.57	6635 (36.9%)
w/ ESA	-	8.02/12.97	7104 (39.5%)

Fig. 6. MemoNet+ESA and MemoNet comparison on SDD. Blue: past; green: ground-truth; red: predictions. (Color figure online)

memory we can reach similar results to MemoNet, which instead requires 14,652 elements in memory (46% difference).

In MemoNet, memory is initialized by writing all past and intention features available in the training data and then a filtering algorithm erases redundant memory instances. The algorithm removes all those elements with similar starting and ending points. The metric used is the L2-norm and the proximity threshold is determined by two configurable parameters, θ_{past} and θ_{int}, related to starting point and destination distance respectively. In our MemoNet+ESA model we retain the classic memory controller used with key-value memory augmented networks [7,13,32,33], which directly optimizes redundancies with a task loss and does not require memory filtering. Our final memory has a size of 7104 samples. In MemoNet instead, as the ratio of the filtering algorithm thresholds varies, the results change significantly. With a memory size similar to ours ($\theta_{past}/\theta_{int} = 10$, memory size of 6635) FDE is about 17% lower. At the same time, MemoNet requires twice the samples in memory compared to MemoNet+ESA to obtain a comparable error rate. In Fig. 6 we show a qualitative comparison on SDD between MemoNet and MemoNet+ESA.

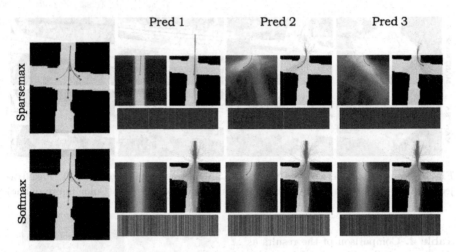

Fig. 7. Explainability analysis on Argoverse. The first column contains the predictions generated with sparsemax and softmax. For each prediction we show the attention vector over memory locations (bottom) and we plot the sum of the semantic maps (left) and the trajectories (right), both weighed by ESA attention.

5.3   Explainability

Providing explainable outputs is a fundamental and crucial aspect for autonomous driving. Since predictors are to be deployed in safety-critical systems, what causes a behavior must be interpretable and observable. One of the advantages of using memory-based trajectory predictors is to explicitly have a link between the memory features used to generate the prediction and the associated training samples. Indeed, by design we can identify the specific samples that allow the generation of a given future trajectory. This in general is not possible with a neural network, where knowledge is distilled in its weights during training. In our work, thanks to the ESA controller, we improve the quality of the explanations. Thanks to the sparsemax activation, each future feature fed to the decoder is a linear combination of a small but significant subset of memory elements. In fact, sparsemax allows the model to generate sparse attentions over the memory. On the contrary, with a softmax activation function, the attention would be smoother and identifying individual sample responsibilities would be harder.

By using the sparsemax activation, we want there to be an evident cause-effect relationship between what is read from memory and the generated output. The quality of the explanation however depends on the architecture of the prediction and in particular on what kind of information is stored in memory. In the following we show how predictions can be interpreted using the MANTRA-M+ESA and MemoNet+ESA models. In Fig. 7 we show an example from the Argoverse dataset, comparing the usage of sparsemax and softmax in the MANTRA-M+ESA model. In the figure, we show attention over memory cells, as well as past, future and semantic maps of the samples retrieved from memory. For each prediction, we represent the attention as a heatmap, normal-

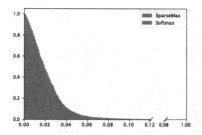

Fig. 8. Normalized histograms of attention values, binned in 0.001-width intervals (x-axis).

Table 5. Quantative analysis of attention values generated by the ESA controller with softmax and sparsemax

Att. Values	Softmax	Sparsemax
>0 (%)	100%	2.37%
>0 (mean)	2328	55
Max	0.21	0.47
Mean	0.0004	0.0180

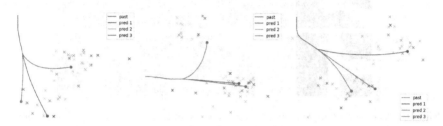

Fig. 9. Explainability for MemoNet+ESA. X-dots are attention-weighted memory intentions associated with predictions of the same color.

ized by its maximum value. We also plot directly on the map all past and future trajectories in memory, weighing their intensity with the respective normalized attention value. In a similar way, we show a semantic heatmap generated by a weighted sum of all maps. Interestingly, the predictions generated by the model using the two activations are very similar. The substantial difference lies in which memory samples are used to generate such predictions. With sparsemax, most attention values are equal to 0 (blue lines in the attention heatmap). Maps and trajectories corresponding to positive attentions can be interpreted as a scenario consistent with the generated prediction. Instead, using softmax, we have a soft attention vector and no element in memory is clearly identifiable as responsible of the prediction. The semantic heatmaps appear similar for all the futures.

We can make a quantitative analysis of this behavior. In Table 5 we report the average number of attention values greater than 0, the maximum attention value and the average attention value. Using sparsemax, only 2.37% of the attention values is positive with an average of 55 elements for each example. On the other hand, with softmax, all memory attention values are positive, the maximum attention value is halved and on average attention values are 45 times lower. This demonstrates that sparsemax allows the model to focus only on relevant memory elements, thus providing interpretable insights about the model's behavior. The same conclusion can be drawn by looking at Fig. 8. We show the normalized histogram of attention values, binned in intervals of width 0.001. Attention values with softmax concentrate close to 0, while with sparsemax are more spreaded.

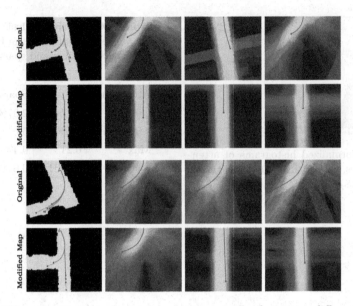

Fig. 10. When changing the observed map, the controller focuses on different memory samples. Past trajectory is in blue, ground-truth in green and the predictions in red. (Color figure online)

A similar analysis can be done for MemoNet+ESA, which stores intentions as endpoint coordinates instead of future trajectories and maps. In Fig. 9, we show for each prediction the intentions retrieved from memory, weighed by attention. The final position of the generated trajectories is always in a neighborhood of the intentions considered to be relevant by the ESA controller.

In addition, to verify the robustness of the model and its explainability, we perform an ablation study on MANTRA-M+ESA. We manually perturb the input and observe how this affects both the predictions and the explainability. In particular, we change the feature of the semantic map, leaving the past unchanged and observe which are the elements in memory that the model focuses on. As we can observe in Fig. 10, different trajectories are generated which are coherent with the new map. The ESA controller also focuses on memory instances that are related to the new semantic map.

6   Conclusions

We proposed ESA a novel reading controller based on explainable sparse attention for Memory-based Trajectory Predictors. Differently from the prior work, ESA allows to generate predictions based on different combinations of the elements in memory, leading to better generalization and robustness. Furthermore, thanks to the sparsemax activation function, it is possible to identify a small subset of samples relevant to generate the output. We tested ESA on top of state of the art Memory-based Trajectory Predictors obtaining considerable improvements and demonstrated the explainability of the predictions.

Acknowledgements. This work was supported by the European Commission under European Horizon 2020 Programme, grant number 951911 - AI4Media.

References

1. Alahi, A., Goel, K., Ramanathan, V., Robicquet, A., Fei-Fei, L., Savarese, S.: Social LSTM: human trajectory prediction in crowded spaces. In: Proceedings of the IEEE Conference on Computer Vision and Pattern Recognition, pp. 961–971 (2016)
2. Berlincioni, L., Becattini, F., Galteri, L., Seidenari, L., Bimbo, A.D.: Road layout understanding by generative adversarial inpainting. In: Escalera, S., Ayache, S., Wan, J., Madadi, M., Güçlü, U., Baró, X. (eds.) Inpainting and Denoising Challenges. TSSCML, pp. 111–128. Springer, Cham (2019). https://doi.org/10.1007/978-3-030-25614-2_10
3. Berlincioni, L., Becattini, F., Seidenari, L., Del Bimbo, A.: Multiple future prediction leveraging synthetic trajectories (2020)
4. Bhattacharyya, A., Hanselmann, M., Fritz, M., Schiele, B., Straehle, C.N.: Conditional flow variational autoencoders for structured sequence prediction (2020)
5. Caesar, H., et al.: nuScenes: a multimodal dataset for autonomous driving. In: Proceedings of the IEEE/CVF Conference on Computer Vision and Pattern Recognition, pp. 11621–11631 (2020)
6. Chang, M.F., et al.: Argoverse: 3D tracking and forecasting with rich maps. In: Proceedings of the IEEE Conference on Computer Vision and Pattern Recognition, pp. 8748–8757 (2019)
7. De Divitiis, L., Becattini, F., Baecchi, C., Bimbo, A.D.: Disentangling features for fashion recommendation. ACM Trans. Multimedia Comput. Commun. Appl. (TOMM) (2022)
8. De Divitiis, L., Becattini, F., Baecchi, C., Del Bimbo, A.: Style-based outfit recommendation. In: 2021 International Conference on Content-Based Multimedia Indexing (CBMI), pp. 1–4. IEEE (2021)
9. Dendorfer, P., Osep, A., Leal-Taixe, L.: Goal-GAN: multimodal trajectory prediction based on goal position estimation. In: Proceedings of the Asian Conference on Computer Vision (ACCV) (2020)
10. Deo, N., Trivedi, M.M.: Multi-modal trajectory prediction of surrounding vehicles with maneuver based LSTMS. In: 2018 IEEE Intelligent Vehicles Symposium (IV), pp. 1179–1184. IEEE (2018)
11. Deo, N., Trivedi, M.M.: Trajectory forecasts in unknown environments conditioned on grid-based plans. arXiv preprint arXiv:2001.00735 (2020)
12. Devlin, J., Chang, M.W., Lee, K., Toutanova, K.: Bert: pre-training of deep bidirectional transformers for language understanding. arXiv preprint arXiv:1810.04805 (2018)
13. De Divitiis, L., Becattini, F., Baecchi, C., Del Bimbo, A.: Garment recommendation with memory augmented neural networks. In: Del Bimbo, A., et al. (eds.) ICPR 2021. LNCS, vol. 12662, pp. 282–295. Springer, Cham (2021). https://doi.org/10.1007/978-3-030-68790-8_23

14. Geiger, A., Lenz, P., Urtasun, R.: Are we ready for autonomous driving? The kitti vision benchmark suite. In: 2012 IEEE Conference on Computer Vision and Pattern Recognition, pp. 3354–3361. IEEE (2012)
15. Giuliari, F., Hasan, I., Cristani, M., Galasso, F.: Transformer networks for trajectory forecasting. In: 2020 25th International Conference on Pattern Recognition (ICPR), pp. 10335–10342. IEEE (2021)
16. Graves, A., Wayne, G., Danihelka, I.: Neural turing machines. arXiv preprint arXiv:1410.5401 (2014)
17. Graves, A., et al.: Hybrid computing using a neural network with dynamic external memory. Nature **538**(7626), 471–476 (2016)
18. Gupta, A., Johnson, J., Fei-Fei, L., Savarese, S., Alahi, A.: Social GAN: socially acceptable trajectories with generative adversarial networks. In: Proceedings of the IEEE Conference on Computer Vision and Pattern Recognition, pp. 2255–2264 (2018)
19. He, Z., Wildes, R.P.: Where are you heading? Dynamic trajectory prediction with expert goal examples. In: Proceedings of the International Conference on Computer Vision (ICCV) (2021)
20. Helbing, D., Molnar, P.: Social force model for pedestrian dynamics. Phys. Rev. E **51**(5), 4282 (1995)
21. Ivanovic, B., Pavone, M.: The trajectron: probabilistic multi-agent trajectory modeling with dynamic spatiotemporal graphs. In: Proceedings of the IEEE International Conference on Computer Vision, pp. 2375–2384 (2019)
22. Kosaraju, V., Sadeghian, A., Martin-Martin, R., Reid, I., Rezatofighi, H., Savarese, S.: Social-bigat: multimodal trajectory forecasting using bicycle-GAN and graph attention networks. In: Advances in Neural Information Processing Systems, vol. 32. Curran Associates, Inc. (2019)
23. Kothari, P., Kreiss, S., Alahi, A.: Human trajectory forecasting in crowds: a deep learning perspective. arXiv preprint arXiv:2007.03639 (2020)
24. Kumar, A., et al.: Ask me anything: dynamic memory networks for natural language processing. In: International Conference on Machine Learning, pp. 1378–1387 (2016)
25. Lee, N., et al.: Desire: distant future prediction in dynamic scenes with interacting agents. In: Proceedings of the IEEE Conference on Computer Vision and Pattern Recognition, pp. 336–345 (2017)
26. Li, J., Yang, F., Tomizuka, M., Choi, C.: Evolvegraph: multi-agent trajectory prediction with dynamic relational reasoning. In: Proceedings of the Neural Information Processing Systems (NeurIPS) (2020)
27. Liang, J., Jiang, L., Hauptmann, A.: *SimAug*: learning robust representations from simulation for trajectory prediction. In: Vedaldi, A., Bischof, H., Brox, T., Frahm, J.-M. (eds.) ECCV 2020. LNCS, vol. 12358, pp. 275–292. Springer, Cham (2020). https://doi.org/10.1007/978-3-030-58601-0_17
28. Lisotto, M., Coscia, P., Ballan, L.: Social and scene-aware trajectory prediction in crowded spaces. In: Proceedings of the IEEE International Conference on Computer Vision Workshops (2019)
29. Ma, C., et al.: Visual question answering with memory-augmented networks. In: Proceedings of the IEEE Conference on Computer Vision and Pattern Recognition, pp. 6975–6984 (2018)
30. Ma, Y., Zhu, X., Zhang, S., Yang, R., Wang, W., Manocha, D.: Trafficpredict: trajectory prediction for heterogeneous traffic-agents. In: Proceedings of the AAAI Conference on Artificial Intelligence, vol. 33, pp. 6120–6127 (2019)

31. Mangalam, K., et al.: It is not the journey but the destination: endpoint conditioned trajectory prediction. arXiv preprint arXiv:2004.02025 (2020)
32. Marchetti, F., Becattini, F., Seidenari, L., Del Bimbo, A.: Mantra: memory augmented networks for multiple trajectory prediction. In: Proceedings of the IEEE Conference on Computer Vision and Pattern Recognition (2020)
33. Marchetti, F., Becattini, F., Seidenari, L., Del Bimbo, A.: Multiple trajectory prediction of moving agents with memory augmented networks. IEEE Trans. Pattern Anal. Mach. Intell. (2020)
34. Marchetti, F., Becattini, F., Seidenari, L., Del Bimbo, A.: Smemo: social memory for trajectory forecasting. arXiv preprint arXiv:2203.12446 (2022)
35. Martins, A., Astudillo, R.: From softmax to sparsemax: a sparse model of attention and multi-label classification. In: International Conference on Machine Learning, pp. 1614–1623. PMLR (2016)
36. Mohamed, A., Qian, K., Elhoseiny, M., Claudel, C.: Social-STGCNN: a social spatio-temporal graph convolutional neural network for human trajectory prediction. In: Proceedings of the IEEE/CVF Conference on Computer Vision and Pattern Recognition, pp. 14424–14432 (2020)
37. Pang, B., Zhao, T., Xie, X., Wu, Y.N.: Trajectory prediction with latent belief energy-based model. In: Proceedings of the IEEE/CVF Conference on Computer Vision and Pattern Recognition, pp. 11814–11824 (2021)
38. Pellegrini, S., Ess, A., Schindler, K., Van Gool, L.: You'll never walk alone: modeling social behavior for multi-target tracking. In: 2009 IEEE 12th International Conference on Computer Vision, pp. 261–268. IEEE (2009)
39. Pernici, F., Bruni, M., Del Bimbo, A.: Self-supervised on-line cumulative learning from video streams. Comput. Vis. Image Underst. **197**, 102983 (2020)
40. Rebuffi, S.A., Kolesnikov, A., Sperl, G., Lampert, C.H.: ICARL: incremental classifier and representation learning. In: Proceedings of the IEEE Conference on Computer Vision and Pattern Recognition, pp. 2001–2010 (2017)
41. Ridel, D., Deo, N., Wolf, D., Trivedi, M.: Scene compliant trajectory forecast with agent-centric spatio-temporal grids. IEEE Robot. Autom. Lett. **5**(2), 2816–2823 (2020). https://doi.org/10.1109/LRA.2020.2974393
42. Robicquet, A., Sadeghian, A., Alahi, A., Savarese, S.: Learning social etiquette: human trajectory understanding in crowded scenes. In: Leibe, B., Matas, J., Sebe, N., Welling, M. (eds.) ECCV 2016. LNCS, vol. 9912, pp. 549–565. Springer, Cham (2016). https://doi.org/10.1007/978-3-319-46484-8_33
43. Sadeghian, A., Kosaraju, V., Gupta, A., Savarese, S., Alahi, A.: Trajnet: towards a benchmark for human trajectory prediction. arXiv preprint (2018)
44. Sadeghian, A., Kosaraju, V., Sadeghian, A., Hirose, N., Rezatofighi, H., Savarese, S.: Sophie: an attentive GAN for predicting paths compliant to social and physical constraints. In: Proceedings of the IEEE Conference on Computer Vision and Pattern Recognition, pp. 1349–1358 (2019)
45. Salzmann, T., Ivanovic, B., Chakravarty, P., Pavone, M.: Trajectron++: multi-agent generative trajectory forecasting with heterogeneous data for control. arXiv preprint arXiv:2001.03093 (2020)
46. Santoro, A., Bartunov, S., Botvinick, M., Wierstra, D., Lillicrap, T.: Meta-learning with memory-augmented neural networks. In: International Conference on Machine Learning, pp. 1842–1850 (2016)
47. Shafiee, N., Padir, T., Elhamifar, E.: Introvert: human trajectory prediction via conditional 3D attention. In: Proceedings of the IEEE/CVF Conference on Computer Vision and Pattern Recognition, pp. 16815–16825 (2021)

48. Shi, L., et al.: SGCN: sparse graph convolution network for pedestrian trajectory prediction. In: Proceedings of the IEEE/CVF Conference on Computer Vision and Pattern Recognition, pp. 8994–9003 (2021)
49. Srikanth, S., Ansari, J.A., Sharma, S., et al.: INFER: intermediate representations for future prediction. In: IEEE/RSJ International Conference on Intelligent Robots and Systems (IROS 2019) (2019)
50. Sukhbaatar, S., Weston, J., Fergus, R., et al.: End-to-end memory networks. In: Advances in Neural Information Processing Systems, pp. 2440–2448 (2015)
51. Sun, J., Li, Y., Fang, H.S., Lu, C.: Three steps to multimodal trajectory prediction: Modality clustering, classification and synthesis. arXiv preprint arXiv:2103.07854 (2021)
52. Vaswani, A., et al.: Attention is all you need. In: Advances in Neural Information Processing Systems, vol. 30 (2017)
53. Weston, J., Chopra, S., Bordes, A.: Memory networks. arXiv preprint arXiv:1410.3916 (2014)
54. Xu, C., Mao, W., Zhang, W., Chen, S.: Remember intentions: retrospective-memory-based trajectory prediction. In: Proceedings of the IEEE/CVF Conference on Computer Vision and Pattern Recognition, pp. 6488–6497 (2022)
55. Yuan, Y., Weng, X., Ou, Y., Kitani, K.: Agentformer: agent-aware transformers for socio-temporal multi-agent forecasting. arXiv preprint arXiv:2103.14023 (2021)
56. Zhao, H., et al.: TNT: target-driven trajectory prediction. arXiv abs/2008.08294 (2020)

Cycle-Consistent World Models for Domain Independent Latent Imagination

Sidney Bender[1], Tim Joseph[2(✉)], and J. Marius Zöllner[2,3]

[1] TU Berlin (TUB), Berlin, Germany
s.bender@tu-berlin.de
[2] FZI Research Center for Information Technology, Karlsruhe, Germany
{joseph,zoellner}@fzi.de
[3] Karlsruhe Institute of Technology (KIT), Karlsruhe, Germany
marius.zoellner@kit.edu

Abstract. End-to-end autonomous driving seeks to solve the perception, decision, and control problems in an integrated way, which can be easier to generalize at scale and be more adapting to new scenarios. However, high costs and risks make it very hard to train autonomous cars in the real world. Simulations can therefore be a powerful tool to enable training. Due to slightly different observations, agents trained and evaluated solely in simulation often perform well there but have difficulties in real-world environments. To tackle this problem, we propose a novel model-based reinforcement learning approach called Cycle-consistent World Models. Contrary to related approaches, our model can embed two modalities in a shared latent space and thereby learn from samples in one modality (e.g., simulated data) and be used for inference in different domain (e.g., real-world data). Our experiments using different modalities in the CARLA simulator showed that this enables CCWM to outperform state-of-the-art domain adaptation approaches. Furthermore, we show that CCWM can decode a given latent representation into semantically coherent observations in both modalities.

Keywords: Domain adaption · Reinforcement learning · World models · Cycle-GAN

1 Introduction

Many real-world problems, in our case autonomous driving, can be modeled as high-dimensional control problems. In recent years, there has been much research effort to solve such problems in an end-to-end fashion. While solutions based on imitation learning try to mimic the behavior of an expert, approaches based on reinforcement learning try to learn new behavior to maximize the expected future cumulative reward given at each step by a reward function. In a wide range of areas, reinforcement learning agents can achieve super-human performance [25,28,33] and outperform imitation learning approaches [32].

L. Karlinsky et al. (Eds.): ECCV 2022 Workshops, LNCS 13805, pp. 561–574, 2023.
https://doi.org/10.1007/978-3-031-25072-9_38

However, for high-dimensional observation spaces many reinforcement learning algorithms that are considered state-of-the-art learn slowly or fail to solve the given task at all. Moreover, when the agent fails to achieve satisfactory performance for a given task, it is hard to analyze the agent for possible sources of failure. Model-based reinforcement learning promises to improve upon these aspects. Recent work has shown that model-based RL algorithms can be a magnitude more data-efficient on some problems [7,11,13,14,19,21]. Additionally, since a predictive world model is learned, one can analyze the agent's perception of the world [5].

Still, such agents are mostly trained in simulations [1,4,31] since interaction with the real world can be costly (for example, the cost for a fleet of robots or the cost to label the data). Some situations should be encountered to learn, but must never be experienced outside of simulation (e.g., crashing an autonomous vehicle). While simulations allow generating many interactions, there can be a substantial mismatch between the observations generated by the simulator and the observations that the agent will perceive when deployed to the real world. Furthermore, observations from simulation and reality are mostly unaligned, i.e., there is no one-to-one correspondence between them. This mismatch is often called the domain gap [9] between the real and simulated domain. When the domain gap is not taken into account, the behavior of an agent can become unpredictable as it may encounter observations in reality that have never been seen before in simulation.

One family of approaches to reduce this gap is based on the shared-latent space assumption [22]. The main idea is that the semantics of an observation are located in a latent space from which a simulated and an aligned real observation can be reconstructed. Approaches grounded on this assumption have recently been able to achieve impressive results in areas such as style transfers [16] and imitation learning [2].

Inspired by this, we propose adopting the idea of a shared latent space to model-based reinforcement learning by constructing a sequential shared-latent variable model. Our main idea is to create a model that allows to plan via latent imagination independently of the observation domain. The model is trained to project observation sequences from either domain into a shared latent space and to predict the future development in this latent space. By repeatedly rolling out the model one can then plan or train a policy based on low-dimensional state trajectories.

Our contributions can be summarized as follows: 1. We present a novel cycle-consistent world model (CCWM) that can embed two similar partially observable Markov decision processes that primarily differ in their observation modality into a shared latent space without the need for aligned data. 2. We show that observation trajectories of one domain can be encoded into a latent space from which CCWM can decode an aligned trajectory in the other domain. This can be used as a mechanism to make the agent interpretable. 3. We test our model in a toy environment and train a policy via latent imagination first and then evaluate and show that it is also able to learn a shared latent representation for observations from a more complex environment based on the CARLA simulator.

2 Preliminaries

Sequential Latent Variable Models. In contrast to model-free reinforcement learning (RL), model-based RL explicitly learns an approximate transition model of the environment to predict the next observation x_{t+1} from the current observation x_t and the chosen action a_t [30]. The model is used to rollout imagined trajectories $x_{t+1}, a_{t+1}, x_{t+2}, a_{t+2}, ...$ which can be either used to find the best future actions or to train a policy without the need to interact with the real environment. A problem with such a model is that rollouts become computationally expensive for high-dimensional observation spaces. For this reason, many recent model-based RL algorithms make use of sequential latent variable models. Instead of learning a transition function in observation space $X \subseteq \mathbb{R}^{d_X}$, observations are first projected into a lower-dimensional latent space $S \subseteq \mathbb{R}^{d_S}$ with $d_S \ll d_X$. Then a latent transition function can be used to rollout trajectories of latent states $s_{t+1}, a_{t+1}, s_{t+2}, a_{t+2}, ...$ computationally efficient [12,15]. Since naive learning of latent variable models is intractable, a prevailing way to train such models is by variational inference [20]. The resulting model consists of the following components:

- Dynamics models: prior $p_\theta(s_t|s_{t-1}, a_{t-1})$ and posterior $q_\theta(s_t|s_{t-1}, a_{t-1}, x_t)$
- Observation model: $p_\theta(x_t|s_t)$

Furthermore, at each time step the resulting loss function encourages the ability to reconstruct observations from the latent states while at the same time enforcing to be able to predict the future states from past observations. This loss function is also known as the negative of the evidence lower bound (ELBO):

$$L_t = \underbrace{-\mathbb{E}[p_\theta(x_t|s_t)]}_{\substack{q_\theta(s_t|x_{\leq t}, a_{\leq t}) \\ \text{reconstruction loss} \\ L_{\text{recon}}}} + \underbrace{\mathbb{E}[KL(q_\theta(s_t|s_{t-1}, a_{t-1}, x_t) \parallel p_\theta(s_t|s_{t-1}, a_{t-1}))]}_{\substack{q_\theta(s_{t-1}|x_{\leq t-1}, a_{\leq t-1}) \\ \text{regularization loss} \\ L_{\text{reg}}}} \quad (1)$$

Shared Latent Space Models. We want to enable our model to jointly embed unaligned observation from two different modalities of the same partially observable Markov decision process into the same latent space. Let X_A and X_B be two observation domains (e.g., image domains with one containing RGB images and the other one containing semantically segmented images). In aligned domain translation, we are given samples (x_B, x_B) drawn from a joint distribution $P_{X_A, X_B}(x_A, x_B)$. In unaligned domain translation, we are given samples drawn from the marginal distributions $P_{X_A}(x_A)$ and $P_{X_B}(x_B)$. Since an infinite set of possible joint distributions can yield the given marginal distributions, it is impossible to learn the actual joint distribution from samples of the marginals without additional assumptions.

A common assumption is the shared-latent space assumption [23,24]. It postulates that for any given pair of samples $(x_A, x_B) \sim P_{X_A, X_B}(x_A, x_B)$ there exists a shared latent code s in a shared-latent space such that both samples can be generated from this code, and that this code can be computed from any of the two samples. In other words, we assume that there exists a function with $s = E^{A \to S}(x_A)$ that maps from domain X_A to a latent space S and a function with $x_A = G^{S \to A}(s)$ that maps back to the observation domain. Similarly, the functions $s = E^{B \to S}(x_B)$ and $x_B = G^{S \to B}$ must exist and map to/from to the same latent state.

Directly from these assumptions follows that observations of domain A can be translated to domain B via encoding and decoding and the same must hold for the opposite direction:

$$G^{S \to B}(E^{A \to S}(x_A)) \in X_B$$
$$G^{S \to A}(E^{B \to S}(x_A)) \in X_A \tag{2}$$

Another implication of the shared latent space assumption is that observations from one domain can be translated the other one and back to the original domain (**cycle-consistency** [37]):

$$E^{A \to S}(x_a) = E^{B \to S}(G^{S \to B}(E^{A \to S}(x_A)))$$
$$E^{B \to S}(x_b) = E^{A \to S}(G^{S \to A}(E^{B \to S}(x_B))) \tag{3}$$

The fundamental idea is that by enforcing both of them on semantically similar input domains, the model embeds semantically similar samples close to each other in the same latent space.

3 Cycle-Consistent World Models

In this section, we present our cycle-consistent world model (CCWM). Considering the structure of sequential latent variable models and the constraints resulting from the shared latent space assumption, we show how both can be integrated into a single unified model. In the following, we explain the model architecture and the associated loss terms (Fig. 1).

Architecture. As our model is a sequential latent variable model, it includes all the components that have been presented in Sect. 2, namely a prior transition model $p_\theta(s_t|s_{t-1}, a_{t-1})$, a posterior transition model $q_\theta(s_t|s_{t-1}, a_{t-1}, h_t)$ and an observation model $p_\theta^A(x_t|s_t)$ with $Dec_A(s_t) = \text{mode}(p_\theta^A(x_t|s_t))$. Additionally, we define a feature extractor with $h_t = Enc_A(x_t)$ and a reward model $p_\theta^A(r_t|s_t)$. So far, this model can be used as the basis of an RL-agent that acts on a single domain by first building up the current latent representation s_t using the feature extractor and posterior and then rolling out future trajectories $s_{t+1}, s_{t+2}, \ldots$ with their associated rewards with the prior dynamics and the reward model. To project to and from another domain X_B into the same latent space S we add another feature extractor $Enc_B(x_t)$ and observation model $p_\theta^B(x_t|s_t)$ with $Dec_B(s_t) = \text{mode}(p_\theta^B(x_t|s_t))$. Both are similar to their domain X_A counterparts but do not share any weights. The prior dynamics model is shared since it does not depend on observation. In contrast, we need another posterior dynamics

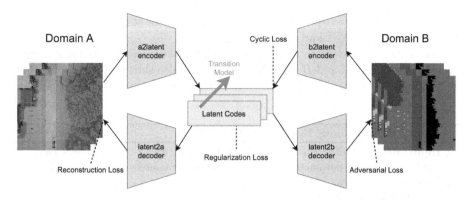

Fig. 1. Cycle-consistent world model. In the pictured situation, a sequence of top camera images is used a the input. The images are encoded frame-wise into latent states and forward predicted by the transition model. From these latent codes, reconstructed top camera images and images translated to semantic top camera images are calculated. From the translated images, cyclic latent codes are calculated. Finally, the four losses can be calculated, which enforce Eqs. (2) and (3).

model for domain B, but since we let it share weights with its domain A counterpart, we effectively only have a single posterior dynamics model. Additionally, we add a reward model $p_\theta(r_t|s_t)$ that also is shared between both domains so that latent trajectories can be rolled out independently of the observation domain. A major advantage of this approach is that we can train a policy with our model without regard to the observation domains.

Finally, for training only, we need two discriminators Dis_ϕ^A and Dis_ϕ^B to distinguish between real and generated samples for each domain. It is important to note that the discriminators have a separate set of parameters ϕ.

Losses. Given a sequence of actions and observations $\{a_t, x_t\}_{t=k}^{k+H} \sim D_A$ from a dataset D_A collected in a single domain X_A, we first roll out the sequential latent variable model using the posterior to receive an estimate for the posterior distribution $q(s_t|s_{t-1}, a_{t-1}, x_t)$ and the prior distribution $q(s_t|s_{t-1}, a_{t-1}, x_t)$ for each time step. We can then calculate the following losses: L_{recon} is the reconstruction loss of the sequential latent variable model and $L_{\mathrm{reg}}(q, p) = KL(q \parallel p)$ is the regularization loss that enforces predictability of futures state as shown in Eq. 1. $L_{\mathrm{adv}}(x) = Dis_B(x)$ is an adversarial loss that penalizes translations from domain X_A to X_B via S that are outside of domain X_B to enforce Eq. 2 of the shared latent space assumption. Here, Dis_B is a PatchGAN [17] based discriminator that is trained alongside our model to differentiate between real and generated observations. The cycle loss $L_{\mathrm{cyc}}(q, p) = KL(q \parallel p)$ is derived from the cycle constraints of Eq. 3 and calculates the KL-divergence between the posterior state distributions conditioned on observations and states from domain A and conditioned on observations and states that have been translated to domain B, i.e. $x_t \rightarrow s_t \rightarrow x_t^{\mathrm{trans}} \rightarrow s_t^{\mathrm{cyc}}$ (see Algorithm 1; line 7, 8 and 12). To calculate the cyclic loss it is necessary to roll out a second set of state trajectories using the cyclic encoding h_t^{cyc} and the cyclic state s_t^{cyc}.

For sequences of domain B, we train with the same loss functions, but with every occurrence of A and B interchanged. This is also shown in Algorithm 1 line 26 and line 28.

Algorithm 1: Training Routine of the CCWM

---

1 **Input:** Replay Buffers D_A and D_B, Encoders Enc_a and Enc_b, Decoders Dec_a and Dec_b, Model parameters Θ, Discriminator parameters Φ;

2 **Function** $L_{gen}(Enc_1, Dec_1, Enc_2, Dec_2, x_{1:T})$:

3   **foreach** $t \in T$ **do**

4          $h_t \leftarrow Enc_1(x_t)$;

5          $s_t' \sim q(s_t|s_{t-1}', h_t)$;

6          $x_t^{\text{recon}} \leftarrow Dec_1(s_t')$;

7          $x_t^{\text{trans}} \leftarrow Dec_2(s_t')$;

8          $h_t^{\text{cyc}} \leftarrow Enc_2(x_t^{\text{trans}})$;

9          $s_t^{\text{cyc}} \sim q(s_t^{\text{cyc}}|s_{t-1}^{\text{cyc}}, h_t^{\text{cyc}})$;

10         $L_{\text{ret}} \mathrel{+}= L_{\text{recon}}(x_t, x_t^{\text{recon}})$;

11         $L_{\text{ret}} \mathrel{+}= L_{\text{adv}}(x_t^{\text{trans}})$;

12         $L_{\text{ret}} \mathrel{+}= L_{\text{reg}}(q(s_t|s_{t-1}', h_t), p(s_t|s_{t-1}'))$;

13         $L_{\text{ret}} \mathrel{+}= L_{\text{cyc}}(q(s_t|s_{t-1}', h_t), q(s_t|s_{t-1}^{\text{cyc}}, h_t^{\text{cyc}}))$;

14   **end**

15   **return** L_{ret};

16 **Function** $L_{dis}(Enc_1, Dec_2, x^1, x^2)$:

17   **foreach** $t \in T$ **do**

18          $h_t \leftarrow Enc_1(x_t^1)$;

19          $s_t \sim q(s_t|s_{t-1}, h_t)$;

20          $x_t^{\text{trans}} \leftarrow Dec_2(s_t)$;

21          $L_{\text{ret}} \mathrel{+}= L_{\text{adv}}(x_t^2) + (1 - L_{\text{adv}}(x_t^{\text{trans}}))$;

22   **end**

23   **return** L_{ret};

24 **while** *not converged* **do**

25   Draw sequence of $x_{a,1:T} \sim D_A$;

26   Draw sequence of $x_{b,1:T} \sim D_B$;

27      $L_{gen} = L_{gen}(Enc_a, Dec_a, Enc_b, Dec_b, x_{a,1:T}) +$
         $L_{gen}(Enc_b, Dec_b, Enc_a, Dec_a, x_{b,1:T})$;

28   Update Model parameters $\Theta \leftarrow \Theta + \Delta L_{gen}$;

29      $L_{dis} = L_{dis}(Enc_a, Dec_b, x_{a,1:T}, x_{b,1:T}) + L_{dis}(Enc_b, Dec_a, x_{b,1:T}, x_{a,1:T})$;

30   Update Discriminator parameters $\Phi \leftarrow \Phi + \Delta L_{dis}$;

31 **end**

---

4   Related Work

Control with Latent Dynamics. World Models [10] learn latent dynamics in a two-stage process to evolve linear controllers in imagination. PlaNet [15] learns them jointly and solves visual locomotion tasks by latent online planning. Furthermore, Dreamer [12,14] extends PlaNet by replacing the online planner with a learned policy that is trained by back-propagating gradients through the transition function of the world model. MuZero [28] learns task-specific reward and value models to solve challenging tasks but requires large amounts of experience. While all these approaches achieve impressive results, they are limited to their training domain and have no inherent way to adapt to another domain.

Domain Randomization. James et al. [18] introduce a novel approach to cross the visual reality gap, called Randomized-to-Canonical Adaptation Networks (RCANs), that uses no real-world data. RCAN learns to translate randomized rendered images into their equivalent non-randomized, canonical versions. In turn, this allows for real images to be translated into canonical simulated images. Xu et al. [36] showed that random convolutions (RC) as data augmentation could greatly improve the robustness of neural networks. Random convolutions are approximately shape-preserving and may distort local textures. RC outperformed related approaches like [26,34,35] by a wide margin and is thereby considered state-of-the-art by us.

Unsupervised Domain Adaptation. The original Cycle-GAN [37] learn to translate images from one domain to another by including a cycle loss and an adversarial loss into training. Liu et al. [23] extend this idea with weight sharing of the inner layers and a normalization loss in the latent state, which enables it to embed images of semantically similar domains into the same latent space. *Learning to drive* [3] uses this idea to train an imitation learning agent in simulation and successfully drive in reality. In *RL-Cycle-GAN* [27], a Cycle-GAN with an RL scene consistency loss is used, and the authors show that even without the RL scene consistency loss, RCAN [18] was outperformed by a wide margin. RL-Cycle-GAN is state-of-the-art for unsupervised domain adaptation to the best of our knowledge.

5   Experiments

First, we will demonstrate our model in a small toy environment. Then we will show its potential in a more realistic setting related to autonomous driving based on the CARLA simulator [8].

Implementation. Our prior and posterior transition models are implemented as recurrent state-space models (RSSM) [15]. In the RSSM, we exchanged the GRU [6] with a convolutional GRU [29]. A challenge of integrating the ideas of a world model and a shared latent space assumption is that it is easier to enforce a shared latent space on a large three-dimensional tensor-shaped latent space. In contrast, most world models use a low-dimensional vector latent space. A bigger latent space makes it easier to embed and align both modalities, but it leads to a less informative self-supervised encoding for the downstream heads, such as the reward model. As we show in our ablation study choosing the right height and width of the latent space is crucial for successful learning.

Proof of Concept. Reinforcement learning environments are often very complex, so that evaluation and model analysis can become hard for complex models such as ours. Additionally, domain adaptation complicates evaluation even more. For this reason, we first construct a toy environment that we call ArtificialV0 to show that our idea is working in principle. ArtificialV0 is constructed as follows: A state of ArtificialV0 is the position of a red and a blue dot. Its state space is a box $[-1, 1] \times [-1, 1]$. As observations, we use images of the red and the blue dot on a white background. The goal is to move the red dot towards the blue dot. The actions are steps by the red dot with an action space of $[-0.2, 0.2] \times [-0.2, 0.2]$. The negative euclidean distance between the blue and the red dot is used as a reward. An episode terminates as soon as the absolute Euclidean distance is smaller than 0.1. The other modality is constructed the same, but the observation images are inverted. Advantages of ArtificialV0 are that the actions and observations are easy to interpret and the optimal policy as a reference benchmark is easy to implement. The optimal policy brings the red dot on a straight line towards the blue dot and achieves an average return of -2.97. We find that CCWM achieves a similar average return after 30K environment steps in an online setting in both modalities, despite us only giving it access to a small offline dataset of 5000 disjunct observations from the reversed modality without downstream information. In Fig. 2, one can see that a trajectory can be started in the inversed modality and successfully continued in both modalities. This indicates that the model is capable of embedding both modalities into a shared latent space.

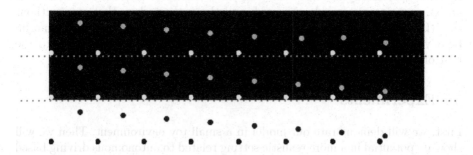

Fig. 2. Qualitative results on ArtificialV0. The top row shows the observations recorded from the environment if one observation is given to the model and the policy is rolled out. It shows that the model can learn the optimal policy (bringing the red/turquoise dot towards the blue/yellow dot on a straight line) only with downstream information from the original modality but also works in the reversed modality. The second row is the prediction of our CCWM back into the domain from that the agent retrieved the initial observation. The last row is the cross-modality prediction. (Color figure online)

Table 1. Comparison with the state-of-the-art. We measured the quality of the reward prediction with the relative squared error against predicting the mean reward to show that something better than predicting the mean is learned. Furthermore, we determined how well the different models can predict the next states based on the peak signal-to-noise ratio (PSNR) between the real future observations and the predicted observations. We can see that all domain adaptation methods can transfer the reward predictions while only using one modality. Our CCWM achieved the best reward transfer and the best video prediction. It is worth mentioning that the cross-modality reward predictions with only one modality and with RC were unstable, varying strongly over time steps depending on the initialization. Since single modality and RC are fast, while Cycle-GAN and CCWM are slow, we show the results after training approximately 24 h on an NVIDIA GTX1080TI to keep the comparison fair.

Approach	Reward RSE	Reward RSE cross-modality	PSNR
Single modality	0.25	3.86	10.21
RC	0.31	0.49	11.39
CycleGAN	0.28	0.57	12.28
Ours	0.23	**0.48**	**13.91**

Table 2. Ablation study on the size of the latent space. The models are identical except that the convolutional GRU is used at different downsampling scales of the network. We can see that latent spaces smaller than 4×4 are having trouble minimizing all objectives at once, and the reward RSE is not falling significantly below simply predicting the mean.

Latent space size	Reward RSE	Reward RSE cross-modality	PSNR
1×1	0.92	1.18	13.00
2×2	0.95	1.10	13.80
4×4	0.57	0.57	13.81
8×8	0.23	**0.48**	**13.91**

Experiment Setup. To show the potential of our approach in a more realistic environment, we also evaluate our model in the CARLA simulator. We choose to use images from a semantic camera as the first modality and images from an RGB camera as the second modality. Both look down onto the cars from a birds-eye-view point.

For an even more realistic setting, one could replace the top view RGB camera with an RGB surround camera in a real car and the schematic top view with an RGB surround-view camera from in simulation. However, since we do not have access to a real car with such sensors and we are restricted in computational resources, we simplified the problem for now. Arguably, the visual difference between the RGB camera from the simulation and the real world RGB camera is smaller than the visual difference between the RGB camera in the simulation and the schematic view of the simulation, so there is reason to believe that a transfer from the RGB camera of the simulation to the RGB camera of the real world would work as well.

Comparsion with the State-of-the-Art. To show that the constructed domain gap is not trivial and our model is outperforming current domain adaptation methods, we compare our model with 1) no adaptation to the other modality at all, 2) the random convolutions (RC) [36] approach, which we regard as being state of the art in domain randomization, and 3) the RL-CycleGan [27], which we consider to be the start of the art in unsupervised domain adaptation. All models are reimplemented and integrated into our codebase. They are apart from their core idea as similar as possible regarding network structure, network size, and other hyperparameters. The performance of a world model rises and falls with two factors: 1) How well the model can predict the current reward based on the current state and 2) how accurate the prediction of the next states is. We recorded three disjunct offline datasets with the CARLA roaming agent (an agent controlled by the CARLA traffic manager). The first contains trajectories of observations of the semantic view with downstream information. The second contains trajectories of observations of the RGB camera without downstream information. The third contains aligned semantic and RGB camera trajectories and downstream information. The first and the second dataset are used for training the model, and the third is used for evaluation. The model without any domain adaptation is trained on the first dataset in the regular dreamer style for the model training. The RC model is trained on the first dataset with randomized inputs. The RL-Cycle-GAN model is trained by first learning a Cycle-GAN-based translation from the first modality to the second modality. Then the model is trained on the translated observations of the first dataset. CCWM is trained as described in the previous section on the first and the second dataset.

Results. All models are evaluated on the third dataset in the following ways: First, we qualitatively analyze the predictive power for the next states of the model. We warm up the model by feeding it some observations and then predict the next observations of the target domain, as shown in Fig. 3. A general advantage of CCWM noteworthy to mention is that it can predict into both modalities simultaneously since both have a shared latent representation, which might be practical for error search. Besides the qualitative analysis of the state predictions based on the predicted observations, we also compare the predictions quantitatively by calculating the PSNR between the predicted and the real trajectory, as seen in the Table 1. Furthermore, we compare the reward prediction in the target domain where no downstream information was available. Both in qualitative and quantitative comparison, one can see that our model outperforms the other approaches.

Analysis. The advantage of our approach over RC is that RC generalizes random distortions of the input image that RC can emulate with a random convolution layer, which might include the semantic segmentation mask, but will also include many other distributions, making it less directed despite its simplicity.

Fig. 3. Qualitative Results on Carla. The first row shows the ground truth of the semantic top camera sampled from dataset 3, and the second row the baseline of what would happen if the dreamer was trained in one modality and rolls out the other modality now. Row 3 and 4 show the state-of-the-art comparison with random convolutions and a preprocessing input with a Cycle-GAN. Both were also only trained on with the RGB top camera. The 5th and the 6th row shows our model rolled out aligned in both modalities. The previous 19 frames and the first frame of the ground truth are fed into the model for all models, and then the model is rolled out for fifteen time steps (every second is shown).

Pre-translating with Cycle-GAN follows a more directed approach but is not able to train the whole network end-to-end. Furthermore, it first encodes a training image, then decodes it to a different domain, and then encodes it again to derive downstream information and predict future states. This is a longer path than encoding it only once like CCWM and leaves room for well-known problems with adversarial nets like artifacts in the image, hindering training progress.

Ablation Study. Although probabilistic graphic models and reinforcement learning approaches are generally susceptible to hyperparameters, the size of the latent space has shown to be especially significant. As shown in Table 2 a 1×1 latent space like it is common in many model-based RL approaches performs poorly, while bigger latent spaces provide much better performance. Our explanation for this is twofold. Firstly, related approaches such as UNIT [23] cannot translate images well with a tiny latent space and instead use huge latent spaces. Secondly, in autonomous driving, it might not be beneficial to compress the whole complicated scene with multiple cars that all have their own location, direction, speed, etc. into one vector, but give the network inductive bias to represent each of them in a single vector and calculate the dynamics through the convolutional GRU with its suiting local inductive bias. Another important consideration is the weights for the different losses, which need to be carefully chosen. The reward loss tends to get stuck around the mean since its signal is relatively weak, so it should be chosen relatively high. The KL based losses in the latent space can get very high and destroy the whole model with a single step. On the other hand, a high normalization loss leads to bad predictive capabilities, and a high cyclic loss leads to a bad alignment of the modalities.

6   Conclusion

In this work, we introduced cycle-consistent world models, a world model for model-based reinforcement learning that is capable of embedding two modalities into the same latent space. We developed a procedure to train our model and showed its performance in a small toy environment and a more complex environment based on the CARLA simulator. Furthermore, we compared it in an offline setting with two state-of-the-art approaches in domain adaptation, namely RC and RL-Cycle-GAN. We outperformed RC by being more directed and Cycle-GAN by training end-to-end without the necessity to encode twice. For the future we plan to extend our model by training a full model-based RL agent that is able to learn to control a vehicle in simulation and generalize to reality given only offline data from reality without any reward information.

Acknowledgement. The research leading to these results is funded by the German Federal Ministry for Economic Affairs and Climate Action" within the project "KI Delta Learning"(Förderkennzeichen 19A19013L). The authors would like to thank the consortium for the successful cooperation.

References

1. Bellemare, M.G., Naddaf, Y., Veness, J., Bowling, M.: The arcade learning environment: an evaluation platform for general agents. J. Artif. Intell. Res. **47**, 253–279 (2013)
2. Bewley, A., et al.: Learning to Drive from Simulation without Real World Labels. CoRR abs/1812.03823 (2018). http://arxiv.org/abs/1812.03823
3. Bewley, A., et al.: Learning to drive from simulation without real world labels. In: 2019 International Conference on Robotics and Automation (ICRA), pp. 4818–4824. IEEE (2019)
4. Brockman, G., et al.: OpenAI Gym. CoRR abs/1606.01540 (2016). http://arxiv.org/abs/1606.01540
5. Chen, J., Li, S.E., Tomizuka, M.: Interpretable end-to-end urban autonomous driving with latent deep reinforcement learning. IEEE Trans. Intell. Transp. Syst. 1–11 (2021). https://doi.org/10.1109/TITS.2020.3046646
6. Cho, K., van Merrienboer, B., Bahdanau, D., Bengio, Y.: On the properties of neural machine translation: Encoder-decoder approaches. CoRR abs/1409.1259 (2014). http://arxiv.org/abs/1409.1259

7. Chua, K., Calandra, R., McAllister, R., Levine, S.: Deep reinforcement learning in a handful of trials using probabilistic dynamics models. In: Bengio, S., Wallach, H., Larochelle, H., Grauman, K., Cesa-Bianchi, N., Garnett, R. (eds.) Advances in Neural Information Processing Systems, vol. 31. Curran Associates, Inc. (2018). https://proceedings.neurips.cc/paper/2018/file/3de568f8597b94bda53149c7d7f595 8c-Paper.pdf

8. Dosovitskiy, A., Ros, G., Codevilla, F., Lopez, A., Koltun, V.: Carla: an open urban driving simulator. In: Conference on Robot Learning, pp. 1–16. PMLR (2017)

9. Ganin, Y., Lempitsky, V.: Unsupervised domain adaptation by backpropagation. In: Bach, F., Blei, D. (eds.) Proceedings of the 32nd International Conference on Machine Learning, Proceedings of Machine Learning Research, Lille, France, vol. 37, pp. 1180–1189. PMLR (2015). http://proceedings.mlr.press/v37/ganin15.html

10. Ha, D., Schmidhuber, J.: World models. arXiv preprint arXiv:1803.10122 (2018)

11. Hafner, D., Lillicrap, T., Ba, J., Norouzi, M.: Dream to control: learning behaviors by latent imagination. arXiv preprint arXiv:1912.01603 (2019)

12. Hafner, D., Lillicrap, T., Ba, J., Norouzi, M.: Dream to Control: learning Behaviors by Latent Imagination. In: International Conference on Learning Representations (2020). https://openreview.net/forum?id=S1lOTC4tDS

13. Hafner, D., et al.: Learning latent dynamics for planning from pixels. In: International Conference on Machine Learning, pp. 2555–2565. PMLR (2019)

14. Hafner, D., Lillicrap, T., Norouzi, M., Ba, J.: Mastering atari with discrete world models. arXiv preprint arXiv:2010.02193 (2020)

15. Hafner, D., et al.: Learning Latent Dynamics for Planning from Pixels. CoRR abs/1811.04551 (2018). http://arxiv.org/abs/1811.04551

16. Huang, X., Liu, M.Y., Belongie, S., Kautz, J.: Multimodal unsupervised image-to-image translation. In: Proceedings of the European Conference on Computer Vision (ECCV), pp. 172–189 (2018)

17. Isola, P., Zhu, J.Y., Zhou, T., Efros, A.A.: Image-to-image translation with conditional adversarial networks. In: Proceedings of the IEEE Conference on Computer Vision and Pattern Recognition, pp. 1125–1134 (2017)

18. James, S., et al.: Sim-to-real via sim-to-sim: data-efficient robotic grasping via randomized-to-canonical adaptation networks. In: Proceedings of the IEEE/CVF Conference on Computer Vision and Pattern Recognition, pp. 12627–12637 (2019)

19. Janner, M., Fu, J., Zhang, M., Levine, S.: When to trust your model: model-based policy optimization. In: Wallach, H., Larochelle, H., Beygelzimer, A., d Alché-Buc, F., Fox, E., Garnett, R. (eds.) Advances in Neural Information Processing Systems, vol. 32. Curran Associates, Inc. (2019). https://proceedings.neurips.cc/paper/2019/file/5faf461eff3099671ad63c6f3f094f7f-Paper.pdf

20. Kingma, D.P., Welling, M.: Auto-Encoding Variational Bayes (2014)

21. Kurutach, T., Clavera, I., Duan, Y., Tamar, A., Abbeel, P.: Model-Ensemble Trust-Region Policy Optimization. In: International Conference on Learning Representations (2018). https://openreview.net/forum?id=SJJinbWRZ

22. Liu, M.Y., Breuel, T., Kautz, J.: Unsupervised image-to-image translation networks. arXiv preprint arXiv:1703.00848 (2017)

23. Liu, M.Y., Breuel, T., Kautz, J.: Unsupervised image-to-image translation networks. In: Guyon, I., et al. (eds.) Advances in Neural Information Processing Systems, vol. 30. Curran Associates, Inc. (2017). https://proceedings.neurips.cc/paper/2017/file/dc6a6489640ca02b0d42dabeb8e46bb7-Paper.pdf

24. Liu, M.Y., Tuzel, O.: Coupled generative adversarial networks. In: Lee, D., Sugiyama, M., Luxburg, U., Guyon, I., Garnett, R. (eds.) Advances in Neural Information Processing Systems, vol. 29. Curran Associates, Inc. (2016). https://proce edings.neurips.cc/paper/2016/file/502e4a16930e414107ee22b6198c578f-Paper.pdf
25. Mnih, V., et al.: Playing atari with deep reinforcement learning. arXiv preprint arXiv:1312.5602 (2013)
26. Qiao, F., Zhao, L., Peng, X.: Learning to learn single domain generalization. In: Proceedings of the IEEE/CVF Conference on Computer Vision and Pattern Recognition, pp. 12556–12565 (2020)
27. Rao, K., Harris, C., Irpan, A., Levine, S., Ibarz, J., Khansari, M.: RL-CycleGAN: reinforcement learning aware simulation-to-real. In: Proceedings of the IEEE/CVF Conference on Computer Vision and Pattern Recognition, pp. 11157–11166 (2020)
28. Schrittwieser, J., et al.: Mastering atari, go, chess and shogi by planning with a learned model. Nature **588**(7839), 604–609 (2020)
29. Siam, M., Valipour, S., Jagersand, M., Ray, N.: Convolutional gated recurrent networks for video segmentation. In: 2017 IEEE International Conference on Image Processing (ICIP), pp. 3090–3094. IEEE (2017)
30. Sutton, R.S., Barto, A.G.: Reinforcement Learning: An Introduction. A Bradford Book, Cambridge (2018). ISBN 0262039249
31. Tassa, Y., et al.: DeepMind Control Suite. CoRR abs/1801.00690 (2018). http://arxiv.org/abs/1801.00690
32. Toromanoff, M., Wirbel, E., Moutarde, F.: End-to-end model-free reinforcement learning for urban driving using implicit affordances. In: Proceedings of the IEEE/CVF Conference on Computer Vision and Pattern Recognition, pp. 7153–7162 (2020)
33. Vinyals, O., et al.: Grandmaster level in StarCraft II using multi-agent reinforcement learning. Nature **575**(7782), 350–354 (2019)
34. Volpi, R., Namkoong, H., Sener, O., Duchi, J., Murino, V., Savarese, S.: Generalizing to unseen domains via adversarial data augmentation. arXiv preprint arXiv:1805.12018 (2018)
35. Wang, H., Ge, S., Xing, E.P., Lipton, Z.C.: Learning robust global representations by penalizing local predictive power. arXiv preprint arXiv:1905.13549 (2019)
36. Xu, Z., Liu, D., Yang, J., Raffel, C., Niethammer, M.: Robust and generalizable visual representation learning via random convolutions. arXiv preprint arXiv:2007.13003 (2020)
37. Zhu, J.Y., Park, T., Isola, P., Efros, A.A.: Unpaired image-to-image translation using cycle-consistent adversarial networks. In: Proceedings of the IEEE International Conference on Computer Vision, pp. 2223–2232 (2017)

W21 - Real-World Surveillance: Applications and Challenges

W21 - Real-World Surveillance: Applications and Challenges

This was the second edition of the RWS workshop, focusing on identifying and dealing with challenges of deploying machine learning models in real-world applications. The workshop also included a challenge on thermal object detection on the largest thermal dataset that was annotated for this challenge.

October 2022

Kamal Nasrollahi
Sergio Escalera
Radu Tudor Ionescu
Fahad Shahbaz Khan
Thomas B. Moeslund
Anthony Hoogs
Shmuel Peleg
Mubarak Shah

W21 - Real-World Surveillance:
Applications and Challenges

Strengthening Skeletal Action Recognizers via Leveraging Temporal Patterns

Zhenyue Qin[1(✉)], Pan Ji[2], Dongwoo Kim[3], Yang Liu[1,4], Saeed Anwar[1,4], and Tom Gedeon[5]

[1] Australian National University, Canberra, Australia
zhenyue.qin@anu.edu.au
[2] OPPO US Research, Palo Alto, USA
[3] GSAI POSTECH, Pohang, South Korea
[4] Data61-CSIRO, Sydney, Australia
[5] Curtin University, Perth, Australia

Abstract. Skeleton sequences are compact and lightweight. Numerous skeleton-based action recognizers have been proposed to classify human behaviors. In this work, we aim to incorporate components that are compatible with existing models and further improve their accuracy. To this end, we design two temporal accessories: discrete cosine encoding (DCE) and chronological loss (CRL). DCE facilitates models to analyze motion patterns from the frequency domain and meanwhile alleviates the influence of signal noise. CRL guides networks to explicitly capture the sequence's chronological order. These two components consistently endow many recently-proposed action recognizers with accuracy boosts, achieving new state-of-the-art (SOTA) accuracy on two large datasets.

1 Introduction

Accurately recognizing human actions is essential for many applications such as human-robot interaction [8], sports analysis [34], and smart healthcare services [28]. Recently, skeleton-based action recognition has attracted increasing attention [4,7,16]. Technically, skeleton data are concise, free from environmental noises (such as background clutter, and clothing), and also lightweight, enabling fast processing speed on edge devices [4]. Ethically, skeletonized humans are privacy-protective and racial-unbiased, promoting fair training and inference.

Recent approaches of skeleton-based action recognition primarily concentrate on devising new network architectures to more comprehensively extract motion patterns from the skeleton trajectories [2,39,41]. Most designs focus on proposing various graph operations to capture spatial interactions. On the other hand, the temporal extractors of [2,30,41] inherit [39]. In contrast, in this paper, we propose two components that are compatible with existing skeletal action recognizers, to more robustly and adequately extract *temporal* movement features.

Supplementary Information The online version contains supplementary material available at https://doi.org/10.1007/978-3-031-25072-9_39.

L. Karlinsky et al. (Eds.): ECCV 2022 Workshops, LNCS 13805, pp. 577–593, 2023.
https://doi.org/10.1007/978-3-031-25072-9_39

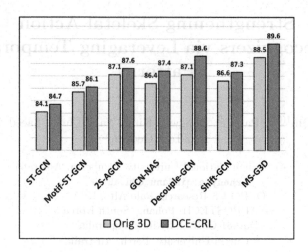

Fig. 1. Our proposed discrete cosine encoding and chronological loss (DCE-CRL) consistently improve the accuracy of numerous state-of-the-art models. The results are evaluated on NTU60 with the joint feature.

We first present the discrete cosine embedding, which promotes extraction of a motion trajectory's frequency information for a robust skeleton-based action recognizer. To achieve this, we do not explicitly transform the action sequence into the frequency domain; instead, we enrich the sequence by concatenating it with additional encoding. As such, the action recognizer's temporal convolutions on this extra encoding absorb frequency information in the downstream process. A primary advantage of the proposed encoding is that it is compatible with most skeleton-based classifiers since most of these models take temporal sequences as input rather than frequency input. Furthermore, this encoding is followed by temporal convolutions of different receptive field sizes, so that frequency features of various temporal resolutions can be effectively learned. Moreover, it also allows for flexible control of the concatenated encoding's frequency. Since temporal noisy pieces of an action (*e.g*, joints in wrongly estimated locations and their sudden absence in some frames) turn out to contain more high-frequency signals than normal parts, we can emphasize the action sequence's low-frequency components to alleviate the noise's adverse influence.

Second, we propose a loss function that facilitates a skeleton-based action recognizer to explicitly capture information about the action sequence's chronological order. The chronological order represents the relative temporal order between two frames, *i.e.*, this information depicts whether a frame appears earlier or later than the other. This information offers rich clues, and sometimes is indispensable, for reliably differentiating between actions. For example, it is intuitive that distinguishing between *putting on shoes* and *taking off shoes* is almost impossible without knowing the chronological order of frames.

We summarize our **contributions** as follows:

i) We propose discrete cosine embedding. It allows a skeleton-based action recognizer to analyze actions in the frequency domain, and also supports disentangling the signal into different frequency components. We can thus alleviate noise's adverse influence by highlighting low-frequency components.

ii) We present the chronological loss function. It directly guides an action recognizer to explicitly capture the action sequence's temporal information, promoting differentiating temporally confusing action pairs such as *putting on shoes* vs *taking off shoes* and *wearing jacket* vs *taking off jacket*.

iii) The proposed two accessories are widely compatible with existing skeleton-based action recognizers. We experimentally show that the two components consistently boost the accuracy of a number of recent models.

iv) Equipped with these two accessories, a simple model is competitive with more complex models in accuracy and consumes substantially fewer parameters and less inference time. When combining the two components with a more complicated network, we achieve new state-of-the-art accuracy.

2 Related Work

2.1 Recognizing Skeleton-Based Actions

In early work, the joint coordinates in the entire sequence were encoded into a feature embedding for skeleton-based action recognition [36]. Since these models did not explore the internal correlations between frames, they lost information about the action trajectories, resulting in low classification accuracy. The advent of convolutional neural networks (CNNs) fortified capturing the correlations between joints and boost the performance of skeleton-based action recognition models [16, 24, 37]. However, CNNs cannot model the skeleton's topology graph, thus missing the topological relationships between joints.

Graph convolution networks (GCNs) were introduced to model these topological relations by representing skeletons as graphs, in which nodes and edges denote joints and bones. ST-GCN [39] aggregated joint features spatially with graph convolution layers and extracted temporal variations between consecutive frames with convolutions. In later work, Li *et al.* proposed AS-GCN [19] to use learnable adjacency matrices to substitute for the fixed skeleton graph, further improving recognition accuracy. Subsequently, Si *et al.* designed AGC-LSTM [32] to integrate graph convolution layers into LSTM as gate operations to learn long-range temporal dependencies in action sequences. Then, the 2s-AGCN model [31] utilized bone features and learnable residual masks to increase skeletons' topological flexibility and ensembled the models trained separately with joints and bones to boost the classification accuracy. Recently, more innovative techniques, such as self-attention [41], shift-convolution [3,4] and graph-based dropout [2], have been introduced into GCNs. MS-G3D [26] employed graph 3D convolutions to extract features from the spatial-temporal domain simultaneously.

2.2 Capturing Chronological Information

Extracting temporal features has been extensively studied for skeleton-based action recognition. Many of these methods are proposed to more adequately capture motion patterns within a local time window (the adjacent several frames) [19,26]. Instead, we investigate learning the video-wide temporal evolution.

Recent skeleton-based action recognition work used the graph neural networks as backbones. They usually did not directly capture the entire motion sequence's relative temporal order. Instead, they gradually increase the sizes of temporal receptive fields to progressively capture temporal information of longer sequences as the networks go deeper. In general action recognition, one of the pioneering works that captures the entire video's temporal information was using a linear ranking machine, named rank pooling, to model the temporal order of a video [10]. Then, Fernando *et al.* proposed the hierarchical rank pooling to more completely capture the action sequence's complex and discriminative dynamics [9]. At a similar time, [12] stated that utilizing subsequences provides more features that are specifically about temporal evolution than merely using single frames. Subsequently, Fernando *et al.* showed how to use deep learning to learn the parameters of rank pooling [11]. Later, Cherian *et al.* further generalized rank pooling's form with subspaces and formulated the objective as a Riemannian optimization problem [5,6].

The approaches along the direction mentioned above usually modeled learning the video-wide temporal order as an optimization problem. The training was generally not end-to-end with the exception of [11]. It first transformed the length-varying video to a fixed-length sequence; then, the chronological information was learned on the new representation and required expensively computing the gradients of a bi-level optimization, resulting in slow convergence.

In this work, based on the spirit of rank pooling, we propose a loss function that facilitates models to explicitly capture the entire sequence-wide chronological information, referred to as the chronological loss.

2.3   Noise Alleviation

In the literature, work exists that aims to alleviate the noise of skeleton-based action sequences, *e.g*, [22] proposed a gate in a recurrent neural network to address noise and [15] partitioned the skeleton of a human body into different parts and used a part's representation to reduce the noise from a single joint.

In human motion prediction, Mao *et al.* applied a discrete cosine transform (DCT) to reduce the noise within the motion trajectories [27]. Similarly, our proposed encoding also receives inspirations from [27] and DCTs; however, our approach is different. In [27], to separate signals of different frequencies, the motion trajectory was firstly mapped to the frequency domain. Then, the graph model processed the signals in the frequency domain. In contrast, skeleton-based action recognizers are usually designed to process motion trajectories in the temporal domain. We evaluated that first converting the signals from the temporal to the frequency domain, followed by using the transformed signals, leads to a plunge of accuracy, reducing more than 10% according to our experiment.

We instead propose an encoding mechanism that facilitates disentangling sequences into different frequencies. At the same time, the encoding is still in the

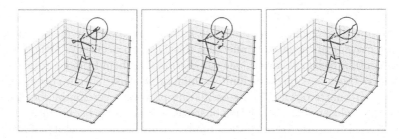

Fig. 2. Illustration of the noisy coordinates highlighted by the red circles. The hand joints show jittering if observing the hand parts of the three figures. In the second figure, the right elbow bends at an unrealistic angle. (Color figure online)

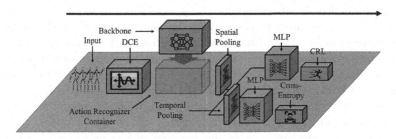

Fig. 3. Illustration of the framework incorporating the two temporal accessories. Read left to right. The skeleton action sequence is first preprocessed by the DCE unit. Then it passes through the deep learning backbone, which is flexibly replaceable with most existing skeleton-based action recognizers. The features then undertake spatial pooling. Finally, the features diverge and go through two parallel branches, one for CRL and the other for classification.

temporal domain rather than in the frequency domain. The proposed encodings are thus compatible with many existing models.

3 Proposed Approach

Our core motivation is to strengthen the capability of skeleton-based action recognizers to more comprehensively and adequately extract temporal features. To this end, we propose two temporal accessories that can be compatibly inserted into existing models. One is the discrete cosine encoding that supports mitigating the undesired impact of noise. The other is the chronological loss function that explicitly guides the model in learning relative temporal orders. In the following, we explain the two components in detail on how they enhance extracting temporal features. Then, we describe the framework that simultaneously integrates the two proposed accessories in the end of this section.

3.1 Discrete Cosine Encoding

In skeleton-based action recognition literature, the input features are usually 3D coordinate-based action movement trajectories along the temporal axis. As in

Fig. 2, we often observe noise in coordinates, resulting in a wobbling of nodes in the temporal dimension. Action recognizers may unexpectedly misinterpret this as a fast movement, negatively affecting classification accuracy. Although some actions contain fast movements, one can distinguish them from coordinate noise because (i) there is a physical limit to human actions' motion speed, and (ii) noise in coordinates often exhibits oscillating patterns over spatial locations.

If we look closer at the oscillation of spatial locations over consecutive frames, we find that the center of oscillation tends to follow the trajectory of the original action, with a small displacement over the frames. The underlying action trajectory is usually reflected in a low-frequency movement signal. In contrast, the oscillation can generally be viewed as a high-frequency signal that does not affect the actors' main trajectories. We can thus mitigate the noise by suppressing high-frequency signals.

To this end, we need to decompose the signals into representations over different frequencies. Meanwhile, the decomposed signals remain suitable for existing action recognizers. It is non-trivial to design decomposition approaches that simultaneously satisfy the two requirements. Naively, we may resort to standard methods that map signals in the temporal domain to the frequency domain before feeding them into skeleton-based action recognizers. Nevertheless, three downsides emerge if we follow this path.

Firstly, the actions like *hopping* and *jumping up* contain fast movements that are intrinsically high-frequency information. Blindly removing all high-frequency components leads to unexpected loss of necessary high-frequency information that are distinct cues for action recognition. To verify this, we mapped original sequences to the frequency domain, and evenly divided the signals into the lower and higher frequencies. Then, we abandoned the higher half and reverted the rest of the signals back to the temporal domain. Compared to the original signals, using these processed sequences dropped the accuracy from 87% to 72%.

Secondly, when an action involves more than one person, the temporal co-occurrences between the two peoples' interacting patterns provide rich clues. However, information about temporal locality inevitably reduces as a result of transferring into the frequency domain. Although some approaches like wavelet transform preserve partial temporal locality, the sizes of temporal windows need to be manually determined, which undesirably limits the flexibility of learning comprehensive features.

Thirdly, existing action recognizers are usually *not* designed for processing frequency-domain signals; hence, if we simply feed the frequency signals into such an action recognizer or concatenate them with the original time-domain sequence, the physical meaning will be illogical.

In sum, we aim to design an encoding mechanism that satisfies two requirements: (i) it supports disentangling high- and low-frequency components of a sequence, enabling mitigating noise; (ii) it can be meaningfully concatenated with the original sequence to prevent the unexpected loss of necessary high-frequency information that are distinct cues for action recognition and ensure compatibility with existing recognizers.

Our key idea for addressing the above-mentioned three requirements is inspired by the following observation. We notice that many existing action recognizers extract movement features with temporal convolutions. Temporal convolutions conduct weighted sums of features within a local time window.

Hence, we intend to leverage temporal convolutions to transform the temporal sequences into signals within the frequency domain. The exact procedures are described as follows. We concatenate original sequence $\mathbf{x} = [x_0, \cdots, x_{T-1}]^\top \in \mathbb{R}^T$ (T is the frame number) with the following encoding:

$$[\mathbf{b}_0 \odot \mathbf{x}, \cdots, \mathbf{b}_k \odot \mathbf{x}, \cdots, \mathbf{b}_{K-1} \odot \mathbf{x}] \in \mathbb{R}^{K \times T}. \tag{1}$$

We define $\mathbf{B} = [\mathbf{b}_0, \cdots, \mathbf{b}_k, \cdots, \mathbf{b}_{K-1}] \in \mathbb{R}^{K \times T}$ as a collection of basis sequences. K is a hyperparameter controlling the complexity of $\mathbf{B}$, and $\odot$ represents the Hadamard (element-wise) product. To be more specific, the form of a single basis sequence can be written as:

$$\mathbf{b}_k = \left[\cos\left(\frac{\pi}{T}(0 + \frac{1}{2})k\right), \cdots, \cos\left(\frac{\pi}{T}(T - 1 + \frac{1}{2})k\right)\right]^\top.$$

A larger k corresponds to a basis sequence $\mathbf{b}_k$ with a higher frequency. Thus, selecting K close to T leads to the encoding preserving more high-frequency information of $\mathbf{x}$, and vice versa. During implementation, K is set to be 8, and the encoding is applied to each joint's 3D coordinates separately and then concatenated with the existing features. Larger K leads to even slightly better performance. A more detailed ablation study on K is in the supplementary materials. The shape of the tensor for an action sequence is changed from $\mathbb{R}^{C \times T \times N \times M}$ to $\mathbb{R}^{(K+1)C \times T \times N \times M}$, where C is the feature channel number, T is the sequence length, N is the joint number, M is the subject number.

One may raise the concern that concatenating with the original sequence seems to preserve the noise. However, our empirical studies show that concatenating with the original signal leads to slightly higher accuracy than discarding the original signal. We conjecture that the network may use the DCE as references to proactively mitigate the noise of the original sequence. More detailed experimental results are in the supplementary materials.

The motivation behind applying such an encoding mechanism is to directly guide the action recognizers to easily capture the signal's frequency information. The basis sequence is from the type-2 discrete cosine transform. Hence, a simple summation of the encoding along the temporal direction reveals a sequence's frequency distribution. To facilitate more flexible learning of the frequency information, we replace the summation with a neural network as:

$$\mathbf{d}_\phi = f_\phi(\mathbf{b}_k \odot \mathbf{x}), \tag{2}$$

where f_ϕ is a neural network parameterized by ϕ. Consequently, $\mathbf{d}_\phi$ can extract frequency-related information on an arbitrary subsequence of $\mathbf{x}$ via temporal convolution. For example, convolution over $\mathbf{b}_k \odot \mathbf{x}$ with a 1×3 filter can learn how much the signals from three consecutive frames are aligned with a given frequency k. If we stack multiple temporal convolution layers, the model can then extrapolate frequency information of longer subsequences. Since temporal convolutional layers are commonly employed in skeleton action recognition models, the encoding can be seamlessly integrated into these models to improve the performance of most existing models.

We summarize the encoding's advantages as follows:

i) The encoding disentangles the action sequence into multiple sequences corresponding to different frequencies. We can mitigate the high-frequency ones to lighten the noise's adverse influence on recognizing actions.
ii) The encoding is in the temporal domain. It thus can be meaningfully concatenated with the original sequence and fed into standard skeletal recognizers.
iii) Since temporal convolutions comprise receptive fields of various sizes, they capture frequency information with various time ranges.

3.2   Chronological Loss Function

We also aim to design a loss function that enables the action recognizer to correctly capture the chronological order of an action sequence. The order depicts whether a frame appears earlier or later than another frame. To facilitate learning the chronological order, we have to find solutions to the two challenges below:

i) What functions express the relative order between two arbitrary frames?
ii) What is the loss function to direct the action recognizer to accurately capture the chronological order?

In the following, we describe the approaches to tackle the two problems and present the chronological loss that simultaneously resolves the two challenges.

Modeling Relative Order Between Frames: In this part, we explain how to mathematically describe the relative order between two frames. To this end, we require formulating three chronological relationships: a frame appears earlier or later than another frame, or the temporal order does not exist between them. Inspired by [10], we assign each frame at time t with a value $f_\phi(\mathbf{h}_t)$, where f is a function (we use a multi-layer perceptron), $\mathbf{h}_t$ is the feature of the frame at time t, and ϕ is the learnable parameter. Then, we use the difference between $f_\phi(\mathbf{h}_t)$ and $f_\phi(\mathbf{h}_{t'})$ to indicate the temporal order between the frames at time t and t'. That is,

- $f_\phi(\mathbf{h}_i) < f_\phi(\mathbf{h}_j)$ iff frame i appears earlier than frame j;
- $f_\phi(\mathbf{h}_i) > f_\phi(\mathbf{h}_j)$ iff frame i appears later than frame j;
- $f_\phi(\mathbf{h}_i) = f_\phi(\mathbf{h}_j)$ iff the action performer stays idle during the period between frames i and j.

We refer to the sequence of values:

$$[f_\phi(\mathbf{h}_1), f_\phi(\mathbf{h}_2), ..., f_\phi(\mathbf{h}_T)]$$

as the chronological values with sequence length T.

Formulating the Chronological Loss Function: We have developed a mathematical form that reflects the temporal order between two frames. Now, we aim to devise a loss function that accurately guides the chronological values to indicate the correct temporal order of an action sequence. That is, the chronological values monotonically increase along with the temporal evolution:

$$\forall t : f_\phi(\mathbf{h}_t) \le f_\phi(\mathbf{h}_{t+1}). \tag{3}$$

Creating a loss function to guide the model to satisfy the above objective is non-trivial. Poor design causes fluctuations in the chronological values within the sequence. For instance, to let $f_\phi(\mathbf{h}_t)$ be smaller than $f_\phi(\mathbf{h}_{t+1})$, we may directly add their difference for every two adjacent frames as the loss:

$$\sum_{t=1}^{T-1} f_\phi(\mathbf{h}_t) - f_\phi(\mathbf{h}_{t+1}). \tag{4}$$

However, optimizing the above loss function unintentionally compromises the desired objective specified in Eq. 3. The problem roots in the negative loss when $f_\phi(\mathbf{h}_t) \le f_\phi(\mathbf{h}_{t+1})$. It can neutralize the positive loss, resulting in undesirably overlooking the cases of $f_\phi(\mathbf{h}_t) > f_\phi(\mathbf{h}_{t+1})$ for some t within the sequence.

Motivated by the above failure, we notice that to create a loss function that properly guides the model for fully satisfying the condition outlined in Eq. 3, we have to ensure every local decrease of the chronological values is tackled by the loss. To this end, we optimize the loss function in Eq. 4 by leveraging the ReLU function to set the negative loss to zero. The loss function then becomes:

$$\sum_{t=1}^{T-1} \text{ReLU} \left(f_\phi(\mathbf{h}_t) - f_\phi(\mathbf{h}_{t+1}) \right). \tag{5}$$

As a result, if an earlier frame's chronological value is unintended larger than that of the next frame, a positive loss value will emerge, and not be neutralized. This loss will encourage the model to eliminate the local decrease. The loss value is zero if and only if the chronological values monotonically increase along with the temporal evolutionary direction of the skeleton sequence.

3.3   Framework

The framework incorporating the two temporal accessories is displayed in Fig. 3[1]. It is integrally composed of five components. DCE is the first part to preprocess the skeleton action sequence before feeding it to the second unit for deep learning, which is flexibly replaceable with arbitrary skeleton-based action recognizers. The third piece is a spatial pooling layer. The processing pathway is then divided into two parallel branches to the remaining two components. One is for CRL, and the other is for classification. The dual loss functions are jointly applied during training.

---

[1] Some icons in the paper's illustrations are from flaticon.

Table 1. Comparison of using CRL and DCE individually and their ensemble. We include the top-1 accuracy and the total number of parameters (#params) in the networks. #Ens is the number of models used in an ensemble. BSL is baseline. Joint and Bone denote the use of joint and bone features, respectively. We highlight accuracy improvement with magenta bold.

Methods	# Ens	NTU60				NTU120				# Params (M)	GFlops
		X-Sub	Acc ↑	X-View	Acc↑	X-Sub	Acc↑	X-Set	Acc↑		
BSL (Joint)	1	87.2	-	93.7	-	81.9	-	83.7	-	1.44	19.4
CRL (Joint)	1	88.2	1.0	95.2	1.5	82.5	0.6	84.3	0.6	1.77	19.6
DCE (Joint)	1	88.5	1.3	95.0	1.3	83.1	1.2	85.1	1.4	1.46	19.6
Ens: DCE-CRL (Joint)	2	89.7	2.5	96.0	2.3	84.6	2.7	86.3	2.6	3.21	39.2
BSL (Bone)	1	88.2	-	93.6	-	84.0	-	85.7	-	1.44	19.4
CRL (Bone)	1	89.4	1.2	95.3	1.7	84.7	0.7	86.6	0.9	1.77	19.6
DCE (Bone)	1	90.0	1.8	95.1	1.5	85.5	1.5	86.6	0.9	1.46	19.6
Ens: DCE-CRL (Bone)	2	90.7	2.5	96.2	2.6	87.0	3.0	88.3	2.6	3.21	39.2
Ens: BSL (Joint+Bone)	2	89.3	-	94.7	-	85.9	-	87.4	-	2.88	38.8
Ens: CRL (Joint+Bone)	2	90.6	1.3	96.2	1.5	86.1	0.2	87.8	0.4	3.54	39.2
Ens: DCE (Joint+Bone)	2	90.6	1.3	96.6	1.9	86.9	1.0	88.4	1.0	2.92	38.9

4   Experiments

4.1   Datasets

NTU60 [29]. As a widely-used benchmark skeleton dataset, NTU60 contains 56,000 videos collected in a laboratory environment by Microsoft Kinect V2. This dataset is challenging because of different viewing angles of skeletons, various subject skeleton sizes, varying speeds of action performance, similar trajectories between several actions, and limited hand joints to capture detailed hand actions. NTU60 has two subtasks, *i.e.*, cross-subject, and cross-view. For cross-subject, data of half of the subjects are used for training, and those of the rest are for validation. For the cross-view task, various cameras are placed in different locations, half of the setups are to train, and the others are to validate.

NTU120 [20]: An extension of NTU60 with more subjects, more camera positions and angles, resulting in a more challenging dataset with 113,945 videos.

4.2   Experimental Setups

Our models employ PyTorch as the framework and are trained on one NVIDIA 3090 GPU for 60 epochs. Stochastic gradient descent (SGD) with momentum 0.9 is used as the optimizer. The learning rate is initialized as 0.05 and decays to be 10% at epochs 28, 36, 44, and 52. In the footsteps of [30], we preprocess the skeleton data with normalization and translation, padding each video clip to be of the same length of 300 frames by repeating the actions. Models are trained independently for NTU60 and NTU120.

Our DCE-CRL is generally compatible with existing models. We choose a modified MS-G3D [26] as our baseline architecture for studying the proposed

Table 2. Comparison of action recognition on two benchmark datasets. We compare the recognition accuracy and the total number of parameters (#params) as well as the inference time (in GFlops) in the networks. #Ens is the number of models used in an ensemble. J and B denote the use of joint and bone features, respectively. BSL represents the utilized backbone model without applying the DCE-CRL. The top accuracy is highlighted in red bold.

(a) NTU60.

Methods	# Ens	NTU60 X-Sub	Acc ↑	X-View	Acc↑	# Para (M)	GFlops
Lie Group [35]	1	50.1	-	52.8	-	-	
STA-LSTM [33]	1	73.4	-	81.2	-	-	-
VA-LSTM [40]	1	79.2	-	87.7	-	-	-
HCN [18]	1	86.5	-	91.1	-	-	-
MAN [38]	1	82.7	-	93.2	-	-	-
ST-GCN [39]	1	81.5	-	88.3	-	2.91	16.4
AS-GCN [19]	1	86.8	-	94.2	-	7.17	35.5
AGC-LSTM [32]	2	89.2	-	95.0	-	-	-
2s-AGCN [31]	4	88.5	-	95.1	-	6.72	37.2
DGNN [30]	4	89.9	-	96.1	-	8.06	71.1
Bayes-GCN [42]	1	81.8	-	92.4	-	-	-
Our method							
Ens: DCE-CRL (J+B)	2	90.6	8.8	96.6	4.2	2.92	39.2

(b) NTU120.

Methods	# Ens	NTU120 X-Sub	Acc ↑	X-View	Acc↑	# Para (M)	GFlops
PA LSTM [29]	1	25.5	-	26.3	-	-	-
Soft RNN [14]	1	36.3	-	44.9	-	-	-
Dynamic [13]	1	50.8	-	54.7	-	-	-
ST LSTM [22]	1	55.7	-	57.9	-	-	-
GCA-LSTM [23]	1	58.3	-	59.2	-	-	-
Multi-Task Net [16]	1	58.4	-	57.9	-	-	-
FSNet [21]	1	59.9	-	62.4	-	-	-
Multi CNN [17]	1	62.2	-	61.8	-	-	-
Pose Evo Map [25]	1	64.6	-	66.9	-	-	-
SkeleMotion [1]	1	67.7	-	66.9	-	-	-
2s-AGCN [31]	1	82.9	-	84.9	-	6.72	37.2
Our method							
Ens: DCE-CRL (J+B)	2	87.0	4.1	88.4	3.5	2.92	39.2

methods. We omit the G3D module because 3D graph convolutions are time-consuming, doubling the training time if employing it, according to our empirical evaluation. We refer to this simplified model as BKB, standing for the backbone. More details about BKB are in the supplementary materials. We will release the code once the paper is published.

4.3 Ablation Studies

In Table 1, we report the results of evaluating the influence of CRL and DCE. The assessments include applying the two components individually to compare against training without the proposed methods. Existing approaches reveal great accuracy improvement resulting from ensembling multiple models [2,3,26]. Hence, we also evaluate ensembling two models trained with the components separately.

We have three observations: First, both CRL and DCE substantially improve the baseline action recognizer's accuracy over all four datasets. For most cases, DCE enhances an action recognizer more than CRL, reflected in the higher improved accuracy. This implies that DCE may provide more complementary information to the recognizer. Second, previous work indicates that using the bone features leads to higher accuracy than applying the joint coordinates. We discover that both CTR and DCE improve accuracy more in the bone domain than the joint domain, implying that bones are not only more suitable representations for skeleton-based action recognition, but also they have richer potential features to be further exploited. Third, ensembling the two models trained with CRL and DCE further largely boosts the accuracy, suggesting that CRL and DCE provide complementary information to each other. We conjecture that CRL focuses on capturing the chronological order, while DCE alleviates the noise.

588   Z. Qin et al.

Table 3. Comparison with and without the DCE-CRL on the confusing actions. The "Action" column shows the ground truth labels, and the "Similar Action" column shows the predictions from the model (with/without the DCE and/or CRL). The accuracy improvements highlighted in red are the substantially increased ones (Acc↑ ≥ 5%) due to using the DCE and/or CRL. The similar actions in cyan show the change of predictions after employing DCE and/or CRL.

(a) Examples of actions whose confusing ones temporally evolve oppositely.

| Action | Joint | | Joint with DCE | | |
	Acc (%)	Similar action	Acc (%)	Acc↑ (%)	Similar action
Putting on bag	91.0	Taking off jacket	93.9	2.9	Wearing jacket
Taking off a shoe	81.0	Wearing a shoe	83.2	2.2	Wearing a shoe
Putting on headphone	85.6	Taking off headphone	87.7	2.1	Sniff

(b) Actions whose accuracy gets most improved by the CRL.

| Action | Joint | | Joint with CRL | | |
	Acc (%)	Similar action	Acc (%)	Acc↑ (%)	Similar action
Opening a box	66.4	Folding paper	73.9	7.5	Folding paper
Making a phone call	78.2	Playing with phone	84.4	6.2	Playing with phone
Playing with phone	60.0	Stapling book	64.7	4.7	Stapling book
Typing on a keyboard	64.4	Writing	69.1	4.7	Writing
Balling up paper	65.7	Folding paper	69.9	4.2	Folding paper

(c) Actions whose accuracy gets most improved by the DCE.

| Action | Joint | | Joint with DCE | | |
	Acc (%)	Similar action	Acc (%)	Acc↑ (%)	Similar action
Yawn	63.5	Hush (quiet)	72.2	8.7	Hush (quiet)
Balling up paper	65.7	Folding paper	73.9	8.2	Playing magic cube
Wipe	83.0	Touching head	90.9	7.9	Touching head
Playing magic cube	61.7	Playing with phone	68.7	7.0	Counting money
Making ok sign	38.1	Making victory sign	44.0	5.9	Making victory sign

One may assume that DCE's boost is due to the dimension increase rather than the encoding. To answer this, we also evaluate the BKB's accuracy using other two approaches. These two extend the dimensionality to be the same as the sequence with the DCE: 1) The extended dimensional values are the elementwise product of the randomly generated values between −1 and 1 and the original sequence; the range between −1 and 1 coincides with the cosine function's amplitude. 2) The enlarged dimensions are the simple repetition of the original skeleton 3D coordinates.

We observe that feeding either of the above two inputs to the BKB consequently leads to an accuracy decrease. Using the joint features of the NTU120 cross-subject setting, the accuracy slightly drops from 81.9% to 80.7% and 81.2% respectively for the first and the second encoding mechanisms mentioned above. Therefore, the DCE's boost in accuracy is due to the unique properties of the proposed components rather than dimension growth.

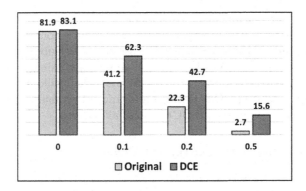

Fig. 4. Recognition accuracy on the perturbed validation set with varying noise levels. DCE stands for encoding the skeleton action sequence with DCE.

4.4   Improvement Analysis

To get insights into the working mechanisms of DCE and CRL, we quantitatively compare each action's recognition accuracy before and after applying the two proposed accessories respectively to the BKB model.

For CRL, its design intention is to dispel the confusion between two actions where one is visually like the temporal inverse of the other by explicitly encoding the chronological order of a skeleton sequence. For example, the hand joints of *opening a box* move in the opposite direction along time to those of *folding a paper*. We observe that the chaos in such extensive pairs of actions has been greatly reduced. We list three action examples in Table 3a. Note that as a result of applying CRL, the most confusing action accordingly changes from the ones with the opposite execution direction along time to those with similar motion trajectories, which implies that the previously missing information on temporal evolution has been sufficiently filled.

For more general actions, CRL also greatly enhances the accuracy of the other actions apart from the above ones, as Table 3b exhibits. This phenomenon indicates that information about temporal evolution is universally beneficial for accurately recognizing general actions. Such information is insufficiently captured without CRL. On the other hand, as Table 3c depicts, DCE also broadly boosts the recognizer's accuracy over many actions. We hypothesize this is because most skeleton action sequences contain a level of noise, and our DCE effectively mitigates its adverse influence on skeleton-based action recognition.

4.5   Noise Alleviation

The proposed DCE decomposes the original skeleton action sequence into different frequency components. We select the lower ones with the intention of alleviating noise in the signal. Here, we empirically verify the effectiveness of noise mitigation. To this end, for each instance in the validation set, we randomly sample noise from a spherical Gaussian distribution: $\mathbf{n} \sim \mathcal{N}(\mathbf{0}, \mathbf{I})$ with the same shape as the sequence. This is followed by adding it to the sequence as: $\mathbf{x} := \mathbf{x} + \epsilon \cdot \mathbf{n}$, where ϵ is a hyperparameter to manually control the intensity of noise. Larger ϵ imposes stronger noise. We use the cross-subject setting and the joint features of NTU120. As Fig. 4 shows, we see that using DCE consistently resists more against the accuracy drop on the perturbed validation set compared with that of not utilizing DCE, testing on a range of ϵ values, directly verifying the effectiveness of the proposed DCE's noise alleviation.

4.6 Compatibility with Existing Models

The proposed DCE-CRL is not merely applicable to the baseline model. The two components are widely compatible with existing networks. We desire to know the quantitative accuracy that DCE-CRL improves for an action recognizer. To this end, we train recently proposed models with and without DCE-CRL. The model pair is fairly trained and compared by setting the same hyperparameters. The utilized dataset is NTU60 cross-subject using the joint features. We summarize the results in Fig. 1. Out of the seven selected action recognizers, DCE-CRL consistently increases the action recognizer's accuracy, indicating the DCE-CRL's great compatibility.

4.7 Comparison with SOTA Accuracy

We show in Table 2 that extending the simple baseline with DCE-CRL outperforms the current competitive state-of-the-art action recognizers; even the number of parameters and the inference time of the utilized model (BKB) are respectively far smaller and shorter than the SOTA networks. Note that DCE-CRL is generally compatible with existing models. Hence, replacing the utilized simple backbone with a more complicated model can achieve even higher accuracy. To illustrate, we equip MSG3D with DCE-CRL and obtain the accuracy of **87.5%** and **89.1%**, respectively on the cross-subject and cross-setup of NTU120. As a comparison, the original MSG3D's accuracy results on these two settings without using the proposed components are 86.9% and 88.4%, respectively. Similarly, this combination achieves **91.8%** and **96.8%** on the cross-subject and cross-view of NTU60, achieving new SOTA accuracy.

5   Conclusion

We propose two temporal accessories that are compatible with existing skeleton-based action recognizers for boosting their accuracy. On the one hand, discrete cosine encoding (DCE) facilitates a model to comprehensively capture information from the frequency domain. Additionally, we can proactively alleviate signal noise by highlighting the low-frequency components of an action sequence. On the other hand, the proposed chronological loss directly leads a network to be aware of the action sequence's temporal direction. Our experimental results on benchmark datasets show that incorporating the two components consistently improves existing skeleton-based action recognizers' performance.

References

1. Caetano, C., Sena, J., Brémond, F., Dos Santos, J.A., Schwartz, W.R.: Skelemotion: a new representation of skeleton joint sequences based on motion information for 3d action recognition. In: 2019 16th IEEE International Conference on Advanced Video and Signal Based Surveillance (AVSS), pp. 1–8. IEEE (2019)
2. Cheng, K., Zhang, Y., Cao, C., Shi, L., Cheng, J., Lu, H.: Decoupling GCN with dropgraph module for skeleton-based action recognition. In: Vedaldi, A., Bischof, H., Brox, T., Frahm, J.-M. (eds.) ECCV 2020. LNCS, vol. 12369, pp. 536–553. Springer, Cham (2020). https://doi.org/10.1007/978-3-030-58586-0_32
3. Cheng, K., Zhang, Y., He, X., Chen, W., Cheng, J., Lu, H.: Skeleton-based action recognition with shift graph convolutional network. In: Proceedings of the IEEE/CVF Conference on Computer Vision and Pattern Recognition, pp. 183–192 (2020)
4. Cheng, K., Zhang, Y., He, X., Cheng, J., Lu, H.: Extremely lightweight skeleton-based action recognition with ShiftGCN++. IEEE Trans. Image Process. **30**, 7333–7348 (2021)
5. Cherian, A., Fernando, B., Harandi, M., Gould, S.: Generalized rank pooling for activity recognition. In: Proceedings of the IEEE Conference On Computer Vision and Pattern Recognition, pp. 3222–3231 (2017)
6. Cherian, A., Sra, S., Gould, S., Hartley, R.: Non-linear temporal subspace representations for activity recognition. In: Proceedings of the IEEE Conference on Computer Vision and Pattern Recognition, pp. 2197–2206 (2018)
7. Du, Y., Wang, W., Wang, L.: Hierarchical recurrent neural network for skeleton based action recognition. In: Proceedings of the IEEE Conference On Computer Vision and Pattern Recognition, pp. 1110–1118 (2015)
8. Fanello, S.R., Gori, I., Metta, G., Odone, F.: Keep it simple and sparse: real-time action recognition. J. Mach. Learn. Res. **14**, 2617–2640 (2013)
9. Fernando, B., Anderson, P., Hutter, M., Gould, S.: Discriminative hierarchical rank pooling for activity recognition. In: Proceedings of the IEEE Conference on Computer Vision and Pattern Recognition, pp. 1924–1932 (2016)
10. Fernando, B., Gavves, E., Oramas, J., Ghodrati, A., Tuytelaars, T.: Rank pooling for action recognition. IEEE Trans. Pattern Anal. Mach. Intell. **39**(4), 773–787 (2016)
11. Fernando, B., Gould, S.: Learning end-to-end video classification with rank-pooling. In: International Conference on Machine Learning, pp. 1187–1196. PMLR (2016)
12. Fernando, B., Gould, S.: Discriminatively learned hierarchical rank pooling networks. Int. J. Comput. Vision **124**(3), 335–355 (2017)
13. Hu, J.F., Zheng, W.S., Lai, J., Zhang, J.: Jointly learning heterogeneous features for RGB-D activity recognition. In: Proceedings of the IEEE Conference On Computer Vision and Pattern Recognition, pp. 5344–5352 (2015)
14. Hu, J.F., Zheng, W.S., Ma, L., Wang, G., Lai, J., Zhang, J.: Early action prediction by soft regression. IEEE Trans. Pattern Anal. Mach. Intell. **41**(11), 2568–2583 (2018)
15. Ji, X., Cheng, J., Feng, W., Tao, D.: Skeleton embedded motion body partition for human action recognition using depth sequences. Signal Process. **143**, 56–68 (2018)
16. Ke, Q., Bennamoun, M., An, S., Sohel, F., Boussaid, F.: A new representation of skeleton sequences for 3d action recognition. In: IEEE/CVF Conference on Computer Vision and Pattern Recognition (CVPR), pp. 3288–3297 (2017)

17. Ke, Q., Bennamoun, M., An, S., Sohel, F., Boussaid, F.: Learning clip representations for skeleton-based 3d action recognition. IEEE Trans. Image Process. **27**(6), 2842–2855 (2018)

18. Li, C., Zhong, Q., Xie, D., Pu, S.: Co-occurrence feature learning from skeleton data for action recognition and detection with hierarchical aggregation. In: International Joint Conference on Artificial Intelligence (IJCAI) (2018)

19. Li, M., Chen, S., Chen, X., Zhang, Y., Wang, Y., Tian, Q.: Actional-structural graph convolutional networks for skeleton-based action recognition. In: Conference on Computer Vision and Pattern Recognition (CVPR), pp. 3595–3603 (2019)

20. Liu, J., Shahroudy, A., Perez, M., Wang, G., Duan, L.Y., Kot, A.C.: NTU RGB+D 120: a large-scale benchmark for 3d human activity understanding. IEEE Trans. Pattern Anal. Mach. Intell. **42**(10), 2684–2701 (2019)

21. Liu, J., Shahroudy, A., Wang, G., Duan, L.Y., Kot, A.C.: Skeleton-based online action prediction using scale selection network. IEEE Trans. Pattern Anal. Mach. Intell. **42**(6), 1453–1467 (2019)

22. Liu, J., Shahroudy, A., Xu, D., Wang, G.: Spatio-temporal LSTM with trust gates for 3d human action recognition. In: Leibe, B., Matas, J., Sebe, N., Welling, M. (eds.) ECCV 2016. LNCS, vol. 9907, pp. 816–833. Springer, Cham (2016). https://doi.org/10.1007/978-3-319-46487-9_50

23. Liu, J., Wang, G., Hu, P., Duan, L.Y., Kot, A.C.: Global context-aware attention LSTM networks for 3d action recognition. In: Proceedings of the IEEE Conference on Computer Vision and Pattern Recognition, pp. 1647–1656 (2017)

24. Liu, M., Liu, H., Chen, C.: Enhanced skeleton visualization for view invariant human action recognition. Pattern Recogn. **68**, 346–362 (2017)

25. Liu, M., Yuan, J.: Recognizing human actions as the evolution of pose estimation maps. In: Proceedings of the IEEE Conference on Computer Vision and Pattern Recognition, pp. 1159–1168 (2018)

26. Liu, Z., Zhang, H., Chen, Z., Wang, Z., Ouyang, W.: Disentangling and unifying graph convolutions for skeleton-based action recognition. In: Proceedings of the IEEE/CVF Conference on Computer Vision and Pattern Recognition, pp. 143–152 (2020)

27. Mao, W., Liu, M., Salzmann, M., Li, H.: Learning trajectory dependencies for human motion prediction. In: Proceedings of the IEEE/CVF International Conference on Computer Vision, pp. 9489–9497 (2019)

28. Saggese, A., Strisciuglio, N., Vento, M., Petkov, N.: Learning skeleton representations for human action recognition. Pattern Recogn. Lett. **118**, 23–31 (2019)

29. Shahroudy, A., Liu, J., Ng, T.T., Wang, G.: NTU RGB+ D: A large scale dataset for 3d human activity analysis. In: Proceedings of the IEEE Conference on Computer Vision and Pattern Recognition, pp. 1010–1019 (2016)

30. Shi, L., Zhang, Y., Cheng, J., Lu, H.: Skeleton-based action recognition with directed graph neural networks. In: Proceedings of the IEEE/CVF Conference on Computer Vision and Pattern Recognition, pp. 7912–7921 (2019)

31. Shi, L., Zhang, Y., Cheng, J., Lu, H.: Two-stream adaptive graph convolutional networks for skeleton-based action recognition. In: Proceedings of the IEEE/CVF conference on computer vision and pattern recognition, pp. 12026–12035 (2019)

32. Si, C., Chen, W., Wang, W., Wang, L., Tan, T.: An attention enhanced graph convolutional LSTM network for skeleton-based action recognition. In: Proceedings of the IEEE/CVF Conference on Computer Vision and Pattern Recognition, pp. 1227–1236 (2019)

33. Song, S., Lan, C., Xing, J., Zeng, W., Liu, J.: An end-to-end spatio-temporal attention model for human action recognition from skeleton data. In: Proceedings of the AAAI Conference on Artificial Intelligence, vol. 31 (2017)
34. Tran, D., Wang, H., Torresani, L., Ray, J., LeCun, Y., Paluri, M.: A closer look at spatiotemporal convolutions for action recognition. In: Proceedings of the IEEE Conference on Computer Vision and Pattern Recognition, pp. 6450–6459 (2018)
35. Vemulapalli, R., Arrate, F., Chellappa, R.: Human action recognition by representing 3d skeletons as points in a lie group. In: Proceedings of the IEEE Conference on Computer Vision and Pattern Recognition, pp. 588–595 (2014)
36. Wang, L., Huynh, D.Q., Koniusz, P.: A comparative review of recent kinect-based action recognition algorithms. IEEE Trans. Image Process. (TIP) **29**, 15–28 (2020). https://doi.org/10.1109/TIP.2019.2925285
37. Wang, L., Koniusz, P., Huynh, D.: Hallucinating IDT descriptors and i3d optical flow features for action recognition with CNNs. In: Proceedings of the IEEE/CVF International Conference on Computer Vision (ICCV) (2019). https://doi.org/10.1109/ICCV.2019.00879
38. Xie, C., Li, C., Zhang, B., Chen, C., Han, J., Liu, J.: Memory attention networks for skeleton-based action recognition. In: International Joint Conference on Artificial Intelligence (IJCAI) (2018)
39. Yan, S., Xiong, Y., Lin, D.: Spatial temporal graph convolutional networks for skeleton-based action recognition. In: Thirty-second AAAI Conference on Artificial Intelligence (2018)
40. Zhang, P., Lan, C., Xing, J., Zeng, W., Xue, J., Zheng, N.: View adaptive recurrent neural networks for high performance human action recognition from skeleton data. In: Proceedings of the IEEE International Conference on Computer Vision, pp. 2117–2126 (2017)
41. Zhang, P., Lan, C., Zeng, W., Xing, J., Xue, J., Zheng, N.: Semantics-guided neural networks for efficient skeleton-based human action recognition. In: Proceedings of the IEEE/CVF Conference on Computer Vision and Pattern Recognition, pp. 1112–1121 (2020)
42. Zhao, R., Wang, K., Su, H., Ji, Q.: Bayesian graph convolution LSTM for skeleton based action recognition. In: Proceedings of the IEEE/CVF International Conference on Computer Vision (ICCV) (2019)

Which Expert Knows Best? Modulating Soft Learning with Online Batch Confidence for Domain Adaptive Person Re-Identification

Andrea Zunino[1]([envelope]) [iD], Christopher Murray[1] [iD], Richard Blythman[1] [iD], and Vittorio Murino[1,2,3] [iD]

[1] Ireland Research Center, Huawei Technologies Co. Ltd., Dublin, Ireland
{andrea.zunino,vittorio.murino}@iit.it,
{christopher.murray,richard.blythman}@huawei.com
[2] Pattern Analysis and Computer Vision (PAVIS), Istituto Italiano di Tecnologia, Genova, Italy
[3] University of Verona, Verona, Italy

Abstract. Deploying a person re-identification (Re-ID) system in a real scenario requires adapting a model trained on one labeled dataset to a different environment, with no person identity information. This poses an evident challenge that can be faced by unsupervised domain adaptation approaches. Recent state-of-the-art methods adopt architectures composed of multiple models (a.k.a. *experts*), and transfer the learned knowledge from the source domain by clustering and assigning hard pseudo-labels to unlabeled target data. While this approach achieves outstanding accuracy, the clustering procedure is typically sub-optimal, and the experts are simply combined to learn in a collaborative way, thus limiting the final performance. In order to mitigate the effects of noisy pseudo-labels and better exploit experts' knowledge, we propose to combine soft supervision techniques in a novel multi-expert domain adaptation framework. We introduce a novel weighting mechanism for soft supervisory learning, named Online Batch Confidence, which takes into account expert reliability in an online per-batch basis. We conduct experiments across popular cross-domain Re-ID benchmarks proving that our model outperforms the current state-of-the-art results.

Keywords: Human re-identification · Unsupervised domain adaptation · Clustering · Soft learning · Mutual mean teaching

1 Introduction

Person re-identification (Re-ID) targets the identification of a person across non-overlapping cameras. Given a query image of a person, the goal is to recognize the same subject in images acquired by different cameras by analysing and extracting appearance information only (without using biometric cues).

© The Author(s), under exclusive license to Springer Nature Switzerland AG 2023
L. Karlinsky et al. (Eds.): ECCV 2022 Workshops, LNCS 13805, pp. 594–607, 2023.
https://doi.org/10.1007/978-3-031-25072-9_40

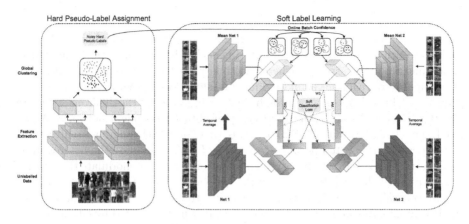

Fig. 1. Our proposed multi-expert domain adaptation technique for Re-ID consisting of: (left) global Hard Pseudo-Label Assignment, and (right) online Soft Label Learning. At each epoch, two multi-branch teachers extract features that are globally clustered, providing the hard pseudo-labels to the data. This label assignment is then locally exploited during the soft loss computations. Each soft contribution is weighted according to how well a teacher-expert locally groups the data in the batch. Our method assigns higher weight to the experts that map global pseudo-labels to more compact feature groups in the online batch of data. Two weighted guidances are further generated from each teacher-expert (lighter yellow and red), providing soft labels to the student-experts (darker yellow and red) of both Net 1 and Net 2 in an asymmetric-branch configuration. (Color figure online)

Unsupervised domain adaptation (UDA) introduces more challenging issues in the Re-ID task. UDA aims to refine a model trained in a source domain (*i.e.*, using labeled data taken from a fixed set of cameras) by using unlabeled target images acquired in a different domain (*e.g.*, acquired by different environmental conditions). Recent state-of-the-art UDA Re-ID methods assign hard pseudo-labels to unlabeled target data [1,7,15,20–24] in order to re-cast the problem as a supervised one. To generate these pseudo-labels, a clustering algorithm is typically used that exploits feature similarity to assign informative initial labels, which are subsequently refined offline once per epoch. However, the pseudo-labels are generally very noisy, and this is due to the limited ability to transfer the source-domain features to the target domain, to difficulties with estimating the unknown number of identities, and the sub-optimal behavior of the clustering algorithm.

To increase the discrimination capability of a Re-ID system and also to mitigate the negative impact of noisy labels, recent works have adopted an ensemble of multiple networks (a.k.a. *experts*) [7,23], which can generate better predictions for unlabelled samples than the output of a single network at the most recent training epoch. The distribution of the training data is roughly captured after the networks are trained with hard pseudo-labels, and thus the class predictions of such networks can serve as soft labels for training in a teacher-student framework [13,17]. A number of recent Re-ID studies have found great success using such approaches. For example, the mutual mean-teaching (MMT) framework [7] combines the ideas of using the temporal average of a network for generating predictions [16], and deep mutual learning [25], where two identical networks with different initializations are soft supervising each other. Chen *et al.* [1] introduced a self-mean teacher strategy where multiple branches in a single network (and its temporal average) act as individual sub-networks to generate distinct predictions. Each branch gets supervision from its peer branch of the different structure, which enhances the divergence between paired teacher-student networks. Multiple Expert Brainstorming (MEB) uses three networks with different architectures, *i.e.* experts and their temporal averages [23]. To accommodate the heterogeneity of experts, MEB-Net introduces a regularization scheme to modulate the experts' authority according to the quality of the generated clusters in the target domain.

In general, the research trend is leaning towards larger ensembles with a higher number of soft supervisory signals that can be provided by either (i) temporal averages, (ii) a collaboration of two or more individual networks, or (iii) collaborative sub-networks. We suggest that these approaches are complementary. In this work, we explore the benefits of incorporating all of these methods within a novel soft classification loss. By doing this, we are able to put together the advantages of feature learning that were exploited separately on prior art, in order to be more robust to noisy labels. Inspired by the MMT framework [7], our approach exploits a multi-branch architecture where the individual branches act as experts. We first take advantage of the availability of multiple teachers, *i.e.* self and mutual, during the adaptation and combine the soft guidance. As the number of predictions acting as soft supervision is increased, it becomes important to consider the confidence of the prediction generated by a particular approach. Most existing frameworks neglect the confidence of the various networks that take part in online peer-teaching, or adopt weighting mechanisms that are not flexible during the epochs (*e.g.*, authority regularization [23]). These strategies in fact disregard the reliability of each expert on smaller subsets of data observed during each training iteration, impacting the learning process negatively. To counterbalance this downside, we design a novel online batch confidence score where the inter- and intra-cluster metrics of each expert are computed on-the-fly for each individual batch of samples. Our method does not require additional local clustering, but exploits the hard pseudo-label assignment provided by the initial global clustering in each training iteration (see Fig. 1).

By estimating a tailored weight to each batch, we can better assess the expert knowledge, thus allowing a better adaptation to the data and increasing the reliability of the training process of each expert and, hence, overall. We evaluate our proposed model on the popular cross-domain Re-ID benchmarks, where we achieve state-of-the-art results for Re-ID in UDA scenario.

In summary, the contributions of our work are:

- We propose a new multi-expert domain adaptation technique for Re-ID introducing a novel unified soft loss to contrast the noise in the pseudo-labels' assignment. The unified soft loss incorporates multiple supervisory signals from (i) models' temporal averages, (ii) collaboration of individual networks, and (iii) collaborative sub-networks, in a unified self, mutual, and multiple-branch learning approach.

- We also introduce a novel online weighting mechanism for the soft-learning strategy that takes into account dynamically the expert confidence in grouping the images in the batches, resulting in a further better overall robustness to label noise.
- By exploiting the final ensemble of expert networks, we obtain state-of-the-art results in popular domain-adaptive Re-ID settings.

2 Related Work

We identified two classes of approaches in relation to our proposed work:

Domain Translation-Based Methods. Domain translation-based methods [2,4,5,9,19,29] provide an effective way to make use of source domain images and their valuable ground-truth identities. These methods generate source-to-target translated images and fine-tune the Re-ID model with those samples. Maintaining the original inter-sample relations and distributions of source-domain data during translation is critical for generating informative training samples, and the generated images should preserve the original identities. To do this, ID-based regularizations are adopted on either pixel level [19] or feature level [2,4,5] by means of contrastive loss [4] or classification loss [2,5]. Ge *et al.* [9] focus on carefully translating the source-domain images to the target domain and they introduce online relation-consistency regularization to maintain inter-sample relations during the training. Zhong *et al.* [29] propose ECN+, a framework equipped with a memory module that can effectively exploit the sample relations over both source and target images. Different from these methods, our model is first pre-trained on the source images, and then the source domain is not further utilized in the pipeline, *i.e.* the model is adapted by only considering the unlabeled target samples, which makes it effective to be used in source (data) free scenarios too.

Pseudo Labelling-Based Methods. Pseudo labelling-based methods [1,7,15, 20–24] focus only on target-domain data and they aim at learning the distribution of the unlabeled samples. In general, they demonstrate better performance in the UDA Re-ID problem by adopting a preliminary clustering phase where pseudo-labels are assigned to unlabelled target samples. UDAP [15] was one of the first works to use pseudo-label generation for UDA, but in a self-teaching scenario. PAST [24] proposed a self-training strategy that progressively augments the model capability during adaptation by adopting conservative and promoting stages alternately. SSG [6] extracted features of all persons in target dataset and group them by three different cues, whole bodies, upper parts and lower parts assigning an independent set of pseudo-labels to each group. MMT [7] incorporated mutual learning between a pair of networks via pseudo-label generation. Along with the use of hard pseudo-labels generated during the clustering phase, each mean teacher network provides online soft labels to supervise the mutual network during domain adaptation. ABMT [1] builds on this idea by promoting diversity between the two experts by modeling them as asymmetric branches that share the same backbone, instead of using separate identical networks. AWB [18] enforces the mutual experts to learn different representations by asymmetrically amplifying feature maps in their proposed Attentive WaveBlock module. ACT [20] diversified the learning process between its two networks by filtering only inliers for one of the experts, while the other learns from data that is as diverse as possible. Opting for varying architectures across three experts, MEB-Net [23] proposed an authority regularization scheme for weighing the contribution of each expert to the mutual learning process using the global clustering metrics of each expert. SpCL [8] proposed a self-paced contrastive learning framework with hybrid memory which gradually creates more reliable clusters and learning target distribution. It fully exploits all available data, source and target, with hybrid memory for joint feature learning.

Our method differs from the above approaches as we combine self-teaching with mutual teaching, while also introducing a novel mechanism for weighting the experts contribution during the soft losses computation based on online batch-level confidence. This regularization exploits the target data only during adaptation and captures more efficiently their finer differences thanks to more reliable experts' guidance during the training process. In this way, we put more emphasis on the expert that has better knowledge on the batch sample identities, as well as leverage a dynamic mechanism allowing the weighting of the experts to change throughout each epoch.

3   Method

Our approach adopts a standard two-stage training scheme consisting of supervised learning in a source domain and unsupervised adaptation to a target domain, with some modifications. During the supervised learning phase, two similar architectures are pre-trained with different initializations on the source dataset. The architectures are composed of multiple branches, where each branch represents an expert. When the pre-trained networks are adapted to the target domain in the second phase, we perform clustering on target samples at each training epoch in order to assign hard pseudo-labels, which are then utilized to fine-tune the two collaborative networks by self and mutual learning. In addition to the hard pseudo-labels, our approach exploits online soft labels produced by specific branch network predictions. In practice, each branch of the network learns from two other experts (see Fig. 1). Moreover, a novel weighting mechanism is employed to enhance the online reliability of each single expert during adaptation on a small batch of data. The knowledge from multiple experts is then mixed for target domain adaptation.

3.1   Supervised Pre-training on Source Domain

In the first stage, we train the multi-branch network in the fully supervised way on the source domain. We adopt the two-branch backbone of [1], which considers average and max pooling branches (we will refer to them in the following using subscripts a and m, respectively). The network is equipped with two branches of different structure, and thus have different levels of discrimination capability. We define these branches as experts. Given a source sample $x_i \in X_s$ (subscript s denotes the source domain) and its ground truth identity $y_i \in Y_s$, the two-branch network (with parameters θ) produces two feature representations: $F_a(x_i|\theta)$ and $F_m(x_i|\theta)$. The representations are then encoded to produce two predictions $P_a(x_i|\theta)$ and $P_m(x_i|\theta)$, respectively. For simplicity, let us denote $F_e(x_i|\theta)$ and $P_e(x_i|\theta)$ as the feature representation and predictions of the expert $e \in E = \{a, m\}$ of the network. Each single branch of the network is optimized with cross-entropy L_{ce} and batch hard triplet [10] L_{tri} losses independently. Assuming y_i one-hot encoded, the cross entropy loss is defined as:

$$L_{ce} = - \sum_{x_i \in X_s} \sum_{j=1}^{Y_s} y_i^j log P_e^j(x_i|\theta). \tag{1}$$

For the triplet loss, we consider the positive and hard negative batch choice [10]. We generate batches by randomly sampling C classes of human identities, and H images for each class. The triplet loss is given by:

$$L_{tri} = \sum_{j=1}^{C} \sum_{i=1}^{H} max(0, D_{max}(F_e(x_i^j|\theta), F_e(x_p^j|\theta)) + \\ + m_{tri} - D_{min}(F_e(x_i^j|\theta), F_e(x_n^w|\theta))), \tag{2}$$

where x_i^j represents a data sample corresponding to the i-th image of the j-th person in the batch, x_p^j and x_n^w indicate the positive and negative samples in each batch, m_{tri} denotes the triplet distance margin, and D_{max} and D_{min} are defined to consider the hard pairs as the maximum and minimum cosine distances between the representations F_e (i.e., F_a or F_m) of x_i^j and the positive and negative samples in the batch, respectively.

The whole network is trained in the source domain with a combination of both losses:

$$L_s = \sum_{e \in E} (L_{ce}(P_e(X_s|\theta), Y_s) + L_{tri}(F_e(X_s|\theta), Y_s)). \tag{3}$$

3.2 Unsupervised Adaptation on Target Domain

We pre-train two networks using Eq. 3 with different initializations. Our approach uses an MMT-based [7] framework where two pairs of networks are trained during the adaptation. In each training iteration, we assume the k-th pre-trained network is parameterized by θ^k, $k \in K = \{1, 2\}$. Such networks will be referred to as student networks. For robustness, we create a supervisory network parameterized by the temporal average of the weights and biases of each student network, which will be referred to as teacher networks. The parameters of the k-th teacher at current iteration T are denoted as Θ_T^k, which is updated as:

$$\Theta_T^k = \alpha\Theta_{T-1}^k + (1-\alpha)\theta^k, \tag{4}$$

where $\alpha \in [0,1]$ is the scale factor, and the initial temporal average parameters are $\Theta_0^k = \theta^k$.

Our unsupervised adaptation process is based on three main stages: (1) clustering and hard-label learning, (2) soft-label learning, and (3) soft losses weighting.

Clustering and Hard Pseudo-Label Learning. At the beginning of each epoch, we need to assign IDs to the unlabelled target training data (see left of Fig. 1). For each target sample $x_i \in X_t$, the multi-branch teacher is predicting two feature representations $F_a(x_i|\Theta^k)$ and $F_m(x_i|\Theta^k)$, which are concatenated to provide a unique set of features $F_{a,m}(x_i|\Theta^k)$. These features are further combined as ensemble representations between the teacher networks to obtain $F(x_i) = \sum_{k\in K} F_{a,m}(x_i|\Theta^k)$, which will be used in the clustering phase to obtain a universal hard pseudo-label set. Mini-batch k-means clustering [7] is performed to cluster all target-domain features $F(x_i)$ and assign them pseudo-IDs. The produced pseudo-IDs are then used as hard pseudo-labels $\tilde{y}_i \in \tilde{Y}_t$ for the target training samples x_i. Much like the source-domain pre-training, each k-th student network is trained independently on clustered target samples with cross-entropy loss:

$$L_{t,ce}^k = \sum_{e\in E} L_{ce}(P_e(X_t|\theta^k), \tilde{Y}_t), \tag{5}$$

and triplet loss:

$$L_{t,tri}^k = \sum_{e\in E} L_{tri}(F_e(X_t|\theta^k), \tilde{Y}_t), \tag{6}$$

where $\tilde{Y}_t$ is the hard pseudo-label set, $P_e(X_t|\theta^k)$, and $F_e(X_t|\theta^k)$ are the predictions and the feature representations of the target samples by the k-th student network, respectively.

Soft Pseudo-Label Learning. Following recent methods [7,23], our approach leverages the temporally-averaged model of each network, generating more reliable soft pseudo-labels to supervise the student networks. Specifically, in the computation of soft losses we exploit the multi-branch nature of the networks to have two teachers (i.e. the self and the mutual mean-teachers) available for each student. ABMT [1] is the first work to tackle UDA for Re-ID with a multi-branch architecture. In that work, better performance is demonstrated when asymmetric-branch guidance is adopted between the self mean-teacher and the student. Following this idea, we propose to guide each student branch with the predictions and feature representations of its own self and mutual mean-teachers. The predictions from the self and mutual teachers supervise the k-th student network with a soft cross-entropy loss [11] in a asymmetric-branch configuration. This can be formulated as:

$$L_{t,sce}^k = - \sum_{n \in K} \sum_{\substack{j,w \in E \\ j \neq w}} \sum_{x_i \in X_t} (P_j(x_i|\Theta_T^n) \log(P_w(x_i|\theta^k))). \tag{7}$$

Similarly, to further enhance the teacher-student networks' discriminative capacity, the feature representations in the teachers supervise those of the student with a soft triplet loss:

$$L_{t,stri}^k = - \sum_{n \in K} \sum_{\substack{j,w \in E \\ j \neq w}} \sum_{x_i \in X_t} (T_j(x_i|\Theta_T^n) \log(T_w(x_i|\theta^k))), \tag{8}$$

where $T(x_i|\theta^k) = \frac{exp(\|F_e(x_i|\theta^k) - F_e(x_n|\theta^k)\|_2)}{exp(\|F_e(x_i|\theta^k) - F_e(x_p|\theta^k)\|_2) + exp(\|F_e(x_i|\theta^k) - F_e(x_n|\theta^k)\|_2)}$ is the softmax triplet distance of the sample x_i, its hardest positive x_p and its hardest negative x_n in a mini-batch. By minimizing Eq. 8, the softmax triplet distance between student and teachers is encouraged to be as small as possible.

Weighting of Soft Losses. The assigned hard pseudo-labels (as described above) have provided a unified label set on which to compare the reliability of the different experts. To accommodate the heterogeneity of experts, we propose a weighting mechanism inside the soft losses, which modulates the reliability of the different experts according to how well the identities of the samples in a given batch are in agreement to the global hard pseudo-labels (see right of Fig. 1). Following the same batch selection as before, we sample C assigned-identities and n instances per identity for a given batch of N target samples. For each expert e in each network k, we denote its Online Batch Confidence $\mathcal{M}_e^k$ as:

$$\mathcal{M}_e^k = \frac{\sum_{i=1}^C n\|\mu_i - \mu\|^2}{\sum_{i=1}^C \sum_{x \in \mathbb{C}_i} \|F_e(x|\Theta_T^k) - \mu_i\|^2}, \tag{9}$$

where $\mu_i = \sum_{x \in \mathbb{C}_i} F_e(x|\Theta_T^k)/n$ and $\mu = \sum_{i=1}^N F_e(x_i|\Theta_T^k)/N$ are the average feature of $\mathbb{C}_i$, *i.e.* the group denoting the *i-th* identity, and the average feature of all batch target samples, respectively. The numerator computes a distance between the centroid of each group and the global batch centroid. It does it for all the groups and sum over them. In practice, it computes how the groups, on average, are far from the global batch-mean. The denominator computes a cumulative distance between the centroid of a group and the features assigned to that group and, analogously, it does it for all the groups and sum over them: it computes how compact are the samples in the same groups. A larger $\mathcal{M}_e^k$ means better discrimination capability over the small batch of data.

Before the computation of losses for each batch, we calculate $\mathcal{M}_e^k$ for each expert and define a set of four weights $w_a^1, w_m^1, w_a^2, w_m^2$ to be used in the soft losses computation as:

$$w_e^k = \frac{|K||E|\mathcal{M}_e^k}{\sum_{n \in K} \sum_{j \in E} \mathcal{M}_j^n}, \tag{10}$$

where $|K|$ and $|E|$ are the number of student networks, and branches per student network participating in the learning process, respectively. With this Online

Batch Confidence, our weighting mechanism modulates the reliability of experts to facilitate better discrimination in the target domain. Such regularization is efficient and the set of weights changes throughout the epoch, as well as the variability in each expert's discriminative ability from batch to batch. We then re-define the soft cross entropy loss in Eq. 7 as the following weighted sum:

$$L_{t,sce}^k = -\sum_{n \in K} \sum_{\substack{j,w \in E \\ j \neq w}} w_j^n \sum_{x_i \in X_t} (P_j(x_i|\Theta_T^n) \log(P_w(x_i|\theta^k))), \quad (11)$$

and the soft triplet losses in Eq. 8 as the following weighted sum:

$$L_{t,stri}^k = -\sum_{n \in K} \sum_{\substack{j,w \in E \\ j \neq w}} w_j^n \sum_{x_i \in X_t} (\mathcal{T}_j(x_i|\Theta_T^n) \log(\mathcal{T}_w(x_i|\theta^k))). \quad (12)$$

The teacher-student networks are trained end-to-end with Eqs. 5, 6, 11, 12 leading to the overall loss for target adaptation as:

$$
\begin{aligned}
L_t = (1 - \lambda_{ce})(L_{t,ce}^1 + L_{t,ce}^2) + \lambda_{ce}(L_{t,sce}^1 + L_{t,sce}^2) \\
+ (1 - \lambda_{tri})(L_{t,tri}^1 + L_{t,tri}^2) + \lambda_{tri}(L_{t,stri}^1 + L_{t,stri}^2).
\end{aligned} \quad (13)
$$

Once the training is over, we consider for prediction the feature ensemble extracted by the two trained teacher networks as $\sum_{k \in K} F_{a,m}(\Theta^k)$.

4   Experiments

We first introduce the Re-ID datasets used in our experiments. Then, the adopted architecture and setup are described. Finally, the results are discussed along with an ablation study.

4.1   Datasets and Evaluation

We consider the most popular ReID benchmark datasets following recent works [1,7,23].

DukeMTMC-reID [27]: This dataset contains 36,411 images of 1,812 identities from 8 high-resolution cameras. It is divided into a training set of 16,522 images belonging to 702 identities that are randomly selected from the overall images, and the testing set comprises the other 2,228 query images and 17,661 gallery ones.

Market-1501 [26]: This dataset contains 32,668 labeled images of 1,501 individuals in total acquired by 6 different cameras. The dataset is divided into a training set of 12,936 images of 751 individuals, and a test set of 19,732 images of 750 people (with 3,368 query images and 16,364 gallery images).

MSMT17 [19]: This dataset includes 126,441 images of 4,101 identities captured by 15 different cameras, considering both outdoor and indoor scenarios. It is divided into a training set of 32,621 images of 1,041 individuals, and a

Table 1. Comparison with state-of-the-art UDA methods during doman adaptation on popular datasets. The **top** and the second-top results are highlighted in bold and underlined, respectively.

Methods	Duke→ Market		Market→ Duke		Market → MSMT		Duke → MSMT	
	mAP	Rank-1	mAP	Rank-1	mAP	Rank-1	mAP	Rank-1
UDAP [15]	53.7	75.8	49.0	68.4	-	-	-	-
PAST [24]	54.6	78.4	54.3	72.3	-	-	-	-
ACT [20]	60.6	80.5	54.5	72.4	-	-	-	-
ECN+ [29]	63.8	84.1	54.4	74.0	15.2	40.4	16.0	42.5
SSG [6]	68.7	86.2	60.3	76.0	16.6	37.6	18.3	41.6
MMT [7] (ResNet-50)	71.2	87.7	65.1	78.0	21.6	46.1	23.5	50.0
ABMT [1] (ResNet-50)	<u>78.3</u>	<u>92.5</u>	<u>69.1</u>	<u>82.0</u>	23.2	49.2	<u>26.5</u>	<u>54.3</u>
SpCL [8]	-	-	-	-	**26.8**	**53.7**	-	-
Ours (ResNet-50)	**81.1**	**93.7**	**70.9**	**83.9**	<u>25.1</u>	<u>50.9</u>	**28.2**	**55.4**
MEB-Net [23]	76.0	89.9	66.1	79.6	-	-	-	-
MMT [7] (IBN-ResNet-50)	76.5	90.9	68.7	81.8	26.3	52.5	29.7	58.8
ABMT [1] (IBN-ResNet-50)	80.4	93.0	70.8	83.3	27.8	55.5	<u>33.0</u>	<u>61.8</u>
AWB [18]	<u>81.0</u>	<u>93.5</u>	<u>70.9</u>	<u>83.8</u>	<u>29.0</u>	<u>57.3</u>	29.5	61.0
Ours (IBN-ResNet-50)	**84.1**	**94.4**	**73.0**	**84.2**	**29.1**	**57.7**	**33.3**	**62.0**

test set of 93,820 images of 3,060 people (with 11,659 query images and 82,161 gallery images).

Following common practice in the Re-ID problem, we use the mean average precision (mAP) and the cumulative matching characteristics (CMC) at Rank-1 to evaluate the performance of our proposed method. In evaluations, we use DukeMTMC-reID or Market-1501 datasets as the source domain and the others as target domains.

4.2 Architectures and Setup

Networks. We test our approach using ResNet-50 and IBN-ResNet-50 [14] backbones pre-trained on ImageNet [3]. Following [1], layer 4 of the network is duplicated to create the multi-branch structure. The first branch remains unchanged from the one used in the original backbone: 3 bottlenecks and global average pooling (GAP). The second branch is composed of 4 bottlenecks and global max pooling (GMP). The two branches of the teacher networks act as experts.

Source Domain Supervised Pre-training. The models are trained for 80 epochs, and 64 images of 16 identities are selected for each iteration. For each epoch, the number of iterations is set to 200. The input image size is set to 256×128 pixels. We adopt the Adam optimizer [12] to train all models with a weight decay of 5×10^{-4}. The initial learning rate is set to 3.5×10^{-4}, which is reduced to 3.5×10^{-5} and 3.5×10^{-6} after 40 and 70 epochs.

Target Domain Unsupervised Adaptation. The models are adapted for 40 epochs. For each epoch, the networks are trained for 800 iterations with a

Table 2. Ablation study for the single modules of our method, using IBN-ResNet-50.

Variants	Duke→ Market		Market→ Duke	
	mAP	Rank-1	mAP	Rank-1
0 - Direct Transfer	38.3	67.9	36.7	57.0
1 - Baseline single-branch + MMT [7]	76.5	90.9	68.7	81.8
2 - Baseline multi-branch + SMT [1]	80.4	93.0	70.8	83.3
3 - Baseline multi-branch + MMT	82.6	93.6	70.9	83.3
4 - Baseline multi-branch + MMT + SMT	82.8	93.7	71.5	83.6
5 - Baseline multi-branch + MMT + SMT + weighting	83.2	93.9	72.2	83.9
6 - Ours	**84.1**	**94.4**	**73.0**	**84.2**

fixed learning rate 3.5×10^{-4}. For each iteration, the batch size is set to 64 with a random selection of 16 clustering-based pseudo-identities (cluster label) and 4 images for each pseudo-identity. The images are resized to 256×128 pixel resolution and augmented with random horizontal flipping and random erasing [28]. We adopt the Adam optimizer [12] with a weight decay of 5×10^{-4}. The mean-teacher network is initialized and updated as in Eq. 4 with a smoothing coefficient $\alpha = 0.999$. We set the losses' coefficients as $\lambda_{ce} = 0.5$ and $\lambda_{tri} = 0.8$. When we perform clustering, we select the optimal k values of k-means following [7,18], *i.e.* 500 for Duke→Market, 700 for Market→Duke, 1500 for Duke→MSMT and Market→MSMT.

4.3 Comparison with State-of-the-Art Methods

We compare the performance of our proposed method with recent state-of-the-art (SOTA) UDA methods in Table 1. We choose the popular UDA settings Duke→Market, Market→Duke, Duke→MSMT17, Market→MSMT17. We report results for both the simpler and the more powerful backbones, ResNet-50 and IBN-ResNet-50 models. The first part of Table 1 shows ReID performance of methods using the ResNet-50 network while the second part of the Table shows the comparison between methods adopting more powerful backbones. In general, our approach achieves SOTA results for both mAP and Rank-1 metrics for all cross-dataset settings using both ResNet-50 or IBN-ResNet-50 networks. Thus, an improvement in performance is observed regardless of the use of a specific backbone, which further validates the robustness of our proposed method. Our approach only performs at second place compared to SpCL [8] in Market→MSMT17 scenario, the only scenario for which we can compare the 2 methods. SpCL adopts a contrastive learning strategy and augmented supervision since it makes use of both source and target data during adaptation while our approach is more flexible since it can be applied whenever the source data is no more available for target adaptation.

4.4 Ablation Study

We analyse several possible learning structures to validate the effectiveness of our proposed method. We perform an ablation study considering Duke→Market and Market→Duke UDA settings, and report results in Table 2 for the different variants of our approach. After the baseline performance in line (0), showing the accuracies without applying any adaptation, in line (1), we report the performance by a standard backbone IBN-ResNet-50 inside a mutual mean-teaching framework [7]. In line (2), we consider the multi-branch backbone proposed in [1] in a self mean teaching framework. The latter backbone can easily be adopted in the MMT framework, line (3), where the asymmetric-branch guidance is coming from the mutual teacher, or coming from both the mutual and self teacher in an equal manner, line (4). By weighting the soft losses with the two teacher experts guiding the predictions and feature representations of the student experts, we further improve performance, line (5). We finally leverage the two trained teachers in an ensemble to showcase our complete method, line (6). We note that the different contributions consistently and progressively improve the baselines. From the simpler *single-branch + MMT* version to the complete method, we see significant gains of mAP (namely, 7.6–4.3% for the two addressed cases). We highlight that the proposed strategy applies on top of the expert models, so a similar behavior can also be noticed when the backbone architecture is different (*e.g.*, ResNet-50, which is not so different from the IBN-ResNet-50 used in our analysis).

5 Conclusions

In this work, we addressed unsupervised domain adaptation for person Re-ID. We proposed a novel multi-expert domain adaptation framework to tackle the noisy labelling effect, resulting from the sub-optimal clustering stage typically adopted in the recent approaches. We combined the benefits of soft supervision techniques and introduced a novel local weighting mechanism for guiding the expert knowledge in the soft loss computation. Our results demonstrated state-of-the art performance for cross-domain Re-ID scenarios.

References

1. Chen, H., Lagadec, B., Bremond, F.: Enhancing diversity in teacher-student networks via asymmetric branches for unsupervised person re-identification. In: Proceedings of the IEEE/CVF Winter Conference on Applications of Computer Vision (2021)
2. Chen, Y., Zhu, X., Gong, S.: Instance-guided context rendering for cross-domain person re-identification. In: International Conference on Computer Vision (2019)
3. Deng, J., Dong, W., Socher, R., Li, L.J., Li, K., Fei-Fei, L.: Imagenet: a large-scale hierarchical image database. In: IEEE Conference on Computer Vision and Pattern Recognition (2009)

4. Deng, W., Zheng, L., Ye, Q., Kang, G., Yang, Y., Jiao, J.: Image-image domain adaptation with preserved self-similarity and domain-dissimilarity for person re-identification. In: IEEE Conference on Computer Vision and Pattern Recognition (2018)

5. Deng, W., Zheng, L., Ye, Q., Yang, Y., Jiao, J.: Similarity-preserving image-image domain adaptation for person re-identification. arXiv preprint arXiv:1811.10551 (2018)

6. Fu, Y., Wei, Y., Wang, G., Zhou, Y., Shi, H., Huang, T.S.: Self-similarity grouping: a simple unsupervised cross domain adaptation approach for person re-identification. In: International Conference on Computer Vision (2019)

7. Ge, Y., Chen, D., Li, H.: Mutual mean-teaching: pseudo label refinery for unsupervised domain adaptation on person re-identification. In: International Conference on Learning Representations (2019)

8. Ge, Y., Zhu, F., Chen, D., Zhao, R., Li, H.: Self-paced contrastive learning with hybrid memory for domain adaptive object re-id. In: Advances in Neural Information Processing Systems (2020)

9. Ge, Y., Zhu, F., Zhao, R., Li, H.: Structured domain adaptation with online relation regularization for unsupervised person re-id. arXiv e-prints pp. arXiv-2003 (2020)

10. Hermans, A., Beyer, L., Leibe, B.: In defense of the triplet loss for person re-identification. arXiv preprint arXiv:1703.07737 (2017)

11. Hinton, G., Vinyals, O., Dean, J.: Distilling the knowledge in a neural network. arXiv preprint arXiv:1503.02531 (2015)

12. Kingma, D.P., Ba, J.: Adam: a method for stochastic optimization. arXiv preprint arXiv:1412.6980 (2014)

13. Lopez-Paz, D., Bottou, L., Schölkopf, B., Vapnik, V.: Unifying distillation and privileged information. In: International Conference on Learning Representations (2016)

14. Pan, X., Luo, P., Shi, J., Tang, X.: Two at once: enhancing learning and generalization capacities via ibn-net. In: European Conference on Computer Vision (2018)

15. Song, L., et al.: Unsupervised domain adaptive re-identification: theory and practice. Pattern Recogn. **102**, 107173 (2020)

16. Tarvainen, A., Valpola, H.: Mean teachers are better role models: weight-averaged consistency targets improve semi-supervised deep learning results. arXiv preprint arXiv:1703.01780 (2017)

17. Wang, L., Yoon, K.J.: Knowledge distillation and student-teacher learning for visual intelligence: a review and new outlooks. IEEE Trans. Pattern Anal. Mach. Intell. (2021)

18. Wang, W., Zhao, F., Liao, S., Shao, L.: Attentive waveblock: complementarity-enhanced mutual networks for unsupervised domain adaptation in person re-identification. arXiv preprint arXiv:2006.06525 (2020)

19. Wei, L., Zhang, S., Gao, W., Tian, Q.: Person transfer GAN to bridge domain gap for person re-identification. In: IEEE Conference on Computer Vision and Pattern Recognition (2018)

20. Yang, F., et al.: Asymmetric co-teaching for unsupervised cross-domain person re-identification. In: AAAI (2020)

21. Yu, H.X., Zheng, W.S., Wu, A., Guo, X., Gong, S., Lai, J.H.: Unsupervised person re-identification by soft multilabel learning. In: IEEE Conference on Computer Vision and Pattern Recognition (2019)

22. Zhai, Y., et al.: Ad-cluster: augmented discriminative clustering for domain adaptive person re-identification. In: IEEE Conference on Computer Vision and Pattern Recognition (2020)
23. Zhai, Y., Ye, Q., Lu, S., Jia, M., Ji, R., Tian, Y.: Multiple expert brainstorming for domain adaptive person re-identification. In: Vedaldi, A., Bischof, H., Brox, T., Frahm, J.-M. (eds.) ECCV 2020. LNCS, vol. 12352, pp. 594–611. Springer, Cham (2020). https://doi.org/10.1007/978-3-030-58571-6_35
24. Zhang, X., Cao, J., Shen, C., You, M.: Self-training with progressive augmentation for unsupervised cross-domain person re-identification. In: International Conference on Computer Vision (2019)
25. Zhang, Y., Xiang, T., Hospedales, T.M., Lu, H.: Deep mutual learning. In: IEEE Conference on Computer Vision and Pattern Recognition (2018)
26. Zheng, L., Shen, L., Tian, L., Wang, S., Wang, J., Tian, Q.: Scalable person re-identification: a benchmark. In: International Conference on Computer Vision (2015)
27. Zheng, Z., Zheng, L., Yang, Y.: Unlabeled samples generated by GAN improve the person re-identification baseline in vitro. In: IEEE Conference on Computer Vision and Pattern Recognition (2017)
28. Zhong, Z., Zheng, L., Kang, G., Li, S., Yang, Y.: Random erasing data augmentation. In: AAAI (2020)
29. Zhong, Z., Zheng, L., Luo, Z., Li, S., Yang, Y.: Learning to adapt invariance in memory for person re-identification. IEEE Trans. Pattern Anal. Mach. Intell. (2020)

Cross-Modality Attention and Multimodal Fusion Transformer for Pedestrian Detection

Wei-Yu Lee[✉], Ljubomir Jovanov, and Wilfried Philips

TELIN-IPI, Ghent University-imec, Gent, Belgium
{Weiyu.Lee,Ljubomir.Jovanov,Wilfried.Philips}@ugent.be

Abstract. Pedestrian detection is an important challenge in computer vision due to its various applications. To achieve more accurate results, thermal images have been widely exploited as complementary information to assist conventional RGB-based detection. Although existing methods have developed numerous fusion strategies to utilize the complementary features, research that focuses on exploring features exclusive to each modality is limited. On this account, the features specific to one modality cannot be fully utilized and the fusion results could be easily dominated by the other modality, which limits the upper bound of discrimination ability. Hence, we propose the Cross-modality Attention Transformer (CAT) to explore the potential of modality-specific features. Further, we introduce the Multimodal Fusion Transformer (MFT) to identify the correlations between the modality data and perform feature fusion. In addition, a content-aware objective function is proposed to learn better feature representations. The experiments show that our method can achieve state-of-the-art detection performance on public datasets. The ablation studies also show the effectiveness of the proposed components.

Keywords: Cross-modality fusion · Multimodal pedestrian detection · Transformer

1 Introduction

Pedestrian detection is one of the most important challenges in computer vision due to its various applications, including autonomous driving, robotics, drones, and video surveillance. For achieving more accurate and robust results, thermal images have been widely exploited as the complementary information to solve the challenging problems that impede conventional RGB-based detection, such as background clutter, occlusions, or adverse lighting conditions. Radiated heat of pedestrians contains sufficient features to differentiate shape of humans from the background but lose visual appearance details in thermal images. On the other

Supplementary Information The online version contains supplementary material available at https://doi.org/10.1007/978-3-031-25072-9_41.

hand, RGB cameras can capture fine visual characteristics of pedestrians (e.g., texture). Hence, designing a fusion scheme to effectively utilize the different visual cues from thermal and RGB modalities has become a popular research interest.

In the existing methods, numerous fusion strategies have been exploited to utilize complementary features from color and thermal images [9,28,32]. In addition to simple feature concatenation [9], semantic information [13] and attention mechanism [29] are also introduced to improve the detection performance. Furthermore, in order to better exploit the characteristics of different modalities, the illumination condition is also considered during feature fusion [6,14].

However, most previous studies only focus on performing feature fusion and exploiting the fused features instead of exploring features specific to each modality (i.e. modality-specific features) before fusion [9,11,13,20,31]. Specifically, most methods from the literature put emphasis on performing detection after the features from the color and thermal images are fused, while the features exclusive for each modality are not entirely utilized in the fusion process.

As a consequence, some of the features specific to one modality cannot be fully utilized. For instance, the texture or color of RGB images in bad illumination conditions might not be properly explored due to the domination of strong features from the other modality in the fusion process. However, in conventional RGB-based detection, the texture of the objects provides important cues that make the targets distinct from the background clutter. Without fully considering specific features, the fused information becomes the main discriminative cues, which limits the upper bound of discrimination ability.

In this paper, we propose a novel network architecture for multimodal pedestrian detection based on exploring the potential of modality-specific features to boost the detection performance.

The first key idea of this paper is better exploitation of modality-specific features by cross-referencing the complementary modality data in order to obtain more discriminative details. Instead of extracting features from modalities by independent feature extractors, we suggest that the aligned thermal-visible image pairs could act as a "consultant" for each other to discover potential specific features. The argumentation for such reasoning can be found in the set theory. Fused features are in fact represented as an intersection of features from thermal and RGB images, while modality-specific features remain unused in disjunctive parts of the feature sets of each modality.

This shows that a large amount of features remains unused. While pixel level fusion resembles finding intersection of two sets, we would like to introduce a union of **features** present in both modalities. This is accomplished by a unique multimodal transformer with a novel cross-referencing self-attention mechanism, called the Cross-modality Attention Transformer (CAT), to consider cross-modality information as keys to compute the weights on the current modality values.

Furthermore, after extracting modality-specific features, we propose a Multimodal Fusion Transformer (MFT) to perform the fusion process on every pair of ROIs. Our MFT further improves the detection performance by merging the multimodal features simultaneously with the help of the self-attention mechanism. Moreover, in order to learn distinct feature representations between foreground (pedestrian) and background, a novel content-aware objective function is proposed to guide the model, which shortens the intra-class variation and expands the inter-class variation.

Our main contributions can be summarized as follows:

- We have identified the disadvantages of the recent techniques relying on fused features for multimodal pedestrian detection, and introduced a new modality fusion scheme, which can effectively utilize modality-specific information of each modality.
- To our best knowledge, we are among the first to propose a Transformer-based network to enhance modality-specific features by cross-referencing the other modality. In feature extraction, we rely on the Cross-Modality Attention Transformer (CAT), which identifies related features in both modalities and consults the second modality in order to perform information aggregation based on the union of sets instead of the intersection type of fusion.
- Thanks to the ability of transformer networks to identify correlations between heterogeneous data, in this case RGB and thermal ROIs, our proposed Multimodal Fusion Transformer (MFT) performs association between detected regions in a more efficient way, compared to classical CNN methods.
- In our experiments, we qualitatively and quantitatively verify the performance of our model against recent relevant methods and achieve comparable or better detection results on the *KAIST* and *CVC-14* datasets.

2   Related Work

2.1   Multimodal Pedestrian Detection

Although deep learning methods have significantly advanced and dominated conventional RGB-based detection, detecting pedestrians in adverse weather and lighting conditions, background clutter or occlusions, is still a non-trivial problem. Motivated by this, numerous researchers have developed different fusion schemes relying on an additional modality to improve detection performance. Hwang et al. [7] have proposed a widely used pedestrian dataset with synchronized color and thermal image pairs. Next, Liu et al. compared various fusion architectures and proposed an important baseline model based on the halfway fusion [9] and Faster R-CNN [22]. Then, in the papers of Konig et al. [11] and Park et al. [20], the fusion models based on Faster R-CNN were also proposed. Moreover, in order to distinguish pedestrian instances from hard negatives, additional attributes have been introduced. For instance, Li et al. [13] leveraged the auxiliary semantic segmentation task to boost pedestrian detection results. Zhang et al. [29] also utilized the weak semantic labels to learn an attentive mask helping the model to focus on the pedestrians. In addition, the illumination condition of the scenes is also studied to improve the fusion results. For example, Guan et al. [6] and Li et al. [14] estimate the lighting condition of the images to determine the weights between thermal and color features. Furthermore, the misaligned and unpaired problems between the modalities have been investigated by Zhang et al. [31] and Kim et al. [10]. However, in most of the aforementioned methods, pedestrian detection is usually performed after the features have been fused. The discussion about exploring or enhancing the specific features of each modality is quite limited.

2.2   Multimodal Transformers

The self-attention mechanism of transformers has shown its advantages in many natural language processing and computer vision tasks [3,25], and recently, it has been also applied to various multimodal fusion problems, such as image and video retrieval [1,4], image/video captioning [8,15,18,24], visual tracking [27] and autonomous driving [21].

Typically, multimodal inputs to transformers are allowed free attention flow between different modalities [19]. For instance, spatial regions in the image and audio spectrum would be considered and aggregated simultaneously without limitation. However, unlike audiovisual learning tasks, aligned thermal and color image pairs have more features in common, such as the shape of objects. Directly applying traditional self-attention to the image pairs might have difficulties to extract useful features due to the redundant information. In addition, the literature on fusing thermal and color images relying on transformers for pedestrian detection is quite limited. Hence, the strategies for utilization of shared information and exploring specific cues of each modality remains an important issue.

3   Proposed Method

The objective of our proposed fusion scheme is to explore potential modality-specific features and perform multimodal pedestrian detection using thermal and color image pairs as input. Our model consist of three main parts: (1) two-stream feature extractor with cross-modality attention, (2) modality-specific region proposals, and (3) multimodal fusion. As illustrated in Fig. 1, we rely on two independent feature extractors to obtain the features from color and thermal image pair I_c and I_t.

Simultaneously with the extraction, we feed feature maps from feature extractors to our proposed Cross-modality Attention Transformer (CAT), for considering cross-modality information used to enhance modality-specific features. The attended results are concatenated with each original input feature maps and forwarded to the corresponding region proposal network RPN_c, and RPN_t to find modality-specific ROIs. Finally, the proposed Multimodal Fusion Transformer (MFT) merges the ROI pair R_c and R_t from ROI pooling module and output classification and bounding box predictions. In the following subsections, we explain the details of each contribution (Fig. 2).

3.1   Cross-Modality Attention Transformer

In previous studies, numerous fusion methods have been proposed to merge and utilize the attributes of each modality in a proper manner. However, the

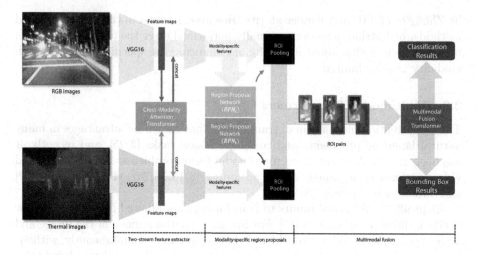

Fig. 1. Our proposed multimodal pedestrian detection network. We propose a two-stream feature extractor with Cross-modality Attention Transformer (CAT) to extract modality-specific features. After fetching the modality-specific region proposals, Multimodal Fusion Transformer (MFT) is introduced to merge the ROI pairs for class and bounding box predictions.

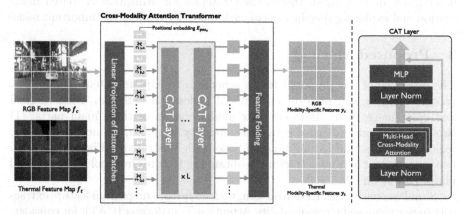

Fig. 2. Our proposed Cross-modality Attention Transformer (CAT). We divide feature maps f_c and f_t from the two feature extractors into patches as input, and fold the output patches to form the attended modality-specific feature maps y_c and y_t. Different from previous works, we introduce cross-modality attention to utilize the other modality information for encouraging the model to focus on the regions ignored by the current modality but highlighted by the other.

discussion about enhancing modality-specific features before fusion is quite limited. Modality-specific information plays an important role in single modality detection, such as textures in color images. Therefore, in this part, we aim at enhancing and preserving the modality-specific features before fusion by our

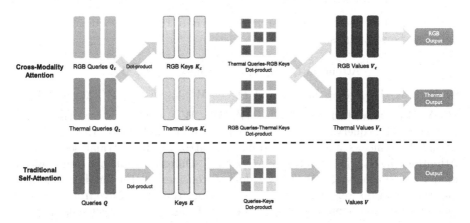

Fig. 3. Illustration of cross-modality attention. Instead of computing the queries with all modality keys, we introduce a novel way to only reference the complementary keys for introducing new perspectives from the other modality.

proposed cross-modality attention mechanism. Specifically, we argue that with our enhanced modality-specific features it is possible to achieve more accurate region proposals.

Cross-Modality Attention. As we have already learned, fused features usually cannot fully exploit modality-specific information and become dominated by strong cues from one of the modalities. In order to explore specific features of each modality, we introduce a novel attention mechanism to reference cross-modality data and to enhance our feature extraction. The main idea behind this design is to utilize the cross-modality information in order to encourage the model to focus on the regions ignored by the current modality but highlighted by the other. For this purpose, we use cross-modality attention transformers, which significantly outperform conventional CNN in discovering subtle and distant correlations in different modalities.

As illustrated in Fig. 3, instead of computing the scaled dot-product of the query with all modality keys, we only compute the scaled dot-product with the complementary keys. To be more specific, following [3]'s standard ViT architecture, we divide the input feature map pairs f_c^i and f_t^i from a certain layer i of the feature extractors into N_x patches for each modality, and then we flatten the patches and project them into N_x d_x-dimensional input sequences as $(x_c^1, ..., x_c^{N_x})$ and $(x_t^1, ..., x_t^{N_x})$.

Further, we add N_x d_x-dimensional learnable positional embeddings E_{pos_x} to each modality, and define the new matrices as X_c and X_t of size $\mathbb{R}^{N_x \times d_x}$. Different from the traditional transformer layers in ViT [3], we introduce CAT layers containing cross-modality attention module to utilize the other modality information. Our cross-modality attention operates on the queries and values from color patches as Q_c, V_c, keys from thermal patches as K_t. Hence, in every

CAT layer, the cross-modality attention matrix of the color sequence for single head can be written as:

$$\text{Attention}(Q_c, K_t, V_c) = \text{softmax}(\alpha Q_c K_t^T)V_c, \tag{1}$$

where α is the scaling factor, and $Q_c = \boldsymbol{W}_q X_c$, $K_t = \boldsymbol{W}_k X_t$, and $V_c = \boldsymbol{W}_v X_c$ are linear transformations. $\boldsymbol{W}_q, \boldsymbol{W}_k, \boldsymbol{W}_v \in \mathbb{R}^{d_x \times \frac{d_x}{N_h}}$ are the weight parameters for query, key, and value projections and N_h is the number of heads. As same as above, the attention matrix of the thermal sequence is: $\text{Attention}(Q_t, K_c, V_t) = \text{softmax}(\alpha Q_t K_c^T)V_t$. Different from the other multimodal transformers [16,21,26], we do not consider all the modality keys to find the attention weights. Instead, we introduce the perspective from the other modality by cross-referencing the keys to see if there is any target sensed by the other sensor and enhance the current sensor's features.

The output of the cross-modality attention module is then passed into Layer Normalization and MLP layers to get the attended features for color and thermal modalities. Next, we repeat several CAT layers and fold the attended output patches to form the feature maps y_c^i and y_t^i, which represent the additional modality-specific features learned from the other modality. Note that we still use the values from each modality to form the outputs, which means we do not directly fuse the multimodal features here. Then, we concatenate the output feature maps to the corresponding input features. Network-in-Network (NIN) [17] is applied to reduce the dimension and merge them with the input features. Furthermore, we apply our CAT on three different scales for considering coarse to fine-grained modality-specific features in practice.

Modality-Specific Region Proposal. In order to fully exploit the modality-specific features, we propose two independent region proposal networks to find the proposals separately. Different from the previous works [9,11,13,20,31], we suggest that using the fused features to perform region proposal might limit the discrimination ability because the fused features might be dominated by one modality without considering the other. Therefore, in our work, we perform region proposal separately relying on our enhanced modality-specific features to explore the potential candidates. Afterwards, we use IoU threshold to match the proposals from the two modalities to fetch ROI pairs. In order to maximize the recall rate, we form the union of the proposals to involve all the possible candidates and to avoid mismatches. In other words, if a proposal from thermal sensor is not matched, we still use the bounding box to fetch the ROI from the RGB sensor to form a ROI pair.

3.2 Multimodal Fusion

In order to optimally use the modality-specific cues from thermal and color images to perform detection, we propose a Multimodal Fusion Transformer (MFT) to perform the fusion of the features. Instead of considering the whole

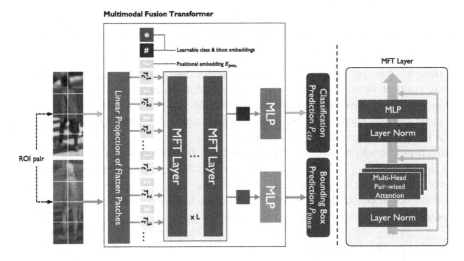

Fig. 4. Our proposed Multimodal Fusion Transformer (MFT). We divide ROI pairs from the two modalities into patches and prepend *class* and *bbox* for learning image representations. Different from CAT, we treat all the modality input sequences equally to apply pair-wised attention for computing attention weights.

image pair with background clutter, our fusion scheme only focuses on combining each ROI pair separately for lower computation time. In addition, to learn more discriminative feature representations, we introduce content-aware loss to enforce MFT to group the features based on the content.

Modality-Specific Features Fusion. As illustrated in Fig. 4, we divide each ROI into N_r patches and project the flatten patches into N_r d_r-dimensional input sequences as $(r_c^1, ..., r_c^{N_r})$ and $(r_t^1, ..., r_t^{N_r})$. In addition, we also add N_r d_r-dimensional learnable positional embeddings E_{pos_r}, and similar to [3], we introduce two d_r-dimensional learnable embeddings *class* and *bbox*, whose output state serve as the image representation for classification and bounding box predictions. Then, we concatenate input sequences with the embeddings as input and feed them into the MFT layers.

Different from the traditional ViT [3], we propose pair-wised attention module inside the MFT layers to apply the self-attention to all the input sequences to perform prediction. In particular, for merging the ROI pairs, we need to consider all the pair-wised modality patches to transfer the complementary information. During the pair-wised attention, we mix all the paired sequences and compute the scaled dot-product of queries with **all** the modality keys to fuse the features from the two sensors. After several MFT layers, finally, we forward the output of *class* and *bbox* token to two independent MLP layers for the classification prediction P_{cls} and bounding box prediction P_{bbox}. We argue the differences from the previous studies [6,14] as follows: without designing external network to learn balancing parameters, we utilize the self-attention mechanism to find the attention weightings and perform information fusion on each ROI pair.

Content-Aware Loss. For learning better feature representations, we propose content-aware loss to utilize the label information of each ROI pair. We impose content-aware loss $\mathcal{L}_{ca}$ on each output state of *class* token $Y_{cls,c}$, where c indicates the class label, 0 for background and 1 for foreground, to maximize the inter-class discrepancy and minimize intra-class distinctness. Specifically, for each input batch, we average the all the $Y_{cls,1}$ to fetch the representative feature of foreground: $Y_{fg} = \sum_{batch} Y_{cls,1}/N_f$, where N_f represents the number of outputs with pedestrian annotation. As the same way, we can fetch representative feature of background Y_{bg}. Then, the distance between each $Y_{cls,c}$ and the corresponding center feature can be calculated as: $d_{bg} = \sum_{batch} \|Y_{cls,0} - Y_{bg}\|, d_{fg} = \sum_{batch} \|Y_{cls,1} - Y_{fg}\|$. With the above definitions, the proposed content-aware loss $\mathcal{L}_{ca}$ can be written as:

$$\mathcal{L}_{ca} = \max\left(d_{bg} + d_{fg} - \|Y_{bg} - Y_{fg}\| + m, 0\right), \tag{2}$$

where $m > 0$ is the margin enforcing the separation between foreground and background features. By this way, we shorten the intra-class feature distance and expand the inter-class distance for better discrimination ability. For each training iteration, we optimize the following objective function:

$$\mathcal{L}_{total} = \mathcal{L}_{RPN_c} + \mathcal{L}_{RPN_t} + \mathcal{L}_{cls} + \mathcal{L}_{bbox} + \lambda\mathcal{L}_{ca}, \tag{3}$$

where $\mathcal{L}_{RPN_c}$ and $\mathcal{L}_{RPN_t}$ represent the loss from region proposal networks [22] of color and thermal modality, and $\mathcal{L}_{ca}$ is weighted by a balancing parameter λ. Similar to the Faster R-CNN [22], we use cross entropy and smooth L1 loss as the classification $\mathcal{L}_{cls}$ and bounding box regression loss $\mathcal{L}_{bbox}$ of MFT. Moreover, for better understanding, we also show the pseudo-code of our proposed method in Algorithm 1 to illustrate the whole process.

4   Experiments

In order to evaluate the performance of our proposed method, we conduct several experiments on the KAIST [7] and CVC-14 [5] datasets to compare with the previous methods. Furthermore, we also conduct ablation studies to demonstrate the impact of the proposed components. In all the experiments, we follow [31]'s settings and use log-average Miss Rate over the range of $[10^{-2}, 10^0]$ false positive per image (FPPI) as the main metric to compare the performance.

4.1   Dataset and Implementation Details

KAIST Dataset. The KAIST dataset [7] consists of $95,328$ visible-thermal image pairs captured in urban environment. The annotation includes $1,182$ unique pedestrians with $103,128$ bounding boxes. After applying the standard

---

Algorithm 1: Multimodal Fusion for Pedestrian Detection

Input   : Color Image: I_c, Thermal Image: I_t
Output : Classification and bounding box predictions P_{cls}, P_{bbox}
// Step 1: Feature extraction
for three different scales i during feature extraction **do**

> Extract color and thermal feature maps f_c^i and f_t^i from $\mathrm{VGG_c}(I_c)$ and $\mathrm{VGG_t}(I_t)$.
> // Enhance modality-specific features by Cross-modality Attention Transformer (CAT)
> $y_c^i = \mathrm{CAT}(f_c^i)$, $y_t^i = \mathrm{CAT}(f_t^i)$
> // Concatenate with f_c and f_t, and reduce the dimension with NIN [17]
> $f_c^{i+1} = \mathrm{NIN}(f_c^i, y_c^i), f_t^{i+1} = \mathrm{NIN}(f_t^i, y_t^i)$

end

// Step 2: Modality-specific region proposals
// Use feature extractor outputs F_c and F_t as inputs
$R_c = \mathrm{ROIpooling}(\mathrm{RPN_c}(F_c)), R_t = \mathrm{ROIpooling}(\mathrm{RPN_t}(F_t))$

// Step 3: ROI matching
$R_p = \mathrm{ROImatch}(R_c, R_t)$

// Step 4: Multimodal Fusion
// Perform feature fusion and prediction by Multimodal Fusion Transformer (MFT)
forall ROI pairs R_p^k and learnable embeddings *class* and *bbox* **do**

> $P_{cls}^k, P_{bbox}^k = \mathrm{MFT}(R_p^k, class, bbox)$

end

---

criterion [7], there are $7,601$ training image pairs and $2,252$ testing pairs. We train our model on the paired annotations released by [31] and apply horizontal flipping with single scale 600 for data augmentation. For fair comparison with the reference state-of-the-art methods, we evaluate the performance on "reasonable" day, night, and all-day subsets defined by [7] with sanitized labels [13].

CVC-14 Dataset. The CVC-14 dataset [5] consist of $7,085$ and $1,433$ visible (grey) and thermal frames captured in various scenes at day and night for training and testing. Different from the KAIST dataset, the field of views of the thermal and visible image pairs are not fully overlapped and calibrated well. The authors provided separated annotations for each modality and cropped image pairs to make thermal and visible images share the same field of view. In our experiments, we use the cropped image pairs and annotations to evaluate our method. The data augmentation strategy is as same as the KAIST dataset.

Implementation. In our proposed method, we use two independent VGG-16 [23] pretrained on ImageNet [12] to extract color and thermal modality feature. Subsequently, we apply our proposed Cross-modality Attention Transformer (CAT) on the last three different scales with patch size 16, 3, and 3 without overlap to reference the other modality. Each of the transformer contains 3 CAT layers. For the proposed Multimodal Fusion Transformer (MFT), we also use 3 MFT layers with patch size 3, and each input patch dimension is $1,024$. Except the attention modules, we follow [3]'s architecture to design CAT and MFT layers. The λ parameter for $\mathcal{L}_{ca}$ is 0.001. For more details, please refer to the Supplementary Materials.

Table 1. Comparisons with the state-of-the-art methods on the KAIST dataset.

Methods	Feature extractor	Miss Rate (IoU = 0.5)		
		All	Day	Night
ACF [7]	-	47.32%	42.57%	56.17%
Halfway Fusion [9]	VGG-16	25.75%	24.88%	26.59%
Fusion RPN + BF [11]	VGG-16	18.29%	19.57%	16.27%
IAF + RCNN [14]	VGG-16	15.73%	14.55%	18.26%
IATDNN + IASS [6]	VGG-16	14.95%	14.67%	15.72%
CIAN [30]	VGG-16	14.12%	14.77%	11.13%
MSDS-RCNN [13]	VGG-16	11.34%	10.53%	12.94%
AR-CNN [31]	VGG-16	9.34%	9.94%	8.38%
MBNet [32]	ResNet-50	8.13%	8.28%	7.86%
MLPD [10]	VGG-16	7.58%	7.95%	6.95%
Ours	VGG-16	**7.03%**	**7.51%**	**6.53%**

Table 2. Comparisons with the state-of-the-art methods on the CVC-14 dataset.

Methods	Feature extractor	Miss Rate (IoU = 0.5)		
		All	Day	Night
MACF [20]	-	69.71%	72.63%	65.43%
Choi et al. [2]	VGG-16	63.34%	63.39%	63.99%
Halfway Fusion [20]	VGG-16	31.99%	36.29%	26.29%
Park et al. [20]	VGG-16	26.29%	28.67%	23.48%
AR-CNN [31]	VGG-16	22.1%	24.7%	18.1%
MBNet [32]	VGG-16	21.1%	24.7%	13.5%
MLPD [10]	VGG-16	21.33%	24.18%	17.97%
Ours	VGG-16	**20.58%**	**23.97%**	**12.85%**

4.2   Quantitative Results

Evaluation on the KAIST Dataset. As illustrated in Table 1, we evaluate our method and compare it with other recent related methods. Our method achieves 7.03% MR, 7.51% MR, and 6.53% MR on day, night, and all-day subsets under the 0.5 IoU threshold. This table clearly shows that our method can achieve superior performance on all the subsets.

Evaluation on the CVC-14 Dataset. As for the KAIST dataset, we show the evaluations of our method and compare it with other state-of-the-art models in Table 2. In this table, we follow the setting introduced in [20] to conduct the experiment. Our method achieves 20.58% MR, 23.97% MR, and 12.85% MR on

Table 3. Ablation studies of proposed cross-modality attention and content-aware loss on the KAIST dataset. The baseline model is Halfway fusion [9], and we evaluate the model with or without the proposed components to verify the improvements.

Methods	CAT		MFT	Miss Rate (IoU = 0.5)		
	Cross-modality attention	Pair-wised attention	Content-aware loss	All	Day	Night
Baseline	-	-	-	25.75%	24.88%	26.59%
Ours	-	✓	-	13.54%	14.87%	13.01%
	-	✓	✓	12.42%	13.75%	12.11%
	✓	-	-	10.14%	10.87%	9.74%
	✓	-	✓	**7.03%**	**7.51%**	**6.53%**

day, night, and all-day subsets under the 0.5 IoU threshold. We can observe that our method outperforms the other method under all the subsets.

4.3   Ablation Study

Effects of Cross-Modality Attention. For further analysis of the effect of our proposed cross-modality attention, we conduct an experiment to compare the results with or without the cross-modality attention mechanism. In Table 3 we list four models to demonstrate the advantages of our method. Instead of using cross-modality attention in CAT, we use pair-wised attention to allow free attention flow between the modalities to compute the attention matrix (computing queries with all the modality keys) and to compare it with our proposed method. We find that cross-modality attention improves the performance of the reference models by a large margin. The performance can be improved by referencing the complementary information from the other modality rather than allowing free attention flow and directly fusing all the modality data in the early stage.

In addition, we also show the feature maps of the models with and without cross-modality attention in Fig. 5 to demonstrate our advantages qualitatively. Heat maps in this figure are generated by averaging the final convolution layer of RGB feature maps and superimposing them on the RGB image. We observe that with our proposed cross-modality attention, the model can correctly identify the pedestrians. In contrast, without cross-modality attention, the model can hardly focus on the targets.

Effects of Content-Aware Loss. In Table 3, we also compare the results with and without the content-aware loss to discuss the effect of content-aware loss. We can see that our proposed loss further improves the detection performance irrespective of the cross-modality attention or pair-wised attention. Especially when the content-aware loss is applied to the model with cross-modality attention, the result shows the best performance and largest improvement among all the combinations, which can demonstrate the effectiveness this component.

(a) Detection results w/o CAT   (b) Detection results w/ CAT   (c) Features w/o CAT   (d) Features w/ CAT

Fig. 5. The effects of cross-modality attention transformer. The left column (a) and (b) show the RGB image with detection results comparison, and the right column (c) and (d) show the feature maps with/without our proposed cross-modality attention transformer. The green boxes represent the correct detection results, and the red and orange boxes represent false negative and false positive samples respectively. (Color figure online)

Effects of Proposed Components. In order to demonstrate the contributions of all the proposed components, we evaluate the model with/without the components to verify the improvements. In Table 4, we use Halfway fusion [9] as our baseline model, and then apply the proposed components gradually to show the performance difference.

First, we only apply CAT to enhance modality-specific features and concatenate the features for single region proposal network. There is only a marginal improvement because the potential ROIs might not be fully explored. In addition, for different proposals, an advanced fusion process is also required to handle

Table 4. Ablation studies of our proposed components on the KAIST dataset. The baseline model is Halfway fusion [9], and we evaluate the model with/without the proposed components to verify the improvements. The results show that the proposed transformer significantly improves the detection performance and outperforms the previous models to achieve the best performance among all the combinations.

Methods	CAT	Modality-specific RPNs	MFT	Miss Rate (IoU = 0.5)		
				All	Day	Night
Baseline	-	-	-	25.75%	24.88%	26.59%
Ours	✓	-	-	20.45%	21.66%	20.47%
	✓	✓	-	13.12%	14.45%	12.87%
	✓	✓	✓	**7.03%**	**7.51%**	**6.53%**

various illumination scenes. Secondly, we apply two independent region proposal networks for each modality to find proposals and merge the ROI pairs by concatenation. This leads to a larger improvement, demonstrating that the enhanced modality-specific features and independent RPNs can truly help the model to find more potential proposals.

Furthermore, we apply MFT to fuse the ROI pairs instead of feature concatenation to verify the effect of fusion by the attention mechanism. The results show that the proposed transformer significantly improves the detection performance and outperforms the previous models to achieve the best performance among all the combinations.

5   Conclusions

In this paper, we propose a novel fusion scheme to combine visible and thermal image pairs to perform multimodal pedestrian detection. We introduce the Cross-modality Attention Transformer (CAT) to reference complementary information from the other modality during feature extraction to investigate the potential of modality-specific features to improve detection performance. Instead of directly fusing all the modality data, by our proposed cross-modality attention, we can extract more discriminative details. Moreover, we propose modality-specific region proposal networks to explore the potential candidates and merge the modality features pair-wisely by our proposed Multimodal Fusion Transformer (MFT) to make better predictions. Finally, a novel content-aware loss is proposed to separate the foreground and background features to increase the discrimination ability. The experiment results on the public KAIST and CVC-14 datasets confirm that our method can achieve state-of-the-art performance, and the ablation studies also clarify the effectiveness of the proposed components.

Acknowledgments. This work was funded by EU Horizon 2020 ECSEL JU research and innovation programme under grant agreement 876487 (NextPerception) and by the Flemish Government (AI Research Program).

References

1. Bain, M., Nagrani, A., Varol, G., Zisserman, A.: Frozen in time: a joint video and image encoder for end-to-end retrieval. In: Proceedings of the IEEE/CVF International Conference on Computer Vision (ICCV) (2021)
2. Choi, H., Kim, S., Park, K., Sohn, K.: Multi-spectral pedestrian detection based on accumulated object proposal with fully convolutional networks. In: International Conference on Pattern Recognition (ICPR) (2016)
3. Dosovitskiy, A., et al.: An image is worth 16x16 words: transformers for image recognition at scale. In: ICLR (2021)
4. Gabeur, V., Sun, C., Alahari, K., Schmid, C.: Multi-modal transformer for video retrieval. In: Vedaldi, A., Bischof, H., Brox, T., Frahm, J.-M. (eds.) ECCV 2020. LNCS, vol. 12349, pp. 214–229. Springer, Cham (2020). https://doi.org/10.1007/978-3-030-58548-8_13
5. González, A., et al.: Pedestrian detection at day/night time with visible and fir cameras: a comparison. Sensors **16**(6), 820 (2016)
6. Guan, D., Cao, Y., Yang, J., Cao, Y., Yang, M.Y.: Fusion of multispectral data through illumination-aware deep neural networks for pedestrian detection. Inf. Fusion **50**, 148–157 (2019)
7. Hwang, S., Park, J., Kim, N., Choi, Y., Kweon, I.S.: Multispectral pedestrian detection: benchmark dataset and baselines. In: Proceedings of the IEEE/CVF Conference on Computer Vision and Pattern Recognition (CVPR) (2015)
8. Iashin, V., Rahtu, E.: Multi-modal dense video captioning. In: Proceedings of the IEEE/CVF Conference on Computer Vision and Pattern Recognition Workshops (CVPRW) (2020)
9. Jingjing Liu, Shaoting Zhang, S.W., Metaxas, D.: Multispectral deep neural networks for pedestrian detection. In: Proceedings of the British Machine Vision Conference (BMVC) (2016)
10. Kim, J., Kim, H., Kim, T., Kim, N., Choi, Y.: MLPD: multi-label pedestrian detector in multispectral domain. IEEE Robot. Autom. Lett. **6**(4), 7846–7853 (2021)
11. Konig, D., Adam, M., Jarvers, C., Layher, G., Neumann, H., Teutsch, M.: Fully convolutional region proposal networks for multispectral person detection. In: Proceedings of the IEEE/CVF Conference on Computer Vision and Pattern Recognition Workshops (CVPRW) (2017)
12. Krizhevsky, A., Sutskever, I., Hinton, G.E.: Imagenet classification with deep convolutional neural networks. In: Advances in Neural Information Processing Systems (2012)
13. Li, C., Song, D., Tong, R., Tang, M.: Multispectral pedestrian detection via simultaneous detection and segmentation. In: Proceedings of the British Machine Vision Conference (BMVC) (2018)
14. Li, C., Song, D., Tong, R., Tang, M.: Illumination-aware faster R-CNN for robust multispectral pedestrian detection. Pattern Recognit. **85**, 161–171 (2019)
15. Li, G., Zhu, L., Liu, P., Yang, Y.: Entangled transformer for image captioning. In: Proceedings of the IEEE/CVF International Conference on Computer Vision (ICCV) (2019)
16. Li, Z., Li, Z., Zhang, J., Feng, Y., Zhou, J.: Bridging text and video: a universal multimodal transformer for audio-visual scene-aware dialog. IEEE/ACM Trans. Audio Speech Lang. Process. **29**, 2476–2483 (2021)
17. Lin, M., Chen, Q., Yan, S.: Network in network. arXiv preprint arXiv:1312.4400 (2013)

18. Lu, J., Batra, D., Parikh, D., Lee, S.: ViLBERT: pretraining task-agnostic visiolinguistic representations for vision-and-language tasks. In: Advances in Neural Information Processing Systems (2019)
19. Nagrani, A., Yang, S., Arnab, A., Jansen, A., Schmid, C., Sun, C.: Attention bottlenecks for multimodal fusion. In: Advances in Neural Information Processing Systems (2021)
20. Park, K., Kim, S., Sohn, K.: Unified multi-spectral pedestrian detection based on probabilistic fusion networks. Pattern Recognit. **80**, 143–155 (2018)
21. Prakash, A., Chitta, K., Geiger, A.: Multi-modal fusion transformer for end-to-end autonomous driving. In: Proceedings of the IEEE/CVF Conference on Computer Vision and Pattern Recognition (CVPR) (2021)
22. Ren, S., He, K., Girshick, R., Sun, J.: Faster R-CNN: towards real-time object detection with region proposal networks. In: Advances in Neural Information Processing Systems (2015)
23. Simonyan, K., Zisserman, A.: Very deep convolutional networks for large-scale image recognition. arXiv preprint arXiv:1409.1556 (2014)
24. Sun, C., Myers, A., Vondrick, C., Murphy, K., Schmid, C.: VideoBERT: a joint model for video and language representation learning. In: Proceedings of the IEEE/CVF International Conference on Computer Vision (ICCV) (2019)
25. Vaswani, A., et al.: Attention is all you need. In: Advances in Neural Information Processing Systems (2017)
26. Wang, W., Chen, C., Ding, M., Yu, H., Zha, S., Li, J.: TransBTS: multimodal brain tumor segmentation using transformer. In: International Conference on Medical Image Computing and Computer-Assisted Intervention (2021)
27. Xiao, Y., Yang, M., Li, C., Liu, L., Tang, J.: Attribute-based progressive fusion network for RGBT tracking (2022)
28. Xu, D., Ouyang, W., Ricci, E., Wang, X., Sebe, N.: Learning cross-modal deep representations for robust pedestrian detection. In: Proceedings of the IEEE/CVF Conference on Computer Vision and Pattern Recognition (CVPR) (2017)
29. Zhang, H., Fromont, E., Lefèvre, S., Avignon, B.: Guided attentive feature fusion for multispectral pedestrian detection. In: Proceedings of the IEEE/CVF Winter Conference on Applications of Computer Vision (WACV) (2021)
30. Zhang, L., et al.: Cross-modality interactive attention network for multispectral pedestrian detection. Inf. Fusion **50**, 20–29 (2019)
31. Zhang, L., Zhu, X., Chen, X., Yang, X., Lei, Z., Liu, Z.: Weakly aligned cross-modal learning for multispectral pedestrian detection. In: Proceedings of the IEEE/CVF International Conference on Computer Vision (ICCV) (2019)
32. Zhou, K., Chen, L., Cao, X.: Improving multispectral pedestrian detection by addressing modality imbalance problems. In: Vedaldi, A., Bischof, H., Brox, T., Frahm, J.-M. (eds.) ECCV 2020. LNCS, vol. 12363, pp. 787–803. Springer, Cham (2020). https://doi.org/10.1007/978-3-030-58523-5_46

See Finer, See More: Implicit Modality Alignment for Text-Based Person Retrieval

Xiujun Shu[1], Wei Wen[1], Haoqian Wu[1], Keyu Chen[1], Yiran Song[1,3], Ruizhi Qiao[1(✉)], Bo Ren[2], and Xiao Wang[4(✉)]

[1] Tencent Youtu Lab, Shenzhen, China
{xiujunshu,jawnrwen,linuswu,yolochen,ruizhiqiao}@tencent.com,
songyiran@sjtu.edu.cn
[2] Tencent Youtu Lab, Hefei, China
timren@tencent.com
[3] Shanghai Jiao Tong University, Shanghai, China
[4] Anhui University, Hefei, China
wangxiaocvpr@foxmail.com

Abstract. Text-based person retrieval aims to find the query person based on a textual description. The key is to learn a common latent space mapping between visual-textual modalities. To achieve this goal, existing works employ segmentation to obtain explicitly cross-modal alignments or utilize attention to explore salient alignments. These methods have two shortcomings: 1) Labeling cross-modal alignments are time-consuming. 2) Attention methods can explore salient cross-modal alignments but may ignore some subtle and valuable pairs. To relieve these issues, we introduce an **I**mplicit **V**isual-**T**extual (**IVT**) framework for text-based person retrieval. Different from previous models, IVT utilizes a single network to learn representation for both modalities, which contributes to the visual-textual interaction. To explore the fine-grained alignment, we further propose two implicit semantic alignment paradigms: multi-level alignment (MLA) and bidirectional mask modeling (BMM). The MLA module explores **finer** matching at sentence, phrase, and word levels, while the BMM module aims to mine **more** semantic alignments between visual and textual modalities. Extensive experiments are carried out to evaluate the proposed IVT on public datasets, *i.e.*, CUHK-PEDES, RST-PReID, and ICFG-PEDES. Even without explicit body part alignment, our approach still achieves state-of-the-art performance. Code is available at: https://github.com/TencentYoutuResearch/PersonRetrieval-IVT.

Keywords: Text-based person retrieval · Person search by language · Cross-model retrieval

1 Introduction

Person re-identification (re-ID) has many applications, *e.g.,* finding suspects or lost children in surveillance, and tracking customers in supermarkets. As a sub-

---

X. Shu and W. Wen—Equal contribution.

ⓒ The Author(s), under exclusive license to Springer Nature Switzerland AG 2023
L. Karlinsky et al. (Eds.): ECCV 2022 Workshops, LNCS 13805, pp. 624–641, 2023.
https://doi.org/10.1007/978-3-031-25072-9_42

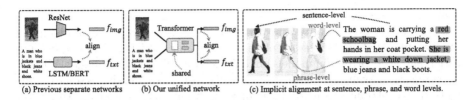

(a) Previous separate networks   (b) Our unified network   (c) Implicit alignment at sentence, phrase, and word levels.

Fig. 1. Illustration of the key idea of IVT. Previous methods use separate models to extract features, while we utilize a single network for both modalities. The shared parameters contribute to learning a common space mapping. Besides, we explore semantic alignment using three-level matchings. Not only see finer, but also see more semantic alignments.

task of person re-ID, text-based person retrieval (TPR) has attracted remarkable attention in recent years [23,39,48]. This is due to the fact that textual descriptions are easily accessible and can describe more details in a natural way. For example, police officers usually access surveillance videos and take the deposition from witnesses. Textual descriptions can provide complementary information and even are critical in scenes where images are missing.

Text-based person retrieval needs to process visual and textual modalities, and its core is to learn a common latent space mapping between them. To achieve this goal, current works [16,42] firstly utilize different models to extract features, *i.e.,* ResNet50 [14] for visual modality, and LSTM [46] or BERT [17] for textual modality. Then they are devoted to exploring visual-textual part pairs for semantic alignment. However, these methods have at least two drawbacks that may lead to suboptimal cross-modal matching. **First**, separate models lack modality interaction. Each model usually contains many layers with a large number of parameters, and it is difficult to achieve full interaction just using a matching loss at the end. To relieve this issue, some works [21,45] on general image-text pre-training use cross-attention to conduct interaction. However, they require to encode all possible image-text pairs to compute similarity scores, which leads to quadratic time complexity at the inference stage. How to design a more suitable network for the TPR task still needs profound thinking. **Second**, labeling visual-textual part pairs, *e.g.,* head, upper body, and lower body, is time-consuming, and some pairs may be missing due to the variability of textual descriptions. For example, some text contains descriptions of hairstyles and pants, but others do not contain this information. Some researchers begin to explore implicit local alignment to mine part matching [39,48]. To ensure reliability, partial local matchings with high confidence are usually selected. However, these parts usually belong to salient regions that can be easily mined by global alignment, *i.e.,* they do not bring additional information gain. According to our observation, local semantic matching should not only see **finer**, but also see **more**. Some subtle visual-textual cues, *e.g.,* hairstyle and logo on clothes, maybe easily ignored but could be complementary to global matching.

To solve the above problems, we first introduce an **Implicit Visual-Textual (IVT)** framework, which can learn representation for both modalities only using a single network (See Fig. 1(b)). This benefits from the merit that Transformer can operate on any modality that can be tokenized. To avoid the shortcomings of separate models and cross-attention, *i.e.,* separate models lack modality interaction and cross-attention models are quite slow at the inference stage, IVT supports separate feature extracting to ensure retrieval speed and shares some parameters that contribute to learning a common latent space mapping. To explore fine-grained modality matching, we further propose two implicit semantic alignment paradigms: multi-level alignment (MLA) and bidirectional mask modeling (BMM). The two paradigms do not require extra manual labeling and can be easily implemented. Specifically, as shown in Fig. 1(c), MLA aims to explore fine-grained alignment by using sentence, phrase, and word-level matchings. BMM shares a similar idea with MAE [13] and BEIT [2] in that both learn better representation through random masking. The difference is that the latter two aim at single-modal autoencoding-style reconstruction, while BMM does not reconstruct images but focuses on learning cross-modal matching. By masking a certain percentage of visual and textual tokens, BMM forces the model to mine more useful matching cues. The proposed two paradigms could not only see finer but also see more semantic alignments. Extensive experiments demonstrate the effectiveness on the TPR task.

Our contributions can be summarized as three folds: (1) We propose to tackle the modality alignment from the perspective of backbone network and introduce an Implicit Visual-Textual (IVT) framework. This is the first unified framework for text-based person retrieval. (2) We propose two implicit semantic alignment paradigms, *i.e.,* MLA and BMM, which enable the model to mine finer and more semantic alignments. (3) We conduct extensive experiments and analyses on three public datasets. Experimental results show that our approach achieves state-of-the-art performance.

2   Related Work

Text-Based Person Retrieval. Considering the great potential economic and social value of text-based person retrieval, Li et al. propose the first benchmark dataset CUHK-PEDES [23] in 2017, and also build a baseline, *i.e.,* GNA-RNN, based on LSTM network. Early works utilize ResNet50 and LSTM to learn representations for visual-textual modalities, and then utilize matching loss to align them. For example, CMPM [47] associates the representations across different modalities using KL divergence. Besides aligning the features, some works study the identity cue [22], which helps learn discriminative representations. Since text-based person retrieval requires fine-grained recognition of human bodies, later works start to explore global and local associations. Some works [4,5] utilize visual-textual similarity to mine part alignments. ViTAA [42] segments the human body and utilizes k-reciprocal sampling to associate the visual and textual attributes. Surbhi et al. [1] propose to create semantic-preserving embeddings through attribute prediction. Since visual and textual attributes require pre-processing, more works attempt to use attention mechanisms to explore fine-grained alignment. PMA [16] proposes a pose-guided multi-granularity attention network. HGAN [48] splits the images into multi-scale stripes and utilizes attention to select top-k part alignments. Other works include adversarial learning and relation modeling. TIMAM [30] learns modality-invariant feature representations using adversarial and cross-modal matching objectives. A-GANet [26] introduces the textual and visual scene graphs consisting of object properties and relationships. In summary, most current works learn modality alignment by exploiting local alignments. In this work, we study the modality alignment from different perspectives, in particular, how to obtain full modality interaction and how to achieve local alignment simply. The proposed framework can effectively address these issues and achieve satisfying performance.

Transformer and Image-Text Pre-training. Transformer [38] is firstly proposed for machine translation in the natural language processing (NLP) community. After that, many follow-up works are proposed and set new state-of-the-art one after another, such as BERT [17], GPT series [3,34,35]. The research on Transformer-based representations in computer vision is also becoming a hot spot. Early works like ViT [9] and Swin Transformer [27] take the dividing patches as input, like the discrete tokens in NLP, and achieve state-of-the-art performance on many downstream tasks. Benefiting from the merit that Transformer can operate on any modality that can be tokenized, it has been utilized in the multi- or cross-modal tasks intuitively [41]. Lu et al. [29] propose the ViLBERT to process both visual and textual inputs in separate streams that interact through co-attentional Transformer layers. Oscar [24] and VinVL [45] take the image, text, and category tags as inputs and find that the category information and stronger object detector can bring better results. Many recent works study which architecture is better for multi- or cross-modal tasks, *e.g.*, UNITER [6], ALBEF [21], and ViLT [18]. These works generally employ several training objectives to support multiple downstream vision-language tasks. The most relevant downstream task for us is image-text retrieval. To obtain modality interaction, these works generally employ cross-attention in the fusion blocks. However, they have a very slow retrieval speed at the inference stage because they need to predict the similarity of all possible image-text pairs. The ALIGN [15] and CLIP [33] are large-scale vision-language models pre-trained using only contrastive matching. These separate models are suitable for image-text retrieval, but they generally achieve satisfying performance in zero-shot settings. Some works, *e.g.*, switch Transformers [10], VLMO [40], attempt to optimize the network structure so that both retrieval and other visual-language tasks can be supported. This paper is greatly inspired by these works and aims to relieve the modality alignment for text-based person retrieval.

3   Methodology

3.1   Overview

To tackle the modality alignment in text-based person retrieval, we propose an Implicit Visual-Textual (IVT) framework as shown in Fig. 2. It consists of a unified visual-textual network and two implicit semantic alignment paradigms, *i.e.*,

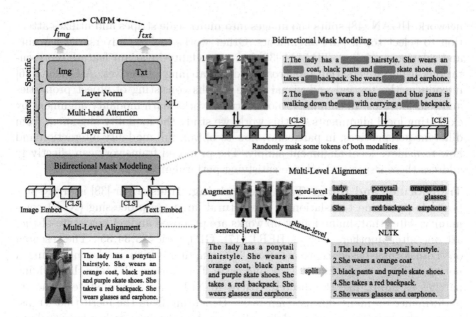

Fig. 2. Architecture of the proposed IVT Framework. It consists of a unified visual-textual network and two implicit semantic alignment paradigms, *i.e.*, multi-level alignment (MLA) and bidirectional mask modeling (BMM). The unified visual-textual network contains parameter-shared and specific modules, which contribute to learning common space mapping. MLA aims to see "finer" by exploring local and global alignments from three-level matchings. BMM aims to see "more" by mining more semantic alignments from random masking for both modalities.

multi-level alignment (MLA) and bidirectional mask modeling (BMM). One key idea of IVT lies in tackling the modality alignment using a unified network. By sharing some modules, *e.g.*, the layer normalization and multi-head attention, the unified network contributes to learning common space mapping between visual and textual modalities. It can also learn modality-specific cues using different modules. The two implicit semantic alignment paradigms, *i.e.*, MLA and BMM, are proposed to explore fine-grained alignment. Different from previous methods that use manually processed parts or select salient parts from attention, the two paradigms could mine not only finer but also more semantic alignments, which is another key idea of our proposed IVT.

3.2 Unified Visual-Textual Network

Embedding. As shown in Fig. 2, the input are image-text pairs, which provide the appearance characteristics of a person from visual and textual modalities. Let the image-text pairs denoted as $\{x_i, t_i, y_i\}|_{i=1}^{N}$, where x_i, t_i, y_i denote the image, text, and identity label, respectively. N is the total number of samples. For an input image $x_i \in \mathbb{R}^{H \times W \times C}$, it is firstly split into $K = H \cdot W / P^2$ patches, where

P denotes the patch size, and then linearly projected into patch embeddings $\{f_k^v\}|_{k=1}^K$. This operation can be realized using a single convolutional layer. The patch embeddings are then prepended with a learnable class token f_{cls}^v, and added with a learnable position embedding f_{pos}^v and a type embedding f_{type}^v.

$$\mathbf{f}^v = [f_{cls}^v, f_1^v, ..., f_K^v] + f_{pos}^v + f_{type}^v. \tag{1}$$

For the input text t_i, it generally consists of one or several sentences and each sentence has a sequence of words. The pretrained word embedding is leveraged to project words into token vectors:

$$\mathbf{f}^t = [f_{cls}^t, f_1^t, ..., f_M^t, f_{sep}^t] + f_{pos}^t + f_{type}^t, \tag{2}$$

where f_{cls}^t and f_{sep}^t denote the start and end tokens. M indicates the length of tokenized subword units. f_{pos}^t is the position embedding and f_{type}^t is the type embedding.

Visual-Textual Encoder. Current works on the TPR task utilize separate models, which lack full modality interaction. Some recent works on general image-text pre-training attempt to utilize cross-attention to achieve modality interaction. However, cross-attention requires encoding all possible image-text pairs at the inference stage, which leads to a very slow retrieval speed. Based on these observations, we propose to take the unified visual-textual network for the TPR task. The unified network has a quick retrieval speed and supports modality interaction.

As shown in Fig. 2, the network follows a standard architecture of ViT [9] and stacks L blocks in total. In each block, two modalities share layer normalization (LN) and multi-head self-attention (MSA), which contribute to learning common space mapping between visual and textual modalities. It is because the shared parameters help learn common data statistics. For example, LN would calculate the mean and standard deviation of input token embeddings, and shared LN would learn the statistically common values of both modalities. This can be regarded as a "modality interaction" from a data-level perspective. As the visual and textual modalities are not the same, each block has modality-specific feed-forward layers, *i.e.*, the "Img" and "Txt" modules in Fig. 2. They are used to capture modality-specific information by switching to visual or textual inputs. The complete processing in each block can be denoted as follows:

$$\mathbf{f}_i^{v/t} = \mathrm{MSA}(\mathrm{LN}(\mathbf{f}_{i-1}^{v/t})) + \mathbf{f}_{i-1}^{v/t}, \tag{3}$$

$$\mathbf{f}_i^{v/t} = \mathrm{MLP}_{\mathrm{img/txt}}(\mathrm{LN}(\mathbf{f}_i^{v/t})) + \mathbf{f}_i^{v/t}, \tag{4}$$

where LN denotes the layer normalization and MSA denotes the multi-head attention. i is the index of the blocks. $\mathbf{f}_{i-1}^{v/t}$ is the visual or textual output of the $(i-1)^{th}$ block and also the input of the ith block. $\mathrm{MLP}_{\mathrm{img/txt}}$ denotes the modality-specific feed-forward layers. $\mathbf{f}_i^{v/t}$ is the output of the ith block.

Output. The class token of the last block serves as the global representation, *i.e.*, f_{img} and f_{txt} in Fig. 2. The dimension of both feature vectors are 768, and the outputs are normalized using the LN layer.

3.3   Implicit Semantic Alignment

Multi-level Alignment. Fine-grained alignment has been demonstrated to be the key to achieving performance improvement, such as segmented attributes [42] and stripe-based parts [11,48]. These methods can be regarded as explicit part alignment, namely telling the model which visual-textual parts should be aligned. In this work, we propose an implicit alignment method, *i.e.*, multi-level alignment, which is intuitive but very effective.

As shown in Fig. 2, the input image is firstly augmented to get three types of augmented ones, *e.g.*, horizontal flipping and random cropping. The input text generally consists of one or several sentences. We split them into more short sentences according to periods and commas. These short sentences are regarded as "phrase-level" representation, which describe partial appearance characteristics of the human body. To mine finer parts, we further utilize the Natural Language Toolkit (NLTK) [28] to extract nouns and adjectives, which describe specific local characteristics, *e.g.*, bag, clothes, or pants. The three-level textual descriptions, *i.e.*, sentence-level, phrase-level, and word-level, correspond to the three types of augmented images. The three-level image-text pairs are randomly generated at each iteration. In this way, we construct a matching process that gradually refines from global to local, forcing the model to mine finer semantic alignments.

The major difference between our approach and previous work is that we do not explicitly define visual semantic parts, instead, automatically explore aligned visual parts guided by three-level textual descriptions. This is due to the following observations that inspire our TPR framework: 1) Previous explicit aligned methods lack inconsistency between the training and inference phases. During training, previous methods utilize an unsupervised way to explore local alignments, e.g., select the top-k salient alignments based on similarities. During the inference stage, only the global embeddings for each modality would be used, resulting in the inconsistent issue. 2) The explicit local alignment makes the training easier but makes it harder at the inference stage. Oversimpilified task design leads to worse generalization performance of the model. Even though we do not provide visual parts, but rather the full images, the model tries hard to mine local alignment at the training stage, thus achieving better performance at the inference stage due to the consistency.

Bidirectional Mask Modeling. To automatically mine local alignment, recent methods [11,48] split images into stripes and utilize attention to select top-k part alignments. However, they ignore the fact that top-k part alignments are usually salient cues, which may have been mined by the global alignment. Therefore, these parts may bring limited information gains. We argue that local alignments should not only be finer, but also more diverse. Some subtle visual-textual cues may be complementary to global alignment.

As shown in Fig. 2, we propose a bidirectional mask modeling (BMM) method to mine more semantic alignments. For the image and text tokens, we randomly mask some percentage of them and then force the visual and textual outputs to keep in alignment. In general, the masked tokens correspond to specific patches of the image or words of the text. If specific patches or words are masked, the model would try to mine useful alignment cues from other patches or words. Let us take the lady in Fig. 2 as an example, if the salient words "orange coat" and "black pants" are masked, the model would pay more attention to other words, e.g., ponytail hairstyle, red backpack. In this way, more subtle visual-textual alignments can be explored. At the training stage, this method makes it more difficult for the model to align image and text but helps it to mine more semantic alignments at the inference stage.

The above method shares a similar idea to Random Erasing [50], MAE [13], and BEIT [2]. However, Random Erasing masks only one region. MAE and BEIT aim at image reconstruction using an autoencoding style. The proposed BMM does not reconstruct images but focuses on cross-modal matching.

3.4 Loss Function

We utilize the commonly used cross-modal projection matching (CMPM) loss [47] to learn visual-textual alignment, which is defined as follows:

$$\mathcal{L}_{cmpm} = \frac{1}{B} \sum_{i=1}^{B} \sum_{j=1}^{B} \left(p_{i,j} \cdot \log \frac{p_{i,j}}{q_{i,j} + \epsilon} \right), \tag{5}$$

$$p_{i,j} = \frac{exp(f_i^T \cdot f_j)}{\sum_{k=1}^{B} exp(f_i^T \cdot f_k)}, \qquad q_{i,j} = \frac{y_{i,j}}{\sum_{k=1}^{B} y_{i,k}}, \tag{6}$$

where $p_{i,j}$ denotes the matching probability. f_i and f_j denote the global features of different modalities. B is the mini-batch size. $q_{i,j}$ denotes the normalized true matching probability. ϵ is a small number to avoid numerical problems.

The CMPM loss represents the KL divergence from distribution $\mathbf{q}$ to $\mathbf{p}$. Following previous work [47], the matching loss is computed in two directions, i.e., image-to-text and text-to-image. The total loss can be denoted as follows:

$$\mathcal{L} = \mathcal{L}_{cmpm}^{t2v} + \mathcal{L}_{cmpm}^{v2t}. \tag{7}$$

4 Experiment

4.1 Experimental Setup

Datasets. We evaluate our approach on three benchmark datasets, i.e., **CUHK-PEDES** [23], **RSTPReid** [51], and **ICFG-PEDES** [8]. Specifically,

Table 1. Comparison with SOTA methods on CUHK-PEDES.

Method	R1	R5	R10
CNN-RNN [36]	8.07	-	32.47
GNA-RNN [23]	19.05	-	53.64
PWM-ATH [5]	27.14	49.45	61.02
GLA [4]	43.58	66.93	76.26
MIA [31]	53.10	75.00	82.90
A-GANet [26]	53.14	74.03	81.95
ViTAA [42]	55.97	75.84	83.52
IMG-Net [43]	56.48	76.89	85.01
CMAAM [1]	56.68	77.18	84.86
HGAN [48]	59.00	79.49	86.60
NAFS [11]	59.94	79.86	86.70
DSSL [51]	59.98	80.41	87.56
MGEL [39]	60.27	80.01	86.74
SSAN [8]	61.37	80.15	86.73
NAFS [11]	61.50	81.19	87.51
TBPS [12]	61.65	80.98	86.78
TIPCB [7]	63.63	82.82	89.01
Baseline (Ours)	55.75	75.68	84.13
IVT (Ours)	**65.59**	**83.11**	**89.21**

Table 2. Comparison with SOTA methods on RSTPReid.

Method	R1	R5	R10
DSSL [51]	32.43	55.08	63.19
Baseline (Ours)	37.40	60.90	70.80
IVT (Ours)	**46.70**	**70.00**	**78.80**

Table 3. Comparison with SOTA methods on ICFG-PEDES.

Method	R1	R5	R10
Dual Path [49]	38.99	59.44	68.41
CMPM+CMPC [47]	43.51	65.44	74.26
MIA [31]	46.49	67.14	75.18
SCAN [20]	50.05	69.65	77.21
ViTAA [42]	50.98	68.79	75.78
SSAN [8]	54.23	72.63	79.53
Baseline (Ours)	44.43	63.50	71.00
IVT (Ours)	**56.04**	**73.60**	**80.22**

CUHK-PEDES [23] contains 40,206 images of 13,003 persons and 80,440 description sentences. It is splitted into a training set with 34,054 images and 68,126 description sentences, a validation set with 3,078 images and 6,158 description sentences, and a testing set with 3,074 images and 6,156 description sentences. RSTPReid [51] is collected from MSMT [44] and contains 20,505 images of 4,101 persons. Each image has two sentences and each sentence is no shorter than 23 words. More in detail, the training, validation, and testing sets have 3,701, 200, and 200 identities, respectively. ICFG-PEDES [8] is also collected from MSMT17 [44] and contains 54,522 images of 4,102 persons. Each image has one description sentence with an average of 37.2 words. The training and testing subsets contain 34,674 image-text pairs for 3,102 persons, and 19,848 image-text pairs for the remaining 1,000 persons, respectively. We also explore pre-training on four image captioning datasets: Conceptual Captions (CC) [37], SBU Captions [32], COCO [25] and Visual Genome (VG) [19] datasets. There are about 4M image-text pairs in total.

Evaluation Metric. The cumulative matching characteristic (CMC) curve is a precision curve that provides recognition precision for each rank. Following previous works, R1, R5, and R10 are reported when compared with state-of-the-art (SOTA) models. The mean average precision (mAP) is the average precision across all queries, which is also reported in ablation studies for future comparison.

Implementation Details. The proposed framework follows the standard architecture of ViT-Base [9]. The patch size is set as 16 × 16 and the dimensions of both visual and textual features are 768. The input images are resized to 384 × 128. We use horizontal flipping and random cropping as data augmenting. At the pre-training stage, we utilize 64 Nvidia Tesla V100 GPUs with FP16 training. At the fine-tuning stage, we employ four V100 GPUs and set the mini-batch size as 28 per GPU. The SGD is used as the optimizer with the weight decay of 1e-4. The learning rate is initialized as 5e-3 with cosine learning rate decay.

4.2 Comparison with State-of-the-Art Methods

In this section, we report our experimental results and compare with other SOTA methods on CUHK-PEDES [23], RSTPReid [51] and ICFG-PEDES [8]. Note that, the Baseline in Table 1, Table 2 and Table 3, denotes the vanilla IVT without pre-training, bidirectional mask modeling (BMM) and multi-level alignment (MLA) components. We employ BERT [17] to initialize the "txt" module and ImageNet pre-trained model from [9] to initialize other modules.

Table 4. Ablation results of components on CUHK-PEDES. "Base" denotes our baseline method. "Pre" is short of pre-training.

No.	Base	Pre	BMM	MLA	R1	R5	R10	mAP
1	✓				55.75	75.68	84.13	53.36
2	✓	✓			60.06	78.56	85.22	56.64
3	✓		✓		60.43	79.55	86.19	56.65
4	✓			✓	61.00	80.60	87.23	56.88
5	✓	✓	✓		62.88	81.60	87.54	59.34
6	✓	✓		✓	63.87	82.67	88.42	59.52
7	✓		✓	✓	64.00	82.72	88.95	58.99
8	✓	✓	✓	✓	**65.59**	**83.11**	**89.21**	**60.66**

Table 5. The computational efficiency terms of several methods. "Time" denotes the retrieval time for testing CUHK-PEDES on a Tesla V100 GPU.

Method	Architecture	Para (M)	Time (s)
ViLT [18]	Transformer	96.50	103,320
ALBEF [21]	Transformer	209.56	12,240
NAFS [11]	ResNet + BERT	189.00	78
SSAN [8]	ResNet + LSTM	97.86	31
TBPS [12]	ResNet + BiGRU	84.83	26
IVT	Transformer	166.45	42

Results on CUHK-PEDES. As shown in Table 1, our baseline method achieves 55.75%, 75.68%, 84.13% on R1, R5, and R10, respectively. It already achieves comparable or even better performance compared with many works proposed in recent years, e.g., MIA [31], ViTAA [42], CMAAM [1]. These experiments demonstrate the effectiveness of the unified visual-textual network for text-based person retrieval. In contrast, our proposed IVT obtains 65.59%, 83.11%, and 89.20% on these metrics, which are significantly better than our baseline method. Specifically, these results have improved considerably, i.e., +9.84%, +7.43%, +5.07%, respectively. It should be noted that many recent SOTA algorithms have taken complex operations, e.g., segmentation, attention, or adversarial learning. Even though, our approach outperforms existing SOTA algorithms, e.g., DSSL [51] and NAFS [11], and can also be easily implemented. These results fully validate the effectiveness of our approach for text-based person retrieval.

Results on RSTPReid. As RSTPReid is newly released, only DSSL [51] has reported the results on it. As shown in Table 2, DSSL [51] achieves 32.43%, 55.08%, 63.19% on the R1, R5, R10, respectively. In contrast, the proposed method achieves 46.50%, 70.20% and 79.70%, which exceed the DSSL [51] by a large margin, *i.e.*, +14.27%, +14.92%, and +15.61%. It should be noted that our baseline method still exceeds DSSL, which benefits from our unified visual-textual network. Besides, our IVT outperforms the baseline with a large margin. The above experiments fully validate the advantages of our proposed modules.

Results on ICFG-PEDES. The experimental results on the ICFG-PEDES dataset are reported in Table 3. We can find that the baseline method achieves 44.43%, 63.50%, and 71.00% on the R1, R5, and R10, respectively. Meanwhile, our proposed IVT achieves 56.04%, 73.60%, and 80.22% on these metrics, which also fully validate the effectiveness of our proposed modules for the TPR task. Compared with other SOTA algorithms, *e.g.*, SSAN [8], ViTAA [42], our results are also better than them significantly. In contrast, even without complex operations to mine local alignments, IVT still achieves SOTA performance.

In summary, our IVT yields the best performance in terms of all metrics on three benchmark datasets. The superior performance is not only due to the well-designed unified visual-textual network, but also owing to the effective implicit semantic alignment paradigms. We hope our work can bring new insights to the text-based person retrieval community.

4.3   Ablation Study

To better understand the contributions of each component in our framework, we conduct a comprehensive empirical analysis in this section. Specifically, the results of different components of our framework on the CUHK-PEDES [23] dataset are shown in Table 4.

Effectiveness of Multi-level Alignment. As shown in Table 4, the Baseline achieves 55.75%, 53.36% on R1 and mAP, respectively. After introducing the MLA module, the overall performance has been improved to 61.00% and 56.88%. The improvements up to +5.25% and +3.52%, respectively. The results demonstrate the effectiveness of our proposed MLA strategy. Further analysis shows that MLA enables the model to mine fine-grained matching through sentence, phrase, and word-level alignments, which in turn improves the visual and textual representations.

Effectiveness of Bidirectional Mask Modeling. The BMM strategy also plays an important role in our framework, as shown in Table 4. By comparing No. 1 and No. 3, we can find that R1 and mAP have been improved from 55.75%, 53.36% to 60.43%, 56.65%. The improvements up to +4.68%, +3.29% on R1 and mAP, respectively. The experimental results fully validate the effectiveness of the BMM strategy for text-based person retrieval.

Effectiveness of Pre-training. Even without pre-training, IVT achieves 64% on R1 accuracy (see No. 7), outperforming current SOTA methods, *e.g.*, NAFS

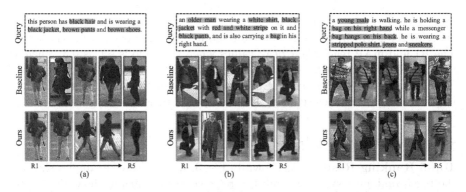

Fig. 3. The top-5 ranking results. The green/red boxes denote the true/false results. (Color figure online)

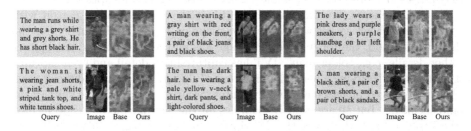

Fig. 4. Comparison of heat maps between the baseline method and IVT.

(61.50%), TBPS (61.65%). To obtain better-generalized features, we pre-train our model using a large-scale image-text corpus. As illustrated in Table 4, we can find that the overall performance can also be improved significantly. Specifically, the R1 and mAP are improved from 55.75%, 53.36% to 60.06%, 56.64% with the pre-training (see No. 1 and No. 2). In addition, it can also be found that pre-training improves the final results by comparing the No. 5/No. 7 and No. 8 in Table 4. Therefore, we can draw the conclusion that pre-training indeed brings more generalized features, which further boost the final matching accuracy.

Effects of Masking Ratio. The BMM method requires setting the masking ratio. This section studies its effect on final performance. The ablation results are shown in Fig. 5. The experiments are conducted on CUHK-PEDES [23] and only the "Baseline+BMM" method is utilized. As shown in Fig. 5, the performance has been

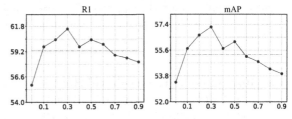

Fig. 5. Ablation results of masking ratios on CUHK-PEDES.

improved gradually as the masking ratio increases. When the masking ratio is set to zero, which equals the baseline method, the performance is the worst. The R1 and mAP reach their peak values when the ratio is set as 0.3. Then as the masking ratio continues to increase, the performance gradually decreases. This is because the model cannot mine enough semantic alignments with a too large masking ratio, thus reducing the final performance.

4.4 Qualitative Results

Top-5 Ranking Results. As shown in Fig. 3, we give three examples showing the top-5 ranking results. Overall, the retrieved top-5 images show high correlations between the visual attributes and the textual descriptions, even for the false matching results. Compared with the baseline method, our proposed IVT has retrieved more positive samples. This is because it can capture more fine-grained alignments. For example, Fig. 3(b) needs to search for a person with "white shirt, black jacket with red white stripe on it and black pants". For the baseline method, the top-1 retrieved image has all these attributes, but ignores other details, *e.g.,* "older man". IVT has captured this detail and even captures the attribute "carrying a bag in his right hand". Figure 3(c) shows a difficult case. All the retrieved images have the attributes "wearing a striped polo shirt, jeans and sneakers", but all are negative samples for the baseline method. Specifically, the baseline method ignores the description "holding a bag on his right hand" while our IVT has captured this detail and retrieved more positive samples.

Fig. 6. Visualization of part alignment between visual and textual modalities. (a) Upper Body, (b) Packbag, (c) Lower Body, (d) Shoes. We compute the similarities of word-level text and all image patches. The brighter the part, the higher the similarity.

Visualization of Sentence-Level Heat Map. To better understand the visual and textual alignment, we give some visualizations of sentence-level heat maps in Fig. 4. The heat maps are obtained by visualizing the similarities between textual [CLS] tokens and all visual tokens. The brighter the image is, the more similar it is to the text. In general, textual descriptions can correspond to human bodies, demonstrating that the model has learned the semantic relevance of visual and textual modalities. Compared with the baseline method, IVT could focus more attributes of human bodies described by the text. For example, the man (1st

row, 1st column) is wearing grey shorts. The baseline method has ignored the attribute, but IVT has captured it. Hence, the proposed IVT can responds to more accurate and diverse person attributes than the baseline method.

Visualization of Local Alignment. To validate the ability of fine-grained alignment, we further conduct word-level alignment experiments. As shown in Fig. 6, we give four types of human attributes, *i.e.,* upper body, packbag, lower body, and shoes. Each row in Fig. 6 shares the same attribute. The brighter the area in the image, the more similar it is to the given textual attribute descriptions. As shown in Fig. 6, our method can recognize not only salient body regions, *e.g.,* clothes and pants, but also some subtle parts, *e.g.,* handbags and shoes. These visualization results show us that our model can focus on exactly the correct body parts, given the word-level attribute description. It indicates that our approach is capable of exploring fine-grained alignments even without explicit visual-textual part alignments. This benefits from the proposed two implicit semantic alignment paradigms, *i.e.,* MLA and BMM. Therefore, we can draw the conclusion that our proposed method indeed achieves **See Finer** and **See More** for text-based person retrieval.

4.5 Computational Efficiency Analysis

In this section, we analyze the parameters and retrieval time at the inference stage. As shown in Table 5, we mainly compare recent methods in the TPR field, *e.g.,* NAFS [11], SSAN [8], TBPS [12], and typical methods in general image-text retrieval, *e.g.,* ViLT [18], ALBEF [21]. Since the parameters of LSTM/BiGRU are less than Transformer, our IVT has more parameters than SSAN and TBPS, but comparable retrieval time. Besides, the performance of these methods would be limited by the text-modeling ability of LSTM. Due to the utilization of Non-local attention, NAFS has 189M parameters and its retrieval time reaches 78s, both exceeding our IVT. Compared with general image-text retrieval methods, *e.g.,* ViLT and ALBEF, our IVT has a significant advantage in retrieval time. Specifically, ViLT needs 103,320 s to test CUHK-PEDES, but our IVT only needs 42 s. This is because they need to encode all possible image-text pairs, other than just extracting features only once. Overall, our method is competitive enough in terms of both parameters and retrieval efficiency.

5 Discussion

From the badcases in Fig. 3, we find two characteristics for the TPR task. First, the text description is usually not comprehensive, which corresponds to only part of the visual features. Second, the textual representation tends to ignore subtle features, especially for a relatively long description. By conducting extensive experiments, we got two valuable conclusions: 1) Unified network is effective for the TPR task. It maybe regarded as the backbone network in the future. 2) More subtle part alignments should be mined, other than only the salient part pairs. Even without complex operations, the proposed approaches can still mine fine-grained semantic alignments and achieve satisfying performance. We hope they can bring new insights to the TPR community.

6   Conclusion

This paper proposes to tackle the modality alignment from two perspectives: backbone network and implicit semantic alignment. First, an Implicit Visual-Textual (IVT) framework is introduced for text-based person retrieval. It can learn visual and textual representations using a single network. Benefiting from the architecture, *i.e.,* shared and specific modules, it is possible to guarantee both the retrieval speed and modality interaction. Second, two implicit semantic alignment paradigms, *i.e.,* BMM and MLA, are proposed to explore fine-grained alignment. The two paradigms could see "finer" using three-level matchings and see "more" by mining more semantic alignments. Extensive experimental results on three public datasets have demonstrated the effectiveness of our proposed IVT framework on text-based person retrieval.

7   Broader Impact

Text-based person retrieval has many potential applications in surveillance, *e.g.,* finding suspects, lost children, or elderly people. This technology can enhance the safety of the cities we live in. This work demonstrates the effectiveness of unified network and implicit alignments for the TPR task. The potential negative impact lies in that surveillance data about pedestrians may cause privacy breaches. Hence, the data collection process should be consented to by the pedestrian and the data utilization should be regulated.

Acknowledgement. This work is supported by National Natural Science Foundation of China (No. 62102205).

References

1. Aggarwal, S., Radhakrishnan, V.B., Chakraborty, A.: Text-based person search via attribute-aided matching. In: Proceedings of the IEEE/CVF Winter Conference on Applications of Computer Vision (WACV), pp. 2617–2625 (2020)
2. Bao, H., Dong, L., Wei, F.: BEit: BERT pre-training of image transformers. In: International Conference on Learning Representations (ICLR) (2022)
3. Brown, T., et al.: Language models are few-shot learners. In: Advances in Neural Information Processing Systems (NeurIPS), vol. 33, pp. 1877–1901 (2020)
4. Chen, D., et al.: Improving deep visual representation for person re-identification by global and local image-language association. In: Proceedings of the European Conference on Computer Vision (ECCV), pp. 54–70 (2018)
5. Chen, T., Xu, C., Luo, J.: Improving text-based person search by spatial matching and adaptive threshold. In: 2018 IEEE Winter Conference on Applications of Computer Vision (WACV), pp. 1879–1887 (2018)

6. Chen, Y.-C., et al.: UNITER: UNiversal image-TExt representation learning. In: Vedaldi, A., Bischof, H., Brox, T., Frahm, J.-M. (eds.) ECCV 2020. LNCS, vol. 12375, pp. 104–120. Springer, Cham (2020). https://doi.org/10.1007/978-3-030-58577-8_7

7. Chen, Y., Zhang, G., Lu, Y., Wang, Z., Zheng, Y.: TIPCB: a simple but effective part-based convolutional baseline for text-based person search. Neurocomputing **494**, 171–181 (2022)

8. Ding, Z., Ding, C., Shao, Z., Tao, D.: Semantically self-aligned network for text-to-image part-aware person re-identification. arXiv preprint arXiv:2107.12666 (2021)

9. Dosovitskiy, A., et al.: An image is worth 16x16 words: transformers for image recognition at scale. In: International Conference on Learning Representations (ICLR) (2020)

10. Fedus, W., Zoph, B., Shazeer, N.: Switch transformers: scaling to trillion parameter models with simple and efficient sparsity. arXiv preprint arXiv:2101.03961 (2021)

11. Gao, C., et al.: Contextual non-local alignment over full-scale representation for text-based person search. arXiv preprint arXiv:2101.03036 (2021)

12. Han, X., He, S., Zhang, L., Xiang, T.: Text-based person search with limited data. In: The British Machine Vision Conference (BMVC) (2021)

13. He, K., Chen, X., Xie, S., Li, Y., Dollár, P., Girshick, R.: Masked autoencoders are scalable vision learners. In: Proceedings of the IEEE Conference on Computer Vision and Pattern Recognition (CVPR) (2022)

14. He, K., Zhang, X., Ren, S., Sun, J.: Deep residual learning for image recognition. In: Proceedings of the IEEE Conference on Computer Vision and Pattern Recognition (CVPR), pp. 770–778 (2016)

15. Jia, C., et al.: Scaling up visual and vision-language representation learning with noisy text supervision. In: International Conference on Machine Learning (ICML), pp. 4904–4916. PMLR (2021)

16. Jing, Y., Si, C., Wang, J., Wang, W., Wang, L., Tan, T.: Pose-guided multi-granularity attention network for text-based person search. In: Proceedings of the AAAI Conference on Artificial Intelligence (AAAI), vol. 34, pp. 11189–11196 (2020)

17. Kenton, J.D.M.W.C., Toutanova, L.K.: Bert: pre-training of deep bidirectional transformers for language understanding. In: Annual Conference of the North American Chapter of the Association for Computational Linguistics: Human Language Technologies (NAACL-HLT), pp. 4171–4186 (2019)

18. Kim, W., Son, B., Kim, I.: ViLT: vision-and-language transformer without convolution or region supervision. In: International Conference on Machine Learning (ICML), pp. 5583–5594 (2021)

19. Krishna, R., et al.: Visual genome: connecting language and vision using crowd-sourced dense image annotations. Int. J. Comput. Vis. (IJCV) **123**(1), 32–73 (2017)

20. Lee, K.H., Chen, X., Hua, G., Hu, H., He, X.: Stacked cross attention for image-text matching. In: Proceedings of the European Conference on Computer Vision (ECCV), pp. 201–216 (2018)

21. Li, J., Selvaraju, R., Gotmare, A., Joty, S., Xiong, C., Hoi, S.C.H.: Align before fuse: vision and language representation learning with momentum distillation. In: Advances in Neural Information Processing Systems (NeurIPS), vol. 34 (2021)

22. Li, S., Xiao, T., Li, H., Yang, W., Wang, X.: Identity-aware textual-visual matching with latent co-attention. In: Proceedings of the IEEE International Conference on Computer Vision (ICCV), pp. 1890–1899 (2017)

23. Li, S., Xiao, T., Li, H., Zhou, B., Yue, D., Wang, X.: Person search with natural language description. In: Proceedings of the IEEE Conference on Computer Vision and Pattern Recognition (CVPR), pp. 1970–1979 (2017)

24. Li, X., et al.: OSCAR: object-semantics aligned pre-training for vision-language tasks. In: Vedaldi, A., Bischof, H., Brox, T., Frahm, J.-M. (eds.) ECCV 2020. LNCS, vol. 12375, pp. 121–137. Springer, Cham (2020). https://doi.org/10.1007/978-3-030-58577-8_8

25. Lin, T.-Y., et al.: Microsoft COCO: common objects in context. In: Fleet, D., Pajdla, T., Schiele, B., Tuytelaars, T. (eds.) ECCV 2014. LNCS, vol. 8693, pp. 740–755. Springer, Cham (2014). https://doi.org/10.1007/978-3-319-10602-1_48

26. Liu, J., Zha, Z.J., Hong, R., Wang, M., Zhang, Y.: Deep adversarial graph attention convolution network for text-based person search. In: Proceedings of the 27th ACM International Conference on Multimedia (MM), pp. 665–673 (2019)

27. Liu, Z., et al.: Swin transformer: hierarchical vision transformer using shifted windows. In: Proceedings of the IEEE/CVF International Conference on Computer Vision (ICCV), pp. 10012–10022 (2021)

28. Loper, E., Bird, S.: NLTK: the natural language toolkit. arXiv preprint cs/0205028 (2002)

29. Lu, J., Batra, D., Parikh, D., Lee, S.: ViLBERT: pretraining task-agnostic visiolinguistic representations for vision-and-language tasks. In: Advances in Neural Information Processing Systems (NeurIPS), vol. 32 (2019)

30. Sarafianos, N., Xu, X., Kakadiaris, I.A.: Adversarial representation learning for text-to-image matching. In: Proceedings of the IEEE/CVF International Conference on Computer Vision (ICCV), pp. 5813–5823 (2019)

31. Niu, K., Huang, Y., Ouyang, W., Wang, L.: Improving description-based person re-identification by multi-granularity image-text alignments. IEEE Trans. Image Process. (TIP) **29**, 5542–5556 (2020)

32. Ordonez, V., Kulkarni, G., Berg, T.: Im2text: describing images using 1 million captioned photographs. In: Advances in Neural Information Processing Systems (NeurIPS), vol. 24 (2011)

33. Radford, A., et al.: Learning transferable visual models from natural language supervision. In: International Conference on Machine Learning (ICML), pp. 8748–8763. PMLR (2021)

34. Radford, A., Narasimhan, K., Salimans, T., Sutskever, I.: Improving language understanding by generative pre-training (2018)

35. Radford, A., Wu, J., Child, R., Luan, D., Amodei, D., Sutskever, I., et al.: Language models are unsupervised multitask learners. OpenAI Blog **1**(8), 9 (2019)

36. Reed, S., Akata, Z., Lee, H., Schiele, B.: Learning deep representations of fine-grained visual descriptions. In: Proceedings of the IEEE Conference on Computer Vision and Pattern Recognition (CVPR), pp. 49–58 (2016)

37. Sharma, P., Ding, N., Goodman, S., Soricut, R.: Conceptual captions: a cleaned, hypernymed, image alt-text dataset for automatic image captioning. In: Proceedings of the 56th Annual Meeting of the Association for Computational Linguistics (ACL), pp. 2556–2565 (2018)

38. Vaswani, A., et al.: Attention is all you need. In: Advances in Neural Information Processing Systems (NeurIPS), vol. 30 (2017)

39. Wang, C., Luo, Z., Lin, Y., Li, S.: Text-based person search via multi-granularity embedding learning. In: International Joint Conference on Artificial Intelligence (IJCAI), pp. 1068–1074 (2021)

40. Wang, W., Bao, H., Dong, L., Wei, F.: VLMo: unified vision-language pre-training with mixture-of-modality-experts. arXiv preprint arXiv:2111.02358 (2021)

41. Wang, X., et al.: Large-scale multi-modal pre-trained models: a comprehensive survey (2022). https://github.com/wangxiao5791509/MultiModal_BigModels_Survey

42. Wang, Z., Fang, Z., Wang, J., Yang, Y.: *ViTAA*: visual-textual attributes alignment in person search by natural language. In: Vedaldi, A., Bischof, H., Brox, T., Frahm, J.-M. (eds.) ECCV 2020. LNCS, vol. 12357, pp. 402–420. Springer, Cham (2020). https://doi.org/10.1007/978-3-030-58610-2_24
43. Wang, Z., Zhu, A., Zheng, Z., Jin, J., Xue, Z., Hua, G.: Img-net: inner-cross-modal attentional multigranular network for description-based person re-identification. J. Electron. Imaging (JEI) **29**(4), 043028 (2020)
44. Wei, L., Zhang, S., Gao, W., Tian, Q.: Person transfer GAN to bridge domain gap for person re-identification. In: Proceedings of the IEEE Conference on Computer Vision and Pattern Recognition (CVPR), pp. 79–88 (2018)
45. Zhang, P., et al.: VinVL: revisiting visual representations in vision-language models. In: Proceedings of the IEEE/CVF Conference on Computer Vision and Pattern Recognition (CVPR), pp. 5579–5588 (2021)
46. Zhang, S., Zheng, D., Hu, X., Yang, M.: Bidirectional long short-term memory networks for relation classification. In: Proceedings of the 29th Pacific Asia Conference on Language, Information and Computation (PACLIC), pp. 73–78 (2015)
47. Zhang, Y., Lu, H.: Deep cross-modal projection learning for image-text matching. In: Proceedings of the European Conference on Computer Vision (ECCV), pp. 686–701 (2018)
48. Zheng, K., Liu, W., Liu, J., Zha, Z.J., Mei, T.: Hierarchical gumbel att ention network for text-based person search. In: Proceedings of the 28th ACM International Conference on Multimedia (MM), pp. 3441–3449 (2020)
49. Zheng, Z., Zheng, L., Garrett, M., Yang, Y., Xu, M., Shen, Y.D.: Dual-path convolutional image-text embeddings with instance loss. ACM Trans. Multimedia Comput. Commun. Appl. (TOMM) **16**(2), 1–23 (2020)
50. Zhong, Z., Zheng, L., Kang, G., Li, S., Yang, Y.: Random erasing data augmentation. In: Proceedings of the AAAI Conference on Artificial Intelligence (AAAI), vol. 34, pp. 13001–13008 (2020)
51. Zhu, A., et al.: DSSL: deep surroundings-person separation learning for text-based person retrieval. In: Proceedings of the 29th ACM International Conference on Multimedia (MM), pp. 209–217 (2021)

Look at Adjacent Frames: Video Anomaly Detection Without Offline Training

Yuqi Ouyang(✉), Guodong Shen, and Victor Sanchez

University of Warwick, Coventry, UK
{yuqi.ouyang,guodong.shen,v.f.sanchez-silva}@warwick.ac.uk

Abstract. We propose a solution to detect anomalous events in videos without the need to train a model offline. Specifically, our solution is based on a randomly-initialized multilayer perceptron that is optimized online to reconstruct video frames, pixel-by-pixel, from their frequency information. Based on the information shifts between adjacent frames, an incremental learner is used to update parameters of the multilayer perceptron after observing each frame, thus allowing to detect anomalous events along the video stream. Traditional solutions that require no offline training are limited to operating on videos with only a few abnormal frames. Our solution breaks this limit and achieves strong performance on benchmark datasets.

Keywords: Video anomaly detection · Unsupervised · Offline training · Online learning · Multilayer perceptron · Discrete wavelet transform

1 Introduction

Video anomaly detection (VAD) aims at detecting anomalous events in a video scene. Since it is hard to define all possible anomalous events a priori and moreover, these anomalies may occur infrequently, VAD is rarely solved by supervised learning. It is then common to exclusively rely on normal video data to train a model for the detection of anomalous events. Hence, the lack of examples of abnormal events during training defines the inherent challenges of this task. However, real-life experiences show that a person could react to a biker moving among pedestrians because the biker moves distinctively, without knowing that the biker is an abnormality in this context. Enlightened by such human intelligence, VAD can also be solved without any knowledge learned from an offline training process by analyzing spatio-temporal differences between adjacent frames since these differences are usually significant where anomalous events occur [10,15,20]. The working mechanism of this type of VAD solution, which requires no offline training and is hereinafter referred to as online VAD[1], is illustrated in Fig. 1 along with the working mechanism of offline VAD. Note that instead of training the model offline, online VAD updates the model sequentially after observing each video frame, thus requiring no training data.

---

[1] Note that in this paper, online VAD does not refer to VAD solutions that are trained offline but operate at high frame rates.

© The Author(s), under exclusive license to Springer Nature Switzerland AG 2023
L. Karlinsky et al. (Eds.): ECCV 2022 Workshops, LNCS 13805, pp. 642–658, 2023.
https://doi.org/10.1007/978-3-031-25072-9_43

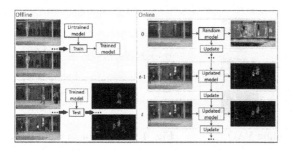 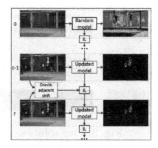

Fig. 1. Working mechanisms of offline and online VAD.   **Fig. 2.** Idea of online VAD.

One main advantage of online VAD is that a model can learn beyond the data used for offline training and adapt much better to the data source during operation, which eases the issue of concept drift caused by, e.g., any data distribution differences between the offline and online data. In offline VAD, a model is trained first to understand the patterns of videos depicting normal events. Once trained, it is used on new videos depicting normal events and possibly, abnormal events. In this case, the concept drift due to the distribution differences between the offline and online data depicting normal events may result in poor performance during operation. To deal with such a concept drift, one can use human support to regularly identify those normal frames on which the model performs poorly, and then retrain the model to recognize these frames correctly. However, such an approach undoubtedly comes with an extra workload.

Despite the fact that the performance of VAD solutions has improved recently [9,35], online VAD solutions are still scarce [28,36]. Moreover, the existing ones are usually flawed. For example, [10,15,20], which are pioneers in online VAD, attain poor performance on videos with a large number of anomalous events. Motivated by these limitations, we present our idea for online VAD in Fig. 2. Our idea focuses on the spatio-temporal differences between adjacent frames, hereinafter referred to as adjacent shifts. It uses an incremental learner (IL) that accounts for the adjacent shifts and sequentially updates the model from a set of randomly-initialized parameters. The IL is expected to easily adapt to gradual adjacent shifts, which usually exist between normal frames and should result in no anomaly detections. Conversely, the IL is expected to encounter difficulties in adapting to drastic adjacent shifts, which usually exist between abnormal frames or between a normal and an abnormal frame, thus resulting in anomaly detections. More specifically, as illustrated in Fig. 3, our solution generates error maps as the detection results by reconstructing spatio-temporal information and pixel coordinates into pixel values and comparing the reconstructed frames with the original ones. In this work, we use the discrete wavelet transform (DWT) to summarize the spatio-temporal information of a video sequence and a Multilayer Perceptron (MLP) for the frame reconstruction task. The MLP is set to be updated by the IL while adapting to the adjacent shifts, after observing each frame. The contributions of this work are summarized as follows:

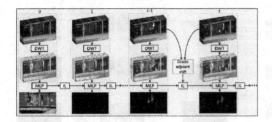 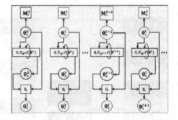

Fig. 3. Our solution for online VAD.   **Fig. 4.** Our problem definition.

- We introduce the first MLP-based model that uses frequency information to produce pixel-level VAD results.
- More importantly, we design a novel solution for online VAD, i.e., detecting anomalous events with **no** offline training, where the network parameters are optimized sequentially after observing each frame starting from random initialization.
- We achieve state-of-the-art performance on benchmark datasets regardless of the temporal order or the number of abnormal frames in a video.

2   Related Work

Offline VAD. State-of-the-art offline VAD solutions use deep-learning techniques. These models can be classified as unsupervised, weakly-supervised, or self-supervised depending on the training data and how these data are used. Unsupervised models exclusively employ normal videos for training, with data mapping via an encoder-decoder structure being the most frequently used strategy. The baseline approaches in this context are based on data reconstruction or prediction, where anomalous events are detected based on the reconstruction or prediction errors, respectively [12,21,24,29,38,49,52]. Adversarial losses can also be applied in such approaches, where the generator is adversarially trained to perform the data mapping so that abnormal videos lead to unrealistic outputs with large reconstruction or prediction errors [19,30,37,45,51]. If the features extracted from the normal and abnormal videos form two distinct and compact clusters, anomalous events can be detected in a latent subspace. In this type of approach, two ways are usually used to train a model. Either separately training the feature extraction and clustering steps [13,14,32,41,48], or jointly training both steps end-to-end [1–3,11,27,34].

Weakly-supervised models use a small number of abnormal frames with labels for training. Based on the triplet loss [18] or multiple instance learning strategy [7,35,40,44,53,54], such models can learn to increase the inter-class distance between normal and abnormal data. Weakly-supervised models outperform unsupervised models but they require abnormal data with ground truth labels [7,44,50].

Self-supervised models are becoming increasingly popular [33,46]. These models are trained with both normal and abnormal videos but without ground truth labels. They are usually trained iteratively by simulating labels through decision-making and then optimizing parameters based on the simulated labels. Data collection is easier for these models because they do not require ground truth labels. However, because these models may use some of the test videos in the training stage, their results should be interpreted with caution.

Online VAD. Compared to offline VAD solutions, online models are limited both in number and performance [28,36]. Three main models [10,15,20] have been designed to tackle VAD online. The seminal work in [10] detects anomalous events by aggregating anomaly values from several frame shuffles, where the anomaly value in each shuffle is calculated by measuring the similarity between all previous frames and the frames in a sliding window. Although that particular model cannot work with video streams, it sets an important precedent for VAD online. The follow-up works in [15,20] propose analyzing adjacent frame batches, where anomalies are defined as abrupt differences in spatio-temporal features between two adjacent batches. However, instead of randomly initializing a model and continuously optimizing it along the video stream, these two works repeat the random initialization of parameters every time a new frame is observed, ignoring the fact that frames may share common information, such as the scene background. Moreover, the performance of these three solutions, i.e., [10,15,20], degrades when used on videos with a large number of abnormal frames.[2] The reason for this is that those models assume a video only contains a few abnormal frames, thus a large number of abnormal frames violate their assumptions regarding the spatio-temporal distinctiveness of abnormal frames in a video.

Offline VAD with Further Optimization. Recently, several offline VAD solutions that are further optimized online have been proposed [4–6]. Specifically, these models are first trained offline and then refined online to reduce false positives. Namely, the model processes each frame online and adds potential normal frames to a learning set where false positives identified manually are also included [4]. Such a learning set is then used to further optimize the model so that it can more accurately learn the normal patterns beyond the offline training data. Scene adaptation has also been recently addressed in VAD [23,25] based on recent advance in meta-learning [8]. Specifically, the model is first trained offline to produce acceptable results for a variety of scenes, and is then further optimized with only a few frames from a specific scene of interest, thus rapidly adapting to that scene.

3 Proposed Solution

Problem Definition. Based on our objective of generating pixel-level detections online in the form of error maps, we start by providing the problem definition:

---

[2] Videos with more than *50%* of frames being abnormal.

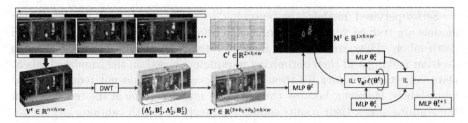

Fig. 5. The workflow of our solution at timestep t. b_1 and b_2 respectively indicate the number of high-frequency maps in tensors $\mathbf{B}_1^t$ and $\mathbf{B}_2^t$ of the temporal DWT.

- At current timestep t, based on the network parameters $\boldsymbol{\theta}_\circ^t$, compute the error map $\mathbf{M}_\circ^t$, where anomalies are indicated at the pixel-level.
- Define the loss $\ell(\boldsymbol{\theta}^t)$ to optimize $\boldsymbol{\theta}_\circ^t$ into $\boldsymbol{\theta}_\bullet^t$.
- Use $\boldsymbol{\theta}_\circ^t$ and $\boldsymbol{\theta}_\bullet^t$ to define $\boldsymbol{\theta}_\circ^{t+1}$ for computing an error map for timestep $t+1$.

Note that we use two types of subscripts, i.e., a clear circle $\circ$ and a black circle $\bullet$, where $\circ$ ($\bullet$) denotes the initial (final) network parameters or the detection results before (after) the optimization process. We illustrate our problem definition in Fig. 4 starting from the first frame of a video stream, i.e., $t=0$, where the IL is in charge of updating the network parameters from $\boldsymbol{\theta}_\circ^t$ to $\boldsymbol{\theta}_\bullet^t$ and subsequently computing $\boldsymbol{\theta}_\circ^{t+1}$.

3.1 Workflow

Figure 5 illustrates the workflow of our solution at the current timestep t.[3] It first applies a temporal DWT to a set of n frames of size $h \times w$, denoted by $\mathbf{V}^t$, to generate DWT coefficients, where the last frame of $\mathbf{V}^t$ is the current frame $\mathbf{I}^t$. The DWT coefficients are used in conjunction with the pixel coordinates $\mathbf{C}^t$ as the input $\mathbf{T}^t$ of a pixel-level MLP to reconstruct $\mathbf{I}^t$, thus leading to an error map $\mathbf{M}^t$ by comparing the reconstruction results with the ground truth frame. $\mathbf{M}^t$ is then used by the IL to define the loss $\ell(\boldsymbol{\theta}^t)$. Based on $\ell(\boldsymbol{\theta}^t)$, the IL updates the network parameters from $\boldsymbol{\theta}_\circ^t$ to $\boldsymbol{\theta}_\bullet^t$, and then calculates $\boldsymbol{\theta}_\circ^{t+1}$ for the next timestep.

Discrete Wavelet Transform. In this work, we use the temporal DWT to summarize the spatio-temporal information of a video sequence because it has been shown to provide motion information that fits the human visual system [16]. As shown in Fig. 6 (left), two levels of temporal DWT are performed on the set of frames $\mathbf{V}^t$. The temporal DWT results in four tensors of sub-bands. Specifically, two low-frequency tensors, $\mathbf{A}_1^t$ and $\mathbf{A}_2^t$, and two high-frequency tensors, $\mathbf{B}_1^t$ and $\mathbf{B}_2^t$. Note that the temporal DWT does not change the spatial dimensions. The input tensor $\mathbf{T}^t$ of our MLP is defined as:

$$\mathbf{T}^t = \left[\mathbf{C}^t, \mathbf{A}^t, \mathbf{B}_1^t, \mathbf{B}_2^t \right], \tag{1}$$

---

[3] In Figs. 5 and 6, the channel dimensions of the data are omitted.

Fig. 6. Details of our temporal DWT (left) and our MLP (right). a_1 and a_2 (b_1 and b_2) respectively indicate numbers of low-frequency (high-frequency) maps in tensors $\mathbf{A}_1^t$ and $\mathbf{A}_2^t$ ($\mathbf{B}_1^t$ and $\mathbf{B}_2^t$) of the temporal DWT. Layers of the MLP are depicted with the dimensions of their outputs. The error map is computed by comparing two grayscale images.

where $\mathbf{A}^t \in \mathbb{R}^{1 \times h \times w}$ is the last low-frequency map of the tensor $\mathbf{A}_1^t$. Here $\mathbf{T}^t$ is formed by stacking the coordinate tensor $\mathbf{C}^t$, the low-frequency map $\mathbf{A}^t$ (appearance information), the first-level high-frequency tensor $\mathbf{B}_1^t$ (sparse motion information) and the second-level high-frequency tensor $\mathbf{B}_2^t$ (dense motion information) along the first dimension.

Multilayer Perceptron. We use an MLP to reconstruct frames because it has been shown to effectively map pixel coordinates into pixel values [39,42,43]. As illustrated in Fig. 6 (right), our MLP maps $\mathbf{T}^t$ into the reconstructed frame $\hat{\mathbf{I}}^t$. Given such an MLP with parameters $\boldsymbol{\theta}^t$, the resulting error map, $\mathbf{M}^t$, and its mean squared error (MSE), ϵ^t, are respectively computed as follows:

$$\mathbf{M}^t = \left(\mathbf{I}^t - f_{\boldsymbol{\theta}^t}(\mathbf{T}^t) \right)^{\odot^2}, \tag{2}$$

$$\epsilon^t = \frac{1}{hw} \left\| \mathbf{M}^t \right\|_1, \tag{3}$$

where $f_{\boldsymbol{\theta}^t}(\cdot)$ denotes the mapping function of the MLP, $\odot^2$ indicates element-wise square, and $\| \cdot \|_1$ indicates ℓ_1-norm.[4] The MLP is optimized by an IL, as detailed next.

3.2 Incremental Learner

As illustrated in Fig. 7 (left), at the current timestep t, the IL comprises a comparator that defines the loss $\ell(\boldsymbol{\theta}^t)$ from the MLP results, an adapter that optimizes the parameters of the MLP based on $\ell(\boldsymbol{\theta}^t)$, and a clipper that calculates the initial parameters of the MLP for the next timestep, i.e., $\boldsymbol{\theta}_\circ^{t+1}$.

The Comparator. This component compares the reconstruction results of frame $\mathbf{I}^{t-1}$, as reconstructed by the MLP with parameters $\boldsymbol{\theta}_\bullet^{t-1}$, with those of frame $\mathbf{I}^t$, as reconstructed by the MLP with parameters $\boldsymbol{\theta}^t$. Based on this

---

[4] To compute the error map and the MSE when optimizing the MLP, please follow Eqs. (2) and (3) but replace $\boldsymbol{\theta}^t$ with current parameter set involved in the optimization; e.g., $\mathbf{M}_\circ^t = \left(\mathbf{I}^t - f_{\boldsymbol{\theta}_\circ^t}(\mathbf{T}^t) \right)^{\odot^2}$, $\epsilon_\circ^t = \frac{1}{hw} \left\| \mathbf{M}_\circ^t \right\|_1$.

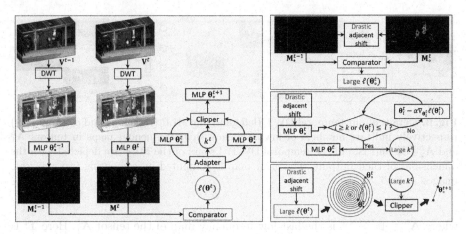

Fig. 7. The incremental learner at the current timestep t (left). An example of a drastic adjacent shift that results in a large value for the loss function (top-right). The iterative process used to update the parameters of the MLP (middle-right). Computation of the initial parameters of the MLP for timestep $t+1$ (bottom-right).

comparison, it defines the loss $\ell(\boldsymbol{\theta}^t)$ for the MLP at timestep t. This loss is based on MSE values:

$$\ell(\boldsymbol{\theta}^t) = \text{Relu}(\epsilon^t - \epsilon_\bullet^{t-1}), \quad \text{if } t \geqslant 1. \tag{4}$$

The rationale behind the loss in Eq. (4) is that such comparison accounts for the adjacent shift between frames $\mathbf{I}^{t-1}$ and $\mathbf{I}^t$. Figure 7 (top-right) shows an example, where a drastic adjacent shift allows a well-fitted MLP ($\boldsymbol{\theta}_\bullet^{t-1}$) to accurately reconstruct frame $\mathbf{I}^{t-1}$, while preventing the precedent MLP ($\boldsymbol{\theta}_\circ^t$) from accurately reconstructing frame $\mathbf{I}^t$. Hence, Eq. (4) implicitly measures the adjacent shift by resulting in a large loss value, i.e., large $\ell(\boldsymbol{\theta}_\circ^t)$.[5]

For the first frame of a video sequence, i.e., timestep $t=0$, note that there is no previous information for the comparator to define the loss in Eq. (4). We say that, in this case, the randomly-initialized MLP is under a cold start and may not generate accurate reconstruction results. Hence, we adjust Eq. (4) for $t=0$ as follows:

$$\ell(\boldsymbol{\theta}^t) = \text{Relu}(\epsilon^t - \bar{\epsilon}), \quad \text{if } t=0, \tag{5}$$

where $\bar{\epsilon}$ is a user-defined MSE value that defines a target value to be achieved for an accurate reconstruction.[6]

---

[5] To compute the loss when optimizing the MLP, please follow Eq. (4) or Eq. (5) but replace $\boldsymbol{\theta}^t$ with current parameter set involved in the optimization; e.g., $\ell(\boldsymbol{\theta}_\circ^t) = \text{Relu}(\epsilon_\circ^t - \epsilon_\bullet^{t-1})$, (if $t \geqslant 1$).

[6] We set $\epsilon_\bullet^{t-1} = \max(\epsilon_\bullet^{t-1}, \bar{\epsilon})$ in Eq. (4) to prevent our MLP from being overfitted towards a MSE value lower than the lower-bound $\bar{\epsilon}$.

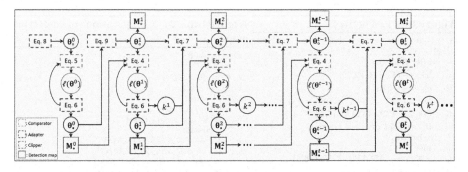

Fig. 8. The mathematical operations of the IL along a video stream.

The Adapter. Based on the loss $\ell(\theta^t)$, the adapter adapts to the adjacent shift by optimizing the parameters of the MLP over iterations of gradient descent (GD), the $(i+1)$th GD iteration is defined as:

$$\theta_i^t \xrightarrow{-\nabla_{\theta_i^t}\ell(\theta_i^t),\ \text{if } \ell(\theta_i^t) > \bar{l} \text{ or } i < \bar{k}} \theta_{i+1}^t, \tag{6}$$

where $i \in [0, \bar{k}]$. Here we use two user-defined parameters, i.e., $\bar{k}$ and $\bar{l}$, to control the end of the iterative optimization: the adapter stops at the current GD iteration if the number of GD iterations is large enough $(i \geqslant \bar{k})$ or if the loss is small enough $(\ell(\theta_i^t) \leqslant \bar{l})$. This iterative process is illustrated in Fig. 7 (middle-right), where the adapter optimizes the parameters from $\theta_\circ^t$ to $\theta_\bullet^t$.[7]

Let k^t denote the number of iterations of GD used by the adapter at timestep t. A large loss $\ell(\theta^t)$, potentially caused by the drastic adjacent shift, will force the adapter to spend many iterations of GD, i.e., a large k^t, to optimize the network parameters.

The Clipper. According to Fig. 7 (left), the loss $\ell(\theta^{t+1})$ at the <u>next</u> timestep $t+1$, is computed by comparing $\mathbf{M}_\bullet^t$, as computed under the parameter set $\theta_\bullet^t$, and $\mathbf{M}^{t+1}$, as computed under the parameter set θ^{t+1}. Since the adapter generates $\theta_\bullet^t$, we need to define the initial parameter set $\theta_\circ^{t+1}$. This is accomplished by the clipper.

As illustrated in Fig. 7 (middle-right), a large value of k^t, i.e., the number of iterations of GD used by the adapter at timestep t, may imply a drastic shift between frames $\mathbf{I}^{t-1}$ and $\mathbf{I}^t$. Such a drastic shift may indicate that the current frame $\mathbf{I}^t$ is abnormal. Hence, the MLP may have been well-fitted to reconstruct the abnormal frame $\mathbf{I}^t$, and such a well-fitted parameter set $\theta_\bullet^t$ may not be appropriate to initialize $\theta_\circ^{t+1}$. Based on these observations, $\theta_\circ^{t+1}$ is defined based on knowledge transfer [31] by clipping between $\theta_\circ^t$ and $\theta_\bullet^t$ as follows:

$$\theta_\circ^{t+1} = \theta_\circ^t + (k^t)^{-\frac{1}{2}}\left(\theta_\bullet^t - \theta_\circ^t\right), \quad \text{if } t \geqslant 1. \tag{7}$$

---

[7] Note that $\theta_\circ^t = \theta_{i=0}^t$.

Algorithm 1. The IL	**Algorithm 2.** Computation of detection maps
1: $i \leftarrow 0$	
2: $\theta_\circ^t \leftarrow \theta_\circ^t$	1: $\mathbb{M}_{\text{det}} \leftarrow \emptyset$
3: **while** *True* **do**	2: $t \leftarrow 0$
4: $\mathbf{M}_i^t \leftarrow (\mathbf{I}^t - f_{\theta_i^t}(\mathbf{T}^t))^{\odot 2}$	3: $\theta_\circ^0 \leftarrow \mathcal{RAND}()$
5: $e_i^t \leftarrow \frac{1}{hw}\|\mathbf{M}_i^t\|_1$	4: $\epsilon_\bullet^{-1} \leftarrow null$
6: $\mathbf{M}_\circ^t \leftarrow \mathbf{M}_i^t,\ \epsilon_\circ^t \leftarrow \epsilon_0^t$ **when** $i=0$	5: $k^{-1} \leftarrow null$
7: $\ell(\theta_i^t) \leftarrow \mathcal{COMPARE}(\bar{\epsilon}, e_\bullet^{t-1}, e_i^t)$	6: **while** *True* **do**
8: **if** $\ell(\theta_i^t) \leqslant \bar{l}$ or $i \geqslant \bar{k}$ **then**	7: **if** $\mathbf{V}^t$ *not exists* **then**
9: **break**	8: **break**
10: **else**	9: $\mathbf{V}^t \leftarrow \mathcal{TRANSFORM}(\mathbf{V}^t)$
11: $\theta_{i+1}^t \leftarrow \mathcal{ADAPT}(\alpha, \nabla_{\theta_i^t}\ell(\theta_i^t))$	10: $\mathbf{C}^t \leftarrow \mathcal{TRANSFORM}(\mathbf{C}^t)$
12: $i \leftarrow i+1$	11: $\mathbf{A}_1^t, \mathbf{B}_1^t, \mathbf{A}_2^t, \mathbf{B}_2^t \leftarrow \mathcal{DWT}(\mathbf{V}^t)$
13: $\theta_\bullet^t \leftarrow \theta_i^t,\ \mathbf{M}_\bullet^t \leftarrow \mathbf{M}_i^t,\ \epsilon_\bullet^t \leftarrow \epsilon_i^t$	12: $\mathbf{T}^t \leftarrow [\mathbf{C}^t, [\mathbf{A}_1^t]_{a_1,:,:}, \mathbf{B}_1^t, \mathbf{B}_2^t]$
14: $\theta_\circ^{t+1} \leftarrow \mathcal{CLIP}(\theta_\circ^t, \theta_\bullet^t, k^{t-1})$	13: $\theta_\circ^{t+1}, \epsilon_\bullet^t, k^t, \mathbf{M}_{\text{det}} \leftarrow \mathcal{IL}(\mathbf{I}^t, \mathbf{T}^t, \theta_\circ^t, \epsilon_\bullet^{t-1}, k^{t-1})$
15: $k^t \leftarrow$ (if $i=0$ then 1 else i end)	14: $\mathbb{M}_{\text{det}} \leftarrow \mathbb{M}_{\text{det}} \bigcup \{\mathbf{M}_\bullet^t\}$
16: $\mathbf{M}_{\text{det}}^t \leftarrow$ (if $t=0$ then $\mathbf{M}_\bullet^t$ else $\mathbf{M}_\circ^t$ end)	15: $t \leftarrow t+1$
17: $\mathcal{RETURN}(\theta_\circ^{t+1}, \epsilon_\bullet^t, k^t, \mathbf{M}_{\text{det}}^t)$	

For a large k^t, potentially caused by a drastic adjacent shift, $\theta_\circ^{t+1}$ is set to be close to $\theta_\circ^t$, thus rejecting the learned knowledge $\theta_\bullet^t$ acquired from a potential abnormal frame $\mathbf{I}^t$. This is illustrated in Fig. 7 (bottom-right).

Since there is no previously learned knowledge at $t=0$, the MLP is randomly-initialized for this first frame as:

$$\theta_\circ^0 = \widetilde{\theta}, \tag{8}$$

where $\widetilde{\theta}$ indicates a random set of parameters. At $t=1$, the MLP is initialized without clipping parameters to avoid using the randomly-initialized parameters $\theta_\circ^0$ as:

$$\theta_\circ^1 = \theta_\bullet^0. \tag{9}$$

Figure 8 illustrates the complete functionality of the IL along a video stream, while Algorithm 1 summarizes it.

Detection and Anomaly Inference. As depicted in Fig. 8, at the current timestep t, the MLP with initial parameters $\theta_\circ^t$ results in an error map $\mathbf{M}_\circ^t$ as the detection results, hereinafter called detection maps. For all observed frames, these detection maps are:

$$\mathbb{M}_{\text{det}} = \{\mathbf{M}_\bullet^0, \mathbf{M}_\circ^1, \mathbf{M}_\circ^2, ..., \mathbf{M}_\circ^t\}, \tag{10}$$

where at $t=0$, we use the detection map $\mathbf{M}_\bullet^0$ computed after the optimization.[8] The detection maps can be visually displayed as heatmaps that depict the abnormal pixels,[9] or alternately, one can calculate the MSE values associated with these maps to numerically quantify the anomalies. Algorithm 2 summarizes the process to compute the detection maps.

---

[8] Because $\mathbf{M}_\circ^0$ is a noisy error map generated by the MLP under random-initialization.

[9] See examples of detection maps in Fig. 11.

4   Experiments

Datasets. We test our model on three benchmark datasets. The UCSD Ped2 [26] dataset, which is a single-scene dataset depicting a pedestrian walkway where the anomalous events are individuals cycling, driving or skateboarding. The CUHK Avenue [22] dataset, which is also a single-scene dataset depicting a subway entrance with various types of anomalous events, such as individuals throwing papers and dancing. And the ShanghaiTech [24] dataset, which is more difficult to analyze as it has complex anomalous events across multiple scenes.[10]

Non-continuous Videos. In each benchmark dataset, videos depicting the same scene may not be continuously shot by one camera. Hence, our model regards these videos as different video streams. However, since such video streams may share common information, e.g., the background information, instead of randomly initializing our model on the first frame of these video streams, as specified in Eq. (8), our model initializes the parameters by transferring the knowledge learned on the last frame of the previous video stream. Note that for continuous videos depicting the same scene and shot by one camera, our model computes the initial parameters as specified by Fig. 8.

Implementation Details. All frames are transformed into gray-scale images and respectively re-sized to 230×410 and 240×428 for the CUHK Avenue and ShanghaiTech datasets. All pixel coordinates and pixel values are re-scaled to the range $[-0.5, 0.5]$. *Daubechies 2* is used as the filter for the temporal DWT on each set of video frames $\mathbf{V}^t$ of length $n = 16$. We randomly initialize the parameters of our MLP based on the settings provided in [39].[11] Adam [17] is used with learning rates 1×10^{-4} and 1×10^{-5}, respectively, on the first frame and subsequent frames of all videos. The two user-defined parameters, $\bar{\epsilon}$ and $\bar{l}$, are respectively set to 1×10^{-4} and 1×10^{-6} on all videos. $\bar{k}$ is set to 500 on the first 5 frames of each video, and then to 100 on the remaining frames.

Evaluation Metrics. We use the frame-level Area Under the Curve (AUC) of the Receiver Operating Characteristic (ROC) curve as the evaluation metric. Higher AUC values indicate better model performance.

4.1   Performance

In Table 1, we tabulate the performance of our model and other baseline models, where we highlight the best performing model in **bold** and underline the second best performing model. The left part of Table 1 tabulates the performance of online VAD solutions, while the right part compares ours with offline VAD solutions. Note that since the other three online VAD solutions cannot work on videos with a large number of abnormal frames while our model does not have

---

[10] The dataset has *13* scenes but there are no test videos for the last scene, thus we only focus on the first *12* scenes.

[11] All activation functions are *Sine* and only the last layer has no activation function.

Table 1. Comparisons in frame-level AUC values (%) on three benchmark datasets with online VAD models (left) and offline VAD models (right).

	Giorno [10]	Iones-cu [15]	Liu [20]	Ours		Chang [3]	Lu [23]	Wang [47]	Sun [41]	Cai [2]	Lv [25]	George-scu [9]	Liu [21]	Ours
UCSD Ped2	–	82.2	87.5	**96.5**	UCSD Ped2	96.5	96.2	–	–	96.6	96.9	**99.8**	99.3	96.5
CUHK Avenue	78.3	80.6	84.4	**90.2**	CUHK Avenue	86.0	85.8	87.0	89.6	86.6	89.5	**92.8**	91.1	90.2
Shang-haiTech	–	–	–	**83.1**	Shang-haiTech	73.3	77.9	79.3	74.7	73.7	73.8	**90.2**	76.2	83.1

that restriction, substantially more videos are tested by our solution. Despite more videos being analyzed, our model outperforms the best model previously reported in [20] by *9.0%* and *5.8%* AUC, respectively, on the UCSD Ped2 and CUHK Avenue datasets. Note that we report the first result computed by an online VAD solution on the ShanghaiTech dataset, i.e., an AUC value of *83.1%*. Compared to offline VAD solutions, the performance of our solution is very competitive, outperforming most of the models on the ShanghaiTech dataset with an AUC value of *83.1%*, which confirms our robustness to deal with complex scenes. The best performing offline solution is [9], which relies on multiple tasks where several networks are separately trained offline to jointly detect anomalies. Our solution attains competitive results by detecting video anomalies on-the-fly without offline training under a single framework based on frame reconstruction.

4.2 Further Studies

Abnormal Start. As specified in Eq. (5), our solution optimizes the MLP on the first frame of a video stream for accurate reconstruction results even if that frame may be abnormal. However, even in such a case, the detection of anomalies at subsequent abnormal frames is not affected. We illustrate this in Fig. 9, which depicts the anomaly values computed on the UCSD Ped2 test video *008* as normalized MSE values (see the blue curve). In this sequence, all frames are abnormal. Notice that even after optimizing the MLP on the first abnormal frame to achieve a low reconstruction error, which leads to an anomaly value close to *0*, the anomaly values increase rather than remaining low for subsequent frames. These values eventually become quite informative to detect anomalous events. The reason for this behavior is that our solution exploits the drastic adjacent shifts that exist between abnormal frames. Hence, despite being fine-tuned on the first abnormal frame, our MLP still leads to high MSE values on subsequent abnormal frames.

Unlimited Amount of Anomalies. Figure 9 also confirms that our solution works very well even on a video containing only abnormal frames, which is a major improvement from the previous online VAD solutions [10,15,20], whose performance degrades on videos with a large number of abnormal frames (i.e., more than *50%* of the frames depicting anomalous events). As mentioned before,

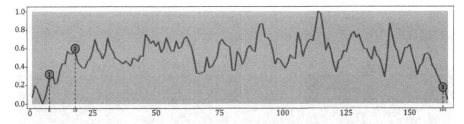

Fig. 9. Anomaly values of the UCSD Ped2 video *008*, where frames *23*, *33* and *179* are pinned with a green circle. (Color figure online)

Table 2. Frame-level AUC values (%) when testing with and without the clipper.

	Ours w/o clipper	Ours
CUHK avenue	86.3	90.2
Shang-haiTech	79.8	83.1

Fig. 10. Frame *23*, *33* and *179* of the UCSD Ped2 video *008* and their corresponding detection maps.

a large number of abnormal frames in a video violate their assumptions that a video only contains a few abnormal frames, i.e., the spatio-temporal distinctiveness of abnormal frames in a video. Conversely, our solution makes no such assumptions, thus being capable of working on videos containing plenty of abnormal frames, or even on videos with abnormal frames only.

Pixel-Level Detection. In Fig. 9, we select three frames, i.e., frames *23*, *33* and *179* (pinned with green circles),[12] and illustrate these frames and their detection maps in Fig. 10. By examining these detection maps, one can see that since the cycling event is detected in all three frames, the skateboarding event should be responsible for the changes in anomaly values in Fig. 9. Specifically, its presence and disappearance, respectively, lead to an increase in the anomaly value of the second frame (frame *33*) and a decrease in the third frame (frame *179*). Moreover, when the skateboarder gradually enters (leaves) the scene, our solution detects it with an increasing (decreasing) trend on anomaly values (see the blue curve before frame *33*, and before frame *179* in Fig. 9), which shows that our solution can detect anomalous events at the frame boundaries where the anomalies are usually not evident enough.

---

[12] Index numbers in the figure are *8*, *18* and *164*, respectively, since our model begins analyzing this sequence at the *16*th frame.

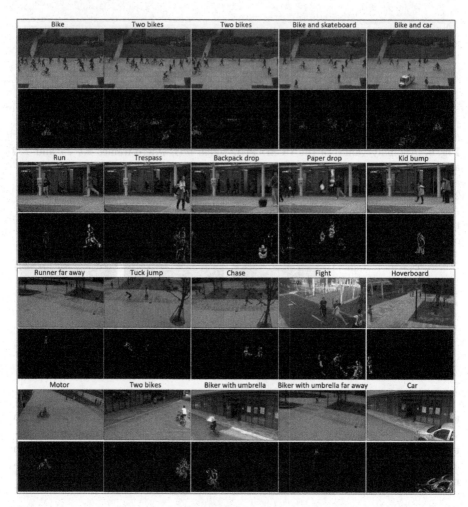

Fig. 11. Examples of frames and their detection maps of the UCSD Ped2 (rows *1–2*), CUHK Avenue (rows *3–4*) and ShanghaiTech (rows *5–8*) datasets, where the type of anomalous event is indicated above each example. Our solution accurately detects various types of anomalous events at the pixel-level.

Figure 11 shows examples of detection maps and their corresponding frames[13]. These sample results confirm the advantages of our pixel-level detections by demonstrating that our solution can identify video anomalies at a fine granularity level. For example, it can detect the umbrellas held by bikers (see the example in the last row, third column), small abnormal objects located in the background (see the examples at the sixth row, first column, and in the last row, fourth column). The visual results in Fig. 11 also show that our solution

---

[13] We use *matplotlib rainbow* as the colormap, with its colors being replaced by black at low values to increase the contrast.

Fig. 12. Anomaly values based on normalized MSE from frames *620* to *715* of the CUHK Avenue dataset video *006* (left). Frame *625* (row *1*) and *631* (row *2*) and their detection maps computed without (column *2*) and with (column *3*) the clipper (right).

accurately reconstructs the scene background and foreground where pixel errors may be hardly recognizable. This demonstrates that the IL used by our solution successfully transfers common information along the video stream.

The Clipper. We perform an ablation study to showcase the functionality of the clipper. First, we use our solution on two benchmark datasets without using the clipper, i.e., directly setting $\theta_\circ^t = \theta_\bullet^{t-1}$, and then tabulate its performance in terms of AUC values with that attained by using the clipper (see Table 2). These results show that by using the clipper, the performance increases, respectively, by *3.9%* and *3.3%* on the CUHK Avenue and ShanghaiTech datasets.

To further understand the effect of the clipper on performance, we focus on frames *620* to *715* of the CUHK Avenue test video *006*, which depicts the end of an anomalous event, i.e., a person trespasses and walks out of the scene. The anomaly values computed with and without the clipper are plotted in Fig. 12 (left) in blue and red curves, respectively. Frames *625* and *631* and their detection maps are depicted in Fig. 12 (right).[14] Regardless of whether the clipper is used, our solution produces informative detection maps to detect the anomalous event in frame *625* (first row). For frame *631* (second row), which is the first frame after the anomalous event ends, one can see that when the clipper is not used, our solution produces a noisier detection exactly in the region where the person leaves the scene. The pixels erroneously marked as abnormal in this error map are the result of overfitting the model on the previous abnormal frames. As depicted by the red curve in Fig. 12 (left), such overfitted model generates higher anomaly values in subsequent normal frames, which results in a poor performance. Hence, without clipping the knowledge learned from previous frames, the model may not be appropriate to be used on subsequent frames.

---

[14] Index numbers in the figure are *610* and *616*, respectively, since our model begins analyzing this sequence at the *16*th frame.

5   Conclusion

In this paper, we proposed a solution for online VAD where offline training is no longer required. Our solution is based on a pixel-level MLP that reconstructs frames from pixel coordinates and DWT coefficients. Based on the information shifts between adjacent frames, an incremental learner is used to optimize the MLP online to produce detection results along a video stream. Our solution accurately detects anomalous events at the pixel-level, achieves strong performance on benchmark datasets, and surpasses other online VAD models by being capable to work with any number of abnormal frames in a video. Our future work focuses on improving performance by reducing the number of false positive detections.

References

1. Abati, D., Porrello, A., Calderara, S., Cucchiara, R.: Latent space autoregression for novelty detection. In: CVPR, pp. 481–490 (2019)
2. Cai, R., Zhang, H., Liu, W., Gao, S., Hao, Z.: Appearance-motion memory consistency network for video anomaly detection. AAAI **35**(2), 938–946 (2021)
3. Chang, Y., Tu, Z., Xie, W., Yuan, J.: Clustering driven deep autoencoder for video anomaly detection. In: Vedaldi, A., Bischof, H., Brox, T., Frahm, J.-M. (eds.) ECCV 2020. LNCS, vol. 12360, pp. 329–345. Springer, Cham (2020). https://doi.org/10.1007/978-3-030-58555-6_20
4. Doshi, K., Yilmaz, Y.: Continual learning for anomaly detection in surveillance videos. In: CVPRW, pp. 1025–1034 (2020)
5. Doshi, K., Yilmaz, Y.: A modular and unified framework for detecting and localizing video anomalies. In: WACV, pp. 3982–3991 (2022)
6. Doshi, K., Yilmaz, Y.: Rethinking video anomaly detection - a continual learning approach. In: WACV, pp. 3961–3970 (2022)
7. Feng, J.C., Hong, F.T., Zheng, W.S.: MIST: multiple instance self-training framework for video anomaly detection. In: CVPR, pp. 14009–14018 (2021)
8. Finn, C., Abbeel, P., Levine, S.: Model-agnostic meta-learning for fast adaptation of deep networks. In: ICML, pp. 1126–1135 (2017)
9. Georgescu, M.I., Barbalau, A., Ionescu, R.T., Khan, F.S., Popescu, M., Shah, M.: Anomaly detection in video via self-supervised and multi-task learning. In: CVPR, pp. 12742–12752 (2021)
10. Del Giorno, A., Bagnell, J.A., Hebert, M.: A discriminative framework for anomaly detection in large videos. In: Leibe, B., Matas, J., Sebe, N., Welling, M. (eds.) ECCV 2016. LNCS, vol. 9909, pp. 334–349. Springer, Cham (2016). https://doi.org/10.1007/978-3-319-46454-1_21
11. Gong, D., et al.: Memorizing normality to detect anomaly: memory-augmented deep autoencoder for unsupervised anomaly detection. In: ICCV, pp. 1705–1714 (2019)
12. Hasan, M., Choi, J., Neumann, J., Roy-Chowdhury, A.K., Davis, L.S.: Learning temporal regularity in video sequences. In: CVPR, pp. 733–742 (2016)
13. Hinami, R., Mei, T., Satoh, S.: Joint detection and recounting of abnormal events by learning deep generic knowledge. In: ICCV, pp. 3639–3647 (2017)

14. Ionescu, R.T., Khan, F.S., Georgescu, M., Shao, L.: Object-centric auto-encoders and dummy anomalies for abnormal event detection in video. In: CVPR, pp. 7834–7843 (2019)
15. Ionescu, R.T., Smeureanu, S., Alexe, B., Popescu, M.: Unmasking the abnormal events in video. In: ICCV, pp. 2914–2922 (2017)
16. Jin, B., et al.: Exploring spatial-temporal multi-frequency analysis for high-fidelity and temporal-consistency video prediction. In: CVPR, pp. 4553–4562 (2020)
17. Kingma, D.P., Ba, J.: Adam: a method for stochastic optimization. In: ICLR (2015)
18. Liu, W., Luo, W., Li, Z., Zhao, P., Gao, S.: Margin learning embedded prediction for video anomaly detection with a few anomalies. In: IJCAI, pp. 3023–3030 (2019)
19. Liu, W., Luo, W., Lian, D., Gao, S.: Future frame prediction for anomaly detection - a new baseline. In: CVPR, pp. 6536–6545 (2018)
20. Liu, Y., Li, C., Póczos, B.: Classifier two sample test for video anomaly detections. In: BMVC, p. 71 (2018)
21. Liu, Z., Nie, Y., Long, C., Zhang, Q., Li, G.: A hybrid video anomaly detection framework via memory-augmented flow reconstruction and flow-guided frame prediction. In: ICCV, pp. 13588–13597 (2021)
22. Lu, C., Shi, J., Jia, J.: Abnormal event detection at 150 FPS in MATLAB. In: ICCV, pp. 2720–2727 (2013)
23. Lu, Y., Yu, F., Reddy, M.K.K., Wang, Y.: Few-shot scene-adaptive anomaly detection. In: Vedaldi, A., Bischof, H., Brox, T., Frahm, J.-M. (eds.) ECCV 2020. LNCS, vol. 12350, pp. 125–141. Springer, Cham (2020). https://doi.org/10.1007/978-3-030-58558-7_8
24. Luo, W., Liu, W., Gao, S.: A revisit of sparse coding based anomaly detection in stacked RNN framework. In: ICCV, pp. 341–349 (2017)
25. Lv, H., Chen, C., Cui, Z., Xu, C., Li, Y., Yang, J.: Learning normal dynamics in videos with meta prototype network. In: CVPR, pp. 15425–15434 (2021)
26. Mahadevan, V., Li, W., Bhalodia, V., Vasconcelos, N.: Anomaly detection in crowded scenes. In: CVPR, pp. 1975–1981 (2010)
27. Markovitz, A., Sharir, G., Friedman, I., Zelnik-Manor, L., Avidan, S.: Graph embedded pose clustering for anomaly detection. In: CVPR, pp. 10536–10544 (2020)
28. Mohammadi, B., Fathy, M., Sabokrou, M.: Image/video deep anomaly detection: a survey (2021)
29. Morais, R., Le, V., Tran, T., Saha, B., Mansour, M., Venkatesh, S.: Learning regularity in skeleton trajectories for anomaly detection in videos. In: CVPR, pp. 11988–11996 (2019)
30. Nguyen, T.N., Meunier, J.: Anomaly detection in video sequence with appearance-motion correspondence. In: ICCV, pp. 1273–1283 (2019)
31. Nichol, A., Achiam, J., Schulman, J.: On first-order meta-learning algorithms (2018)
32. Ouyang, Y., Sanchez, V.: Video anomaly detection by estimating likelihood of representations. In: ICPR, pp. 8984–8991 (2021)
33. Pang, G., Yan, C., Shen, C., Hengel, A.v.d., Bai, X.: Self-trained deep ordinal regression for end-to-end video anomaly detection. In: CVPR, pp. 12170–12179 (2020)
34. Park, H., Noh, J., Ham, B.: Learning memory-guided normality for anomaly detection. In: CVPR, pp. 14360–14369 (2020)
35. Purwanto, D., Chen, Y.T., Fang, W.H.: Dance with self-attention: a new look of conditional random fields on anomaly detection in videos. In: ICCV, pp. 173–183 (2021)

36. Ramachandra, B., Jones, M.J., Vatsavai, R.R.: A survey of single-scene video anomaly detection (2020)
37. Ravanbakhsh, M., Nabi, M., Sangineto, E., Marcenaro, L., Regazzoni, C., Sebe, N.: Abnormal event detection in videos using generative adversarial nets. In: ICIP, pp. 1577–1581 (2017)
38. Shen, G., Ouyang, Y., Sanchez, V.: Video anomaly detection via prediction network with enhanced spatio-temporal memory exchange. In: ICASSP, pp. 3728–3732 (2022)
39. Sitzmann, V., Martel, J.N., Bergman, A.W., Lindell, D.B., Wetzstein, G.: Implicit neural representations with periodic activation functions. In: NeurIPS (2020)
40. Sultani, W., Chen, C., Shah, M.: Real-world anomaly detection in surveillance videos. In: CVPR, pp. 6479–6488 (2018)
41. Sun, C., Jia, Y., Hu, Y., Wu, Y.: Scene-aware context reasoning for unsupervised abnormal event detection in videos. In: ACM MM, pp. 184–192 (2020)
42. Tancik, M., et al.: Learned initializations for optimizing coordinate-based neural representations. In: CVPR, pp. 2846–2855 (2021)
43. Tancik, M., et al.: Fourier features let networks learn high frequency functions in low dimensional domains. In: NeurIPS (2020)
44. Tian, Y., Pang, G., Chen, Y., Singh, R., Verjans, J.W., Carneiro, G.: Weakly-supervised video anomaly detection with robust temporal feature magnitude learning. In: ICCV, pp. 4975–4986 (2021)
45. Vu, H., Nguyen, T.D., Le, T., Luo, W., Phung, D.: Robust anomaly detection in videos using multilevel representations. AAAI **33**(01), 5216–5223 (2019)
46. Wang, S., Zeng, Y., Liu, Q., Zhu, C., Zhu, E., Yin, J.: Detecting abnormality without knowing normality: a two-stage approach for unsupervised video abnormal event detection. In: ACM MM, pp. 636–644 (2018)
47. Wang, Z., Zou, Y., Zhang, Z.: Cluster attention contrast for video anomaly detection. In: ACM MM, pp. 2463–2471 (2020)
48. Xu, D., Ricci, E., Yan, Y., Song, J., Sebe, N.: Learning deep representations of appearance and motion for anomalous event detection. In: BMVC, pp. 8.1–8.12 (2015)
49. Ye, M., Peng, X., Gan, W., Wu, W., Qiao, Y.: AnoPCN: video anomaly detection via deep predictive coding network. In: ACM MM, pp. 1805–1813 (2019)
50. Zaheer, M.Z., Mahmood, A., Astrid, M., Lee, S.-I.: CLAWS: clustering assisted weakly supervised learning with normalcy suppression for anomalous event detection. In: Vedaldi, A., Bischof, H., Brox, T., Frahm, J.-M. (eds.) ECCV 2020. LNCS, vol. 12367, pp. 358–376. Springer, Cham (2020). https://doi.org/10.1007/978-3-030-58542-6_22
51. Zaigham Zaheer, M., Lee, J.H., Astrid, M., Lee, S.I.: Old is gold: redefining the adversarially learned one-class classifier training paradigm. In: CVPR, pp. 14171–14181 (2020)
52. Zhao, Y., Deng, B., Shen, C., Liu, Y., Lu, H., Hua, X.S.: Spatio-temporal autoencoder for video anomaly detection. In: ACM MM, pp. 1933–1941 (2017)
53. Zhong, J.X., Li, N., Kong, W., Liu, S., Li, T.H., Li, G.: Graph convolutional label noise cleaner: train a plug-and-play action classifier for anomaly detection. In: CVPR, pp. 1237–1246 (2019)
54. Zhu, Y., Newsam, S.D.: Motion-aware feature for improved video anomaly detection. In: BMVC, p. 270 (2019)

SOMPT22: A Surveillance Oriented Multi-pedestrian Tracking Dataset

Fatih Emre Simsek[1,2]([envelope]) [ID], Cevahir Cigla[1] [ID], and Koray Kayabol[2] [ID]

[1] Aselsan Inc., Ankara, Turkey
{fesimsek,ccigla}@aselsan.com.tr
[2] Gebze Technical University, Kocaeli, Turkey
koray.kayabol@gtu.edu.tr

Abstract. Multi-object tracking (MOT) has been dominated by the use of track by detection approaches due to the success of convolutional neural networks (CNNs) on detection in the last decade. As the datasets and bench-marking sites are published, research direction has shifted towards yielding best accuracy on generic scenarios including re-identification (reID) of objects while tracking. In this study, we narrow the scope of MOT for surveillance by providing a dedicated dataset of pedestrians and focus on in-depth analyses of well performing multi-object trackers to observe the weak and strong sides of state-of-the-art (SOTA) techniques for real-world applications. For this purpose, we introduce *SOMPT22* dataset; a new set for multi person tracking with annotated short videos captured from static cameras located on poles with 6-8 m in height positioned for city surveillance. This provides a more focused and specific benchmarking of MOT for outdoor surveillance compared to public MOT datasets. We analyze MOT trackers classified as one-shot and two-stage with respect to the way of use of detection and reID networks on this new dataset. The experimental results of our new dataset indicate that SOTA is still far from high efficiency, and single-shot trackers are good candidates to unify fast execution and accuracy with competitive performance. The dataset will be available at: sompt22.github.io.

Keywords: Multiple object tracking · Pedestrian detection · Surveillance · Dataset

L. Karlinsky et al. (Eds.): ECCV 2022 Workshops, LNCS 13805, pp. 659–675, 2023.
https://doi.org/10.1007/978-3-031-25072-9_44

1   Introduction

Multi-object tracking (MOT) is a popular computer vision problem that focuses on tracking objects and extracting their trajectories under different scenarios that are later used for various purposes. The use cases of MOT can be various including but not limited to autonomous cars, video surveillance, missile systems, and sports. The output of MOT could be predicting the next positions of the objects, prevent collusion or extract some statistics about the scene and object behaviours. Detection is the first and perhaps the most crucial stage in MOT to identify the object or target and define the purpose of the track. There are different well-defined types of objects such as pedestrians, faces, vehicles, animals, air-crafts, blood cells, stars or any change as a result of motion with respect to background that are subject to the need of tracking in several scenarios. In real world, monitored objects can alter visibility within time due to environmental conditions, physical properties of the object/motion or crowdedness. The change in visibility as well as existence of multiple objects yield occlusion, object/background similarities, interactions between instances. These effects combined with environmental factors such as illumination change, introduce common challenges and difficulties in MOT literature both in detection and tracking stages.

In recent years, various algorithms and approaches have been proposed via CNNs to attack aforementioned problems. The research on MOT has also gained popularity with the appearance of the first MOTChallenge [7] in 2014. This challenge contains annotated videos, with object (pedestrian) bounding boxes among consecutive frames that are captured through moving or static cameras. In the MOT challenge, the variety of data is large, including eye-level videos captured on moving vehicles, high-level moving, or stationary surveillance cameras at various heights. This variation of the data limits the understanding of the capabilities of state-of-the-art (SOTA) techniques such that the positions of the cameras and the types of scene objects change the scenarios, as mentioned previously. These challenges require MOT approaches to be generic and capable of tracking vehicles or pedestrians under varying conditions. However, MOT techniques, as well as any other technique, should be optimized per scenario to get maximum efficiency and low false alarms for real-world usage.

In this paper, we focus on MOT challenge on static and single cameras positioned for the sake of surveillance of pedestrians at around 6–8 m in height. In this manner, we create a new dataset and analyze the well-known MOT algorithms in this dataset to understand the performance of SOTA for surveillance. This approach is expected to optimize MOT algorithms by further analyzing motion behaviour and object variations for semi-crowded scenes in the limited scope. Public MOT datasets challenge the detection and tracking performances of MOT algorithms by increasing the density of the frames. On the other hand, we try to challenge MOT algorithms in terms of long term tracking by keeping sequences longer with less tracks. The paper continues with the discussion of the related works in the following section that summarizes recent and popular approaches and datasets for MOT. Section 3 is devoted to the description of the MOT problem in terms of a surveillance perspective, discussing challenges and use cases. Section 4 presents the details of the newly introduced dataset as well as the evaluation approach utilized that clearly analyzes the detection and tracking steps. Section 5 shows the experimental results that evaluate the performance of well-known MOT approaches in the proposed dataset. Section 6 concludes the paper with a discussion of the pros and cons of the MOT techniques as well as future comments.

2   Related Work

MOT is a common problem in most vision systems that enable relating objects among consecutive frames. This temporal relation provides an ID for each individual in a scene that is utilized to extend additional information about the attributes and behaviors of the objects and the scene statistics. In that manner, assigning an ID to an object and tracking it correctly are the first steps in gathering high-level inference about the scenes. In this section, we summarize the methods (in the scope of pedestrian tracking) proposed for MOT and the datasets that play a key role in the training and relatively high performance of modern CNNs.

2.1   MOT Methods

MOT techniques involve two main stages: detection and association, a.k.a., tracking. The detection determines the main purpose of the tracking by indicating the existence of an object type within a scene. Once objects are detected, next step is the association of objects. Throughout the MOT literature, various detection methodologies have been utilized involving moving object detection [47], blob detection [1], feature detection [22], predefined object detection [5]. Until the last decade, hand-crafted features and rules have been utilized to detect objects in a scene. CNNs have dominated smart object detection with the help of a large amount of annotated objects and computational power, yielding wide application areas for machine learning.

Once the objects are detected in a frame, association along the consecutive frames can be provided via two different approaches: assign a single object tracking per object or optimize a global cost function that relates all the objects detected in both frames. The first approach utilizes representation of the detected objects by defining a bounding box, foreground mask or sparse features. Then, these representations are searched along the next frame within a search region that is defined by the characteristics of the motion of the objects. In these types of methods, object detection is not required for each frame; instead, it is performed with low frequency to update the tracks with new observations. Feature matching [15], Kalman filters [16], correlation-based matching [13] are the main tools for tracking representations among consecutive frames.

The second type of tracking exploits detection for each frame and relating the objects based on their similarities defined by several constraints such as position, shape, appearance, etc. The Joint Probabilistic Data Association Filter (JPDAF) [46] and the Hungarian algorithm [17] are the most widely used approaches to provide one-to-one matching along consecutive frames. In this way, independently extracted objects per frame are matched based on the similarity criteria. As CNNs improve, reID networks have also been utilized to yield robust similarities along objects in consecutive frames. The MOT literature has recently focused on the second approach, tracking-by-detection, where CNNs are utilized to detect objects [21,28,49] in each frame, and object similarities are extracted using different approaches [2,4,9,37,43] that are fed into a matrix. The matrix representation involves the objects in the rows and columns, one side is devoted to track objects, the other is devoted to the new comers. Hungarian algorithm performed on the similarity matrix yields an optimum match of one to one. Methods are mostly differentiated according to the similarity formulation of the matrix, whereas the correspondence search is mostly achieved by a Hungarian algorithm.TransTrack [33], TrackFormer [25], and MOTR [40] have made attempts to use the attention mechanism in tracking objects in movies more recently with the current focus of utilizing transformers [34] in vision tasks.In these works, the query to associate the same objects across frames is transmitted to the following frames using the features of prior tracklets.To maintain tracklet consistency, the appearance information in the query is also crucial.

2.2   Datasets

Datasets with ground-truth annotations are important for object detection and reID networks, which form the fundamental steps of modern MOT techniques. In this manner, we briefly summarize the existing person detection and multi-object tracking datasets which are within the scope of this study. In both sets, bounding boxes are utilized to define objects with a label that indicates the type of object, such as face, human, or vehicle. On the other hand, there are apparent differences between object detection and multi-object tracking datasets. Firstly, there is no temporal and spatial relationship between adjacent frames in object detection datasets. Second, there is no unique identification number of objects in object detection datasets. These differences make creation of multi-object tracking datasets more challenging than object detection datasets.

Table 1. A comparison of person detection datasets

	Caltech [8]	KITTI [12]	COCOPersons [20]	CityPersons [41]	CrowdHuman [30]	VisDrone [50]	EuroCityPersons [3]	WiderPerson [42]	Panda [35]
# countries	1	1	-	3	-	1	12	-	1
# cities	1	1	-	27	-	1	31	-	1
# seasons	1	1	-	3	-	1	4	4	1
# images (day/night)	249884/-	14999/-	64115/-	5000/-	15000/-	10209/-	40219/7118	8000/-	45/-
# persons (day/night)	289395/-	9400/-	257252	31514/-	339565/-	54200/-	183182/35323	236073/-	122100/-
# density(person/image)	1	0.6	4	6.3	22.6	5.3	4.5	29.9	2713.8
Resolution	640 × 480	1240 × 376	-	2048 × 1024	-	2000 × 1500	1920 × 1024	1400 × 800	> 25k × 14k
Weather	Dry	Dry	-	Dry	-	-	Dry, wet	Dry, wet	-
Year	2009	2013	2014	2017	2018	2018	2018	2019	2020

2.3   Person Detection Datasets

The widespread use of person detection datasets dates back to 2005 with the INRIA dataset [5]. Then after two more datasets emerge to serve the detection community in 2009 TudBrussels [36] and DAIMLER [10]. These three datasets increase structured progress of detection problem. However, as algorithms performance improves, these datasets are replaced by more diverse and dense datasets, e.g. Caltech [8] and KITTI [12]. CityPersons [41] and EuroCityPersons [3] datasets come to the fore with different country, city, climate and weather conditions. Despite the prevalence of these datasets, they all suffer from a low-density problem (person per frame); no more than 7. Crowd scenes are significantly underrepresented. CrowdHuman [30] and WiderPerson [42] datasets attack this deficiency and increase the density to 22. Recently, the Panda [35] dataset has been released, which is a very high resolution (25k × 15k) human-oriented detection and tracking dataset where relative object sizes are very small w.r.t. full image. This dataset focuses on very wide-angle surveillance through powerful processors by merging multiple high-resolution images. Another common surveillance dataset is Visdrone [50], which includes 11 different object types that differentiate humans and various vehicles. This dataset is captured by drones where view-points are much higher than surveillance, the platforms are moving and bird-eye views are observed. A summary of the datasets with various statistics is shown in Table 1.

Table 2. A comparison of *SOMPT22* with other popular multi-object tracking datasets

	PETS [11]	KITTI [12]	CUHK-SYSU [38]	PRW [45]	PathTrack [24]	MOT17 [26]	MOT20 [6]	BDD100K [39]	DanceTrack [32]	SOMPT22
# images	795	8000	18184	6112	255000	11235	13410	318000	105855	21602
# persons	4476	47000	96143	19127	2552000	300373	1.6M	3.3M	1M	801341
# density(ped/img)	5.6	4.78	4.2	3.1	10	26	120	10.3	10	37
# tracks	19	917	8432	450	16287	1666	3456	131000	990	997
Camera angle	High view	Eye-level	Eye-level	Eye-level	Unstructured	Eye-level	High view	Eye-level	Eye-level	High view
Fps	7	-	-	-	25	30	25	30	20	30
Camera state	Static	Moving	Moving	Moving	Moving	Moving/static	Moving/static	Moving/static	Static	Static
Resolution	768 × 576	1240 × 376	600 × 800	1920 × 1080	1280 × 720	1920 × 1080	1920 × 1080	1280 × 720	1920 × 1080	1920 × 1080
Year	2009	2013	2016	2016	2017	2017	2020	2020	2021	2022

Multi-object Tracking Datasets. There is a corpus of multi-object tracking datasets that involve different scenarios for pedestrians. In terms of autonomous driving, the pioneering MOT benchmark is the KITTI suite [12] that provides labels for object detection and tracking in the form of bounding boxes. Visual surveillance-centric datasets focus on dense scenarios where people interact and often occluding each other and other objects. PETS [11] is one of the first datasets in this application area. The MOTChallenge suite [7] played a central role in the benchmark of multi-object tracking methods. This challenge provides consistently labeled crowded tracking sequences [6,18,26]. MOT20 [6] boosts complexity of challenge by increasing density of the frames. MOT20 introduces a large amount of bounding boxes; however, the scenes are over-crowded and motion directions are that various to correspond real surveillance scenarios including squares and intersections. The recently published BDD100K [39] dataset covers more than 100K videos with different environmental, weather, and geographic circumstances under unconstrained scenarios. In addition to these, the CUHK-SYSU [38], PRW [45], PathTrack [24] and DanceTrack [32] datasets serve variety for multi-object tracking. These datasets are diverse in terms of static/moving camera; eye-level, high view angle and low/high resolution as shown in Table 2.

3 Problem Description

As shown in Table 2 the existing benchmarks mostly tackle MOT problem within autonomous driving perspective (eye level capture) due to the advances of vehicle technology. On the other hand, surveillance is one of the fundamental applications of video analysis that serves city-facility security, law enforcement, and smart city applications. As in most outdoor surveillance applications, the cameras are located high to cover large areas to observe and analyze. Stationary high-view cameras involve different content characteristics compared to common datasets provided in MOT, including severe projective geometry effects, larger area coverage, and longer but slower object motion. Therefore, it is beneficial to narrow the scope of MOT and analyze-optimize existing approaches for the surveillance problem.

It is also important to consider the object type to be tracked during the scope limitation. In surveillance, the are two main object types under the spot, pedestrians and vehicles. Pedestrians have unpredictable motion patterns, interact in different ways with the other individuals, yielding various occlusion types, while showing 2D object characteristics due to thin structure. On the other hand, vehicles move faster along predefined roads with predictable (constant velocity-constant acceleration) motion models, interfere with other vehicles with certain rules defined by traffic enforcement, and suffer from object view point changes due to thickness in all 3 dimensions. In that manner, there are significant differences between the motion patterns and object view-point changes of pedestrians and vehicles that have influence on identification features and tracking constraints that definitely require careful attention. That is the main reason, the challenges differentiate by the type of objects as in [27] and [7].

As a result of wide area coverage and high-view camera location, the viewpoints of pedestrians change significantly, where the distant objects can be observed frontally, and closer objects get high tilt angles w.r.t. camera. Besides, objects' slow relative motion yield longer tracks within the scene that requires tracking to be robust against various view changes for long periods of time. In this type of scene capture, object sizes in terms of image resolution get smaller, and the number of appearance-based tempters (human-like structures such as tree trunk, poles, seats, etc.) increases, while the videos involve less motion compared to eye level scene capture. Thus, object detection becomes more difficult and requires special attention to outliers as well as appearance changes in long tracks. In addition, consistent camera positioning enables the use of 3D geometry cues in terms of projective imaging, where several assumptions can be exploited such as objects closer to the camera being the occluders on a planar scene. Apart from the challenges in detection, especially observation of a longer object motion introduces new problems for tracking, mostly the change of floor illumination and object view point due to the object motion or occlusions.

Constraining the MOT problem from a surveillance perspective, we propose a new annotated dataset and try to diagnose the capability of state-of-the-art MOT algorithms on pedestrians. As mentioned previously, MOT is achieved by two steps, detection and tracking where we also base our evaluation on the recent popular techniques for both steps. SOTA one-shot object detectors can be grouped into two: anchor-based, e.g., Yolov3 [28] and anchor-free, e.g., CenterNet [49]. When we analyzed the 20 best performing MOT algorithms in the MOTChallenge [7] benchmarks, we observed that these algorithms were based upon either CenterNet/FairMOT or Yolo algorithms. Therefore, we decided to build experiments around these basic algorithms to assess the success of detection and tracking in the *SOMPT22* dataset. FairMOT [44] and CenterTrack [48] are two one-shot multi-object trackers based on the CenterNet algorithm. FairMOT adds a reID head on top of the backbone to extract people embeddings. CenterTrack adds a displacement head to predict the next position of the centers of people. Two of the most common association methods are SORT [2] and DeepSORT [37]. SORT uses IOU(Intersection over union) and Kalman filter [16] as the criterion for linking detection to tracking. DeepSORT incorporates deep features of the detected candidates for linking detection to tracking. Three one-shot multi-object trackers (CenterTrack, FairMOT and Yolov5 & SORT) and one two-stage multi-object tracker (Yolov5 & DeepSORT) were trained to benchmark tracking performances.

4   *SOMPT22* Dataset

4.1   Dataset Construction

Video Collection. In order to obtain surveillance videos for MOT evaluation, 7/24 static cameras publicly streaming located at 6–8 m high poles are chosen all around the globe. Some of the chosen countries are Italy, Spain, Taiwan, the United States, and Romania. These cameras mostly observe squares and road intersections where pedestrians have multiple moving directions. Videos are recorded for around one minute at different times of the day to produce a variety of environmental conditions. In total, 14 videos were collected, by default using 9 videos as a training set and 5 as a test set. It is important to note that faces of the pedestrians are blurred to be anonymous in a way that does not affect pedestrian detection and reID features significantly. We conducted the object detection test with and without face blurring and have not observed any difference for base algorithms.

Annotation. Intel's open-source annotation tool CVAT [14] is used to annotate the collected videos. Annotation is achieved by first applying a pre-trained model to have rough detection and tracking labels, which are then fine tuned by human annotators. The annotated labels include bounding boxes and identifiers

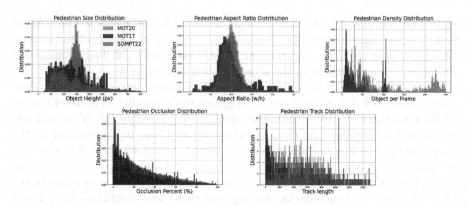

Fig. 1. Statistics of MOT17, MOT20 and proposed *SOMPT22* for persons of the training datasets (histogram of height, aspect ratio, density, occlusion and track length)

(unique track ID) of each person within MOTChallenge [7] format. The file format is a CSV text file that contains one instance of an object per line. Each line contains information that includes *frameID*, *trackID*, *top-left corner*, *width and height*. In order to enable track continuity, partially and fully occluded objects are also annotated as long as they appear within the video again. Bounding boxes annotated with dimensions that overflow the screen size are trimmed to keep inside the image. Bounding boxes also include the part of the person who is occluded.

4.2   Dataset Statistics

Table 2 presents some important statistics of the existing and proposed datasets. *SOMPT22* has density of 37 pedestrians per frame that is between MOT17 and MOT20. MOT20 is a big step up in terms of number of person in MOTChallenge datasets. It is the most dense dataset right now. On the other hand, MOT17 and MOT20 are not mainly surveillance-oriented datasets rather they are proposed to challenge algorithms in terms of detection and occlusion. Especially in MOT20, videos are recorded during crowded events or at a metro station while people get off the train. The motion patterns of the pedestrians are not variable in these videos; each video includes a dominant direction with much fewer different directions. This is not a natural pattern of motion for people on surveillance cameras. On the contrary, people in *SOMPT22* dataset act more spontaneously through city squares covering almost every direction. Figure 1 shows statistical benchmarking of MOT17, MOT20 and the proposed dataset. Although *SOMPT22* has more images compared to MOT17 and MOT20, the number of tracks is the least. This shows that MOT17 and MOT20 have shorter tracklets with shorter sequences than *SOMPT22*. A tracklet is a section of a track that a moving object is following as built by an image recognition system. This is an expected result, where surveillance oriented cameras cover larger field-of-views that enable longer observation of each individual. *SOMPT22* provides a high-view dataset which is lacking in MOT datasets. In this manner, the *SOMPT22* dataset challenges algorithms in terms of long-term detection, recognition, and tracking that require robust adaptation to changes in the scale and view points of pedestrians. We trained a YoloV5 [28] with a medium model backbone on the *SOMPT22* training sequences, obtaining the detection results presented in Table 8. A detailed breakdown of the detection bounding boxes on individual sequences and annotation statistics are shown in Table 3.

Table 3. *SOMPT22* Dataset & detection bounding box statistics

Annotation								Detection			
Sequence	Resolution	Length(sec)	FPS	Frames	Boxes	Tracks	Density	nDet.	nDet/fr.	Min Height	Max Height
Train											
SOMPT22-02	1280 × 720	20	30	600	16021	45	28	15281	25	75	300
SOMPT22-04	1920 × 1080	60	30	1800	35853	44	20	35697	19	15	363
SOMPT22-05	1920 × 1080	60	30	1800	55857	40	31	58764	32	18	212
SOMPT22-07	1920 × 1080	60	30	1800	82525	94	46	81835	45	17	330
SOMPT22-08	1920 × 1080	60	30	1800	121708	138	68	124880	69	32	400
SOMPT22-10	1920 × 1080	60	30	1800	102077	139	57	104928	58	31	492
SOMPT22-11	1920 × 1080	60	30	1800	41285	58	23	42706	23	22	417
SOMPT22-12	1920 × 1080	60	30	1800	72334	74	40	72987	40	16	506
SOMPT22-13	1920 × 1080	30	30	900	8244	38	9	8528	9	28	400
Test											
SOMPT22-01	1920 × 1080	60	30	1800	49970	75	28	59134	32	24	355
SOMPT22-03	1280 × 720	40	30	1200	19527	42	16	17249	14	45	270
SOMPT22-06	1920 × 1080	60	30	1800	122225	121	68	104148	57	29	321
SOMPT22-09	1920 × 1080	60	30	1800	59092	57	33	50442	28	18	311
SOMPT22-14	1920 × 1080	30	30	902	14623	32	16	12781	14	29	439
Total		12 min			21602	801341	997	37	789360	36	15	506

4.3 Evaluation Metrics

The multi-object tracking community used MOTA [31] as the main metric to benchmark for a long time. This measure combines three sources of error: false positives, missing targets, and identity switching. However, recent results have revealed that this metric places too much weight on detection rather than association quality, which owes too much weight to the quality of detection instead. Higher Order Tracking Accuracy (HOTA) [23] is proposed to correct this historical bias since then. HOTA is geometric mean of detection accuracy and association accuracy which is averaged across localization thresholds. In our benchmarking, we use HOTA as the main performance metric. We also use the AssA and IDF1 [29] scores to measure association performance. AssA is the Jaccard association index averaged over all matching detections and then averaged over localization thresholds. IDF1 is the ratio of correctly identified detections to the average number of ground-truth and computed detections. We use DetA and MOTA for detection quality. DetA is the Jaccard index of detection averaged above the localization thresholds. In this paper, we propose a new multi-pedestrian tracking dataset called *SOMPT22*. This dataset contains surveillance oriented video sequences that captured on publicly streaming city cameras. The motivation is to reveal the bias in existing datasets that tend to be captured either on eye-level for autonomous driving systems or in high view and in crowded scenes. We believe that the ability to analyze the complex motion patterns of people in daily life on a well constrained surveillance scenario is necessary for building a more robust and intelligent trackers. *SOMPT22* serves such a platform to encourage future work on this idea. localization thresholds. ID switch is the number of Identity Switches (ID switch ratio = #ID switches/recall). The complexity of the algorithms is measured according to the processing cost (fps), including only the tracking step. The fps values may or may not be provided by the authors with non-standard hardware configurations. MOTChallenge [7] does not officially take the reported fps of the algorithms into account in the evaluation process.

5   Experiments

5.1   Experiment Setup

Table 4 depicts the experimental configurations of the object detectors, multi-object trackers, and association algorithms. As can be seen in Table 1, Crowd-Human is a recent person detection dataset with a large volume and density of images. CenterTrack was pre-trained on CrowdHuman dataset [30] by us. Fair-MOT and YoloV5 had already been pre-trained on it by the respective authors. The model parameters pre-trained on this dataset were utilized to initialize the detectors and trackers. Then, we fine-tune (transfer learning) CenterTrack, Fair-MOT and YoloV5 on the proposed *SOMPT22* train dataset via 240, 90 and 90 epochs correspondingly. For the sake of fairness, we keep the network input resolution fixed for all detectors and trackers during the training and inference phase. We followed the presented training protocols of the detectors and trackers in their respective source codes. Therefore, each object detector and tracker has its own pre-processing techniques, data augmentation procedures, hyperparameter tuning processes as well as accepted dataset annotation formats e.g. yolo [28] for YoloV5, MOTChallenge [7] for FairMOT and COCO [20] for CenterTrack. DeepSORT algorithm has CNN based feature extractor module. This module is pre-trained on the Market1501 [19] public reID dataset. All algorithms are implemented and executed on the PyTorch framework with Python, some of which are provided by the corresponding authors. The inference experiments were conducted on an Intel i7-8700k CPU PC with Nvidia GTX1080ti (11GB) GPU.

As following tracking-by-detection technique, detection is performed for each frame independently. We use the Kalman filter [16] and bounding box intersection over union as initial stages for all trackers to associate the detection results along consecutive frames. Further details of the experimented trackers are given in Table 5. YoloV5 object detector constructs a two-stage multi-object tracker in collaboration with the DeepSORT algorithm. YoloV5 & SORT, CenterTrack and

FairMOT algorithms are three one-shot trackers that have only one backbone to extract deep features from objects. The multi-object tracker formed by cascading YoloV5 and SORT algorithms is classified as one-shot tracker due to the fact that association is performed purely on CPU. These three multi-object trackers are trained in end-to-end fashion. DeepSORT association algorithm and Fair-MOT benefit from reID features, while CenterTrack and SORT only accomplish the association task without reID features. CenterTrack exploits an additional head within detection framework that provides prediction of the displacement.

Table 4. Algorithms & Specs (* not trained by us)

Algorithms	Function	Type	Backbone	Resolution	Train Dataset
CenterTrack [48]	Multi-object tracker	One-shot	DLA34	640 × 640	CH / + SOMPT22
FairMOT [44]	Multi-object tracker	One-shot	DLA34	640 × 640	CH* / + SOMPT22
YoloV5 [28]	Object Detector	One-shot	Darknet	640 × 640	CH* / + SOMPT22
DeepSORT [37]	Associator	-	CNN	64 × 128	Market1501* [19]
SORT [2]	Associator	-	-	64 × 128	-

Table 5. Association methods of MOT algorithms

Methods	Box IOU	Re-ID Features	Kalman Filter	Displacement
CenterTrack [48]	✓		✓	✓
FairMOT [44]	✓	✓	✓	
YoloV5 & DeepSORT [4]	✓	✓	✓	
YoloV5 [28] & SORT [2]	✓		✓	

5.2 Benchmark Results

In this section, we will compare and contrast the performance of the four aforementioned trackers according to the HOTA [23] and CLEAR [31] metrics and the inference speed, as shown in Table 6. We can observe that the detection performance (DetA) of CenterTrack is better than that of FairMOT, which may be due to the displacement head that improves the localization of people. On the other hand, the association performance (AssA) of FairMOT is better than that of CenterTrack with the help of the reID head that adds a strong cue to the association process. CenterTrack requires less computation sources compared to FairMOT. According to the HOTA score which is the geometric mean of DetA and AssA, combination of YoloV5 and SORT variants outperform the other techniques significantly. The key role in this result belongs to the detection accuracy, where YoloV5 improves the detection by at least 10%. DeepSORT and SORT approaches perform on similar to each other with certain expected deviations. DeepSORT adds reID based matching of object patches over SORT and decreases ID switch by 80% while increasing the computational complexity by x2.5. However, the use of reID representations introduces some decrease in association accuracy (AssA), which is probably due to long tracklets that change appearance significantly. YoloV5 is an anchor-based object detector. Fine selection of anchor combination seems to lead better detection performance on surveillance cameras than methods that utilize anchor-free methods e.g. CenterTrack and FairMOT. This reveals the importance of detection during the track-by-detect paradigm that is the most common approach in MOT literature. MOT20 is the most similar public dataset to *SOMPT22* in the literature in terms of camera perspective. Therefore, we repeated same experiments to observe contribution of *SOMPT22* to MOT algorithms on MOT20 dataset as well. Table 7 shows the benchmark of the MOT algorithms in the MOT20 train set. The comparative results are parallel to those given in Table 6, where the YoloV5 and (Deep)SORT methods outperform. It is also clear that transfer learning in our proposed dataset *SOMPT22* increases performance.

Table 6. MOT algorithms performance on *SOMPT22* Test Set before and after fine-tuning on *SOMPT22* (↑: higher is better, ↓: lower is better)

Methods	HOTA ↑	DetA ↑	AssA ↑	MOTA ↑	IDF1 ↑	IDsw ↓	FPS ↑
CenterTrack [48]	23.4/32.9	31.2/42.3	18.0/26.1	35.6/43.5	23.4/35.3	5426/3843	57.7
FairMOT [44]	36.3/37.7	38.2/34.6	35.5/41.8	41.0/43.8	44.8/42.9	1789/1350	12
YoloV5 & DeepSORT [4]	39.5/43.2	48.8/**47.2**	32.4/39.8	58.4/ 55.9	47.1/53.3	162/**152**	26.4
YoloV5 [28] & SORT [2]	44.5/**45.1**	47.8/47.1	41.7/**43.5**	59.4/56	55.9/**55.1**	747/822	**70.2**

Table 7. MOT algorithms performance on MOT20 Train Set before and after fine-tuning on *SOMPT22* (↑: higher is better, ↓: lower is better)

Methods	HOTA ↑	DetA ↑	AssA ↑	MOTA ↑	IDF1 ↑	IDsw ↓
CenterTrack [48]	13.2/18.8	12.7/28.8	14.2/12.5	14.3/33.4	12.1/20.7	64/1318
FairMOT [44]	15.4/18.1	7.9/16.8	30.9/22.8	3.3/−21.1	11.4/15.1	232/**153**
YoloV5 & DeepSORT [4]	26.8/41.5	35.8/**52.5**	20.4/33.0	47.3/**67.5**	33.2/51.0	345/5859
YoloV5 [28] & SORT [2]	32.1/**48.0**	32.7/50.4	31.7/**45.9**	42.7/63.2	44.1/**62.6**	4788/3013

As mentioned, combination of SORT based associator and anchor-based object detector YoloV5 performs better than one-shot MOT algorithms. In addition, detection performance is critical for overall tracking. To observe the contribution of the *SOMPT22* dataset to the performance of object detection solely, we evaluated the YoloV5 object detector on the test set of *SOMPT22* after fine tuning with the same dataset. Detection scores are shown in Table 9. The precision and recall measurements are calculated as 0.89 and 0.68 respectively, which indicates that there is still room for improvement of detection under surveillance scenario. The problem in surveillance is the wide field of view that results in small objects which are rather difficult to detect. We provide these public detections as a baseline for the tracking challenge so that the trackers can be trained and tested.

Table 8. Overview of Performance of YoloV5 detector trained on the *SOMPT22* Training Dataset

Sequence	Precision	Recall	AP@.5	AP@.5:.95
Train				
SOMPT22-02	0.98	0.89	0.96	0.61
SOMPT22-04	0.98	0.94	0.98	0.73
SOMPT22-05	0.97	0.92	0.97	0.73
SOMPT22-07	0.96	0.85	0.93	0.66
SOMPT22-08	0.96	0.82	0.93	0.66
SOMPT22-10	0.97	0.84	0.94	0.68
SOMPT22-11	0.98	0.92	0.97	0.74
SOMPT22-12	0.99	0.95	0.98	0.83
SOMPT22-13	0.98	0.93	0.95	0.67
Test				
SOMPT22-01	0.85	0.72	0.79	0.41
SOMPT22-03	0.95	0.72	0.80	0.42
SOMPT22-06	0.89	0.66	0.77	0.43
SOMPT22-09	0.87	0.51	0.60	0.33
SOMPT22-14	0.94	0.70	0.80	0.50

Table 9. The performance of YoloV5 [28] object detector evaluated on the test set of *SOMPT22*

Train set	Precision	Recall	AP@.5	AP@.5:.95
CH/+*SOMPT22*	0.88/0.89	0.65/0.68	0.74/0.79	0.40/0.42

Figure 2 shows some success and failure cases of YoloV5 & SORT method on *SOMPT22* dataset where green color indicates successful detection and tracks and red vice-versa. The detectors fail along cluttered regions where pedestrians are occluded or pedestrian-like structures show off, on the other hand, trackers fail. Detector failure leads to ID switch, trajectory fragmentation and losing long-term tracking. The experimental results on the proposed dataset indicate that SOTA is still not very efficient. On the other hand, detection plays a key role in the overall tracking performance. Thus, enrichment of detectors with additional attributes and representations that support association of objects seem be a competitive alternative in terms of lower computational complexity and higher

Fig. 2. Some success/failure cases of YoloV5 & SORT MOT algorithm on *SOMPT22*. First row shows true detections in green and false detections in red. Second row shows true trajectories in green and fragmented trajectories in red

performance. Two-stage approaches perform obviously better and provide room for improvement such that they are fast and enable a low number of ID switches with a much more accurate association.

6   Conclusion and Future Work

In this paper, we propose a new multi-pedestrian tracking dataset: *SOMPT22*. This dataset contains surveillance oriented video sequences that captured on publicly streaming city cameras. The motivation is to reveal the bias in existing datasets that tend to be captured either on eye-level for autonomous driving systems or in high view and in crowded scenes. We believe that the ability to analyze the complex motion patterns of people in daily life on a well constrained surveillance scenario is necessary for building a more robust and intelligent trackers. *SOMPT22* serves such a platform to encourage future work on this idea.

The benchmark of four most common tracking approaches in *SOMPT22* demonstrates that the multi-object tracking problem is still far from being solved with at most 48 % HOTA score and requires fundamental modifications prior to usage of heuristics in specific scenarios.

FairMOT and CenterTrack multi-object trackers show complementary performance for the detection and association part of the tracking task. On the other hand, improved detection with YoloV5 followed by SORT based trackers outperform joint trackers. Moreover, SORT provides higher tracking scores, apart from ID switch compared to DeepSORT. These indicate that detection is the key to better tracking, and reID features require special attention to be incorporated within the SORT framework.

References

1. Alberola-López, C., Casar-Corredera, J.R., Ruiz-Alzola, J.: A comparison of CFAR strategies for blob detection in textured images. In: 1996 8th European Signal Processing Conference (EUSIPCO 1996), pp. 1–4 (1996)

2. Bewley, A., Ge, Z., Ott, L., Ramos, F., Upcroft, B.: Simple online and realtime tracking. In: 2016 IEEE International Conference on Image Processing (ICIP). IEEE (2016). https://doi.org/10.1109/icip.2016.7533003

3. Braun, M., Krebs, S., Flohr, F., Gavrila, D.M.: EuroCity persons: a novel benchmark for person detection in traffic scenes. IEEE Trans. Pattern Anal. Mach. Intell. **41**(8), 1844–1861 (2019). https://doi.org/10.1109/tpami.2019.2897684

4. Broström, M.: Real-time multi-object tracker using YOLOv5 and deep sort with OSNet (2022). https://github.com/mikel-brostrom/Yolov5_DeepSort_OSNet

5. Dalal, N., Triggs, B.: Histograms of oriented gradients for human detection. In: 2005 IEEE Computer Society Conference on Computer Vision and Pattern Recognition (CVPR 2005), vol. 1, pp. 886–893 (2005). https://doi.org/10.1109/CVPR.2005. 177

6. Dendorfer, P., et al.: MOT20: a benchmark for multi object tracking in crowded scenes. arXiv:2003.09003 (2020). http://arxiv.org/abs/1906.04567

7. Dendorfer, P., et al.: Motchallenge: a benchmark for single-camera multiple target tracking. Int. J. Comput. Vis. **129**(4), 845–881 (2020)

8. Dollar, P., Wojek, C., Schiele, B., Perona, P.: Pedestrian detection: an evaluation of the state of the art. IEEE Trans. Pattern Anal. Mach. Intell. **34**(4), 743–761 (2012). https://doi.org/10.1109/TPAMI.2011.155

9. Du, Y., Song, Y., Yang, B., Zhao, Y.: Strongsort: make deepsort great again (2022). https://doi.org/10.48550/arxiv.2202.13514

10. Enzweiler, M., Gavrila, D.M.: Monocular pedestrian detection: survey and experiments. IEEE Trans. Pattern Anal. Mach. Intell. **31**(12), 2179–2195 (2009). https://doi.org/10.1109/TPAMI.2008.260

11. Ferryman, J., Shahrokni, A.: Pets 2009: dataset and challenge (2009)

12. Geiger, A., Lenz, P., Urtasun, R.: Are we ready for autonomous driving? The kitti vision benchmark suite. In: Conference on Computer Vision and Pattern Recognition (CVPR) (2012)

13. Henriques, J.F., Caseiro, R., Martins, P., Batista, J.: High-speed tracking with kernelized correlation filters. IEEE Trans. Pattern Anal. Mach. Intell. **37**(3), 583–596 (2015). https://doi.org/10.1109/TPAMI.2014.2345390

14. INTEL: Cvat. https://openvinotoolkit.github.io/cvat/docs/

15. Jabar, F., Farokhi, S., Sheikh, U.U.: Object tracking using SIFT and KLT tracker for UAV-based applications. In: 2015 IEEE International Symposium on Robotics and Intelligent Sensors (IRIS), pp. 65–68 (2015). https://doi.org/10.1109/IRIS. 2015.7451588

16. Kalman, R.E.: A new approach to linear filtering and prediction problems. Trans. ASME-J. Basic Eng. **82**, 35–45 (1960). https://doi.org/10.1109/CVPR.2005.177

17. Kuhn, H.W.: Variants of the Hungarian method for assignment problems (1956)

18. Leal-Taixé, L., Milan, A., Reid, I., Roth, S., Schindler, K.: MOTChallenge 2015: towards a benchmark for multi-target tracking. arXiv:1504.01942 (2015). http://arxiv.org/abs/1504.01942

19. Zheng, L., Shen, L., Tian, L., Wang, S., Wang, J., Tian, Q.: Scalable person re-identification: a benchmark (2015)

20. Lin, T.Y., et al.: Microsoft coco: common objects in context (2014). https://doi. org/10.48550/arxiv.1405.0312

21. Liu, W., et al.: SSD: single shot MultiBox detector. In: Leibe, B., Matas, J., Sebe, N., Welling, M. (eds.) ECCV 2016. LNCS, vol. 9905, pp. 21–37. Springer, Cham (2016). https://doi.org/10.1007/978-3-319-46448-0_2

22. Lowe, D.G.: Distinctive image features from scale-invariant keypoints. Int. J. Comput. Vision **60**, 91–110 (2004)

23. Luiten, J., et al.: HOTA: a higher order metric for evaluating multi-object tracking. Int. J. Comput. Vision **129**(2), 548–578 (2020)
24. Manen, S., Gygli, M., Dai, D., Van Gool, L.: Pathtrack: fast trajectory annotation with path supervision (2017). https://doi.org/10.48550/arxiv.1703.02437
25. Meinhardt, T., Kirillov, A., Leal-Taixe, L., Feichtenhofer, C.: Trackformer: multi-object tracking with transformers (2021). https://doi.org/10.48550/arxiv.2101.02702
26. Milan, A., Leal-Taixé, L., Reid, I., Roth, S., Schindler, K.: MOT16: a benchmark for multi-object tracking. arXiv:1603.00831 (2016). http://arxiv.org/abs/1603.00831
27. Naphade, M., et al.: The 5th AI city challenge. In: The IEEE Conference on Computer Vision and Pattern Recognition (CVPR) Workshops (2021)
28. Redmon, J., Farhadi, A.: Yolov3: an incremental improvement (2018). https://doi.org/10.48550/arxiv.1804.02767
29. Ristani, E., Solera, F., Zou, R.S., Cucchiara, R., Tomasi, C.: Performance measures and a data set for multi-target, multi-camera tracking (2016). https://doi.org/10.48550/arxiv.1609.01775
30. Shao, S., et al.: Crowdhuman: a benchmark for detecting human in a crowd (2018). https://doi.org/10.48550/arxiv.1805.00123
31. Stiefelhagen, R., Bernardin, K., Bowers, R., Rose, R., Michel, M., Garofolo, J.: The clear 2007 evaluation (2007). https://doi.org/10.1007/978-3-540-68585-2_1
32. Sun, P., et al.: Dancetrack: multi-object tracking in uniform appearance and diverse motion (2021). https://doi.org/10.48550/arxiv.2111.14690
33. Sun, P., et al.: Transtrack: multiple object tracking with transformer (2020). https://doi.org/10.48550/arxiv.2012.15460
34. Vaswani, A., et al.: Attention is all you need (2017). https://doi.org/10.48550/arxiv.1706.03762
35. Wang, X., et al.: Panda: a gigapixel-level human-centric video dataset. In: 2020 IEEE International Conference on Computer Vision and Pattern Recognition (CVPR). IEEE (2020)
36. Wojek, C., Walk, S., Schiele, B.: Multi-cue onboard pedestrian detection. In: 2009 IEEE Conference on Computer Vision and Pattern Recognition, pp. 794–801 (2009). https://doi.org/10.1109/CVPR.2009.5206638
37. Wojke, N., Bewley, A., Paulus, D.: Simple online and realtime tracking with a deep association metric (2017). https://doi.org/10.48550/arxiv.1703.07402
38. Xiao, T., Li, S., Wang, B., Lin, L., Wang, X.: Joint detection and identification feature learning for person search (2016). https://doi.org/10.48550/arxiv.1604.01850
39. Yu, F., et al.: BDD100K: a diverse driving dataset for heterogeneous multitask learning (2018). https://doi.org/10.48550/arxiv.1805.04687
40. Zeng, F., Dong, B., Zhang, Y., Wang, T., Zhang, X., Wei, Y.: MOTR: end-to-end multiple-object tracking with transformer (2021). https://doi.org/10.48550/arxiv.2105.03247
41. Zhang, S., Benenson, R., Schiele, B.: Citypersons: a diverse dataset for pedestrian detection (2017). https://doi.org/10.48550/arxiv.1702.05693
42. Zhang, S., Xie, Y., Wan, J., Xia, H., Li, S.Z., Guo, G.: Widerperson: a diverse dataset for dense pedestrian detection ,in the wild (2019). https://doi.org/10.48550/arxiv.1909.12118
43. Zhang, Y., et al.: Bytetrack: multi-object tracking by associating every detection box (2021). https://doi.org/10.48550/arxiv.2110.06864
44. Zhang, Y., Wang, C., Wang, X., Zeng, W., Liu, W.: FairMOT: on the fairness of detection and re-identification in multiple object tracking. Int. J. Comput. Vision **129**(11), 3069–3087 (2021). https://doi.org/10.1007/s11263-021-01513-4

45. Zheng, L., Zhang, H., Sun, S., Chandraker, M., Yang, Y., Tian, Q.: Person re-identification in the wild (2016). https://doi.org/10.48550/arxiv.1604.02531
46. Zhou, B., Bose, N.: An efficient algorithm for data association in multitarget tracking. IEEE Trans. Aerosp. Electron. Syst. **31**(1), 458–468 (1995). https://doi.org/10.1109/7.366327
47. Zhou, D., Zhang, H.: Modified GMM background modeling and optical flow for detection of moving objects. In: 2005 IEEE International Conference on Systems, Man and Cybernetics, vol. 3, pp. 2224–2229 (2005). https://doi.org/10.1109/ICSMC.2005.1571479
48. Zhou, X., Koltun, V., Krähenbühl, P.: Tracking objects as points (2020). https://doi.org/10.48550/arxiv.2004.01177
49. Zhou, X., Wang, D., Krähenbühl, P.: Objects as points (2019). https://doi.org/10.48550/arxiv.1904.07850
50. Zhu, P., et al.: Detection and tracking meet drones challenge. IEEE Trans. Pattern Anal. Mach. Intell. (2021). https://doi.org/10.1109/TPAMI.2021.3119563

Detection of Fights in Videos: A Comparison Study of Anomaly Detection and Action Recognition

Weijun Tan[1,2]([envelope]) [ID] and Jingfeng Liu[1]

[1] Jovision-Deepcam Research Institute, Shenzhen, China
{sz.twj,sz.ljf}@jovision.com, {weijun.tan,jingfeng.liu}@deepcam.com
[2] LinkSprite Technology, Longmont, CO 80503, USA
weijun.tan@linksprite.com

Abstract. Detection of fights is an important surveillance application in videos. Most existing methods use supervised binary action recognition. Since frame-level annotations are very hard to get for anomaly detection, weakly supervised learning using multiple instance learning is widely used. This paper explores the detection of fights in videos as one special type of anomaly detection and binary action recognition. We use the UBI-Fight and NTU-CCTV-Fight datasets for most of the study since they have frame-level annotations. We find that anomaly detection performs similarly or even better than action recognition. Furthermore, we study to use anomaly detection as a toolbox to generate training datasets for action recognition in an iterative way conditioned on the performance of the anomaly detection. Experiment results should show that we achieve state-of-the-art performance on three fight detection datasets.

Keywords: Fight detection · Video analysis · 3D CNN · I3D · Anomaly detection · Action recognition

1 Introduction

Surveillance cameras are widely used in public places for safety purposes. Empowered by machine learning and artificial intelligence, surveillance cameras become smarter using automatic object or event detection and recognition. Video anomaly detection is to identify the time and space of abnormal objects or events in videos. Examples include industrial anomaly detection and security anomaly detection, and more.

This paper studies a special type of anomaly detection - single-type abnormal event detection in videos. Specifically we study the detection of fights in public places [1,4,19,21,27,29], but the algorithms can be easily extended to other single-type event detection, e.g., gun event detection. A fight is a common event that needs the attention of safety personnel to prevent its escalation and to become more destructive or even deadly.

© The Author(s), under exclusive license to Springer Nature Switzerland AG 2023
L. Karlinsky et al. (Eds.): ECCV 2022 Workshops, LNCS 13805, pp. 676–688, 2023.
https://doi.org/10.1007/978-3-031-25072-9_45

In video anomaly detection, more attention is given to the detection of general abnormal events. It is fair since the algorithms for the detection of general abnormal events can mostly be used for the detection of a single-type abnormal event. There are two major categories of anomaly detection algorithms. The first algorithm is self-supervised learning only using normal videos. A model is learned to represent normal videos. Based on the modeling of the past frames, a prediction of a new frame is expected. An abnormal event is reported if the actual new frame is too different from the predicted frame exceeding a pre-set threshold. In this category, latent-space features or generative models can be used for both hand-crafted and deep learning features. For more details of this algorithm, please refer to [18].

The second is a weakly supervised learning approach, where a weak annotation label of whether an abnormal event is presented in a video is provided. In this case, multiple instance learning (MIL) is used [22]. From a pair of abnormal and normal videos, a positive bag of instances is formed on the abnormal video and a negative bag of instances on the normal video. The model tries to maximize the distance between the maximum scores of two bags.

Anomaly detection is closely related to action recognition. However, action recognition typically needs frame-level annotation, which is very hard to get for large-scale video datasets. That is why the self and weakly supervised learning becomes prevailing. For single-type abnormal event detection, annotating the frame-level labels becomes relatively easier. For fight detection, there are at least two datasets, the UBI-Fight [4] and the NTU-CCTV-Fight [19], which provide frame-level labels. In this paper, we mainly use the UBI-Fight dataset in our study, while other datasets are used to benchmark our algorithms' performance. With the frame-level labels, we can use supervised action detection. It is believed that supervised learning should outperform self or weakly supervised learning. However, this has not been well studied in the literature.

To our knowledge, except for [4], we are the first to explore using anomaly detection for the detection of single-type abnormal events in videos. In this work, we do a comparison study of weakly supervised learning vs. supervised one. Our goal is to suggest which one should be used in terms of performance and resources to achieve it.

Secondly, we study if we can use weakly supervised learning as a data generation tool to generate training data for supervised action recognition. The idea is to find the most reliable snippets out of the abnormal videos and the hardest snippets out of the normal videos. The data generation approach helps the training of the supervised action recognition when an annotated dataset is unavailable. Since we have frame-level annotation labels in the two datasets we use, we can compare the performance of this action recognition with the one using the frame annotation labels.

2   Related Work

In this section we review literature on both the general anomaly detection and the fight detection in videos.

2.1   General Anomaly Detection

Weakly supervised anomaly detection has shown substantially improved performance over the self-supervised approaches by leveraging the available video-level annotations. This annotation only gives a binary label of abnormal or normal for a video. Sultani et al. [22] propose the MIL framework using only video-level labels and introduce the large-scale anomaly detection dataset, UCF-Crime. This work inspires quite a few follow-up studies [4,6,14,17,23,25,26,26,28].

However, in the MIL-based methods, abnormal video labels are not easy to be used effectively. Typically, the classification score is used to tell if a snippet is abnormal or normal. This score is noisy in the positive bag, where a normal snippet can be mistakenly taken as the top abnormal event in an anomaly video. To deal with this problem, Zhong et al. [28] treat this problem as a binary classification under a noisy label problem and use a graph convolution neural (GCN) network to clear the label noise. In RTFM [23], a robust temporal feature magnitude (RTFM) is used to select the most reliable abnormal snippets from the abnormal videos and the normal videos. They unify the representation learning and anomaly score learning by a temporal feature ranking loss, enabling better separation between normal and abnormal feature representations, and improving the exploration of weak labels compared to previous MIL methods. In [14] a multiple sequence learning (MSL) is used. The MSL uses a sequence of multiple instances as the optimization unit instead of one single instance in the MIL. In [4], an iterative weak and self-supervised classification framework is proposed where their key idea is to add new data to the learning set. They use Bayesian classifiers to choose the most reliable segments for the weak and self-supervised paradigms.

There is very little work on supervised learning for anomaly detection since the frame level annotation is very hard to get. Two examples are [15] and [13].

2.2   Fight Detection Using Action Recognition

Detection of fights in videos mostly follows the approach of action recognition. It is simply a binary classification task to classify fight or non-fight actions. Typical methods include 2D CNN feature extraction followed by some types of RNN, or 3D CNN feature extraction [1,19,21,27,29].

Early work uses the Hockey, and Peliculas dataset [2], and others, which are easy tasks. With pretrained 2D or 3D CNN feature extraction plus some feature aggregation techniques, the accuracy of the prediction can be very closed to 100% [21,27,29]. Later a few more realistic datasets from surveillance or mobile cameras are made available, including the one in [1], the NTU-CCTV-Fight dataset [19], and the UBI-Fight dataset [4]. On the NTU-CCTV-Fights dataset, the frame mean average precision (mAP) is only 79.5% [19], and on the UBI-Fights dataset, the frame detection AUC is 81.9% [4].

3   Proposed Methods

In this section, we first compare supervised action recognition and weakly supervised anomaly detection in detecting fights in videos. Our purpose is not to present new action recognition or anomaly detection algorithms but to use existing algorithms to shed light on what directions we should go. This serves as a baseline for how well we can do and motivates the proposed solution.

3.1   Action Recognition of Fights

There are many approaches for action recognition on trimmed video clips, including 2D CNN+RNN, 3D CNN, with or without attention mechanism, transformers, and many variations. Please refer to [11] for a review of action recognition. In this work, we use standard I3D CNN action recognition network [3], R(2+1)D CNN recognition network [24], and the ones with late temporal fusion using BERT [9], as shown in Fig. 1(a). This approach achieves the SOTA performance on the UCF101 [20] and HMDB51 [12] action recognition datasets.

In video action recognition tasks, the videos are already trimmed with the correct action boundary. These videos can be short or long, with tens of or thousands of frames. One important point is how to select the proper frames for the training and testing of the action recognition. A few studies propose to choose smartly which frames to use [8] or use all frames [16]. In typical action recognition networks, 16, 32, or 64 frames are used in a video snippet sample. Too many frames used in a step of training can easily cause out-of-memory problems since these frames need to be remembered in the GPU cache for the gradient calculation in the backpropagation. When the video clip is too long, a typical method is first to cut off the clip into shorter and equal-length segments, then sample a snippet of the given length at a random location in the segment. A problem with this sampling method is that the selected snippet may not be optimum for the action recognition training. We will discuss this problem later in this work with experiment results.

3.2   Anomaly Detection of Fights

To our knowledge, very little work treats the detection of fights as an anomaly detection task except for [4]. This is probably because the annotated dataset is relatively easy to get, so people tend to use action recognition. The second reason is that people do not realize the power of anomaly detection when the action recognition methods are available to use off-shelve. In reality, when we try to deploy such a system into a practical application, we usually find that there are always cases not covered by current datasets. In this case, the weak supervision method makes the work of preparing the new dataset a lot easier than that for the supervised action recognition.

We have reviewed the MIL-based anomaly detection methods in the previous section. We find that the RTMF [23], shown in Fig. 1(b), gives excellent performance using a relatively easy architecture. The key innovation is using a temporal feature magnitude to discriminate abnormal snippets from normal ones. The AUC of the RTMF on the UCF-Crime dataset is 84.03% when only the RGB modal is used with the I3D backbone.

In this work, we modify the RTMF framework [23] as our anomaly detector for fight detection in videos. In addition to the basic binary cross-entropy loss, the temporal smoothness loss and sparsity loss [22], a new loss on the feature magnitude on the top-k selected abnormal and normal segments is introduced,

$$loss_{FM} = \begin{cases} max(0, m - d_k(X_i, X_j)), \; if \; y_i = 1, \; y_j = 0 \\ 0, otherwise \end{cases} \tag{1}$$

where y is the label, X is a snippet, d_k is a distance function that separates the top-k magnitude segment features of the abnormal video y_i and the normal video y_j, and m is a predefined margin. The form of this loss function is similar to the MIL ranking loss in [22], but it is now on the feature magnitude and the top-k magnitude segments.

3.3 Iterative Anomaly Detection and Action Recognition

Other than the extensive comparative study of anomaly detection and action recognition, we propose to use anomaly detection and action recognition iteratively, as shown in Fig. 1. Our novelty is to use anomaly detection to find good-quality training data for the action recognition. Since the UBI-Fight dataset [4] has the frame-level annotation, we use it extensively in our study. When anomaly detection is used, we only use the video-level annotation and ignore (or assume unknown) the frame-level annotations. We use the frame-level annotation in the supervised action recognition.

In the anomaly detection dataset, we know there are at least one or more abnormal snippets in every abnormal positive video. In every normal negative video, we know all snippets are negative. So, intuitive thinking is that we only need to generate positive training data for action recognition. In our study, we find that the negative training data can also be refined. When random data sampling is used, all samples have an equal chance of being picked for a training epoch. However, not all samples contribute equally to the training process. The training prefers to use hard negative samples to increase the model's discrimination capability. They help not only the performance but also the convergence.

In the RTFM training process, every video is divided into 32 equal-length segments, the same as what is done in [22]. The features of all 16-frame snippets in a segment are averaged as the feature for this segment. The top three segments whose feature magnitudes are the largest are picked, and their average magnitude is used in training. At the inference time, the original video snippet is used, and a classification score is generated for every snippet. This score is used to pick the snippets for the action recognition. Please note that this is different from using the classification score in [22], where the score is used to pick the most probable segment in training. In RTFM, the score is used at the inference time after the training is already done.

This score is typically large for an abnormal positive sample, very close to 1.0. We notice that when RTFM converges too fast or overfits, the picked snippets may have some wrong results, meaning that normal snippets are picked as abnormal snippets, which is very detrimental to the action recognition performance. So we use a threshold like 0.995 to pick snippets whose scores are higher than it.

Normal negative samples' scores are typically small but can be large. Our strategy is to use a threshold around 0.5. The samples whose scores are larger than this threshold are hard samples. One other factor we need to consider is to balance the number of generated positive and negative samples. We can adjust this threshold to make them close to each other.

There are two cases in which this iterative method can be used. In anomaly detection, we use the I3D backbone [3] pre-trained on the Imagenet [5] and Kinetics-400 [10] datasets. We use the above method to generate training samples for the action recognition network. In the first case, the backbones of the anomaly detection and the action recognition are the same. In this case, after the first iteration of the action recognition, its updated backbone can be used to regenerate the video features for anomaly detection in the second iteration.

In the second case, the backbone of the action recognition network is different from that of the anomaly detection network. Then after the anomaly detection networks generate training samples for the action recognition, this process can stop. Even though the action recognition network can regenerate video features for anomaly detection, then the anomaly detection becomes a different network. This process is not preferred and is not tested in our work. In this sense, the anomaly detection network is more like a toolbox for generating training data for a different action recognition network.

4   Experiments

4.1   Datasets

A small fight detection dataset in [1] is used. It consists of trimmed video clips good for action recognition. It includes 150 fight videos and 150 normal videos.

UBI-Fight. The UBI-Fights dataset is released in [4]. It provides a wide diversity in fighting scenarios. It includes 80 h of video fully annotated at the frame level consisting of 1000 videos, where 216 videos contain a fight event, and 784 are normal daily life situations.

NTU-CCTV-Fight. The NTU-CCTV-Fight is released in [19]. It includes 1000 videos, each including at least one fight event. No normal videos are included. Frame-level annotations are provided in a JSON format. It mixes two types of cameras - surveillance and mobile cameras, making fight detection hard to detect in this dataset.

Table 1. Action recognition results on three fight detection datasets. Notes [1]: This method uses weakly supervised anomaly detection. We put it here for comparison with our action recognition results. [2]: We use multiple sets of snippets and average the features as input to the classification layers for better performance.

Dataset	Method	Frames	Acc	AUC	mAP
Dataset [1]	2DCNN+Bi-LSTM	5 random	0.72	–	–
Dataset [1]	ours R(2+1)D-BERT	32 random	**0.9562**	–	–
UBI-Fight [4]	Iterative WS/SS[1]	16 avg	–	0.819	–
UBI-Fight	Ours R(2+1)D-BERT	32 random	0.9177	**0.9150**	–
UBI-Fight	Ours I3D	32 random	0.8861	0.9058	–
UBI-Fight	Ours I3D	16 random[2]	0.80	0.8149	–
NTU-CCTV-Fight [19]	RGB+Flow-2DCNN	16	–	–	0.795
NTU-CCTV-Fight	ours R(2+1)D-BERT	32 random	0.8244	0.8715	**0.8275**

4.2 Implementation Details

For anomaly detection, we use the RTMF codebase [23] in PyTorch. We use the I3D with Resnet18 backbone [3] for the video feature generation. A training rate of 1E−3 is used, and the training runs 100 epochs. Two dataset iterators, one for the abnormal data and the other for the normal data, are used. This way, the pairing of abnormal and normal data is random, even when the numbers of abnormal and normal samples are different.

For the action recognition, we also use a codebase in PyTorch. We use the I3D network with Resnet18 backbone [3], and the R(2+1)D network with Resnet34 backbone [24]. We use the BERT similar to [9]. For the I3D network, we use the model pre-trained on the Imagenet [5] and Kinetics-400 [10] datasets. For the R(2+1)D network, we use the model pre-trained on the IG65 dataset [7]. Our start learning rate is 1E-5, and the learning rate reduces by a 0.1 factor on a plateau with 5-epochs patience. The checkpoint with the best validation accuracy is saved for evaluation on the testing dataset.

Evaluation Metric. For action recognition, we use accuracy as the metric. The frame AUC is used for all the results on the UBI-Fight dataset, [4]. While for the results on the NTU-CCTV-Fight dataset [19], the frame mAP is used.

4.3 Action Recognition Results

We first show the action recognition results in accuracy (Acc, 4th column), frame AUC (5th column), and mAP (6th column) on a few fight detection datasets. These results all outperform previous SOTA results. They also serve as a baseline or bound for our iterative anomaly detection and action recognition approach.

Table 2. Anomaly detection results vs. action recognition results on the UBI-Fight dataset. AR means action recognition.

Method	Crop	AUC
I3D16F Action recognition (AR)	1-crop	0.8149
RTFM	10-crop	0.9192
RTFM	1-crop	0.9097
RTFM w/o MSNL	1-crop	0.8692
RTFM-Action Recognition (AR)	1-crop	0.9129
RTFM-AR w/o MSNL	1-crop	0.8708
RTFM-AR w/o MSNL, 1-snippet	1-crop	0.8408

Shown in Table 1 are the results where we test both the standard I3D [3] and the more advanced R(2+1)D-BERT [9]. Other than the backbone and classification network, the number of frames input to the 3D CNN and how they are sampled are also important. For long action recognition samples, an ideal sampling method is to use all available frames in every epoch of the training, similar to what is done in anomaly detection. However, due to the limited memory in GPU, it is very hard to do so if the backbone network is trained. In our test, we find that I3D with 16 frames does not have good performance and realize that the number of fewer frames may be one of the causes. So we sample multiple sets (typically 4 or 8) of 16 frames in an epoch and average the features before it is fed to the classification layers. The accuracy result improves from 0.778 to 0.800 in Table 1.

In this subsection, we focus on the anomaly detection results on the UBI-Fight dataset [4]. We use the anomaly detection method of the RTFM [23]. All videos are first divided into 32 equal-length segments. Then features are extracted for all 16-frame snippets sequentially in every segment, and these features are averaged as the representing feature for this segment. The standard I3D pre-trained on the Imagenet is used. The dimension of the feature is 1024.

Shown in Table 2 are the anomaly detection results with their corresponding action recognition results for reference. Since all snippets use 16 frames, we put the I3D with 16-frames (I3D16F) in the table for comparison. The original RTFM uses 10-crop data augmentation. However, in practice, particularly at inference time, it is impossible to use a 10-crop, so we use the 1-crop result as our baseline, with an AUC = 0.9097. We see that this value is much better than the I3D16F result for a few reasons. First, the RTFM uses a powerful MS-NL aggregation neck before the classification head. When we turn off this aggregation neck, the AUC becomes 0.8692, which is still better than the I3D16F's 0.8149. We do more analysis and find that this is partly due to using all 16-frame snippets in the segment. As explained in Subsect. 3.1, in every epoch of the action recognition training, due to the limit of the GPU memory, only one set of 16 frames is randomly sampled. We do an experiment in Subsect. 4.3, where we use multiple sets of 16-frames and achieve better performance. This effect becomes less noticeable when 32 or 64-frame snippets are used. In anomaly detection, since the backbone is not trained, the features of all snippets can be calculated and used in the training of the classification head.

The second cause is the nature of the MIL in the anomaly detection [22]. The MIL uses the segments whose scores or the feature magnitudes in the RTFM are the maximum instead of randomly chosen ones. Particularly on the negative samples, this has an effect of using the hard samples. We do a test to confirm this effect, where we randomly choose a negative segment, and the performance of the anomaly detection gets worse.

4.4   Anomaly Detection or Action Recognition?

We devise a set of supervised anomaly detection experiments to do our analysis. Instead of using the whole abnormal videos as positive samples in the RTFM, we use the frame-level annotation and extract the ground truth positive portions as the positive samples. These are the same as using trimmed positive samples in the training of action recognition in Subsect. 4.3. There are no changes in the negative samples. We remove all other loss functions and only use the BCE-loss function. We also remove choosing the top-3 feature magnitude segments and replace them with a random choice. Since all the positive segments use ground truth abnormal portions, it cannot make mistakes in choosing normal snippets as abnormal ones. This essentially makes the RTFM a different implementation of action recognition. There are still two differences. The first one is that the backbone is not trained. The second one is that all snippets of a segment are still used for feature calculation. The results are shown in Table 2 denoted by RTFM-AR. The AUCs are now a little bit higher than the ones using weak supervision. The AUC for the one without MSNL is still higher than the I3D16F whose AUC is 0.8149. Going one step further, we remove the feature averaging on all snippets in a segment and use one randomly chosen snippet per segment instead. At the same time, we reduce the number of segments per video and change other settings so that the effective batch number is equivalent to that in the action recognition. The AUC reduces significantly to 0.8408, but still higher than that of the I3D16F. This is very likely because, in the RTMF action recognition, the training runs a lot faster since the I3D backbone is not trained. As more epochs are run, all the snippets have a better chance of being picked. While in the standard I3D16F, since the backbone is trained, the training runs very slowly, and some snippets may never get a chance to be picked.

From these experiments, we notice that with the same backbone and similar classification head, the weakly supervised anomaly detection (RTFM w/o MSNL) and supervised action recognition (RTFM-AR w/o MSNL) can achieve almost the same performance. With additional MSNL neck, anomaly detection can achieve better performance. Since the data preparation and training of anomaly detection are much easier, we conclude that anomaly detection is preferred for the detection of a single-type abnormal event in video.

4.5   Iterative Anomaly Detection and Action Recognition

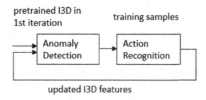

Fig. 1. Pipeline of iterative anomaly detection and action recognition.

In this subsection, we explore the iterative anomaly detection and action recognition, as shown in Fig. 1. The primary purpose is to find out if anomaly detection can be used as a practical toolbox to generate training samples for standard action recognition and achieve the best performance of supervised action recognition.

In the first iteration, we first train the RTFM [23] anomaly detector using the features extracted from an I3D network pre-trained on the Imagenet. To have the best quality snippet selection for the action recognition, we keep the MSNL aggregation neck in the RTFM. After the training is done, we run the training data in the RTFM inference mode. Snippets from the abnormal videos whose scores are larger than 0.995 are selected as positive samples, and snippets from the normal videos whose scores are larger than 0.5 are selected as hard negative samples. Then we train the standard I3D16F action recognition network using these generated training samples. Please note that the validation and test data do not go through the anomaly detection generation. This completes the first iteration of the anomaly detection and action recognition interaction.

In the second iteration, the updated I3D backbone is used to generate a new dataset for anomaly detection. And the same training process for the RTFM follows on the newly generated data. Similarly are the new training data generation for the action recognition and the new action recognition training. This process can keep going for a few iterations.

Table 3. Iterative anomaly detection and action recognition on the UBI-Fight dataset. AR means action recognition.

Dataset	Iter	RTFM	I3D-AR
UBI-Fight	0-th	–	0.8149 (AUC)
UBI-Fight	1st	0.9097	**0.8789**
UBI-Fight	2nd	0.9279	0.8115
NTU-CCTV-Fight	0-th	–	**0.8275**(mAP)
NTU-CCTV-Fight	1st	0.8140	0.8070

Shown in Table 3 are the anomaly detection and action recognition AUCs in the first two iterations. We call the previous standard action recognition the 0-th iteration. It is observed that the I3D16F AUC improves from 0.8149 to 0.8789 in the first iteration. This proves that the RTFM anomaly detector can be used to generate training samples for the I3D16F action recognition. We do some deeper analysis and find that the hard negative samples contribute to this performance improvement. If the I3D16F action recognition can be trained well enough and as many as possible snippets can be used, we believe that the I3D16F itself can also achieve this AUC of 0.8789.

However, the AUC does not improve further in the 2nd iteration. Analysis shows that even though the RTFM AUC improves a little bit from the 1st iteration, the scores' values also get larger. In this sense, the RTFM is overfitted, and more normal video snippets have scores close to 1; therefore are selected erroneously as positive samples for the action recognition. That is why the I3D16F action recognition AUC goes back to 0.8115 in the 2nd iteration.

We repeat this same experiment on the NTU-CCTV-Fight dataset but with I3D32F since 32-frames perform better than 16-frames. Since there are surveillance cameras and mobile cameras mixed in this dataset, the fights are harder to detect in this dataset. In the first iteration, The anomaly detection's AUC is 0.8140, and the I3D32F action recognition's mAP is 0.8070, worse than that in the 0-th iteration, which is 0.8275. Our analysis shows that the anomaly detection performance must be good enough to generate good quality training samples for action recognition. In Table 2, the anomaly detection AUC is 0.9097, while it is only 0.8140 in Table 3. We conjecture that there exists a threshold between 0.8140 and 0.9097. When the anomaly detection's AUC is larger than the threshold, the generated training samples will help the performance of the subsequent action recognition. Since the performance does not improve in the 1st iteration, we do not continue the 2nd iteration. But overall, the best action recognition performance (0.8789 AUC on UBI-Fight) is about the same as that of the anomaly detection (0.8692 RTFM w/o MSNL on UBI-Fight in Table 2).

4.6   Comparison with SOTA Results

Our results in Table 1 are better than existing SOTA results on the three datasets. Furthermore, if we take the best result out of all our studies, the best AUC on the UBI-Fight dataset [4] is 0.9192 in Table 2.

5   Conclusion

In this paper, we do a comparison study of weakly supervised anomaly detection and supervised action recognition. Based on our experiment results, we find that anomaly detection can work as well as or even better than action recognition. Since the weak supervision annotations are a lot easier to get, anomaly detection is a very good choice for detecting single-type abnormal events in videos. In addition to the detection of fights in videos, we also study the detection of gun events in videos and find that the conclusion is the same.

We also find that anomaly detection can be used as a toolbox for the generation of training data for supervised action recognition in some conditions. In this study, the condition is that the anomaly detection's AUC in the first iteration must be good enough. This may not generalize to other different scenarios.

References

1. Aktı, S., Tataroğlu, G., Ekenel, H.: Vision-based fight detection from surveillance cameras. In: IEEE/EURASIP 9th International Conference on Image Processing Theory, Tools and Applications (2019)
2. Bermejo Nievas, E., Deniz Suarez, O., Bueno García, G., Sukthankar, R.: Violence detection in video using computer vision techniques. In: Real, P., Diaz-Pernil, D., Molina-Abril, H., Berciano, A., Kropatsch, W. (eds.) CAIP 2011. LNCS, vol. 6855, pp. 332–339. Springer, Heidelberg (2011). https://doi.org/10.1007/978-3-642-23678-5_39
3. Carreira, J., Zissermana, A.: Quo vadis, action recognition? A new model and the kinetics dataset. In: CVPR (2017)
4. Degardin, B., Proença, H.: Human activity analysis: iterative weak/self-supervised learning frameworks for detecting abnormal events. In: 2020 IEEE International Joint Conference on Biometrics (IJCB), pp. 1–7. IEEE (2019)
5. Deng, J., Dong, W., Socher, R., Li, L.J., Li, K., Fei-Fei, L.: ImageNet: a large-scale hierarchical image database. In: CVPR 2009 (2009)
6. Feng, J.C., Hong, F.T., Zheng, W.S.: MIST: multiple instance self-training framework for video anomaly detection. In: CVPR (2021)
7. Ghadiyaram, D., Feiszli, M., Tran, D., Yan, X., Wang, H., Mahajan, D.: Large-scale weakly-supervised pre-training for video action recognition. In: CVPR (2019)
8. Gowda, S.N., Rohrbach, M., Sevilla-Lara, L.: Towards understanding action recognition smart frame selection for action recognition. In: AAAI (2021)
9. Kalfaoglu, M.E., Kalkan, S., Alatan, A.A.: Late temporal modeling in 3D CNN architectures with BERT for action recognition. In: Bartoli, A., Fusiello, A. (eds.) ECCV 2020. LNCS, vol. 12539, pp. 731–747. Springer, Cham (2020). https://doi.org/10.1007/978-3-030-68238-5_48
10. Kay, W., et al.: The kinetics human action video dataset. arXiv preprint 1705.06950 (2017)
11. Kong, Y., Fu, Y.: Human action recognition and prediction: a survey. Int. J. Comput. Vis. **130**, 1366–1401 (2021)
12. Kuehne, H., Jhuang, H., Garrote, E., Poggio, T., Serre, T.: HMDB: a large video database for human motion recognition. In: International Conference on Computer Vision (ICCV) (2011)
13. Landi, F., Snoek, C.G.M., Gucchiara, R.: Anomaly locality in video surveillance. arXiv preprint 1901(10364), p. 10 (2019). https://doi.org/10.48550/ARXIV.1901.10364, https://arxiv.org/abs/1901.10364
14. Li, S., Liu, F., Jiao, L.: Self-training multi-sequence learning with transformer for weakly supervised video anomaly detection. In: AAAI (2022)
15. Liu, K., Ma, H.: Exploring background-bias for anomaly detection in surveillance videos. In: ACM MM (2019)

16. Liu, X., Pintea, S.L., Nejadasl, F.K., Booij, O.: No frame left behind: Full video action recognition. In: CVPR (2021)

17. Lv, H., Zhou, C., Cui, Z., Xu, C., Li, Y., Yang, J.: Localizing anomalies from weakly-labeled videos. IEEE Trans. Image Process. (TIP) **30**, 4505–4515 (2021)

18. Pang, G., Shen, C., Cao, L., Hengel, A.V.D.: Deep learning for anomaly detection: a review. ACM Comput. Surv. **54**(2), 1–38 (2021)

19. Perez, M., Kot, A.C., Rocha, A.: Detection of real-world fights in surveillance videos. In: IEEE ICASSP (2019)

20. Soomro, K., Zamir, A.R., Shah, M.: UCF101: a dataset of 101 human action classes from videos in the wild. CRCV-TR-12-01 (2012)

21. Sudhakaran, S., Lanz, S.: Learning to detect violent videos using convolutional long short-term memory. In: 2017 14th IEEE International Conference on Advanced Video and Signal Based Surveillance (AVSS), pp. 1–6 (2017). https://doi.org/10.1109/AVSS.2017.8078468

22. Sultani1, W., Chen, C., Shah, M.: Real-world anomaly detection in surveillance videos. In: CVPR (2018)

23. Tian, Y., Pang, G., Chen, Y., Singh, R., Verjans, J., Carneiro, G.: Weakly-supervised video anomaly detection with robust temporal feature magnitude learning. arXiv preprint arXiv:2101.10030 (2021)

24. Tran, D., Wang, H., Torresani, L., Ray, J., LeCun, Y., Paluri, M.: A closer look at spatiotemporal convolutions for action recognition. In: 2018 IEEE/CVF Conference on Computer Vision and Pattern Recognition, pp. 6450–6459 (2018). https://doi.org/10.1109/CVPR.2018.00675

25. Wu, J., et al.: Weakly-supervised spatio-temporal anomaly detection in surveillance video. In: IJCAI (2021)

26. Wu, P., Liu, J.: Learning causal temporal relation and feature discrimination for anomaly detection. IEEE Trans. Image Process. **30**, 3513–3527 (2021). https://doi.org/10.1109/TIP.2021.3062192

27. Xu, Q., See, J., Lin, W.: Localization guided fight action detection in surveillance videos. In: 2019 IEEE International Conference on Multimedia and Expo (ICME), pp. 568–573 (2019). https://doi.org/10.1109/ICME.2019.00104

28. Zhong, J.X., Li, N., Kong, W., Liu, S., Li, T.H., Li, G.: Graph convolutional label noise cleaner: train a plug-and-play action classifier for anomaly detection. In: CVPR (2019)

29. Zhou, P., Ding, Q., Luo, H., Hou, X.: Violent interaction detection in video based on deep learning. J. Phys. Conf. Ser. **844**, 012044 (2017)

Privacy-Preserving Person Detection Using Low-Resolution Infrared Cameras

Thomas Dubail$^{(\boxtimes)}$ ⓘ, Fidel Alejandro Guerrero Peña ⓘ,
Heitor Rapela Medeiros ⓘ, Masih Aminbeidokhti ⓘ, Eric Granger ⓘ,
and Marco Pedersoli ⓘ

Laboratoire d'imagerie, de vision et d'intelligence artificielle (LIVIA), Department of
Systems Engineering, ETS Montreal, Montreal, Canada
{thomas.dubail.1,fidel-alejandro.guerrero-pena.1,
heitor.rapela-medeiros.1,masih.aminbeidokhti.1}@ens.etsmtl.ca,
{eric.granger,marco.pedersoli}@etsmtl.ca

Abstract. In intelligent building management, knowing the number of people and their location in a room are important for better control of its illumination, ventilation, and heating with reduced costs and improved comfort. This is typically achieved by detecting people using compact embedded devices that are installed on the room's ceiling, and that integrate low-resolution infrared camera, which conceals each person's identity. However, for accurate detection, state-of-the-art deep learning models still require supervised training using a large annotated dataset of images. In this paper, we investigate cost-effective methods that are suitable for person detection based on low-resolution infrared images. Results indicate that for such images, we can reduce the amount of supervision and computation, while still achieving a high level of detection accuracy. Going from single-shot detectors that require bounding box annotations of each person in an image, to auto-encoders that only rely on unlabelled images that do not contain people, allows for considerable savings in terms of annotation costs, and for models with lower computational costs. We validate these experimental findings on two challenging top-view datasets with low-resolution infrared images.

Keywords: Deep learning · Privacy-preserving person detection ·
Low-resolution infrared images · Weak supervision · Embedded systems

1 Introduction

Intelligent building management solutions seek to maximize the comfort of occupants, while minimizing energy consumption. These types of solutions are crucial for reducing the use of fossil fuels with a direct impact on the environment. Such energy-saving is usually performed by adaptively controlling lighting, heating, ventilation, and air-conditioning (HVAC) systems based on building occupancy, and in particular the number of people present in a given room. For this, low-cost methods are needed to assess the level of room occupancy, and efficiently control the different systems within the building.

© The Author(s), under exclusive license to Springer Nature Switzerland AG 2023
L. Karlinsky et al. (Eds.): ECCV 2022 Workshops, LNCS 13805, pp. 689–702, 2023.
https://doi.org/10.1007/978-3-031-25072-9_46

Among the different levels of occupancy information that can be extracted in an intelligent building [25], Sun et al. define the location of occupants as the most important and fine-grained for smart building control. Given the recent advances in machine learning and computer vision, most solutions usually rely on deep convolutional neural networks (CNNs) to detect people [7,8]. Despite the high level of accuracy that can be achieved with CNNs for visual object detection based on RGB images, their implementation for real-world video surveillance applications incurs in a high computational complexity, privacy issues, and gender and race biases [4,22]. Finally, building occupancy management solutions are typically implemented on compact embedded devices, rigidly installed on the ceiling or portals of rooms, and integrating inexpensive cameras that can capture low-resolution IR images.

To mitigate these issues, He et al. [9] have proposed a privacy-preserving object detector that blurs people's faces before performing detection. To strengthen the detector against gender/race biases, the same authors proposed a face-swapping variation that also preserves privacy at the cost of increased computational complexity. Regardless of the good performance, their approach does not ensure confidentiality at the acquisition level, relying on RGB sensors to build the solution. Furthermore, their detector was designed for fully annotated settings using COCO [16] as the base dataset. This makes it difficult to generalize to people detection under different capture conditions (like when cameras are located on the ceiling on compact embedded systems), and extreme changes in the environment. In addition, it is difficult to collect and annotate image data to train or fine-tune CNN-based object detectors for a given application, so weakly-supervised or unsupervised training is a promising approach.

In contrast, our work tackles the occupants location problem by detecting people in infrared (IR) images at low resolution, which avoids most of the above-mentioned issues on privacy. Low resolution not only reduces computational complexity but also improves privacy, i.e., a detection on high-resolution infrared images would not be enough as it is possible to re-identify people [32]. More specifically, we analyse people detection with different levels of supervision. In this work, we compare unsupervised, weakly-supervised and fully supervised solutions. This is an essential aspect of the detection pipeline since producing bounding box annotations is very expensive, and there is a lack of good open-source object detection datasets for low-resolution infrared scenarios. In fact, reducing the level of supervision can lead to improved scalability for real applications and reduced computation, which is important considering the use of the proposed algorithms on embedded devices.

The contributions of this work are the following. (i) We propose cost-effective methods for estimating room occupancy under a low-supervision regime based on low-resolution IR images, while preserving users privacy. (ii) We provide an extensive empirical comparison of several cost-effective methods that are suitable for person detection using low-resolution IR cameras. Results indicate that, using top-view low-resolution images, methods that rely on weakly-labeled image data can provide good detection results, and thus save annotation efforts and reduce the required complexity of the detection model. (iii) To investigate the performance of person detecting methods on low-resolution IR images, our results are shown in two challenging datasets – the FIR-Image-Action and Distech-IR datasets. Finally, we provide bounding box annotations for the FIR-Image-Action [31] dataset[1].

2 Related Work

Privacy-preserving methods are of great interest to the scientific community. As a result, multiple approaches have been proposed in the past to circumvent the challenge. Ryoo et al. [21] proposed a method for learning a transformation that obtained multiple low-resolution images from a high-resolution RGB source. The method proved effective for action recognition even when inputs' resolution were down-sampled to 16×12. These findings were validated by others [6,29] using similar downsampling-based techniques to anonymize the people displayed in the images. On the other hand, recent approaches [9,20] have focused on producing blurred or artificial versions of people's faces while preserving the rest of the image intact. These methods usually rely on Generative Adversarial Networks (GANs) to preserve image utility, while producing unidentifiable faces. Specifically, the method of He et al. [9] is one of the first approaches to apply such anonymization in an object detection task. Despite recent advances in the area, these proposals are specialized for RGB images, which are anonymized after the undisclosed acquisition. As an alternative to RGB cameras, others authors [15,26–28] have proposed using low-resolution IR cameras to preserve anonymity at the phase of image acquisition. While most of these works target action recognition task, we focus on people detection as in [5,24].

Complementary to privacy preservation, this work also focuses on studying detection methods with different levels of supervision. Such a study aims to find techniques that reduce annotation costs without a significant performance reduction when compared with fully supervised approaches. In this work, we followed the auto-encoders (AE)-based anomaly-detection method similar to the one proposed by Baur et al. [2]. The technique is used to learn the distribution of typical cases, and applied later to identify abnormal regions within the image. However, different from their proposal, we focus on object detection instead of image segmentation. We also evaluate weakly-supervised detection methods based on Class Activation Maps (CAMs) [23,33] following the authors algorithm. Nonetheless, we use a customized CNN to be consistent with our low-resolution inputs. Finally, we consider also the fully supervised techniques Single Shot Detector [17] and Yolo v5 [11] as upper-bound references, sharing the same backbone as the previously described approaches. The aim of this work is not to use the latest developments for each type of technique. Instead, we aim to achieve a good trade-off between performance and computational complexity with simple and commonly used techniques. In particular, we favor simple

---

[1] https://github.com/ThomasDubail/FIR-Image-Action-Localisation-Dataset.

occupants' detection approaches that can run in embedded devices and with capabilities for handling low-resolution infrared images.

3   Person Detection with Different Levels of Supervision

Several cost-effective methods may be suitable for person detection using low-resolution IR cameras installed on ceilings. The goal of object detection is to find a mapping f_θ such that $f_\theta(x) = z$, where z are the probabilities that a bounding box belongs to each class. Note that such a mapping can be obtained using any level of supervision. In this paper, we seek to compare the detection accuracy of methods that rely on different levels of image annotation, and thereby assess the complexity needed to design embedded person detection systems.

In this work we consider a "fully annotated" dataset of IR images at low resolution $\mathcal{F} = \{(x_0, b_0), (x_1, b_1), ..., (x_N, b_N)\}$, with x an IR image and $b = \{(c_0, d_0, w_0, h_0), ..., (c_B, d_B, w_B, h_B)\}$ rectangular regions enclosing the objects of interest, also known as bounding boxes. Without loss of generality, we use a center pixel representation *(center x, center y, width, height)* [17] for defining the bounding boxes. In the given formulation, all bounding boxes belong to the persons' category. Consequently, a "weakly annotated" IR dataset is defined as $\mathcal{W} = \{(x_0, y_0), ..., (x_N, y_N)\}$ in which $y_i \in \{0, 1\}$ corresponds to an image-level annotation indicating whether a person is present ($y_i = 1$) or not ($y_i = 0$) in the image x_i. Finally, at the lowest level of supervision, an "unlabeled" dataset containing only IR images without annotations is expressed as $\mathcal{U} = \{x_i\}$. Please note that for this study, the datasets $\mathcal{F}$, $\mathcal{W}$, and $\mathcal{U}$ are drawn from the same pool of IR images but with different levels of annotations.

The rest of this section details the methods compared in this paper, each one trained according to a different level of supervision. Here, the backbone of all deep learning based methods remained the same. We focus on low-cost methods that can potentially be implemented on compact embedded devices.

3.1   Detection Through Thresholding

Let x be an IR thermal image from $\mathcal{U}$. The people within the images appear as high-temperature blobs easily distinguishable from the low-temperature background, Fig. 1a. Such a property allows us to directly apply a threshold-based mapping $g_\tau(x) = \Phi([\![x \geq \tau]\!])$ to obtain the persons' location. In this formulation, we use $[\![\cdot]\!]$ to refer to the Iverson bracket notation, which denotes the binarization of the image x according to the threshold τ, Fig. 1b. Here, the value of τ is manually determined according to a validation set, or automatically following Otsu's method [19], hereafter referred as *Threshold* and *Otsu's Threshold* respectively. Finally, a mapping Φ converts the segmentation map into a bounding box by taking the minimum and maximum pixels from each binary blob (see Fig. 1c). This method is more seen as a post-processing step for the following methods, although we have evaluated it to have a lower bound for detection.

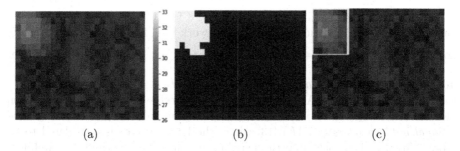

(a)   (b)   (c)

Fig. 1. Example of an IR image (a) that is binarized using the threshold $\tau = 29$ (b), and represented as a bounding box (c).

3.2 Unsupervised Anomaly-Based Detection Using Auto-encoders

For this next approach, people are considered anomalies for the distribution of empty rooms. In this work, we follow the method proposed by Baur et al. [2] to model the background distribution using an auto-encoder (AE) f_θ trained using only empty rooms, $\mathcal{W}^0 = \{x_i \mid y_i = 0\}$. Such an approach acts as a background reconstruction technique whenever an anomaly is present, i.e., the AE will not be able to reconstruct it. Then, we can highlight the anomaly by taking the difference between the input image and the obtained reconstruction, $x - f_\theta(x)$ (see Fig. 2). Finally, the anomaly detection method for person detection is defined as $\Lambda_{\theta,\tau}(x) = g_\tau(x - f_\theta(x))$ where g_τ is the thresholding technique explained in the previous section. Thus, the detection is performed in a two-step process: anomaly boosting and anomaly segmentation-localization, which can be done by setting a threshold τ.

The encoder architecture comprises six convolutional layers with kernels of size 3×3. Max-pooling operations are used every two convolutional layers to increase the field of view. The decoder follows a symmetrical architecture of the encoder using transposed convolution with a stride of 2 as upsampling technique. The bottleneck uses a linear layer with 256 neurons which encode input information as a vector projected onto the latent space. Finally, a reconstruction loss

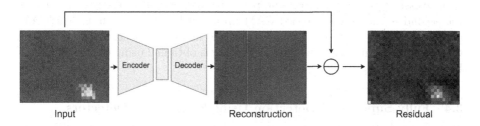

Input   Reconstruction   Residual

Fig. 2. Unsupervised anomaly-based method for person detection from low-resolution IR images.

694   T. Dubail et al.

is used to guide the training process and find a feasible minimizer θ for solving our background reconstruction task. In this work the Mean Square Error (MSE) loss is used, $\mathcal{L}_{MSE}(x,\theta) = \dfrac{1}{|W^0|} \sum_{x_i \in W^0} (x_i - f_\theta(x_i))^2$. In later sections we refer to this approach as *deep auto-encoder* or simply *dAE*.

Besides the classical AE, several types of hourglass architectures have been used for anomaly detection [2]. One of the most popular versions are the *variational auto-encoders* (*dVAE*) [13] where the latent vector is considered to be drawn from a given probabilistic distribution. Here we use the same architecture for both AE-based methods with a distinction for the loss function where the KL-divergence regularization is added to the MSE loss to enforce a normal distribution to the latent space.

3.3 Weakly-Supervised Detection Using Class Activation Mapping

Let (x,y) be a generic tuple from $\mathcal{W}$ where x is an image and $y \in \{0,1\}$ is a category indicating whether a person is present or not in x. A weakly-supervised approach for object localization is such that, by exploiting only the image-level annotation during training, learns a mapping c_φ to retrieve the object location during the evaluation. This kind of approaches based on Class Activation Maps (CAM) techniques [1,14,30] has been widely explored in the literature. The principle is based on the use of the compound function $c_\varphi(x) = (c_\psi^1 \circ c_\phi^0)(x)$ that relies on a feature extractor c_ϕ^0 followed by a binary classifier c_ψ^1. Here c_ϕ^0 is implemented as a CNN and c_ψ^1 as a Multi-layer Perceptron. Then, the following minimization based on cross-entropy loss is performed in order to find the optimal set of weights: $\min_\theta - \sum_{(x,y) \in \mathcal{W}} y \cdot \log c_\varphi(x)$. Once a feasible set of weights is found, a non-parametric transformation function uses the output from the feature extractor c_ϕ^0 to produce an activation map for each category in the task, $M(c_\phi^0(x))$. Note that in this task, the computation of such an activation map is only performed whenever a positive classification is obtained, i.e., $c_\varphi(x) \geq 0.5$. The architecture used for c_ϕ^0 is the same as in the encoder for the *dAE* technique, but without using the Max-pooling layers to avoid losing resolution. In this work, we use three variants for the Global Average Pooling M. The first is the classic weighting approach proposed by [33], hereafter referred to as *CAM*, the second is the gradient-based CAM proposed by [23], known as *GradCAM*, and the last one is the hierarchical approach known as *LayerCAM* [10]. As in the previous techniques, the final localization is obtained using the thresholding-based mapping g_τ.

3.4 Fully-Supervised Detection Using Single Shot Detectors

At the higher level of supervision, we explore the mapping function with bounding box annotations within the cost function. Let h_ϑ be a mapping parameterized over ϑ, which produces bounding box predictions. Among the different types of detectors existing in the literature, we used Single Shot Detector (*SSD*) [17] and *Yolo v5* [11] solutions since they are suitable for low-resolution images and allows using a custom backbone without serious implications for the training process. In this study, we enforce the same architecture for feature extraction as in the AEs and CAMs approaches. However, unlike the previous techniques, such supervised mappings do not require the composition with the thresholding function g_τ. Let $(x, b) \in \mathcal{F}$ be an IR image with its corresponding bounding box annotation. The learning process for both methods solves the optimization problem $\vartheta^* = \arg\min \mathcal{L}(x, b, \vartheta)$, by minimizing the cost of the model $\mathcal{L}$. Despite their differences in terms of representation, in both cases the loss function uses a supervised approach that measures the difference between the output $z = h_\vartheta(x)$ and the expected detections b.

4   Experimental Methodology

4.1   Datasets

In this study, two datasets were used to assess the IR person detection using models trained with different levels of annotation – the public FIR-Image-Action [31] dataset, and our Distech-IR dataset.

1) FIR-Image-Action with bounding box annotations

The FIR-Image-Action [31] dataset includes 110 annotated videos. We randomly selected 36 videos from this pool for the test and the others 74 for training and validation. Furthermore, training and validation sets were separated using a random selection of the frames (70% and 30%, respectively). All the approaches have been trained using the same data partition to ensure comparability.

To the best of our knowledge, there are no low-resolution IR datasets with bounding box annotations for person detection. Therefore, we annotated this dataset at bounding box level. The dataset was created by Haoyu Zhang of Visiongo Inc. for video-based action recognition. Such a dataset offers 126 videos with a total combined duration of approximately 7 h. Since this study aims to evaluate the performance of different techniques for IR-based people localization, we only used the IR images provided by the authors for our experiments. Nevertheless, it is worth mentioning that two modalities are available within the dataset: RGB with a spatial resolution of 320×240 acquired at 24 FPS, and IR with a resolution of 32×24 and sampled at 8 FPS. Although the RGB falls outside the scope of this work, we used them for obtaining bounding box annotations, as described later. As part of this work's contributions, we have publicly created and released the localization annotations for 110 videos out of the 126 for both RGB and IR modalities. Since there is redundancy within neighboring frames and our application does not require video processing, we further sampled the IR dataset obtaining the equivalent of 2 FPS videos.

We used a semi-automated approach to obtain bounding box annotations for the challenging low-resolution IR images in FIR-Image-Action. First, we create bounding box annotations by hand of a randomly selected subset of the RGB frames. We carefully curated these bounding boxes to reduce the impact of the misplacement when decreasing the resolution for the IR modality. Then, an *SSD* detector [17], h_ϑ, was trained over the RGB annotated dataset, being used afterward to obtain pseudo-labels over the remaining unannotated partition of the RGB dataset. A new randomly selected subset is then curated and h_ϑ training is repeated but using a larger partition of the data. This process was repeated three times resulting in a fully annotated version of the RGB dataset.

Finally, bounding box annotations for the IR dataset are obtained by pairing images from both modalities, followed by a coordinate aligning procedure. Since the videos for IR and RGB were out of synchronization, the initial time shift was manually determined using an overlay visualization of both modalities (see Fig. 3c). Such a synchronization was performed individually for every video. The final IR localization annotations were obtained by doing a linear interpolation of the bounding box coordinates from the labeled RGB dataset. The parameters for the alignment were estimated using linear regression. An example of the obtained bounding box annotation for both RGB and IR modalities can be observed in Fig. 3a and b, respectively.

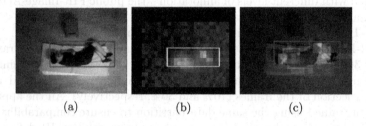

(a)   (b)   (c)

Fig. 3. Example of an RGB image with its ground truth (a), the corresponding IR image with the aligned bounding box (b), and an overlay of RGB and IR modalities (c) from the FIR-Image-Action dataset.

2) Distech-IR
The second dataset, named hereafter Distech-IR, followed the same separation proportions containing 1500 images for training, 500 for validation, and 800 for testing. Such a dataset, similar to FIR-Image-Action, contains two modalities of images (RGB and IR) with their corresponding bounding box level annotations provided by Distech Controls Inc. The dataset reflects the increasing interest by the industry for privacy preserving-based solutions for person localization and constitute an actual use case for this task. The Distech-IR dataset also proved to reflect better real-world scenarios since it is composed of seven rooms with different levels of difficulties, i.e., heat radiating appliances, sun-facing windows, and more than one person per room. For simulating deployment, we used rooms not seen before during training for the test set. Figure 4 shows some examples of images from both datasets.

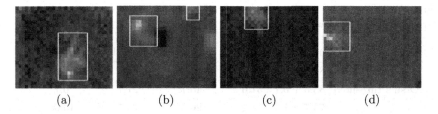

Fig. 4. Examples of IR images with their corresponding ground truth for FIR-Image-Action (a) and Distech-IR (b)-(d) datasets.

4.2 Implementation Details

We used normalization by 50° to ensure small scale within the input map. Adam optimizer [12] was employed with an initial learning rate of 10^{-4} and decay of 0.2 with a patience of ten epochs. Additionally, a 15 epoch patience early-stopping was implemented. Then, the best model according to the validation loss was selected for each case. The time calculations were evaluated on an Intel Xeon CPU at 2.3 Ghz, however the training of the models was done on an NVIDIA Tesla P100 GPU. Each experiment was performed 3 times with different seeds. The validation protocols are presented with each dataset in Sect. 4.1.

4.3 Performance Metrics

In this study, we use the optimal Localization Recall Precision (oLRP) [18] in order to characterize the ability of each method to detect the presence of people and locate them. This metric allows us to evaluate with the same measurement methods that provide bounding box detections without a score associated (such as AEs and CAMs) and methods with a detection score (as SSD and Yolo v5 detectors). Additionally, as stated by the authors, it reflects the localization quality more accurately than other measurements, providing separate measures for the different errors that a detection method can commit. The metric takes values between 0 and 100, with lower values being better. As part of the metric computation, we also calculate the localization (oLRP_{loc}), False Positive (oLRP_{FP}), and False Negative (oLRP_{FN}) components which provide more insights of methods behavior. Finally, execution time on the same hardware has been computed to obtain an approximation of the time complexity.

5 Results and Discussion

Tables 1 and 2 summarize the obtained results for FIR-Image-Action and Distech-IR datasets, respectively. As expected, the fully supervised approaches obtained the best performance for most metrics and comparable results to the other approaches regarding False Negatives. The methods proved useful for locating people in diverse scenarios, even under low-resolution settings. However, they

Table 1. Performance of detection methods on the FIR-Image-Action dataset. All metrics are calculated with an IoU of 0.5.

Model	oLRP↓	oLRP$_{loc}$ ↓	oLRP$_{FP}$ ↓	oLRP$_{FN}$ ↓	Time(ms)↓
Threshold	86.5 ± 0.1	32.3	45.3	44.2	0.4
Otsu's threshold	83.5 ± 0.1	31.6	45.7	27.7	0.7
dVAE	74.7 ± 1.3	31.2	26.6	24.5	13.0
dAE	77.4 ± 1.3	30.4	31.2	29.1	12.3
CAM	85.1 ± 1.1	34.3	41.2	29.0	11.4
GradCAM	85.5 ± 3.2	34.5	43.0	32.1	24.3
LayerCAM	84.8 ± 2.2	34.9	37.2	33.1	25.6
SSD	63.8 ± 2.7	**25.3**	12.6	18.6	46.6
Yolo v5	**56.9** ± 1.8	25.5	**6.3**	**6.2**	45.9

take longer to execute than the second-best performed techniques, which is an important downside to take into account for measuring real-time occupancy levels in intelligent buildings. In particular, the *dAE* approach was 3.7 times faster than *Yolo v5* and 3.8 times faster than *SSD*.

We can refer to the AE-based anomaly detection approaches as second-placed strategies. Both *dAE* and *dVAE* showed similar performance in terms of LRP and efficiency between them. Furthermore, the techniques obtained localization performance comparable to the fully-supervised approaches, especially in more complex situations like the Distech-IR dataset. This result is remarkable considering that localization supervision is not used during AE training, and only empty room images were used (10% of the data in the FIR-Image-Action and 30% in Distech-IR). The methods also showed acceptable execution times for the application.

Table 2. Performance of detection methods on the Distech-IR dataset. All metrics are calculated with an IoU of 0.5.

Model	oLRP↓	oLRP$_{loc}$ ↓	oLRP$_{FP}$ ↓	oLRP$_{FN}$ ↓	Time(ms)↓
Threshold	93.6 ± 2.4	37.1	72.7	54.1	0.4
Otsu's threshold	95.5 ± 1.2	34.2	83.3	50.0	0.7
dVAE	83.3 ± 8.9	33.2	32.7	40.6	13.0
dAE	82.7 ± 9.0	32.4	33.7	40.0	12.3
CAM	93.1 ± 1.9	37.6	59.7	52.3	11.4
GradCAM	91.6 ± 2.3	37.5	50.3	48.8	24.3
LayerCAM	91.1 ± 2.5	37.7	45.5	50.3	25.6
SSD	82.0 ± 7.2	31.1	**26.3**	44.7	46.6
Yolo v5	**80.2** ± 7.7	**30.2**	31.4	**37.4**	45.9

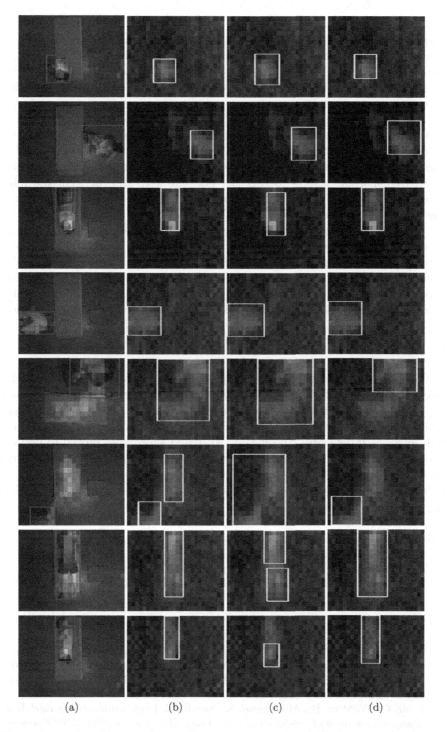

Fig. 5. Examples of low-resolution IR people detection results. Overlay of RGB and IR modalities with their corresponding ground truth (a), along with bounding box predictions of *dVAE* (b), *gradCAM* (c), and *Yolo v5* (d).

CAM methods provide a lower level of performance than other methods, despite having access to class-label annotations for training. Indeed, these methods are known to activate strongly for discriminant regions of an input image (since the backbone CNN is trained to discriminate classes), and be affected by complex image backgrounds [3]. These two factors affect its ability to define precise contours around a person.

As can be seen in the tables, *Otsu's Threshold* provides good person localization. However, it assumes a multi-modal intensity distribution for finding the threshold, which leads to false person localization in empty rooms. This effect can be observed by the high values of $oLRP_{FP}$. The rest of the approaches showed comparable results to this last one but were still far from the fully-supervised process. Figure 5 shows some examples of the obtained result over the FIR-Image-Action for *dVAE*, *gradCAM*, and *Yolo v5* methods.

As expected, real scenarios like those depicted in Distech-IR proved harder to generalize. A decrease in the performance was observed in all levels of supervision with a significant drop of 18.2% oLRP for *SSD* and 23.3% for *Yolo v5*. A smaller decrease was observed for AEs obtaining even closer results to the one from supervised approaches. The primary issue in this dataset was the large number of False Negative which almost doubled the FN obtained for FIR-Image-Action.

6   Conclusions

In this work, we presented a study comprising different methods with increasing levels of supervision for privacy-preserving person localization. Our experimental results over two low-resolution top-view IR datasets showed that reduced image-level supervision is enough for achieving results almost comparable to a fully-supervised detectors. Specifically, AE-based approaches proved to perform similarly to Yolo v5 in real-world scenarios by only using images of empty rooms for training and with 3.7 times less execution time. Such a result is significant for reducing annotation costs and improving the scalability of intelligent building applications. Additionally, we detailed the process for producing bounding box annotations for low-resolution IR images and provided the localization for the publicly available dataset FIR-Image-Action.

Acknowledgements. This work was supported by Distech Controls Inc., and the Natural Sciences and Engineering Research Council of Canada (RGPIN-2018-04825).

References

1. Bae, W., Noh, J., Kim, G.: Rethinking class activation mapping for weakly supervised object localization. In: Vedaldi, A., Bischof, H., Brox, T., Frahm, J.-M. (eds.) ECCV 2020. LNCS, vol. 12360, pp. 618–634. Springer, Cham (2020). https://doi.org/10.1007/978-3-030-58555-6_37
2. Baur, C., Wiestler, B., Albarqouni, S., Navab, N.: Deep autoencoding models for unsupervised anomaly segmentation in brain MR images. arXiv:1804.04488 [cs] 11383, pp. 161–169 (2019)

3. Belharbi, S., Sarraf, A., Pedersoli, M., Ayed, I.B., McCaffrey, L., Granger, E.: F-CAM: full resolution cam via guided parametric upscaling. In: Proceedings of the IEEE/CVF Winter Conference on Applications of Computer Vision, pp. 3490–3499 (2022)
4. Buolamwini, J., Gebru, T.: Gender shades: intersectional accuracy disparities in commercial gender classification. In: Conference on Fairness, Accountability and Transparency, pp. 77–91. PMLR (2018)
5. Cao, J., Sun, L., Odoom, M.G., Luan, F., Song, X.: Counting people by using a single camera without calibration. In: 2016 Chinese Control and Decision Conference (CCDC), pp. 2048–2051. IEEE, May 2016
6. Chen, J., Wu, J., Konrad, J., Ishwar, P.: Semi-coupled two-stream fusion convnets for action recognition at extremely low resolutions. In: 2017 IEEE Winter Conference on Applications of Computer Vision (WACV), pp. 139–147. IEEE (2017)
7. Chen, Z., Wang, Y., Liu, H.: Unobtrusive sensor-based occupancy facing direction detection and tracking using advanced machine learning algorithms. IEEE Sens. J. **18**(15), 6360–6368 (2018)
8. Gao, C., Li, P., Zhang, Y., Liu, J., Wang, L.: People counting based on head detection combining AdaBoost and CNN in crowded surveillance environment. Neurocomputing **208**, 108–116 (2016)
9. He, P., et al.: Privacy-preserving object detection. arXiv preprint arXiv:2103.06587 (2021)
10. Jiang, P.T., Zhang, C.B., Hou, Q., Cheng, M.M., Wei, Y.: LayerCAM: exploring hierarchical class activation maps for localization. IEEE Trans. Image Process. **30**, 5875–5888 (2021)
11. Jocher, G., et al.: ultralytics/yolov5: v6.1 - TensorRT, TensorFlow Edge TPU and OpenVINO Export and Inference, February 2022
12. Kingma, D.P., Ba, J.: Adam: a method for stochastic optimization. arXiv preprint arXiv:1412.6980 (2014)
13. Kingma, D.P., Welling, M.: Auto-Encoding Variational Bayes, May 2014
14. Kitano, H.: Classification and localization of disease with bounding boxes from chest X-ray images, p. 6 (2020)
15. Tao, L., Volonakis, T., Tan, B., Zhang, Z., Jing, Y.: 3D convolutional neural network for home monitoring using low resolution thermal-sensor array. In: 3rd IET International Conference on Technologies for Active and Assisted Living (TechAAL 2019), pp. 1–6. Institution of Engineering and Technology, London, UK (2019)
16. Lin, T.-Y., et al.: Microsoft COCO: common objects in context. In: Fleet, D., Pajdla, T., Schiele, B., Tuytelaars, T. (eds.) ECCV 2014. LNCS, vol. 8693, pp. 740–755. Springer, Cham (2014). https://doi.org/10.1007/978-3-319-10602-1_48
17. Liu, W., et al.: SSD: single shot multibox detector. arXiv:1512.02325 [cs] 9905, pp. 21–37 (2016)
18. Oksuz, K., Cam, B.C., Kalkan, S., Akbas, E.: One metric to measure them all: Localisation Recall Precision (LRP) for evaluating visual detection tasks. arXiv:2011.10772 [cs], November 2021
19. Otsu, N.: A threshold selection method from gray-level histograms. IEEE Trans. Syst. Man Cybern. **9**(1), 62–66 (1979)
20. Ren, Z., Lee, Y.J., Ryoo, M.S.: Learning to anonymize faces for privacy preserving action detection. In: Ferrari, V., Hebert, M., Sminchisescu, C., Weiss, Y. (eds.) ECCV 2018. LNCS, vol. 11205, pp. 639–655. Springer, Cham (2018). https://doi.org/10.1007/978-3-030-01246-5_38

21. Ryoo, M.S., Rothrock, B., Fleming, C., Yang, H.J.: Privacy-preserving human activity recognition from extreme low resolution. In: Thirty-First AAAI Conference on Artificial Intelligence (2017)
22. Schwemmer, C., Knight, C., Bello-Pardo, E.D., Oklobdzija, S., Schoonvelde, M., Lockhart, J.W.: Diagnosing gender bias in image recognition systems. Socius **6**, 2378023120967171 (2020)
23. Selvaraju, R.R., Cogswell, M., Das, A., Vedantam, R., Parikh, D., Batra, D.: Grad-CAM: Visual Explanations from Deep Networks via Gradient-based Localization. Int. J. Comput. Vision **128**(2), 336–359 (2020)
24. Shengsheng, Yu., Chen, X., Sun, W., Xie, D.: A robust method for detecting and counting people. In: 2008 International Conference on Audio. Language and Image Processing, pp. 1545–1549. IEEE, Shanghai, China, July 2008
25. Sun, K., Zhao, Q., Zou, J.: A review of building occupancy measurement systems. Energy Build. **216**, 109965 (2020)
26. Tao, L., Volonakis, T., Tan, B., Jing, Y., Chetty, K., Smith, M.: Home activity monitoring using low resolution infrared sensor. arXiv:1811.05416 [cs], November 2018
27. Tateno, S., Meng, F., Qian, R., Hachiya, Y.: Privacy-preserved fall detection method with three-dimensional convolutional neural network using low-resolution infrared array sensor. Sensors **20**(20), 5957 (2020)
28. Tateno, S., Meng, F., Qian, R., Li, T.: Human motion detection based on low resolution infrared array sensor. In: 2020 59th Annual Conference of the Society of Instrument and Control Engineers of Japan (SICE), pp. 1016–1021. IEEE, Chiang Mai, Thailand, September 2020
29. Wang, Z., Chang, S., Yang, Y., Liu, D., Huang, T.S.: Studying very low resolution recognition using deep networks. In: Proceedings of the IEEE Conference on Computer Vision and Pattern Recognition, pp. 4792–4800 (2016)
30. Yang, S., Kim, Y., Kim, Y., Kim, C.: Combinational class activation maps for weakly supervised object localization. In: 2020 IEEE Winter Conference on Applications of Computer Vision (WACV), pp. 2930–2938. IEEE, Snowmass Village, CO, USA, March 2020
31. Zhang, H.: FIR-Image-Action-Dataset (2020). https://github.com/visiongo-kr/FIR-Image-Action-Dataset#fir-image-action-dataset
32. Zheng, H., Zhong, X., Huang, W., Jiang, K., Liu, W., Wang, Z.: Visible-infrared person re-identification: a comprehensive survey and a new setting. Electronics **11**, 454 (2022)
33. Zhou, B., Khosla, A., Lapedriza, A., Oliva, A., Torralba, A.: Learning deep features for discriminative localization. arXiv:1512.04150 [cs], December 2015

Gait Recognition from Occluded Sequences in Surveillance Sites

Dhritimaan Das[1], Ayush Agarwal[2], and Pratik Chattopadhyay[1(✉)] ⓘ

[1] Indian Institute of Technology (BHU), Varanasi 221005, India
{dhritimaand.cd.eee17,pratik.cse}@iitbhu.ac.in
[2] Motilal Nehru National Institute of Technology, Allahabad 211004, India
ayushagarwal@mnnit.ac.in

Abstract. Gait recognition is a challenging problem in real-world surveillance scenarios where the presence of occlusion causes only fractions of a gait cycle to get captured by the monitoring cameras. The few occlusion handling strategies in gait recognition proposed in recent years fail to perform reliably and robustly in the presence of an incomplete gait cycle. We improve the state-of-the-art by developing novel deep learning-based algorithms to identify the occluded frames in an input sequence and henceforth reconstruct these occluded frames. Specifically, we propose a two-stage pipeline consisting of occlusion detection and reconstruction frameworks, in which occlusion detection is carried out by employing a VGG-16 model, following which an LSTM-based network termed RGait-Net is employed to reconstruct the occluded frames in the sequence. The effectiveness of our method has been evaluated through the reconstruction Dice score as well as through the gait recognition accuracy obtained by computing the Gait Energy Image feature from the reconstructed sequence. Extensive evaluation using public data sets and comparative study with other methods verify the suitability of our approach for potential application in real-life scenarios.

Keywords: Gait recognition · Occlusion detection · Occlusion reconstruction · Real-world surveillance

1 Introduction

Over the past decade, rapid advances in the field of Computer Vision-based gait recognition show that a gait-based biometric identification system can be potentially used to identify suspects in surveillance sites to strengthen public security. The main advantage of gait over any other biometric feature is that it can be captured by a surveillance camera from a distance, and the videos/images captured by the camera are not required to be of very high resolution. The literature on gait recognition shows that any gait feature extraction algorithm work satisfactorily only if a complete gait cycle is available [14,25,36]. However, in most real-life situations, the presence of a complete cycle is not guaranteed. This is

---

D. Das and A. Agarwal—Contributed equally to the work.

© The Author(s), under exclusive license to Springer Nature Switzerland AG 2023
L. Karlinsky et al. (Eds.): ECCV 2022 Workshops, LNCS 13805, pp. 703–719, 2023.
https://doi.org/10.1007/978-3-031-25072-9_47

because static or dynamic occlusion can occur at any time during the video capture phase. Static occlusion may be caused due to the presence of a stationary object such as a pillar, or a standing person. On the contrary, dynamic occlusion is caused by moving objects, such as walking people. Figure 1 provides an example dynamic occlusion scenario in gait sequences using background-subtracted frames. It can be observed from Fig. 1 that due to the presence of occlusion, binary silhouette frames of a target subject also get corrupted resulting in a substantial loss of silhouette-level information that drastically affects the effectiveness of the computed gait features.

Fig. 1. Binary silhouettes extracted from a gait video with dynamic occlusion

Traditional gait recognition techniques are not suited to perform recognition effectively in the presence of occlusion. Only a few techniques in the literature have attempted to provide solutions to this challenging problem. Existing approaches to handling occlusion in gait recognition are mainly of three types: (i) methods that reconstruct (or predict) occluded frames [28,37], (ii) methods that follow a key pose-based recognition approach and compare features corresponding to the available poses in a gait cycle [15,16], and (iii) methods that carry out feature reconstruction instead of frame-by-frame reconstruction [5,6]. However, the effectiveness of these approaches depends on several factors, e.g., the approach in [28] works satisfactorily only if multiple occluded cycles are available, while the methods in [5,6,15,16,37] are effective only for small degrees of occlusion. Dependency on these unrealistic constraints limits the practical applicability of the existing methods in real-world scenarios. In this work, we take up this challenge and propose a fully automated approach for occlusion detection and reconstruction by employing deep neural network models. Specifically, a VGG-16-based model has been used for occlusion detection, and an LSTM-based generator termed *Reconstruct Gait Net* (abbreviated *RGait-Net*) has been employed to reconstruct the occluded frames.

To the best of our knowledge, there exists only one gait data set containing occluded gait sequences in the public domain, namely the TUM-IITKGP gait data set [27]. However, due to the unavailability of clean ground truth sequences corresponding to an occluded sequence, this data cannot be used to train the RGait-Net occlusion reconstruction model. Hence, we consider another popular gait data set of unoccluded sequences, namely the CASIA-B data [46,50], and prepare an extensive dataset for training the RGait-Net by introducing varying degrees of synthetic occlusion in its sequences. Our algorithm can be conveniently integrated with the existing security setup in surveillance sites such as airports, railway stations, and shopping malls to improve public security. The main contributions of the work may be summarized as follows:

- To the best of our knowledge, ours is the first-ever work on appearance-based gait recognition that employs deep neural network-based models to detect occlusion and predict missing frames in a given occluded gait sequence.
- We propose a novel multi-objective loss function to train the RGait-Net model suitably to reconstruct the corrupted frames from any given occluded gait sequence.
- The effectiveness of our approach against varying degrees of occlusion has been established through extensive experimental evaluation and comparison with existing popular gait recognition approaches.
- The pre-trained models will be made publicly available for further comparative studies.

2   Related Work

Initial approaches to gait recognition can be classified as either appearance-based or model-based. The appearance-based techniques aim to utilize and extract meaningful features from the binary silhouettes and then use them for gait recognition. The work in [25] presents the first appearance-based gait feature called Gait Energy Image (GEI) that computes the average of gait features over a complete gait cycle. Following this approach, other approaches such as [36], introduce a pose-based feature called Pose Energy Image (PEI), which is extracted from fractional parts of a gait cycle instead of the complete cycle. Similarly, [14] and [16] also use fractional part of the gait cycle to extract features and then use them for gait recognition using the RGB, depth, and skeleton streams from Kinect. The work in [47] presents a feature termed Active Energy Image (AEI) that captures dynamic information more effectively than GEI. In [44], the GEI features are projected into a lower-dimensional space using Marginal Fisher Analysis and recognition is carried out in this projected space. A viewpoint-invariant gait recognition approach is described in [18], in which cyclic gait analysis is performed to identify key frames present in a sequence. The above approaches consider training and test sequences that are captured from a uniform viewpoint. Later, a few cross-view gait recognition techniques have also been proposed, e.g., [8] that performs recognition by aligning the GEIs [25] from different viewpoints using coupled bi-linear discriminant projection.

While the appearance-based approaches extract gait features from silhouette shape variation over a gait cycle, the model-based methods attempt to fit the kinematics of human motion into a pre-defined model. For example, in [4], an articulated 3D points cylinder has been considered as a gait model. The static gait features are extracted from the height, stride length, and other body-part dimensions of the fitted model, while the kinematics of gait are extracted from the lower part of the body. In [19] also, skeleton-based gait features extracted from the lower parts of the human body are fitted to a simple harmonic motion

model. The work in [9] considers a skeleton model and fits the static body structural parameters such as distance from neck to torso, stride length, etc., to the
model. The work in [20] predicts the body parts and extracts features from the
Fourier Transform of the motion data derived from these body parts and, finally,
uses a KNN classifier to predict the class of a subject. In [31], the silhouettes
are divided into seven segments and an ellipse is fitted to each of these segments and geometric parameters like ratio of major to minor axis length, etc.,
are considered for computing the gait features.

With the advancement of Deep Learning, CNN-based recognition systems
are being widely used in several applications at present. For example, in [29]
and [34], a CNN architecture is trained on sensor data from the accelerometer
and smartphone gyroscope that use the temporal and frequency domain data
to extract features, and then an SVM-based classification is performed. In [38],
the averaged feature computed from an incomplete gait cycle is fed to a CNN
generator to predict the GEI feature corresponding to a full cycle. To compensate for the scarce training data available for gait recognition as well as to avoid
underfitting, the authors in [1] suggest employing a small-scale CNN consisting
of four convolutional layers (with eight feature maps in each layer) and four
pooling layers for gait recognition. CNN-based generators have also been extensively used to perform cross-view gait recognition and handle varying co-variate
conditions during the training and testing phases. For example, the method in
[41] performs recognition by employing a deep Siamese architecture-based feature comparison. This method has been seen to work satisfactorily for large view
angle variation. In [23], a gait recognition algorithm is presented that utilizes a
GAN-based model to transform the gait features with co-variate conditions to
that without co-variate conditions. In [45], the authors propose a method to
handle gait recognition from varying viewpoints and co-variate conditions. This
method uses a GAN to remove co-variate features from any input test sequence
and generates a GEI without co-variate features which is next used for classification. The approach in [26] introduces MGAN that learns view-specific features
and transform gait features from one view to another. However, multiple such
models must be trained for transformations between any arbitrary views, which
is cost-intensive. To handle the shortcomings of [26,49] describes another GAN-
based model termed the View Transformation GAN (VT-GAN) which performs
transformation between any arbitrary pair of views. The approach in [24] also
discusses a view-invariant gait recognition method in which features are learned
in the Cosine space. For effective feature learning and transformation, an angular
loss function and a triplet loss function are employed. Approaches described in
[12,13] introduce a specific type of neural network termed GaitSet that extracts
useful spatio-temporal information from a gait sequence and integrate this information for view transformation. The work in [40] uses knowledge distillation on
GaitSet and introduces a lightweight CNN model that performs equivalently to
the GaitSet model. The GaitPart introduced in [21] is developed as an improvement to GaitSet, in which micromotions from different body parts are encoded
and aggregated to derive features for gait recognition.

With the introduction of RGB-D cameras, such as Kinect, some frontal view gait recognition techniques [14,39] have also been developed. The work in [7] jointly exploits structured data and temporal information using a spatio-temporal neural network model, termed *TGLSTM*, to effectively learn long and short-term dependencies along with the graph structure. Some recent approaches towards handling the problem of occlusion in gait recognition are discussed next. Occlusion reconstruction has been done using a Gaussian process dynamic model in [37]. In [30], the authors proposed an approach that uses Support Vector Machine-based regression to reconstruct the occluded data. These reconstructed data are first projected onto a PCA subspace, following which classification of the projected features has been carried out in this canonical subspace. Three different techniques for the reconstruction of missing frames have been discussed in [32], out of which the first approach uses an interpolation of polynomials, the second one uses auto-regressive prediction, and the last one uses a method involving projection onto a convex set. In [2], an algorithm focusing on tracking pedestrians has been presented in which the results are evaluated on a synthetic data set containing sequences with partial occlusion. Some approaches involving human tracking [3,48] and activity recognition [33,43] techniques have also been developed that handle the challenging problem of occlusion detection and reconstruction. The work in [35] presents modeling and characterization of occlusion in videos. In [5], a deep neural network-based gait recognition approach has been proposed to reconstruct GEI from an input sequence that does not contain all the frames of a complete gait cycle. An improvement to the above work has been proposed in [6] by the same authors in which the transformation from incomplete GEI computed from an incomplete gait cycle to the complete GEI has been performed progressively using an Autoencoder-based neural network. In addition, in [17], a robust gait representation technique termed *Frame Difference Energy Image* (FDEI) is introduced, which accounts for the loss of information due to self-occlusion in a gait cycle.

From the extensive literature survey, it is observed that most existing gait recognition approaches require a complete gait cycle for proper functioning. The few occlusion handling approaches in gait recognition suffer either from lack of robustness or depend on certain unrealistic constraints which are difficult to match with real-world scenarios. Further, to date, no neural network-based approach has been developed to reconstruct the missing/occluded frames in a gait sequence. We improve the state-of-the-art research on gait recognition by proposing a novel deep neural network-based approach for occlusion detection and reconstruction, as detailed in the following section.

3   Proposed Approach

In this section, we describe our proposed gait-based person identification approach in the presence of occlusion. A schematic diagram explaining the steps of the proposed approach is diagrammatically shown using Fig. 2. With reference

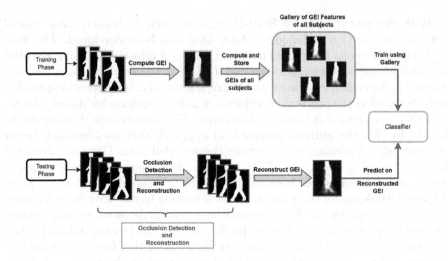

Fig. 2. An overview of the pipeline followed for gait recognition along with the proposed occlusion detection and reconstruction setup.

to the figure, given a set of RGB training sequences, first standard preprocessing steps such as background subtraction, silhouette extraction, and dimension normalization to fixed height and width are applied to all frames of each RGB sequence, as in any other image-based gait recognition approaches [25,36]. From the training sequences, the GEI features are computed, and a database of gallery GEI features is constructed. Now, given an RGB test sequence, similar preprocessing steps are carried out to obtain the normalized silhouette for each frame. Next, an occlusion detection model is employed that scans each frame of an input gait sequence and classifies it as 'Occluded' or 'Unoccluded'. The silhouette classified as occluded is next predicted using our RGait-Net occlusion reconstruction model. Finally, the GEI feature is computed from the above reconstructed sequence and its dimension is reduced by applying Principal Component Analysis (PCA) and retaining only those components that preserve more than *98%* variation present in the data. A Random Forest classifier which is trained on the gallery set of GEIs is then used to predict the class of the reconstructed GEI. The above diagram has been drawn by assuming the training sequences are free from occlusion. The preprocessing and GEI feature computation stages are explained in detail in [25], and hence, we avoid discussions on these topics.

3.1   Occlusion Detection Using VGG-16

Occlusion detection is carried out in an automated manner by employing a deep VGG-16 Convolutional Neural Network as the backbone of the occlusion detection network. This model is trained using an extensive gallery set of occluded and unoccluded binary silhouettes. While preparing this gallery set, the unoccluded silhouettes are taken from both the CASIA-B [46,50] and the TUM-IITKGP

[27] datasets, whereas the occluded silhouettes are considered from the statically/dynamically occluded sequences present in the TUM-IITKGP dataset. The model takes as input a single binary frame and classifies it as either '*Occluded*' or '*Unoccluded*'. For an input frame $\mathcal{F}_i$, let $P_1(i)$ and $P_2(i)$ denote the probability of '*Occluded*' and '*Unccluded*', respectively. Then if y_i denotes the ground truth class label for $\mathcal{F}_i$, the cross-entropy loss $\mathcal{L}_{occ}$ is computed as:

$$\mathcal{L}_{occ} = -y_i log(P_1(i)) - (1 - y_i)log(P_2(i)). \tag{1}$$

The optimal hyper-parameters of this model, i.e., learning rate and dropout rate, are determined through cross-validation and grid-search considering different configurations.

3.2 Occlusion Reconstruction Through RGait-Net

Since the gait pattern of any person can be viewed as a spatio-temporal sequence data, in this work each occluded frame present in a binary gait sequence is predicted by utilizing the information given by a few previous unoccluded (or, reconstructed) frames. We propose to perform this reconstruction by employing a deep neural network, more specifically a Fully Convolutional Long-Short-Term Memory (FC-LSTM) network. The FC-LSTM model used in this work consists of a deep Auto-Encoder with seven convolutional layers and four pooling layers stacked on top of each other and is termed as the Reconstruct Gait Net (RGait-Net). The network architecture with the convolutional layers is shown in Fig. 3.

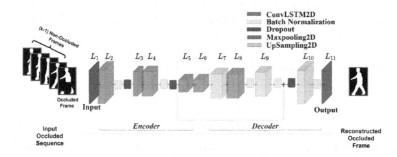

Fig. 3. Architecture of the proposed occlusion reconstruction model

With reference to the figure, each layer of the network is represented by L_i where i denotes the layer number. We use a kernel size of 3 for all the convolutional layers and a kernel size of 2 for the max-pooling and up-sampling operations. *Tanh* activation is used at the output of all convolutional layers. If $(k-1)$ number of previous unoccluded (or, predicted frames) are used to predict the kth frame using RGait-Net, then $(k-1)$ different $256 \times 256 \times 1$ dimensional binary images are stacked one after another before feeding to the input layer L_1 of the network, as shown in the figure. If a gait sequence has multiple occluded frames, we follow the same process, i.e., detection of the occluded frame followed by reconstructing the occluded frame using previously predicted/unoccluded frames through RGait-Net. For regularization, we consider dropout layers and apply batch normalization after each convolutional layer. Furthermore, to improve the robustness of our model against noisy data, we add zero-centered Gaussian noise to the input frames during training. We also use skip connections in our model to achieve better gradient flow and minimize the effect of vanishing gradient.

Let the function learned by the occlusion reconstruction model be represented by T. This function takes as input a sequence of frame-level features from the previous $(k-1)$ frames, denoted by $\mathcal{F}_1$, $\mathcal{F}_2$, ..., $\mathcal{F}_{k-1}$, to predict the feature for the kth frame, i.e., $\widehat{\mathcal{F}_k}$. Mathematically, this can be represented as:

$$\widehat{\mathcal{F}_k} = T(\mathcal{F}_1, \mathcal{F}_2, ..., \mathcal{F}_{k-1}). \tag{2}$$

A multi-objective loss function is used to train the proposed RGait-Net. This loss has two components: (a) a binary cross-entropy loss, and (b) a dice loss. The pixel-wise binary cross-entropy loss ($\mathcal{L}_{rec}$) between the reconstructed image and the ground truth image is computed as follows:

$$\mathcal{L}_{rec} = -\frac{1}{I}\sum_{i=1}^{I}[\mathcal{F}_k(i)log(\widehat{\mathcal{F}_k}(i)) + (1 - \mathcal{F}_k(i))log(1 - \widehat{\mathcal{F}_k}(i))], \tag{3}$$

where I denotes the total number of pixels in each image and $\mathcal{F}_k(i)$ and $\widehat{\mathcal{F}_k}(i)$ respectively denote the pixel intensity of the ith pixel corresponding to the ground truth and the reconstructed binary frames. The dice loss $\mathcal{L}_{Dice}$ is computed as the negative of the Sørensen-Dice similarity score (or, the dice score) [10]. This dice score between two images provides a value in the range $[0,1]$, where a value closer to '1' indicates high overlap, and a value closer to '0' corresponds to insignificant overlap. The dice loss between the ground truth and the reconstructed binary silhouette images is computed as:

$$\mathcal{L}_{Dice} = -\frac{2\widehat{\mathcal{F}_k}^T \mathcal{F}_k}{\widehat{\mathcal{F}_k}^T \widehat{\mathcal{F}_k} + \mathcal{F}_k^T \mathcal{F}_k}. \tag{4}$$

The final objective function is computed as a weighted summation of the binary cross-entropy and the dice coefficient, as shown in (5).

$$\mathcal{L}_{total} = \lambda_1 \mathcal{L}_{rec} + \lambda_2 \mathcal{L}_{Dice}, \tag{5}$$

where λ_1 and λ_2 are positive constants. Given an occluded gait sequence, occlusion detection is first performed to identify the frames that have to be reconstructed. Next, the proposed RGait-Net-based reconstruction model is employed to predict the occluded frames one by one. Once all occluded frames in a sequence are reconstructed, the GEI feature [25] is extracted, its dimension is reduced by applying Principal Component Analysis (PCA) by retaining 98% variance, and a Random Forest classifier with bagging is used to compute the gait recognition accuracy from the reconstructed sequences.

4   Experimental Analysis

The proposed algorithm has been implemented on a system with *96* GB RAM, one i9-18 core processor, along with three GPUs: one Titan Xp with *12* GB RAM, *12* GB frame-buffer memory, and *256* MB BAR1 memory, and two GeForce GTX *1080* Ti with *11* GB RAM, *11* GB frame-buffer memory and *256* MB of BAR1 memory. Evaluation of the proposed approach has been done on two public data sets, namely the CASIA-B [46] and the TUM-IITKGP [27] data sets. Among these, the CASIA B data consists of *106* subjects, and for each of these subjects, the data was captured under the following conditions: (a) six sequences with normal walk (*nm-01* to *nm-06*), (b) two sequences with carrying bag (*bg-01* and *bg-02*), (c) two sequences with coat (*cl-01* and *cl-02*). In the present set of experiments, we use the normal walking sequences (i.e., sequences *nm-01* to *nm-06*) captured from the fronto-parallel view. Among these, four sequences of each subject are considered for training the RGait-Net and the Random Forest with GEI features, while the remaining two are used during the testing phase by introducing synthetic occlusions of varying degrees.

The TUM-IITKGP data set, on the other hand, consists of walking videos of *35* subjects under the following conditions: (a) one video of normal walking without any occlusion, (b) one with carrying bag, (c) one with wearing gown, (d) one with static occlusion, and lastly (e) one with dynamic occlusion. It may be noted that both the CASIA-B and TUM-IITKGP data sets consist of background-subtracted frame sequences extracted from RGB videos. Therefore, the generation of binary silhouettes through background subtraction was not required in our experiments. Only the silhouette cropping and normalization to a fixed height and width have been applied to each frame as a preprocessing step. As explained in Sect. 3, we use unoccluded gait cycles to construct the gallery of GEI features for the *35* subjects for the TUM-IITKGP dataset, while videos with static and dynamic occlusion are used during the testing phase. The sequences present in the TUM-IITKGP data set are sufficiently large, and we extract eight gait cycles from the normal walking video and four sequences from each of the static and dynamic occluded videos. The eight GEIs corresponding to normal walking of each of the *35* subjects results in a total of *280* GEIs, which are used for training the Random Forest classifier to recognize humans from their gait signatures. Evaluation is also done on eight occluded sequences corresponding to the videos with static and dynamic occlusion in this data set. Throughout this section, the test sequences in the TUM-IITKGP data are labeled as *Sequence 1*, *Sequence 2*, ..., *Sequence 8*, respectively.

4.1   Training Details

The VGG-16-based occlusion detection network is trained from scratch by constructing a dataset of *1524* silhouette images consisting of *664* unoccluded silhouettes selected from both the TUM-IITKGP and the CASIA-B gait data sets, along with *860* occluded silhouettes (with both static and dynamic occlusion) from the TUM-IITKGP dataset. RMSprop optimizer is used to train the network till convergence i.e., the loss (computed from (1)) between successive epochs is smaller than a predefined threshold ϵ. Usually, $\epsilon << 1$, and here we choose its value to be $2e-4$. It may be noted that our proposed occlusion detection model is trained with both originally and synthetically occluded silhouettes corresponding to the TUM-IITKGP and CASIA-B datasets, and hence it is expected to perform satisfactorily for real occluded test gait silhouettes also.

The RGait-Net model is trained using synthetically occluded sequences generated from the CASIA-B data [46] along with their unoccluded versions. From the training gait sequence of each of the 106 subjects in this data, we generate four new sequences with different sets of frames occluded in each. In this way, we generate 1696 synthetically occluded sequences and use these as input and their unoccluded versions as the desired outputs while training the proposed RGait-Net. Since the CASIA-B dataset contains sequences of binary silhouettes, synthetic occlusions have been added in this dataset by randomly selecting a few frames from a gait cycle and putting random black patches on the objects present in these frames. For a certain percentage of occlusion (say, $p\%$), the number of frames (N) to be occluded in a gait cycle of length C is computed as $N = pC/100$. An extensive performance evaluation of the RGait-Net model has been performed using the various synthetically occluded sequences generated from the CASIA-B data by varying the value of p from 5 to 50. The network is trained for 62 epochs (up to convergence), and the hyperparameters involved in training the RGait-Net model are the learning rate, dropout rate, and weights λ_1 and λ_2 of the loss function shown in (5). Upon fine-tuning, we obtain the values of these parameters as the learning rate $= 0.001$, the dropout rate $= 0.2$, $\lambda_1 = \lambda_2 = 1$ (refer to (5)).

4.2   Results

In the first experiment, we verify the effectiveness of the trained occlusion detection model on the synthetically occluded test set of the CASIA-B data. We observe that the precision and recall for occlusion detection are 99.53% and 98.72%, respectively and the overall accuracy in identifying the occluded and unoccluded frames correctly is 98.89%. Visually, we observe that the occlusion detection performance for the real-occluded TUM-IITKGP data are also quite accurate. However, due to unavailability of ground-truth unoccluded frames, we could not present the quantitative metrics for occlusion detection using this data. Next, we study the extent to which the performance of the proposed RGait-Net depends on the number of preceding clean/unoccluded frames. Figure 4(a) shows a sample synthetically occluded sequence and Fig. 4(b) shows the reconstruction results using the RGait-Net model qualitatively.

Comparison of the reconstructed frames in Fig. 4(b) with the ground truth shown in Fig. 4(c) reveals that the reconstruction quality of the RGait-Net model is quite good. To quantitatively verify this, dice scores are computed between the GEIs obtained before and after occlusion reconstruction through RGait-Net. The synthetically occluded test sequences of the CASIA-B data have been used for this purpose, and the average dice score obtained after comparing the generated

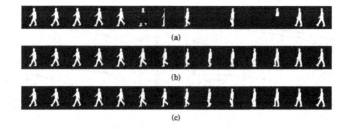

Fig. 4. (a) Sequence with synthetically occluded frames, (b) reconstructed frames, and (c) ground truth frames

frames with the ground-truth is *0.83*, which is significantly high, emphasizing the effectiveness of our proposed RGait-Net.

Table 1. Rank 1 accuracy with different levels of synthetic occlusion in CASIA-B data set with *10* initial unoccluded frames and *5* initial unoccluded frames respectively

Degree of occlusion	Rank 1 accuracy	
	10 initial unoccluded frames	5 initial unoccluded frames
≤10%	93.39%	92.38%
10%–20%	86.66%	85.71%
20%–30%	76.66%	73.33%
30%–40%	63.80%	62.85%
40%–50%	47.37%	40.75%

As discussed in Sect. 3.2, the occlusion reconstruction model RGait-Net needs a few good quality frames at the start of the sequence to predict the occluded frames in the later part of the sequence. The next experiment studies the influence of these initial clean/unoccluded frames on the gait recognition accuracy after reconstructing occluded gait sequences using the proposed RGait-Net. Results are shown in Table 1 by incorporating varying degrees of synthetic occlusion in the test sequences of the CASIA-B data (i.e., sequences in folders *nm-05* and *nm-06*). The first column in the table corresponds to the degree of occlusion and the next two columns show the gait recognition accuracy considering *10* and *5* preceding unoccluded frames, respectively. As a classification model, we employ a Random Forest with bagging algorithm. The optimal number of decision trees in the Random Forest is chosen to be 100, and this value is decided during the cross-validation phase in which we choose the best configuration (i.e., the number of decision tree estimators) based on the cross-validation accuracy on the gait recognition gallery set. It is observed from Table 1 that the Rank 1 accuracy corresponding to any degree of occlusion are closely comparable for both *10* and *5* initial unoccluded frames, which proves that the proposed RGait-Net model works satisfactorily even if a small number of initial unoccluded frames are present in a test gait sequence. Also, the recognition performance is quite good (i.e., more than 85%) if the degree of occlusion is less than 20%.

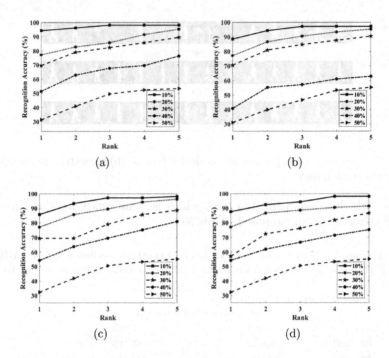

Fig. 5. Sequence in (a) folder nm-05 with 5 initial unoccluded frames, (b) folder nm-05 with 10 initial unoccluded frames, (c) folder nm-06 with 5 initial unoccluded frames, and (d) folder nm-06 with 10 initial unoccluded frames

Next, in Figs. 5(a)–(d), we plot the rank-wise improvement in the accuracy by varying the amount of synthetic occlusion in the sequences *nm-05* and *nm-06* of the CASIA data set. Specifically, Figs. 5(a) and (b) show the Cumulative Match Characteristic (CMC) curves corresponding to the synthetically occluded sequence *nm-05* by setting the number of preceding unoccluded frames to *5* and *10*, respectively. Figures 5(c) and (d) show the corresponding results for the synthetically occluded *nm-06* sequence. It can be seen from the figures that our

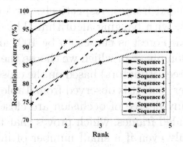

Fig. 6. Rank-wise improvement in the accuracy of our algorithm on the eight occluded sequences present in the TUM-IITKGP data set

method performs significantly accurately (greater than *85%*) if the percentage of frames occluded in a gait cycle is less than *20* and performs decently (with an accuracy of more than *60%* in all cases) if the percentage is less than *30*.

We also perform a similar experiment with the TUM-IITKGP data. The eight GEIs computed from the unoccluded normal walking sequences are used to train the Random Forest classifier with *100* decision trees. Each of the occluded sequences extracted from the data set, namely, *Sequence 1*, *Sequence 2*, ..., *Sequence 8* for each subject, are used during testing. First, occlusion detection and reconstruction are performed using the proposed deep learning models, and the GEIs computed from these reconstructed sequences are input to the trained Random Forest model to determine their appropriate class. Figure 6 presents a rank-wise improvement in the classification accuracy of the eight test occluded sequences using CMC curves as the value of the rank is increased from *1* to *5*. It is seen from the figure that the Rank 1 accuracy of recognition is greater than or equal to *75%* for each of the sequences and the accuracy is more than *80%* for all sequences above Rank 2. Further, the average Rank 1 accuracy obtained from the eight sequences of the TUM-IITKGP data is *87.50%*, while the average Rank 5 accuracy for the same sequences is *96.43%*. The significantly high accurate recognition results for the TUM-IITKGP data emphasizes that the RGait-Net trained using the CASIA-B data performs occlusion reconstruction satisfactorily for the TUM-IITKGP data as well.

4.3   Comparative Analysis

Finally, a comparative study of our work is made with some popular gait recognition techniques (with and without the occlusion handling scheme) and also with a few recent video frame prediction methods in terms of Rank 1 accuracy. The same training and occluded test sequences of the TUM-IITKGP data described for the previous experiment have also been used here. Among the occlusion handling techniques in gait recognition, we consider the methods given in [5,6,37], each of which performs recognition after occlusion reconstruction. Among the traditional methods that do not incorporate occlusion reconstruction, we consider the work in [1,25,36,38,47], while among the video frame prediction methods we consider those in [11,22,42]. Corresponding results are presented in Table 2. It is seen from the table that the proposed approach outperforms each of the other gait recognition approaches in terms of recognition accuracy by a significantly high margin. Although [6] and [5] work well in presence of occlusion(Rank 1 accuracy of *80.00%* and *78.92%*), the effectiveness of these approaches depends on the quality and the number of clean initial frames available so that a reasonable GEI can be constructed. Also, it can be observed that the video frame prediction approaches, namely [11,22,42], perform worse than the proposed approach. This is primarily because the binary silhouettes extracted from the TUM-IITKGP data set are noisy and the Vanilla Autoencoder-based reconstruction followed by these algorithms is not effective enough. The proposed spatio-temporal occlusion reconstruction using RGait-Net is significantly accurate and outperforms each of the other approaches used in the comparative study in terms of average

Table 2. Comparative analysis of the proposed work with existing approaches on the TUM-IITKGP data set in terms of accuracy

Category	Method	Average accuracy (%)
Proposed	**RGait-Net+ [25]**	**91.42**
Methods with occlusion handling mechanism	[5]	78.92
	[6]	80.00
	[37]	53.57
	[17]	77.65
Methods without occlusion handling mechanism	[25]	65.71
	[36]	70.23
	[47]	73.54
	[1]	76.42
	[38]	76.79
Frame prediction methods	[42]	47.66
	[22]	63.33
	[11]	76.66

accuracy. Moreover, our approach performs reconstruction and recognition accurately from only partial gait cycle information, and it does not require a large number of unoccluded frames to be present in an input sequence, which makes it highly suitable for application in real-life surveillance scenarios.

5   Conclusions and Future Work

We propose effective neural approaches to carry out occlusion detection and reconstruction in gait sequences. Our occlusion detection model based on VGG-16 performs with *98.9%* accuracy, and the occlusion reconstruction RGait-Net model also performs quite satisfactorily for low to moderate degrees of occlusion. For the synthetically occluded CASIA-B dataset, the accuracy of gait recognition after RGait-Net-based sequence reconstruction has been seen to be more than *92%* if the degree of occlusion is less than *10%*, and the accuracy is more than *73%* if the degree of occlusion is between *20%* and *30%*. Our complete gait recognition approach has also been seen to outperform several other traditional and occlusion handling approaches in gait recognition as well as some video frame prediction approaches in terms of accuracy. The proposed occlusion reconstruction model RGait-Net can be conveniently integrated with any other suitable gait recognition model. Although, RGait-Net is quite effective for online frame reconstruction, for offline purposes bi-directional LSTM models may help in making even better frame predictions, which can be studied in the future. Additionally, knowledge distillation-based network squeezing may be applied to carry out reconstruction on platforms with less computational power.

Acknowledgments. The authors would like to thank NVIDIA for supporting their work with a TITAN Xp GPU. The authors also acknowledge KLA Corporation for assisting them in attending the conference with an international travel grant.

References

1. Alotaibi, M., Mahmood, A.: Improved gait recognition based on specialized deep convolutional neural network. Comput. Vis. Image Underst. **164**, 103–110 (2017)
2. Aly, S.: Partially occluded pedestrian classification using histogram of oriented gradients and local weighted linear Kernel support vector machine. IET Comput. Vision **8**(6), 620–628 (2014)
3. Andriluka, M., Roth, S., Schiele, B.: People-tracking-by-detection and people-detection-by-tracking. In: Proceedings of the IEEE Conference on Computer Vision and Pattern Recognition, pp. 1–8. IEEE (2008)
4. Ariyanto, G., Nixon, M.S.: Model-based 3D gait biometrics. In: Proceedings of the IEEE International Joint Conference on Biometrics, pp. 11–13, October 2011
5. Babaee, M., Li, L., Rigoll, G.: Gait recognition from incomplete gait cycle. In: Proceedings of the 25th IEEE International Conference on Image Processing, pp. 768–772. IEEE (2018)
6. Babaee, M., Li, L., Rigoll, G.: Person identification from partial gait cycle using fully convolutional neural networks. Neurocomputing **338**, 116–125 (2019)
7. Battistone, F., Petrosino, A.: TGLSTM: a time based graph deep learning approach to gait recognition. Pattern Recogn. Lett. **126**, 132–138 (2019)
8. Ben, X., Gong, C., Zhang, P., Yan, R., Wu, Q., Meng, W.: Coupled bilinear discriminant projection for cross-view gait recognition. IEEE Trans. Circ. Syst. Video Technol. **30**(3), 734–747 (2019)
9. Bobick, A.F., Johnson, A.Y.: Gait recognition using static, activity-specific parameters. In: Proceedings of the IEEE Computer Society Conference on Computer Vision and Pattern Recognition, pp. 423–430, December 2001
10. Carass, A., et al.: Evaluating white matter lesion segmentations with refined SØRensen-Dice analysis. Sci. Rep. **10**(1), 1–19 (2020)
11. Chang, Z., et al.: MAU: a motion-aware unit for video prediction and beyond. In: Beygelzimer, A., Dauphin, Y., Liang, P., Vaughan, J.W. (eds.) Proceedings of the Advances in Neural Information Processing Systems (2021)
12. Chao, H., He, Y., Zhang, J., Feng, J.: GaitSet: regarding gait as a set for cross-view gait recognition. In: Proceedings of the AAAI Conference on Artificial Intelligence, vol. 33, pp. 8126–8133 (2019)
13. Chao, H., Wang, K., He, Y., Zhang, J., Feng, J.: GaitSet: cross-view gait recognition through utilizing gait as a deep set. IEEE Trans. Pattern Anal. Mach. Intell. **44**, 3467–3478 (2021)
14. Chattopadhyay, P., Roy, A., Sural, S., Mukhopadhyay, J.: Pose depth volume extraction from RGB-D streams for frontal gait recognition. J. Vis. Commun. Image Represent. **25**(1), 53–63 (2014)
15. Chattopadhyay, P., Sural, S., Mukherjee, J.: Exploiting pose information for gait recognition from depth streams. In: Agapito, L., Bronstein, M.M., Rother, C. (eds.) ECCV 2014. LNCS, vol. 8925, pp. 341–355. Springer, Cham (2015). https://doi.org/10.1007/978-3-319-16178-5_24
16. Chattopadhyay, P., Sural, S., Mukherjee, J.: Frontal gait recognition from occluded scenes. Pattern Recogn. Lett. **63**, 9–15 (2015)

17. Chen, C., Liang, J., Zhao, H., Hu, H., Tian, J.: Frame difference energy image for gait recognition with incomplete silhouettes. Pattern Recogn. Lett. **30**(11), 977–984 (2009)
18. Collins, R.T., Gross, R., Shi, J.: Silhouette-based human identification from body shape and gait. In: Proceedings of 5th IEEE International Conference on Automatic Face Gesture Recognition, pp. 366–371. IEEE (2002)
19. Cunado, D., Nixon, M.S., Carter, J.N.: Using gait as a biometric, via phase-weighted magnitude spectra. In: Proceedings of the 1st International Conference on Audio and Video-Based Biometric Person Authentication, pp. 93–102, March 1997
20. Cunado, D., Nixon, M.S., Carter, J.N.: Automatic extraction and description of human gait models for recognition purposes. Comput. Vis. Image Underst. **90**(1), 1–41 (2003)
21. Fan, C., et al.: GaitPart: temporal part-based model for gait recognition. In: Proceedings of the IEEE/CVF Conference on Computer Vision and Pattern Recognition, pp. 14225–14233 (2020)
22. Guen, V.L., Thome, N.: Disentangling physical dynamics from unknown factors for unsupervised video prediction. In: Proceedings of the IEEE/CVF Conference on Computer Vision and Pattern Recognition, pp. 11474–11484 (2020)
23. Gupta, S.K., Chattopadhyay, P.: Gait recognition in the presence of co-variate conditions. Neurocomputing **454**, 76–87 (2021)
24. Han, F., Li, X., Zhao, J., Shen, F.: A unified perspective of classification-based loss and distance-based loss for cross-view gait recognition. Pattern Recogn. **125**, 108519 (2022)
25. Han, J., Bhanu, B.: Individual recognition using gait energy image. IEEE Trans. Pattern Anal. Mach. Intell. **28**(2), 316–322 (2005)
26. He, Y., Zhang, J., Shan, H., Wang, L.: Multi-task GANs for view-specific feature learning in gait recognition. IEEE Trans. Inf. Forensics Secur. **14**(1), 102–113 (2018)
27. Hofmann, M., Sural, S., Rigoll, G.: Gait recognition in the presence of occlusion: a new dataset and baseline algorithm. In: Proceedings of the 19th International Conference on Computer Graphics, Visualization and Computer Vision (2011)
28. Hofmann, M., Wolf, D., Rigoll, G.: Identification and reconstruction of complete gait cycles for person identification in crowded scenes. In: Proceedings of the International Conference on Computer Vision Theory and Applications (2011)
29. Hu, H., Li, Y., Zhu, Z., Zhou, G.: CNNAuth: continuous authentication via two-stream convolutional neural networks. In: IEEE International Conference on networking, Architecture and Storage, pp. 1–9. IEEE (2018)
30. Isa, W.N.M., Alam, M.J., Eswaran, C.: Gait recognition using occluded data. In: Proceedings of the IEEE Asia Pacific Conference on Circuits and Systems, pp. 344–347. IEEE (2010)
31. Lee, L., Grimson, W.E.L.: Gait Analysis for Recognition and Classification. In: Proceedings of the 5th IEEE International Conference on Automatic Face and Gesture Recognition, pp. 155–162, May 2002
32. Lee, T.K., Belkhatir, M., Sanei, S.: Coping with full occlusion in fronto-normal gait by using missing data theory. In: Proceedings of the 7th International Conference on Information, Communications and Signal Processing, pp. 1–5. IEEE (2009)
33. de León, R.D., Sucar, L.E.: Continuous activity recognition with missing data. In: Proceedings of the International Conference on Pattern Recognition: Object Recognition Supported by User Interaction for Service Robots, vol. 1, pp. 439–442. IEEE (2002)

34. Li, Y., Hu, H., Zhu, Z., Zhou, G.: SCANet: sensor-based continuous authentication with two-stream convolutional neural networks. ACM Trans. Sens. Netw. (TOSN) **16**(3), 1–27 (2020)
35. Roy, A., Chattopadhyay, P., Sural, S., Mukherjee, J., Rigoll, G.: Modelling, synthesis and characterisation of occlusion in videos. IET Comput. Vision **9**(6), 821–830 (2015)
36. Roy, A., Sural, S., Mukherjee, J.: Gait recognition using pose kinematics and pose energy image. Signal Process. **92**(3), 780–792 (2012)
37. Roy, A., Sural, S., Mukherjee, J., Rigoll, G.: Occlusion detection and gait silhouette reconstruction from degraded scenes. SIViP **5**(4), 415 (2011)
38. Shiraga, K., Makihara, Y., Muramatsu, D., Echigo, T., Yagi, Y.: GEINet: view-invariant gait recognition using a convolutional neural network. In: Proceedings of the International Conference on Biometrics, pp. 1–8. IEEE (2016)
39. Sivapalan, S., Chen, D., Denman, S., Sridharan, S., Fookes, C.: Gait energy volumes and frontal gait recognition using depth images. In: Proceedings of the International Joint Conference on Biometrics, pp. 1–6. IEEE (2011)
40. Song, X., Huang, Y., Shan, C., Wang, J., Chen, Y.: Distilled light GaitSet: towards scalable gait recognition. Pattern Recogn. Lett. **157**, 27–34 (2022)
41. Takemura, N., Makihara, Y., Muramatsu, D., Echigo, T., Yagi, Y.: On input/output architectures for convolutional neural network-based cross-view gait recognition. IEEE Trans. Circuits Syst. Video Technol. **29**(9), 2708–2719 (2017)
42. Wang, Y., Jiang, L., Yang, M.H., Li, L.J., Long, M., Fei-Fei, L.: Eidetic 3DLSTM: a model for video prediction and beyond. In: Proceedings of the International Conference on Learning Representations (2019)
43. Weinland, D., Özuysal, M., Fua, P.: Making action recognition robust to occlusions and viewpoint changes. In: Daniilidis, K., Maragos, P., Paragios, N. (eds.) ECCV 2010. LNCS, vol. 6313, pp. 635–648. Springer, Heidelberg (2010). https://doi.org/10.1007/978-3-642-15558-1_46
44. Xu, D., Yan, S., Tao, D., Lin, S., Zhang, H.J.: Marginal fisher analysis and its variants for human gait recognition and content-based image retrieval. IEEE Trans. Image Process. **16**(11), 2811–2821 (2007)
45. Yu, S., Chen, H., Garcia Reyes, E.B., Poh, N.: GaitGAN: invariant gait feature extraction using generative adversarial networks. In: Proceedings of the Conference on Computer Vision and Pattern Recognition Workshops, pp. 30–37 (2017)
46. Yu, S., Tan, D., Tan, T.: A framework for evaluating the effect of view angle, clothing and carrying condition on gait recognition. In: Proceedings of the 18th International Conference on Pattern Recognition, vol. 4, pp. 441–444. IEEE (2006)
47. Zhang, E., Zhao, Y., Xiong, W.: Active energy image plus 2DLPP for gait recognition. Signal Process. **90**(7), 2295–2302 (2010)
48. Zhang, J., Sun, H., Guan, W., Wang, J., Xie, Y., Shang, B.: Robust human tracking algorithm applied for occlusion handling. In: Proceedings of the 5th International Conference on Frontier of Computer Science and Technology, pp. 546–551. IEEE (2010)
49. Zhang, P., Wu, Q., Xu, J.: VT-GAN: view transformation GAN for gait recognition across views. In: Proceedings of the International Joint Conference on Neural Networks, pp. 1–8 (2019)
50. Zheng, S., Zhang, J., Huang, K., He, R., Tan, T.: Robust view transformation model for gait recognition. In: Proceedings of the 18th IEEE International Conference on Image Processing, pp. 2073–2076. IEEE (2011)

Visible-Infrared Person Re-Identification Using Privileged Intermediate Information

Mahdi Alehdaghi[✉][iD], Arthur Josi[iD], Rafael M. O. Cruz[iD], and Eric Granger[iD]

Laboratoire d'imagerie, de vision et d'intelligence artificielle (LIVIA), Department of Systems Engineering, ETS Montreal, Montreal, Canada
{mahdi.alehdaghi.1,arthur.josi.1}@ens.etsmtl.ca,
{rafael.menelau-cruz,eric.granger}@etsmtl.ca

Abstract. Visible-infrared person re-identification (ReID) aims to recognize a same person of interest across a network of RGB and IR cameras. Some deep learning (DL) models have directly incorporated both modalities to discriminate persons in a joint representation space. However, this cross-modal ReID problem remains challenging due to the large domain shift in data distributions between RGB and IR modalities. This paper introduces a novel approach for a creating intermediate virtual domain that acts as bridges between the two main domains (i.e., RGB and IR modalities) during training. This intermediate domain is considered as privileged information (PI) that is unavailable at test time, and allows formulating this cross-modal matching task as a problem in learning under privileged information (LUPI). We devised a new method to generate images between visible and infrared domains that provide additional information to train a deep ReID model through an intermediate domain adaptation. In particular, by employing color-free and multi-step triplet loss objectives during training, our method provides common feature representation spaces that are robust to large visible-infrared domain shifts. Experimental results on challenging visible-infrared ReID datasets indicate that our proposed approach consistently improves matching accuracy, without any computational overhead at test time. The code is available at: https://github.com/alehdaghi/Cross-Modal-Re-ID-via-LUPI.

Keywords: Person re-identification · Cross-modal recognition · Multi-step domain adaptation · Learning under privileged information

1 Introduction

Person re-identification (ReID) has a growing interest in several real-world computer vision applications, such as video monitoring and surveillance, search and retrieval, and pedestrian tracking for autonomous driving. It involves matching

Supplementary Information The online version contains supplementary material available at https://doi.org/10.1007/978-3-031-25072-9_48.

images or videos to retrieve individuals captured across a distributed network of non-overlapping video cameras. The non-rigid structure of human bodies, variations in capture conditions in the wild, due to changes in illumination, motion blur, resolution, pose, view point, and occlusion and background clutter leads to high intra-person and low inter-person variability [18,27]. Despite these considerable challenges, recent progress in deep learning (DL) has allowed to develop state-of-art person ReID models that can achieve a high level of accuracy when trained on large fully-labeled image datasets. However, the domain shift typically associated with diverse operational capture conditions (e.g., camera viewpoints and lighting) may translate to a significant decline in performance. Most image-based ReID methods are formulated as a single-modality retrieval problem, where features extracted from input query images are matched against reference gallery images. For example, where all the images are captured by visible cameras, encouraging results have been observed [15]. Visible cameras, however, are somewhat limited in real-world applications because of insufficient discrimination under poor lighting conditions, e.g., in night time surveillance.

Most modern surveillance systems operate in dual modes, i.e., in the visible mode during the day and switch to the infrared mode at night. In this way, a supplementary task is also developed alongside visual ReID: Given a visible target image, the goal is to match it with the infrared image of the corresponding individual. This cross-modal image matching task is named Visible-Infrared person re-identification (VI-ReID). Alongside Intra-class variations (e.g., viewpoint, pose, illumination, and background clutter) and noisy samples (e.g., misalignment and occlusion), the cross-modal discrepancies between visible and infrared images make VI-ReID extremely challenging.

VI-ReID currently comprises two components: feature extraction and distance measurement. An imperative objective of feature extraction is to extract discriminating features to separate different individuals captured by several cameras. For example, in [16,24,28,39,40] two separated CNN are used as backbone for feature extraction for infrared and visible images. Based on the extracted features, distance measurement seeks to measure the similarity between two images and then optimize the feature extractor backbones, aiming to increase the similarity for pairs with the same individual while decreasing it otherwise. Considering the cross-modal task, the model needs to leverage and extract the shared information between two modalities. Such information should be modality invariant and also can be disentangled from modality-specific ones, for example, Yang et al. [43], Choi et al. [5], and Kniaz et al. [19] used GANs models to translate visible and infrared images. Since there are significant differences in the visual attributes and statistics of visible/infrared images [40], one special challenge lies in bridging the modality gap between the visible and infrared images. Multiple approaches [9,40,49] focus on spectrum mining between visual image colors and infrared and make gray images the bridge. These approaches consider the bridge as one static domain between infrared and visible. For example, the method proposed in [49] used gray images as the middle modality, and [46] used augmentation, selecting one channel randomly from the color spectrum of optic inputs. In fact, making proper intermediate images for tackling the large gap between infrared and visible is crucial for the model training process. But, using gray scale only or one channel from colored images is not perfectly able to link between colored and infrared images. For example, a person with same colored clothes, may have different infrared view in different environment. So, it motivates us to use more effective way for such intermediate domains.

In this paper, an effective way is proposed for bridging RGB and IR images by selecting different amounts of the visible spectrum. The learning strategy is based on the learning under privileged information (LUPI) paradigm [31], where image data from an intermediate domain is generated for training DL models for cross-modal person ReID. The intermediate domain allow the model to extract a common yet discriminant feature representation for visible and infrared modalities, and reduce the domain shift between the distribution of these modalities (as analysed Sect. 4.3). Besides LUPI, our approach is also inspired by methods in domain adaptation (DA), where a model (trained on source data) is adapted to a target domain by learning a common source-target representation space. Given a large domain shift, multi-task DA methods adapt models through one or more intermediate domains [36].

Our training model is comprised of 2 CNN backbones – two students for the main modalities, and a teacher sub-model for the intermediate modality. Instead of learning the infrared model from visible images (and vice versa), we generated intermediate images from visible inputs with less visual difference from infrared inputs, while preserving discriminating information from visible images. We also introduce a simple but effective spatial image augmentation mechanism intended to force the main branches of our model to focus on learning from the PI branch and consequently avoid focusing on modality-specific information. We show that common semantic information between visible and infrared images is, to a greater extent, dependent on textures. Therefore, intermediate images with the same texture pattern as visible images that do not contain color information may be a suitable intermediate step in linking to RGB and IR. During testing, our proposed model uses only one of the branches for the cross-modal ReID, so the generated intermediate domain does not impact the inference time. During training, a one-channel image is created from visual inputs with random color combinations. There are two reasons for this. First, they are not dependent on color information, making the feature representation of visual inputs less color discriminative. Second, the visual difference between infrared and grayscale images is smaller than between color and such color-free images.

The main contributions of this paper are summarized as follows. (1) The cross-modal VI-ReID problem is reformulated according to the LUPI paradigm. To address the large domain shifts between RGB and IR data distributions, we propose leveraging related PI as intermediate domains to train the CNN backbone. (2) An effective method is introduced to create intermediate virtual domains – the PI – from visible images based on a random linear combination of color channels. This allows bridging the RGB and IR modality domains during training. (3) Multi-step triplet and color-free losses are proposed to learn common feature representations for visible and infrared images, and also provide color-independent intermediate images as an effective bridge between modalities.

In addition, the cross-modal distance between modalities is minimized according to intermediate features. (4) An extensive set of experiments on the challenging SYSU-MM01 [39] and RegDB [28] datasets show that our proposed approach can significantly outperform state-of-art methods for cross-modal visible-infrared person ReID, yet incur no computational overhead at test time.

2 Related Work

Cross-Modality Person Re-ID: Visible-Infrared Person ReID (VI-ReID) focuses on matching visible daytime images against infrared nighttime images of a person for cross-modal identification problems [2,7,10,12,30,42,45,52]. A general view of pipelines to train DL models for cross-modal matching is categorized as [37]: (1) Representation: How to represent and summarize multimodal data to better exploit the modality-invariant and complementarity or even redundancy of multiple modalities. (2) Translation: How to map data from one modality to another, which could be heterogeneous, and the relationship between them is often open-ended or subjective. (3) Co-learning: How to transfer knowledge between modalities, their representation, and their predictive models.

A challenging problems encountered in cross-modal matching is the large discrepancies between visible and infrared modalities. Wu Ancong *et al.* [40], for example, attempts to reduce the discrepancies in the spectrum level between color channels and infrared values by using a grayscale version of colored images. Fan *et al.* [9] used an image created from a combination of infrared, visible and Gray channels information, as using gray only may result in the omission of specific information in the visible spectrum. [25,49] used the grayscale modality as a bridge between the infrared and visible ones in a triple CNN model training. [49] used gray images, produced from the visible images, and constrained the last convolution layer of the Gray and visible backbones to produce similar features. Using gray alongside visible and infrared improves the performance [25,49], but it is limited to a specific version of colored images. Part-based representation models learn to extract part/region aggregated features, making them robust against misalignment or occlusion. [52] introduced horizontal-divided region features at the end of a deep model to learn and represent local sub-feature. To capture the relations across multiple body parts, Intra-modality Weighted-Part Aggregation was presented in [47] to mine the contextual information in local parts and formulate an enhanced part-aggregated representation. Finally, some models have been proposed to extract modality-invariant information from each modality, and then training their CNN backbone in loss functions in the feature space. In order to learn shared features across modalities, a dual-constrained top-ranking loss, and a two-stream network have been developed in [44]. Adversarial approaches to create a common feature space are used in [7] and [12], where an embedding model tries to fool a modality discriminator model.

Generative Methods for Cross-Modality ReID: ThermalGAN [19] used an autoencoder to translate visible images to infrared in pixel-level similarity. AlignGAN [35] has two encoder-decoders to translate infrared to visible, and vice versa from shared encoded feature space, which is used for matching in the testing stage. [5,34,43] contain two specific encoders for infrared and visible, as well as a shared encoder to separate the modality-invariant information from the modality-specific one. By using GAN-based methods, synthetic images are generated and are especially conditioned by a modality-invariant representation which works at diminishing the pixel-level modality discrepancy. However, these methods are complex, and generated images do not possess enough quality to bridge the modality gap between synthetic and target data. To avoid using GAN models for generation of intermediate images, our approach instead relies on spectrum information, making our approach much faster.

Learning Under Privileged Information: Traditionally, supervised learning focuses on finding a decision function that minimizes the generalization error on a labeled training data. In the case of easy learning problems and with larger training datasets, the function found by the learner typically converges rapidly to the optimal value. On the other hand, if the learning problem is challenging and the learner's space of decision functions is large, the convergence rate (or learning rate) will be slow [31]. The idea of LUPI was first proposed by Vapnik & Vashist in [33] to increase the convergence rate of Support Vector Machines (SVMs). Hoffman *et al.* [14] was the first to introduce network hallucination, aiming to introduce PI knowledge through CNN by using an additional CNN stream. This additional stream learns to extract depth-related features (PI) from the RGB modality thanks to knowledge distillation.

The LUPI approach had been applied with success in various domains, and its benefits go beyond faster convergence [4,17,20,22,32]. Instead of hallucinating the modality through another network, Pande [29] learned a GAN to generate the PI-related features at inference time. Crasto et al. [6] directly distilled the PI knowledge (optical flow) to the current RGB stream, learning with both the cross-entropy and MSE losses, and avoiding the extra inference parameters. [3] recently extended the approach to Visual Question Answering (VQA) method, playing with multiple teachers. To the best of our knowledge, our work is the first application of the LUPI paradigm in the context of cross-modal person VI-ReID, where each modality is considered as a PI for the other. Moreover, we introduce knowledge from the PI through new losses and layers with shared parameters. It has the advantage of avoiding the extra parameters related to LUPI. Our innovative method is suitable for a wide range of cross-modal ReID frameworks.

3   Proposed Method

In cross-modality matching, the main learning objective is to allow for extraction of image features that are invariant across modalities. Let $V = \{v_i\}_1^n$ be the visible and $T = \{t_i\}_{n+1}^{n+m}$ be the infrared or thermal images of several persons. Also, $Y = \{y_i\}_1^{n+m}$ contains the identity of each person in dataset. The multi-modal dataset is represented as $D = \{D_{tr}, D_{te}\}$, where D_{tr} denotes the training data and D_{te} denotes the testing data, in which $D_{tr} = \{V_{tr}, T_{tr}\}$ and $D_{te} = \{V_{te}, T_{te}\}$. The learning objective is to optimize the relation between the extracted features $\mathbf{f}_i^v = \varphi(v_i; \Theta)$ of visible images, space and the features $\mathbf{f}_j^t = \varphi(t_j; \Theta)$ of near-infrared images, denoted by:

$$\mathcal{L} = \sum l(\mathbf{f}_i^v, \mathbf{f}_j^t, y_i, y_j) \tag{1}$$

where y_i and y_j are the annotated training labels for each image and $\mathbf{f}_i^v, \mathbf{f}_j^t \in \mathbb{R}^d$, d being the features dimension.

3.1   LUPI Framework

With LUPI [21], the DL model has access to PI, $x^\star$, only during the training while main information x is available during both training and inference phases. When using the cross-modal representative model, two modalities should be used to represent inputs during the training phase so that the models can learn to use the cross-modal information, whereas only individual inputs will be fed to the model during the inference phase. That is, the missing modality at inference can be seen as a PI.

Two methods were provided to incorporate Privilege Information (PI) through the training process. The first approach uses the infrared modality as PI to train the visible models, as shown in Fig. 1.a and the first part of Eq. 2, or vice versa, as shown in Fig. 1.b and the second part of Eq. 2.

$$\begin{aligned} \min_{\Theta_1} \mathbf{E}_{v,t^\star,y\sim p(v,t^\star,y)}[\mathcal{L}(\varphi_1^+(v, t^\star; \Theta_1), y)], \\ \min_{\Theta_2} \mathbf{E}_{t,v^\star,y\sim p(t,v^\star,y)}[\mathcal{L}(\varphi_2^+(t, v^\star; \Theta_2), y)]. \end{aligned} \tag{2}$$

where $\mathcal{L}$ is the loss function. For the second approach, since the domain discrepancies between infrared and visible are numerous [49], we employ a mutual (intermediate) modality while learning a joint model for visible-infrared sensor data (Fig. 1.c and the Eq. 3). The intermediate modality allows bridging the cross-modality gap by making each main modality benefit from the other more easily, since mutual domain has more information in common with both main domains than each main domains.

$$\min_{\Theta} \mathbf{E}_{v,t,z^\star,y\sim p(v,t,z^\star,y)}[\mathcal{L}(\varphi^+(v, t, z^\star; \Theta), y)]. \tag{3}$$

where $z^\star$ is intermediate domain, Θ is parameter set of the model.

For these approaches, the teacher's knowledge (related to PI knowledge) is transferred to the student(s) thanks to shared parameter layers after the first convolutional blocks. For instance, the intermediate modality backbone shares all its parameters with the Visible and Infrared modality streams except for the first convolutional block. In addition, specific losses are used to help the knowledge transfer from our specific intermediate modality.

3.2   Intermediate Domain Generation

Employing an intermediate domain from which the data from the main domains (RGB and IR) have a lower domain shift, can help the model learn DL embeddings with shared discriminative features. The most appropriate choice for an

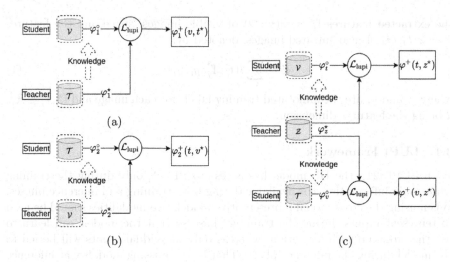

Fig. 1. Summary of methodology to train a model under privileged information: (a) learning under privileged infrared domain, (b) learning under privileged visual domain, and (c) learning under privileged intermediate domain, between infrared and visual.

intermediate domain in the context of infrared-visible person ReID appears to be grayscale images, since they can be derived directly from the visible images, but also because it intuitively considers the nature of each modality. Infrared cameras have one sensor which measures the wavelength of 1 mm to the nominal red edge of the visible spectrum, while visible images have three channels showing the wavelength of the red, green, and blue spectrum on persons. So, in a classic cross-modal person-ReID model, the DL model must to mine shared abstraction between these pieces of information to re-identify the same person from three visible channels to one infrared channel, or vice-versa. As a motivating example, Fig. 2.a and 2.b show Visible and Infrared images of the same person, where a large shift can be visually observed. By generating one channel images from a random selection through the visible channels (see Fig. 2.c), the domain shift appears to be reduced significantly.

The semantic features extracted by the model from the synthetic images, which contain one modality, such as infrared, but have pose information from the visible images, may be one solution for modality-independent features between visible and infrared since they rely less on color. Based on such information, synthetic images in the training phase can be seen as the PI, z^* in Eq. 3. We trained the model shown in Fig. 1.c with the one-channel images used as teacher, and showed that the random selection of images from the color channel (see Eq. 4) can be an relevant intermediate domain between infrared and visible:

$$G_s = \frac{\alpha R + \beta G + \gamma B}{\alpha + \beta + \gamma} \tag{4}$$

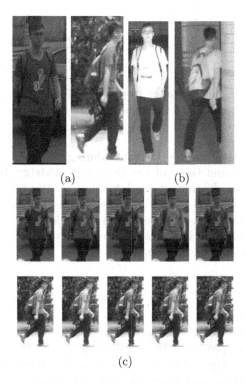

Fig. 2. Illustration domain shift between visible and infrared images: (a) visual images, (b) infrared images, and (c) a selected random linear combination of RGB channels.

where $\alpha, \beta,$ and $\gamma \sim \mathbb{U}(0, 1)$, a uniform random distribution generator, and R, G, and B are one channel color images. Instead of being limited to a specific version of colored images as in [25, 49], our model uses color-free images, without any color information from visible modality, to bridge the gap between visible and infrared image modalities.

3.3 Loss Functions

(1) Intermediate Dual Triplet Loss. The ultimate goal of deep cross-modal person ReID is to encode an embedding where the same person captured with different camera modalities are more similar in the feature space. In such space, the feature vectors of different individuals should also be as far as possible from each other. Triplet loss is one objective constraint usually used for the model to achieve this goal. This loss compares a reference input (called the anchor) with a matching input person (called positive) and a different input person (called negative). With the intermediate domain, we apply such loss in dual mode for comparing each main domain with the intermediate:

$$\mathcal{L}_{\text{tri}} = \mathbb{E}_{(\mathbf{f}_a,\mathbf{f}_p,\mathbf{f}_n)\in(\mathcal{F}^v,\mathcal{F}^t,\mathcal{F}^z)}[M - D_{\mathbf{f}_a,\mathbf{f}_p} + D_{\mathbf{f}_a,\mathbf{f}_n}]$$
$$+\mathbb{E}_{(\mathbf{f}_a,\mathbf{f}_p,\mathbf{f}_n)\in(\mathcal{F}^t,\mathcal{F}^z,\mathcal{F}^v)}[M - D_{\mathbf{f}_a,\mathbf{f}_p} + D_{\mathbf{f}_a,\mathbf{f}_n}] \qquad (5)$$
$$+\mathbb{E}_{(\mathbf{f}_a,\mathbf{f}_p,\mathbf{f}_n)\in(\mathcal{F}^z,\mathcal{F}^v,\mathcal{F}^t)}[M - D_{\mathbf{f}_a,\mathbf{f}_p} + D_{\mathbf{f}_a,\mathbf{f}_n}]$$

where subscript a, p, and n stands respectively for anchor, positive and negative indicator. The superscript t, v, and z select visible, infrared and color-free modalities features.

(2) Color-Free Loss. Feature invariance during extraction of the original visible images features and those of the generated modality, has been shown to be beneficial [49]. In this regard, we introduce a color-invariant loss function between visible images and virtual color-free images to enhance the robustness of feature representations against variations in modality. This color-free modality teaches the visible branch not rely on the colors spectrum. The color-free loss to train the CNN backbone is defined by:

$$\mathcal{L}_c = \mathbb{E}_{(\mathbf{f}_i^v,\mathbf{f}_i^z)\in(\mathcal{F}^v,\mathcal{F}^z)}[\begin{cases} |\mathbf{f}_i^v - \mathbf{f}_i^z|, & \text{if } |\mathbf{f}_i^v - \mathbf{f}_i^z| > 0.5 \\ \alpha_c KL(\mathcal{C}(\mathbf{f}_i^v),\mathcal{C}(\mathbf{f}_i^z)), & \text{otherwise.} \end{cases}] \qquad (6)$$

where $\mathbf{f}_i^z$ is the feature vector extracted from generated images z_i, $\mathcal{C}$ is the fully-connected layer that produces the classification probability for each identify of image i, and $KL(\cdot)$ is the Kullback-Leibler divergence between two probabilities. Since the augmentation mechanism uses different distortions for RGB and intermediate images, the model may not extract similar features from these types of images, so we use the distance of identification probability between them.

(3) Overall Loss for LUPI. The loss proposed to leverage a privileged intermediate domain $\mathcal{L}_{\text{lupi}}$ is defined by combining $\mathcal{L}_c$ and $\mathcal{L}_{\text{tri}}$:

$$\mathcal{L}_{\text{lupi}} = \mathcal{L}_{\text{tri}} + \lambda\mathcal{L}_c, \qquad (7)$$

where λ is a predefined trade off parameter to balance the color-free loss. Also, to extract features that are discriminant according to person identity, the total loss $\mathcal{L}$ in Eq. 3 includes an identity loss:

$$\mathcal{L} = \mathcal{L}_{\text{lupi}} + \mathcal{L}_{\text{id}}. \qquad (8)$$

By optimizing DL model parameters with identity supervision, using an $\mathcal{L}_{\text{id}}$ like cross-entropy [48] allows for feature representations that are identity-variant.

4   Results and Discussion

4.1   Experimental Methodology

Datasets: The two datasets RegDB [28], and SYSU-MM01 [39] have been widely used in cross-modal research. SYSU-MM01 is a large-scale dataset containing 22,258 visible images and 11,909 infrared images (captured from four

RBG and two near-infrared cameras) of 491 individuals. Following its authors, 395 identities were used as the training set, and 96 identities were used as the testing set for fairness comparisons with other papers. It contains two evaluation modes in SYSU-MM01 based on images in the gallery: single-shot and multi-shot. The former setting allows us to select a random image from each person's identity within the gallery, while the latter setting allows us to select ten images from each identity. The RegDB dataset consists of 4,120 visible and 4,120 infrared images representing 412 identities, which are collected from a single visible and an infrared camera co-located. Each identity contains ten visible images and ten infrared images from consecutive clips. Additionally, ten trail configurations randomly divide the images into two identical sets (206 identifiers) for training and testing. Two different test settings are used – matching infrared (query) with visible (gallery), and vice versa.

Performance Measures: We use the Cumulative Matching Characteristics (CMC), and Mean Average Precision (mAP) as evaluation metrics.

Implementation Details: Our proposed approach is implemented using the PyTorch framework. For fair comparison with the other methods, the ResNet50 [13] CNN was used as backbone for feature extraction, and pre-trained on ImageNet [8]. Our training process and model was implemented on the code released by [48], following the same hyper-parameter configuration. For Sect. 3.2, we generate three uniform random numbers, and then normalize them for each channel of visible images to create color-free images. At first, all the input images are resized to 288×144, then cropped randomly, and filled with zero-padding or with the mean of pixels as a data augmentation approach. A stochastic gradient descent (SGD) optimizer is used in the optimization process. The used margin used for the triplet loss is the same as [48], and the warming-up strategy is applied in the first ten epochs. In order to perform the triplet loss in each mini-batch during training, the group must contain at least two distinct individuals for whom at least two images (one for anchor and one for positive) have to be included. So, in each training batch, we select $b_s = 8$ persons from the dataset and $n_p = 4$ positive images for each person, still following [48] parameters. The $\alpha_c = 0.5$ and $\lambda = 10$ is set.

4.2   Comparison with the State-of-Art

Tables 1 and 2 compare the results with our proposed approach against state-of-the-art cross-modal VI-ReID methods published in the past two years, on the SYSU-MM01 and RegDB datasets. The results show that our method outperforms these methods in various settings. Our method does not require additional and costly image generation, conversion from visible to infrared, or adversarial learning process. It is very to generate random channel images using matrix dot production, which is a simple mathematics operation with low computational overhead. Also, our model learns a robust feature representation against different cross-modality matching settings. Notably, on the large-scale SYSU-MM01

Table 1. Performance of state-of-art techniques for cross-modal person ReID on the SYSU-MM01 dataset.

Family	Method	All search			Indoor search		
		R1(%)	R10(%)	mAP(%)	R1(%)	R10(%)	mAP(%)
Representation	DZP [40]	14.8	54.1	16.0	20.58	68.38	26.92
	CDP [9]	38.0	82.3	38.4	–	–	–
	SFANET [25]	65.74	92.98	60.83	71.60	96.60	80.05
	HAT [49]	55.29	92.14	53.89	–	–	–
	xModal [23]	49.92	89.79	50.73	62.1	95.75	69.37
	CAJ [46]	69.88	–	66.89	76.26	97.88	80.37
	EDFL [24]	36.94	84.52	40.77	–	–	–
	BDTR [44]	27.82	67.34	28.42	32.46	77.42	42.46
	AGW [48]	47.50	–	47.65	54.17	91.14	62.97
	cmGAN [7]	26.97	67.51	31.49	31.63	77.23	42.19
	MANN [12]	30.04	74.34	32.18	–	–	–
	TSLFN [52]	62.09	93.74	48.02	59.74	92.07	64.91
	DDAG [47]	54.75	90.39	53.02	61.02	94.06	67.98
	JFLN [1]	66.11	95.69	64.93	–	–	–
	MPANet [41]	70.58	96.21	**68.24**	76.74	98.21	80.95
Translation	AGAN [35]	42.4	85.0	40.7	45.9	92.7	45.3
	D^2RL [38]	28.9	70.6	29.2	–	–	–
	JSIA [34]	45.01	85.7	29.5	43.8	86.2	52.9
	Hi-CMD [5]	34.94	77.58	35.94	–	–	–
	TS-GAN [51]	58.3	87.8	55.1	62.1	90.8	71.3
Co-learning	HCML [45]	14.32	53.16	16.16	–	–	–
	DHML [50]	60.08	90.89	47.65	56.30	91.46	63.70
	SSFT [26]	63.4	91.2	62.0	70.50	94.90	72.60
	SIM [15]	56.93	–	60.88	–	–	–
	LbA [30]	55.41		54.14	61.02	–	66.33
	CM-NAS [10]	61.99	92.87	60.02	67.01	97.02	72.95
Ours		**71.08**	**96.42**	67.56	**82.35**	**98.3**	**82.73**

dataset, our model improves the Rank-1 accuracy and the mAP score by 23.58% and 19.11%, respectively, over the AGW model [48] (our baseline). Recall that our approach can be used on top of any cross-modal approach, so the performances obtained working over the baseline are promising. On SYSU-MM01, our model shows an improvement of 5.61% in terms of Rank-1 and of 1.78% mAP in comparison to the second best approach considering Indoor Search scenario. For RegDB dataset, from Visible to Thermal ReID, our model improves the best R1 and mAP by respectively 2.92% and 1.80 %. Similar improvement can be observed from the RegDB Thermal to Visible setting. Another advantage of our approach is that it can be combined with other state-of-art models without any overhead during inference, since it only triggers training process to force visible features such that they are less discriminative on colors.

Table 2. Performance of state-of-art techniques for cross-modal person ReID on the RegDB dataset.

Family	Method	Visible → Thermal			Thermal → Visible		
		R1(%)	R10(%)	mAP(%)	R1(%)	R10(%)	mAP(%)
Representation	DZP [40]	17.75	34.21	18.90	16.63	34.68	17.82
	CDP [9]	65.3	84.5	62.1	64.40	84.50	61.50
	SFANET [25]	76.31	91.02	68.00	70.15	85.24	63.77
	HAT [49]	71.83	87.16	67.56	70.02	86.45	66.30
	xModal [23]	62.21	83.13	60.18	–	–	–
	CAJ [46]	85.03	95.49	79.14	84.75	95.33	77.82
	EDFL [24]	48.43	70.32	48.67	51.89	72.09	52.13
	BDTR [44]	33.47	58.96	31.83	34.21	58.74	32.49
	AGW [48]	70.05	–	50.19	70.49	87.21	65.90
	MANN [12]	48.67	71.55	41.11	38.68	60.82	32.61
	DDAG [47]	69.34	86.19	63.46	68.86	85.15	61.80
	MPANet [41]	83.70	–	80.9	82.8	–	80.7
Translation	AGAN [35]	57.9	–	53.6	56.3	–	53.4
	D^{2}RL [38]	43.4	66.1	44.1	–	–	–
	JSIA [34]	48.5	–	49.3	48.1	–	48.9
	Hi-CMD [5]	70.93	86.39	66.04	–	–	–
	TS-GAN [51]	68.2	–	69.4	–	–	–
Co-learning	HCML [45]	24.44	47.53	20.80	21.70	45.02	22.24
	SSFT [26]	71.0	–	71.7	–	–	–
	SIM [15]	74.47	–	75.29	75.24	–	78.30
	LbA [30]	74.17	–	67.64	72.43	–	65.64
	CM-NAS [10]	84.54	95.18	80.32	82.57	94.51	78.31
Ours		**87.95**	**98.3**	**82.73**	**86.80**	**96.02**	**81.26**

4.3 Ablation Study

In this section, impact on performance of each component in our model is analyzed. At first, it begins with domain shift between visible and infrared images, then comparing the loss functions on training process. Also the qualitative results are discussed in the supplementary.

Domain Shift between Visual, Infrared and Gray. In a way to determine the domain shift between domains, we explored two distinct strategies, the first Maximum Mean Discrepancy (MMD) [11] which is done in supplementary and then measuring and comparing mAP and Rank-1 directly. We compared such metrics to measure the domain shift on model performance, by playing with different modalities as query and gallery in the inference phase while training only from each modality (three first columns in Table 3). To do so, we used the AGW model [48] with its default settings since its architecture corresponds to our final model, except we used only a single branch here by which we fed the query and gallery images. Indeed, AGW used two Resnet50 with shared parameters between the visible and infrared branches for the cross-modal ReID

Table 3. Results of uni-modal and multi-modal training on Visual (V), Infrared (I), and Grayscale (G) modality images of the SYSU-MM01 dataset under the multi-shot setting. G images are grayscale of V images. During testing, query and gallery images are processed by the model separately. Note that when the query and gallery pairs are in the same modality, the same cameras are excluded to avoid the existence of images from in the same scene.

Testing	Training modalities									
	V		I		G		I-V		I-V-G	
Query → Gallery	R1	mAP	R1	mAP	R1	mAP	R1	mAP	R1	mAP
V → V	**97.95**	91.67	60.61	20.24	95.69	78.60	97.40	**91.82**	97.22	89.22
I → I	49.76	21.97	94.90	80.38	56.93	27.05	95.96	80.49	**96.92**	**82.78**
G → G	88.36	54.26	49.30	16.30	**96.10**	80.06	90.16	60.35	95.94	**84.16**
I → V	9.86	6.83	15.72	5.92	14.20	9.01	58.69	41.57	**65.16**	**46.94**
I → G	10.89	6.50	13.09	5.71	16.15	9.22	50.85	27.79	**64.98**	**45.01**

purpose, which should not be used this way here since, in the inference phase, we need to feed the cross-modal images from one branch.

In practice, we observe that the model trained with only gray images performs better on other modalities. For instance, the mAP of the **G**ray model is 5% better than the **V**isible only model when working from infrared images (Respectively 27.05% and 21.97%). Similarly, the **G**ray model is ahead of the **I**nfrared one by a large margin when performing from visible images, respectively 78.60% and 20.24% mAP. Those results confirm the intermediate position of the gray domain in between the two main domains. Moreover, this model has better performance to retrieve the best candidate (R1 Accuracy) when the Query comes from **I**nfrared and Gallery from **V**isual in comparison to model trained only with **V**isuals (11.2 to 9.86).

Effect of Random Gray Image Generation. As the visual differences between gray and infrared images are much lower than between visible and infrared images in our eyes, we take advantage of such phenomena by learning our model with the gray version of colored images alongside infrared and visible images. In fact, instead of directly optimizing the distances between infrared and visible, the model utilizes gray images as a bridge, and tries to represent infrared and visible features so that they are similar to the gray ones. The result are shown Table 3. The main result in the table concerns the ability of the models to well re-identify from **I**nfrared to **V**isible. Here, the gray model version (**I-V-G** training) is successfully ahead of the model trained with infrared and visible images only (**I-V**), reaching 46.94% against 41.57% mAP respectively. Also, the gray model is much better to translate from and to the gray domain, which was expected but is of good omen regarding the intermediate characterisation of this domain. Finally, we can also observe pretty equivalent performances under **V** to **V** and **I** to **I** testing settings, making the gray version the way to go for a robust person ReID. Additionally, in comparison with **V** training and **I** training

Table 4. Evaluation of each component on the SYSU-MM01 dataset under single- and multi-shot settings.

Model	R1(%)		mAP(%)		Memory
	Single-Shot	Multi-Shot	Single-Shot	Multi-Shot	
Pre-trained	2.26	3.31	3.61	1.71	23.5M
Visible-trained	7.41	9.86	8.83	6.83	23.5M
Baseline	49.41	58.69	41.55	38.17	23.5M
Grayscale	61.45	65.16	59.46	46.94	23.7M
RandG	64.50	69.01	61.05	50.84	23.7M
RandG + Aug	**71.08**	**77.23**	**66.76**	**61.53**	23.7M
Infrared (train-test)	90.34	94.90	87.45	80.38	23.5M

Table 5. Impact on performance of loss functions for the SYSU-MM01 dataset.

Strategy	$\mathcal{L}_c$	$\mathcal{L}_{tri}$	R1 (%)	R10(%)	mAP (%)
Baseline	×	×	42.89	83.51	43.51
Triplet Loss	×	✓	55.56	90.40	52.96
Color Loss	✓	×	53.97	89.32	51.68
Ours	✓	✓	64.50	95.19	61.05

(Table 3), which are respectively lower and upper-bound, the performance of our approach is close to the upper-bound, showing that the model can benefit from the privileged missed modality in LUPI paradigm (Table 4).

Effect of Intermediate Dual Triplet Loss. Combined with the dual triplet loss, it has been demonstrated that the bridging cross-modality triplet loss provides strong supervision to improve discrimination in the testing phase. Using such regularization, as shown in Table 5, improves the mAP by 9 and rank-1 by 12%.

Effect of Color-Free Loss. A comparison of the results of our model with and without that loss function is conducted in order to verify the effectiveness of the proposed loss function in focusing on color-independent discriminative parts of visible images. Table 5 shows that when the model uses such regularizing loss, it improves the mAP by 8 and rank-1 by 11 percentile points.

5   Conclusions

In this paper, we formulated the cross-modal VI-ReID as a problem in learning under privileged information problem, by considering PI as an intermediate domain between infrared and visible modalities. We devised a new method to generate images between visible and infrared domains that provides additional information to train a deep ReID model through intermediate domain adaptation. Furthermore, we show that using randomly generated grayscale images as PI can significantly improve cross-modality recognition. Future directions involve using adaptive image generation based on images from the main modalities, and an end-to-end learning process to find the one channel images as infrared version of visible. Our experiments on the challenging SYSU-MM01 [39] and RegDB [28] datasets show that our proposed approach can significantly outperform state-of-art methods specialized for cross-modal VI-ReID. This does not require extra parameters, nor additional computation during inference.

Acknowledgements. This work was supported by Nuvoola AI Inc., and the Natural Sciences and Engineering Research Council of Canada.

References

1. Chen, K., Pan, Z., Wang, J., Jiao, S., Zeng, Z., Miao, Z.: Joint feature learning network for visible-infrared person re-identification. In: Peng, Y., et al. (eds.) PRCV 2020. LNCS, vol. 12306, pp. 652–663. Springer, Cham (2020). https://doi.org/10.1007/978-3-030-60639-8_54
2. Chen, Y., Wan, L., Li, Z., Jing, Q., Sun, Z.: Neural feature search for RGB-infrared person re-identification. In: Proceedings of the IEEE/CVF Conference on Computer Vision and Pattern Recognition, pp. 587–597 (2021)
3. Cho, J.W., Kim, D.J., Choi, J., Jung, Y., Kweon, I.S.: Dealing with missing modalities in the visual question answer-difference prediction task through knowledge distillation. In: Proceedings of the IEEE/CVF Conference on Computer Vision and Pattern Recognition, pp. 1592–1601 (2021)
4. Choi, C., Kim, S., Ramani, K.: Learning hand articulations by hallucinating heat distribution. In: Proceedings of the IEEE International Conference on Computer Vision, pp. 3104–3113 (2017)
5. Choi, S., Lee, S., Kim, Y., Kim, T., Kim, C.: HI-CMD: hierarchical cross-modality disentanglement for visible-infrared person re-identification. In: Proceedings of the IEEE/CVF Conference on Computer Vision and Pattern Recognition, pp. 10257–10266 (2020)
6. Crasto, N., Weinzaepfel, P., Alahari, K., Schmid, C.: Mars: motion-augmented RGB stream for action recognition. In: Proceedings of the IEEE/CVF Conference on Computer Vision and Pattern Recognition, pp. 7882–7891 (2019)
7. Dai, P., Ji, R., Wang, H., Wu, Q., Huang, Y.: Cross-modality person re-identification with generative adversarial training. In: IJCAI, vol. 1, p. 2 (2018)
8. Deng, J., Dong, W., Socher, R., Li, L.J., Li, K., Fei-Fei, L.: ImageNet: a large-scale hierarchical image database. In: 2009 IEEE Conference on Computer Vision and Pattern Recognition, pp. 248–255. IEEE (2009)
9. Fan, X., Luo, H., Zhang, C., Jiang, W.: Cross-spectrum dual-subspace pairing for RGB-infrared cross-modality person re-identification. ArXiv abs/2003.00213 (2020)
10. Fu, C., Hu, Y., Wu, X., Shi, H., Mei, T., He, R.: CM-NAS: cross-modality neural architecture search for visible-infrared person re-identification. In: Proceedings of the IEEE/CVF International Conference on Computer Vision, pp. 11823–11832 (2021)

11. Gretton, A., Borgwardt, K.M., Rasch, M.J., Schölkopf, B., Smola, A.: A kernel two-sample test. J. Mach. Learn. Res. **13**(1), 723–773 (2012)
12. Hao, Y., Li, J., Wang, N., Gao, X.: Modality adversarial neural network for visible-thermal person re-identification. Pattern Recogn. **107**, 107533 (2020)
13. He, K., Zhang, X., Ren, S., Sun, J.: Deep residual learning for image recognition. In: Proceedings of the IEEE Conference on Computer Vision and Pattern Recognition, pp. 770–778 (2016)
14. Hoffman, J., Gupta, S., Darrell, T.: Learning with side information through modality hallucination. In: Proceedings of the IEEE Conference on Computer Vision and Pattern Recognition, pp. 826–834 (2016)
15. Jia, M., Zhai, Y., Lu, S., Ma, S., Zhang, J.: A similarity inference metric for RGB-infrared cross-modality person re-identification. arXiv preprint arXiv:2007.01504 (2020)
16. Jiang, J., Jin, K., Qi, M., Wang, Q., Wu, J., Chen, C.: A cross-modal multi-granularity attention network for RGB-IR person re-identification. Neurocomputing **406**, 59–67 (2020)
17. Kampffmeyer, M., Salberg, A.B., Jenssen, R.: Urban land cover classification with missing data modalities using deep convolutional neural networks. IEEE J. Sel. Topics Appl. Earth Observ. Remote Sens. **11**(6), 1758–1768 (2018)
18. Kiran, M., Praveen, R.G., Nguyen-Meidine, L.T., Belharbi, S., Blais-Morin, L.A., Granger, E.: Holistic guidance for occluded person re-identification. In: British Machine Vision Conference (BMVC) (2021)
19. Kniaz, V.V., Knyaz, V.A., Hladůvka, J., Kropatsch, W.G., Mizginov, V.: Thermal-GAN: multimodal color-to-thermal image translation for person re-identification in multispectral dataset. In: Leal-Taixé, L., Roth, S. (eds.) ECCV 2018. LNCS, vol. 11134, pp. 606–624. Springer, Cham (2019). https://doi.org/10.1007/978-3-030-11024-6_46
20. Kumar, S., Banerjee, B., Chaudhuri, S.: Improved landcover classification using online spectral data hallucination. Neurocomputing **439**, 316–326 (2021)
21. Lambert, J., Sener, O., Savarese, S.: Deep learning under privileged information using heteroscedastic dropout. In: Proceedings of the IEEE Conference on Computer Vision and Pattern Recognition, pp. 8886–8895 (2018)
22. Lezama, J., Qiu, Q., Sapiro, G.: Not afraid of the dark: NIR-VIS face recognition via cross-spectral hallucination and low-rank embedding. In: Proceedings of the IEEE Conference on Computer Vision and Pattern Recognition, pp. 6628–6637 (2017)
23. Li, D., Wei, X., Hong, X., Gong, Y.: Infrared-visible cross-modal person re-identification with an X modality. In: Proceedings of the AAAI Conference on Artificial Intelligence, vol. 34, pp. 4610–4617 (2020)
24. Liu, H., Cheng, J.: Enhancing the discriminative feature learning for visible-thermal cross-modality person re-identification. CoRR abs/1907.09659 (2019). https://arxiv.org/abs/1907.09659
25. Liu, H., Ma, S., Xia, D., Li, S.: SFANet: a spectrum-aware feature augmentation network for visible-infrared person re-identification. arXiv preprint arXiv:2102.12137 (2021)
26. Lu, Y., et al.: Cross-modality person re-identification with shared-specific feature transfer. In: Proceedings of the IEEE/CVF Conference on Computer Vision and Pattern Recognition, pp. 13379–13389 (2020)
27. Mekhazni, D., Bhuiyan, A., Ekladious, G., Granger, E.: Unsupervised domain adaptation in the dissimilarity space for person re-identification. In: Vedaldi, A.,

Bischof, H., Brox, T., Frahm, J.-M. (eds.) ECCV 2020. LNCS, vol. 12372, pp. 159–174. Springer, Cham (2020). https://doi.org/10.1007/978-3-030-58583-9_10

28. Nguyen, D.T., Hong, H.G., Kim, K.W., Park, K.R.: Person recognition system based on a combination of body images from visible light and thermal cameras. Sensors **17**(3), 605 (2017)

29. Pande, S., Banerjee, A., Kumar, S., Banerjee, B., Chaudhuri, S.: An adversarial approach to discriminative modality distillation for remote sensing image classification. In: Proceedings of the IEEE/CVF International Conference on Computer Vision Workshops (2019)

30. Park, H., Lee, S., Lee, J., Ham, B.: Learning by aligning: visible-infrared person re-identification using cross-modal correspondences. In: Proceedings of the IEEE/CVF International Conference on Computer Vision, pp. 12046–12055 (2021)

31. Pechyony, D., Vapnik, V.: On the theory of learning with privileged information. In: Advances in Neural Information Processing Systems, vol. 23 (2010)

32. Saputra, M.R.U., et al.: DeepTIO: a deep thermal-inertial odometry with visual hallucination. IEEE Robot. Autom. Lett. **5**(2), 1672–1679 (2020)

33. Vapnik, V., Vashist, A.: A new learning paradigm: learning using privileged information. Neural Netw. **22**(5–6), 544–557 (2009)

34. Wang, G.A., et al.: Cross-modality paired-images generation for RGB-infrared person re-identification. In: Proceedings of the AAAI Conference on Artificial Intelligence, vol. 34, pp. 12144–12151 (2020)

35. Wang, G., Zhang, T., Cheng, J., Liu, S., Yang, Y., Hou, Z.: RGB-infrared cross-modality person re-identification via joint pixel and feature alignment. In: The IEEE International Conference on Computer Vision (ICCV), October 2019

36. Wang, M., Deng, W.: Deep visual domain adaptation: a survey. Neurocomputing **312**, 135–153 (2018)

37. Wang, Z., Wang, Z., Zheng, Y., Wu, Y., Zeng, W., Satoh, S.: Beyond intra-modality: a survey of heterogeneous person re-identification. arXiv preprint arXiv:1905.10048 (2019)

38. Wang, Z., Wang, Z., Zheng, Y., Chuang, Y.Y., Satoh, S.: Learning to reduce dual-level discrepancy for infrared-visible person re-identification. In: Proceedings of the IEEE Conference on Computer Vision and Pattern Recognition, pp. 618–626 (2019)

39. Wu, A., Zheng, W.S., Gong, S., Lai, J.: RGB-IR person re-identification by cross-modality similarity preservation. Int. J. Comput. Vision **128**(6), 1765–1785 (2020)

40. Wu, A., Zheng, W.S., Yu, H.X., Gong, S., Lai, J.: RGB-infrared cross-modality person re-identification. In: Proceedings of the IEEE International Conference on Computer Vision, pp. 5380–5389 (2017)

41. Wu, Q., et al.: Discover cross-modality nuances for visible-infrared person re-identification. In: Proceedings of the IEEE/CVF Conference on Computer Vision and Pattern Recognition, pp. 4330–4339 (2021)

42. Xu, X., Wu, S., Liu, S., Xiao, G.: Cross-modal based person re-identification via channel exchange and adversarial learning. In: Mantoro, T., Lee, M., Ayu, M.A., Wong, K.W., Hidayanto, A.N. (eds.) ICONIP 2021. LNCS, vol. 13108, pp. 500–511. Springer, Cham (2021). https://doi.org/10.1007/978-3-030-92185-9_41

43. Yang, Y., Zhang, T., Cheng, J., Hou, Z., Tiwari, P., Pandey, H.M., et al.: Cross-modality paired-images generation and augmentation for RGB-infrared person re-identification. Neural Networks **128**, 294–304 (2020)

44. Ye, M., Lan, X., Wang, Z., Yuen, P.C.: Bi-directional center-constrained top-ranking for visible thermal person re-identification. IEEE Trans. Inf. Forensics Secur. **15**, 407–419 (2020)

45. Ye, M., Lan, X., Li, J., Yuen, P.: Hierarchical discriminative learning for visible thermal person re-identification. In: Proceedings of the AAAI Conference on Artificial Intelligence, vol. 32 (2018)
46. Ye, M., Ruan, W., Du, B., Shou, M.Z.: Channel augmented joint learning for visible-infrared recognition. In: Proceedings of the IEEE/CVF International Conference on Computer Vision, pp. 13567–13576 (2021)
47. Ye, M., Shen, J., J. Crandall, D., Shao, L., Luo, J.: Dynamic dual-attentive aggregation learning for visible-infrared person re-identification. In: Vedaldi, A., Bischof, H., Brox, T., Frahm, J.-M. (eds.) ECCV 2020. LNCS, vol. 12362, pp. 229–247. Springer, Cham (2020). https://doi.org/10.1007/978-3-030-58520-4_14
48. Ye, M., Shen, J., Lin, G., Xiang, T., Shao, L., Hoi, S.C.H.: Deep learning for person re-identification: a survey and outlook. arXiv preprint arXiv:2001.04193 (2020)
49. Ye, M., Shen, J., Shao, L.: Visible-infrared person re-identification via homogeneous augmented tri-modal learning. IEEE Trans. Inf. Forensics Secur. **16**, 728–739 (2020)
50. Zhang, Q., Cheng, H., Lai, J., Xie, X.: DHML: deep heterogeneous metric learning for VIS-NIR person re-identification. In: Sun, Z., He, R., Feng, J., Shan, S., Guo, Z. (eds.) CCBR 2019. LNCS, vol. 11818, pp. 455–465. Springer, Cham (2019). https://doi.org/10.1007/978-3-030-31456-9_50
51. Zhang, Z., Jiang, S., Huang, C., Li, Y., Da, X., R.Y.: RGB-IR cross-modality person ReID based on teacher-student GAN model. Pattern Recogn. Lett. **150**, 155–161 (2021)
52. Zhu, Y., Yang, Z., Wang, L., Zhao, S., Hu, X., Tao, D.: Hetero-center loss for cross-modality person re-identification. Neurocomputing **386**, 97–109 (2020)

Video in 10 Bits: Few-Bit VideoQA for Efficiency and Privacy

Shiyuan Huang[1(✉)], Robinson Piramuthu[2], Shih-Fu Chang[1],
and Gunnar A. Sigurdsson[2]

[1] Columbia University, New York, USA
shiyuan.h@columbia.edu
[2] Amazon Alexa AI, Cambridge, USA

Abstract. In Video Question Answering (VideoQA), answering general questions about a video requires its visual information. Yet, video often contains redundant information irrelevant to the VideoQA task. For example, if the task is only to answer questions similar to "Is someone laughing in the video?", then all other information can be discarded. This paper investigates how many bits are really needed from the video in order to do VideoQA by introducing a novel *Few-Bit VideoQA* problem, where the goal is to accomplish VideoQA with few bits of video information (e.g., 10 bits). We propose a simple yet effective task-specific feature compression approach to solve this problem. Specifically, we insert a lightweight **Feat**ure **Comp**ression Module (**FeatComp**) into a VideoQA model which learns to extract task-specific tiny features as little as 10 bits, which are optimal for answering certain types of questions. We demonstrate more than 100,000-fold storage efficiency over MPEG4-encoded videos and 1,000-fold over regular floating point features, with just 2.0–6.6% absolute loss in accuracy, which is a surprising and novel finding. Finally, we analyze what the learned tiny features capture and demonstrate that they have eliminated most of the non-task-specific information, and introduce a Bit Activation Map to visualize what information is being stored. This decreases the privacy risk of data by providing k-anonymity and robustness to feature-inversion techniques, which can influence the machine learning community, allowing us to store data with privacy guarantees while still performing the task effectively.

Keywords: Video question answering · Feature compression · Privacy · Applications

1 Introduction

Video data exemplifies various challenges with machine learning data: It is large and has privacy issues (*e.g.* faces and license plates). For example, an autonomous robot driving around may collect gigabytes of video data every hour, quickly filling up all available storage with potentially privacy-sensitive information. Yet, video is useful for a variety of computer vision applications, *e.g.* action detection [25], object tracking [31], and video question answering (VideoQA) [19,44].

L. Karlinsky et al. (Eds.): ECCV 2022 Workshops, LNCS 13805, pp. 738–754, 2023.
https://doi.org/10.1007/978-3-031-25072-9_49

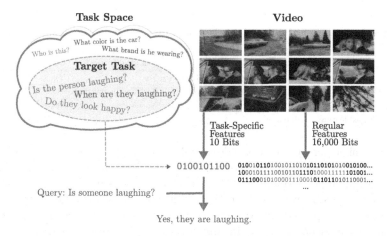

Fig. 1. We introduce the problem of *Few-Bit VideoQA* where only few bits of video information is allowed to do VideoQA tasks. Our proposed method compresses the features inside a neural network to an extreme degree: A video input of 1 MB can be compressed down to 10 bits, and still solve common question-answering tasks with a high degree of accuracy, such as "Is someone laughing?". This has both storage and privacy implications. Here the *Target Task* is all questions related to *laughing*, a subset of all possible tasks in the *Task Space*. *Regular Features* have additional task-irrelevant information visualized with questions and bits of corresponding color. *Regular Features* corresponds to 512 floating point features from a state-of-the-art video network, such as [40].

With increasing adoption of computer vision applications, and mobile devices, efficiently storing video data becomes an increasingly important problem.

VideoQA [19,44] is a general machine learning task that requires analyzing video content to answer a general question. As such, it contains elements of multiple video problems, such as classification, retrieval, and localization. The questions in VideoQA are not all available beforehand, which means that the model needs to extract general semantic information that is still useful enough to answer a diverse set of questions, such as any question about the task "Human Activities", for example. Concretely, the system may have been trained on queries including "Is the person laughing?", and at inference time, the system needs to record and store a video, and may later receive the query "Do they look happy?" which was not seen at training time. In contrast with classification problems where the system could potentially store the answer to all possible queries, a VideoQA system needs to keep enough information to answer any query about the video.

While early video analysis used hand-crafted visual descriptors to represent videos, recent advances in deep neural networks are able to learn highly abstracted features [3,30,32]. These features still contain more information than what is actually needed in a specific task. For example, a popular video encoder network ResNet3D [40] extracts a 512-dim floating number feature, which requires 16,384 bits. If the task is to only answer questions about the

Table 1. Overview of different levels of feature storage and privacy limitations.

	Data amount	Can identify user?	Can discriminate user?
Original data	1,000,000 bits (e.g. MPEG4 video)	Yes	Yes
Regular feature	16,000 bits (e.g. 512 floats)	Yes (Can be inverted)	Yes
Regular compressed feature	100 bits	No	Yes
Task-compressed feature	10 bits	No	No (k-Anonymity [33])

happiness of a person, such as "do people look happy", we theoretically only need 1-bit of information to accomplish that task by encoding the existence of person + laughing. A single 16,384-bit feature encodes much more information than required to answer this question. Moreover, it is hard to interpret what information is captured by continuous features, and hence hard to guarantee that the features do not contain any sensitive information, which may carry privacy concerns.

Much of recent work on VideoQA focuses on learning stronger vision features, improved architectures, or designing better multi-modal interaction [8,20,22,29, 36]. This paper instead investigates how many bits are really needed from video in current VideoQA tasks. To this end, we introduce a novel *"Few-Bit VideoQA"* problem, which aims to accomplish VideoQA where only few bits of information from video are allowed. To our knowledge, the study of few-bit features is an understudied problem, with applications to storing and cataloging large amounts of data for use by machine learning applications.

We provide a simple yet effective task-specific compression approach towards this problem. Our method is inspired by recent learning-based image and video coding [4,7,9,13,14,24,27,38,39,43], which learns low-level compression with the goal of optimizing visual quality. In contrast, our method looks at compressing high-level features, as shown in Fig. 1. Given a video understanding task, we compress the deep video features into few bits (e.g., 10 bits) to accomplish the task. Specifically, we utilize a generic **Feat**ure **Comp**ression module (**FeatComp**), which can be inserted into neural networks for end-to-end training. FeatComp learns compressed binarized features that are optimized towards the target task. In this way, our task-specific feature compression can achieve a high compression ratio and also address the issue of privacy. This approach can store large amounts of features on-device or in the cloud, limiting privacy issues for stored features, or transmitting only privacy-robust features from a device. Note that this work is orthogonal to improvements made by improved architectures, shows insights into analyzing how much video data is needed for a given VideoQA dataset, and provides a novel way to significantly optimize storage and privacy risks in machine learning applications.

10-bits is a concrete number we use throughout this paper to demonstrate the advantage of tiny features. While 10-bits seems like too little to do anything useful, we surprisingly find that predicting these bits can narrow down the solution space enough such that the model can correctly pick among different answers in VideoQA [19,44]. Different tasks require different number of bits, and we see in

Sect. 4 that there are different losses of accuracy for different tasks at the same number of bits. Storing only 10 bits, we can assure that the stored data does not contain any classes of sensitive information that would require more than 10 bits to be stored. For example, we can use the threshold of 33 bits (8 billion unique values) as a rule-of-thumb threshold where the features stop being able to discriminate between people in the world. In Table 1 we show different levels of features and how they compare to this threshold. At the highest level, *Original Data*, we would have the full image or video with all their privacy limitations.[1] Even with feature extraction, various techniques exist to invert the model and reconstruct the original data [10,28]. Using *Regular Compressed Features* (*i.e.* off-the-shelf compression on features in Sect. 5.1 or task-independent compression in Sect. 4.1) would still pose some privacy threats. Our *Task-Compressed Features*, provide privacy guarantees from their minimal size.

We experiment on public VideoQA datasets to analyze how many bits are needed for VideoQA tasks. In our experiments, the "**Task**" is typically question types in a specific dataset, such as TGIF-Action [17], but in a practical system this could be a small set of the most common queries, or even a single type of question such as "do people look happy in the video?". Note that while the "Task" of all questions in a single dataset might seem very diverse and difficult to compress, the task is much more narrow than any possible question about any information in the video, such as "What color is the car?", "What actor is in the video?", etc. In summary, our contributions can be summarized as:

1. We introduce a novel Few-Bit VideoQA problem, where only few bits of video information is used for VideoQA; and we propose a simple yet effective task-specific feature compression approach that learns to extract task-specific tiny features of very few bits.
2. Extensive study of how many bits of information are needed for VideoQA. We demonstrate that we lose just 2.0%–6.6% in accuracy using only 10 bits of data. It provides a new perspective of understanding how much visual information helps in VideoQA.
3. We validate and analyze the task specificity of the learned tiny features, and demonstrate their storage efficiency and privacy advantages.

The outline of the paper is as follows. In Sect. 2, we review related work in VideoQA, image/video coding and deep video representations. In Sect. 3 we introduce the Few-Bit VideoQA problem, and our simple solution with feature compression. In Sect. 4 we discuss several experiments to analyze and validate our approach. Finally, in Sect. 5 we demonstrate applications of this novel problem, including distributing tiny versions of popular datasets and privacy.

2 Related Work

Video Question Answering. VideoQA is a challenging task that requires the system to output answers given a video and a related question [17,21,37,44,45].

---

[1] In the MSRVTT-QA dataset the videos are 630 KB on average, which consists of 320 × 240-resolution videos, 15 s on average.

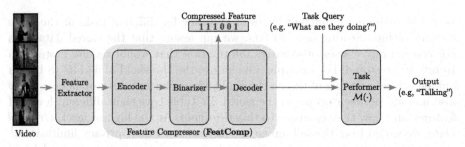

Fig. 2. Pipeline of our generic feature compression approach (**FeatComp**) towards Few-Bit VideoQA, which follows the procedure of encoding, binarization and decoding. It has learned to only encode information relevant to the questions of interest through task-specific training.

Recent approaches include multi-modal transformer models [20,22] and graph convolutional networks [15]. We instead look at VideoQA under limited bits, which shares some philosophy with work that has looked at how much the visual content is needed for Visual Question Answering [11].

Image and Video Coding. Image and video coding is a widely studied problem which aims to compress image/video data with minimum loss of human perceptual quality. In the past decades, standard video codecs like HEVC [35], AVC [42], image codes like JPEG [41] have been used for compressing image/video data. More recently, learning-based image/video compression [4,6,7,9,12–14,24,27] has been proposed to replace the codec components with deep neural networks that optimize the entire coding framework end-to-end, to achieve better compression ratios. All of these existing approaches compress with the goal of pixel-level reconstruction. Our approach is inspired by these works but is applied under a different context—we aim to solve a novel Few-bit VideoQA problem.

Deep Video Representation. Deep neural networks have been shown effective to learn compact video representation, which is now a favored way to store video data for machine learning applications. With the emergence of large-scale video datasets like Kinetics [3] and HowTo100M [30], recent advances in representation learning [8,20,22,29,36] extract continuous video features which contain rich semantic information. Such pre-computed video features can be successfully applied to video understanding tasks like action detection, action segmentation, video question answering, etc. [20,22,29]. However, deep video features could still contain more information than what's actually needed by a specific task. In this work, we instead focus on learning tiny video features, where we aim to use few bits of data to accomplish the target task.

3 Few-Bit VideoQA

In this section, we first establish the problem of Few-Bit VideoQA; then provide a simple and generic task-specific compression solution; finally we introduce our simple implementation based on a state-of-the-art VideoQA model.

3.1 Problem Formulation

In a standard VideoQA framework, a feature extractor (e.g., ResNet3D [40]) is applied to a video sequence to extract the video embedding x, which is a high-level and compact representation of the video, normally a vector composed of floating point numbers. We write $\mathcal{M}(\cdot)$ as the VideoQA task performer, and the output from $\mathcal{M}(x, q)$ is the predicted answer in text, where q refers to the text embedding of the question. In our context, we assume $\mathcal{M}(\cdot)$ is a neural network that can be trained with an associated task objective function $\mathcal{L}_{\text{task}}$.

Though the compact feature x already has a much smaller size compared to the original pixel data, it still raises storage and privacy concerns as discussed in the introduction. To this end, we introduce a novel problem of Few-Bit VideoQA, where the goal is to accomplish VideoQA tasks with only few bits of visual information, i.e., we want to perform $\mathcal{M}(x, q)$ with the size of x less than N bits, where N is small (e.g., $N = 10$).

3.2 Approach: Task-Specific Feature Compression

We propose a simple yet effective approach towards the problem of Few-Bit VideoQA. As shown in Fig. 2, we insert a feature compression bottleneck (**FeatComp**) between the video feature extractor and task performer $\mathcal{M}$ to compress x. We borrow ideas from compression approaches in image/video coding [38,39,43]. But rather than learning pixel-level reconstruction, we train the compression module solely from a video task loss, which has not been explored before to our knowledge. In fact, FeatComp is a generic module that could be applied to other machine learning tasks as well.

Encoding and Decoding. FeatComp follows the encode-binarize-decode procedure to transform floating-point features into binary, and then decode back to floating-point that can be fed into the task performer. In encoding, we first project x to the target dimension, $x' = f_{\text{enc}}(x)$, where $x' \in \mathbb{R}^N$ and N is the predefined bit level to compress into. Binarization is applied directly on feature values; so we map the feature values to a fixed range. We use batch normalization (BN) [16] to encourage bit variance, and then a hyperbolic function $tanh(\cdot)$ to convert all the values to $[-1, 1]$. Binarization is inherently a non-differentiable operation, in order to incorporate it in the learning process, we use stochastic binarization [38,39,43] during training. The final equation for the encode-binarize-decode procedure is:

$$x_{\text{dec}} = \text{FeatComp}(x) = (f_{\text{dec}} \circ \text{BIN} \circ \tanh \circ \text{BN} \circ f_{\text{enc}})\,(x) \qquad (1)$$

where $\circ$ denotes function composition. The output after the binarization step is $x_{\text{bin}} \in \{0,1\}^N$, which is the N-bit compressed feature to be stored.

Learning Task-Specific Compression. Our FeatComp is generic and can be inserted between any usual feature extractor and task performer $\mathcal{M}(\cdot)$. The task performer will instead take the decoded feature to operate the task: $\mathcal{M}(x_{\text{dec}})$ or $\mathcal{M}(x_{\text{dec}}, q)$. To compress in a task-specific way, FeatComp is trained along with $\mathcal{M}(\cdot)$ with the objective $\mathcal{L} = \mathcal{L}_{\text{task}}$, where $\mathcal{L}_{\text{task}}$ is the target task objective.

Simple Implementation. Our FeatComp can be easily implemented into any VideoQA models. As an instantiation, we choose a recent state-of-the-art model ClipBERT [20] as our baseline, and add our compression module to study the number of bits required for VideoQA. Specifically, ClipBERT follows the similar pipeline as in Fig. 2. We then insert FeatComp after feature extractor. For encoding and decoding, we use a fully connected layer where $f_{\text{enc}} \colon \mathbb{R}^{(T \times h \times w \times D)} \mapsto \mathbb{R}^N$ and $f_{\text{dec}} \colon \{0,1\}^N \mapsto \mathbb{R}^{(T \times h \times w \times D)}$. x is flattened to a single vector to be encoded and binarized, and the decoded x_{dec} can be reshaped to the original size. Then the answer is predicted by $\mathcal{M}(x_{\text{dec}}, q)$. For FeatComp with N-bit compression, we write it as FeatComp-N. This implementation is simple and generic, and as we see in Sect. 4 already works surprisingly well. Various other architectures for encoding, decoding, and binarization could be explored in future research.

Intuition of FeatComp. To learn a compression of the features, various methods could be explored. We could cluster the videos or the most common answers into 1024 clusters, and encode the cluster ID in 10 bits, etc. For a task such as video action classification, where only the top class prediction is needed, directly encoding the final answer may work. However, in a general video-language task such as VideoQA, the number of possible questions is much more than 1024, even for questions about a limited topic. Our method can be interpreted as learning 2^N clusters end-to-end, that are predictable, and useful to answering any questions related to the task.

4   Experiments

In this section, we show the experimental results on Few-Bit VideoQA, study how much visual information is needed for different VideoQA tasks, and analyze what the bits capture.

Datasets. We consider two public VideoQA datasets: 1) TGIF-QA [17] consists 72K GIF videos, 3.0 s on average, and 165K QA pairs. We experiment on 3 TGIF-QA tasks—Action (*e.g.* "What does the woman do 5 times?"), Transition (*e.g.* "What does the man do after talking?"), which are multiple-choice questions with 5 candidate answers; and FrameQA (*e.g.* What does an airplane drop which bursts into flames?"), which contains general questions with single-word answers. 2) MSRVTT-QA [44] consists of 10k videos of duration 10–30 s each and 254K general QA pairs, *e.g.* "What are three people sitting on?".

Implementation Details. We leverage the ClipBERT model pre-trained on COCO Captions [5] and Visual Genome Captions [18], and train on each VideoQA dataset separately, following [20]. During training, we randomly initialize FeatComp, and finetune the rest of the network; we randomly sample $T = 1$ clip for TGIF-QA and $T = 4$ clips for MSRVTT-QA, where in each clip we only sample the middle frame. We fix the ResNet backbone and set the learning rate to 5×10^{-5} for FeatComp, and 10^{-6} for $\mathcal{M}$. We use the same VideoQA objective and AdamW [26] optimizer as in [20]. During inference, we uniformly sample T_{test} clips to predict answer, where $T_{\text{test}} = T$, unless noted otherwise. We set $N = 1, 2, 4, 10, 100, 1000$ for different bit levels. Code will be made available.

Evaluation Metrics. QA accuracy at different bit levels.

Baselines. To our knowledge, there is no prior work directly comparable; hence we define the following baselines:

- *Floats*: the original floating-point-based ClipBERT; it provides performance upper bound using enough bits of information.[2]
- *Q-only*: answer prediction solely from question.
- *Random Guess*: randomly choose a candidate/English word for multiple-choice/single-word-answer QA.

Additionally, since our approach follows an autoencoder-style design, we also study whether an objective of feature reconstruction helps with Few-Bit VideoQA. We add the following two approaches for comparison:

- *R-only*: learn FeatComp solely with a reconstruction loss, $\mathcal{L}_R = \text{MSE}(x, x_{\text{dec}})$, then finetune $\mathcal{M}$ using learned compressed features with task objective.
- *FeatComp+R*: learn FeatComp from both task and reconstruction objectives, i.e., $\mathcal{L} = \mathcal{L}_{task} + \mathcal{L}_R$.

We also study how test-time sampling affects the results with:

- *4×FeatComp*: test-time sampling $T_{test} = 4T$.

4.1   Few-Bit VideoQA Results

We demonstrate our Few-Bit VideoQA results in Fig. 3 and compare to the baselines. As expected, more bits yields accuracy improvements. Notably, for our *FeatComp*, even a 1-bit video encoding yields a 7.2% improvement over *Q-only* on TGIF-Action. On all datasets, at 1000-bit, we maintain a performance drop $<2.8\%$ while compressing more than >1000 times. At 10-bit, the drop is within 2.0–6.6% while compressing over $>100,000$ times, demonstrating 10-bit visual information already provides significant aid in VideoQA tasks.

We also compute *supervision size* as a naive upper bound of bits required to accomplish the task. Theoretically, the compressed features should be able to

---

[2] ClipBERT uses 16-bit precision, we use this for calculation.

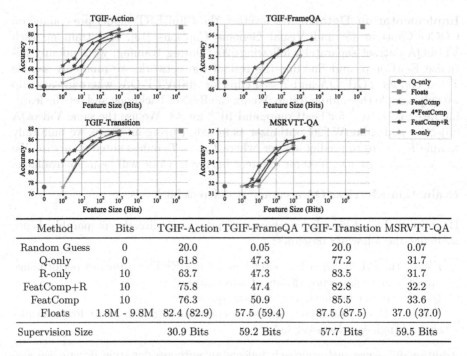

Method	Bits	TGIF-Action	TGIF-FrameQA	TGIF-Transition	MSRVTT-QA
Random Guess	0	20.0	0.05	20.0	0.07
Q-only	0	61.8	47.3	77.2	31.7
R-only	10	63.7	47.3	83.5	31.7
FeatComp+R	10	75.8	47.4	82.8	32.2
FeatComp	10	76.3	50.9	85.5	33.6
Floats	1.8M - 9.8M	82.4 (82.9)	57.5 (59.4)	87.5 (87.5)	37.0 (37.0)
Supervision Size		30.9 Bits	59.2 Bits	57.7 Bits	59.5 Bits

Fig. 3. Analysis of how bit size affects VideoQA accuracy on MSRVTT-QA and TGIF-QA. *Floats* refers to the original ClipBERT model that serves as an upper bound of the performance without bit constraints. We both report the number we reproduced and cite paper results in parenthesis. With our simple approach *FeatComp*, we can reach high performance using only a few bits. Notably, at 10-bit level (see table), we get only 2.0–6.6% absolute loss in accuracy.

differentiate the texts in all QA pairs. Hence, for each task, we use bzip2[3] to compress the text file for the training QA pairs, whose size is used as the upper bound. We observe correspondence between supervision size and compression difficulty. TGIF-Action and TGIF-Transition are easier to compress, requiring less bits to get substantial performance gains; and they also appear to require less bits in supervision size. Instead TGIF-FrameQA and MSRVTT-QA are harder to compress, also reflected by their relatively larger supervision size.

Role of Reconstruction Loss. Our approach FeatComp is learned in a task-specific way in order to remove any unnecessary information. On the other hand, it is also natural to compress with the goal of recovering full data information. Here we investigate whether direct feature supervision helps learn better compressed features from *FeatComp+R* and *R-only*. We can see that integrating feature reconstruction harms the accuracy at every bit level. In fact, R-only can be considered as a traditional lossy data compression that tries to recover full data values. It implies that recovering feature values requires more informa-

---

[3] https://www.sourceware.org/bzip2.

Table 2. Task-specificity of FeatComp-10. Compression learned from an irrelevant source task is less helpful for a target task, while compression from the same task gives the best performance.

Source task	Target task		
	Action	FrameQA	Transition
Action	**76.9**	47.6	83.8
FrameQA	62.3	**51.7**	77.1
Transition	76.0	47.3	**85.0**
Q-Only	61.8	47.3	77.2

tion than performing VideoQA task; hence feature reconstruction loss may bring task-irrelevant information that hurts the task performance.

Role of Temporal Context. For longer videos, frame sampling is often crucial for task accuracy. We study the impact in Fig. 3 (*FeatComp* vs. *4 × FeatComp*). We can see denser sampling benefits compression ratio especially at higher bits on longer videos like MSRVTT-QA. For example, 400-bit compression with $T_{\text{test}} = 16$ outperforms 1000-bit compression with $T_{\text{test}} = 4$. However, on short video dataset TGIF-QA, it does not yield improvements, which implies TGIF-QA videos can be well understood with single frames.

How Task-Specific are the Features? Here we evaluate the task-specificity by analyzing how much information is removed by the compression. We use FeatComp-10 learned on a source TGIF-QA task to extract the compressed features x_{bin}; then train a new VideoQA model for different target tasks on top of x_{bin}. For fair comparison, we use the same $\mathcal{M}$ and decoding layers and follow the previous practice to initialize $\mathcal{M}$ from the model pre-trained on COCO Captions and Visual Genome, and randomly initialize the decoding layers. Then we train the network with learning rate 5×10^{-5} for 20 epochs. Table 2 shows the results, where we also cite *Q-only* from Fig. 3. Compression from the same task gives the best result as expected. Note that *Action* and *Transition* are similar tasks in that they query for similar actions but *Transition* asks change of actions over time. *FrameQA* instead queries for objects, which has minimal similarity with *Action/Transition*. Compression from a highly relevant source task (*Action* vs. *Transition*) gives pretty high performance. However, compression from an irrelevant source task (*Action/Transition* vs. *FrameQA*) yields similar performance as *Q-only* (0-bits), which again implies that the learned compression discards any information unnecessary for its source task.

4.2 Qualitative Analysis

Here we study qualitatively what the learned bits are capturing. We apply Grad-CAM [34] directly on the compressed features x_{bin} to find the salient regions

Table 3. Top-7 closest words from FeatComp-10 on TGIF-Action.

Word	Most similar words
head	bob neck stroke fingers raise cigarette open
wave	touch man bob hands hand stroke raise
spin	around ice pants jump foot kick knee
step	walk spin around ice pants jump kick
guitar	object who keys black a strum bang
eyes	blink smile laugh cigarette lip tilt sleeve

over frames. Grad-CAM is originally a tool for localizing the regions sensitive to class prediction. It computes the weighted average of feature maps based on its gradients from the target class, which localizes the regions that contribute most to that class prediction. Here we instead construct *Bit Activation Map* (BAM) by treating each binary bit x_{bin}^i as a "class": 1 is a positive class while 0 is negative. We calculate BAM at the last convolutional layer of ResNet. We average BAM for all bits where for $x_{bin}^i = 0$ we multiply them with -1. Note that no class annotations, labels or predictions are used for BAM. Figure 4a shows heat map visualization results on TGIF-Action for FeatComp-10. The learned compression is capturing salient regions (e.g., eyes, lips, legs) which align with human perception. We also study in Fig. 4b how important temporal signals are captured, where we average the BAM over frames. The frames with high BAM scores tend to capture more important semantic information across time that is relevant to the task.

Additionally, we investigated whether the bits capture task-specific information by analyzing the correspondences between VideoQA vocabularies and compressed features. We calculate a word feature as the averaged FeatComp-10 bit features over the videos whose associated QAs contain this word, and find its top-7 closest words based on Euclidean distance. Table 3 shows some sample results on TGIF-Action. Neighbor words tend to demonstrate semantic associations. E.g., 'eyes' is closely related to 'blink', 'smile' and 'laugh'; 'step' is associated with its similar actions like 'walk' and 'jump', implying that the compression captures task-relevant semantic information.

5   Applications of Few-Bit VideoQA

The problem of Few-Bit VideoQA has practical applications to data storage efficiency and privacy. Below we discuss how our approach can compress a video dataset into a tiny dataset and use that to achieve the same task (Sect. 5.1); then we demonstrate the privacy advantages of few-bit features (Sect. 5.2).

5.1   Tiny Datasets

With our task-specific compression approach, we can represent a video using only a few bits, which allows extreme compression of gigabytes of video datasets into

(a) Bit Activation Map w.r.t. FeatComp-10 compressed features x_{bin} on TGIF-Action. The learned tiny feature is able to capture salient regions useful for the source task of FeatComp-10.

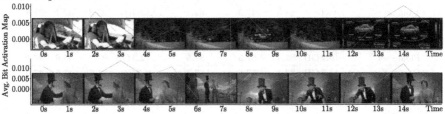

(b) Averaged Bit Activation Map over video frames on MSRVTT-QA. We can see that learned tiny feature can capture scene changes — the bit activation peaks appear when there is a different scene useful for the source task of FeatComp-10.

Fig. 4. Qualitative visualization examples of bit activation map w.r.t FeatComp-10 compressed features x_{bin}. Note that no question, answer, prediction, or class label was used for these visualizations—This is visualizing what the network used to generically compress the video.

almost nothing. For example, TGIF-QA contains 72K videos whose MPEG4-encoded format takes up about 125 GB of storage; with FeatComp-10, we end up with a 90 KB dataset. We follow Sect. 4.1 to train a model using stored compressed features x_{bin}, with source and target tasks being the same, to evaluate the feature quality. Table 4 compares the compression size and testing accuracy on TGIF-QA and MSRVTT-QA tasks. Our tiny datasets achieve similar performance with Fig. 3 at all bit levels. In addition to MPEG4 video and uncompressed Floats, we also report the size of Floats compressed with the off-the-shelf lossless compression standard ZIP.[4] The result implies that the original features indeed contain a lot of information not easily compressible, and we are learning meaningful compression.

5.2 Privacy Advantages from Tiny Features

Here we demonstrate how our tiny features offer privacy advantages.

Advantages of Data Minimization. Intuitively, we expect 10 bits of information to not contain very much sensitive information. The principle of Shannon Information[5], or the pigeonhole principle, tells us that 10 bits of information

---

[4] numpy.savez_compressed.

[5] en.wikipedia.org/wiki/Information_content.

Table 4. VideoQA accuracies with our compressed datasets at different bit levels. Looking at 10-bit, we can get 100,000-fold storage efficiency, while maintaining good performance.

Datasets	TGIF-QA				MSRVTT-QA			
	Size	Task Accs.			Size	Task Accs.		
		Action	FrameQA	Transition		What	Who	All
1-bit set	9 KB	68.3	47.3	82.4	1.3 KB	24.8	37.7	31.8
10-bit set	90 KB	76.9	51.7	85.0	13 KB	27.4	44.1	33.7
1000-bit set	9 MB	80.8	54.4	87.2	1.3 MB	28.3	46.0	34.9
Floats set	16.2 GB	82.4	57.5	87.5	12.3 GB	31.7	45.6	37.0
Comp. floats set	14.0 GB	82.4	57.5	87.5	9.5 GB	31.7	45.6	37.0
MPEG4 set	125 GB	82.4	57.5	87.5	6.3 GB	31.7	45.6	37.0

Table 5. What sensitive information can be stored in 10 bits?

Impossible (bits)	Possible (bits)
Credit card num. ($\approx$53)	Gender ($\approx$2)
Social security num. ($\approx$30)	Skin Color ($\approx$3)
Street address ($\approx$28)	Social Class ($\approx$3)
License plate ($\approx$36)	...
Personal image ($\approx$100,000)	
Phone number ($\approx$33)	
...	

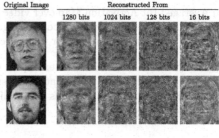

Fig. 5. Feature inversion with increasingly compressed features.

cannot contain a full image (approximately 10 KB). Using 10 bits as an example, we can divide sensitive information into two groups based on if it can be captured by 10 bits or not in Table 5.

Note we assume the data was not in the training set as otherwise the model could be used to extract that potentially sensitive information (training networks with differential privacy can relax this assumption [1]). By capturing only 10 bits, we can assure a user that the stored data does not contain any classes of sensitive information that require more bits to be stored.

Furthermore, considering that there are 8B unique people in the world, the identity of a person in the world can be captured in 33 bits, so we cannot reconstruct the identity of a person from 10 bits. The identity can be identifying information, photographic identity, biometrics, etc. This is consistent with the following experiment where feature-inversion techniques do not seem to work, whereas they work on regular compressed 16,384 bit features (as in Table 1). This effectively de-identifies the data.

Robustness to Feature Inversion. Storing only a few bits also has applications to preventing reconstruction of input from stored features (*i.e.* Feature Inversion). Various methods exist for inverting features [10,28] typically by optimizing an input to match the feature output, with an optional regularization. To evaluate this, we started with a re-implementation[6] of the Frederikson *et al.* attack [10] for model inversion, but instead of class probability, we used a MSE loss between the target feature and the model feature output before the last linear layer, or after the binarizer. We used a two layer neural network trained on the AT&T Faces Dataset [2]. In Fig. 5 we demonstrate how the inversion deteriorates as the model uses fewer bits. The 1,280 bits result has no compression (40 32-bit numbers) whereas the rest of the results have increasing compression. All models were trained until at least 97.5% accuracy on the training set.

Feature Quantization. In Frederikson *et al.* [10] the authors note that rounding the floating point numbers at the 0.01 level seems to make reconstruction difficult and offer that as mitigation strategy. This suggests that compression of features will likely help defend against model inversion and we verify this here. For comparison, rounding a 32-bit float (less than 1) at the 0.01 level corresponds to 200 different numbers which can be encoded in 8 bits, a 4-fold compression. Our method can compress 512 32-bit floats (16,384 bits) into 10 bits, a 1,638-fold compression, while having a defined performance on the original task.

K-Anonymity of Tiny Features. For a system that has N users and stores 10 bits (1,024 unique values), we can assure the user that:

> Any data that is stored is indistinguishable from the data from approximately $N/1,024$ other users.

This ensures privacy by implementing k-anonymity [33] assuming the 10 bits are uniformly distributed, for example if $N = 10^7$, $k \approx 10,000$. Finally, we can use a variable number of stored bits to ensure any stored bits are sufficiently non-unique, similar to hash-based k-anonymity for password checking [23].

6   Conclusion

In this work, we introduce a novel Few-Bit VideoQA problem, which aims to do VideoQA with only few bits of video information used; we propose a simple and generic FeatComp approach that can be used to learn task-specific tiny features. We experiment over VideoQA benchmarks and demonstrate a surprising finding that a video can be effectively compressed using as little as 10 bits towards accomplishing a task, providing a new perspective of understanding how much visual information helps in VideoQA. Furthermore, we demonstrate the storage, efficiency, and privacy advantages of task-specific tiny features—tiny features ensure no sensitive information is contained while still offering good task performance. We hope these results will influence the community by offering insight into how much visual information is used by question answering systems, and

---

[6] github.com/Koukyosyumei/secure_ml

752 S. Huang et al.

opening up new applications for storing large amounts of features on-device or in the cloud, limiting privacy issues for stored features, or transmitting only privacy-robust features from a device.

Acknowledgement. We would like to thank Vicente Ordonez for his valuable feedback on the manuscript.

References

1. Abadi, M., et al.: Deep learning with differential privacy. In: Proceedings of the 2016 ACM SIGSAC Conference on Computer and Communications Security (2016)
2. Cambridge, A.L.: The orl database. https://www.cl.cam.ac.uk/research/dtg/attarchive/facedatabase.html
3. Carreira, J., Zisserman, A.: Quo Vadis, action recognition? A new model and the kinetics dataset. In: CVPR (2017)
4. Chadha, A., Andreopoulos, Y.: Deep perceptual preprocessing for video coding. In: CVPR (2021)
5. Chen, X., et al.: Microsoft coco captions: data collection and evaluation server. arXiv (2015)
6. Choi, J., Han, B.: Task-aware quantization network for JPEG image compression. In: Vedaldi, A., Bischof, H., Brox, T., Frahm, J.-M. (eds.) ECCV 2020. LNCS, vol. 12365, pp. 309–324. Springer, Cham (2020). https://doi.org/10.1007/978-3-030-58565-5_19
7. Djelouah, A., Campos, J., Schaub-Meyer, S., Schroers, C.: Neural inter-frame compression for video coding. In: ICCV (2019)
8. Feichtenhofer, C., Fan, H., Malik, J., He, K.: SlowFast networks for video recognition. In: ICCV (2019)
9. Feng, R., Wu, Y., Guo, Z., Zhang, Z., Chen, Z.: Learned video compression with feature-level residuals. In: CVPR Workshops (2020)
10. Fredrikson, M., Jha, S., Ristenpart, T.: Model inversion attacks that exploit confidence information and basic countermeasures. In: Proceedings of the 22nd ACM SIGSAC Conference on Computer and Communications Security (2015)
11. Goyal, Y., Khot, T., Summers-Stay, D., Batra, D., Parikh, D.: Making the V in VQA matter: elevating the role of image understanding in visual question answering. In: CVPR, July 2017
12. He, T., Sun, S., Guo, Z., Chen, Z.: Beyond coding: detection-driven image compression with semantically structured bit-stream. In: PCS (2019)
13. Hu, Z., Chen, Z., Xu, D., Lu, G., Ouyang, W., Gu, S.: Improving deep video compression by resolution-adaptive flow coding. In: Vedaldi, A., Bischof, H., Brox, T., Frahm, J.-M. (eds.) ECCV 2020. LNCS, vol. 12347, pp. 193–209. Springer, Cham (2020). https://doi.org/10.1007/978-3-030-58536-5_12
14. Hu, Z., Lu, G., Xu, D.: FVC: a new framework towards deep video compression in feature space. In: CVPR (2021)
15. Huang, D., Chen, P., Zeng, R., Du, Q., Tan, M., Gan, C.: Location-aware graph convolutional networks for video question answering. In: AAAI (2020)
16. Ioffe, S., Szegedy, C.: Batch normalization: accelerating deep network training by reducing internal covariate shift. In: ICML (2015)
17. Jang, Y., Song, Y., Yu, Y., Kim, Y., Kim, G.: TGIF-QA: toward spatio-temporal reasoning in visual question answering. In: CVPR (2017)

18. Krishna, R., et al.: Visual genome: connecting language and vision using crowd-sourced dense image annotations. IJCV **123**, 32–73 (2017)
19. Le, T.M., Le, V., Venkatesh, S., Tran, T.: Hierarchical conditional relation networks for video question answering. In: CVPR (2020)
20. Lei, J., et al.: Less is more: ClipBERT for video-and-language learning via sparse sampling. In: CVPR (2021)
21. Lei, J., Yu, L., Bansal, M., Berg, T.L.: TVQA: localized, compositional video question answering. arXiv preprint arXiv:1809.01696 (2018)
22. Li, L., Chen, Y.C., Cheng, Y., Gan, Z., Yu, L., Liu, J.: Hero: Hierarchical encoder for video+language omni-representation pre-training. In: EMNLP (2020)
23. Li, L., Pal, B., Ali, J., Sullivan, N., Chatterjee, R., Ristenpart, T.: Protocols for checking compromised credentials. In: Proceedings of the 2019 ACM SIGSAC Conference on Computer and Communications Security (2019)
24. Lin, J., Liu, D., Li, H., Wu, F.: M-LVC: multiple frames prediction for learned video compression. In: CVPR (2020)
25. Lin, T., Liu, X., Li, X., Ding, E., Wen, S.: BMN: boundary-matching network for temporal action proposal generation. In: ECCV (2019)
26. Loshchilov, I., Hutter, F.: Decoupled weight decay regularization. arXiv preprint arXiv:1711.05101 (2017)
27. Lu, G., Ouyang, W., Xu, D., Zhang, X., Cai, C., Gao, Z.: DVC: an end-to-end deep video compression framework. In: CVPR (2019)
28. Mahendran, A., Vedaldi, A.: Understanding deep image representations by inverting them. In: CVPR (2015)
29. Miech, A., Alayrac, J.B., Smaira, L., Laptev, I., Sivic, J., Zisserman, A.: End-to-end learning of visual representations from uncurated instructional videos. In: CVPR (2020)
30. Miech, A., Zhukov, D., Alayrac, J.B., Tapaswi, M., Laptev, I., Sivic, J.: HowTo100M: learning a text-video embedding by watching hundred million narrated video clips. In: ICCV (2019)
31. Minderer, M., Sun, C., Villegas, R., Cole, F., Murphy, K., Lee, H.: Unsupervised learning of object structure and dynamics from videos. arXiv preprint arXiv:1906.07889 (2019)
32. Russakovsky, O., et al.: ImageNet large scale visual recognition challenge. IJCV **115**(3), 211–252 (2015)
33. Samarati, P., Sweeney, L.: Protecting privacy when disclosing information: k-anonymity and its enforcement through generalization and suppression (1998)
34. Selvaraju, R.R., Cogswell, M., Das, A., Vedantam, R., Parikh, D., Batra, D.: Grad-CAM: visual explanations from deep networks via gradient-based localization. In: ICCV (2017)
35. Sullivan, G.J., Ohm, J.R., Han, W.J., Wiegand, T.: Overview of the high efficiency video coding (HEVC) standard. IEEE Trans. Circ. Syst. Video Technol. **22**(12), 1649–1668 (2012)
36. Sun, C., Myers, A., Vondrick, C., Murphy, K., Schmid, C.: VideoBERT: a joint model for video and language representation learning. In: ICCV (2019)
37. Tapaswi, M., Zhu, Y., Stiefelhagen, R., Torralba, A., Urtasun, R., Fidler, S.: MovieQA: understanding stories in movies through question-answering. In: CVPR (2016)
38. Toderici, G., et al.: Variable rate image compression with recurrent neural networks. arXiv preprint arXiv:1511.06085 (2015)
39. Toderici, G., et al.: Full resolution image compression with recurrent neural networks. In: CVPR (2017)

40. Tran, D., Wang, H., Torresani, L., Ray, J., LeCun, Y., Paluri, M.: A closer look at spatiotemporal convolutions for action recognition. In: CVPR (2018)
41. Wallace, G.K.: The JPEG still picture compression standard. IEEE Trans. Consum. Electron. **38**, 18–34 (1992)
42. Wiegand, T., Sullivan, G.J., Bjontegaard, G., Luthra, A.: Overview of the H.264/AVC video coding standard. IEEE Trans. Circ. Syst. Video Technol. **13**, 560–576 (2003)
43. Wu, C.-Y., Singhal, N., Krähenbühl, P.: Video compression through image interpolation. In: Ferrari, V., Hebert, M., Sminchisescu, C., Weiss, Y. (eds.) ECCV 2018. LNCS, vol. 11212, pp. 425–440. Springer, Cham (2018). https://doi.org/10.1007/978-3-030-01237-3_26
44. Xu, D., et al.: Video question answering via gradually refined attention over appearance and motion. In: ACM MM (2017)
45. Zhu, L., Xu, Z., Yang, Y., Hauptmann, A.: Uncovering the temporal context for video question answering. IJCV **124**, 409–421 (2017)

ChaLearn LAP Seasons in Drift Challenge: Dataset, Design and Results

Anders Skaarup Johansen[1]([✉]) [ID], Julio C. S. Jacques Junior[2] [ID],
Kamal Nasrollahi[1,3] [ID], Sergio Escalera[2,4] [ID], and Thomas B. Moeslund[1] [ID]

[1] Aalborg University, Copenhagen, Denmark
{asjo,kn,tmb}@create.aau.dk
[2] Computer Vision Center, Barcelona, Spain
jjacques@cvc.uab.cat
[3] Milestone Systems, Brondby, Denmark
kna@milestone.dk
[4] University of Barcelona, Barcelona, Spain
sergio@maia.ub.es

Abstract. In thermal video security monitoring the reliability of deployed systems rely on having varied training data that can effectively generalize and have consistent performance in the deployed context. However, for security monitoring of an outdoor environment the amount of variation introduced to the imaging system would require extensive annotated data to fully cover for training and evaluation. To this end we designed and ran a challenge to stimulate research towards alleviating the impact of concept drift on object detection performance. We used an extension of the Long-Term Thermal Imaging Dataset, composed of thermal data acquired from 14th May 2020 to 30th of April 2021, with a total of 1689 2-min clips with bounding-box annotations for 4 different categories. The data covers a wide range of different weather conditions and object densities with the goal of measuring the thermal drift over time, from the coldest day/week/month of the dataset. The challenge attracted 184 registered participants, which was considered a success from the perspective of the organizers. While participants managed to achieve higher mAP when compared to a baseline, concept drift remains a strongly impactful factor. This work describes the challenge design, the adopted dataset and obtained results, as well as discuss top-winning solutions and future directions on the topic.

1 Introduction

In the context of thermal video security monitoring the sensor type that is responsible of quantifying the observed infrared-radiation as a thermograph can be split into two groups: sensors that produce relative thermographs and sensors that produce absolute thermographs. Absolute thermographs can correlate the observed radiation directly with temperature, whereas relative thermographs produce observations relative to the "coldest" and "warmest" radiation. In security monitoring contexts the absolute temperature readings produced by an absolute thermograph are not necessary and can potentially suppress thermal details

© The Author(s), under exclusive license to Springer Nature Switzerland AG 2023
L. Karlinsky et al. (Eds.): ECCV 2022 Workshops, LNCS 13805, pp. 755–769, 2023.
https://doi.org/10.1007/978-3-031-25072-9_50

when observing thermally uniform environment. Furthermore the price of absolute thermal cameras are much higher than their relative counterpart.

When performing image recognition tasks the visual appearance of objects and their surroundings is very important, and in an outdoor context that is subjected to changes in temperature, weather, sun-radiation, among others, the visual appearance of objects and their surroundings change quite drastically. This is further expanded by societal factors like the recent pandemic which could introduce mandatory masks. This is known as "Concept Drift" where objects remain the same however the concept definition which is observed through representation changes. While in theory it could be possible to collect a large enough dataset encompassing the weather conditions, the actors, usually people, within the context also dress and act differently. Furthermore the cost of producing such a dataset would be quite extensive as potentially years worth of data would have to be annotated. Typically deployment of object detectors would have a pretrained baseline, and the model would have to be retrained when the observed context drifts too far away from the training context. The reliability in such a system is questionable as deployed algorithms tend not to have a way to quantify the performance during deployment and extra data would have to be routinely annotated to verify that the system is still performing as expected. To address this issue and foster more research into long-term reliability of deployed learning based object detectors a benchmark for classifying the impact of concept drift could greatly benefit the field.

The ECCV 2022 ChaLearn LAP Seasons in Drift Challenge aims to propose a setting for evaluating the impact of concept drift at a month to month basis and evaluating the impact of concept drift in a weighted manner. The problem of concept drift is exacerbated with limited training data, particularly when the distribution of the visual appearance in the data is similar. To explore the consistency of performance across varied levels of concept drift particularly of object detection algorithms, an extended set of frames were annotated spanning several months. The challenge attracted a total of 184 participants on its different tracks. With a total of 691 submissions at the different challenge stages and tracks, from over 180 participants, the challenge managed to successfully establish a benchmark for thermal concept drift. Top-wining solutions outperformed the baseline by a large margin following distinct strategies, detailed in Sect. 4.

The rest of the paper is organized as follows. In Sect. 2 we present the related work. The Challenge design, which includes a short description of the adopted dataset, evaluation protocol and baseline are detailed in Sect. 3. Challenge results and top-winning solutions are discussed in Sect. 4. Finally, conclusion and suggestions for future research directions are drawn in Sect. 5.

2 Related Work

Popular thermal detection and segmentation datasets, such as KAIST [13] and FLIR-ADAS [24], provide thermal and visible images. The focus of a large part of academic research have been focused on leveraging a multi-modal input [10, 16,29,30] or using the aligned visible/thermal pairs as a way to do unsupervised domain adaptation between the visible and thermal [7,10,25,28]. Approaches that leverage the multi-modal input directly typically use siamese style networks to perform modality specific feature extraction, subsequently leveraging a fusion scheme to combine the information in a learned manner [16,25,29], alternatively simple concatenation or addition is performed after initial feature extraction [10, 30]. In contrast, a network can be optimized to be domain agnostic. HeatNet [25] and DANNet [28] leverage an adversarial approach to guide the network to extract domain agnostic features.

It has been proven that in security monitoring contexts fusion of visible and thermal images outperforms any modality alone [14,17], however in a real-world scenario camera setups tend to be single sensor setups. While thermal cameras are robust to changes in weather and lighting conditions, they still struggle with the change of visual appearance of objects due to the change of scene temperature [6,8,9,15,17]. Early work [9] leveraged edges to highlight objects, making detection possible robust to the variation when the relative contrast between objects and their surroundings were consistent. Recent studies leverage research in the visible imaging domain, and directly apply it to the thermal domain [6,17]. Until recently thermal specific detection methods have been a rarity and recently it was proven that contextual information is important to increase robustness to day/night variation [15,23] for thermal only object detection. By employing a conditioning of the latent representation guided by an auxiliary day/night classification head, the accuracy of day and night accuracy can be significantly increased [15]. Similar increase in performance can also be gained with a combination of a shallow feature-extractor and residual FPN-style connections [8]. Most notably the residual connections are leverage during training to enforce learning of discriminative features throughout the network, and serve no purpose during inference, and as such can be removed.

3   Challenge Design

The ECCV 2022 Seasons in Drift Challenge[1] aimed to spotlight the problem of concept drift in a security monitoring context and highlight the challenges and limitations of existing methods, as well as to provide a direction of research for the future. The challenge used an extension of the LTD Dataset [21] which consists of thermal footage that spans multiple seasons, detailed in Sec. 3.1. The challenge was split into 3 different tracks associated with thermal object detection. Each track having the same evaluation criteria/data but varying the amount of train data as well as the time span of the data, as detailed next.

– **Track 1 - Detection at day level:** Train on a predefined and single day data and evaluate concept drift across time[2]. The day is the 13th of February 2020 as it is the coldest day in the recorded data, due to the relative thermal appearance of objects being the least varied in colder environments this is our starting point.

---

[1] Challenge - https://chalearnlap.cvc.uab.cat/challenge/51/description/.

[2] Track 1 (on Codalab) - https://codalab.lisn.upsaclay.fr/competitions/4272.

- **Track 2 - Detection at week level:** Train on a predefined and single week data and evaluate concept drift across time[3]. The week selected is the week of the 13th - 20th of February 2020 - (i.e. expanding from our starting point)
- **Track 3 - Detection at month level:** Train on a predefined and single month data and evaluate concept drift across time[4]. The selected month is the entire month of February.

The training data is chosen by selecting the coldest day, and surrounding data as cold environments introduce the least amount of concept drift. Each track aims at evaluating how robust a given detection method is to concept drift, by training on limited data from a specific time period (day, week, month in February) and evaluating performance across time, by validating and testing performance on months of unseen data (Jan., Mar., Apr., May., Jun., Jul., Aug. and Sep.). The February data is only present in the training set and the remaining months are equally split between validation and test.

Each track is composed of two phases, i.e., development and test phase. At the development phase, public train data was released and participants needed to submit their predictions with respect to a validation set. At the test (final) phase, participants needed to submit their results with respect to the test data, which was released just a few days before the end of the challenge. Participants were ranked, at the end of the challenge, using the test data. It is important to note that this competition involved the submission of results (and not code). Therefore, participants were required to share their codes and trained models after the end of the challenge so that the organizers could reproduce the results submitted at the test phase, in a "code verification stage". At the end of the challenge, top ranked methods that pass the code verification stage were considered as valid submissions.

3.1 The Dataset

The dataset used in the challenge is an extension of the Long-Term Thermal Imaging [21] dataset, and spans 188 days in the period of 14th May 2020 to 30th of April 2021, with a total of 1689 2-minute clips sampled at 1fps with associated bounding box annotations for 4 classes (Human, Bicycle, Motorcycle, Vehicle). The collection of this dataset has included data from all hours of the day in a wide array of weather conditions overlooking the harborfront of Aalborg, Denmark. In this dataset depicts the drastic changes of appearance of the objects of interest as well as the scene over time in a static security monitoring context to develop robust algorithms for real-world deployment. Figure 1 illustrates the camera setup and two annotated frames of the dataset, obtained at different time intervals.

For a detailed explination of the datasets weather contents, an overview can be found in the original dataset paper [21]. As for the extended annotations

---

[3] Track 2 (on Codalab) - https://codalab.lisn.upsaclay.fr/competitions/4273.
[4] Track 3 (on Codalab) - https://codalab.lisn.upsaclay.fr/competitions/4276.

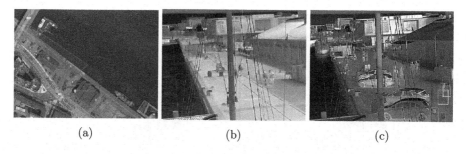

(a)   (b)   (c)

Fig. 1. Illustration of the camera setup (a) and two annotated frames of the dataset, captured at different time intervals (b-c).

provided with this challenge, we can observe that the distribution of classes is heavily skewed towards the classes that are most commonly observed in the context. As can be seen in Table 1 the total number of occourances of each class is heavily scewed towards the *Person* class. Furthermore, as can be seen in Fig. 2, each class follows roughly the same trend in terms of the density of which they occur. While the most common for all classes is a single count of the given object present in a given image is 1, the range of occurrences are greater for the *Person* category.

The camera used for recording the dataset was elevated above the observed area, and objects often appear very distant with regards to the camera, in combination with the resolution of the camera most objects appear very small in the image (see Fig. 1). Table 1 summarizes the amount of objects from each class pertaining to each size category. The size is classified using the same scheme as used in the COCO dataset [19], where objects with areas $area < 32^2$, $32^2 < area < 96^2$ and $area > 96^2$ are considered small, medium and large respectively. The density of object sizes are also illustrated in Fig. 3, where it can be more clearly seen that the vast majority of objects fall within the small category for classes. This holds true for classes *Person*, *Bicycle* and *Motorcycle*, where as the *Vehicle* class more evenly covers all size categories. This is a result of larger vehicles only being allowed to drive in the area closest to the camera.

Table 1. Object frequency observed for each COCO-style size category.

	Class			
Size	Person	Bicycle	Motorcycle	Vehicle
Small	5.663.804	288.081	27.153	113.552
Medium	454	7	0	37.007
Large	176.881	5.192	5.240	550.696
Total	5.841.139	293.280	32.393	701.255

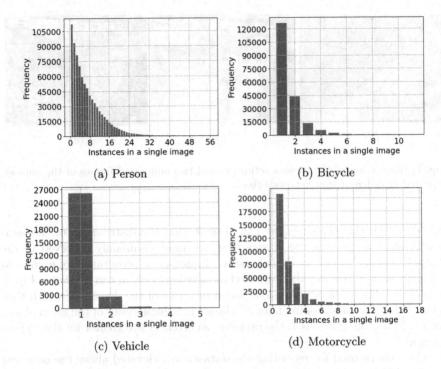

Fig. 2. Histogram of object density, across the dataset, density of objects (x-axis) and occurrences (y-axis).

3.2 Evaluation Protocol

The challenge followed the COCO evaluation[5] scheme for mAP. The primary metric is, mAP across 10 different IoU thresholds (ranging from 0.5 to 0.95 at 0.05 increments). This is calculated for each month in the validation/test set and the model is then ranked based on a weighted average of each month (more distant months having a larger weight as more concept drift is present), referred to as mAP_w in the analysis of the results (Table 2). The evaluation is performed leveraging the official COCO evaluation tools[6].

3.3 The Baseline

The baseline is a YOLOv5 with the default configuration from the Ultralytics[7] repository, including augmentations. It was trained with a batch size of 64 for 300 epochs, with an input image size of 384×288 and the best performing model is chosen. Naturally, the labels were converted to the normalized yolo format ([cls] [c_x] [c_y] [w] [ht]) for both training and evaluation. For submission on the

---

[5] https://cocodataset.org/#detection-eval.
[6] https://github.com/cocodataset/cocoapi.
[7] https://github.com/ultralytics/yolov5.

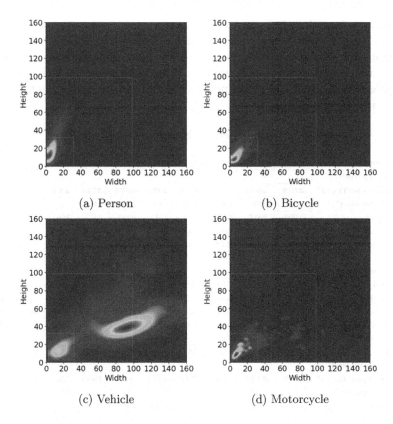

Fig. 3. Illustration of object size (height×width, in pixels) across the dataset. The white outlines seperate the areas that would be labeled as small, medium and large following COCO standards.

Codalab platform they were converted back to the ([cls] [tl$_x$] [tl$_y$] [br$_x$] [br$_y$]) coordinates. The models were all trained on the same machine with 2x Nvidia RTX 3090 GPUs, all training is also conducted as multi GPU training using the pytorch distributed learning module.

4   Challenge Results and Winning Methods

The challenge ran from 25 April 2022 to 24 June 2022 through Codalab[8] [22], a powerful open source framework for running competitions that involve result or code submission. It attracted a total of 184 registered participants, 82, 52 and 50 on track 1, 2 and 3, respectively. During development phase we received 267 submissions from 17 active teams in track 1, 117 submissions from 6 teams in track 2, and 96 submissions from 4 teams in track 3. At the test (final) phase, we

---

[8] Codalab - https://codalab.lisn.upsaclay.fr.

received 84 submissions from 23 active teams in track 1, 55 submissions from 22 teams in track 2, and 72 submissions from 24 teams in track 3. The reduction in the number of submissions from the development to the test phase is explained by the fact that the maximum number of submissions per participant on the final phase was limited to 3, to minimize the change of participants to improve their results by trial and error.

Table 2. Codalab leaderboards* at the test (final) phase.

Participant	mAP_w	mAP	Jan	Mar	Apr	May	Jun	Jul	Aug	Sep
Track 1 (day level)										
Team GroundTruth*	**.2798**	**.2832**	.3048	**.3021**	**.3073**	**.2674**	**.2748**	**.2306**	**.2829**	**.2955**
Team heboyong*	.2400	.2434	**.3063**	.2952	.2905	.2295	.2318	.1901	.2615	.1419
Team BDD	.2386	.2417	.2611	.2775	.2744	.2383	.2371	.1961	.2365	.2122
Team Charles	.2382	.2404	.2676	.2848	.2794	.2388	.2416	.2035	.2446	.1630
Team Relax	.2279	.2311	.2510	.2642	.2556	.2138	.2336	.1856	.2214	.2235
Baseline*	.0870	.0911	.1552	.1432	.1150	.0669	.0563	.0641	.0835	.0442
Track 2 (week level)										
Team GroundTruth*	**.3236**	**.3305**	.3708	.3502	**.3323**	.2774	**.2924**	**.2506**	**.3162**	.4542
Team heboyong*	.3226	.3301	.3691	.3548	.3279	**.2827**	.2856	.2435	.3112	.4662
Team Hby	.3218	.3296	.3722	.3556	.3256	.2806	.2818	.2432	.3067	**.4714**
Team PZH	.3087	.3156	**.3999**	**.3588**	.3212	.2596	.2744	.2502	.3013	.3592
Team BDD	.3007	.3072	.3557	.3367	.3141	.2562	.2735	.2338	.2936	.3942
Baseline*	.1585	.1669	.2960	.2554	.2014	.1228	.0982	.1043	.1454	.1118
Track 3 (month level)										
Team GroundTruth*	**.3376**	**.3464**	**.4142**	**.3729**	**.3414**	**.3032**	**.2933**	**.2567**	.3112	**.4779**
Team heboyong*	.3241	.3316	.3671	.3538	.3289	.2838	.2864	.2458	**.3132**	.4735
Team BDD	.3121	.3186	.3681	.3445	.3248	.2680	.2843	.2450	.3062	.4076
Team PZH	.3087	.3156	.3999	.3588	.3212	.2596	.2744	.2502	.3013	.3592
Team BingDwenDwen	.2986	.3054	.3565	.3477	.3241	.2702	.2707	.2337	.2808	.3598
Baseline*	.1964	.2033	.3068	.2849	.2044	.1559	.1535	.1441	.1944	.1827

Top solutions are highlighted in bold, and solutions that passed the "code verification stage" are marked with a *.

4.1 The Leaderboard

The leaderboards at the test phase for the different tracks are shown in Table 2. Note that we only show here the top-5 solutions (per track), in addition to the baseline results. Top solutions that passed the "code verification stage" are highlighted in bold. The full leaderbord of each track can be found in the respective Codalab competition webpage.

As expected, Table 2 shows that overall better results are obtained with more train data. That is, a model trained at the month level is overall more accurate than the same model trained at the week level, which is overall more accurate than the one trained at the day level. Therefore, the differences in performance improvement when training the model at the month level (compared to week level) are smaller than those obtained when training the model at the week level (compared to day level), particularly when a large shift in time is observed (e.g., from Jun. to Sep.), suggesting that the increase of train data from week to month level may have a small impact when large shifts are observed. This was also observed by the *Team heboyong* (described in Sec. 4.3), which reported to have only used week level data to train their model (i.e., on Tracks 2 and 3), based on the observation that using more data was not improving the final result. This raises an interesting point in that even for winning approaches the variation of the training data is much more important than the amount of training data, a further analysis of what causes the loss of mAP across will be discussed in 4.4.

Table 3 shows some general information about the top winning approaches. As it can be seen from Table 3, common strategies employed by top-winning solutions are the use of pre-trained models combined with data augmentation. Next, we briefly introduce the top-winning solutions that passed the code verification stage based on the information provided by the authors. For a detailed information, we refer the reader to the associated fact sheets, available for download in the challenge webpage(see footnote 1). Two participants (i.e., *Team GroundTruth* and *Team heboyong*) ranked best on all tracks. Each participant applied the same method on all tracks, but trained at day, week or month level, detailed as follows.

Table 3. General information about the top winning approaches.

	Top-1 *Team GroundTruth*	Top-2 *Team heboyong*
Pre-trained model	✓	✓
External data	✗	✗
Data augmentation	✓	✓
Use of the provided validation set as part of the training set at the final phase	✗	✗
Handcrafted features	✗	✗
Spatio-temporal feature extraction	✗	✗
Object tracking	✗	✗
Leverage timestamp information	✗	✗
Use of empty frames present in the dataset	✗	✗
Construct any type of prior to condition for visual variety	✗	✗

4.2 Top-1: *Team GroundTruth*

The *Team GroundTruth* proposed to take benefit of temporal and contextual information to improve object detection performance. Based on Scaled-YOLOv4 [26], they first perform sparse sampling at the input. The best sampling setting is defined based on experiments given different sampling methods (i.e., average sampling, random sampling, and active sampling). Mosaic [1] data augmentation is then used to improve the detector's recognition ability and robustness to small objects. To obtain a more accurate and robust model at inference stage, they adopt Model Soups [27] for model integration, given the results obtained by Scaled-YOLOv4p6 and Scaled-YOLOv4p7 detectors trained using different hyperparameters, also combined with horizontal flip data augmentation to further improve the detection performance. Given a video sequence of region proposals and their corresponding class scores, Seq-NMS [12] associates bounding boxes in adjacent frames using a simple overlap criterion. It then selects boxes to maximize a sequence score. Those boxes are used to suppress overlapping boxes in their respective frames and are subsequently re-scored to boost weaker detections. Thus, Seq-NMS [12] is applied as post-processing to improve the performance further. An overview of the proposed pipeline is illustrated in Fig. 4.

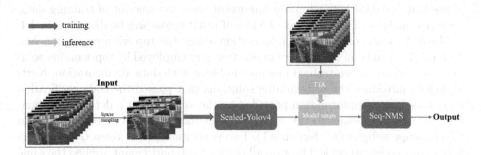

Fig. 4. Top-1 winning solution pipeline: *Team GroundTruth.*

4.3 Top-2: *Team Heboyong*

The *Team heboyong* employed Cascade RCNN [4], a two-stage object detection algorithm, as the main architecture for object detection, with Swin Transformer [20] as backbone. According to the authors, Swin Transformer gives better results when compared with other CNN-based backbones. CBNetv2 [18] is used to enhance the Swin Transformer to further improve accuracy. MMdetection [5] is adopted as the main framework. During training, only 30% of the train data is randomly sampled, to reduce overfitting, combined with different data augmentation methods, such as Large Scale Jitter, Random Crop, MixUp [31], Albu Augmentation [3] and CopyPaste [11]. At inference stage, they use Soft-NMS [2] and flip augmentation to further enhance the results. An overview of the proposed pipeline is illustrated in Fig. 5. They also reported to have not addressed well the long-tail problem caused by the extreme sparsity of the bicycle and motorcycle categories, which resulted in low mAP for these two categories.

4.4 What Challenge the Models the Most?

In this section we analyze the performance of the baseline, Team GroundTruths and Team heboyongs models on the test set. Particularly, we inspect the performance of each model with regards to temperature, humidity object area and object density. Temperature and humidity are chosen as they were discovered that these two factors have the highest correlation with visual concept drift [21]. Additionally, because of the uneven distribution of object densities across dataset, the impact of the object density is also investigated.

Impact of Temperature can be observed in Fig. 6, as the temperature increases the performance of the model degrades. This is expected as the available training data has been picked from the coldest month and as such warmer scenes are not properly represented in the training data, and as mentioned in 3 this is deliberately done as temperature is one of the most impactfull factors of concept drift in thermal images [21]. The performance of the baseline model shows severe degradation when compared to the winner and *Team heboyong*, while the performance consistently degrades for all models. Interestingly, *Team heboyong* method is distinctly more sensitive to concept drift with the smaller training set, while the winning solutions seems to perform consistently regardless of the amount of data trained on.

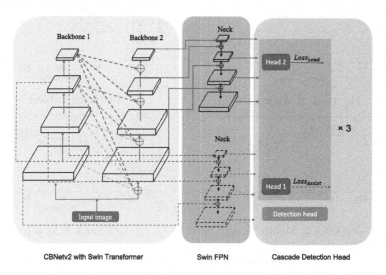

Fig. 5. Top-2 winning solution pipeline: *Team heboyong.*

Impact of Humidity. According to the initial paper [21], humidity is one of the most impactfull factors of concept drift, as it tends to correlate positively with the different types of weather. This leads to a quite interesting observation, which can be made across all tracks with regards to the impact of humidity. As can be observed in Fig. 7, the mAP of detectors increases with the humidity across all tracks. This could be because higher humidity tends to correlate with the level of rain-clouds, which would explain partially cloudy being more difficult for the detectors as the visual appearance in the image is less uniform.

Impact of Object Size. As would be expected the models converge towards fitting bounding-boxes to the most dominant object size of the training data (see Table 1). As shown in Fig. 8, the models obtain very good performance on the

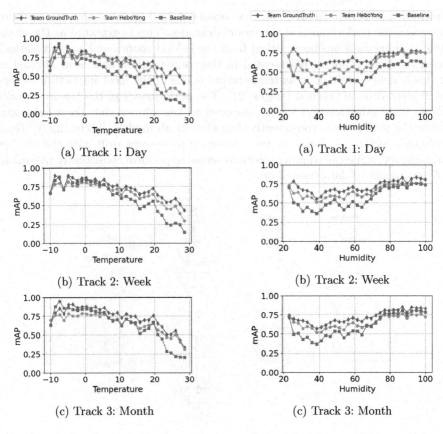

(a) Track 1: Day

(b) Track 2: Week

(c) Track 3: Month

(a) Track 1: Day

(b) Track 2: Week

(c) Track 3: Month

Fig. 6. Overview of performance with samples separated with regards to the temperature recorded for the given frame.

Fig. 7. Overview of performance with samples separated with regards to the humidity recorded for the given frame.

most common of object sizes and struggle with objects as they increase in size and rarity. In this case the participants see strong improvement over baseline, and also manage to become more robust towards rarer cases. As can also be observed in the figure this problem is increasingly alleviated with the increase of training data.

Impact of Object Density. As shown in Fig. 2, the density of objects for the majority of the images is towards the lower end, as such one would expect the detectors' mAP to degrade when a scene becomes more crowded and the individual objects become more difficult to detect due to occlusions. However what is observed is the mAP of highlighted methods are consistent as density increases, while the performance across densities also correlate to the amount of training data.

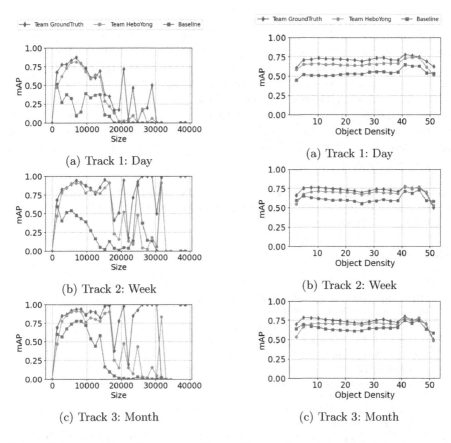

(a) Track 1: Day

(b) Track 2: Week

(c) Track 3: Month

Fig. 8. Overview of performance with samples separated with regards the size of objects bounding-box

(a) Track 1: Day

(b) Track 2: Week

(c) Track 3: Month

Fig. 9. Overview of performance with samples separated with regards to the object density of the frame

5   Conclusions

The Seasons in Drift challenge attracted over 180 participants whom made 480 submissions during validation and 211 submissions for test set and a potential place on the finale leaderboard. While the concept of measuring the impact of thermal drift on detection performance in a security monitoring context is a very understudied field, a lot of people participated. Many of the participants managed to beat the proposed baseline by quite a large margin, especially with limited training data, and achieved more robust solutions when compared to the degradation of the baseline in terms of performance with respect to drift. Allthough great improvements can be observed, the problem of concept drift still negatively affects the performance of participating methods. Interestingly while the winner and *Team heboyong* methods use different architectures, the impact of concept drift seems to transcend the choice of SotA object detectors. This lends merit investigating methods that could condition layers of the network given the input image, and introduce a venue for the model to learn an adaptable approach as opposed to learning a generalized model specific to the thermal conditions of the training context. As can be observed in Figs. 8 and 9 the size of the observed objects seem to be a more challenging factor than the density of which they occour in. Detection of small objects is a known and well documented problem, and despite the nature of thermal cameras, still persist as an issue in the thermal domain. Further research could be done to learn more scale invariant object detectors or rely entirely on other methods than an RPN or Anchors to produce object proposals.

Acknowledgements. This work has been partially supported by Milestone Research Program at AAU, the Spanish project PID2019-105093GB-I00 and by ICREA under the ICREA Academia programme.

References

1. Bochkovskiy, A., Wang, C., Liao, H.M.: Yolov4: optimal speed and accuracy of object detection. CoRR abs/2004.10934 (2020)
2. Bodla, N., Singh, B., Chellappa, R., Davis, L.S.: Soft-NMS - improving object detection with one line of code. In: ICCV (2017)
3. Buslaev, A.V., Iglovikov, V.I., Khvedchenya, E., Parinov, A., Druzhinin, M., Kalinin, A.A.: Albumentations: fast and flexible image augmentations. Information **11**(2), 125 (2020)
4. Cai, Z., Vasconcelos, N.: Cascade R-CNN: delving into high quality object detection. In: CVPR (2018)
5. Chen, K., et al.: MMDetection: open MMLab detection toolbox and benchmark. CoRR abs/1906.07155 (2019)
6. Chen, Y.Y., Jhong, S.Y., Li, G.Y., Chen, P.H.: Thermal-based pedestrian detection using faster r-cnn and region decomposition branch. In: ISPACS (2019)
7. Dai, D., Van Gool, L.: Dark model adaptation: semantic image segmentation from daytime to nighttime. In: ITSC (2018)
8. Dai, X., Yuan, X., Wei, X.: Tirnet: object detection in thermal infrared images for autonomous driving. Appl. Intell. **51**(3), 1244–1261 (2021)
9. Davis, J.W., Keck, M.A.: A two-stage template approach to person detection in thermal imagery. In: WACV-W (2005)
10. Devaguptapu, C., Akolekar, N., M Sharma, M., N Balasubramanian, V.: Borrow from anywhere: pseudo multi-modal object detection in thermal imagery. In: CVPR-W (2019)
11. Ghiasi, G., et al.: Simple Copy-Paste is a strong data augmentation method for instance segmentation. In: CVPR (2021)
12. Han, W., et al.: Seq-NMS for video object detection. CoRR abs/1602.08465 (2016)
13. Hwang, S., Park, J., Kim, N., Choi, Y., So Kweon, I.: Multispectral pedestrian detection: benchmark dataset and baseline. In: CVPR (2015)
14. Jia, X., Zhu, C., Li, M., Tang, W., Zhou, W.: Llvip: a visible-infrared paired dataset for low-light vision. In: ICCV (2021)
15. Kieu, M., Bagdanov, A.D., Bertini, M., del Bimbo, A.: Task-conditioned domain adaptation for pedestrian detection in thermal imagery. In: Vedaldi, A., Bischof, H., Brox, T., Frahm, J.-M. (eds.) ECCV 2020. LNCS, vol. 12367, pp. 546–562. Springer, Cham (2020). https://doi.org/10.1007/978-3-030-58542-6_33

16. Kim, J., Kim, H., Kim, T., Kim, N., Choi, Y.: Mlpd: multi-label pedestrian detector in multispectral domain. Rob. Autom. Lett. **6**(4), 7846–7853 (2021)
17. Krišto, M., Ivasic-Kos, M., Pobar, M.: Thermal object detection in difficult weather conditions using yolo. IEEE Access **8**, 125459–125476 (2020)
18. Liang, T., et al.: Cbnetv2: a composite backbone network architecture for object detection. CoRR abs/2107.00420 (2021)
19. Lin, T.-Y., et al.: Microsoft COCO: common objects in context. In: Fleet, D., Pajdla, T., Schiele, B., Tuytelaars, T. (eds.) ECCV 2014. LNCS, vol. 8693, pp. 740–755. Springer, Cham (2014). https://doi.org/10.1007/978-3-319-10602-1_48
20. Liu, Z., et al.: Swin transformer: hierarchical vision transformer using shifted windows. In: ICCV (2021)
21. Nikolov, I., et al.: Seasons in drift: a long-term thermal imaging dataset for studying concept drift. In: NeurIPS (2021)
22. Pavao, A., et al.: CodaLab Competitions: An open source platform to organize scientific challenges. Ph.D. thesis, Université Paris-Saclay, FRA (2022)
23. Siris, A., Jiao, J., Tam, G.K., Xie, X., Lau, R.W.: Scene context-aware salient object detection. In: ICCV (2021)
24. Telodyne: FLIR AADAS Dataset. https://www.flir.com/oem/adas/adas-dataset-form/
25. Vertens, J., Zürn, J., Burgard, W.: Heatnet: bridging the day-night domain gap in semantic segmentation with thermal images. In: IROS (2020)
26. Wang, C., Bochkovskiy, A., Liao, H.: Scaled-YOLOv4: scaling cross stage partial network. In: CVPR (2021)
27. Wortsman, M., et al.: Model soups: averaging weights of multiple fine-tuned models improves accuracy without increasing inference time. CoRR abs/2203.05482 (2022)
28. Wu, X., Wu, Z., Guo, H., Ju, L., Wang, S.: Dannet: a one-stage domain adaptation network for unsupervised nighttime semantic segmentation. In: CVPR (2021)
29. Zhang, H., Fromont, E., Lefevre, S., Avignon, B.: Multispectral fusion for object detection with cyclic fuse-and-refine blocks. In: ICIP (2020)
30. Zhang, H., Fromont, E., Lefèvre, S., Avignon, B.: Guided attentive feature fusion for multispectral pedestrian detection. In: WACV (2021)
31. Zhang, H., Cissé, M., Dauphin, Y.N., Lopez-Paz, D.: mixup: beyond empirical risk minimization. CoRR abs/1710.09412 (2017)

Author Index